Handbook of Deontic Logic and Normative Systems

Handbook of Deontic Logic and Normative Systems

Edited by

Dov Gabbay

John Horty

Xavier Parent

Ron van der Meyden

Leendert van der Torre

© Individual author and College Publications 2013. All rights reserved.

ISBN 978-1-84890-132-2

College Publications
Scientific Director: Dov Gabbay
Managing Director: Jane Spurr

http://www.collegepublications.co.uk

Cover produced by Laraine Welch
Printed by Lightning Source, Milton Keynes, UK

All rights reserved. No part of this publication may be reproduced, stored in a retrieval system or transmitted in any form, or by any means, electronic, mechanical, photocopying, recording or otherwise without prior permission, in writing, from the publisher.

CONTENTS

Preface — vii

PART I   BACKGROUND — 1

RISTO HILPINEN AND PAUL MCNAMARA
Deontic Logic: A Historical Survey and Introduction — 3

JÖRG HANSEN
Imperative Logic and Its Problems — 137

PART II   CONCEPTS AND PROBLEMS — 193

SVEN OVE HANSSON
The Varieties of Permission — 195

LOU GOBLE
Prima Facie Norms, Normative Conflicts, and Dilemmas — 241

MAREK SERGOT
Normative Positions — 353

DAVIDE GROSSI AND ANDREW J. I. JONES
Constitutive Norms and Counts-as Conditionals — 407

PART III   NEW FRAMEWORKS — 443

SVEN OVE HANSSON
Alternative Semantics for Deontic Logic — 445

XAVIER PARENT AND LEENDERT VAN DER TORRE
Input/output Logic — 499

LARS LINDAHL AND JAN ODELSTAD
The Theory of Joining-Systems — 545

Preface

This Handbook presents a detailed overview of the main lines of research on contemporary deontic logic and related topics. Although building on decades of previous work in the field, it is the first collection to take into account the significant changes in the landscape of deontic logic that have occurred in the past twenty years. These changes have resulted largely, though not entirely, from the interaction of deontic logic with a variety of other fields, including computer science, legal theory, organizational theory, and economics.

As editors, we have been guided by four ideas. First, although the Handbook contains important historical work, we have tried to highlight new developments, and new prospects for deontic logic. Second, we have tried to combat the impression that deontic logic exists only as a collection of abstract formal systems, sometimes lacking in motivation. Instead, we wanted to emphasize the real problems that give rise to the formalisms developed by deontic logicians, as well the potential for real applications in a variety of fields. Third, we have made every effort to provide authors with the freedom to present their material in depth, sometimes resulting in chapters of monographic length and scope, containing the first comprehensive treatments of their subjects. Finally, we wanted the work to be affordable for individual researchers, not simply for those institutions willing to pay the exhorbitant prices charged by commercial publishers, and even by certain commercial ventures masking as university presses. For this reason, we chose to work with College Publications, a non-profit publisher run by academics and for academics. We recommend this service to others.

The Handbook is more than a set of individual chapters. It is a community project. The need for the Handbook was first identified at the 2008 Conference on Deontic Logic in Computer Science, held in Luxembourg. Each selected author was invited to present an outline of his or her chapter at the 2009 Augustus de Morgan Workshop, again in Luxembourg, where the material was discussed and coordinated among a group of authors, editors, and other experts. Chapters were then written, evaluated by independent readers, and first drafts were presented and discussed by authors and editors at yet another meeting co-located with the 2010 Conference on Deontic Logic in Computer Science, held in Florence. As a result of this discussion, drafts were revised, again sent to readers, discussed among editors and revised further until, at last, we are now ready to publish the first volume of the Handbook. A second volume, and perhaps a third, are under way, and we hope to set up a web site where revisions and additions can appear.

Throughout the process, we have relied heavily on the expertise and time of our readers. For help with the current volume, we would particularly like to thank the following individuals: Jörg Hansen, Sven Ove Hansson, Loes Olde Loohuis, Shyam Nair, Jan Odelstad, Antonio Rotolo, and Giovanni Sartor.

The first volume of the Handbook is divided into three parts: Background, Concepts and Problems, and New Frameworks.

The chapters in the first part concentrate on historical background, while laying the foundations for later developments. Chapter 1 provides a historical and philosophical introduction to core developments in deontic logic. Chapter 2 provides an analysis of the historical debate over the possibility of imperative logic, a topic that was closely intertwined with early developments in deontic logic.

The chapters in the second part discuss some key normative concepts and problems that have become prominent in the literature. Chapter 3 focuses on the notion of permission as a normative concept in its own right, not simply the dual of obligation. Chapter 4 is about the motivation for and design of deontic logics allowing conflicts among obligations, a possibility that is ruled out by standard systems. Chapter 5 is devoted to the theory of normative positions, whose main purpose is to lay the foundations for a general theory of rights, duties, and other normative relations among individuals and institutions. While the previous chapters deal with the so-called regulative norms of obligation and permission, Chapter 6 explores the logic of constitutive norms—the system of norms that allow us to view certain human actions as, for example, establishing binding contracts within a particular legal system, or as scoring touchdowns within the game of American football.

The chapters in the third part describe three new logical frameworks that have now become part of the mainstream literature on deontic logic. Chapter 7 discusses the limitations of the traditional approach, and explores, as an alternative, preference-based deontic logics. Although this chapter is still tied to the usual possible worlds framework, the next two chapters stray further. Chapter 8 introduces input/output logic, an operational theory, originally devised for reasoning with conditional norms, which provides a fine-grained analysis of normative reasoning by imposing different constraints on the process of detachment. Chapter 9 is devoted to the algebraic theory of joining systems, which focuses special attention on the normative role of intermediate, or bridge, concepts.

For future volumes, we currently have commitments from authors for chapters on at least the following topics: contrary to duty obligations, supererogation and allied normative concepts, deontic logic and actions, normative system change, deontic logic and changing preferences, reactive approaches to deontic logic, the formalization of practical reasoning, deontic logic and natural language semantics, deontic logic and legal reasoning, deontic logic based on imperatives, modalities for access control, distributed systems management, obligations for usage control, privacy and epistemic obligations, epistemic norms, and compliance with rules and norms.

Dov Gabbay
John Horty
Xavier Parent
Ron van der Meyden
Leendert van der Torre

PART I

BACKGROUND

1
Deontic Logic: A Historical Survey and Introduction

RISTO HILPINEN AND PAUL MCNAMARA

ABSTRACT. This chapter provides both a historical overview and an introduction to core developments in deontic logic up to the end of the 20th century. The presentation becomes more systematic for the last half century covered, but continues to convey historical developments. In particular, we present some key developments from the Middle Ages through the 19th century, then turn to Meinong and Mally's contributions near the early part of the last century, followed by the full emergence of modern deontic logic with von Wright's work in the early 1950's. We next cover the emergence of the so-called "standard" systems of deontic logic in the 1960s, including the emergence of formal semantics for those systems. Then we cover a wide array of objections to, and limitations of, the standard systems, while also often indicating various lines of response to these challenges. Finally, we turn to the issue of the representation of action and agency in deontic contexts. Supplements to some sections or sub-sections provide the reader with the option of more details on a given topic.

1	Early developments. From medieval deontic logic to the 19th century	5
2	Alexius Meinong on normative concepts and value concepts .	9
3	Ernst Mally's *Deontik*	15
4	On the interpretation of deontic logic	20
5	G. H. von Wright's deontic interpretation of modal logic . . .	31
6	The standard system of deontic logic (SDL) and close cousins	36
	6.1 SDL	36
	6.2 The Leibnizian-Kangerian-Andersonian reduction . . .	39
7	The semantics of Standard Deontic Logic and close cousins .	45
	7.1 The semantics for SDL	45
	7.2 Semantics for the Leibnizian-Kangerian-Andersonian reduction	47
	7.3 A generalization of SDL: VW logics and their non-standard semantics	49
	7.4 Classical quantification and SDL semantics	51

8		Problems and paradoxes regarding the standard systems . . .	58
	8.1	A puzzle with the very idea of deontic logic	58
	8.2	A problem centering around OT	59
	8.3	Puzzles centering around the rule RMD	60
	8.4	Puzzles centering around DD and OD	66
	8.5	Puzzles centering around deontic conditionals	82
	8.6	Some further expressive inadequacies of the standard systems	91
	8.7	A problem calling for attention to action and agency in deontic contexts	96
9		Actions and agency in deontic logic	97

The following handbook entry is primarily historical, aimed at providing orientation to the newcomer to deontic logic by conveying a sense of the sweep of themes associated with core material in deontic logic, while at the same time occasionally offering something of interest to those who have already entered the field. We aim to cover the history of deontic logic through roughly the end of the last millennium.[1] Once we turn to the emergence of deontic logic as a full-fledged active area within symbolic logic in the second half of the last century (in Sections 6-9) with the emergence of the so-called "standard" systems of deontic logic, the approach shifts to a more systematic orientation, but with historical information included in the process. We also then must become more selective, and that leads to a focus on core areas and ideas, familiarity with which we think is often presupposed by those actively working in this area. Although there is a narrative structure, especially in the earlier half of this chapter, most sections of the chapter (even the earlier ones) are relatively self-contained despite occasional references backward. Section 8 in particular, which catalogs an array of puzzles and challenges that the standard systems faced, puzzles that often served (some still do) as catalysts for new work, is designed so that the reader can dip into even one of its sub-sections and read about one puzzle more or less independently of the rest. Overall, we think the chapter allows for the acquisition of a narrative sense of the historical core of deontic logic, but without paying the price "all or nothing".

The authors are quite aware that we have had to make choices, and the result falls short of all we had hoped to include, especially as we approach the end of the last millennium; still, it is our hope that such shortcomings

---

[1] With the exception of reference to some work by Islamic philosophers, the main focus is on the Western tradition. We are not aware of work of relevance in other traditions (although we did make some preliminary inquiries of experts, e.g. about South Asian literature), and we would not be competent to discuss such material unless it was already covered in secondary sources, which we did not succeed in finding.

will be partially overcome by the fine chapters that follow ours on particular areas and issues in deontic logic that our colleagues have written.[2]

Introduction: What is deontic logic?

Deontic logic is an area of logic which investigates normative concepts (deontic concepts), along with closely associated evaluative concepts, norms and norm systems, and normative reasoning. The word 'deontic' is derived from the Greek expression 'déon' δέον, which means 'what is binding' or 'proper'. Jeremy Bentham used the word 'deontology' for "the science of morality" [Bentham, 1983], and the Austrian philosopher [Mally, 1971], who developed in the 1920's a system of the "fundamental principles the logic of ought", called his theory "Deontik". Normative concepts include the concepts of obligation (duty, requirement, the concept of ought), permission (permissibility, 'may'), prohibition ('may not', 'forbidden'), and related notions, for example, those expressed by the words 'right', 'optional' (normatively contingent), 'claim', 'power', 'immunity', and 'supererogatory'. Deontic logic is also concerned with the relations among normative concepts, axiological concepts (value concepts, e.g., 'good' and 'better than'), and agent-evaluative concepts (e.g., 'blameworthy' and 'praiseworthy'). Thus the formal languages of deontic logic contain, in addition to propositional connectives and quantifiers, logical constants for normative concepts, and in some cases operators representing axiological concepts, praxeological concepts (for agency and action), prohairetic concepts (for preference and interest), aretaic concepts (for agent evaluation) and perhaps other modalities. The concepts of agency, action, and preference connect deontic logic to the logic of practical reasoning.

1 Early developments. From medieval deontic logic to the 19th century

In his manuscript *Elementa iuris naturalis* [1930] Gottfried Wilhelm Leibniz called the deontic categories of the obligatory (*debitum*), the permitted (*licitum*) and the prohibited (*illicitum*) "modalities of law" (*iuris modalia*), and observed that important basic principles of alethic modal logic hold for these legal modalities. Much of the development of deontic logic during the past half century, and especially in the founding decade of the 1950s, has been based on just such modal analogies, and thus deontic logic has often been studied as a branch of modal logic.[3] In other words, the con-

---

[2]The authors of this chapter in particular would welcome any corrections or suggestions for improvement.

[3]Deontic necessity was taken to represent what was morally obligatory *all things considered*, and not merely *prima facia* obligatory, a distinction stressed explicitly in

cept of obligation has been studied as normative (deontic) necessity, and the concept of permission or permissibility has been construed as normative (deontic) possibility. Moreover, Leibniz suggested that legal (or deontic) modalities can be defined in terms of (or "reduced to") the alethic modalities of necessity and possibility, and one evaluative notion, in a way that is reminiscent of recent approaches in virtue ethics to defining deontic concepts.[4] According to Leibniz, the permitted is "what is possible for a good person to do", and the obligatory is "what is necessary for a good person to do". (Cf. [Hruschka, 1986, pp. 35-6]) As it turns out, one strain of the early developments in deontic logic emerging in the mid-Twentieth Century also concurred with the sort of "reductive" approach to the deontic operators endorsed here by Leibniz. Leibniz might thus be said to be prescient regarding some of the first main lines of approach to deontic logic as an area of symbolic logic emerging solidly in the 1950s. It is interesting to note here that five hundred years before Leibniz, Peter Abelard (1079-1144) and other early medieval philosophers often endorsed an inverted form of Leibniz' reduction by defining alethic modal concepts by means of normative or quasi-normative concepts. According to this characterization, necessity is taken to be what nature demands, possibility is identified with what nature allows, and impossibility with what nature forbids. [Knuuttila, 1993, p. 182] Thus analogical links between deontic logic and alethic modal logic have a long and rich history before their widespread reemergence a half century ago in symbolic deontic logic.

In particular, formal analogies between deontic notions and "pure" (alethic) modalities (necessary, possible and impossible) were studied by many 14th century philosophers who regarded deontic logic as a branch of modal logic. They presupposed and used the following equivalences in their discussions on normative concepts: (**O** stands here for the concept of obligation (*obligatum*), **P** for permission, and **F** for prohibition.)

(1.1) (i) $\mathbf{P}p \leftrightarrow \neg\mathbf{O}\neg p$,
 (ii) $\mathbf{O}p \leftrightarrow \neg\mathbf{P}\neg p$,
 (iii) $\mathbf{O}p \leftrightarrow \mathbf{F}\neg p$, and
 (iv) $\mathbf{F}p \leftrightarrow \mathbf{O}\neg p$.

The interest in deontic modalities in late medieval philosophy, especially in the 14th century, was related to the attempts to systematize modal theory and overcome the observed inadequacies in Aristotle's account of modal syl-

---

W. D. Ross' seminal [Ross, 1939].

[4][Zagzebski, 1996; Slote, 1997; Hursthouse, 1999] are representative of this recent approach to analyzing deontic concepts via virtue theoretic concepts in ethical theory.

logisms. Many logicians thought that the logic of alethic modalities (modalities of truth or being), such as those of necessity and possibility, could be used as a formal model for other concepts which show apparent similarities to modal concepts, such as knowing, believing, having an opinion, doubting, appearing, and being obligatory, permitted, or forbidden. These interpretations of modal logic led to the development of the elements of epistemic and deontic logic in the fourteenth century [Boh, 1985; Boh, 1993; Knuuttila, 1981], and to critical discussions of the applicability of the basic principles of modal logic to epistemic and normative concepts. The principles investigated include the following inference patterns

(1.2) $$\frac{\mathbf{N}(p \rightarrow q)}{\mathbf{N}p \rightarrow \mathbf{N}q}$$

and

(1.3) $$\frac{\mathbf{N}(p \rightarrow q)}{\mathbf{M}p \rightarrow \mathbf{M}q}$$

where $\mathbf{N}$ and $\mathbf{M}$ represent the concepts of alethic necessity and possibility. (1.2) is equivalent to the principle K of contemporary modal logic, and is a fundamental principle for normal modal logics. (See [Chellas, 1980, p. 114]) A number of medieval philosophers discussed the epistemic and deontic variants of these principles, and concluded that the following rules do not hold for deontic (nor epistemic concepts) without restrictions:

(1.4) $$\frac{\mathbf{N}(p \rightarrow q)}{\mathbf{O}p \rightarrow \mathbf{O}q}$$

and

(1.5) $$\frac{\mathbf{N}(p \rightarrow q)}{\mathbf{P}p \rightarrow \mathbf{P}q}$$

Principles (1.4) and (1.5) were discussed already in the 12th century as principles concerning the logic of will, and the counterexamples to them were formulated in terms of the concept of willing. Peter of Poitiers (1130-1205) gave the following example: If a sinner repents of a sin, he is guilty of sin, but if a sinner wills to repent of a sin, it does not follow that he wills to be guilty of sin. The example can be expressed in normative terms as follows: Necessarily, if a person R repents of a sin, R is guilty of sin, but it does not follow that R ought to be guilty of sin if he ought to repent of a sin. Stephen Langton's (1150-1228) counter-example was similar: Necessarily, if a man visits his sick father, the father is sick. But it does not follow that if this man wills to visit his sick father, then he wills the father to be sick. If the concept of willing is replaced by the concept of *ought*, we get the following counterexample to (1.4): Necessarily, if this man visits his sick father, the father is sick. But it does not follow that if this man ought to visit his sick father, then his father ought to be sick. For surely it ought not to be the case that this man's father is sick. (See [Knuuttila and Hallamaa, 1995, p. 77].) The 14the century philosopher Roger Roseth reformulated Peter of Poitiers's example as follows:

> "There are consequences which are good and known to be good the antecedent of which I am permitted to will, without being permitted to will be consequence, For example, this consequence is good and known to be good: I repent of my sin, therefore I am in sin. I am permitted to will the antecedent but I am not permitted to will the consequent, because I am permitted to repent of my sins, but I am not permitted to will to be in sin."
> [Knuuttila and Hallamaa, 1995, p. 77]

This example serves as a counterexample to principle (1.5); a permission to repent of one's sins does not entail a permission to sin. These medieval authors in effect argued that deontic logic is not a *normal* modal logic, that is, a logic satisfying, among other conditions, the rules

(1.6) $$\frac{p \models q}{\mathbf{O}p \models \mathbf{O}q}$$

and

(1.7) $$\frac{p \models q}{\mathbf{P}p \models \mathbf{P}q}$$

where $\models$ represents the concept of logical consequence. (1.6) is a deontic version of the modal rule usually called RM (see [Chellas, 1980, p. 114]); (1.6) and (1.7) may be called the consequence rules for $\mathbf{O}$ and $\mathbf{P}$. In the 20th century deontic logic counterexamples to the consequence principle (1.6) reappeared in various forms, as Ross's paradox (the paradox of disjunctive obligation), the paradox of disjunctive permission ("free choice permission"), the paradox of the Good Samaritan, and the paradox of Epistemic Obligation. These paradoxes will be discussed below in Section 8.

In the 17th and 18th century literature on normative discourse and the logic of norms, some authors regarded normative concepts as analogous to modal concepts, like in the medieval literature, as was observed above in the particular case of Leibniz. Thus it was assumed that the concepts of obligation, permission and prohibition were related to each other in the same way as the modal concepts of necessity, possibility and impossibility ([Hruschka, 1986, pp. 39-43] and [Knebel, 1991]). Moreover, deontic concepts were usually applied to actions, thus deontic modalities were regarded as *action modalities*. Like the English word 'action', the expression 'actio' used in the 17th and 18th century literature did not refer only to human actions, but also to events which take place as a result of natural necessity; human actions (or actions in the narrow sense) were distinguished from other actions and events by the attribute 'liber'; thus the philosophers of this period made a distinction between *actio libera* and *actio physice necessaria* [Hruschka, 1986, p. 10]. Normative concepts were regarded as being properly applicable only to the former ('free') actions. The concept of obligatory action (*actio obligatoria*) was usually understood as *legally determined* action; an obligatory action in this sense may be an action which must be performed (*actio praecepta*, a morally necessary action) or an action which must be omitted (*actio prohibita*) [Hruschka, 1986, pp. 17-22]. In this context 'an action' meant an act-type or a kind of action rather than an individual action. These features are present in Leibniz's deontic logic. Leibniz defined the permitted (*licitum*) as "that which is possible for a good man [person] to do", and the obligatory (*debitum*, duty) as "that which is necessary for a good person to do". He called deontic modalities "iuris modalia", "modalities of law", and observed that the basic principles of the Aristotelian modal logic hold for the "iuris modalia" (modalities of law) as well as for the other modalities [Leibniz, 1930, p. 466].

2 Alexius Meinong on normative concepts and value concepts

In his *Psychologisch-ethische Untersuchungen zur Werth-Theorie* [1968b] the Austrian philosopher Alexius Meinong (1853-1920) divided acts from the normative and evaluative point of view into four mutually exclusive "value-classes" (88-93):

(2.1) (a) Meritorious (*verdienstlich*),
  (b) Correct (*correct*),
  (c) (Merely) Excusable (*zulässig*), and

(d) Reprehensible, inexcusable (*verwerflich*).[5]

(See [Chisholm, 1982, pp. 104-5] and [Sajama, 1988, pp. 71-2].)The actions in classes (a) and (b) have a positive value, whereas those in (c) and (d) lie below the zero point of the "value line" (1894/1968, 90) Using what Meinong called the "vulgar expressions" 'good' and 'bad', the actions in categories (a) and (b) might be characterized as good, and those in (c) and (d) as bad [Meinong, 1968b, p. 92], even though "zulässige" acts can be said to be "bad" only in a rather weak sense (cf. [Sajama, 1988, p. 71]). Meinong's value classes can be correlated with a comparative concept of goodness in the following way: If an action is meritorious or correct, it is better to perform it than not perform it, and if it is reprehensible or excusable, omitting the action is preferable to doing it. Meinong's schema does not include (morally or normatively) indifferent actions; they are obviously actions whose commission is neither more nor less preferable than their omission. According to the simple act-utilitarian or optimizing consequentialist account of the concept of obligation, an act is obligatory (required) whenever it is better (or has better consequences) than its omission, but this analysis of the concept of obligation leaves no room for meritorious but optional (that is, supererogatory) actions, and this has often been regarded as one of the weaknesses of act-utilitarianism. In this respect the act-utilitarian account does not agree with our conception of the relations between normative concepts and value concepts. Some attempts to define obligation in terms of 'better' and rely on a logic for the latter to derive a logic for former, are subject to the same difficulty; for example, Lennart Åqvist's [1963, p. 286] definition

(DOÅqv) $\mathbf{O}p \leftrightarrow p\mathbf{B}\neg p.$

Meritorious, correct, and excusable actions are actions an agent may perform, and belong to the deontic category of the permitted (permitted actions); inexcusable actions are not permitted (that is, are prohibited), and Meinong's concept of 'correct' can be regarded as equivalent to the

---

[5]Regarding Meinong's terminology, the intention is clearly that the four categories are to be mutually exclusive. We add "merely" in front of "Excusable" because if (c) is simply the complement of (d), the inexcusable, then it must include (b) and (a), which are also not inexcusable surely; but the intention of (c) seems to be a category of offense or suberogation—permissible yet blameworthy. Regarding (b), surely the meritorious is also "correct", but Meinong seems to intend what is obligatory but not meritorious, else (b) overlaps with (a), again violating mutual exclusivity. (a) seems to be intended for a category like the supererogatory, something that is meritorious and optional. There are a variety of interesting subtleties here we must pass over. See [McNamara, 1996a; McNamara, 1996b; McNamara, 1996c; McNamara, 2011a; McNamara, 2011b] for closely related issues.

deontic concept of obligation or requirement, not of permitted as might be suggested by "correct".

Meinong's deontological-axiological categories are represented in Table 1. The category of indifferent actions has been added here to Meinong's schema. 'A' represents an action type ("generic action"), $V(A)$ is the value of A (in a given situation), $\mathbf{om}(A)$ represents the omission of A, and deontic and axiological concepts are symbolized as follows.

$\mathbf{L}A = A$ is laudable (meritorious, supererogatory),
$\mathbf{O}A = A$ is obligatory (required),
$\mathbf{I}A = A$ is normatively indifferent,
$\mathbf{E}A = A$ is excusable (suberogatory),
$\mathbf{P}A = A$ is permitted, that is, not forbidden, and
$\mathbf{F}A = A$ is forbidden (not permitted, reprehensible).

These categories are represented in Table 1 and linked to corresponding value notions.

$\mathbf{P}A$: A is permitted	$\mathbf{L}A$: A is meritorious (laudable, supererogatory, *verdienstlich*)	A is good, an action with a positive value $V(A) > V(\mathbf{om}A)$
	$\mathbf{O}A$: A is obligatory (required, *correct*)	
	$\mathbf{I}A$: A is indifferent	A is indifferent $V(A) = V(\mathbf{om}A)$
$\mathbf{F}A$: A is forbidden (prohibited)	$\mathbf{E}A$: A is excusable (*zulässig*)	A is bad, an action with a negative value, an undesirable action $V(A) < V(\mathbf{om}A)$
	$\mathbf{F}A$: A is forbidden (reprehensible, *verwerflich*)	

Table 1: Meinong's deontological-axiological action categories

To say that an action has positive value means here that it is preferred to its omission. The value of an action, $V(A)$, need not be regarded as interval measurable; $V(A) < V(B)$ may be taken to mean only that B is strictly preferred to A, $V(A) = V(B)$ means that there is no noticeable value-difference between A and B. The arrangement of the five categories in Table 1 does not mean that supererogatory actions are invariably better (more valuable) than obligatory (required) actions, but it is clear that both are better than normatively indifferent actions.

Meinong formulated a "law of omission" concerning the four main deontological-axiological categories, according to which an action A is meritorious if and only if its omission is excusable, excusable if and only if its omission is meritorious, correct (obligatory) if and only if its omission is reprehensible (forbidden), and reprehensible if and only if its omission is correct. ([1968b, p. 89] and [1968a, p. 32]) These laws are expressed by the following formulas:

(2.2) $\quad\quad\quad\quad\quad\quad\quad$ **L**$A \leftrightarrow$ **Eom**A

(2.3) $\quad\quad\quad\quad\quad\quad\quad$ **E**$A \leftrightarrow$ **Lom**A

(2.4) $\quad\quad\quad\quad\quad\quad\quad$ **O**$A \leftrightarrow$ **Fom**A

(2.5) $\quad\quad\quad\quad\quad\quad\quad$ **F**$A \leftrightarrow$ **Oom**A

Moreover, according to the definability of **P** in terms of **F**, we have

(2.6) $\quad$ **P**$A \leftrightarrow \neg$**F**$A \leftrightarrow \neg$**Oom**A

where '$\neg$' is the sign of propositional negation. (2.4)-(2.6) are analogous to the standard interdefinability principles for deontic operators. If the **om**–operator is formally analogous to negation and satisfies the principle of "double omission (negation)", then (2.3) follows from (2.2) and (2.5) follows from (2.4). Meinong does not accept this principle; according to him, the omission of an act requires an opportunity to perform the act [1968a, pp. 691-2]. However, if the concept of omission is interpreted as not-doing or if we consider only actions which are possible for an agent to perform or not to perform in a given situation, the **om**-operator is analogous to propositional negation, and subject to the principle of "double omission".

According to Meinong, normative and axiological concepts can be defined in terms of the praiseworthiness or blameworthiness of actions and their omissions. The omission of a correct (obligatory, required) act is always forbidden and deserves blame, which can be regarded as a social or moral sanction associated with the action. The performance of a meritorious act is praiseworthy, that is, deserves praise or a reward. An act is excusable if its performance does not deserve blame or praise; in this respect excusable actions do not differ from indifferent actions, but the omission of an excusable action is praiseworthy. Thus Meinong's definitions suggest the following analysis of the main axiological and deontological concepts in terms of a reward (**R**) or sanction or punishment (**S**).

(2.7) $\quad$ **L**$A \leftrightarrow A \ni$ **R**,

(2.8) $\quad$ **O**$A \leftrightarrow$ om$A \ni$ **S**,

(2.9) $\quad$ **E**$A \leftrightarrow$ om$A \ni$ **R**,

and

(2.10) $\mathbf{F}A \leftrightarrow A \ni \mathbf{S}$,

where the letter 'э' (the Cyrillic 'E') signifies the association between an action type and a sanction **S** or reward **R**. If 'A' and '**om**A' are read as the propositions that A is performed or omitted, the sign $\ni$ may be read as a sign of a defeasible or non-defeasible conditional. If 'э' is read as a strict (necessary) conditional and '**om**' is read as a sign for propositional negation, (2.8) becomes equivalent to Alan Ross Anderson's [1967, p. 169-71] and [1958] proposal to reduce deontic logic to alethic modal logic by means of the translation

(2.11) $\mathbf{O}A \leftrightarrow \mathbf{N}(\neg A \to \mathbf{S})$.

According to Meinong's interpretation of (2.1)-(2.4), **S** represents strong negative value-feelings, and **R** stands for positive value-feelings [Meinong, 1968b, pp. 73-4] and [Sajama, 1988, p. 75].

$\mathbf{O}A$ implies that doing A is better than its omission, but the converse does not hold. According to Meinong's analysis of value and obligation, the principle

(2.12) $V(A) \leq V(A') \to (\mathbf{O}A \to \mathbf{O}A')$

does not hold, a meritorious act may be better than an obligatory act. An obligatory or normatively required action need not be the best or optimal action available to the agent in a given situation.[6] (2.12) may be called the principle of value-positivity; Sven Ove Hansson calls it the principle of preference positivity ($\geq$'-positivity, where $\geq$' is the preference relation associated with **O**; see [Hansson, 2001, p. 115]). Hansson gives this counter-example to (2.12): Serving a fine dinner to unexpected guests may be better than offering them something to eat and drink, but the former need not be a moral or social requirement if the latter is so [Hansson, 2001, p. 146]. On the other hand, Meinong's system is consistent with the rule of value-positivity (preference positivity) for the (standard) concept of permission (as defined in Table 1), that is,

(2.13) $V(A) \leq V(B) \to (\mathbf{P}A \to \mathbf{P}B)$.

---

[6] Although the best act will then be one that fulfills the obligation, and so the obligatory act (e.g. providing some help) will be done if the more specific best act (providing maximal help) is done.

Given the definition of **F** as ¬**P**, (2.13) entails the principle of value-negativity (preference-negativity),

(2.14) $V(A) \leq V(B) \to (\mathbf{F}B \to \mathbf{F}A)$.

(2.13) and (2.14) are based on the plausible assumption that a permissible action (including an excusable one) cannot be worse than an inexcusable (prohibited) action.

It is interesting to note here that in the 11th century, the Islamic rationalist philosopher Abd-al-Jabbār (935-1025) presented a schema essentially similar to that proposed by Meinong and distinguished normative categories on the basis of whether a given act or its omission deserves blame or praise. Like Meinong, he distinguished four main normative categories ([Hourani, 1975, p. 132], [Hourani, 1985, pp. 100-1] and [Sajama, 1988, p. 80]):

(2.15) (i) An act A is an act of grace (*tafaddul*) or recommended (*nadb*) if and only if the doer deserves praise, the omitter does not deserve blame.
 (ii) A is obligatory (*wājib*) if and only if the omitter deserves blame.
 (iii) A is merely permissible (optional, *mubāh*) if and only if neither the doer nor the omitter deserves blame or praise.
 (iv) A is evil (*qabīh*) if and only if the doer deserves blame.

The actions in categories (i)-(iii) are described as "good" (*hasan*) actions [Hourani, 1985, p. 101]; the word 'acceptable' might be more suitable. Hourani's use of 'permissible' for category (iii) may be misleading; 'optional' seems more apt. The main difference between (2.15) and Meinong's definitions (2.7)-(2.10) is that (2.15.iii), a "permissible" (i.e., optional) action corresponds to the category of normative indifference (**I**), and the category of excusable actions (**E**) is missing from Abd-al-Jabbār's schema. The omission of a meritorious (laudable) action does not deserve blame, i.e.,

(2.16) $\mathbf{L}A \to \neg(\mathrm{om}A \ni \mathbf{S})$,

(This is obvious if **L** represents supererogatory actions.) An analogous principle holds for the other categories, for example, the commission of an obligatory action does not generally merit a reward. Thus we may adopt the following principles:

(2.17) $A \ni \mathbf{S} \to \neg(\mathrm{om}A \ni \mathbf{R})$
(2.18) $\mathrm{om}A \ni \mathbf{S} \to \neg(A \ni \mathbf{R})$

(2.19) $A \ni \mathbf{R} \to \neg(\mathbf{om}A \ni \mathbf{S})$
(2.20) $\mathbf{om}A \ni \mathbf{R} \to \neg(A \ni \mathbf{S})$

(2.17) does not mean that an agent cannot be rewarded after fulfilling his duty, only that such a reward is contingent, and not associated with the duty by a general rule or custom. According to [Sajama, 1988, pp. 77-8, n. 11], some formulations of the *sharia* law violate (2.18) and (2.17) by defining an obligatory action (a duty) as an action whose performance deserves reward and omission a punishment, and a forbidden action as one whose performance deserves punishment and omission a reward. (Sajama refers to [Hartmann, 1987, p. 60].)

3 Ernst Mally's *Deontik*

Meinong gave a conceptual analysis of some axiological and normative concepts and investigated their interrelations, but apart from the formulation of the Laws of Omission, he did not attempt to develop a systematic logical theory in the field. Such an attempt was made by his student Ernst Mally [1971] who was inspired by the formal axiom systems of logic developed in the early 20th century, especially by that of Russell and Whitehead's *Principia Mathematica*. He wanted to develop a formal logic for the concepts of ought (*Sollen*) and the concept of willing something (*Wollen*).

According to Mally, judging (*Urteilen*) and willing are distinct attitudes towards states of affairs. Classical logic is concerned with judgments, and Mally proposed to develop a parallel logic for the attitude of willing. Willing that a certain state of affairs p should obtain was expressed by sentence of the form 'It ought to be that p' (p *soll sein*); and Mally thought that Deontik, the logic of ought, can also serve the logic of will (or willing). [Mally, 1971, p. 241] The non-logical signs of Mally's system represent (possible) states of affairs, not actions (or action types), thus his logic may be conceived as a logic of the ought-to-be (*Seinsollen*) rather than a logic of the ought-to-do (*Tunsollen*). In the following discussion we shall frequently not make a distinction between the concepts of *ought* and *being obligatory* (or the concept of obligation), even though these terms are often not interchangeable in ordinary speech [McNamara, 1990; McNamara, 1996c]. Occasionally though we will note issues that arise in assuming they are equivalent.

Mally's deontic logic is based on classical propositional logic. Its vocabulary consists of a sign for the concept of ought, the standard truth-functional connectives, and sentence letters. Here we shall use p, q, r, ... as sentence letters, and the usual symbol **O** as the ought-operator (instead of Mally's '!'). In addition, Mally uses propositional constants for the "unconditionally" or "actually" ("tatsächlich") obligatory, here expressed as '**u**'; for the "negation" of **u** ("das Sollenswidrige"), '**n**'; for what is the case (the facts, '**w**'), and for what is not the case ("das Untatsächliche"), here expressed by the letter '**m**' [Mally, 1971, p. 239, pp. 249-50], as well as an existential quantifier over propositions. [Mally, 1971, pp. 249-50]

Mally reads '$p \to \mathbf{O}q$', as 'p requires q' ('p fordert q'), and abbreviates it '$p\boldsymbol{f}q$', that is,

(Df.$\boldsymbol{f}$) $(p\boldsymbol{f}q) \leftrightarrow (p \to \mathbf{O}q)$.

The arrow '$\to$' is a sign for the truth-functional conditional; '$p \to q$' means that "it is not the case that p and not q." [Mally, 1971, p. 243] Mally adopts the following axioms for the **O**-operator [Mally, 1971, pp. 246-50]:

(MA1) $((p \to \mathbf{O}q) \& (q \to r)) \to (p \to \mathbf{O}r)$
(MA2) $((p \to \mathbf{O}q) \& (p \to \mathbf{O}r)) \to (p \to \mathbf{O}(q\& r))$
(MA3) $(p \to \mathbf{O}q) \leftrightarrow \mathbf{O}(p \to q)$
(MA4) $(\exists \mathbf{u})\, \mathbf{Ou}$
(MA5) $\neg(\mathbf{u} \to \mathbf{On})$

Mally takes (MA4) to mean that there is an unconditionally obligatory state of affairs **u** (ibid., 249), but if '**u**' is a constant, the quantifier in (MA4) is superfluous, and it can be simplified to

(MA4')   **Ou**.

Since **n** is a state of affairs logically incompatible with **u**, (MA5) can be expressed as

(MA5') $\neg(\mathbf{u} \to \mathbf{O}\neg\mathbf{u})$.

Mally calls (MA5) the principle of the consistency of the unconditionally (or actually) obligatory. [Mally, 1971, p. 250]. For the constants **w** and **m**, Mally adopts the principles (or schemata) [Mally, 1971, p. 239]:

(3.1)   For any p, $p \to \mathbf{w}$,

and

(3.2)   For any p, $\mathbf{m} \to p$.

Mally's attempt to systematize deontic logic was unsuccessful, as some of his critics were quick to point out. Karl Menger was the first to show that the consequences of Mally's axioms include the theorem

(3.3) $\mathbf{O}p \leftrightarrow p.$

(For proofs of (3.3) from Mally's axioms, see [Føllesdal and Hilpinen, 1971, p. 4], [Lokhorst and Goble, 2004, pp. 45-6] and [Lokhorst, 2008].) Menger observed that because of (3.1), the introduction of the sign $\mathbf{O}$ for the concept of ought is "superfluous in the sense that it may be cancelled or inserted in any formula at any place you please". ([Menger, 1939, p. 58] quoted from [Lokhorst and Goble, 2004, p. 46] and [Lokhorst, 2008].) We also get the theorem

(3.4) $\mathbf{Ow},$

which Mally takes to mean that a state of affairs that actually obtains ought to obtain, or that "the facts are unconditionally required" [Mally, 1971, p. 266]. Menger also pointed out that Mally's system is incompatible with his own informal remarks on the concept of ought, for example, that $\mathbf{O}(p \vee q)$ is not equivalent to $\mathbf{O}p \vee \mathbf{O}q$; the latter entails the former but the converse does not hold. ([Mally, 1971, p. 260] and [Menger, 1939]) Mally himself thought that some consequences of his axioms are counter-intuitive or "strange", including (3.3) and (3.4), but instead of revising the system, he tried to interpret his theory as the theory of "correct willing", willing in accordance with the facts ("richtiges Wollen"; [1971, pp. 286ff.] and [Lokhorst, 2008].) A. Hofstadter and J. C. C. McKinsey's attempt to develop a logic of imperative discourse (1939) was subject to the same problem as Mally's system; in their system an imperative '!p' (where ! is the imperative operator) turned out to be equivalent to 'p', and the imperative sign was logically superfluous [1939, p. 453].

The interpretation of the constants $\mathbf{u}$, $\mathbf{n}$, $\mathbf{w}$, and $\mathbf{m}$ is unclear. Principles (3.1) and (3.2) suggest that $\mathbf{w}$ should be regarded as the constant *Verum* (or $\top$), a sentence which is true in every possible situation, that is, any tautology. According to this interpretation, (3.4) holds in any normal system of modal (deontic) logic, and is not necessarily objectionable or "surprising", since it then says that whatever cannot not be ought to be. On the other hand, as Menger pointed out, (3.3) would reduce deontic logic to classical propositional logic, and consequently trivialize it.

It is easy to see that Mally's axioms (MA1) and (MA3), unlike (MA2), are not intuitively valid. According to (MA3), a wide-scope obligation $\mathbf{O}(p \rightarrow q)$

is equivalent to a narrow-scope obligation ($p \rightarrow \mathbf{O}q$), but this is clearly not the case. $\mathbf{O}q$ follows logically from the assumptions p and $p \rightarrow \mathbf{O}q$ by Modus Ponens, but not from p and $\mathbf{O}(p \rightarrow q)$. For example, rationality presumably requires that if you believe that the world was made in six days, you believe that it was made in less than a week. But if you believe that the world was made in six days, it does not follow that you ought to believe that it was made in less than a week, on the contrary, you ought not to believe that, because it is false, and you ought not to believe what you actually believe, viz., that the world was made in less than a week. (This example is from [Broome, 2004, p. 29].) It is also easy to find counterexamples to Mally's first axiom. For example, assume that Brutus is your neighbor's bad-tempered dog which is sometimes let outside on his yard. Then the following sentences may be assumed to be true:

(3.5) (i) If Brutus is outside, the gate ought to be closed.
(ii) If the gate is closed, I am not afraid of Brutus.[7]

Then, according to (MA1),

(3.5) (iii) If Brutus is outside, I ought not to be afraid of him.

However, (3.5.iii) is false in a situation in which the gate is not closed, that is, if the requirement in the consequent of (3.5.i) is not satisfied. As [Hintikka, 1971, p. 82] has noted, "One can 'escape' the obligation that r [in (MA1)] simply by failing to carry out the duty expressed by $\mathbf{O}q$."

Some of Mally's successors made similar mistakes in judging the validity of deontic formulas. In a short paper on the logic of imperatives and deontic propositions Kurt Grelling [1939, p. 45] put forward the following rule of deontic logic:

(3.6) If r follows from p and q, the conjunction of p and $\mathbf{O}q$ implies $\mathbf{O}r$

If the expression 'follows' (Grelling's "folgt") is interpreted as a truth-functional conditional, (3.6) can be formalized as

(3.7) $((p \mathbin{\&} q) \rightarrow r) \rightarrow ((p \mathbin{\&} \mathbf{O}q) \rightarrow \mathbf{O}r)$,

---

[7]In examples throughout, as well as in textual discussions of examples and in our expositions of positions, we will often use the first person singular "I" rather than the more cumbersome "we". It is an invitation for the reader to identify with the position being discussed or agent in focus as an expositional tool.

that is,

(3.8) $((p \to (q \to r)) \to (p \to (\mathbf{O}q \to \mathbf{O}r))$.

This schema seems invalid. It may be true that if Brutus is outside, I am not afraid of Brutus if the gate is closed, but false that if Brutus is outside, I ought not to be afraid of Brutus if the gate ought to be closed; on the contrary, I ought to be afraid if the gate is in fact not closed even if it ought to be. (Cf. [Hintikka, 1971, p. 83].) In a note on Grelling's paper Karl Reach [1939, p. 72] observed that if q is replaced by $\neg p$ and r by p, (3.7) becomes

(3.9) $((p \,\&\, \neg p) \to p) \to ((p \,\&\, \mathbf{O}\neg p) \to \mathbf{O}p)$.

The antecedent is a logical truth, thus Grelling's rule implies

(3.10) $(p \,\&\, \mathbf{O}\neg p) \to \mathbf{O}p$,

which means that if something that ought not to be the case is the case, it ought to be the case.

The problems with Mally's system are mainly due to his failure to distinguish wide-scope *oughts* (obligations) from narrow scope *oughts*, and partly to a problematic interpretation of conditionals and the word 'implies'. Mally reads the truth-functional conditional '$p \to q$' as 'p implies ("impliziert") q' [1971, p. 238], but this word often means some kind of strict implication. According to Mally's axiom (MA3), $p \to \mathbf{O}q$ is equivalent to $\mathbf{O}(p \to q)$, and it should be possible to express (3.5.i) as

(3.11.i)   It ought to be that if Brutus is outside, the gate is closed,

using the somewhat artificial construction 'It ought to be that' to indicate a wide-scope ought. In the same way, the suggested conclusion may be expressed as

(3.11.iii)   It ought to be that if Brutus is outside, I am not afraid of him.

If the second premise (3.5.ii) is interpreted as a strict (necessary) conditional or as a *deontic* implication of the form $\mathbf{O}(q \to r)$, that is, as

(3.11.iin)   In all possible circumstances (situations), if the gate is closed, then I am not afraid of Brutus, or as

(3.11.iid)   It ought to be that if the gate is closed, then I am not afraid of Brutus.

the conclusion follows from (3.11.i) and (3.11.ii). If things are the way they ought to be, the gate is closed if Brutus is outside, and I need not be afraid of him. Here the possible circumstances should be taken as the situations which differ from the actual situation only with respect to Brutus's location (in the house or outside on the yard), the gate's being open or closed, and my state of fear or lack of fear. The wide-scope ought $\mathbf{O}(p \to q)$ is *prima facie* a more plausible representation (or partial representation) of the normative relation of requirement between possible states of affairs (Mally's '$p\boldsymbol{f}q$') than the narrow-scope ought, and as we have seen, they are not equivalent. (Cf. [Broome, 1999, pp. 401-5] and [Broome, 2004, p. 29].)

4   On the interpretation of deontic logic

Mally's and Grelling's failure to formulate workable principles of the logic of norms may have reinforced the skepticism expressed by some authors in the late 1930's and early 1940's about the very possibility of the logic of norms and imperatives.

In the late 1930's Jørgen Jørgensen and a number of other philosophers considered the following problem concerning the logic of imperatives and directives. According to the standard conception of logical entailment, a conclusion follows logically from certain premises if and only if the conclusion cannot be false if the premises are true. Thus it is essential for logical inference that the premises and the conclusion are sentences which can be true or false. But since imperatives cannot be said to be true or false, they cannot function as the premises or conclusions of logical inferences, and it is therefore in principle impossible to justify an imperative by means of logical reasoning. [Jørgensen, 1938, p. 184] On the other hand, Jørgensen notes that it seems equally evident that there are "inferences in which one or both premises as well as the conclusion are imperative sentences, and yet the conclusion seems just as inescapable as the conclusion in any syllogism containing sentences in the indicative mood only." [Jørgensen, 1937 and 1938, p. 290] Jørgensen gives the following example (*loc. cit.*):

> Love your neighbor as yourself!
> Love yourself!
> (Therefore:) Love your neighbor!

This seems to be an example of logically valid reasoning with imperatives.

Jørgensen's countryman Alf Ross called this problem "Jørgensen's dilemma" [Ross, 1941, p. 55]. The word 'imperative' should be taken here to refer to an imperative speech act or what is expressed by it, for example, a command or a directive, not to the grammatical mood of a sentence. The word may be regarded here as interchangeable with 'directive' or 'command'. (For imperatives and commands, see [Aikhenvald, 2010, pp. 1-16] and [Lyons, 1977].) It is clear that Jørgensen's dilemma concerns normative discourse in general. Norms cannot be said to be true or false, and if deontic logic is defined as the logic of norms, Jørgensen's dilemma is a problem for deontic logic.

This problem continues to engage philosophers. G. H. von Wright published in the 1990's a paper entitled 'Is There a Logic of Norms?' [von Wright, 1996], and [Makinson, 1999, p. 29] has called Jørgensen's dilemma "a fundamental problem of deontic logic". Von Wright formulated the problem as follows: "Since norms are usually thought to lack truth-value, how can logical relations such as contradiction and entailment (logical consequence) obtain between norms?" [1996, p. 35] (but see also [von Wright, 1983, pp. 130-1]). Makinson observed that "there is a singular tension between the philosophy of norms and the formal work of deontic logicians", because "the usual presentations of deontic logic .... treat norms as if they could bear truth-values", but "it makes no sense to describe norms as true or false" [1999, pp. 29-30].

[Jørgensen, 1937 and 1938, p. 290] suggests two possible ways out of this dilemma.

1. We may widen the concept of logical consequence in such a way that it need not be defined in terms of the concept of truth, but some semantic feature of norms or imperatives which can be regarded as analogous to truth. (Cf. [Grue-Sörensen, 1939, p. 197].) According to this proposal, logic can be said to have "a wider reach than truth" [von Wright, 1957, vii].

2. We might also try to solve the puzzle by defining the concept of validity for reasoning about norms (or imperatives) indirectly, in terms of the truth-values of propositions which are related to norms in a suitable way. In this way of dealing with the puzzle, the logical relations among norms and imperatives are regarded as being constituted by relations among certain propositions associated with them.

[Hofstadter and McKinsey, 1939, p. 447] adopted the first approach, and suggested that the concept of *satisfaction* can replace the concept of truth in the definition of validity and inconsistency for imperatives. An imperative or a directive cannot be said to be true or false, but it can be satisfied or not satisfied by the actions of the addressee. An imperative is satisfied if (and only if) what is commanded is the case. G. H. von Wright has made a similar proposal concerning the logic of norms, and suggested that deontic logic can be understood as "a logic of norm-satisfaction" [von Wright, 1983, p. 130, pp. 138-42]. In another variant of this approach, logical relations among directives are defined in terms of the "validity" (or "correctness") of a directive or a norm so that the concept of (norm) validity plays the same role in the analysis of normative reasoning as the concept truth in "indicative" reasoning. To distinguish this use of the word 'valid' from the concept of logical validity used in the evaluation of an argument ('argument validity'), it may be called 'norm validity'. For example, Alf Ross has argued that our conception of logically valid normative reasoning is based on the concept of norm validity: "The logical deduction of [a directive] I_2 from I_1 then means that I_2 has objective validity in case I_1 has objective validity." [Ross, 1941, p. 59] The validity of a norm means its " 'existence' or 'being in force' – however these expressions are to be understood." [Ross, 1968, p. 175] It has also been suggested that it is possible to distinguish two logics of imperatives and norms, the logic of satisfaction and the logic of validity, which are not the same. ([Segerberg, 1990, p. 203] and [Ross, 1944, pp. 39-43].)

Jørgensen prefers the second approach, following a proposal made by Walter Dubislav. According to [Dubislav, 1937, p. 341], every directive ("Forderungssatz") D is related to a certain statement ("Behauptungssatz") s(D) in such a way that our judgments about the logical relations among directives are determined by the logical relations among the corresponding statements. This proposal may be also be expressed by using the word 'proposition': a directive G can be inferred from D is and only if the proposition s(G) associated with G is a logical consequence of s(D). A set of directives or norms is regarded as inconsistent if and only if the set of the corresponding proposition is inconsistent. What we are inclined to take as logical relations among norms or imperatives are really relations among the propositions associated with the norms, or derived from such relations.

The philosophers who have adopted this conception of the logic of norms have understood the relevant proposition associated with a norm in different ways. According to Jørgensen, an imperative (or directive) can be analyzed into two parts, the "imperative factor" and the "indicative factor". The former indicates that something is commanded or requested, and the latter describes what is commanded, the content of the command [Jørgensen, 1937 and 1938, p. 291]. The indicative factor of the directive

(4.1)   Bertie, pinch the cow-creamer!

can be taken to be the proposition that Bertie pinch the cow-creamer. To indicate that a proposition is not asserted, it may be expressed in a subjunctive form or by an infinitive clause:

(4.2)   Bertie to pinch the cow-creamer.

As C. S Peirce's observed, "the proposition in the sentence, 'Socrates est sapiens', strictly expressed, should be 'Socratem sapientem esse'." [Peirce, 1998, p. 312] The content may also be expressed by an indicative sentence; hence the term "indicative factor". If the imperative factor (or directive factor) is expressed by the exclamation mark '!', (4.1) has (according to Jørgensen) the form

(4.3)   !(Bertie to pinch the cow-creamer).

The distinction between the content and the directive factor of a directive is a special case of the distinction between the illocutionary character and the content of a speech act. If $D = !p$, where p is a proposition, p is the "indicative" (the proposition or statement) $s(D)$ which determines the logical relations of D to other directives, that is,

(4.4) $s(!p)=p$.

According to (4.4), imperative reasoning (or reasoning with directives) as reasoning about their propositional contents:

(4.5)   An imperative $!q$ is said to be derivable from $!p$ if and only if the proposition q is derivable from p.

Here "the imperative factor is so to speak put outside the brackets much as the assertion-sign in the ordinary logic [logic of statements], and the logical operations are only performed within the brackets" [Jørgensen, 1937 and 1938, p. 292]. The logic of imperatives is thus reduced to the logic of statements for which the concept of logical consequence can be defined in the usual way, and "there seems to be no reason for, and hardly any possibility of, constructing a special 'logic of imperatives'." [Ross, 1941, p. 57].

In this way of analyzing imperative inference, the "indicative" $s(D)$ associated with a given directive or norm is assumed to express the propositional content of the directive. Instead of the "indicative factor" we may use the term 'semantic component' to refer to the content which determines the logical relations among directives and norms.

According to Jørgensen, the logic of imperatives can also be based on another way of transforming imperatives into indicatives. In this method, imperative sentences are transformed into statements which say that "the ordered actions are to be performed, resp. the wished state of affairs is to be produced." [Jørgensen, 1937 and 1938, p. 292] According to this proposal, the content of the command "Pinch the cow-creamer!" may be expressed as "The cow-creamer is to be stolen." Thus the semantic content of the command (4.1) is expressed by

(4.6)   Bertie is to pinch the cow-creamer.

The sentence (4.6) does not differ much from (4.2), but it is possible to see a significant difference in meaning. Unlike (4.2), (4.6) can be interpreted as having normative (deontic) content, and can be regarded as equivalent to 'Bertie must pinch the cow-creamer' (or 'Bertie ought to pinch the cow-creamer'), where the word 'must' functions as a deontic operator. If the requirement (or obligation) expressed or created by a command is expressed by the deontic O-operator, (4.6) can be written as

(4.7)   **O**(Bertie to pinch the cow-creamer)

According to this construal of the logic of directives,

(4.8) $s(!p) = \mathbf{O}p$,

where $\mathbf{O}p$ is a deontic proposition. This way of correlating norms and directives with deontic sentences (understood as "indicatives") helps to solve Jørgensen's problem if it is supplemented by an account the truth-conditions or "truth-makers" of deontic propositions. (For the concept of a truth-maker, see [Mulligan *et al.*, 1984; Armstrong, 2004].) How should the meaning of such sentences be understood? Like many logical empiricists of his time, Jørgensen formulated this question as a question about the verifiability of deontic sentences: "How is a sentence of the form 'Such and such is to be so and so' to be verified?" [Jørgensen, 1937 and 1938, p. 292]. His answer was that the phrase "is to be etc." describes a "quasi-property" ascribed to an action or a state of affairs when "a person is willing or commanding the action to be performed, resp. the state of affairs to be produced." According to him, the sentence "Such and such action is to be performed" may be regarded as an abbreviation of the sentence form

(4.9) There is a person who is commanding that such and such action is to be performed.

Sentences of this form state only that some normative source is issuing a certain command. According to this proposal, (4.6) corresponds to

(4.10) It is commanded that Bertie pinch the cow-creamer.

However, it is clear that (4.10) is not equivalent to (or synonymous with) (4.6) or (4.7). It does not necessarily have any normative import.

Nevertheless Jørgensen's proposal suggests a possible truth-maker (or "falsity-maker") for deontic propositions: Often certain speech acts or other actions, for example, the actions of legislative bodies, judicial decisions, and contracts between individuals, function as truth-makers of legal ought-sentences. This does not hold for all deontic propositions, for example for moral *oughts* and obligations. In this case different metaethical theories can be regarded as theories about the truth-makers of deontic propositions. A moral realist may hold the view that "there are objective normative facts, existing independently of our conceptualization and thinking" [Tannsjö, 2010, p. 38], or we may say that an agent has a moral obligation to perform an action if and only if its omission would violate the interests of the persons affected by the action. If the word 'ought' (or 'must') is regarded as an expression of practical necessity or of a prudential ought, (4.6) is true if and only if Bertie's theft of the cow-creamer is necessary for satisfying Bertie's or some other person's current interests or the best way of satisfying such interests. The nature of the truth-makers of ought-sentences depends on the kind of ought (or obligation) under consideration.

Deontic logicians have often made a distinction between two interpretations of deontic sentences. It has been suggested that a deontic sentence of the form $\mathbf{O}p$ can be interpreted normatively (or prescriptively) as expressing a mandatory norm, or descriptively as a statement that it is obligatory that p according to some unspecified system of norms. ([von Wright, 1963, viii, pp. 104-5], [Stenius, 1963, pp. 250-1], [Alchourrón, 1969, pp. 243-5], [Alchourrón and Bulygin, 1971, p. 121], [Hansson, 1971, p. 123], [Bulygin, 1982, pp. 127ff.] and [Alchourrón and Bulygin, 1993, p. 285].) According to [von Wright, 1963, viii]:

> "The deontic sentences of ordinary language ... exhibit a characteristic ambiguity. Tokens of the same sentence are used, sometimes to enunciate a certain prescription (*i.e.*, to enjoin, permit, or prohibit a certain action), sometimes again to express a

proposition to the effect that *there is* a prescription enjoining or permitting or prohibiting a certain action."

For example, the deontic sentence 'The Florida Keys must be evacuated' can function as an evacuation order given by the authorities of Monroe County to the inhabitants of the Keys before the arrival of a hurricane, or as a proposition which gives information about current evacuation orders in South Florida. The announcement 'You may return to your homes' can likewise issue the permission abrogating the prior order, or instead report the fact that the order has already been canceled. Von Wright calls propositions of the latter kind *norm-propositions* or *normative statements* [1963, viii, p. 105]. We shall use for this purpose below the term 'norm-statement'. Norm-statements, unlike the norms themselves, can be said to be true or false, and the logical relationships among them can therefore be understood in the usual way in terms of the concept of truth. The descriptive interpretation of deontic sentences and formulas is essentially the same as Jørgensen's second method of associating indicatives with imperatives. This distinction solves Jørgensen's problem for deontic logic if it is regarded as the logic of normative statements, statements about the existence of norms. ([Stenius, 1963, p. 251] and [Hansson, 1971, p. 123].) However, Carlos Alchourrón and Eugenio Bulygin have made a distinction between the logic of norms and the logic of normative statements, and argued that the logic of descriptively interpreted normative statements differs from the logic of norms (deontic logic proper), and therefore cannot serve as a substitute for the latter, nor can the latter be derived from the former. ([Alchourrón, 1969], [Alchourrón and Bulygin, 1971, pp. 121-7] and [Alchourrón and Bulygin, 1993, p. 285].) They distinguish two sets of deontic operators, the "prescriptive" operators O and P, and the "descriptive" operators $\mathbf{O}_\alpha$ an $\mathbf{P}_\alpha$, where α refers to a system of norms and rules, for example, a certain system of legal or moral norms. According to Alchourrón and Bulygin, the principle of consistency

(4.11) $\mathbf{O}_\alpha p \to \neg \mathbf{O}_\alpha \neg p$

does not hold for the descriptive **O**-operator for a norm system α: $\mathbf{O}_\alpha p$ & $\mathbf{O}_\alpha \neg p$ should be regarded as consistent, because a legislator can promulgate two incompatible norms and a norm system may generate norm conflicts; the existence of such a system is not logically impossible [1993, pp. 290-1] On the other hand, the counterpart of (4.11) for the "prescriptive" Ought is valid, because a norm (prescription) commanding that p be the case is inconsistent with a norm commanding that $\neg p$ be the case [1993, p. 283]. If the logic of normative statements cannot serve as the foundation for the logic of norms, the possibility of (apparent) logical relations between norms and directives must be explained in some other way. Alchourrón and Bulygin introduce for this purpose the concept of *norm-lekton* as the content of a possible prescription. A norm-lekton is related to a prescription in the same way as a proposition to an assertion; the content of a possible assertion is a proposition [1993, pp. 275-6]. The consistency and other logical properties of norms are constituted by the logical properties of norm-lekta and the relations between them; thus the logic of norms (deontic logic proper) is, strictly speaking, the logic of norm-lekta. Moreover, consistency is not a necessary condition for the existence of norms, because a norm-authority can promulgate incompatible norms (i.e., norms with mutually inconsistent lekta) or prescriptions which can lead to conflict situations; hence the difference between the logic of norms and the logic of normative statements (statements about norms). [Alchourrón and Bulygin, 1993, pp. 281-2]

The term 'lekton' ($\lambda\epsilon\kappa\tau\acute{o}\nu$) was used in Stoic logic to refer to the sense of an expression or utterance. The lekta were divided into incomplete and complete lekta. The latter were the contents of complete speech acts, and were divided further into propositions, questions, commands (imperatives), and other kinds. [Mates, 1965, pp. 16-9] Alchourrón and Bulygin's notion of a norm-lekton as the semantic content of a norm agrees in this respect with the Stoic account of lekta.

Alchourrón and Bulygin call the view that norms and directives have norm-lekta as their semantic content the "hyletic conception of norms". According to an alternative view, the "expressive conception" [1981, pp. 95-9][1993, pp. 273-4], the normative component of a norm is not part of its semantic content, but indicates only how the content is presented, that is, as a command or prescription rather than a statement about matters of fact; thus there are no special "norm-lekta" distinct from ordinary descriptive propositions. In the discussion of Jørgensen's problem above, proposal (4.4) represents the expressive conception, (4.8) exemplifies the hyletic conception. According to schema (4.8), the content of the directive or "norm" that Bertie must pinch the cow-creamer (or 'Bertie, pinch the cow-creamer!') is expressed by the deontic sentence **O**(Bertie to pinch the cow-creamer). Instead of calling the content expressed by such a sentence a "norm-lekton" we may call it a deontic proposition. Some philosophers have been reluctant to recognize the possibility that such propositions can have truth-conditions or have confused them with norms, and this has given rise to Jørgensen's problem for deontic logic.

As von Wright notes in the passage quoted above, the distinction between the normative and the descriptive "interpretation" of deontic sentences can be understood as a distinction between two ways of *using* such sentences: they can be used normatively, to create norms, or assertorically to inform the hearer about the content of some system of norms. Jeremy Bentham distinguished between *authoritative* and *unauthoritative* books of "expository jurisprudence": a book is authoritative when it is composed by the legislator, and unauthoritative when it is the work of any other author. [Bentham, 1948, pp. 323-4] [Hedenius, 1941, pp. 65-6] makes a similar distinction between "genuine" and "spurious" legal sentences, and [Kelsen, 1967, p. 355] distinguishes an "authentic" interpretation of law by legal organs from a jurisprudential ("nonauthentic") interpretation: only the former can create law.

The distinction between two ways of using norm sentences can be regarded as a distinction between two kinds of *utterances* of sentences; the "tokens" of a sentence mentioned by von Wright in the passage quoted above are utterances or inscriptions of the sentence. Deontic propositions can be uttered either *performatively*, for creating norms (bringing about an obligation or requirement) or assertorically. [Kamp, 1979, pp. 263-4] In the former case the utterance of the proposition in the appropriate circumstances by a proper norm authority has normative force, and is sufficient to make the deontic proposition in question true, but the truth of an assertoric utterance of the same sentence depends on whether it fits a norm system whose content is determined independently of the utterance in question. The utterer of a deontic proposition can make the intended normative force of the utterance evident by expressing the proposition in the (grammatically) imperative mood or by adding to the utterance the word 'hereby', as in 'You are hereby ordered to pinch the cow-creamer.' Adding the word 'hereby' to the utterance does not change its content. In the case of legal norms and directives, normative utterances include the written inscriptions (occurrences) of norm sentences in authoritative legal texts and documents.

The authoritative (performative) utterances of norm sentences determine the truth-conditions of the deontic propositions which constitute the content of a norm system, and the system derives its normative force from the authoritative utterances of norm sentences which identify the system and tie it to reality. The sense of a deontic proposition can be understood independently of the system to which it belongs, and the same deontic proposition can belong to different systems. Sameness of content is not enough to determine the identity of normative systems; even if α_1 and α_2 contain the same deontic propositions, they are distinct systems if they originate from different normative sources.

The purpose of an "unauthoritative" utterance of a deontic sentence is presumably to convey the content of an existing norm to an audience, and to do this, it must express the same deontic proposition as the original "authoritative" utterance. We may also say that if the former is a replica of the latter, it informs the audience about the normative force of the authoritative utterance. There are no performative utterances of the deontic propositions of general morality, and as was noted earlier, their truth-conditions are determined by the morally relevant objective facts, for example, by the interests of moral subjects.

According to this view, $\mathbf{O}p$ is a complete deontic proposition, and its sense can be grasped independently of the system to which it belongs; thus the same deontic proposition can belong to different systems. The present use of the expression 'deontic proposition' differs from von Wright's and Alchourrón and Bulygin' notion of norm-proposition (norm-statement, a proposition about the existence of a norm). We have to distinguish here the following entities and signs:

(4.12)   (i) A norm N (directive, command, imperative).
         (ii) '$\mathbf{O}p$': a deontic proposition (norm-lekton, norm-content).
         (iii) 'According to a norm system α, $\mathbf{O}p$' or '$\mathbf{O}p$ is part of the content of α'; a normative statement; a proposition which states that a certain norm is part of a norm-system, and conveys the content of the norm.

(For an earlier discussion of (i)-(iii), see [Hilpinen, 2006].) In the present paper we shall not discuss the nature or existence of norms, except to remark that there is a clear conceptual difference between a norm and its content (a lekton or a deontic proposition). Norms, unlike deontic propositions (norm-lekta), are temporal entities which come into existence by the establishment of a customary rule or (in the case of many legal norms) by acts of promulgation, and they cease to exist by acts of derogation or by being replaced with new customary rules. (Cf. [Alchourrón and Bulygin, 1993, pp. 276-8].) Norms cannot be said to be true or false, but as noted earlier, deontic propositions can have truth-makers, and are capable of being true or false.

If normative and descriptive utterances of a deontic proposition $\mathbf{O}p$ have the same sense, the reference to a specific norm system α and the truth-value of the proposition are determined by the context of utterance. Deontic propositions are true or false relative to a context of normative utterances which determine the identity of the relevant normative system. The reference to a system α can be added to $\mathbf{O}p$; in this way we get norm-statements of the form (12.iii).

The logic of norms can be understood as the logic of norm-statements in the way suggested by Alchourrón and Bulygin as determining the possibility of the existence of norms and norm systems, but the principles of the logic of norms can also be regarded as conditions of the consistency or rationality of norm systems. According to the latter interpretation, deontic logic (the logic of deontic propositions) functions simultaneously as the logic of norms and the logic of norm-statements. The logic of imperatives can be understood in the same way if the semantic content of an imperative is formulated by (4.8), not by (4.4); understood in this way, the logic imperatives is the same as the logic of deontic propositions. This provides a solution to Jørgensen's problem.

[Kamp, 1979, p. 264] has observed that the assertoric use of deontic sentences depends on their performative use. This is true in the sense that the performative utterances of deontic sentences determine the truth-conditions of deontic propositions and their assertoric utterances. Therefore the proposal that logical relations among norms can be understood by studying statements of the form (4.9) or (4.12.iii) puts the cart before the horse. Performative utterances of deontic propositions constitute their own "truth-makers", and they also constitute the truth-makers of assertoric (descriptive) utterances of the same propositions. In their performative use, the function of O- and F-sentences (obligation and prohibition-sentences) representing all things considered norms, is to restrict the range of normatively acceptable options ("the field of permissibility") available to a norm-subject (the addressee), whereas permission sentences have the opposite effect; they enlarge the set of normatively acceptable possibilities. An O-sentence $\mathbf{O}p$ excludes all possibilities in which p does not hold, and a permissive utterance $\mathbf{P}p$ enlarges the set of acceptable options in such a way that they include some possibilities in which p is true. ([Lewis, 1979, p. 166] and [Kamp, 1979, p. 264].) Often it does not matter whether a deontic sentence is used performatively or assertorically, because a true assertoric utterances convey the normative content of a norm or directive to an audience, and can guide the agent's actions in the same way as their performative utterances. For example, in the case of a permission sentence, "either the utterance is a performative and creates a number of new options, or else it is an assertion; but then if it really is appropriate it must be true; and its truth then guarantees that these very same options already exist" [Kamp, 1979, p. 264]. The two kinds of utterances are informationally equivalent.

According to the view that the logic of norms is the same as the logic of deontic propositions, the validity conditions of norms are the truth-makers of deontic propositions. These conditions depend on the kind of directive under consideration. In some cases we may assume that "saying makes it so" [Lewis, 1979, p. 166], and the utterance of an imperative by an authority, (for example, "Bertie, pinch the cow-creamer!" uttered by Stiffy Byng) is enough to make it valid and the corresponding deontic proposition 'Bertie is required to pinch the cow-creamer' true. In the case of legal norms the mere utterance of a deontic proposition does not ensure the validity of the norm (or directive) if the utterer does not have the competence to issue the norm in question. The question about the validity conditions of legal norms is one of the central questions of legal philosophy, but it is not a question for the logic of norms; in the logic of norms the possible validity of normative utterances and norms is presupposed.

5 G. H. von Wright's deontic interpretation of modal logic

In the early 1950's Georg Henrik von Wright [1951] revived the old conception of normative concepts as modal concepts, and presented a system of deontic logic in which the concepts of obligation (ought) and permission (may) were represented by modal operators.

In von Wright's [1951] system the usual deontic operators **P** (for 'permitted'), and **O** (for 'obligatory' or 'ought') are prefixed, not to propositional expressions, but to expressions for action-types or (in von Wright's terminology) "act-names". His syntax contains no propositional letters. In this respect he interprets deontic operators in the same way as the 17th and 18th century authors mentioned above in Section 1. Such expressions can be predicated of individual acts, and may be called act-predicates. Thus the deontic operators of von Wright's system are expressions which turn act-predicates into deontic propositions. This interpretation has certain syntactical consequences. If deontic operators are prefixed to names of generic acts (or act-predicates), the iteration of the operators is ungrammatical: '**P**A' and '**O**A' are not act-predicates, and therefore '**OP**A' and '**OO**A' are not well-formed formulas. For the same reason "mixed" sentences such as '$A \to \mathbf{P}A$', in which logical connectives are used to combine deontic and non-deontic components, are not well-formed, since A is (in effect) a predicate, not a proposition. However, mixed sentences and unmodalized propositional expressions are needed, for example, for the representation of conditional norms.

The act-predicates of von Wright's system can be simple (atomic) or complex; and he assumes that complex act-properties are built from atomic predicates by "act-connectives" which are analogous to the classical propositional connectives; thus von Wright speaks of the "negation-act of a given act" and "the conjunction-, disjunction-, implication-, and equivalence-act of two given acts". He assumes that act-predicates have "performance values" analogous to the truth-values of propositions, and the performance-value of a complex act-predicate is determined by the performance-values of its constituents in the same way as the value of a complex propositions is determined by the truth-values of its constituent propositions; for example, a "conjunction-act" $A \& B$ has the value *performed* if and only if A and B

each have the value *performed*.

Von Wright uses the same signs for performance-functions and truth-functions, and adopts the standard interdefinability principles (1.1)-(1.6) (see above) for deontic operators, for example,

(DP0) $\mathbf{O}A \leftrightarrow \neg\mathbf{P}\neg A$,

and two additional principles which he calls the *Principle of Deontic Distribution* and the *Principle of Permission*, using **P** as his deontic primitive:

(DP1) $\mathbf{P}(A \vee B) \leftrightarrow \mathbf{P}A \vee \mathbf{P}B$,

and

(DP2) $\mathbf{P}A \vee \mathbf{P}\neg A$.

Schemata analogous to (DP1) and (DP2) hold for the "ordinary" alethic notion of possibility; thus the P-operator may be said to represent the concept of deontic or normative possibility

Moreover, von Wright adopts the standard inference rules of propositional logic and a modal rule he calls the *Rule of Extensionality*:

(DRE)   If A and B are logically equivalent, $\mathbf{P}A$ and $\mathbf{P}B$ are logically equivalent.

According to the customary use of 'intension', logically equivalent expressions have the same intension, not only the same extension; therefore this rule may be called the Rule of Intensionality. In addition, von Wright accepts a third principle which he calls the *Principle of Deontic Contingency*, according to which

(PDC) $\mathbf{O}(A \vee \neg A)$ and $\neg\mathbf{P}(A\&\neg A)$ are not theorems.

If von Wright's system is formulated as an axiomatic system, (PDC) is superfluous, because $\mathbf{O}(A \vee \neg A)$ and $\neg\mathbf{P}(A\&\neg A)$ do not follow from (DP0)-(DP2) by his rules of inference.

Given the restrictions on well-formed formulas in von Wright's system, all well-formed formulas can be written in the form

(5.1) $F(\mathbf{P}A_1, \ldots, \mathbf{P}A_i \ldots, \mathbf{P}A_n)$, where F is a truth-function, all occurrences of '**O**' and '**F**' have been replaced respectively by '$\neg\mathbf{P}\neg$ ' and '$\mathbf{P}\neg$', and each A_i is a complex act-expression. Let $a_1, \ldots, a_m$ be the simple (atomic) act-predicates in $A_1, \ldots, A_n$, let D_i be the perfect disjunctive normal form of A_i in terms of $a_1, \ldots, a_m$, and let $C^i_{1(i)}, \ldots C^i_{k(i)}$ be the conjunctive constituents of each D_i ($i = 1, \ldots, n$). Each C^i_j contains for every atomic act-predicate, either a_i or its negation, but not both, and no other act-predicates. These conjunctions will be called here the C-constituents of A_i. According to the principle of intensionality (DP4), (5.1) is equivalent to

(5.2) $\quad F(\mathbf{P}D_1, \ldots, \mathbf{P}D_i, \ldots, \mathbf{P}D_n),$

and according to deontic distribution principle (DP1), (5.2) is equivalent to

(5.3) $\quad F((\mathbf{P}C^1_1 \vee \ldots \vee \mathbf{P}C^1_{k(1)}), \ldots (\mathbf{P}C^n_1 \vee \ldots \vee \mathbf{P}C^n_{k(n)})).$

The formulas $\mathbf{P}C^i_j$ ($i = 1, 2, \ldots, n;\ j = 1, 2, \ldots, k(n)$) are called the P-constituents of (5.1). The P-constituents of a deontic formula are logically independent of each other, except that according to the principle of permission, not all of them can be false. Given m atomic act-predicates, there are 2^m different P-constituents and $2^{2^m}-1$ possible truth-value distributions over the P-constituents $\mathbf{P}C^i_j$. Every deontic formula is a truth-function of its P-constituents, and the value of any deontic formula can be determined for each value assignment to P-constituents. This gives a decision method for von Wright's system; a formula is logically true if and only if it is true under every assignment of truth-values to its P-constituents.[8]

The C-constituents C^i_j of deontic formulas are essentially maximally informative descriptions of an agent's possible actions, and as action descriptions they are analogous to Carnap's state-descriptions–complete descriptions of

---

[8]This allows for the computation of a semantic value for a compound *relative to any given* assignment of values to the compound's P-constituents, but it provides no representation of how the latter deontic formulas are given those assignments. This portion of the story is a bit like an empty box in a chart for a more complete semantic picture. In typical Kripke and post-Kripke style (model theoretical) semantics for deontic formulas (to be discussed in Section 7), the assignments of all deontic formulas (including analogs to C-constituents and P-constituents) is at once determined by structural elements in the models, along with an assignment of truth values to all (non-deontic) atomic formulas relative to points in the structure of the models. Thus the empty box left for von Wright's P-constituents is filled in, or "pushed down" to the lower level assignment of values to atomic descriptive sentences at points, and the relation of points to others in the structures. Of course, all this also presupposes the shift that took place from construing deontic concepts as predicates of act descriptions to construing them as connectives operating on sentences.

a possible state of the world (cf. [Carnap, 1956, pp. 9-10].)

Von Wright's "act-names" are not names in the ordinary sense, but rather general terms which can be predicated of individual acts. It might be argued that as predicative expressions they should contain an empty place which can be filled by an individual expression which refers to a particular act, an act-individual. If von Wright's "act-names" are seriously regarded as expressions which can be used for the purpose of characterizing (or "qualifying") individual acts, and deontic concepts are construed as operators by means of which such generic act-expressions are transformed into (deontic) statements, deontic operators should be syntactically analogous to quantifiers which turn open sentences into complete quantified propositions. According to this view, the infinitive clause

(5.4)   Bertie to pinch the cow-creamer

in the deontic proposition

(5.5)   Ought(Bertie to pinch the cow-creamer)

should be understood as an act-qualifying expression: it applies to those individual actions which consist in Bertie's pinching the cow-creamer. Thus the expression in the scope of the Ought-operator may be regarded as containing an empty place or a free variable for individual actions. Consequently the sentence (5.5) can be regarded as a complete (or closed) sentence only if the Ought-operator "binds" the free variable in question; this can be made explicit by writing (5.5) as

(5.6) $(\mathbf{O}x)$(Bertie to pinch the cow-creamer(x)),

where 'x' is an act-variable, a variable for individual acts. In this way deontic operators transform predicative expressions into complete sentences in the same way as quantifiers.

Once act-variables are introduced, it is natural to let ordinary quantifiers perform the function of binding variables and turning predicative expressions into complete (closed) sentences. Thus (5.6) should be regarded as an abbreviation of

(5.7) $\mathbf{O}(\exists x)$(Bertie to pinch the cow-creamer(x)),

where 'x' is an act-variable. In (5.7), the expression 'Bertie to pinch the cow-creamer (x)' is an action predicate, and '$(\exists x)$(Bertie to pinch the cow-creamer(x)' is the proposition that Bertie pinches the cow-creamer. The existential quantifier indicates Bertie's performance of the act of pinching. According to (5.7), (5.5) says that Bertie ought to perform an act of pinching the cow-creamer, where the word 'an' functions as an existential quantifier. This is clearly the intended meaning of (5.5).

In deontic logic, this way of treating action sentences and deontic operators was proposed by Jaakko Hintikka ([1957]; see also [Hintikka, 1971, pp. 63-5, pp. 99-101]. The view of action sentences underlying (5.7) has become familiar from Donald Davidson's [1980b] work on action sentences and the logic of action. However, the English legal philosopher John Austin argued already in the early 19th century that legal norms and rules involve quantification over individual actions and not merely over agents [Austin, 1954, pp. 19-24]. According to (5.7), the grammar of deontic operators is similar to that of other modal concepts: they are sentential operators which can be applied to (action) propositions to make deontic propositions.

(5.7) represents a "positive" obligation (or ought), and the schema

(5.8) $\mathbf{O}(\exists x) A(x)$

can be regarded as a general form of such an obligation. According to the standard interdefinability principles of deontic operators, (1.1)-(1.6), actions of type A are prohibited if and only if

(5.9) $\mathbf{F} \exists x A(x) \leftrightarrow \mathbf{O} \neg \exists x A(x)$
$\leftrightarrow \mathbf{O} \forall x \neg A(x)$
$\leftrightarrow \neg \mathbf{P} \exists x A(x),$

Actions of type A are prohibited if and only if it is obligatory that every action to be performed by the agent not be of type A, in other words, and in less contrived language, it is not permitted to perform any action of that kind. Here the word 'any' has a narrow scope, and should be translated as an existential quantifier. An action A is permitted if and only if it is permitted to perform such an action:

(5.10) $\mathbf{P} \exists x A(x).$

In cases of simple obligation, permission and prohibition, the deontic operator has wide scope, for example, it is clear that the proposition that an agent ought to do A does not mean that some particular act ought to be an instance of A. This does not mean that quantifying in is never intelligible for act-variables in deontic contexts; for example, a promise or a contract creates an obligation to fulfill the promise or satisfy the contract, and the general obligation to keep one's promises seems to have the form

(5.11) $\forall x(Cx \to \mathbf{O}(\exists y S(y,x)))$,

where 'Cx' is the predicate of making a promise, and '$S(y,x)$' means that the (individual) act y satisfies the promise x. (5.11) signifies a claim allowing for detachable, "absolute" obligation. [Hintikka, 1971, p. 100] has suggested that the wide-scope ought-proposition

(5.12) $\mathbf{O}\forall x(Cx \to (\exists y S(y,x)))$,

can serve as a representation of a prima facie commitment. (For some of difficulties related to the quantification into deontic contexts with act-variables, see [Makinson, 1981].)

The fact that deontic operators are attached to complete propositions rather than generic action terms does not mean that the logic in question is a theory of the ought-to-be rather than a theory of the ought-to-do. If the propositions and predicates in the scope of deontic operators are action propositions and predicates, the logic can also be regarded as a logic of the ought-to-do. Deontic sentences can be read in both ways.

6 The standard system of deontic logic (SDL) and close cousins

6.1 SDL

Two alterations in von Wright's initial approach in 1951 easily lead to what soon came to be known as *Standard Deontic Logic*, and thus what became the dominant approach to deontic logic soon after his [von Wright, 1951]. First, if deontic concepts are represented by propositional operators, the limitations on well-formed formulas in von Wright's [1951] system can be dropped, and both mixed formulas (e.g., $p \to \mathbf{O}p$) and formulas with embedded operators (e.g., $\mathbf{O}(\mathbf{O}p \to p)$) can be accepted as well-formed. This brings the interpretation of deontic logic closer to other interpretations of modal logics, for example, the alethic and the epistemic interpretations. Such an approach offers the enormous practical convenience of just adding a new modal operator layer on top of well-known systems of propositional logic (e.g., classical truth-functional systems).[9] Von Wright adopted this

---

[9]The first edition of [Prior, 1955] appears to have initiated this shift after [von Wright, 1951]; of course Mally had used propositional variables two decades earlier. See Section 3.

approach himself at times.[10] However, there was also von Wright's principle of deontic contingency (PDC), whereby a logically impossible act–an act-type identified by a contradictory act-description ($A\&\neg A$), is not necessarily prohibited, and an act-type identified by a tautologous act-description ($A \vee \neg A$) is not necessarily obligatory.[11] However, in von Wright's [1951] system, any act B is permitted if $A\&\neg A$ is permitted, and no act B is obligatory unless a tautologous act $A \vee \neg A$ is obligatory. Thus accepting $\mathbf{O}(A \vee \neg A)$ and $\neg\mathbf{P}(A\&\neg A)$ as logically true excludes only empty normative systems, systems according to which everything is permitted and nothing is obligatory. Many felt this was too small a difference to matter. Moreover, the principle (PDC) is inconsistent (for example) with Leibniz's analysis of an obligatory act as an act which is necessary for a good person to perform, because any act a good person, or indeed any person, can perform, satisfies a tautologous act-description $A \vee \neg A$ of the sort von Wright embraces. In the same way, it is not possible for a good person, or any person, to perform an impossible act; thus such acts should be regarded as not permitted from the standpoint of Leibniz's analysis of obligation and permission. For the sake of theoretical simplicity, as well as continuity with classical propositional logic and modal logic, the deontic necessitation rule, applied to propositions, was routinely included in deontic systems:

(RND)   If p is a theorem, $\mathbf{O}p$ is a theorem

These revisions of von Wright's [1951] system transform it into what is usually called "the standard system of deontic logic", abbreviated 'SDL' ([Føllesdal and Hilpinen, 1971, p. 13] and [Hansson, 1971, p. 122]). The propositional SDL is defined by adding to non-modal propositional logic the modal axiom schemata

(KD) $\mathbf{O}(p \rightarrow q) \rightarrow (\mathbf{O}p \rightarrow \mathbf{O}q)$

and

(DD) $\mathbf{O}p \rightarrow \neg\mathbf{O}\neg p$

and the rule of deontic necessitation (RND). Here the letters p and q can be regarded as representing arbitrary formulas. On the basis of the axioms (KD) and (DD), this system may be called the system KD (or simply D). It is a member of the family of *normal* modal logics, all of which contain

---

[10]For example in his key early revisions of his "old system" in [von Wright, 1964; von Wright, 1965]; see also [von Wright, 1968].

[11]Although it is a theorem of his system that all tautologous act-types are *permissible*. $\mathbf{P}(A \vee \neg A)$, or equivalently that no contradictory one is obligatory, $\neg\mathbf{O}(A\&\neg A)$.

(a counterpart of) the rule RND [Chellas, 1980, p. 114]. With its origins in the 14th century, the "Traditional Definitional Scheme" is routinely taken for granted in formulations of SDL:

(TDS) $\mathbf{P}p =_{\text{df}} \neg\mathbf{O}\neg p$ and $\mathbf{F}p =_{\text{df}} \mathbf{O}\neg p$[12]

The theorems of the system include the formulas:

(6.1) $\mathbf{O}(p\,\&\,q) \to (\mathbf{O}p\,\&\,\mathbf{O}q)$ (*Conjunctive Distributivity of* $\mathbf{O}$)
(6.2) $\mathbf{O}p\,\&\,\mathbf{O}q \to \mathbf{O}(p\,\&\,q)$ (*Aggregation for* $\mathbf{O}$)
(6.3) $\mathbf{O}p \to \mathbf{O}(p \lor q)$ (*Weakening*)
(6.4) $\mathbf{O}(p \to q) \to (\mathbf{P}p \to \mathbf{P}q)$
(6.5) $\mathbf{P}p \to \mathbf{P}(p \lor q)$
(6.6) $\mathbf{P}(p \lor q) \to (\mathbf{P}p \lor \mathbf{P}q)$ (*Disjunctive Distributivity of* $\mathbf{P}$)
(6.7) $\mathbf{P}(p\,\&\,q) \to \mathbf{P}p$
(6.8) $\mathbf{O}\top$ (*ON*)
(6.9) $\neg\mathbf{O}\bot$ (*OD*)
(6.10) $\mathbf{O}p \to \mathbf{P}p$ (*DD'*)
(6.11) $(\mathbf{O}p\,\&\,\mathbf{P}q) \to \mathbf{P}(p\,\&\,q)$
(6.12) $\mathbf{O}p \lor (\mathbf{P}p\,\&\,\mathbf{P}\neg p) \lor \mathbf{O}\neg p$ (*Exhaustion*)
(6.13) $\neg(\mathbf{O}p\,\&\,(\mathbf{P}p\,\&\,\mathbf{P}\neg p))\,\&\,\neg(\mathbf{O}\neg p\,\&\,(\mathbf{P}p\,\&\,\mathbf{P}\neg p))\,\&\,\neg(\mathbf{O}p\,\&\,\mathbf{O}\neg p)$[13]

Two important derivable rules of inference are

(RMD) If $p \to q$ is a theorem, then $\mathbf{O}p \to \mathbf{O}q$ is a theorem,

sometimes called the "Inheritance Principle", as well as "RM", and

(RED) If $p \leftrightarrow q$ is a theorem, then $\mathbf{O}p \leftrightarrow \mathbf{O}q$ is a theorem,

a deontic "equivalence rule".

---

[12] Letting **OB**, **PE**, **IM**, **OP**, **OM** stand respectively for it is *obligatory* that, *permissible* that, *impermissible* that, *optional* that, and *omissible* that, the more extended scheme would be: $\mathbf{PE}p =_{\text{df}} \neg\mathbf{OB}\neg p$ and $\mathbf{IM}p =_{\text{df}} \mathbf{OB}\neg p$, $\mathbf{OP}p =_{\text{df}} (\neg\mathbf{OB}p\,\&\,\neg\mathbf{OB}\neg p)$, and $\mathbf{OM}p =_{\text{df}} \neg\mathbf{OB}p$.

[13] (6.13) expresses the exclusiveness of the three classes that (6.12) says are exhaustive. The conjunction of (6.12) and (6.13) expresses the "Traditional Threefold Classification" asserting that every alternative is either obligatory, optional or impermissible, and no more than one of these. Using $\underline{\lor}$ for the *exclusive or*, this can be succinctly expressed as: $\mathbf{OB}p \underline{\lor} \mathbf{IM}p \underline{\lor} \mathbf{OP}p$. See [McNamara, 1990; McNamara, 1996a] on the significance of this feature of the traditional scheme.

As expected, if we recast von Wright's 1951 system, now construing "**P**" and "**O**" as propositional operators (not predicates of act-types), and the variables as propositional variables (not variables for act-types), the key principles and rules mentioned earlier (DP0-DP2 and DRE) are all easily derivable in the SDL system above.

It should be noted that some principles not derivable in SDL were often deemed truths of deontic logic, especially if the "**O**" is interpreted as "It ought to be that". Perhaps the most salient example of this kind is the principle: **O**(**O**p → p). This can be construed as saying (roughly) that it ought to be the case that whatever ought to be is.[14]

It is important to note here that we use throughout "Standard Deontic Logic" and its abbreviation, "SDL", as proper names, not as descriptions. Many think that these are misnomers, not quite as bad as the "Holy Roman Empire", which fails on all three counts, but surely not quite the "standard" the label might suggest. It is indeed probably fair to say that most researchers think there is at least some thesis of SDL that on some prominent interpretation of "**O**" is not a logical truth at all, and furthermore there are a number of somewhat independent such complaints about SDL. So it is hardly a widely popular system of logic with only occasional outliers rejecting it as the title might suggest. Rather, it is the most widely known, well-studied system, and central in the accelerated historical development of the subject over the last 50 or so years. As such, it serves as a historical comparator, where various important developments in the subject were explicit reactions to its perceived shortcomings, and even when not, sometimes can be fruitfully framed as such.

6.2 The Leibnizian-Kangerian-Andersonian reduction

It is easy to see that Leibniz's definition of the concept of obligation (or ought),

(O.Leibniz) p is obligatory for S if and only p is necessary for S's being a good person,

can be seen as supporting the principles of SDL when conjoined with plausibly intended assumptions about the possible instantiation of goodness and about the notion of necessity involved. If the explicit reference to the agent is suppressed, (O.Leibniz) can be expressed in the form

(O.GWL) $\mathbf{O}p \leftrightarrow \mathbf{N}(g \to p)$,

where '**N**' is an alethic necessity operator and 'g' represents a proposition

---

[14]See [Kanger, 1971; Hintikka, 1971].

which expresses the agent's being a good person or, in the case of the ought-to-be, the goodness of the world.

The Leibnizian concept of permission (the concept of may) is defined by

(P.GWL) $\mathbf{P}p \leftrightarrow \mathbf{M}(g \& p)$.

These schemata can be regarded as partial reductions of deontic logic to alethic modal logic.

It should be observed that Leibniz's definition of 'obligation' in terms of 'good' is prima facia not subject to the same difficulty as one main definition of normative concepts in terms of a comparative concept of goodness as applying to any state of affairs that is better than its negation, as with (D.OÅqv).[15] Thus (O.GWL) leaves room for supererogatory actions.[16] If deontic logic is regarded as a theory about the ought-to-be rather than ought-to-do, the Leibnizian interpretation of $\mathbf{O}p$ may be expressed (for example) as 'for things to be best, it is necessary that p' or 'for things to be apt, it is necessary that p'.

If it is assumed that it is possible to be good or that the requirements of morality can be satisfied (if it is possible for things to be in order), that is,

(M.g) $\mathbf{M}g$,

the $\mathbf{O}$-operator defined by (O.GWL) satisfies all the principles of SDL, provided that the $\mathbf{N}$-operator satisfies the axioms of the modal system called T in [Chellas, 1980, p. 131], viz.

(K) $\mathbf{N}(p \to q) \to (\mathbf{N}p \to \mathbf{N}q)$

and

(T) $\mathbf{N}p \to p$

and the modal "rule of necessitation", viz.

(RN)   If p is a theorem, $\mathbf{N}p$ is a theorem

---

[15] There are complexities here. If we interpret "$\mathbf{O}$" as "ought", then (D.OÅqv) is more plausible, since it is plausible that "ought" is some sort of optimizing notion, unlike "must" or "obligatory". So if p is better than not p, then it plausibly does follow that it ought to be that p, even though it need not be a must that p or obligatory that p. The Leibnizian reading of $\mathbf{O}$ is also better for "must" or "obligatory". Notice however that if we assume Leibniz means a "perfectly good" man, we end up again with an optimizing notion, where it is plausible to now see "$\mathbf{O}$" as ought" not "obligatory", and as once again in tension with supererogation. This is sometimes a problem in virtue ethical attempts to analyze permissibility and obligation while allowing for supererogation.

[16] See [McNamara, 1999] for one development along this line.

then (O.GWL) guarantees the validity of all principles of SDL.[17] Axiom (M.G) is needed for proving the consistency principle (DD).

In the 20th century deontic logic, the Leibnizian analysis of the concepts of obligation and permission was rediscovered by the Swedish philosopher Stig Kanger [1971, pp. 53-4]. Kanger interpreted the constant g as "what morality prescribes". According to this interpretation, $\mathbf{O}p$ (it is obligatory that p) means that p follows from the requirements of morality. Alan Ross Anderson [1967] put forward a reduction schema equivalent to Kanger's[18],

(O.S) $\mathbf{O}p \leftrightarrow \mathbf{N}(\neg p \to S),$

where S may be taken to mean the threat of a sanction or simply the proposition that the requirements of law or morality have been violated. If p is an action proposition and the negation sign is understood as the omission of the action expressed by p, (O.S) is equivalent to Meinong's schema (2.8).

It should be noted that in order to validate all of SDL in this "reduction", the principle T above is overkill, the normal modal system K with just an axiom saying G is possible is sufficient to generate all of SDL, and if we do add axiom T, this results in a stronger deontic fragment, namely SDL with the addition of $\mathbf{O}(\mathbf{O}p \to p)$, which we noted above is not derivable in SDL. However, it is also surely correct that the intended reductions of Leibniz, Anderson, and Kanger are ones where the notion of necessity involved is *alethic* necessity, and so ones for which the T thesis above holds. Thus from the standpoint of the Leibnizian, Kangerian, and Andersonian reductions, SDL is *too weak* a system for deontic logic, and needs to be augmented with the addition of $\mathbf{O}(\mathbf{O}p \to p)$. As noted above, others also thought this additional might be needed for independent reasons. Lastly we should note that the resulting "reduction" also allows for mixed modal-deontic formulas, and formulas involving deontic operators and the deontic constant, g. We note a few salient ones that are theses:[19]

(6.14) $\mathbf{O}g$

---

[17]See [Åqvist, 2002; Åqvist, 1987] for proofs of correspondences between SDL and its extensions and normal modal systems employing the Leibnizian-Kangerian-Andersonian style reduction.

[18]The arrow is interpreted as material implication here. Anderson also explored alternative interpretations using non-truth functional conditionals, such as relevant logic conditionals.

[19]Note that if the reduction is treated as offering an analysis of obligation (6.14) can have a meta-obligatory flavor, saying something like it is obligatory that all one's obligations are met; if we read g *à la* Leibniz, it says it is obligatory that what a good person would do is done.

(6.15) $\Box(p \to q) \to (\mathbf{O}p \to \mathbf{O}q)$ (RM′)
(6.16) $\Box p \to \mathbf{O}p$ (RND′)
(6.17) $\mathbf{O}p \to \Diamond p$ (A weak version of "Kant's Law")[20]
(6.18) $\neg\Diamond(\mathbf{O}p \& \mathbf{O}\neg p)$ (DD′)

The supplement to this section provides some more formal details.[21]

Supplement to Section 6: Some formalities

A6.1 The SDL wffs (well-formed formulas)

PV is a set of sentence letters $P_1, ..., P_i, ...$ – where "i" is 1,2,.... There are three primitive propositional (sentential) operators: $\neg$, $\to$, **OB** (for it is **ob**ligatory that); and a pair of parentheses: (,).

Let the set of SDL-wffs be the smallest set such that (lower case "p" and "q" are metavariables ranging over formulas):

1. PV is a subset of the SDL-wffs

2. For any p, p is among the SDL-wffs only if $\neg p$ and **OB**p are as well.

3. For any p and q, p and q are in SDL-wffs only if $(p \to q)$ is in SDL-wffs.

We shift to two letter abbreviations here and in a few other select places to make it easier to express more operators (as well as for mixing these with other operators later on): **OB, PE, IM, OM, OP** for it is *obligatory* that, *permissible* that, *impermissible* that, *omissible* that, and *optional* that, respectively. It will be clear when this shift is in play, so no ambiguity or confusion should result.

We use the following abbreviations for formulas and subformulas of the SDL-wffs:

(DF 1) &, ∨, ↔ as usual.

---

[20] Kant's law is more accurately rendered as involving agential possibility (agential ability), not merely impersonal possibility, although it is possible to move closer to this by reading the modal operators in the reduction as keyed to what is predetermined and possible for a given agent. See [McNamara, 2000]. [Hilpinen, 1969] showed how this sort of relativization can be easily done explicitly for modal logics, though authors often leave it implicit.

[21] We will specify one other logic (VW) in tandem with its semantics, in Section 7, and in the next supplement.

(DF 2)   **PE**$p =_{df}$ ¬**OB**¬p ("it is permissible that").
(DF 3)   **IM**$p =_{df}$ **OB**¬p ("it is forbidden/impermissible that")
(DF 4)   **OM**$p =_{df}$ ¬**OB**p ("it is omissible/gratuitous that"
(DF 5)   **OP**$p =_{df}$ (¬**OB**p & ¬**OB**¬p)   ("it is optional that")

A6.2 SDL and one extension

We assume the language is that specified in 6.1. We provide this standard axiomatization of SDL, where ⊢ before a formula indicates it is a thesis (axiom or theorem) of the relevant system:

(A1)   All tautologies of the language   (TAUT)
(A2)   **OB**$(p \to q) \to ($**OB**$p \to$ **OB**$q)$ (KD)
(A3)   **OB**$p \to$ ¬**OB**¬p (DD)
(MP)   If ⊢ p and ⊢ $p \to q$ then ⊢ q
(RND)   If ⊢ p then ⊢ **OB**p

This is essentially just the normal modal logic D, with a notational variant to indicate the deontic interpretation.

Let's also introduce one extension of SDL, called here contextually, SDL$^+$. SDL$^+$ is the system that results from adding just A4 to SDL:

(A4)   **OB**(**OB**$p \to p$)

A6.3 Two Leibnizian-Kangerian-Andersonian systems

PV is a set of sentence variables $P_1, ..., P_i, ...$ – where "i" is 1,2,... There are three primitive propositional (sentential) operators: ¬, →, □. There is a distinguished propositional constant g, and a pair of parentheses, (and).

One can read g a variety of ways (e.g. as "all normative demands are met"), but we use "g" to honor Leibniz who essentially is the first to analyze the basic operators in "reductive terms", essentially a kind of virtue ethical reduction: in terms of what it is necessary for a *good* person to do. We can think of g this way as expressing the proposition that *what a good person would do is done*.

Let the set of <u>LKA-wffs</u> be the smallest set such that:

1. g is among the LKA-wffs.
2. PV is a subset of the LKA-wff, and
3. For any p, p is among the LKA-wffs only if ¬p and □p are among the LKA-wffs.

4. For any p and q, p and q are in LKA-wffs only if $(p \to q)$ is in LKA-wffs.

We use the following abbreviations for formulas and subformulas of the LKA-wffs:

(DF 1') $\&, \vee, \leftrightarrow$ as usual.
(DF 2') $\Box p =_{df} \neg \Diamond \neg$ ("it is possible that").
(DF 3')   **OB**$p =_{df} \Box(g \to p)$
(DF 4')   **PE**$p =_{df} \Diamond(g \& p)$
(DF 5')   **IM**$p =_{df} \Box(p \to \neg g)$
(DF 6')   **OM**$p =_{df} \Diamond(g \& \neg p)$
(DF 7')   **OP**$p =_{df} (\Diamond(g \& p) \& \Diamond(g \& \neg p))$

The logic LKA$_1$ (essentially the Normal Modal Logic K with (A3') added):

(A1')   All tautologies of the language   (TAUT)
(A2') $\Box(p \to q) \to (\Box p \to \Box q)$ (K)
(A3') $\Diamond g$ $(\Diamond g)$
(R1')   If $\vdash p$ and $\vdash p \to q$ then $\vdash q$ (MP)
(R2')   If $\vdash p$ then $\vdash$ **OB**p (RN)

Metatheorem 6.1 *(Loosely stated) SDL is the strongest pure deontic fragment contained in LKA$_1$.*

The Logic LKA$_2$ (essentially the normal modal logic T with A3 added). LKA$_2$ is the system that results from adding just (A4') to LKA$_1$:

(A4') $\Box p \to p$ (T)

This logic just adds the characteristic T axiom to LKA$_1$, and Leibniz, Kanger, and Anderson all had alethic necessity in mind. The only reason we give the prior K version, is that (LKA$_2$) generates a pure deontic logic stronger than SDL, namely SDL$^+$.

Metatheorem 6.2 *(Loosely stated) SDL$^+$ is the strongest pure deontic fragment contained in LKA$_2$.*

See [Åqvist, 2002; Åqvist, 1987] for the exact statements of the metatheorems, as well as their proofs, which are semantic in nature.

[Lokhorst, 2006] explores the correlation between quantified propositional variable systems reading g as "all normative demands/obligations are met", especially in the context of Anderson's original reduction using relevance logic. [McNamara, 1999] extends this approach to cover action beyond the call of duty and other concepts. The ideas of the reduction have been employed and modified in a variety of contexts, as a perusal of the DEON conference volumes indicates.

7 The semantics of Standard Deontic Logic and close cousins

7.1 The semantics for SDL

The sentences of SDL can be interpreted in terms of possible situations or world states ("possible worlds") in the same way as other normal modalities. A possible worlds model of SDL is a triple $M = \langle W, R, I \rangle$, where W is a universe of possible situations, also called points of the model, R is a binary relation on W, and I is an interpretation function which assigns to each sentence letter of the modal language a subset of W, that is, the points $u \in W$ at which the sentence letter is to be deemed true. The truth of any formula p at u under M is then expressed briefly as '$M, u \models p$', or even more briefly, where the model is left tacit, '$u \models p$', and defined recursively. If p is not true at u, it is false at u ($u \not\models p$). A sentence is called *valid* (logically true) if and only if it is true at every situation $u \in W$ for every model M, and q is a logical consequence of p if and only if there is no model M and world u such that $M, u \models p$ and $M, u \not\models q$. Truth at a world in a model, $\models$, is defined in accord with the usual Boolean conditions which ensure that the truth-functional compounds of simple sentences receive appropriate truth-values at each possible world. The alternativeness relation R is needed for the interpretation of sentences involving the deontic operators. In the semantics of modal logic, necessary truth at a given world u is understood as truth at all points which are possible relative to u or are *alternatives* to u, and possible truth at u means truth at some alternative to u. For the concepts of deontic necessity or obligation (sometimes read as "ought") and deontic possibility or permission (may), these conditions can be formulated as follows:

(CO) $u \models \mathbf{O}p$ if and only if $v \models p$ for every $v \in W$ such Ruv,

and

(CP) $u \models \mathbf{P}p$ if and only if $v \models p$ for some $v \in W$ such Ruv.

To ensure the validity of axiom DD, it is necessary to regard R as a serial relation, in other words,

(CD)   For every $u \in W, Ruv$ for some $v \in W$.

Further assumptions about the structural properties of the R-relation validate different deontic principles, and lead to different systems of deontic logic. For example, it is clear that

(7.1) $\mathbf{O}p \to p$

is not a logical truth as interpreted, and therefore R cannot be assumed to be a reflexive relation, but reading "**O**" as it ought to be that, the principle

(7.2) $\mathbf{O}(\mathbf{O}p \to p)$

seems a valid principle: It ought to be the case that whatever ought to be the case is the case. The validity of (7.2) follows from the assumption that R is secondarily reflexive, in other words,

(COO)   If Ruv for some u, then Rvv.

The semantics sketched above may be termed the "standard semantics" of deontic logic. [Hintikka, 1957; Hintikka, 1971; Kanger, 1971] were among the first philosophers who used an alternativeness relation between possible worlds or situations to formulate the truth-conditions of deontic sentences. It is also more generally referred to as "Kripke semantics" or "Kripke style semantics" in modal logic [Kripke, 1963].[22]

Recalling our earlier discussion of Meinong's scheme and his use of axiological preference ordering concepts, let us briefly note here that a more axiological background semantic picture for SDL, one having affinities with utilitarianism, was often endorsed. Suppose we have a set of relations, one for each world u in W, where $v \geq_u w$ is thought of as indicating that v is ranked at least as high as w relative to u. Suppose further that we assume that for each world, u in W:

1. $v \geq_u v$ (reflexivity),
2. if $v \geq_u w$ and $w \geq_u x$ then $v \geq_u x$ (transitivity),
3. either $v \geq_u w$ or $w \geq_u v$ (connectivity).[23]

---

[22] An excellent source on the history of the emergence of formal semantic frameworks for deontic logic is [Wolenski, 1990].

[23] So each $\geq_u$ is a total pre-ordering of W (sometimes called a complete pre-ordering or total quasi-ordering).

Now suppose we add something else people often assumed, the "Limit Assumption":

(LA) For each u, there is v such that for any $w, v \geq_u w$

That is, relative to any world u, there is always a world ranked at least as high relative to u as any worlds (i.e. there is at least one u-best world). Lastly, we use this framework to provide a truth clause for **O** via bests:

(COB) **O**p is true at u iff p holds in all the u-best worlds

It was widely recognized that this approach will also determine SDL, but the metatheory of SDL and related systems via generalized ordering semantics has not been very widely explored compared to Kripke-style semantics.[24] Essentially, this ordering framework provides a way to *generate* the set of u-acceptable worlds out of the ordering:

v is u-acceptable iff v is a u-best world

So we get a u-acceptability relation for each world, just as is presupposed in the standard Kripke semantic structures. We need only look at what propositions hold at the u-best worlds to interpret the truth-conditions of SDL's deontic operators exactly as we did with the simpler Kripke relational structures. If the reader wonders about how seriality in the Kripke structures is captured, it is guaranteed by the Limit Assumption. For that assumption entails that for each world u, there is a u-best world. As such, the ordering semantics is overkill for SDL and most of its resources go unutilized for SDL; but as we will see a bit later, when people started thinking about how to generalize or adjust SDL to handle more complex deontic concepts, these sorts of ordering structures (and their generalizations) became quite important, as was the recognition that SDL itself could be easily subsumed under such ordering frameworks.

7.2 Semantics for the Leibnizian-Kangerian-Andersonian reduction

We now use the Kripke-style semantics to turn back to the Leibnizian-Kangerian-Andersonian reduction. Assume we have a classical propositional language with a distinguished propositional constant, g, and the modal operator, $\Box$, intended as expressing alethic necessity (with $\Diamond$ defined as $\neg\Box\neg$). Then as with the semantics above, $\langle W, R, I \rangle$ will be a model, with W interpreted as a set of points or worlds, R a binary relation on W, and I

---

[24]But see [Goble, 2003; Åqvist, 1987; Spohn, 1975; Jennings, 1974] and in a slightly different setting, [Lewis, 1973; Lewis, 1974].

an interpretation assigning a subset of W to each sentence letter. Truth in a model at a world is defined just as we did above, where for the necessity operator, we have:

(C□) $u \models \Box p$ if and only if $v \models p$ for every $v \in W$ such Ruv

For the moment, we place no structural constraint on R. How do we interpret g, the only element in the reduction that has a deontic or valuative flavor? As a propositional constant, let's read it as *what a good person would do is done*. This is close to Leibniz. With this in mind, we then interpret g by having I assign it a subset of W, with the intention that these are the worlds where what a good person would do is done. We thus add one more element to the models, $\langle W, R, G, I \rangle$, and add the constraint:

(CG) $G \subseteq W$.

Then g will be true at a world u in a model if and if u is a G-world:

(Cg) $u \models g$ if and only if $u \in G$

We need one structural constraint to validate the axiom $\Diamond g$, to the effect that it is always possible in the models that what the good person would do is done. At the semantic level this amounts to adding an analog to seriality used for the semantics for SDL, namely a constraint to the effect that for every world u, u has accessible to it a world where what the good person would do is done:

For every $u \in W$, there is a v, such that $v \in G$ and Ruv.

With the definitions identified in the last section, this semantical system will validate all the pure deontic principles of SDL, along with other mixed principles such as **O**g and **O**$p \to \Diamond p$. However, given the intended interpretation of the $\Box$ as expressing *alethic* necessity, and thus as supporting the T axiom,

(T) $\Box p \to p$

we would need to add reflexivity,

(Rflx)   For every u $\in W, Ruu$

thus generating a pure deontic fragment stronger than SDL, but that is clearly a consequence of the intended interpretation of the reduction.

7.3 A generalization of SDL: VW logics and their non-standard semantics

We now sketch a slight generalization of the standard framework for SDL, one that allows us to include a weakening of SDL that accords with one key aspect of von Wright's earliest work. Recall that in opening this section, we noted that von Wright interpreted "**O**" and "**P**" as act predicates, and that he also endorsed "Deontic Contingency" (PDC), thereby rejecting **O**⊤ and ¬**P**⊥ and their equivalents as logical truths. The former act predicates issue seems separately motivated, and von Wright himself later flirted with treating "**O**" and "**P**" as propositional operators. These facts raise the interesting question: What might a propositional deontic framework look like which treats "**O**" and "**P**" as propositional operators and is as close to SDL as is consistent with *Deontic Contingency*? It turns out that there is a simple such syntactic and semantic framework, one that is a conservative generalization of that for SDL.[25] The language is that of SDL. In honor of von Wright, let's call the base logic VW:

(A1)   All tautologies of the language   (TAUT)
(A2)   **O**$(p \to q) \to ($**O**$p \to $**O**$q)$ (KD)
(A3)   ¬**O**⊥   (OD)
(MP)   If ⊢ p and ⊢ $p \to q$ then ⊢ q
(RM)   If ⊢ $p \to q$ then ⊢ **O**$p \to $**O**$q$ (RMD)

Although **O**$p \to $ ¬**O**¬p is easily derivable from VW, neither **O**⊤ nor ¬**P**⊥ is derivable, but VW is easily derivable from SDL. If we add **O**⊤ to VW, we get a system equipollent to SDL. But how will the semantics for **O** work? It can't be standard, else **O**⊤ would be validated, and thus it would need to be a theorem for the logic to match the semantics.

The basic idea is simple. The model structures are those for normal modal logics like K and SDL above, with no structural constraints on the accessibility relation. The key difference at the semantic level is that the clause for **O** is non-standard:

(CO') $u \models $ **O**p iff there is a v such that Ruv and for every v
   if Ruv, then $v \models p$

Thus something is obligatory at u iff there is a u-acceptable world to begin with, and all such worlds are p-worlds. Plainly, if there is no u-acceptable

---

[25]This basic orientation appeared in [McNamara, 1990], but the elementary metatheory was done in [McNamara, 1988]. Max Cresswell pointed out in conversation that there are affinities to the [Kripke, 1965] treatment of some non-normal modal logics using "non-normal worlds". See also [Cresswell, 1967].

world, $u \not\models \mathbf{O}p$ for all p (including $\top$), so $\mathbf{O}\top$ is *not* valid. With $\mathbf{P}p$ defined as $\neg\mathbf{O}\neg p$, we get this non-standard clause for $\mathbf{P}$:

(CP') $u \models \mathbf{P}p$ iff either there is no v such that Ruv or
$v \models p$ for some v such that Ruv.[26]

But then if there is no v such that Ruv, $u \models \mathbf{P}p$, for any p (including $\bot$), and so $\neg\mathbf{P}\bot$ is *not* valid.

Notice however that $\mathbf{O}p \to \neg\mathbf{O}\neg p$, which is derivable in VW, is also valid in all models per the clause above. For suppose the antecedent is true at u. Then there is a p-world that is a u-alternative and all u-alternatives are p-worlds, but if so, $\mathbf{O}\neg p$ must be false at u for otherwise there would have to be at least one u-alternative that was both a $\neg p$-world and a p-world.

What then is the role of seriality if $\mathbf{O}p \to \neg\mathbf{O}\neg p$ is already validated without any constraints? Given the non-standard clause for $\mathbf{O}$, if we add seriality as a constraint, "$\mathbf{O}\top$" is then validated, SDL is determined, and all is "back to normal". Adding seriality essentially assures that the first clause in the non-standard truth definition of "$\mathbf{O}$" above is automatically met, and so the conjunctive clause is then equivalent to the standard one (the right conjunct above). Similarly, for the clause for $\mathbf{P}$: the first clause is excluded by seriality, so the disjunctive clause is equivalent to the familiar one (the right disjunct above). Thus the framework is a conservative generalization of the standard one for SDL, but one where von Wright's contingency intuitions can be modeled, and so nothing need be guaranteed obligatory or impermissible in the base logic, since there need not be, as it were, any normative standard at all, though in keeping with von Wright's intuitions, if anything is obligatory at all, then so too will $\top$ be; and semantically, that anything at all is obligatory at u amounts to saying u has some standard, namely a non-empty set of u-acceptable ways things might be. It also seems to us more fitting to frame things this way given the place of von Wright's work in stimulating the emergence of SDL, and the fact that his principle of contingency is really separate from his initial conception of "$\mathbf{O}$" and "$\mathbf{P}$" as predicates. We also note in passing that had propositional deontic logic originally been conceived this way, the base deontic logic playing the role SDL now plays would not have been a normal modal logic (although obviously a close cousin).

---

[26] Similarly for the remaining three operators defined above: $u \models \mathbf{IM}p$ iff there is a v such that Ruv and for every v, if Ruv then $v \models \neg p$; $u \models \mathbf{OM}p$ iff either there is no v such that Ruv or $v \models \neg p$ for some v such that Ruv; $u \models \mathbf{OP}p$ iff either there is no v such that Ruv or both $v \models \neg p$ for some v such that Ruv and $v \models p$ for some v such that Ruv.

7.4 Classical quantification and SDL semantics

Most presentations of deontic logic are restricted to propositional logic. This is a serious and unnecessary limitation; as was observed earlier (in Section 5), some normative propositions and relations can be formalized in a plausible way by combining deontic operators and quantifiers. The semantics outlined above can be extended in an obvious way to quantified deontic logic by adding to our formal language quantifiers, predicative expressions, individual variables, individual parameters (arbitrary names), and functional expressions which can be used for generating complex individual terms from simple terms. The models of quantified SDL are structures $\langle W, R, U, D, I \rangle$, where W is a universe of possible situations ("worlds"), R is a binary alternativeness (accessibility) relation between situations (as with propositional SDL), U is a set of individuals, D is function which assigns a subset of U to each $v \in W$ —the individuals existing in v, $D(v)$, and the interpretation function I assigns to each non-logical expression (individual term, predicative expression, or functional expression) the intension of the expression, that is, a function from possible situations to extensions or referents:

(7.3)   (i) For a simple individual term (parameter) c, $I(c, u) \in D(u)$.
(ii) For each n-place predicate G, $I(G, u)$ is a set of ordered sets of n individuals (n-tuples) $\langle i_1, i_2, ..., i_n \rangle$, where each $i_j \in D(u)$, that is, $I(G, u)$ is a subset of $D(u)^n$.
(iii) For each function symbol f, $I(f, u)$ is an operation on $D(u)$, that is a function which has $D(u)$ as its domain as well as its range of values.

For example, if G is the relation of loving, $I(G, u)$ is the set of all ordered pairs of individuals $c \in D(u)$, $d \in D(u)$ such that c loves d in the situation (world) u.

The truth-conditions of quantified sentences may be defined in some standard way, for example, in terms of variant interpretations (cf. [Bostock, 1997, pp. 85-6] and [Mates, 1965, pp. 54-6]). If M is a model with an interpretation function I, let M/c be a model with an interpretation function I/c, called the *c-variant* of I, which is like I except that it may assign a different individual to the singular term c; thus I/c and M/c differ from I and M at most with respect to the value of c. As long as c does not appear in the formula (open sentence) Φ, c can be regarded as denoting any arbitrary individual in the relevant domain under some variant of I, I/c, and the sentence $\forall x \Phi$ is then true under I if and only if the sentence $\Phi(c/x)$ obtained by substituting c for x in Φ is true under every c-variant of I. Thus the truth-conditions of quantified sentences can be expressed as follows:

(7.4) $M, I, u \models \forall x \Phi$ if and only if $M/c, I/c, u \models \Phi(c/x)$ for every c-variant I/c of I such that $I/c(c) \in D(u)$, where the parameter c does not appear in Φ.

(7.5) $M, I, u \models \exists x \Phi$ if and only if $M/c, I/c, u \models \Phi(c/x)$ for some c-variant I/c of I such that $I/c(c) \in D(u)$, where the parameter c does not appear in Φ.

The truth-conditions of other complex sentences and atomic sentences are defined in the standard way, except that they are relativized to possible situations, and the meanings of deontic operators are defined in the same way as in propositional deontic logic.

The individual variables of quantified deontic logic may be interpreted as variables for individual actions, as in the examples discussed in Section 5, or as variables for agents, and the domains $D(u)$ may be interpreted in the similar ways. (See [Åqvist, 1987, pp. 84-5].)[27] The function $D(u)$ may be a constant function, in which case all situations $u \in W$ have the same domain of individual objects, or the situations may involve different domains. The deontic counterpart of the Barcan formula,

(7.6) $\forall x \mathbf{O} G x \to \mathbf{O} \forall x G x,$

states that if G is obligatory (or required) for everyone, then it ought to be so that everyone satisfies G. It is clear that this inference is not valid in all applications, but it is valid in all models with constant domains. For suppose the antecedent is true at u, then for each individual i at u, at each deontic alternative to u, i satisfies G. But if the domains at each world are the same, then the individuals from u satisfying G at the alternatives are all the individuals there are at the alternatives. So at each u-alternative, everyone individual satisfies G, so the consequent must be true at u. (In ordinary idiomatic English, a sentence like, 'Everyone ought to be happy' may be understood as expressing either a wide-scope (de dicto) proposition to the effect that it ought to be the case that everyone is happy or a narrow scope (de re) proposition stating that each actually existing person ought to be happy. (7.6) does not hold in models in which the domain of individuals may expand when we move from a situation to one of its deontic alternatives. For example, suppose at u there are just 10 people left and that that is perfectly evident to each of those 10 people, so that each ought to believe there are just ten people left; if there are u-alternatives with expanded domains, although it will follow that each of these ten will there believe there are just ten people left, it will not follow that the additional other

---

[27] Although see [Makinson, 1981] for problems with interpreting the variables as ranging over concrete actions.

people there will share that belief. It is easy to formally verify that the Barcan formula does not hold in all variable domains models. On the other hand, the converse of (7.6) holds both in constant domain and in variable domain models in which the domains of all deontic alternatives to a world u must contain every individual that exists at u (and perhaps more), but the conditional

(7.7) $\quad \forall x \mathbf{P} G x \to \mathbf{P} \forall x G x,$

that is,

(7.8) $\quad \mathbf{O} \exists x G x \to \exists x \mathbf{O} G x,$

is invalid in both kinds of models. Everyone is permitted to have a dinner in Casa Paco, a public restaurant, but no situation in which everyone is having dinner in Casa Paco is permitted (normatively acceptable), because the legal seating capacity of the restaurant is 40 customers.[28] In the same way, the sentence 'Someone ought to rescue the cat Gussie from the shelter for abandoned pets' is ambiguous: It can be understood as having the form of the antecedent of (7.8) (a wide-scope ought) or the form of its consequent, and the former interpretation does not mean that some specific person has a (personal) obligation to rescue Gussie. In these respects deontic modalities are logically similar to alethic and epistemic modalities. It should be observed that both interpretations of (7.8) can be regarded as ought-to-do propositions in the sense that the predicate in the scope of the deontic operator may be an action predicate. The failure of (7.7) and (7.8) to hold makes possible the tragedy (or paradox) of the commons and other similar problems. An attempt by everyone to perform in the same situation or at the same time what they take to be a permitted action can have normatively unacceptable consequences. (See [Hardin, 1968] and [McConnell and Brue, 2002, p. 596].) Different assumptions about deontic alternativeness relation and about the domains of individuals lead to different systems of quantified deontic logic. (For quantifiers in modal logic, see [Garson, 2001], [Girle, 2009, pp. 106-125], [Priest, 2008, pp. 308-48] and [Bell et al., 2001, pp. 171-183]; see also [Hintikka, 1957; Makinson, 1981; Goble, 1994; Goble, 1996] on quantifiers in deontic logic.)

The supplement to this section provides some more formal details (for the propositional systems).

---

[28]Or per (7.8), it may be obligatory that someone leave the lifeboat (else no one will be saved), but not that there is some one person such that she is obligated to leave, else there would be no need to draw straws to transform the first situation into one like the second, and it would also mean that at least someone in the boat could not go beyond the call by going overboard voluntarily, since s/he would be obligated to do so by (7.8).

Supplement to Section 7: Some formalities

A7.1 Semantics for SDL and SDL$^+$:

We first define the frames (structures) for modeling SDL.

<u>F is an SDL (or KD) frame</u>: $F = \langle W, A \rangle$ where:

1. W is a non-empty set (the points or worlds)
2. A is a subset of $W \times W$ (the acceptability relation)
3. A is serial: $\forall u \exists v Auv$.

A model is such a frame paired with an assignment function from the sentence letters of the language of SDL to the subsets of W:

<u>M is an SDL model</u>: $M = \langle F, I \rangle$, where F is an SDL Frame and I is a function from the propositional letters to subsets of W in the frame (the "truth sets" for the letters).

We now define truth in a model for all sentences of the language of SDL, where "$M, u \models p$" stands for "p is true at u in model M":

<u>Basic truth-conditions</u>: (Here and occasionally in proofs we will use "PC" as short for truth-functional propositional calculus.)

(PC) $M, u \models p$ iff $u \in I(p)$

             $M, u \models \neg p$ iff $M, u \not\models p$ where p is a sentence letter

             $M, u \models p \rightarrow q$ iff either $M, u \not\models p$ or $M, u \models q$

(OB) $M, u \models \mathbf{OB}p$ iff $\forall v$ (if Auv then $M, v \models p$)

<u>Derivative truth-conditions</u>:
(Truth functional operators as usual.)

(PE) $M, u \models \mathbf{PE}p$ iff $\exists v$ (Auv and $M, v \models p$)

(IM) $M, u \models \mathbf{IM}p$ iff $\neg \exists v$ (Auv and $M, v \models p$)

(OM) $M, u \models \mathbf{OM}p$ iff $\exists v$ (Auv and $M, v \models \neg p$)

(OP) $M, u \models \mathbf{OP}p$ iff $\exists v$ (Auv and $M, v \models p$) &

                              $\exists v$ (Auv and $M, v \models \neg p$)

Truth in a model: $M \models p$ iff p is true at every world in M.
Validity in a class C of models: $C \models p$ iff $M \models p$, for every M in C.

Recall that SDL$^+$ was used for convenience to denote the result of adding A4, **OB**(**OB**$p \to p$), to SDL. We also noted that the semantic constraint associated with this was secondary reflexivity:

(COO) $\quad \forall u \forall v \, (Auv \to Avv)$

We now note two well-known metatheorems:

Metatheorem 7.1 *SDL is determined by the class of all SDL models. That is, any theorem of SDL is valid per this semantics (soundness), and any formula valid per this semantics is a theorem of SDL (completeness).*

Metatheorem 7.2 *SDL$^+$ is determined by the class of all secondary reflexive SDL models.*

For key elements see [Åqvist, 2002]; but for some additional metatheory see [Åqvist, 1987].

A7.2 Semantics for LKA$_1$ and LKA$_2$:

F is an LKA$_1$-frame: $F = \langle W, R, G \rangle$ where:

1. W is a non-empty set
2. R is a subset of $W \times W$ (accessibility relation)
3. G is a subset of W (deontically acceptable worlds)
4. $\forall u \exists v \, (Ruv \,\&\, v \in G)$.

Note that here the acceptable worlds do not vary relative to a world as in the SDL frames.

We then add an assignment function to get a model:

M is an LKA$_1$-model: $M = \langle F, I \rangle$, where F is an LKA$_1$ frame and I is a function from the propositional letters to various subsets of W in the frame.

The truth conditions for formulas can now be given.

Basic truth-conditions at a world, u, in a model, M:

(PC) $\qquad$ same as for SDL
($\Box$) $\qquad$ $M, u \models \Box p$ iff $\forall v$ (if Ruv then $M, v \models p$)
(g) $\qquad$ $M, u \models g$ iff $u \in G$

Derivative truth-conditions:

(Truth functional operators as usual.)

($\Diamond$) $M, u \models \Diamond p$ iff $\exists v \, (Ruv$ and $M, v \models p)$
(**OB**) $M, u \models \mathbf{OB}p$ iff $\forall v$ (if $Ruv \,\&\, v \in G$ then $M, v \models p)$
(**PE**) $M, u \models \mathbf{PE}p$ iff $\exists v \, (Ruv \,\&\, v \in G \,\&\, M, v \models p)$
(**IM**) $M, u \models \mathbf{IM}p$ iff $\forall v$ (if $Ruv \,\&\, v \in G$ then $M, v \models \neg p)$
(**OM**) $M, u \models \mathbf{OM}p$ iff $\exists v \, (Ruv \,\&\, v \in G \,\&\, M, v \models \neg p)$
(**OP**) $M, u \models \mathbf{OP}p$ iff $\exists v \, (Ruv \,\&\, v \in G \,\&\, M, v \models p) \,\&$
                                  $\exists v \, (Ruv \,\&\, v \in G \,\&\, M, v \models \neg p)$

Truth in a model and validity in a class of models is defined as above for SDL.

Recall that LKA_2 was used for convenience to denote the result of adding axiom T, $\Box p \to p$, to LKA_1. The semantic constraint associated with T is reflexivity: $\forall u Ruu$.

Metatheorem 7.3 *LKA_1 is determined by the class of all LKA_1 models.*

Metatheorem 7.4 *LKA_2 is determined by the class of all reflexive LKA_1 models.*

Metatheorem 7.5 *The pure deontic fragment of LKA_1 is SDL.*

Metatheorem 7.6 *The pure deontic fragment of LKA_2 is SDL^+.*

A7.3 Semantics for VW:

We first define the frames (structures) for modeling VW.

F is an VW frame: $F = \langle W, A \rangle$ where:

1. W is a non-empty set (the points or worlds)
2. A is a subset of $W \times W$ (the acceptability relation)

This is the same as the definition of the SDL frames except that seriality is dropped.

An assignment function is added to get a model:

<u>M is an VW model</u>: $M = \langle F, I \rangle$, where F is an SDL frame and I is a function from the propositional letters to subsets of W in the frame.

We now define truth in a model for all sentences of the language of VW:

Basic truth-conditions:

(PC) same as for SDL, including $M, u \not\models \bot$
(OB) $M, u \models \mathbf{OB}p$ iff $\exists v Auv \,\&\, \forall v$ (if Auv then $M, v \models p$)

Derivative truth-conditions:
(Any remaining truth-functional operators as usual, including for $\top$: $M, u \models \top$.)

(**PE**) $M, u \models \mathbf{PE}p$ iff $\neg \exists v Auv \lor \exists v\, (Auv \,\&\, M, v \models p)$
(**IM**) $M, u \models \mathbf{IM}p$ iff $\exists v Auv \,\&\, \neg \exists v\, (Auv \,\&\, M, v \models p)$
(**OM**) $M, u \models \mathbf{OM}p$ iff $\neg \exists v Auv \lor \exists v\, (Auv \,\&\, M, v \models \neg p)$
(**OP**) $M, u \models \mathbf{OP}p$ iff $\neg \exists v Auv \lor [\,\exists v\, (Auv \,\&\, M, v \models p)$
$\&\, \exists v\, (Auv \,\&\, M, v \models \neg p)\,]$

Truth in a model: $M \models p$ iff p is true at every world in M. Validity in a class C of models: $C \models p$ iff $M \models p$, for every M in C.

Recall that SDL is VW+**OB**⊤, and that for the semantic clause for **OB** above, the semantic constraint associated with **OB**⊤ is serialty itself: $\forall u \exists v Auv$.

Recall that DD in this framework is valid without any semantic constraints, but not **O**⊤, thus reversing the situation in the standard SDL semantics.

We now state two metatheorems:

Metatheorem 7.7 ([McNamara, 1988]) *VW is determined by the class of all VW models.*

Metatheorem 7.8 ([McNamara, 1988]) *SDL is determined by the class of all serial VW models.*

8 Problems and paradoxes regarding the standard systems

In this section[29] we will consider some of the "paradoxes" associated with the "standard systems": SDL and suitably similar systems, here to include the two expressively stronger LKA systems that generate SDL, and the logically weaker VW system, which shares the same language as SDL. The use of "paradox" is widespread in discussions of deontic logic, and is consistent with a broad use of the term elsewhere, but "puzzles", "challenges", "problems", "dilemmas" will often be used, and seems less loaded. The number of problems attributed to standard systems is large, and these have often served to fuel new work after the classic period of the 1950s. We will list and briefly describe many of them, grouped under various associated headings (e.g. principles that are often thought to figure centrally in the associated puzzles). There will be both continuity with some of the earlier historical material presented, and occasional repetition of coverage, since we wish this section and its subsections to be something a reader might at times consult without the preceding material in focus; there will also be more detailed coverage of puzzles that have received the most attention or seemed the most challenging.

8.1 A puzzle with the very idea of deontic logic

Jørgensen's dilemma - truth and normative language[30]

(This was discussed in considerable detail in Section 4 above. Here we merely give a very compressed sketch.)

The view that evaluative sentences (e.g. "That is beautiful/ugly", "That is good/bad", "That is wrong/right") are not the sort of sentences that can be either true or false was held by many researchers in the first half of the twentieth century, especially during the heyday of positivism. This leads to a dilemma. Deductive logic involves the study of what follows from what. Truth is essential to deductive consequence, as well as to notions of consistency, entailment, contradiction, etc. But then deontic logic is impossible, since its sentences are among the evaluative ones, and thus neither true nor false. Yet normative sentences of the sort studied in deontic logic do seem to stand in familiar logical relationships to one another, so deontic logic must be possible after all.

A widespread distinction was made between *norms* and *normative propositions*[31]. The idea is that a normative sentences such as "You may enter

---

[29]This section benefits from [McNamara, 2006; McNamara, 2010].
[30][Jørgensen, 1937 and 1938].
[31][Hedenius, 1941; von Wright, 1963; Alchourrón and Bulygin, 1981; Alchourrón and Bulygin, 1971; Makinson, 1999; Stenius, 1963]. [von Wright, 1963] credits Hedenius for

freely" may be used by an authority to provide permission on the spot or it may be used by a passerby to report on an already existing norm (e.g. a standing municipal regulation for free entrance to a museum). Using a normative sentence as in the first example is sometimes referred to as "norming" - it creates a norm by granting permission by the very use of the sentence by the authority. In contrast, the use by the passerby is deemed descriptive: it is used to report that permission to do so is a standing state, not to grant permission. Often the two uses are deemed mutually exclusive, with only the latter use allowing for truth or falsity. Some have challenged the exclusiveness, by appealing to speech-act theory along with semantics. The idea is that the one in authority not only grants the permission by performing the speech act of uttering the relevant sentence "You may enter freely", but also thereby makes what it said true (that you may enter freely).[32] Still, many believe that norms are nonetheless distinct from normative propositions, and that a logic of norms is also needed.

"Input-Output Logic" is a recent robust program to provide a logical framework for norms as non-truth-evaluable items. (See e.g. [Makinson, 1999; Makinson and van der Torre, 2000; Makinson and van der Torre, 2003], and the chapter by Parent and van der Torre in this volume.) In a more general vein, there is the older tradition of developing logics for imperatives, and the debate about whether there can even be such. See Hansen's chapter in this volume on imperatival logic. [Vranas, 2010] provides a recent defense of the possibility of imperatival logic, and [Vranas, 2011] offers a new theory of validity for imperative inference. Let us also mention another tradition, the imperative-based (or norm-based) approach to deontic logic, which focuses not on a logic of imperatives (perhaps even denying that possibility), but on the use of imperatives as a key foundational component in a semantics for logics for truth-evaluable deontic sentences. See for example, [Hansen, 2004; Hansen, 2008; Horty, 1994; Horty, 1997; Horty, 2003] and the seminal [van Fraassen, 1973].

8.2 A problem centering around O⊤

The logical necessity of obligations problem[33]

---

the distinction.

[32]Cf. [Lemmon, 1962b; Kamp, 1974; Kamp, 1979]. [Kempson, 1977] argues that performative utterances often work this way. For example, if a legitimate authority in the right context pronounces two people married, it may not only be the case that the speech act performed renders them married, but the sentence "You are now married" may be a true description at its moment of utterance, the dual character perhaps captured by "You are, hereby, married".

[33]We are unaware of any standard name for this problem.

Consider the apparent possibility that

(8.1) Nothing is obligatory.

A natural representation of this in the language of the standard systems (with quantifiers added) is:

(8.2) $\neg \exists q \mathbf{O} q$.

But RND of SDL entails ON, $\vdash \mathbf{O}\top$, so supplemented with propositional quantification, we would get $\vdash \exists q \mathbf{O} q$. SDL thus seems to imply that it is a truth of logic that something is obligatory—that there could not be a situation with no obligations; yet (8.1) appears to express something not only possible, but plausibly thought to be true at times in the past in our universe, so SDL appears to be too strong [Chellas, 1974]. This holds for all the standard systems except VW. [von Wright, 1951] argues that neither $\mathbf{O}\top$ nor $\mathbf{P}\bot$ nor their denials, are logical truths, so we should opt for their absence as logical truths as a "principle of contingency" for deontic logic. The weakening of SDL described in Section 7.3 above, VW, captures this absence.

Later, in [von Wright, 1963, pp.152-4], von Wright argues that $\mathbf{O}\top$ does not express a real prescription. [Føllesdal and Hilpinen, 1971, p. 13] argue that ON of SDL at best excludes "*empty* normative systems" with no obligations, and that furthermore, since no one can fail to fulfill $\mathbf{O}\top$ anyway, it is not a pressing concern.[34] However, no one can bring it about that $\top$, so it would seem that no one can fulfill $\mathbf{O}\top$, although no one can violate it either. [al Hibri, 1978] discusses various early takes on this problem, rejects ON, and later develops a deontic logic without it. [Jones and Pörn, 1985] explicitly reject ON for "ought" in the system developed there, where the concern is with what people ought to do.

Note that reading $\mathbf{O}$ as "it ought to be the case that", as it often was read, makes it less clear that ON is problematic. "It ought to be that contradictions are false" does not sound jarring, but here there is no longer a clear link to "$\mathbf{O}$" and what agents ought to do or bring about.

8.3 Puzzles centering around the rule RMD
The violability of obligations problem[35]

It is often thought that a central distinguishing mark of obligations is that they are *violable* in principle, unlike purely factual claims. It seems hard to swallow that it is obligatory that the sun will set today, much less that

---

[34] See also [Prior, 1958].
[35] This objection is suggested by remarks in [von Wright, 1963, p. 154].

it is obligatory that either it does set today or it is not the case that it does, something that couldn't be otherwise on purely logical grounds. This suggests the following as a conceptual truth about obligations:

(8.3) If p is logically necessary, it is logically impossible that it is obligatory that p

In SDL, a natural expression of such a violability constraint would seem to be this rule:

(8.4)   If $\vdash p$ then $\vdash \neg \mathbf{O}p$ (Violability)

But Violability immediately yields $\neg \mathbf{O}\top$ directly contradicting theorem ON of SDL. So no SDL (or stronger) system can consistently rule out inviolable obligations.[36]

For systems with the expressive resources of the LKA systems, a stronger violability condition is expressible, to the effect that nothing obligatory is necessary:

(8.5) $\Box p \to \neg \mathbf{O}p$ (Violability')

But this is inconsistent with all LKA systems, since $\Box p \to \mathbf{O}p$ and $\Box \top$ are theses.

Note that even in a system weaker than SDL, one that lacked RND and ON, as long as the rule RMD (if $\vdash p \to q$ then $\vdash \mathbf{O}p \to \mathbf{O}q$) is derivable, then adding the Violability rule above would render the system useless. For by PC, $\vdash p \to \top$, so it follows by RMD that $\vdash \mathbf{O}p \to \mathbf{O}\top$; but since by PC, $\vdash \top$, by Violability, it follows that $\vdash \neg \mathbf{O}\top$, and hence $\vdash \neg \mathbf{O}p$. Thus with Violability added to such a system, we get as a thesis that nothing is obligatory. And this means that although the weaker VW can consistently rule out inviolable obligations, it can do so only at the expense of ruling out all obligations.[37]

Free choice permissions puzzle[38]
Consider:

(8.6) You may either have c̲ake or i̲ce cream.[39]
(8.7) You may have cake and you may have ice cream.

---

[36]VW can, but only at the expense of ruling out violable obligations, and thus all obligations, as well.

[37][Jones and Pörn, 1985] design a system explicitly intended to countenance a violability condition for one of their operators, and this constraint is endorsed in [Carmo and Jones, 2002] as well.

[38][Ross, 1941].

[39]Underlined letters suggest our intended symbolization schemes.

Natural symbolizations of (8.6) and (8.7) in the language of the standard systems are:

(8.8) $\mathbf{P}(c \vee i)$
(8.9) $\mathbf{P}c \,\&\, \mathbf{P}i$

Furthermore, it is also natural to see (8.7) as following from (8.6): if I am permitted to have either, then having each is permissible (though perhaps not both). But (8.9) does not follow from (8.8) in standard systems. This is not a theorem:

(P*) $\mathbf{P}(p \vee q) \to (\mathbf{P}p \,\&\, \mathbf{P}q)$.

Furthermore, if (P*) were added to a system that contained VW (and thus to any that contained SDL), disaster would result. For from RMD and the definition of $\mathbf{P}$, we get $\mathbf{P}p \to \mathbf{P}(p \vee q)$.[40] But then with (P*), it would follow that $\mathbf{P}p \to (\mathbf{P}p \,\&\, \mathbf{P}q)$, for all q, so we easily generate a theorem to the effect that everything is permissible if anything is:

(P**) $\mathbf{P}p \to \mathbf{P}q$,

This is absurd on its face, and in any system containing VW, it would in turn yield as a theorem that nothing is obligatory:

(8.10) $\vdash \neg \mathbf{O}p$.

For reductio, assume $\mathbf{O}p$, for some p. Then by DD (No Conflicts), it follows that $\mathbf{P}p$. But one instance of (P**) is $\mathbf{P}p \to \mathbf{P}\neg p$, so we would then have $\mathbf{P}\neg p$, which by RED (Substitution of Provable Equivalents), PC and the definition of $\mathbf{P}$ generates $\neg \mathbf{O}p$, contradicting our assumption.

This puzzle has led many to conclude that there are two senses of "permissibility" that need to be separated out, and that the language of SDL (and thus VW) represents only one of those. One sense might be the simple absence of a prohibition, and thus expressible in standard systems. The other might be a stronger sense of permission that would support (P*) but without supporting (P**), as suggested in [von Wright, 1968]. [Føllesdal and Hilpinen, 1971] (and [Carmo and Jones, 2002]) wonder if this problem might not be a pseudo-problem calling for no more than SDL's expressive resources, since the conjunctive sense of an "or-permission" can simply be expressed as a conjunction of permitting conjuncts, $\mathbf{P}p \,\&\, \mathbf{P}q$, and the weaker sense as a permitted disjunction, $\mathbf{P}(p \vee q)$. [Kamp, 1974;

---

[40] Since $\vdash \neg(p \vee q) \to \neg p$, $\vdash \mathbf{O}\neg(p \vee q) \to \mathbf{O}\neg p$, and thus $\vdash \neg\mathbf{O}\neg p \to \neg\mathbf{O}\neg(p \vee q)$.

Kamp, 1979] provide nuanced analyses of the semantics and pragmatics of permission statements, including disjunctive ones. Here we have only skimmed the surface. For more on the rich topic of various concepts and analyses of permission, see the chapter in this volume by Hansson on the topic.

Ross's paradox[41]
Consider:

(8.11) It is obligatory that you mail the letter.
(8.12) It is obligatory that you mail the letter or you burn the letter.

Natural renderings in the standard systems are:

(8.13) $\quad\quad\quad\quad\quad\quad\quad\quad\quad$ **O**m
(8.14) $\quad\quad\quad\quad\quad\quad\quad\quad\quad$ **O**$(m \vee b)$

However, $\vdash$ **O**$m \to$ **O**$(m \vee b)$ follows by RMD from $\vdash m \to m \vee b$, so (8.14) follows from (8.13) in any VW system, but arguing from (8.11) to (8.12) seems rather odd. Among other things, it seems to suggest that the obligation expressed in (8.11) to mail the letter automatically generates a distinct obligation that I am able to fulfill by burning the letter. Of course, the latter is presumably forbidden, but it remains odd to think I could plead partial mitigation in failing to mail the letter by burning it instead with "Well, at least I fulfilled my obligation to mail or burn it".

The Good Samaritan paradox[42]
Consider:

(8.15) It is obligatory that Jones help Smith who is being mugged.
(8.16) It is obligatory that Smith is being mugged.

Now the following equivalence appears to be logically true:

(8.17) Jones helps Smith who is being mugged if and only if Jones helps Smith and Smith is being mugged.

But relying on this equivalence, if we then symbolize (8.15) and (8.16) in the language of VW in the most natural way, we get:

(8.18) $\quad\quad\quad\quad\quad\quad\quad\quad\quad$ **O**$(h\&m)$
(8.19) $\quad\quad\quad\quad\quad\quad\quad\quad\quad$ **O**m

---
[41][Ross, 1941].
[42][Prior, 1958].

But $(h\&m) \to m$ follows by truth-functional logic, so by RMD, it follows that $\mathbf{O}(h\&m) \to \mathbf{O}m$, and then we can derive (8.19) from (8.18). But does it really follow from its being obligatory that Jones come to the aid of Smith who is being mugged that Smith's mugging it itself obligatory?

Note that Prior casts this paradox in a prohibition form, using this trivial variant of RMD given the definition of $\mathbf{F}$: If $\vdash p \to q$ then $\vdash \mathbf{F}q \to \mathbf{F}p$, suggesting that the impermissibility of Smith being robbed implies the impermissibility of helping him who is being robbed.

It is also doubtful that the paradox is due to the fact that there are two people involved. It can be recast with just one agent via $\mathbf{F}$ (i.e. $\mathbf{O}\neg$) as "The victim's paradox": the victim of a crime can help herself only if there is a crime, but then, it will follow under similar symbolization that it is impermissible for the victim of the crime to help herself, since the crime is impermissible. Similarly for "The repenter's paradox": the robber repents for his crime only if there is a crime, and so by similar reasoning we might get a symbolization suggesting that repenting is wrong since the crime is wrong. These early variations were used to argue against certain early proposed solutions to the Good Samaritan paradox (e.g. [Nowell Smith and Lemmon, 1960]).

Åqvist's paradox of epistemic obligation[43]

Here is a much discussed variant of the preceding paradox. Consider:

(8.20) The bank is being robbed.

(8.21) It is obligatory that Jones (the guard) knows that the bank is being robbed.

(8.22) It is obligatory that the bank is being robbed.[44]

Let us imagine that we have added a logic for a propositional knowledge operator, K. We can then let "Kr" symbolize "Jones knows that the bank is being robbed", and then a natural way to symbolize (8.20)-(8.22) in a VW system so augmented is:

(8.23) $\qquad\qquad\qquad r$

(8.24) $\qquad\qquad\qquad \mathbf{O}Kr$

(8.25) $\qquad\qquad\qquad \mathbf{O}r$

But a logic for propositional knowledge will presumably support knowledge's entailment of truth (e.g. that if Jones knows that the bank is being robbed then the bank is being robbed). So $Kr \to r$ would be a thesis of any such

---

[43][Åqvist, 1967].

[44](8.20) is inessential but is listed to suggest part of the natural context for (8.21).

augmented VW system. But then it would follow by RMD that $\mathbf{O}Kr \to \mathbf{O}r$ is also a thesis of any such system, and we can then use that thesis to derive (8.25) from (8.24) by MP. Applied to the case above, the logic seems to suggest that (8.22) follows from (8.21), which seems absurd. It seems that it is obligatory for the guard to be in the know about any bank robberies taking place, and so this one, but surely it does not follow that it is thereby also obligatory that the robbery take place. Thus it appears to be in the spirit of the standard systems, and any weaker systems endorsing RMD, to generate fictitious consequences from cases of obligatory knowledge.

It is also worth noting that we should not be misled by the typical examples into thinking of this problem as "such logics can't handle obligatory knowledge of *contrary-to-duty facts*", for the same problem extends to cases of facts that are optional in normative status. If a reformed but known bank robber, Jones, enters the bank, it may also be that it is obligatory that the guard knows Jones is in the bank, and then RMD appears to wrongly entail that it is obligatory that Jones is in the bank, even though it is completely optional that Jones is there.

There have been a variety of responses to these RMD-related paradoxes. One response to these has been to try to explain them away. For example, Ross's paradox is often quickly dispensed with as based on confusion or as not really being a problem when the system in question is properly understood (e.g. [Føllesdal and Hilpinen, 1971; Brown, 1996a]). Others deflect it arguing that it is semantically correct and only pragmatically odd, and reflects features that any adequate theory of the pragmatics of deontic language must predict, so no special problem for deontic logic [Castañeda, 1981]. It is also often suggested that regarding the Good Samaritan paradox, RMD is not the real culprit because, viewed rightly, it is really a conditional obligation paradox [Castañeda, 1981; Tomberlin, 1981]. Others however suggest that things are as they seem and the above paradoxes are of a piece in all genuinely invoking RMD and reflecting RMD's problematic character for genuine deontic reasoning. [Jackson, 1985; Goble, 1990a] are closely related examples of approaches to deontic logic rejecting RMD from a principled philosophical perspective. [Jackson, 1985] links an "ought to be" operator to counterfactuals and informally explores its semantics and logic; whereas [Goble, 1990a] takes a similar approach but generalizes the idea to cover "good" and "bad" as well, with [Goble, 1990b] providing characterization results for the identified logics. Interestingly, their approaches also intersect with the philosophical issue of "actualism" and "possibilism" in ethical theory.[45] [Loewer and Belzer,

---

[45]Possibilists assert that an agent is obligated to bring about any p that is part of the the optimal overall outcome she could achieve by her actions, even when the goodness of

1986] provides an interesting discussion of the traditional puzzle, as well as Forrester's puzzle (below), in terms of their system 3D. [Hansson, 1990; Hansson, 2001] systematically explore systems of deontic logic in terms of general attributes of different preference orderings, using these to classify types of normative predicates or operators as *prohibitive* and *prescriptive* (e.g. a prohibitive status as one where anything worse than it has that status also). His work is predicated on the assumption that RMD is a key source of the main paradoxes of the standard systems, and so he devises non-standard systems intended to not countenance principles such as RMD. In this vein, see Hansson's chapter on alternative semantics for deontic logics in this volume. [Hansson, 2001] is also important in its own right for its extensive and original work on preference logic and preference structures, which, as we have already noted, are used regularly in deontic logic (and elsewhere).[46] Opinions about these puzzles we have grouped together need not be monolithic of course. For example, [Carmo and Jones, 2002] take Ross's puzzle seriously, the free choice permission puzzle to be a pseudo problem, and the Good Samaritan puzzle to be resolvable using the resources needed for resolving puzzles with deontic conditionals.

8.4 Puzzles centering around DD and OD
Sartre's dilemma - conflicting obligations[47]

A *conflict* or *dilemma* is a situation where there are one or more obligations not jointly realizable. The typical case involves a conflict of two obligations. For example, suppose I promised Mary to meet her, and that I promised another friend that I would not meet Mary. It would then seem that I have, by my promises, made the following true:

---

p depends on all sorts of other things that she would not in fact bring about were she to bring about p. Actualists assert that an agent is obligated to bring about any p if that would in fact be better than not doing so, and this of course can crucially depend on what else I would do (optimal or not) were I to bring about p ([Jackson and Pargetter, 1986; Jackson, 1988]; [Greenspan, 1975; Goldman, 1976; Thomason, 1981a] provide early discussions.)

[46][van der Torre, 1997] is a nice general source covering issues surrounding RMD as well, along with much else.

[47][Lemmon, 1962a]. In the original example, a young man is obliged to avenge his brother's death (by leaving home and fighting the Nazi occupation) and he is also obliged to stay home and aid his mother (devastated by the loss of the brother). [von Wright, 1968] talks of "predicaments" and cites the Book of Judges, where Jephthah promises God he will sacrifice the first being he meets on his way home from war, if God gives him victory. God does, and the first being he meets upon his return is his beloved daughter. Note that both of these are plausibly thought of as dual-sourced obligations, but our example, following [Marcus, 1980], reflects the possibility of conflicts generated by a single normative principle (e.g. it is obligatory to keep one's promises), which in turn reinforces the idea that conflicts of obligation can be circumstantial and needn't be generated by normative systems that are supposedly inconsistent (cf. [Williams, 1965]).

(8.26) It is obligatory that I meet Mary (now).
(8.27) It is obligatory that it is not the case that I meet Mary (now).

If so, then I have an explicit conflict of obligations. People generate conflicting appointments easily enough under pressure to please, in forgetful moments, due to errors in our calendar entries, etc. It also appears that they result in conflicting obligations in a perfectly ordinary sense of the term.[48] But a natural first blush representation of these in the language of VW and SDL is:

(8.28) $\mathbf{O}m$
(8.29) $\mathbf{O}\neg m$

But given DD, $\mathbf{O}p \rightarrow \neg\mathbf{O}\neg p$, is a theorem of VW, we are quickly led from the conflict expressed by (8.28) and (8.29), to the contradiction expressed by $O\neg m \& \neg O\neg m$. So we must conclude that (8.28) and (8.29) make an inconsistent pair per VW. Yet, the original seems not only logically coherent but all too familiar.[49]

At the end of this section, there is a supplement where we consider some challenges faced once we decide to develop conflict tolerant logics.

A puzzle surrounding Kant's law

Kant's law typically involves a notion of possibility stronger than that of mere logical or metaphysical possibility. In discussions in ethical theory, where "Kant's law" arose, it is agential:

(KL) Anything morally obligatory for an agent must be *within the agent's ability*.[50]

This principle has been widely advocated in ethical theory, one thought being that, at least for all things considered obligations, the fact that something is not even in an agent's power to do is itself a sufficient consideration

---

[48]These obligations need not be all-things-considered-non-overridden obligations, but that does not entail that these are not obligations (any more than "it's not a brown dog" entails "it's not a dog") nor does it mean that we needn't model them. For a simple framework that allows for such obligations, as well as comparing them, see [Brown, 1996b].

[49][Lemmon, 1962a] argues early on that a conflict of obligations may involve no contradiction. [Williams, 1965] stresses the contingency of conflicting obligations and briefly contrasts this with inconsistency as unrealizability in any world. [Marcus, 1980] argues explicitly for the standard world-theoretic conception of consistency as joint realizability in some world in some model (not in all worlds in all models, as with say the set of tautologies). See also [Marcus, 1996].

[50]It is sometimes used more broadly in deontic logic for a weakened version, one that follows from Kant's stronger version. See next puzzle.

to eliminate it from further consideration in a determination of what is to be all in all required. In an optimizing framework like utilitarianism, (KL) is strongly supported by the standard maxim that one is morally obligated to do the best she *can*. It is also often endorsed in various deontic-agential frameworks (e.g. [Horty, 2001; McNamara, 2000]).

But now consider:

(8.30) I'm obligated to pay you back $100 by tonight.

(8.31) I can't pay you back $100 by tonight (e.g. I just spent it on something shopping).

Let us represent the above sentences in the language of our LKA systems where we have a possibility operator. Although agency is not itself represented in the LKA systems, we can still interpret the possibility operator therein as "what is *consistent with the abilities of* some background agent, Jane Doe", and likewise for the deontic operators we might interpret them as indicating what is obligatory for such a Jane Doe [Brown, 1992; McNamara, 2000].[51] (8.30 and (8.31) might then be naturally symbolized as follows:

(8.32) $\mathbf{O}p$

(8.33) $\neg \Diamond p$

(8.30) and (8.31) appear to be consistent. Alas, people often wind up with financial obligations they cannot fulfill, be it from neglect, unforeseen circumstances, or whatever. So it seems that the notion of an unfulfillable obligation is no contradiction in terms. But in the LKA systems, it is a theorem that $\mathbf{O}p \to \Diamond p$. So from (8.32) and (8.33) we get $\Diamond p \& \neg \Diamond p$, a contradiction, and so (8.32) and (8.33) are inconsistent. Yet (8.30) and (8.31) seem consistent. I can clearly *owe* money I'm unable to pay back, but doesn't that ordinarily entail that I have a financial *obligation* I cannot meet?

One strategy here might be to posit ambiguity or context shift and employ a distinction between deliberative contexts of evaluation and judgmental contexts as suggested by [Thomason, 1981a] and [Thomason, 1981b], bolstered by arguing that we need the distinction elsewhere anyway.[52] In judgmental contexts (or the judgmental sense), evaluations such as (8.30)

---

[51]The puzzle would remain even if agency and agential ability were explicitly represented.

[52][Thomason, 1981a] credits [Greenspan, 1975; Powers, 1967] for stressing the contextuality of oughts. In the distinct context of the contrary-to-duty paradox (discussed below), others have endorsed the contextuality or ambiguity of *oughts* (e.g. [Jones and Pörn, 1985; Prakken and Sergot, 1996; Carmo and Jones, 2002]).

above need not satisfy Kant's law since, roughly, we go back in time and evaluate the present in terms of where things would now be relative to optimal past options that were accessible then but need no longer be; whereas in deliberative contexts, where we are focused on what to do now, we deny that there is an obligation to do what is now undoable. Whether this interesting distinction provides a truly satisfactory solution to the above problem is beyond the scope of this essay, but the puzzle appears to be underexplored in deontic logic.

Conflation of impossible obligations with conflicting obligations[53]

Weaker than Kant's Law is the claim that nothing *logically* impossible is obligatory. In the standard systems, this weaker claim can be expressed as a rule:

(8.34) $\qquad$ If $\vdash \neg p$ then $\vdash \neg \mathbf{O}p$,

(8.34) is a derived rule in any VW system. For suppose $\vdash \neg p$. Then by PC, $\vdash p \leftrightarrow \bot$, and then from OD, $\vdash \neg \mathbf{O}\bot$, we get $\vdash \neg \mathbf{O}p$. This in itself is not necessarily a problem for standard systems. For claiming that, say, I'm obligated to both be home and not be home because I for some reason promised you just this logically impossible thing is less convincing then saying that two separate promises might yield two distinct obligations to keep conflicting appointments, each executable, though not jointly. For it might be maintained that the concept of obligation is such that obligation claims to do the logically impossible are *logically* self-defeating. In either event, the standard systems are better insulated from this sort of objection than from the objection that conflicts of obligation are possible. Assuming we are dealing with a system that has RED (and thus any of the standard systems), the rule above is equivalent to OD, and so we can put the point more simply by saying that the following is a thesis of all standard systems, and is plausible:

(OD) $\qquad \neg \mathbf{O}\bot.$[54]

However all this suggests that there is a clear difference between conflicting obligations and a singular obligation regarding something logically impossible, and this in turn means there is a serious expressive limit in the standard systems. For within them, from a conflict of obligations such as $\mathbf{O}p \& \mathbf{O}\neg p$, we can derive an obligatory logical contradiction, $\mathbf{O}\bot$, and vice versa. In

---

[53][Chellas, 1974]. See also [Chellas, 1980; Schotch and Jennings, 1981].

[54]However [Da Costa and Carnielli, 1986] develops a paraconsistent deontic logic that would at least allow for some contradictory obligations.

any logic with both KD and RMD (and thus RED), the following is a theorem:

(8.35) $\qquad (\mathbf{O}p \& \mathbf{O}\neg p) \leftrightarrow \mathbf{O}\bot \qquad$ (Collapse)

For by KD, $\vdash \mathbf{O}(\neg p \to \bot) \to (\mathbf{O}\neg p \to \mathbf{O}\bot)$, and by RED, $\vdash \mathbf{O}(\neg p \to \bot) \leftrightarrow \mathbf{O}p$, so $\vdash \mathbf{O}p \to (\mathbf{O}\neg p \to \mathbf{O}\bot)$. For the right to left direction, by RMD, $\vdash \mathbf{O}\bot \to \mathbf{O}p$ and $\vdash \mathbf{O}\bot \to \mathbf{O}\neg p$, since $\vdash \bot \to q$, for any q. Yet it seems that one can have a conflict of obligations without it being obligatory that some logically impossible state of affairs obtains. A distinction seems to be lost here.

Separating DD from OD is now quite routine in conflict-allowing deontic logics, and OD is assumed in most deontic logics. [Chellas, 1974; Chellas, 1980; Schotch and Jennings, 1981] contain early discussions of this expressive limit and advocate different non-normal modal logics to handle this problem (among others).[55]

The limit assumption dilemma[56]

Recall our sketch of an alternative ordering semantics for SDL, and the use there of the Limit Assumption:

(LA) $\qquad$ For each u, there is v such that for any $w, v \geq_u w$

Although the limit assumption has often been assumed true in the use of ordering semantics for deontic logic, it is a controversial assumption to make, especially as a matter of logic. It seems at least conceivable that there might be a scenario in which the ordering of the worlds in the purview of some world u has no upper limit on their goodness. Blake Barley gave a nice example in an unpublished paper, "The Deontic Dial", circulated at the University of Massachusetts-Amherst in the early 1980's: you have a dial that can be set anywhere from 0 to 1, where both 0 and 1 yield disaster, but all the numbers, n, between 0 and 1 not only avoid disaster, but yield increasingly more overall good as n grows (cf. [McMichael, 1978]). If we countenance the possibility of such scenarios, and thus drop (LA) in our semantics, we must alter the standard truth clause for $\mathbf{O}$ via bests:

(COB) $\qquad \mathbf{O}p$ is true at u iff p holds in all the u-best worlds.

For in models with no u-best worlds, nothing is obligatory and everything becomes permissible by this clause, but this seems too strong a result. For

---

[55] Chellas employs minimal models (or neighborhood semantics) and Schotch and Jennings generalize Kripke models using multiple accessibility relations.

[56] [Lewis, 1973].

example, in the deontic dial case, it seems clearly obligatory to not turn the dial to 1 or 0, however otherwise perplexing the scenario is.

[Lewis, 1973; Lewis, 1974] argued that the limit assumption's use in deontic logics (and for counterfactual logics) is unjustified, and thus that our clauses for deontic (and counterfactual) operators must be adjusted. Most logicians accept this in principle, and often employ a more complex clause such as:

(COB') $\mathbf{O}p$ is true at u iff p is true from some point on up in the u-ordered worlds.[57]

In models where (LA) holds, (COB') is provably equivalent to the simpler (COB) which assumes there are u-best worlds, so the new clause is conservative. But in models where (LA) fails, it will not follow from (COB') that nothing is obligatory and everything is permissible. For example, in models intended to represent Barley's deontic dial scenario, the new clause does get the result that it is obligatory to not turn the dial to 0 and obligatory to not turn the dial to 1, and thus the dial endpoints are impermissible.

However, the new clause also has some perplexing results. For example, in the case of the dial, it seems for each setting, n, between and including 0 but before 1, it ought to be set *past* n, since there will be an accessible u-world where all such worlds ranked as high as u are worlds where the dial was turned past n. Not only does this raise the question about where to turn it positively (specificity), but the truth of the set consisting of all recommended settings of the form *the dial is turned past n*, for $0 < n \leq 1$, entails the truth of "the dial is turned to 1", which is something you ought *not* do by the same clause. Thus the set of things you ought to do is an inconsistent set. Although no syntactic conflict will show up in the system (no finite set of formulas of the form $\mathbf{O}p_1, ..., \mathbf{O}p_k$ are all true in the model while $p_1 \& ... \& p_k$ is false in all models, we nonetheless can have an infinite set of obligations which cannot be jointly fulfilled. This seems to be a case of *conflicting obligations* - a situation where one's obligations are not jointly realizable, and thus belongs under that heading.[58] So although we are given some clear directions - don't place the dial at either extreme and do place it somewhere in between 0 and 1, it is also the case that anywhere in between that we do place it, we will be wrong for not having placed it closer to 1 than that. [McMichael, 1978] argues that a related problem (called "The Confinement Problem" in [McNamara, 1995]) is a problem for Lewis

---

[57][Lewis, 1973; Lewis, 1974].

[58]As it turns out, SDL is also characterized by COB and COB', whether or not the limit assumption holds, as long as the preference relation is a total preordering of W (connected, reflexive and transitive). Things however become more complex once connectivity is dropped. See [Goble, 2003].

semantics, but [Lewis, 1978] argues that it is a problem for utilitarianism, not deontic logic. However, [Fehige, 1994] suggests that logicians must still make choices here and that, ironically, there is no clear best choice for them either: "...When the best options are lacking, then so are flawless accounts of the lack"[Fehige, 1994, p. 42]. Endorsing (LA) as a matter of logic seems unjustified, yet accommodating its denial seems to lead to its own challenges and puzzles.

Plato's dilemma - deontic defeasibility[59]

Suppose I promised to meet you for dinner, and thereby incurred an obligation to do so, but suppose also that as I am about to leave, my child begins to have an asthmatic attack, and it is clear that he needs me to rush him to the hospital. It would then seem that both of these claims are true:

(8.36) I'm obligated to meet you for dinner (now)
(8.37) I'm obligated to rush my child to the hospital (now).

Here we seem to have an indirect non-explicit conflict of obligations, where it is not practically possible to satisfy both obligations, but neither thing required is logically inconsistent with the other. Note that unlike the earlier case of two conflicting appointment obligations that appeared to be on a par for all we said, here we are immediately inclined to judge that the obligation to help my child *overrides* my obligation to meet you for dinner—the former takes clear *precedence over* the latter. Shifting focus to the weaker obligation in (8.36), we might say that it is *defeated* by that of (8.37). Furthermore, except in extra-ordinary circumstances, we would also judge that no other obligation overrides the obligation to help my child, and thus that this obligation is an *all things considered obligation* (or an *undefeated obligation*), unlike the obligation to meet for dinner. Lastly, we would ordinarily think that my obligation to rescue my child is not only not overridden by any other obligation, but that it strictly overrides any obligations I might have that conflict with it, and thus that it is not only not defeated or overridden, but is *overridingly obligatory* or a *strict obligation*. We are also prone to speak more abstractly and say that there is an *exception* here to the *general obligation* to keep one's appointments (or promises), for the circumstances are extenuating.

It should be noted there is no uniform use of terms such as "dilemma" in deontic logic (or ethical theory); some define a "dilemma" as an *unresolvable conflict*: a conflict of obligations where neither of the conflicting obligations defeats the other (cf. [Sinnott-Armstrong, 1988]). On this use of

---

[59][Lemmon, 1962a]. Plato's dilemma involves returning a weapon when the owner is in a rage and intending to (unjustly) kill someone with it. Our interpretation of the issues raised by Plato's dilemma is a bit different than Lemmon's.

"dilemma", although the earlier case with two appointments, as well as the above case, can be construed as *conflicts* of obligation, the current example is not construed as a *dilemma*, since one of the two obligations does defeat and override the other. Sometimes "predicament" is also used, again either for a conflict or for a conflict that is unresolvable.

We have already indicated that standard systems have no mechanism for representing a conflict of obligations as a logical possibility. So clearly the issues here go beyond their capacity, but it is also important to note that once we set out to represent conflicts of obligation, there is the further issue of representing the logic of relationships between conflicting obligations and statuses of obligations deriving from these relationships, such as one overriding another, one defeating another, one being undefeated by any others and so being an all things considered obligation, one being a general one (e.g. it is obligatory to keep one's promises) that holds by default but not unexceptionally, etc. The issue of conflicting obligations *of different weight*[60] and the *defeasability* of obligations by other obligations (or even by circumstances–an obligation to meet a friend for dinner who now himself can't make it because ill) clearly requires much more than just having a logic that allows for conflicts, although that is a necessary condition.

There have been a variety of approaches to this domain and the associated issues, with considerable intensification in the 1990s. [von Wright, 1968] *informally* proposed minimization of evil as a natural tool for resolving conflicts of obligation, thereby suggesting the aptness of reliance on an ordering relation. [Alchourrón and Makinson, 1981] gives an early formal system for conflict resolution using partial orderings of regulations and regulation sets. [Chisholm, 1964] has been very influential conceptually, as witnessed, for example, by [Loewer and Belzer, 1983]. In ethical theory, the informal conceptual landmark is [Ross, 1939]. [Horty, 1994] is a very influential discussion forging a link between Reiter's default logic developed in AI (see [Brewka, 1989]), and an early influential approach to conflicts of obligation, [van Fraassen, 1973], which combines a preference ordering with an imperatival approach to deontic logic (see also [Horty, 1997; Horty, 2003]). [Prakken, 1996] discusses Horty's approach and an alternative that strictly separates the defeasible component from the deontic component, arguing that handling conflicts should be left to the former component only. See also [Makinson, 1993] for a discussion of defeasiblity and the place of deontic conditionals in this context. Other approaches to defeasibility in deontic logic that have affinities to semantic techniques developed in artificial intelligence for modeling defeasible reasoning about defeasible conditionals generally are [Asher and Bonevac, 1996; Moreau, 1996], both

---

[60]Cf. [Brown, 1996b].

of which attempt to represent W. D. Ross-like notions of prima facia obligation, etc. Earlier related works of interest that were ahead of the curve on some aspects of defeasibility are the influential conceptual framework of [Chisholm, 1964], and in a similar but more formal vein, that of [Loewer and Belzer, 1983; Belzer, 1986; Loewer and Belzer, 1991]. Also notable are the discussions of defeasibility and conditionality in [Alchourrón, 1993; Alchourrón, 1996], where a revision operator (operating on antecedents of conditionals) is relied on in conjunction with a strict implication operator and a strictly monadic deontic operator. [Smith, 1994] contains an interesting informal discussion of conflicting obligations, defeasibility, violability and contrary-to-duty conditionals. Since it is very much a subject of controversy and doubt as to whether deontic notions contribute anything special to defeasible inference relations (as opposed to defeasible conditionals), we leave this issue aside here, and turn to conditionals, and the problem in deontic logic that has received the most concerted attention.[61] See also the chapter by Goble in this volume, which covers many of the topics in this section, including those in the following supplement.

Supplement to 8.4 on some challenges for conflict tolerant logics

A minimal conflict tolerant logic

Two early conflict-tolerant logics are [van Fraassen, 1973; Chellas, 1974]. ([Lewis, 1974] contains a note suggesting Chellas may have circulated his system in 1970, but of course this may be true of van Fraassen as well for all we know. We list both, since so proximate.)

Suppose we want a conflict tolerant logic (and we are not yet concerned with representing the further notions associated with defeat among conflicting obligations). What should we keep from the standard systems and what should we reject? Answers to this question are not easy nor uncontroversial. Here, we cannot possibly consider all the options, much less their comparative merits, so instead we will consider one natural and simple pathway to an elementary conflict tolerant logic, one much like the earliest ones to emerge, and then describe some of the challenges it faces. We hope this will give the reader some flavor for issues and complications that arise in developing conflict tolerant

---

[61][Alchourrón, 1996; Asher and Bonevac, 1996; Moreau, 1996; Prakken, 1996] are all found in *Studia Logica* 57.1, 1996. [Nute, 1997] is dedicated to defeasibility in deontic logic (both CTDs and defeasible deontic consequence) and is an excellent single source with articles by many key players.

logics. (See [Goble, 2009] for a more elaborate discussion of various issues and options, as well as his chapter in this handbook.)

If we will allow conflicts, then minimally, we want a conjunction like the following to be consistent of course:

(EC) $\mathbf{O}p \mathbin{\&} \mathbf{O}\neg p$ (Explicit conflict)

(We ignore here that a general definition of conflicts of obligation does not say anything about explicit conflicts merely that there is a set of two or more obligations not all jointly realizable.)

What else? First and foremost, we want to make sure that unlike with all the standard systems, we cannot generate *Deontic Explosion*— the indiscriminate derivability of all formulas in the face of conflicts:

(DEX) $\mathbf{O}p \mathbin{\&} \mathbf{O}\neg p \to \mathbf{O}q$ (Deontic explosion)

This would render any system that recognized the possibility of conflicts utterly useless in the face of one. Suppose also that we want the logic to reject the possibility of obligatory contradictions—that is, suppose we want to retain (6.9)/(OD) as a thesis:

(OD) $\neg \mathbf{O}\bot$

This is not unreasonable, since we might say that the prospect of an obligatory logical contradiction is immediately logically self-defeating. Assuming so, we will then have to avoid our previously mentioned (8.35) "Collapse":

$$\mathbf{O}p \mathbin{\&} \mathbf{O}\neg p \leftrightarrow \mathbf{O}\bot \quad \text{(Collapse)}$$

For otherwise this will immediately rule out conflicts when conjoined with (OD); and even if we wanted to allow for some special cases where there were obligatory contradictions, we don't want every conflict to generate one. Collapse seems undesirable for any reasonable conflict tolerant logic. What about the consequence principle:

(RMD)   If $\vdash p \to q$ then $\vdash \mathbf{O}p \to \mathbf{O}q$?

Well, we have seen that there are certainly considerations that can be raised against this principle, especially without any restriction (e.g. so that even tautologies are obligatory if anything is); but on the other hand, it is certainly attractive to be able to draw conclusions about what else is obligatory from some of the logical consequences of things that are obligatory. So let's here retain RMD in our exploration of conflict tolerance. However, let's set RND, that any theorem is obligatory, aside as a distraction. What of (6.2) - Aggregation?

(6.2) $\mathbf{O}p \,\&\, \mathbf{O}q \to \mathbf{O}(p\&q)$ (O-aggregation)

Clearly this must be rejected given what we have said already. For consider our explicit conflict above, $\mathbf{O}p\&\mathbf{O}\neg p$. From any such explicit conflict, if we granted (6.2), an obligatory contradiction, $\mathbf{O}(p\&\neg p)$ would be derivable, and we have already said this is not plausible; furthermore, from RMD, since $(p\&\neg p) \to q$, for any q, we would get deontic explosion.

What of SDL's KD:

(KD) $\mathbf{O}(p \to q) \to (\mathbf{O}p \to \mathbf{O}q)$?

We must reject this as well, for even without RM, and with the very plausible and widely endorsed

(RED) If $\vdash p \leftrightarrow q$ then $\vdash \mathbf{O}p \leftrightarrow \mathbf{O}q$,

as the only deontic principle, KD would generate the left to right portion of Collapse, $\mathbf{O}p\&\mathbf{O}\neg p \to \mathbf{O}\bot$, which is surely unacceptable in its own right. (From RED, $\mathbf{O}\neg p \leftrightarrow \mathbf{O}(p \to \bot)$, then along with K, we would get $\mathbf{O}\neg p \to (\mathbf{O}p \to \mathbf{O}\bot)$, viz. $(\mathbf{O}\neg p\&\mathbf{O}p) \to \mathbf{O}\bot$ [McNamara, 2004].) Let's take just what we have so far to be our minimal conflict tolerant logic, the logic EMD [Chellas, 1974; Chellas, 1980]. Assume it has the same language as SDL, but it has just one deontic axiom, OD, and one deontic rule, RMD, along with the power of truth-functional logic for the language:

(PL) Propositional logic
(OD) $\neg \mathbf{O}\bot$
(RMD) If $\vdash p \to q$ then $\vdash \mathbf{O}p \to \mathbf{O}q$

(Regarding "EMD": "M" is Chellas's label for the thesis $\Box(p\&q) \to (\Box p\&\Box q)$, "E" for the rule "If $\vdash p \leftrightarrow q$ then $\vdash \Box p \leftrightarrow \Box q$", and "D" for $\neg\Box\bot$. Given truth-functional logic, RMD is interderivable with the combination of E and M [Chellas, 1980], so EM plus D is equivalent to RM plus D above.)

A simple semantics for this can be easily given in terms of what are called "minimal models" or "neighborhood semantics" [Chellas, 1980] (also called Montague-Scott semantics [Scott, 1970; Montague, 1970]). As in Kripke models, we have a set of worlds, W, and a valuation function, v, assigning sets of worlds to the atomic sentences, but now we replace the Kripke accessibility relation with a function that maps worlds to sets of sets of worlds (often thought of as sets of propositions).

(OB) $\quad OB : W \to \text{Pow}(\text{Pow}(W)) \quad$ i.e. $OB(u) \subseteq \text{Pow}(W)$

So the value of the obligation function for any given world, u, is a set of subsets of W—the propositions the obligation function assigns to u as mandated. The truth condition (relative to a model) for obligation statements is as follows:

(CO') $\quad u \models \mathbf{O}p$ iff $||p|| \in OB(u)$

It is obligatory that p at u (in a model) iff the proposition expressed by p (the set of p-worlds) is among those mandated by OB for u. We then validate OD by stipulating that the empty set (representing the contradictory proposition true at no worlds) is never mandated at a world:

(OB-D) $\quad \emptyset \notin OB(u)$, for every u

Notice that in some models, $OB(u)$ will not contain W, so $\mathbf{O}\top$ is not validated. Similarly, nothing has been said to indicate that if $OB(u)$ contains a set α and a set β, that it thereby must contain their intersection, so Aggregation is invalid, as desired.

With this as our brief framing, we now turn to some puzzles/challenges such a conflict tolerant logic faces.

Van Fraassen's general challenge

In [van Fraassen, 1973], van Fraassen published perhaps the first logical-semantic framework for conflicts of obligation. (Compare also [Chellas, 1974] for a different early conflict tolerant logical and semantic framework.) His approach is to layer it on top of a framework for imperatives, the interesting details of which will not concern us here. (See [Horty, 1994; Horty, 1997] for very influential expositions and explorations of van Fraassen's framework in tandem with developing new conflict tolerant systems inspired by developments in AI. Horty's work has helped to bring the importance of van Fraassen's challenge to the attention of deontic logicians.) The first key point is that he has an initial conflict tolerant system for **O** much like the one above. Van Fraassen then gives a simple example of a prima facia desirable inference that the simple conflict tolerant logic he has endorsed cannot ratify. The example (attributed by van Fraassen to Robert Stalnaker), in its **O**-version (ignoring the underlying imperatives) is :

(vFI) You ought to either honor your father or your mother. You ought not honor your father. So you ought to honor your mother.

Formalized, the inference pattern looks like this:

(vFI') $\mathbf{O}(f \vee m), \mathbf{O}\neg f$, so $\mathbf{O}m$

With **O**-aggregation it is easy to generate the conclusion using RMD: from $\mathbf{O}(f \vee m)$ and $\mathbf{O}\neg f$, we get $\mathbf{O}((f \vee m)\&\neg f)$, and from the latter along with RMD (which generates RED), we easily get $\mathbf{O}m$. However, we have cast **O**-aggregation aside above so as to avoid being able to derive an obligatory contradiction from every explicit conflict of obligations. So we cannot reason like this here. Van Fraassen raises a technical question, explored by Horty in the aforementioned papers, which we also pass over here, and asks a more general question, that Horty also articulates more fully, and we will call it "van Fraassen's challenge":

(vFC) Having accepted the possibility of conflicting obligations, how do we develop a conflict tolerant logic that avoids the two extremes of a logic so anemic that there is virtually no conclusions at all we can draw from joint premises (that don't follow from each premise alone), and a logic that is so strong that it generates deontic explosion (or an equally unacceptable variant thereof)?

Van Fraassen also offers the first proposed solution to this untilrecently neglected problem that he identified, which we set aside for the moment other than to say vaguely that it is a sort of two-level generalized consistent aggregation approach. See [van Fraassen, 1973] and/or the aforementioned references to Horty for details.

Van der Torre's van Fraassen-inspired puzzle

We provide a reconstruction based on [van der Torre, 1997].

Recall that above we pointed out that we needed to reject KD in a conflict tolerant context because it would otherwise generate this part of deontic collapse: $(\mathbf{O}p \& \mathbf{O}\neg p) \to \mathbf{O}\bot$. But now notice that (vFI') above is really a barely disguised instance of KD: $\mathbf{O}(\neg f \to m) \to (\mathbf{O}\neg f \to \mathbf{O}m)$. So the *pattern* of inference (vFI'), despite any initially plausible ring, is unacceptable on reflection, and then so is (vFI), since *logically invalid*—not valid in virtue of its form. What makes the pattern sound plausible, as with the instance (vFI) itself, is that we naturally think of the wffs, $(f \vee m)$ and $\neg m$ *as mutually consistent* [McNamara, 2004]. This suggests endorsing the following principle modestly restricting $\mathbf{O}$-aggregation as a natural solution to van Fraassen's challenge above:

(CA)   If $\not\vdash p \to \neg q$ then $\vdash \mathbf{O}p \& \mathbf{O}q \to \mathbf{O}(p \& q)$ (Consistent aggreg.)

[van der Torre, 1997] attributes "consistent aggregation" to van Fraassen, but it appears on reflection (and in conversation with van der Torre) that this was more likely adopted from [Horty, 1994], who uses "consistent agglomeration" there, attributing the notion to [Brink, 1994].

[Brink, 1994] mentions this principle and gives it a qualified endorsement, and it was endorsed in an earlier draft of [McNamara, 2004] for DEON 2002. It is certainly a natural first amendment to consider in developing a conflict tolerant logic, and plausible at first blush. After all, the main reason for rejecting aggregation is the possibility of aggregating incompatible obligations, so if we know that two obligatory things are mutually compatible, what reason can there be to not go ahead and aggregate them. Van der Torre provides a decisive answer in the context we are exploring, for he shows us that however plausible this may sound, as long as we have (CA) and (RMD), we will quite easily generate, a trivial variant of the very result we said had to be avoided first and foremost in a conflict tolerant logic: Deontic Explosion. Here is the trivial and surely unacceptable variant of (DEX):

(DEX') $\mathbf{O}p \& \mathbf{O}\neg p \to \mathbf{O}q$ for any q such that $\not\vdash \neg q$

In other words, if there is any conflict of obligations, every non-contradiction will be obligatory.

Now call this *van der Torre's thesis*:

(vdTT)   From (RMD) and (CA), (DEX') follows

Suppose we have an explicit conflict, $\mathbf{O}p \& \mathbf{O}\neg p$, and suppose we have some consistent q. Then, either $\neg p \& q$ or $p \& q$ is consistent. First suppose $\neg p \,\&\, q$ is consistent. Then its equivalent, $(p \vee q) \,\&\, \neg p$ is consistent. So by (CA), we have $\mathbf{O}(p \vee q) \,\&\, \mathbf{O}\neg p \to \mathbf{O}((p \vee q) \,\&\, \neg p)$. Now, from $\mathbf{O}p$, by (RMD) we get $\mathbf{O}(p \vee q)$, and along with $\mathbf{O}\neg p$, we get $\mathbf{O}((p \vee q) \,\&\, \neg p)$, and then from this by (RMD) again, $\mathbf{O}q$ follows. Secondly, suppose $p \,\&\, q$ is consistent. Then by precisely parallel reasoning, we get $\mathbf{O}q$. So either way, $\mathbf{O}q$ follows.

So, here we have a prima facia well-motivated restriction on O-aggregation that along with RMD generates a version of explosion no more palatable than the original (especially considering that we have already ruled on the only exception to (DEX')'s explosive scope: $\neg \mathbf{O}\bot$ and its equivalents, so that, in effect, we get that everything that is not already logically ruled out as impossibly obligatory becomes obligatory in the face of any conflict).

So meeting *van Fraassen's challenge* is not as easy as it might at first seem. We are thus left puzzling about what form of aggregation specifically, if any, can be endorsed in allowing us to meet van Fraassen's challenge, given that this natural one fails; or must we whittle away instead at RMD, or follow yet some other path to meet the challenge?

Van der Torre, and others, have referred to this puzzle as "van Fraassen's puzzle", but this appears to be a misnomer. Although no doubt derived by van der Torre's reflection on van Fraassen's rich, compact, and sometimes cryptic remarks at the end of [van Fraassen, 1973], there is no mention of the deontic explosion problem that van der Torre articulates (although explosion is mentioned much earlier by van Fraassen in the article), nor is there more than, at best, a suggestion of the simple consistent aggregation principle above, and no mention of it. However, it is noteworthy that van Fraassen offers a solution to what we have called here "van Fraassen's challenge" that certainly involves a semantic version of a different restricted aggregation principle, and so it is certainly possible that he entertained the simpler consistent aggregation principle.

The idle aggregation puzzle (van Fraassen-Hansen)

Let us now point out that the problem may not be so readily solved merely by restricting RMD either, suggesting the challenge and puzzle of how to solve it is robust. For even if RMD is too strong, we must surely adopt some principles governing practical reasoning allowing us to reason about consequences of what is obligatory for us. Consider the following example adapted from one communicated by Jörg Hansen in 2002:

p: Jones keeps an appointment this morning in New York.
p': Jones travels to New York this morning.
q: Jones keeps an appointment in London this afternoon.
q': Jones travels to London this morning.

([Horty, 2003] provides a similar example where there is an obligation to both attend an event and pre-notify, and likewise for another event, in a different place at the same time. Here we have a conflict between two obligatory conjunctions, but from RMD, it follows that each notificational conjunct is obligatory, and as these are mutually consistent, by consistent aggregation, their conjunction would be obligatory too.)

Imagine, not implausibly, that p practically necessitates p', and q practically necessitates q', and add that given the times and distances, Jones is unable to keep both appointments. Nonetheless it might be that traveling to both places this morning is open to Jones, for example, by driving to JFK airport in New York early this morning (long before his New York meeting) and flying directly from there to London. It seems implausible to conclude that Jones is obligated to both travel to New York and travel to London, which in turn is only achievable through the mad frenzied dash just sketched. The travel obligations derive exclusively from the appointment obligations that conflict and can't be jointly realized, so it is not plausible that a singular conjunctive obligation to travel to both New York and to London follows. Now notice that this problem is not easily solved by just saying RMD is implausible, for p' and q' above are actions in my power that are *practical prerequisites* of p and q respectively. Reasonable restrictions of RMD need to allow us to make inferences like these from premises about practical prerequisite of obligations we have, since this seems to be nothing short of central to practical reason itself.

So it appears that it is more plausible to see the key issue as being about how to properly restrict aggregation beyond (CA), even if there are other independent reasons to want to restrict RMD. A faithful representation of obligations must allow us to derive from our obligations further obligations to realize their practical prerequisites; but at the same time, it seems we must disallow the derivation of idle conjunctive obligations. (This presentation draws on [McNamara, 2004].)

For more on the last three interrelated [van Fraassen, 1973]-inspired issues which have only received their due attention more recently, see for example, [Horty, 1994; van der Torre, 1997; van der Torre and Tan, 2000; Horty, 2003; Hansen, 2004; McNamara, 2004; Goble, 2005; Goble,

2009], as well as Goble's chapter in this volume.

(The above problem was conveyed by Jörg Hansen to McNamara at DEON 2002 and discussed in McNamara 2004; it appears that van Fraassen recognized a very similar problem of potential over-generation of conclusions that served as the inspiration for Hansen's articulation of the problem as it applied to McNamara's earlier system. We benefited from discussion with Lou Goble, Jörg Hansen, John Horty, and Leon van der Torre on these van Fraassen-inspired puzzles.)

8.5 Puzzles centering around deontic conditionals
The paradox of derived obligations[62]

Consider this statement:

(8.38) Bob's promising to meet you commits him to meeting you

Two very natural attempted representations of claims like that in (8.38) in standard systems were suggested:

(8.39) $\mathbf{O}(p \to m)$ ([von Wright, 1951])

(8.40) $p \to \mathbf{O}m$ ([Prior, 1955])[63]

Consider (8.39) first, which was how von Wright first interpreted statements like (8.38). The following are theorems by RMD, and thus in all standard systems: $\mathbf{O}\neg r \to \mathbf{O}(r \to s)$ and $\mathbf{O}s \to \mathbf{O}(r \to s)$. Thus if the logic of (8.38) were correctly realized in SDL by representing it as (8.39), it would follow that anything impermissible commits us to everything, and that for anything obligatory, everything commits us to it. Does (8.40) work better? No. The following are simply tautologies: $\neg r \to (r \to \mathbf{O}s)$ and $\mathbf{O}s \to (r \to \mathbf{O}s)$. So if the logic of (8.38) were correctly realized in standard systems by representing it as (8.40), it would follow that, anything false would commit us to anything whatsoever (e.g. since I did not promise you to meet, it would follow that my promising to meet you commits me to not meeting you) and again, for anything obligatory, everything commits us to it (e.g. if I'm obligated to phone you, then my living in a time with no phones commits me to phoning you). As Prior notes, the problems are reminiscent of the paradoxes of strict implication (reading (8.39) and material implication (reading (8.40), respectively. This raises the question: is it simply beyond the resources of standard systems to properly represent notions of commitment or conditional obligations? The next paradox convinced logicians that indeed it is.

---
[62][Prior, 1954].
[63][Prior, 1955] credits G. E. Hughes for this alternative symbolization.

Chisholm's contrary-to-duty paradox[64]

Here is Chisholm's famous quartet:[65]

(8.41) It ought to be that Jones goes to the assistance of his neighbors

(8.42) It ought to be that if Jones goes to the assistance of his neighbors, then he tells them he is coming

(8.43) If Jones doesn't go to the assistance of his neighbors, then he ought not tell them he is coming

(8.44) Jones does not go to their assistance

It is widely thought that (8.41)-(8.44) constitute a *mutually consistent* and *logically independent* set of sentences: all four might be true at once, and none is a deductive consequence of the others. We will treat these as central desiderata: a correct representation of the logic of (8.41)-(8.44) must be consistent with these two constraints.[66] The problem, in a nutshell and from a high altitude, is that it is not at all as easy as it might seem to faithfully represent scenarios like those in the quartet and still meet the above two constraints, and it proved to be a real shortcoming of the standard systems as people quickly came to realize that they could not be represented there. On the positive side, it has been a catalyst for distinctive and expansive work in deontic logic. It is perhaps the most important puzzle in the history of 20th century deontic logic, and so we will spend some more time on it. Here we will briefly characterize the problem for the standard systems. The supplement to 8.5 provides more detail about some attempted solutions to this puzzle.

First we provide some terminology that has emerged regarding the taxonomy of the puzzle ingredients. Since (8.41) tells us what Jones ought to do unconditionally, it is a *primary obligation*, the only one in this context.[67] (8.43) is a *contrary-to-duty obligation* (a CTD), an instance of the type of claim after which the puzzle is named. In the context of (8.41), (8.43) says what Jones ought to do on the condition that he *violates* (or at least does not fulfill) his primary obligation in (8.41). In contrast, (8.42) says what

---

[64][Chisholm, 1963].

[65]We give Chisholm's original example since the piece is so seminal in deontic logic, but this also means that (8.42) and (8.43) have a different form with *ought* having a different surface scope. The difference between (8.42) and (8.43) in Chisholm's original formulation is largely seen as a distracting artifact, and in many presentations of the puzzle, (8.42) is adjusted to follow the form of (8.43), which form is thought to be the more challenging one to represent, and the most central to contrary-to-duty conditionals.

[66]Others will be alluded to in passing below. [Carmo and Jones, 2002] argues for no less than seven desiderata for any solution.

[67]We will follow tradition here in sloughing over the differences between an obligation and what ought to be and what one ought to do, since we believe the puzzle reappears as we shift across these three distinct notions.

else Jones ought to do on the condition that Jones *fulfills* his primary obligation, and so (8.42) could be called a "compliant-with-duty obligation" in this context. Finally, (8.44) is just a factual claim, which conjoined with (8.41), implies that Jones violates (or at least does not fulfill) his primary obligation. The relativization to context for both labels is crucial, since in another context, (8.42) could be a contrary-to-duty instead (e.g. advanced notice is not important, and Jones agreed with a friend that they would both surprise the neighbors with their help) and (8.42) might be compliant with duty in the right context (e.g. due to character defects of the neighbors in question, if they knew Jones was coming to their assistance they would not make the vital efforts now essential to Jones not being too late to help at all). Thus the taxonomy involves tracking the relationships between the normative and factual claims across a piece of discourse.

How might we represent the Chisholm quarter in the standard systems?[68] The most natural first stab appears to be:

(8.41') $\mathbf{O}g$

(8.42') $\mathbf{O}(g \rightarrow t)$

(8.43') $\neg g \rightarrow \mathbf{O}\neg t$

(8.44') $\neg g$

Here we read (8.42) with $\mathbf{O}$ having wide scope, and (8.41) with $\mathbf{O}$ having narrow scope, following the surface of the original. Chisholm noted that by principle KD, we get $\mathbf{O}g \rightarrow \mathbf{O}t$ from (8.42'), and then $\mathbf{O}t$ from (8.41') by MP. In turn, from (8.43') and (8.44'), we get $\mathbf{O}\neg t$ by MP alone. But the combination of $\mathbf{O}t$ and $\mathbf{O}\neg t$ contradicts DD ($\mathbf{O}t \rightarrow \mathbf{O}\neg t$). Thus (8.41')-(8.44') is an inconsistent quartet in any of the standard systems, unlike the original whose logical form they are alleged to represent. Various other representations in the standard systems have similar shortcomings. For example, we might try reading the second and third premises uniformly either on the model of (8.42') or on the model of (8.43'). After all, it is not clear what motivates framing them differently in the original quartet (8.42)-(8.43), and this oddity is often dropped in contemporary discussions. If we use

(8.43") $\mathbf{O}(\neg g \rightarrow \neg t)$

instead of (8.43'), we lose independence since (8.43") is derivable from (8.41') in the standard systems by RMD (as we saw in discussing the *paradox*

---

[68] We use primes here to make it easier to keep track of the various correlated statements in the different quartets.

of derived obligation). Likewise if we use

(8.42") $\quad\quad g \rightarrow \mathbf{O}\neg t$

instead of (8.42'), independence is again lost, for this is derivable from (8.44') by PC alone. So again, we end up with unfaithful representations of the logic of the original quartet.

We can sum up the problems with these three ways to interpret (8.41)-(8.44) in the standard systems in the following table:

	First	Second	Third
	$\mathbf{O}g$	$\mathbf{O}g$	$\mathbf{O}g$
	$\mathbf{O}(g \rightarrow t)$	$\mathbf{O}(g \rightarrow t)$	$g \rightarrow \mathbf{O}t$
	$\neg g \rightarrow \mathbf{O}\neg t$	$\mathbf{O}(\neg g \rightarrow \neg t)$	$\neg g \rightarrow \mathbf{O}\neg t$
	$\neg g$	$\neg g$	$\neg g$
Problem	$\therefore \bot$	$\mathbf{O}g \vdash \mathbf{O}(\neg g \rightarrow \neg t)$	$\neg g \vdash g \rightarrow \mathbf{O}t$

Our first attempt yields a contradiction—the set is rendered inconsistent. The second and third attempts lose independence, since one of the four follows from the others in each case (in fact from just one premise, as indicated). The only remaining apparent combination would replace (8.42') with (8.42") $g \rightarrow \mathbf{O}\neg t$ and (8.43') with (8.43") $\mathbf{O}(\neg g \rightarrow \neg t)$, but that just combines the loss of independence in the second and third attempts, so it is rarely mentioned.

Given the extreme simplicity of the semantics offered for the standard systems, it is not surprising that it is not capable of representing complex normative situations in a satisfactory way, and it is not difficult to see why Chisholm's example in particular cannot be represented in a satisfactory way. As was observed above, the semantics of SDL is based on a division of worlds (situations) into normatively acceptable and unacceptable ones, with the O-sentences defined so that they describe how things are in the deontically acceptable situations. But CTD sentences do not describe how things are in deontically acceptable worlds; instead they tell what is to be done or how things ought to be under deontically *unacceptable* conditions, and specific ones at that (e.g. in worlds where Jones does not go to the assistance of his neighbors). For that, we need a way to pick out not only worlds where things have gone wrong, but where things have gone wrong in some specific way indicated by the clause of the CDT that expresses the violation of the primary obligation; and then we must go on to select propositions that are *relatively-acceptable*—acceptable relative to the assumption that the worlds will be those unacceptable ones where the specific violation conditions hold.[69] So it is no wonder that SDL and kin cannot express

---

[69] Cf. [Lewis, 1974; Jones and Pörn, 1985], although note that Lewis, followed by many

CTDs.

One difference with these conditional obligations or ought-statements that was noted early on was that they are *defeasible* in the sense that they do not satisfy the principle of *strengthening the antecedent*, which of course does hold for material implication:

(SA) $\qquad (p \to q) \to ((p \& r) \to q)$ $\quad$ (Strengthening the Antecedent)

The corresponding thesis for the deontic conditionals in focus in Chisholm's puzzle was virtually universally recognized to be invalid:

(8.45) $\qquad \mathbf{O}(q/p) \to \mathbf{O}(q/(p \& r))$

Even if Jones ought to tell if he will help, it will not follow that it is also true that he ought to tell if he will help and telling will cause some disaster.

It is now virtually universally acknowledged that the Chisholm Paradox shows that the sort of deontic conditional expressed in (8.43) above can't be faithfully represented in SDL, or even as a composite of some sort of a unary deontic operator and a *material* conditional. Here is one of the key places where deontic logicians are in full agreement.

By giving pride of place to contrary-to-duty requirements, the puzzle also brings into relief a crucial feature of (most if not all) normative requirements: their violability.[70] Since we are quite imperfect creatures, the possibility of violation is hardly idle. It is crucial for us to know not only what is to be done, but also what to do in turn when what ought to be done in the first place is not done. Consider the role of apologies in repairing the torn social fabric, or statements in contract law about what is owed in reparation if one party fails to provide what is owed in the primary clauses of the contract (e.g. amazon.com owes you a refund if the wrong item arrives). We would have nuclear meltdowns without emergency clauses about what a crew is to do at a nuclear power plant when the crew has failed to do something required and things have started to go wrong. When things do go wrong, thankfully, they often have not gone as wrong as they can go; we can take adjusting actions mitigating the harm. Damage control is vital,

---

others, concludes that more is needed, namely a preference ordering of worlds allowing for a selection of the "best of the bad" compatible with any particular given violation, whereas Jones and Pörn argue that this is not necessary, a representation of what holds in the non-acceptable worlds (along with what holds in the acceptable worlds, and thus in both classes) will suffice.

[70][Jones, 1990] argues for the importance of violability to legal knowledge representation and consequentially for the importance of deontic logic for such knowledge representation, stressing particularly the issue of representing contrary-to-duty contexts; See also [Jones and Sergot, 1993] which argues that violability (and the possibility thereof) is what gives deontic logic much of its importance.

and in turn, the ability to reason accordingly is vital. The puzzle also obviously places deontic *conditional* constructions at center stage, inviting us to ponder: What is the correct logic behind reasoning with deontic conditionals generally, and particularly, in contexts where the conditionals appear to tell us what we are obligated to do if we violate some other obligation? Lastly, the puzzle also raises the question of how to track fulfillment, non-fulfillment, and particularly violation of obligations, along with conditional obligations, across a set of statements (a piece of extended discourse), as indicated by our initial context-relative taxonomy of the Chisholm quartet.

Among the things contested regarding this paradox are whether or not what is needed is some special *primitive dyadic deontic conditional* operator or just some *non-material* conditional conjoined to a monadic deontic operator, as well as the more general question of what essentially is needed to faithfully represent the logic of deontic conditionals like those in the puzzle. For the interested reader, we explore these further in the *Appendix on Chisholm's puzzle & conditional norms*. Next, we introduce a closely related puzzle.

Forrester's paradox[71]

Here is a version very close to the original:[72]

(8.46) Smith ought not kill Jones
(8.47) If Smith will <u>k</u>ill Jones, then Smith ought to kill Jones <u>gently</u>
(8.48) Smith will kill Jones

As with the kindred Chisholm puzzle, this triplet appears to express a mutually consistent set of claims, with each claim independent of the remainder.

Here is a natural way to symbolize (8.46)-(8.48) in our standard systems:

(8.46') $\mathbf{O}\neg k$
(8.47') $k \to \mathbf{O}g$
(8.48') k

---

[71][Forrester, 1984]. This is also called the "Gentle Murder Paradox".
[72]As the title, "Gentle Murder or the Adverbial Samaritan" indicates, Forrester introduced the puzzle as "the most powerful version yet" of the *Good Samaritan puzzle*, one intended to rule out prior scope solutions targeting the original, and as might then be expected, he opts to drop RMD. However, Forrester's puzzle was cast in terms of deontic conditionals much like those above (with (8.47) above as a key auxiliary premise, and its wide scope analogue as main premise), and it is thus often construed instead as a variant of *Chisholm's paradox*, pointing again to the challenge of modeling deontic conditionals that seem to be telling us what to do if wrong will be done. We thus place it here, although RMD will also be invoked in showing that SDL and ilk are inadequate. It has features of both puzzles.

Now from (8.47') and (8.48') by MP, it follows that

(8.49) $\mathbf{O}q$

But the following seems to be a natural language logical truth:

(8.50) Smith kills Jones gently only if Smith kills Jones

Assuming so, let's imagine an augmented standard deontic system where (8.50) is a formal logical truth, which we will symbolize here for simplicity as

(8.51) $g \to k$

From this, by RMD, it will follow in such an augmented system that

(8.52) $\mathbf{O}g \to \mathbf{O}k$

and so by MP again, we get

(8.53) $\mathbf{O}k$

But now with (8.53) added to (8.46'), we have conflicting obligations, in contradiction with DD of the standard systems. So it looks like the standard systems cannot coherently represent Forrester's triplet.

Although (8.46)-(8.48) seem like they could all be true, it seems difficult to swallow that (8.47) and (8.48) entail that Smith is obligated simpliciter to kill Jones gently (e.g. it seems we can consistently add that Jones has not the slightest justification for harming Smith at all).[73] On the other hand, if we side with those favoring interpreting the deontic conditional in (8.47) as non-material but still subject to a version of modus ponens, we must then accept the inference from (8.47) and (8.48) to the informal analog of $\mathbf{O}g$.[74] So it appears that unless we reject some principle of SDL such as RMD, we will still generate a contradiction.[75] Note also that simply opting

---

[73] In the Appendix on Chisholm's paradox at the end of this chapter, the reasoning here will be discussed in in the context of motivating the position of the friends of *"deontic detachment"*.

[74] This corresponds to the position favored by friends of *"factual detachment"*, also discussed in the Appendix on Chisholm's paradox.

[75] Some have suggested this is still a problem stemming from scope difficulties, others have argued that the problem is that RMD is in fact invalid, and rejecting it solves the problem. [Sinnott-Armstrong, 1985] argues for a scope solution; [Goble, 1991] criticizes the scope solution approach, and argues instead for rejecting RMD. We have listed this puzzle here rather than under the *Good Samaritan puzzle* (and thus under puzzles associated with RMD) since, unlike the standard Good Samaritan, this puzzle seems to

to reject DD as a response is not a natural avenue, since it does not seem very plausible to say that the problem is simply that we have a conflict of obligations, the obligation to kill Jones and the obligation to not kill Jones (or the obligation to kill Jones gently and to not kill Jones gently). If this is right, then it supports the contention that what is at issue ultimately in this puzzle, and with the Chisholm puzzle, pertains to these particularly troubling CTDs.

There is a vast literature on this subject, and here we can only sketch a fragment. Some of this material is briefly discussed in the Appendix to 8.5. In a brief note responding to Prior's paradox of derived obligation, [von Wright, 1956] introduced the often-used undefined dyadic operator approach to the syntax of conditional obligations, $O(q/p)$. See also [von Wright, 1964; von Wright, 1965] for further developments of his approach, and an explicit recognition of the importance of Chisholm's paradox. Von Wright's approach is primarily syntactic and axiomatic. [Danielsson, 1968; Hansson, 1971], followed a bit later by [Lewis, 1973; Lewis, 1974], provide formal semantics for conditional obligation, using preference orderings of worlds to model CTDs construed via an undefined dyadic operator of von Wright's sort. [Åqvist, 2002; Åqvist, 1987] provide systematic presentations of this sort of approach, as well as analogue systems for deontic conditionals in the Leibniz-Kangerian-Andersonian vein (and discussions of other paradoxes). [van Fraassen, 1972; Loewer and Belzer, 1983; Jones and Pörn, 1985] give important and influential alternative models for CTDs, each offering some interesting variants of the former more standard picture. [al Hibri, 1978] contains an early important survey of a number of these approaches to CTDs (and other puzzles), as well as a defense and development of her own system. In addition to [Lewis, 1973; Lewis, 1974], an important recent presentation and development of the metatheory of standard and near-standard monadic and dyadic deontic logics via classic and near-classic ordering structures is [Goble, 2003]. [Mott, 1973; Chellas, 1974; Chellas, 1980; Goble, 1990a] offer influential alternative approaches that do not use an undefined dyadic deontic operator. Instead they opt for representing deontic conditionals using a non-material conditional, $\Rightarrow$, along with a unary deontic operator to generate a genuine compound sentence form,

---

crucially involve a contrary-to-duty conditional, and so it is often assumed that a solution to the Chisholm paradox should be a solution to this puzzle as well (and vice versa). Alternatively, one might see the puzzle as one where we end up obligated to kill our mother gently because of our decision to kill her (via factual detachment), and then by RMD, we would appear obligated to kill her, which has no plausibility by anyone's lights, and thus the puzzle calls for rejecting RMD. However, this would still include a stance on contrary-to-duty conditionals and detachment.

$p \Rightarrow \mathbf{O}q$. [DeCew, 1981] contains an important early critical discussion of this sort of approach. [Tomberlin, 1981] is a very influential informal discussion of various approaches to deontic conditionals. [Bonevac, 1998] is a recent argument against the dyadic approach to conditional obligation, suggesting that defeasible reasoning techniques developed in AI (see [Brewka, 1989]) can handle the problems with CTDs. In contrast, [Smith, 1993; Smith, 1994] stress the difference between violability and defeasibility, and the relevance of the former rather than the latter to CTDs. [Åqvist and Hoepelman, 1981; van Eck, 1982; Loewer and Belzer, 1983] are early approaches to solving the puzzle (or versions thereof) by incorporating temporal notions. [Jones, 1990], as well as [Prakken and Sergot, 1994; Prakken and Sergot, 1996], are influential for their arguments that temporal notions are not essential to the Chisholm's paradox, so the solution cannot lie there. [Castañeda, 1981] argued that by distinguishing between propositions and "practitions" (roughly actions), most puzzles for deontic logic could be solved, including Chisholm's paradox; [Meyer, 1988] takes a similar approach but employing techniques from dynamic logic to represent actions, their combinations, and deontic notions. [Prakken and Sergot, 1996] is influential for arguing that action is inessential to the Chisholm paradox, so that the solution cannot lie there. For some work on CTDs in a branching time framework see [Horty, 1996; Horty, 2001; Bartha, 1999] and Bartha's chapter 11 in [Belnap et al., 2001]. [Carmo and Jones, 2002] is an important recent handbook chapter reviewing various approaches to deontic conditionals in detail, as well as proposing and defending a solution of their own. [Nute, 1997] is a collection dedicated to defeasible deontic logic with a number of essays on Chisholm's puzzle (see especially [van der Torre and Tan, 1997; Prakken and Sergot, 1997]).

We should also mention the important and influential topic of the *counts as conditional*, which we regret not being able to discuss but briefly here. In short, a conditional is introduced to represent the idea of one proposition's realization counting as or constituting another's realization. For example, as mentioned in Section 8.1 and earlier sections, a performative act may constitute the realizing conditions of some other act. For example, in the right institutional settings, my raising my hand counts as my voting on the measure, the proposition that "You are hereby married", uttered by a magistrate counts as a truth-maker for the proposition that you are married, and a delegated person in a business signing an agreement for an order counts as a truth-maker for the business itself being obligated to meet the terms of the agreement. Since such constructions can be regimented into relations between propositions as we did in the middle example just above, they can be treated as a form of conditional, and logics devised accordingly.

Such logics can then be integrated with various agential notions, to in turn represent various institutional transactional phenomena like delegation, authorization, and the trigger of changes in the normative position of various agents in institutions, and even of the institutions themselves, as in our last example. The locus classicus on this topic is [Jones and Sergot, 1996]. See also the chapter in this volume by Jones and Grossi covering this topic.

Let us also note here the not altogether unrelated accumulating work of Lindahl and Odelstad on formal representations of the role of "intermediate concepts" in law. If you are accused of a crime such a burglary, a variety of legally (and stipulatively) defined concepts will be invoked, such as "forced entry", "property", "theft", "person", etc. Although these terms are familiar, appropriated for a legal system, they are defined, and do not always track their normal use precisely. For example judges and lawyers need to be familiar with the exact legal definitions of these terms in order to be competent in adjudicating a legal accusation. In turn, such terms are often defined explicitly (eventually) in terms of extra-legal or "natural" terms that are not encoded in any legal definitions, such as "entering", "object", "transport", "human being", "building", etc. These serve as the "grounds" for the applicability of the intermediate legal concept. The intermediate concepts in turn might be associated with various normative consequences such as sanctions, loss of rights, etc. There is a sense in which, roughly, the stipulatively defined concepts function as intermediaries between the extra-legal grounds they are defined in terms of and the higher level legal-normative consequences that they are linked to (e.g. that burglary is punishable by imprisonment for up to ten years, that the convicted burglar can be held accountable for damages with loss of some of his property for reparations, etc.). See [Lindahl and Odelstad, 2000] for a concise overview, as well as the chapter by Lindahl and Odelstad in this volume for a more comprehensive account.

8.6 Some further expressive inadequacies of the standard systems

We have already noted the apparent expressive inadequacies in standard systems regarding Chisholm's paradox and deontic conditionals. In this section we turn instead to some monadic normative notions that appear to be inexpressible in the languages for VW, SDL and LKA_1 and LKA_2. In a number of cases, it appears that these notions were at least tacitly targeted for representation in standard systems although not actually expressible in them.

Urmson's puzzle - Indifference versus optionality[76]

Consider:

(8.68) It is optional that Jones helps Smith, but not a matter of indifference

We routinely assume that optional matters are not thereby matters of indifference. Yet deontic logicians and ethicists routinely read the condition "$\neg \mathbf{O}p \& \neg \mathbf{O} \neg p$" as "It is *indifferent* that p" ($\mathbf{IN}p$) rather than as "It is optional that p" ($\mathbf{OP}p$).[77] But then it would seem to follow trivially that $(\neg \mathbf{O}p \& \neg \mathbf{O} \neg p) \to \mathbf{IN}p$: if anything is neither obligatory nor prohibited then it is indifferent, that is, neither obligatory nor prohibited. So in the standard systems, the best we can do in symbolizing (8.68) is by way of a tautological contradiction:

(8.68') $(\neg \mathbf{O}p \& \neg \mathbf{O} \neg p) \& \neg (\neg \mathbf{O}p \& \neg \mathbf{O} \neg p)$

Many alternative actions, including heroic ones, are neither obligatory, prohibited, nor matters of indifference. Urmson implores us to not conflate these two concepts and thereby indirectly rule out many cases of moral heroism that are often morally exemplary and optional, and to instead develop logical schemes that ratify the following constraint (Urmson's Constraint):[78]

(UC) $\mathbf{IN}p \to \mathbf{OP}p$, but not $\mathbf{OP}p \to \mathbf{IN}p$

But the standard systems can only represent optionality at best; they lack the expressive resources to carve up the optional zone into the *indifferent* and the *optional but non-indifferent*. Yet indifference was tacitly an early target for representation in as much as it was thought that this was aptly represented in the standard systems.

The problem of action beyond the call of duty[79]

Some alternatives are beyond the call of duty (**BC**) or supererogatory (e.g. volunteering to take on a challenging project for your department having already "paid your dues"). The standard systems have no resources to represent this notion, since they can say nothing more fine-grained about

---

[76] [Urmson, 1958].

[77] Beginning with [von Wright, 1951], and recurring pervasively. Note that because we will begin to discuss distinct notions often conflated with one another in ethical theory and deontic logic, we will employ two letter abbreviations for operators to represent additional concepts more transparently, as we did above in A6.1.

[78] See [McNamara, 1996a] for further discussion.

[79] [Urmson, 1958]. It may be that supererogation and action beyond the call are subtly distinct [McNamara, 2011a; McNamara, 2011b]. We slough over this issue here.

them than that they are optional (neither obligatory nor impermissible), but although the optionality of what is beyond the call.

(BC-OP) $\mathbf{BC}p \to \mathbf{OP}p$

is desirable, its converse, $\mathbf{OP}p \to \mathbf{BC}p$, surely is not. As Urmson's Constraint above (UC) indicates, matters of indifference are optional.[80] So representations of this asymmetry in standard systems will end up being trivial or incoherent (e.g. $\vdash \mathbf{OP}p \to \mathbf{OP}p$, $\nvdash \mathbf{OP}p \to \mathbf{OP}p$, respectively). Also, note that $\mathbf{BC}p \to \neg\mathbf{IN}p$ is also desirable, so to represent this notion fully, we need distinct representations of optionality and indifference.

The must versus ought dilemma[81]

Consider:

(8.69) Although you can skip this meeting, you ought to attend (and you must attend either this one or the next).

We routinely make such distinctions in situations where no conflicting obligations are present. (8.69) appears to properly entail that it is optional that you attend - that you *can* attend and that you *can* also not attend, although preferable to attend. In context, the latter two uses of "can" paradigmatically express permissibility. Yet "ought" is routinely the reading authors give for deontic necessity in deontic logic and in ethical theory, and "permissibility" is routinely presented as its dual. But this suggests the following symbolization of the first two conjuncts of (8.69):

(8.70) $\mathbf{PE}\neg p\,\&\,\mathbf{OB}p$

But (8.70) is equivalent to $\neg\mathbf{OB}p\,\&\,\mathbf{OB}p$ (by RED and TDS), which contradicts DD. It is much more plausible to construe the "can" of permissibility ($\mathbf{PE}$) as the dual of "must" ($\mathbf{MU}$) than as the dual of "ought" ($\mathbf{OU}$). It appears that $\mathbf{MU}p \to \mathbf{OU}p$ is desirable but not $\mathbf{OU}p \to \mathbf{MU}p$, and $\mathbf{MU}p \to \neg\mathbf{PE}\neg p$ is desirable, but not $\mathbf{OU}p \to \neg\mathbf{PE}\neg p$. This suggests a dilemma for the standard systems (and most work in deontic logic):

> Either permissibility is represented in the standard systems, but "ought" is inexpressible in it (despite the widespread assumption otherwise) or "ought" is represented in those systems, but permissibility and impermissibility are inexpressible in them despite the widespread assumption otherwise. You can't have it both ways at the same time.

---

[80][Chisholm, 1963] is a landmark here, and as we've seen, these issues clearly overlap with things Meinong was beginning to explore much earlier (as Chisholm notes).

[81][McNamara, 1990].

That the dual of permissibility is expressed by "ought" is a problematic but pervasive "Bipartisan Presupposition" in both deontic logic and ethical theory.[82] [McNamara, 1996c], and in more detail [McNamara, 1990], argues that there is also very strong pressure from the use of the modal auxiliaries "must" and "ought" in non-deontic contexts to a) distinguish them, b) to take the former to properly entail the latter, and c) to not posit that there is an ambiguity in "ought" for the purpose of saying that there is one sense in which it means the same as "must".[83]

The least you can do problem[84]

Consider

(8.71) You ought to have been on time; the least you could have done was called, and you didn't do even that

Although there has been lots of attention to constructions like that in the first clause of (8.71), the construction in the second clause has been almost totally ignored in deontic logic and ethical theory. Yet it is familiar and widely used, and it appears to entail that there was some minimally acceptable alternative that included calling (to say you would be late), whereas the first clause suggest there was also an acceptable but preferred alternative, which was to just be on time (and so not call to say you would be late). The third clause suggests the criticism that even though you had permissible options of different ranks, you did less than even the minimally acceptable option, and thus you comported yourself impermissibly. Presumably, we want things such as $\mathbf{OB}p \to \mathbf{LE}p$, $\mathbf{LE}p \to \mathbf{PE}p$, but not $\mathbf{LE}p \to \mathbf{OB}p$, etc. This rich notion of what is minimally acceptable among the permissible options is plainly not expressible in the standard systems.

As the reader might surmise, the set of notions above appear to be part of an underexplored interlocking family of normative notions.[85]

---

[82]It is a merit of [Jones and Pörn, 1986] that it recognizes there is a clear difference between deontic uses of "ought" and "must", and it provides an early attempt to distinguish the two (in a logical system with a formal semantics). However, "must" ends up being modeled as something akin to *practical* necessity in their system (whatever obtains in all scenarios - permissible or not) rather than deontic necessity (whatever holds in all *permissible* scenarios). For a cumulative case argument that "must" is the dual of permissibility, not "ought", and thus that it is "must", not "ought", that tracks the traditional concern in ethical theory and deontic logic with what is permissible, impermissible, and obligatory, see [McNamara, 1990; McNamara, 1996c].

[83]Note that c) is of limited interest as a reply anyway, since even granting it for sake of argument, surely the important task then is to analyze the sense of "ought" that is *not* equivalent to the sense of "must", and to integrate these two in one logic.

[84][McNamara, 1990].

[85]For attempts to begin to address the last four problems by the simplifying ploy of extending familiar standard or near standard systems, see [McNamara, 1996b;

The challenge of normative gaps[86]

As we say in discussing Jørgensen above, in some normative contexts, explicit permissions, prohibitions and requirements are issued by some normative authority. But then we must allow for a type of gap: cases where p is neither *explicitly* obligatory, impermissible, nor permissible in a normative system because it is not *explicitly* commanded, prohibited or permitted. Yet in all the standard systems,

(6.12) $\mathbf{O}p \vee (\mathbf{P}p \,\&\, \mathbf{P}\neg p) \vee \mathbf{F}p$ (Exhaustion)

is a thesis. In fact, it is nearly *tautological* given the TDS (Traditional Definitional Scheme). For in primitive notation, it amounts to just this: $\mathbf{O}p \vee (\neg\mathbf{O}\neg p \,\&\, \neg\mathbf{O}\neg\neg p) \vee \mathbf{O}\neg p$, and so only RED is needed to replace $\neg\mathbf{O}\neg\neg p$ with $\neg\mathbf{O}p$ with the result saying essentially "it is this, that, or neither" (where "neither" always gets cashed out as entailing permissibility). So any system endorsing just the language and definitional scheme of SDL that includes truth-functional logic and just the one deontic rule, RED, will ratify Exhaustion. Then, for any proposition p, p will either have the status of being impermissible or obligatory or permissible (since optional). This precludes **P**, **O**, and **F** in the standard systems from being normative notions that allow for gaps.

The problem of the directionality of obligations[87]

In the standard systems, the bearers of obligations, if any are intended, go unrepresented. Furthermore very often obligations are obligations to a specific person or institution: *Jones* is obligated *to Smith* that p be the case. For example, I am obligated to you (by contract say) to paint your fence, and you in turn (upon completion) are obligated to me to pay me. Directed obligations are also related to rights. If I am obligated to you to paint your fence, then you have a claim on me to do so, and if you are obligated to pay me for the paint job, then (upon completion), I have a claim on you to pay me. Notice also that, typically, no one else, other than perhaps representatives of the law, have any claim on me to paint your fence, or on you to pay me for having done so. [Herrestad and Krogh, 1995] argues that an explicit representation of the directionality of obligations (and prohibitions and permissions) is not only needed to represent one important aspect of many (if not most) obligations, but that it also facilitates a better representation of relations between claims and obligations in the tradition

---

McNamara, 1996c; Mares and McNamara, 1997], which provide a cumulative case argument for the broad outlines of a solution to these representational problems.

[86][von Wright, 1968]. See also [Alchourrón and Bulygin, 1971] for another key early source.

[87][Herrestad and Krogh, 1995].

of [Hohfeld, 1919], and in the logical work on normative positions inspired by Hohfeld's work, beginning with Kanger's seminal work (e.g. [Kanger, 1971]). See also Sergot's chapter in this volume on normative positions.

8.7 A problem calling for attention to action and agency in deontic contexts

The jurisdictional problem and the need for the representation of agency[88]

Consider the following claims:

(8.72) Jeeves is obligated to not bring it about that Bertie's teeth are brushed

(8.73) Jeeves is obligated to not bring it about that Bertie's teeth are not brushed

There are limits to Jeeves duties and his rights as valet for Bertie, and Bertie's teeth-brushing is out of his jurisdiction - he is required to not interfere in that area, and thus to neither bring it about that Bertie's teeth are brushed (e.g. by forcibly doing so), nor to bring it about that they are not brushed (e.g. by pinching Bertie's tooth brush). Can we represent these in the standard systems? Many, following [von Wright, 1964; von Wright, 1965], have freely read "**O**" as "Smith is obligated to bring it about that__" (for some mock agent, Smith) or as "Smith is obligated to see to it that __". Even ignoring the complex integration of agential and deontic notions in this reading, and letting advocates of this reading have it, can we represent (8.72) and (8.73)? It does not seem we can do any better than this:

(8.74) $\neg \mathbf{O} p$
(8.75) $\neg \mathbf{O} \neg p$

Together, (8.74) and (8.75) simply provide the conditions for optionality - Jeeves is not obligated to bring it about that Bertie's teeth are brushed and not obligated to bring it about that they are not brushed. But that is not what (8.72) and (8.73) are saying. They decidedly entail that it is *not* an optional matter whether or not Jeeves brings it about that Bertie's teeth are brushed. Put another way, (8.74) and (8.75) could be true even if (8.72) and (8.73) are both false. For example, that is the situation if Jeeves is permitted to assure that Bertie's teeth are brushed and permitted to

---

[88]The first reference we have found coming close to explicitly formulating this problem is [Lindahl, 1977, p. 94], where the "none of your business" terminology is invoked, but it was recognized by Kanger, since essentially presupposed in his analysis of rights-related notions in his seminal [Kanger, 1971]. See also [von Wright, 1968].

assure that they are not brushed, since Bertie has had recent gum surgery, and Jeeves gets to decide what is apt. Shifting the negation signs inward, thereby creating conflicting obligations, is no help either.

If we want to represent scenarios like the one above, it seems we must allow for the negations to be able to operate on the agency itself, so reading "**O**" with agency built in will not serve. To adequately represent these situations, we need to represent agency separately from the deontic operators, and then explore their interactions. Action, agency, and deontic operators will be taken up in the next section.

9 Actions and agency in deontic logic

Philosophers have made a distinction between two kinds of ought, the ought-to-be (Seinsollen) and the ought-to-do (Tunsollen) [Castañeda, 1970], and it has been suggested that since the deontic operators of SDL are propositional operators, the standard deontic logic and the extensions and revisions discussed above should be regarded as theories of the ought-to-be rather than theories of the ought-to-do. However, as was observed earlier, the propositions in question may be action propositions, propositions to the effect that an agent does something or that an action of a certain kind is performed or not performed (omitted). In this approach, deontic concepts are not applied to generic actions or act-types, as in von Wright's 1951 system (see Section 5), but to propositions about individual actions. Another alternative that resides between the impersonal reading and the personal and agential reading of **O** is reading **O** as specifying that it is obligatory *for Smith* that it be the case that p (personal, but not agential), and then the agential form is a special case - it is obligatory for Smith that it be the case that Smith brings it about that q. (See [Krogh and Herrestad, 1996; McNamara, 2004].)

G. H. von Wright has observed that actions (or acts) usually involve changes in the world:

> "Many acts may ... be described as the bringing about or effecting ('at will') of a change. To act is, in a sense, to interfere with the 'course of nature'." [von Wright, 1963, p. 36]

Von Wright analyzes actions in terms of three world-states or occasions: (i) the *initial state* or *origin* which the agent changes or which would have changed if the agent had not been active (had not interfered with the course of nature), (ii) the *end-state* or the *result-state* which results from the action [von Wright, 1963, p. 28], and (iii) the *counter-state* which would have resulted from the initial state without the agent's interference, in other words, the state which would have resulted from the agent's passivity. The counter-state is needed for expressing the "counterfactual element" [von Wright, 1968, pp. 43-4] or *sine qua non* condition of action (cf. [Hart and Honoré, 1959, pp. 103-22]).

The characterization of acts by means of three states or occasions makes it possible to distinguish $2^3 = 8$ different modes of action with respect to a single state of affairs p. These modes of action may be defined as follows: Let $W = \{u, v, w...\}$ be a set of possible world-states or occasions, and let us assume that the agent can be either active or passive in a given state. Let d be a function which assigns to each $u \in W$ a state which results from the agent's activity at u, and let e be a function which assigns to each $u \in W$ the corresponding counter-state. The truth-value of p at u is denoted by '$V(p, u)$', and as usual, '$V(p, u) = 1$' (where '1' means the value *true*) will be abbreviated '$u \models p$'. For example, if $u \models \neg p$, $d(u) \models p$ and $e(u) \models \neg p$, we can say that the agent brings it about that p or produces the state of affairs that p. In this case p becomes true as a result of the agent's action: without the agent's action it would have remained false that p. The falsity of p at the initial state and at the counter-state constitute an opportunity for the agent to bring it about that p. On the other hand, if p is false at $d(u)$ (the end-state) under otherwise similar circumstances, we can say that the agent omits to bring it about that p. In this way we obtain the action possibilities presented in Table 2. Here '**BA**' abbreviates 'bring it about that' and '**SS**' stands for 'sustain (the state that)'. For the sake of brevity,

	u	$d(u)$	$e(u)$	Mode of action	Rendering
Act 1	$\neg p$	p	$\neg p$	Bringing it about that p	**BA**p
Act 2	p	p	$\neg p$	Sustaining the state that p	**SS**p
Act 3	$\neg p$	$\neg p$	$\neg p$	Letting it remain the case that $\neg p$	**omBA**p
Act 4	p	$\neg p$	$\neg p$	Letting it become the case that $\neg p$	**omSS**p
Act 5	p	$\neg p$	p	Bringing it about that $\neg p$	**BA**$\neg p$
Act 6	$\neg p$	$\neg p$	p	Sustaining the state that $\neg p$	**SS**$\neg p$
Act 7	p	p	p	Letting it remain the case that p	**omBA**$\neg p$
Act 8	$\neg p$	p	p	Letting it become the case that p	**omSS**$\neg p$

Table 2: The main action-types according to von Wright

we shall use below the expression 'the agent brings about (or produces) p', instead of saying that an agent brings it about that p. In this simplified terminology, we can say that Act1 is an act of *producing* p, Act2 is an act of *sustaining (preserving)* p, Act5 is an act of *destroying* p, and A6 is an act of *preventing* p.

If $V(p, d(u)) \neq V(p, e(u))$, the truth-value of p depends on the agent's activity; in this case the agent is active with respect to p; otherwise the agent may be said to be passive with respect to p. The action-types in which $V(p, d(u)) = V(p, e(u))$ are omissions (abbreviated '**om**'). As was observed earlier, an omission in the proper sense should be distinguished from the non-performance of an act: an agent can omit an act only in a situation in which he has an opportunity to perform the act in question; thus an omission entails non-performance, but not conversely. If $V(p, d(u)) \neq V(p, e(u))$ and $V(p, d(u)) \neq V(p, u)$, the action in question is a productive or a destructive act, but if $V(p, d(u)) \neq V(p, e(u))$ and $V(p, d(u)) = V(p, u)$, the action is an act of sustaining or preserving some state of affairs.

In von Wright's analysis, actions are characterized by means of propositional expressions which refer to the result-state, the initial state, and the counter-state of the action, and the propositions which describe the states are transformed into action propositions by means of the praxeological operators **BA**, **SS**, and **om**. In many recent systems of the ought-to-do, action propositions are formed in this way, and simple action descriptions are given the form '**Do**(a, p)', where '**Do**' is a modal (praxeological) operator for action or agency and p is a propositional expression. The **Do**-operator is usually read 'a brings it about that' or 'a sees to it that'. This analysis of action sentences goes back to the 11th century philosopher St. Anselm, who investigated the meaning of the Latin phrases 'facere esse' (to bring it about that'), 'facere non esse', 'non facere esse', and 'non facere non esse'. (Cf. [Henry, 1967, pp. 123-9] and [Segerberg, 1992, pp. 348-51].) The logical relations among these concepts can be represented as a square analogous to the square of modalities, and this suggests that they can be treated for logical purposes as modal concepts, as praxeological modalities.

[Kanger, 2001] has presented an analysis of action and agency in terms of the concept of *seeing to it that p*. He regarded a statement of the form 'a sees to it that p', '**Do**(a, p)', as a conjunction

(CDo) $\mathbf{Do}(a, p) \leftrightarrow \mathbf{Ds}(a, p) \,\&\, \mathbf{Dn}(a, p)$,

where '**Ds**' may be said to represent the *sufficient condition aspect* of agency and '**Dn**' stands for the *necessary condition aspect* of agency. (Cf. [Hilpinen, 1974, p. 170] Kanger reads '**Ds**(a, p)' as

p is necessary for something a does

and '**Dn**(a, p)' as

p is sufficient for something a does

These readings are equivalent to

(9.1) **Ds**(a, p): Something a does is sufficient for p

and

(9.2) **Dn**(a,p): Something a does is necessary for p

Kanger interpreted the agency operators **Ds** and **Dn** in terms of two alternativeness relations on possible "universes" [Kanger, 2001, p. 152, 159]. In a simplified form, Kanger's conditions may be expressed as follows:

(CDoS) $u \models \mathbf{Ds}(a,p)$ iff $w \models p$ for every w such that $S^s(u,w)$

and

(CDoN) $u \models \mathbf{Dn}(a,p)$ iff $w \models \neg p$ for every w such that $S^n(u,w)$

Kanger's phrase "something a does" may be paraphrased as "some action D performed by a"; thus (9.1) and (9.2) may be rewritten as

(9.3) **Ds**(a,p): Some action D performed by a is sufficient for p

and

(9.4) **Dn**(a,p): Some action D performed by a is necessary for p

where 'D' is a variable for action types. Reformulated in this way, it is clear that strictly speaking, Kanger's theory is not an analysis of action, but an analysis of the concept of seeing to it that. The concept of action ("something a does") is part of *analysans*.

The praxeological action (or agency) operator is sometimes read 'see to it that', sometimes 'bring it about that'. In so far as these expressions are used in ordinary discourse, they do not have the same meaning. An agent a can "see to that p" either by bringing it about that p or sustaining the state that p, that is, by making sure that p is not "destroyed"; thus seeing to it that p does not entail bringing it about that p. According to this interpretation, **Do**(a,p) is equivalent to **BA**p ∨ **SS**p in von Wright's schema. The modality of the action does not depend on the initial state (situation), and we get the four action modalities distinguished by St. Anselm:

(9.5) (i) **Do**(a,p): a sees to it that p
  (ii) ¬**Do**(a,p): a does not see to it that p
  (iii) **Do**(a,¬p): a sees to it that ¬p
  (iv) ¬**Do**(a,¬p): a does not see to it that ¬p

A common feature of von Wright's and Kanger's analyses is that both analyze action/agency in terms of two conditions. Kanger's first condition, the **Ds**-condition, may be termed the *positive* condition, and the second condition, the **Dn**-condition, may be termed the *negative* condition of agency.

(Cf. [Belnap, 1991, p.792]) The latter condition corresponds to von Wright's counterfactual condition of agency. It states that if the agent had not acted the way he did, p would not have been the case. Some philosophers have disagreed about the formulation of the negative condition. [Pörn, 1977] has argued that we should accept instead of Kanger's $\mathbf{Dn}$-condition only a weaker negative requirement, viz. '$\neg \mathbf{Dn}(a, \neg p)$', abbreviated here '$\mathbf{Cn}(a,p)$':

(ACN) $u \models \mathbf{Cn}(a,p)$ iff $w \models \neg p$ for *some* w such that $S^n(u,w)$

This condition can be read: but for a's action it might not have been the case that p [Pörn, 1977, p. 7]; that is, it was not unavoidable for a that p. [Åqvist, 1974, p. 81] has accepted a similar weak form of the counterfactual condition. According to Pörn and Åqvist, the negative condition should be formulated as a might-conditional, not as a would-conditional. Other versions of the analysis of agency by means of a positive and a negative condition have been [Lindahl, 1977; Åqvist and Mullock, 1989], and Nuel Belnap, John Horty, Michael Perloff, and others. (For discussion of such approaches, see [Belnap *et al.*, 2001; Horty, 2001]; for different forms of the positive and the negative condition, see [Hilpinen, 1997, pp. 11-20].)

There is also a morally and legally relevant concept of bringing it about with a might-conditional as a positive condition and a would-conditional as a negative condition:

(9.6) $\mathbf{BA}^\star(a,p) \to \mathbf{Cs}(a,p) \,\&\, \mathbf{Dn}(a,p)$

where '$\mathbf{Cs}(a,p)$' means that something a does makes p possible or enables (contributes to) p. In cases of this kind, a's actions are a sine qua non-condition of p, and a may be regarded as a contributing agent of the state of affairs p, and held at be least partly responsible for it.

According to von Wright's, formulation of the counterfactual (*sine qua non*) aspect of action, the agent's "passivity" at any given world-state or occasion (situation) u would lead to a single world-state (counter-state) $e(u)$. The values of the functions d and e are assumed to be world-states or situations, not sets of world-states. This means that the counterfactuals underlying von Wright's analysis satisfy the principle of Conditional Excluded Middle:

(9.7) Either: if the agent had been passive, it would have been the case that q, or: if the agent had been passive, it would have been the case that not-q

more generally,

(CEM) $(p \Rightarrow q) \lor (p \Rightarrow \neg q)$,

where $\Rightarrow$ is a sign for a counterfactual or subjunctive conditional. (CEM) dos not always hold because sometimes q might or might not be the case if it were the case that p. (Cf. [Lewis, 1973, p. 79]) Thus we should revise von Wright's analysis by assuming that the agent's passivity in a situation u might lead to various alternative world-states, depending on how u might change without the agent's interference, for example, as a result of the actions of other agents. This can be represented by means of a function which has as its value the set of those world-states which could result from the agent's passivity. Such a representation agrees with the analysis of counterfactuals based on set selection functions given above in the Appendix to 8.5. In the same way, an action whose initial state or origin is u is representable by a function which assigns to u the set of possible world-states which could result from the action.

Von Wright formulates the counterfactual condition in terms of the agent's passivity, or what might be called the *zero action*. Such an account is inapplicable to many action situations which do not include a clear alternative of passivity. If D is the action of bringing it about that p or seeing to it that p in a certain way, we may define the counterfactual aspect of D in terms of the omission of D, or not doing D, or doing something else instead of D. Von Wright's analysis can be enriched in the same way as Kanger's theory, by assuming that the agent can change the initial situation u in different ways by undertaking different actions or by performing some action in different ways, in other words, we may assume that the agent can perform in a given situation various actions $A_1, ..., A_n$, each of which is represented by means a function which assigns to each situation u the set of world-states to which the action might lead the agent from u. In this way von Wright's analysis, applied to the concept of seeing to it that, assumes the form

(9.8)   **Do**(a,p) if and only if a performs some action D such that
        (i) if a were to do D, p would be the case, and
        (ii) if a did not do D, it would not be the case that p

According to (9.8), an agent a may be said to see to it that p if and only if p's being the case is counterfactually dependent on something a does.

According to von Wright, the truth-values of sentences, including those of action sentences, are relative to occasions or world-states [von Wright, 1963, p. 23]: occasions are the points of evaluation of sentences (or propositions). As we have seen, an action proposition involves three occasions, the initial state, the end-state, and a possible counter-state. Is an action sentence regarded as true or false in the initial state or in the end-state; in other words, on which occasion does the agent perform the action? This question is closely related to the question about the time of an action (cf. [Thomson,

1971]). In his [1983] paper von Wright argues that the sentence

(9.9) $\mathbf{BA}p \to p$

is not a logical truth on the ground that

> "[$\mathbf{BA}p \to p$] would say that if a state is produced on some occasion then it is (already) there on this occasion. But this is logically false." [von Wright, 1983, pp.195-6]

This suggests that if action sentences are evaluated with respect to occasions or world-states, we should regard the initial occasion as the point of evaluation. (If an agent brings it about that p, p is false on the initial occasion.) Thus we should define (for example) the truth of '$\mathbf{BA}p$' as follows:

(9.10) $u \models \mathbf{BA}p$ iff $u \models \neg p, d(u) \models p$ and $e(u) \models \neg p$

According to (9.10), sentence (9.9) is logically false, whereas

(9.11) $\mathbf{BA}p \to \neg p$

is logically true. (Cf. [Segerberg, 1992, p. 358])

Condition (9.10) is problematic if '$\mathbf{BA}p$' is read 'the agent brings it about that p', that is, if '$\mathbf{BA}p$' is regarded as a genuine action proposition which says that the agent *does* something. According to von Wright, an action involves changing a situation or a state in some respect or keeping it unchanged, and the state (or 'world') u is understood here as the situation which either is or is not changed by the agent's action. We cannot assume that '$\mathbf{BA}p$' is part of the description of the very situation which is changed (or kept unchanged) by that action. It is natural to say that the agent chooses to perform an action at the initial state u: u is the state from which the action 'originates', but the sentences '$\mathbf{BA}p$', '$\mathbf{SS}p$', '$\mathbf{omBA}p$' and '$\mathbf{omSS}p$' cannot be regarded as true or false at u if they are understood as genuine action sentences. It would be better to say that the agent does something *to* the initial state, that is, changes it or keeps it unchanged, than to say that the action is performed *at* the initial state. Von Wright's view seems to be supported by Nero Wolfe, who has remarked:

> "The average murder, I would guess, consumes ten or fifteen seconds at the outside. In cases of slow poison and similar ingenuities death of course is lingering, but the act of murder is commonly quite brief." [Stout, 1980, p. 16]

According to Wolfe (and Donald Davidson, see [Davidson, 1980a]), a poisoner kills the victim, that is, brings it about that the victim is dead, in a situation in which the victim is not dead; the death may occur much later. However, according to Wolfe (and Davidson), the act of bringing about the death of the victim consists in pouring the poison in his drink, and the initial situation changed by that action is a situation in which the poison is still safely in the little bottle in the poisoner's hand. The act of pouring the poison cannot be said to be performed in such a situation.

Many other authors who have analyzed the concept of action as a praxeological modality have accepted the success principle analogous to (9.9),

(9.12) $\mathbf{Do}(a,p) \to p$

as a valid principle for the concept of seeing to it that p. For example, Brian Chellas, who uses '$\Delta_a p$' for 'a sees to it that p', says about (9.12):

> "This is perhaps the most minimal substantive axiom for Δ. One can see to it that such-and-such is, or be responsible for such-and-such's being, the case only if such-and-such is the case." [Chellas, 1969, p. 66] (See also [Kanger, 2001, pp. 149-50].)

Many subsequent theories of action and agency have followed Chellas's example in this respect. (See [Belnap, 1991; Belnap and Perloff, 1988; Belnap and Perloff, 1992; Elgesem, 1993; Elgesem, 1997; Sandu and Tuomela, 1996; Belnap et al., 2001; Horty, 2001].) It is clear that one can be responsible only for what is in fact the case or what has actually happened, but it is not equally clear that one can "see to it that p" only if it is the case that p. This does not hold in von Wright's theory of action. It is misleading to say that one can see to it that p only if it is the case that p: as von Wright has pointed out, a person can bring it about that p only if it is *not* the case that p, and bringing it about that p may be a case of seeing to it that p. We can say, of course, that an agent *has seen* to it (or *has brought* it about) that p, and is held responsible for p, only if it is the case that p. Statements about (causal) responsibility are evaluated only at the end-states of actions. Thus we have to distinguish here between (present tense) action sentences and statements about agency. A person is an agent of a certain result only if he *has done* something which has caused (or will cause) the result.[89] (In Nero Wolfe's example of killing by poisoning, and in other similar cases in which the outcome can be known beforehand with certainty, we may hold

---

[89]This fits the intention, if not the reading, of what Belnap calls an *"achievement* stit" operator ("stit" for "sees to it that"): Smith achievement-sees to it that p just in case p *now* holds and *was* guaranteed by a *prior* choice of Smith's. Thus p must now hold for this compound sentence to be true, but as a result of some past action that was instrumental in p's now being the case (there is a negative might condition as well as a positive condition).

the agent (the poisoner) responsible for what will happen, even though a court of law would not find him guilty of murder before the victim is dead.)

[Segerberg, 1992, p. 373] has observed that Chellas's action semantics provides no picture of action itself and suggested that this failure may be related to the validity of the T-principle mentioned above. But von Wright's rejection of the T-principle of modal logic does not make his theory superior to Chellas's theory in this respect; on the contrary, as we have seen, Chellas's theory can be given a reasonable (re)interpretation as a theory of agency statements, but von Wright's choice of the initial states as the circumstances of evaluation of action sentences excludes such an interpretation. If Chellas's theory is understood in this way, the lack of a counterfactual condition seems to be a weakness, but such a condition can of course be added to his analysis. ([Hilpinen, 1997, p. 17].)

One potential source of confusion here is the possibility of understanding the expression 'possible world' in two different ways. It can mean either temporary world-state (a moment) or a world-history, that is, a sequence of world-states. In von Wright's approach, a possible world is understood in the former way; it is a possible state of the world at a given moment, a world-state. If events are regarded as changes (or world-state transformations) and an action is regarded as the bringing about of a change, we obviously cannot assume that action propositions are interpreted as sets of possible worlds: actions do not take place *within* possible worlds. On the other hand, if possible worlds are understood as histories or courses of events, we can say that an agent performs an action in a possible world.

Von Wright's analysis of action in terms of alternative successions of world-states suggests an integration of these two conceptions into a semantic representation based on a branching frame of moments (states of the world, situations) and transitions between moments, that is, a structure $(W, <)$, where the elements of W represent moments (situations, world-states), and $<$ is a treelike partial ordering such that for any u, v, and $w \in W$, if $u < w$ and $v < w$, then either $u < v$ or $v < u$ or $u = v$. (Cf. [von Wright, 1968, pp. 38-57].) The moments $u \in W$ can be interpreted as possible choice situations or the initial world-states which the agent may change by his actions, and some of the successors of u in the ordering are the situations which may result from his action, that is, the possible end-states of the action. In this model, an action A can be represented by a set of ordered pairs (u, w), with u as the initial state and w as a possible end-state or result-state of A, in other words, actions are regarded as binary relations on W. (See [Åqvist, 1974, p. 77] and [Czelakowski, 1997, p. 50].) Von Wright suggests this model of action when he observes that when a state of affairs either begins (or ceases) to exist as a result of an agent's action, the "occasion" on which the action takes place should be regarded as consisting of two "phases", one in which the state of affairs is absent (present), and another phase in which the state of affairs is present (absent). ([von Wright, 1983, p. 174, pp. 195-6]; see also [von Wright, 1968, p. 65].) Many philosophers have characterized actions in ways which fit this model. For example, [Apostel, 1982, p. 104] has observed that "an action is a transformation of nature in order to realize a purpose", and in his "action-state semantics" for imperatives C. L. Hamblin has analyzed actions or deeds in terms successive world-states [Hamblin, 1987, pp. 137-166]. According to [Weinberger, 1985, p. 314], "an action is a transformation of states within the flow of time" involving a subject (an agent), who "has at his disposal a range for action, i.e., at least two states of affairs which are possible continuations of a given trajectory in the system of states."

This way of representing actions and world-states requires two kinds of predicates and propositional expressions, expressions which describe possible states of the world (for example, 'the door is closed'), and action terms (predicates) and propositions which describe the way in which an agent changes the world (for example, 'to open the door', 'Bertie opens the door'). The former are true or false at the states $u \in W$, and the latter characterize the transitions (u, w) in W. An action term becomes a propositional sign when it is completed by an indexical sign which indicates an agent (or agents). Let $p, r, s, \ldots$ be propositional symbols, and let $F, G, H, \ldots$ be action terms or action descriptions. Action terms can be simple or complex: the latter are formed from simple action terms by act-connectives, some of which are analogous to propositional connectives. For example, if F and G are action terms, the following expressions are also action terms:

(ActT1) $\quad F + G :$ doing A or B

(ActT2) $\quad F \wedge G :$ doing A and B together

$F + G$ represents a choice between the actions F and G. It is also convenient to have an expression for the omission of an act, in the sense of doing something instead of F:

(ActT3) $\quad \sim F :$ omitting F

'$\sim F$' is applicable to all individual actions (world state transitions) which fail to exemplify F. Systems of dynamic deontic logic usually also contain act-connectives which have no counterparts in propositional logic, for example [Segerberg, 1990, pp. 205-6] and [Segerberg, 1992, p. 376]:

(ActT4) $\quad F; G : F$ followed by G

(ActT5) F^* : doing F a finite number of times

The ordered pairs of states assigned to an action sentence A may be called the possible *performances* of A. A world-state w is said to be possible relative to u or accessible from u if and only if it is possible for some action or sequence of actions to lead from u to w. Let us denote this accessibility relation by POS, and let POS/u be the set of transitions which originate from u, briefly expressed, 'u-transitions'. In the following, the expression 'c does A at u', where c is an agent, is used to refer to an action which has u as its initial state, that is, that is, a set of transitions from u to various possible outcome states. Normative concepts can be defined in this framework by dividing world state transitions into normatively acceptable (legal, permitted, right) and deontically unacceptable (illegal, forbidden, wrong) transitions (cf. [Segerberg, 1982, pp. 270-1, 276-80]). Let LEG/u be the set of legal transitions which originate from u, and let ILL/u be set of illegal u-transitions. The following conditions express the assumptions that any possible transition from u is either legal or illegal, and no transition is both legal and illegal:

(DDet) $\text{LEG}/u \cup \text{ILL}/u = \text{POS}/u$

and

(DCons) $\text{LEG}/u \cap \text{ILL}/u = \emptyset$

(DDet) may be called the principle of deontic determinacy. If it holds for any situation u, the normative system in question has no gaps. (Cf. [von Wright, 1996, p. 47].) According to (DCons), no transition can be both legal and illegal. The assumption that there is some normatively acceptable way out of every situation, in other words,

(DactD)   For every $u \in W, \text{LEG}/u \neq \emptyset$

corresponds to principle (D) of SDL, that is, the postulate that every world (situation) has some deontic alternative.

Let I be an interpretation function which assigns to each action A its possible performances (a subset of $W \times W$), and let $I/u(A)$ be the performances of A which originate from u; thus $I/u(A) \subseteq \text{POS}/u$. The basic normative concepts of prohibition, permission (may), and obligation (ought) - deontic action modalities - can be defined by the following truth-conditions:

(CF.act) $u \models \mathbf{F}A$ iff $I/u(A) \subseteq \text{ILL}/u$
(CP.act) $u \models \mathbf{P}A$ iff $I/u(A) \cap \text{LEG}/u \neq \emptyset$

and

(CO.act) $u \models \mathbf{O}A$ iff $\mathrm{I}/u(\sim A) \subseteq \mathrm{ILL}/u$

where '$\sim A$' means that the agent does at u something incompatible with A, i.e., does not do A. These definitions are variants of the truth-conditions of normative propositions in SDL. According to (CF.act), an act A is prohibited in a given situation if every possible performance of A at that situation is illegal, and A is permitted if and only if it can be performed in a legal way. (cf. [Czelakowski, 1997, p. 60].) According to (CO.act), A is obligatory at u if only if the failure to do A would be illegal.

According to (CP.act), the permissibility of an action A means that some possible performances of A (at a given moment u) are deontically acceptable. For example, A may be permitted in this sense if it can be only performed together with some other acts, or performed in a legal way. This is a "weak" concept of permission which corresponds to that defined in SDL. In the present framework it is possible to define another concept of permission which may be termed a "strong permission". When we say that an act A is permitted in a given situation, we often mean that A itself is not illegal, in other words, that no sanction is attached A, and not only that some (possible) performances of A would be deontically acceptable in the situation. This sense of 'permission' can also be expressed in the form of a conditional: If the agent were to do A, he would not do anything illegal. The truth-conditions of such a conditional can be formulated by means of a selection function f which selects from $\mathrm{I}/u(A)$ the transitions which exemplify A but change the original situation u in other respects in a "minimal" way. Such transitions may often be described by saying that the agent does *only* A. This concept of "strong" permission is defined as follows:

(CP^s.act) $u \models \mathbf{P}^s A$ iff $f(\mathrm{I}/u(A), u) \subseteq \mathrm{LEG}/u$

We might say that the f-function selects from $\mathrm{I}/u(A)$ the *minimal* performances of A. For example, if Bertie's Aunt Agatha gives him permission to take one scone, it means that the action of taking one scone is acceptable, in other words, that Aunt Agatha would not reprimand Bertie if he were to take one scone and do nothing else. On the other hand, it is permitted for a driver to flash her right turn signal - but only if she is going to make a right turn as well. The latter action is an example of a weakly permitted action (assuming that making a right turn is permitted), whereas the former action (taking a cookie) is strongly permitted. The formulation (CP^s.act) is analogous to one of the standard ways of expressing the truth-conditions of conditionals by means of a selection function $f(I(p), u)$ which selects, for each proposition $I(p)$ and a situation u, the p-situations closest to u (as close to u as the truth of p permits): a conditional $p \Rightarrow q$ is true at u if and only if the consequent q is true at all selected p-worlds (i.e., the worlds in which p is true). Thus (CP^s.act) fits the most natural reading of a strong permission to do A: if you were to do only A, you would not be doing anything illegal. (Cf. [Dignum et al., 1996, pp. 200-3].) The selection function f used in (CP^s.act) selects the "minimal" performances of A from the set of all possible performances of A, just as the truth of a conditional $p \Rightarrow q$ is determined by the selection of the p-worlds minimally different from the actual situation (or the situation where the conditional is being evaluated) [Hilpinen, 1993, p. 309].

If the disjunctive permission 'You may do F or G' is interpreted as a strong permission in the sense defined by (CP^s.act),

(9.13) $\mathbf{P}^s(F+G) \to \mathbf{P}^s F \,\&\, \mathbf{P}^s G$

if and only if

(9.14) $f(\mathrm{I}/u(F), u) \cup f(\mathrm{I}/u(G), u) \subseteq f(\mathrm{I}/u(F+G), u)$

i.e., if the minimal performances of a disjunctive act includes the minimal performances of both disjuncts. This need not always be the case; for example, assume that Aunt Dahlia has ordered Bertie to wear black socks, and then gives the following permission:

> Bertie, you may also wear grey socks or purple socks, but you should consult Jeeves before wearing purple socks.

(See [Kamp, 1979, p. 271].) If this sentence is used normatively (performatively), Aunt Dahlia makes a disjunctive action permitted for Bertie, but refers to Jeeves's authority for the determination of the permissibility of one of the disjuncts. Therefore (9.13) is not a logical truth, but it may hold in many situations, and for pragmatic reasons it may be assumed to hold in situations in which a permission sentence is used performatively, if it would not be otherwise clear what has been permitted, that is, which performances of $F+G$ have been made normatively acceptable.

In this conceptualization of action and action propositions, the expression 'a sees to it that p', where p is an "ordinary" proposition which describes a state of the world, can be taken to mean that (i) a performs some action F which is sufficient to transform the initial state into one in which p holds, or if p is already the case, is sufficient to sustain p, and (ii) there is an alternative action G such that if a had performed G instead of F, p might have been false in the result state. This notion of 'seeing to it that' can be formally expressed by means of a modal operator $[F]$ which we define as follows first:

(9.15) $\quad u \models [F]p$ iff $f(u, F) \subseteq I(p)$,

i.e., $[F]p$ means that any possible performance of F at u would lead to a situation in which p is true, and f here is a selection function mapping a world and an action to a set of p-worlds. The [ ]-operator is a necessity operator relativized to the action F. In general, a necessity operator relativized to the antecedent of a conditional can be used to express the meaning of a subjunctive conditional; thus the left-hand side of (9.15) may be read: if an agent were to do F, p would be true. The corresponding possibility operator is defined by

(9.16) $\quad u \models \langle F \rangle$ iff $f(u, F) \cap I(p) \neq \emptyset$

Now in turn, 'a sees to it that p', as $\mathbf{Do}(a, p)$, can be defined as follows:

(9.17) $\quad (u, w) \models \mathbf{Do}(a, p)$ iff there is an action F (with a as the agent) such that
 (i) $u \models [F]p$, and $(u, w) \in I/u(F)$, and
 (ii) F has in u an alternative G such that $u \models \langle G \rangle \neg p$

(i) expresses here the sufficient condition aspect of 'seeing to it that', and (ii) is a weak form of the necessary condition aspect.

In many systems of the logic of the ought-to-do developed in the 1980's and 1990's, simple action descriptions are not regarded as primitive terms, as in the approach outlined above, but are obtained from propositional expressions by means of an action operator similar to the Do-operators considered earlier, which turns propositional expressions into action propositions, usually read 'a sees to it that' or 'a brings it about that'. As was noted earlier, such representations do not give as good an analysis of the concept of action, but can be regarded as representations of different forms of agency. An analysis of that kind has become widely employed in the recent work on the logic of agency and deontic logic. (See [Belnap and Perloff, 1988; Xu, 1995; Brown, 1996a; Bartha, 1999; Belnap et al., 2001; Horty, 2001].)

The combination of different modes of agency with deontic concepts makes it possible to represent several types of obligation and permission and different legal or deontic relations between individuals and groups. For example, consider a state of affairs involving two persons, $F(a, b)$. According to [Kanger, 1971; Kanger and Kanger, 1966], a suitable agency operator $\mathbf{Do}(x, p)$ can be combined with deontic operators to distinguish four basic types of right (or different basic senses of the expression 'right'):

(R1) $\quad\quad\quad\quad \mathbf{ODo}(b, F(a, b))$

(R2) $\neg \mathbf{O}\mathbf{Do}(a, \neg F(a,b)) \leftrightarrow \mathbf{P}\neg \mathbf{Do}(a, \neg F(a,b))$
(R3) $\neg \mathbf{O}\neg \mathbf{Do}(a, \neg F(a,b)) \leftrightarrow \mathbf{P}\mathbf{Do}(a, F(a,b))$
(R4) $\mathbf{O}\neg \mathbf{Do}(b, F(a,b))$

(R1)-(R4) define four basic normative relations between a and b which from a's perspective can be regarded as different relational concepts of right. In (R1), b has a duty to see to it that $F(a,b)$; this is equivalent to a's *claim* in relation to b that $F(a,b)$. (R2) can be described as a's freedom (or *privilege*) in relation to b that $F(a,b)$; this means that a has no obligation to see to it that $\neg F(a,b)$. Kanger called (R3) a's *power* in relation to b that $F(a,b)$, and (R4) a's *immunity* in relation to b that $F(a,b)$. The replacement of the state of affairs $F(a,b)$ by its opposite $\neg F(a,b)$ yields four additional concepts of right which [Kanger and Kanger, 2001, pp. 121-2] called counter-claim (R1'), counter-freedom (R2'), counter-power (R3'), and counter-immunity (R4'). Kanger and Kanger called the 8 relations defined by (R1)-(R4) and their negative analogs *simple* types of right. The normative relationship between any two individuals with respect to a state of affairs p can be characterized completely by means of the conjunctions of the eight simple types of right or their negations. There are $2^8 = 256$ such conjunctions, but the simple types of right are not logically independent of each other: according to the logic of the deontic $\mathbf{O}$-operator and the agency operator $\mathbf{Do}$, only 26 combinations of the simple types of right or their negations are logically consistent. [Kanger and Kanger, 2001, pp. 126-7] called these 26 relations the "atomic types of right". The atomic types provide a complete characterization of the possible legal relationships between two persons with respect to a single state of affairs. It is perhaps misleading to call these 26 relations "types of right", because they include as their constituents duties as well as claims and freedoms. Thus Kanger's theory of normative relations can be regarded as a theory of duties as well as rights [Lindahl, 2001].

Kanger's concepts (R1)-(R4) seem to correspond to the four ways of using the word 'right' (or four concepts of a right) distinguished by W. N. Hohfeld (1919), and he adopted the expressions 'privilege', 'power' and 'immunity' from Hohfeld. Kanger apparently intended (R1)-(R4) as approximate explications of Hohfeld's notions. However, Kanger's concepts of power and immunity differ from Hohfeld's concepts. According to Kanger, both power and freedom are permissions: a power consists in the permissibility of actively seeing to it that something is the case, whereas freedom means that there is no obligation to see to it that the opposite state of affairs should be the case. [Lindahl, 1977, pp. 193-211] and others have argued that Hohfeld's concept of power should be analyzed as a legal *ability* rather than a permission (a *can* rather than *may*). (See [Lindahl, 2001; Bulygin, 1992;

Makinson, 1986])

An agency operator such as the **Do**-operator considered above can be iterated, and it can therefore be used to form sentences which contain several nested occurrences of various modal operators: deontic operators, praxeological operators (for various forms of action and agency), and epistemic operators, which can be relativized to different agents. This feature makes it possible to apply deontic logic and the logic of agency to the analysis of complex social and normative phenomena, for example, the analysis of different concepts of right and other normative relations [Kanger, 1984; Makinson, 1986; Lindahl, 2001], governmental structures and the concept of parliamentarism [Kanger and Kanger, 2001], normative positions and normative change [Lindahl, 1977; Jones and Sergot, 1993; Sergot, 1999], and the study of social control, influence, and responsibility [Pörn, 1989; Santos and Carmo, 1996].

As even this brief exposition of Kanger's analysis of legal relations might suggest, the specification of such relations lends itself rather well to computational techniques, as demonstrated rather explicitly by the work of Sergot in extending the theory of normative positions. (See [Sergot, 1999], as well as his chapter in this volume.) This is only one among the many rich ways in which computer science and deontic logic have developed a fruitful relationship. For a locus classicus on this, see most of the chapters (including the chapter two overview) in [Meyer and Weiringer, 1993], the first volume of papers drawn from the inauguration of the *Deontic Logic in Computer Science* series of binannual conferences (DEONs), the preeminent conference forums (with associated publications) for work in deontic logic.

Acknowledgements

We would like to thank the editors, Dov Gabbay, Jeff Horty, Ron van der Meyden, Xavier Parent, and Leendert van der Torre for overseeing the handbook project, and for their patience, with special thanks going to Xavier Parent for his careful reading and helpful comments on final drafts of this chapter. Guglielmo Feis has our thanks for noting a number of typos in an earlier draft.

Appendix to 8.5 on Chisholm's puzzle and conditional norms

Consider the key inferences generating the Chisholm paradox: the inference from (8.41') and (8.42') to **O**t, and the inference from (8.43') and (8.44') to **O**$\neg t$. Each involves "detachment" of an **O**-statement from a pair of premises, one being a deontic conditional. Let us explore this by introducing the following symbolism,

(NS) $\mathbf{O}(q/p)$,

taken here merely as a neutral shorthand for a natural language conditional obligation or ought statement such as (8.43) above.[90] $\mathbf{O}(q/p)$ is then to be read as "if p, then it ought to be that q". We will also assume that monadic obligations are necessarily equivalent to special dyadic obligations, per the following fairly standard analysis of unconditional obligations:

(UCO) $\mathbf{O}p =_{\mathrm{df}} \mathbf{O}(p/\top)$

That is, it is obligatory (simpliciter) that p if and only if it is obligatory that p if tautological conditions hold (which they always do of course).[91]

Two types of "detachment principles" [Greenspan, 1975] emerged quickly in the literature on Chisholm's paradox:

(FDt) $(p \,\&\, \mathbf{O}(q/p)) \to \mathbf{O}q$ (Factual detachment)

(DDt) $(\mathbf{O}p \,\&\, \mathbf{O}(q/p)) \to \mathbf{O}q$ (Deontic detachment)[92]

(FDt) says it is a logical truth that given both if p then it ought to be that q and p *itself*, then it ought to be that q. (DDt) says that given both if p then it ought to be that q and *it ought to be that p*, then it ought to be that q. As the principles' names indicate, given the same deontic conditional ($\mathbf{O}(q/p)$), the main difference is that per (FDt), it is the factual claim (p) that allows us to detach the deontic conclusion ($\mathbf{O}q$); whereas per (DDt), it is the deontic claim ($\mathbf{O}p$) that allows us to detach that conclusion.[93]

---

[90] The logical differences between "obligation" and "ought" will not matter here, so we will use them interchangeably here.

[91] This definition has been widely endorsed and employed, but not universally so (e.g. [Alchourrón, 1993; Carmo and Jones, 2002] reject it). But see also [Parent, 2012] on some difficulties with some alternatives to UCO.

[92] We add "t"'s to the labels so that references to deontic detachment will not be confused with those to DD (SDL's no conflicts principle).

[93] Those who followed [von Wright, 1956] in viewing deontic conditionals as sui generis and not definable via a monadic operator and any non-evaluative conditional notion rejected (FDt), even if we shift to a non-material conditional. (8.42) above follows Chisholm's original example in having the conditional explicitly in the scope of the English "it ought to be that" construction, so it is not a "deontic conditional" as just characterized. For that, we would have to add that (8.42) is logically equivalent to the non-wide-scope construction: "if Jones does go, then he ought to tell them he is coming". Although this is hardly obvious, as mentioned above, the difference between (8.42) and (8.43) in Chisholm's original formulation is largely seen as inessential, so that "purified" presentations of premises in the role of (8.42) and (8.43) would both match each other in superficial form, and usually that of "if ..., then it ought to be that..." as in (8.43).

Regarding the standard systems, if we were to interpret a deontic conditional with "$\mathbf{O}$" having narrow scope, that is, as a material conditional with an obligatory consequent (i.e. $(p \to \mathbf{O}q)$, as in (8.43') above), (FDt) would be derivable by MP, but (DDt) would not be derivable (e.g. the T axiom is not a thesis). Conversely, if we interpret a deontic conditional with "$\mathbf{O}$" having wide scope, that is, as an obligatory material conditional (i.e. $\mathbf{O}(p \to q)$, as in (8.42') above), (DDt), but not (FDt), is derivable by principle KD.[94]

Earlier we saw that neither of these two interpretations of natural language deontic conditionals via *material* conditionals is at all tenable, but the fact that the two interpretations require an acceptance of one form of detachment and a rejection of the other reflects an important fact: endorsement of both types of detachment (without some restriction) is only plausible if it is plausible to conclude that the Chisholm scenario involves an outright conflict of obligations. For if both detachment forms are endorsed, we end up both obligated to tell (the neighbor we are coming) and also obligated to not tell them we are coming. Most have thought that this is not a case with conflicting obligations, and that something else generates the puzzle. As a result, researchers tended to divide up into two camps according to which principle they took to be deductively valid [Loewer and Belzer, 1983]. We can thus think of the two emerging positions as refinements on the failed narrow scope and wide scope readings of deontic conditionals via material conditionals. For as we can see above, the second premise needed in addition to the reinterpreted deontic conditional, in each case, parallels the narrow scope and wide scope readings via material conditionals: p itself is needed for (FDt); $\mathbf{O}p$ is needed for (DDt).

Deontic logicians who favored (FDt) typically held that deontic conditionals like those in (8.43) involve a *non-material* conditional, such as a subjunctive conditional, but otherwise things are just as they appear. The logical form matches the surface grammatical form: the main operator is deemed to be a conditional that has a consequent in the scope of a monadic

---

Either way, the inference from (8.41) and (8.42) - or the relevant analog to (8.42) - to "it ought to be that Jones tells" is still called "deontic detachment", and likewise for their formal analogues in the standard systems, where KD validates the inference from (8.41') and (8.42') to $\mathbf{O}t$. The crucial thing is the deontic character of the simpler premise in deontic detachment from a deontic conditional.

[94][Smith, 1994] notes that if $\mathbf{O}(q/p)$ is interpreted as $\mathbf{O}(p \to q)$ and we add factual detachment to SDL, we get Mally's collapse: $\vdash \mathbf{O}p \leftrightarrow p$. The part from right to left follows from (FDt) by RND since $\mathbf{O}(p \to p)$ is a thesis. Of greater interest, Smith, crediting Andrew Jones, points out that even a minimal deontic logic that contains merely RED and OD will generate $\mathbf{O}p \to p$. From (FDt), we get $\vdash p \,\&\, \mathbf{O}(p \to \bot) \to \mathbf{O}\bot$, and then from OD, that yields $\vdash \neg(p \,\&\, \mathbf{O}(p \to \bot))$, and then from RED we get $\vdash \neg(p \,\&\, \mathbf{O}\neg p)$, and $\vdash \mathbf{O}p \to p$.

deontic operator, and an ordinary antecedent - the result being an analysis of such conditionals as genuine conditional-deontic *compounds*:

(CDC) $\mathbf{O}(q/p) =_{df} p \Rightarrow \mathbf{O}q$, for some independent conditional $\Rightarrow$

Non-classical "closest antecedent worlds" conditionals of the sort made famous by Stalnaker and Lewis predominated. The logics of such deontic conditional compounds will then derive from the logic for the non-deontic conditional operator and the logic for the monadic deontic operator. Typically, the conditionals offered, along with the truth of their antecedents, would entail their consequents (a version of modus ponens would hold for the non-material conditional), so (FDt) would hold under this sort of analysis. Its truth-conditions might be formulated in a natural way by means of a selection function $f(I(p), u)$ which selects for each world u and proposition $I(p)$ (a set of possible worlds) presented for consideration by the protasis 'if p'; the apodosis then states that q is true in all situations selected by the protasis:

(8.54) $u \models p \Rightarrow q$ iff $f(I(p), u) \subseteq I(q)$,

where $I(q)$ is the set of possible situations in which q is true.[95] It is important to note here that the set of situations (the proposition) selected by the f-function (i.e., selected by the protasis) depends on the situation u about which the conditional statement is made; thus the antecedent may be said to express different propositions in different situations, and the conditionals defined by (8.54) can be said to be "variably strict" rather than strictly necessary conditionals. It is typically assumed that the selection function satisfies the following condition:

(8.55) if $u \models p$ then $u \in f(I(p), u)$

According to (8.54) and (8.55), the following is valid:

(8.56) $(p \Rightarrow q) \rightarrow (p \rightarrow q)$

So a version of modus ponens applies: given p and $p \Rightarrow q$, q follows.[96]

Those favoring this sort of approach usually typically justify their rejection of (DDt) on the following grounds. Conditional obligations like those in (8.42) tell us only what to do in *ideal* circumstances where we keep our primary obligations like those in (8.41); but they thus do not provide guidance or "cues" for action in circumstances where the primary obligation is

---

[95] Cf. [Mott, 1973; Chellas, 1974; Chellas, 1980].
[96] For other conditions for the f-function and other semantic models for conditionals, see [Lewis, 1973].

not met.⁹⁷ In the Chisholm scenarios, combinations like (8.44) and (8.41) entail that the primary obligation, whose execution is hypothesized in the first clause of the conditional obligation (8.42), has been violated. Thus that Jones ought to tell is not entailed by the fact that he ought to go and he ought to tell *if* he goes. If he tells and doesn't go, he makes things worse than if he merely doesn't go. Perhaps the most we can say is that *ideally* he ought to tell. But since on its face, a version of Modus Ponens holds for the conditional in (8.43), if he does not go (8.44) then it follows that Jones ought to not tell.⁹⁸

In contrast, those favoring (DDt) over (FDt) might object by citing a conditional reminiscent of Forrester's, such as "If Jones will kill his rich aunt now (for the inheritance), then he ought to shoot her to death" (his only immediate means being strangulation or a nearby hunting rifle, say). They then might explore how the picture of those favoring (FDt) holds up for such an example as follows. Suppose Jones will kill his rich aunt as a matter of contingent fact, although he could refrain. Then those favoring (FDt), by parity of reasoning, would have to say that although Jones is obligated to not kill his aunt, nonetheless, because he in fact will do so, he is obligated to shoot her to death, and at most only *ideally* ought to not do so. But the idea that Jones' obligation to not shoot his aunt to death merely expresses an ideal obligation, not an actual obligation, is not easy to accept. Similarly, if it is unqualifiedly obligatory that Jones not kill his aunt, as the friends of (FDt) agree, then it must be impermissible to kill her, and so impermissible to do so by any particular means.⁹⁹

The suggestion then is that *unrestricted* factual detachment seems to allow the mere fact that Jones will do something avoidable and terribly wrong to generate an actual obligation to do something also terribly wrong, though less wrong (even if only infinitesimally less wrong).

It is also to be noted that accounts that allow for factual detachment risks entailing "the pragmatic oddity" [Prakken and Sergot, 1994; Prakken and Sergot, 1996]. Using the Chisholm's quartet, suppose one's analysis

---

⁹⁷See [van Eck, 1982] for this idea of "cues" for action.

⁹⁸[DeCew, 1981] was influential in arguing that despite the importance of subjunctive conditionals in deontic contexts, the Chisholm puzzle involves a different special deontic conditional not expressible by this means.

⁹⁹In Chisholm's example it is not intuitively clear what is involved—helping the neighbors with a fire, load their moving truck, help them in with the groceries. It is thus easier to accept that letting them know you will help is merely ideal, but not required, since helping might be something you ought to do, but not something you must do or are obligated to do. By default, examples like the one above immediately rule out this sort of "recommended but not required" interpretation. Furthermore, it is not a case where the apparent consequent entails the antecedent as in the case above where what ought to be done if... is a way of doing what is hypothetically posited in the antecedent.

countenances the conclusion that "Jones ought to not tell his neighbors he is coming" (from factual detachment applied to (8.43) and (8.44), as well as countenacing the truth of the primary obligation (8.41), "Jones ought to go to his neighbor's assistance"? If the theory allows for aggregation of these two obligations, as all the standard systems do (by RMD and KD), we get the conclusion that "Jones ought to go to his neighbor's assistance and not tell them he is coming", which certainly sounds odd, if not false. The original surely does not have this consequence, and yet it looks like any account that embraces factual detachment and aggregation will generate this oddity. Prakken and Sergot suggest that this secondary puzzle places pressure on assuming univocality for the "oughts" in Chisholm's paradox; [Carmo and Jones, 2002] make avoiding the pragmatic oddity a desiderata of any adequate account of Chisholm's paradox.

Many who rejected factual detachment represented conditional obligations via a primitive *dyadic* obligation operator (reminiscent of the syntax of a conditional probability operator). They rejected CDC ($\mathbf{O}(q/p) =_{df} p \Rightarrow \mathbf{O}q$). They believed the logical form of such conditionals was hidden by the surface grammar: the meaning of the compound is not a straightforward function of the meaning of the apparent parts. The underlying intuition is that even if Jones will violate his obligation, that doesn't get him off the hook from obligations that derive from the one he will violate. If he must go help and he must inform his neighbors if he will go, then he must inform them as well, and the fact that he will violate the primary obligation does not block the derivative obligation any more than it does the primary one itself. He is still an agent subject to both constraints.

A "best of the antecedent worlds" semantic picture for the latter approach quickly emerged with Hansson's seminal work:

(BAW) $\mathbf{O}(q/p)$ is true at a world u iff the u-best p-worlds are all q-worlds.[100]

It follows from this by the standard analysis of the monadic operator (UCO) in dyadic contexts that

(8.57) $\mathbf{O}q$ is true iff $\mathbf{O}(q/\top)$, so iff all the unqualifiedly u-best worlds are q-worlds.

This approach, which relies on preference-orderings for the semantics of dyadic conditional obligations, became a widespread trend. (Structurally,

---

[100][Hansson, 1971]. See [Spohn, 1975] for a weak completeness theorem for one key system DSDL3 for which Hansson provided a semantics, and see [Parent, 2008] for a strong completeness proof for DSDL3, as well as [Parent, 2010] for such a proof for another system proposed by Hansson, DSDL2.

this ordering semantics approach was also a forerunner of a variety of approaches (to different phenomena) employing what [Makinson, 1993] characterizes as "the notion of minimality under a relation", as in that for defeasible conditionals such as "if p, normally q" that became so central in AI.) Factual detachment does not hold on this picture, since even if our world is one where Jones does not go to the assistance of his neighbors, and the best among those worlds are ones where he doesn't tell them he is coming, it does not follow that the *unqualifiedly* best worlds are ones where he doesn't tell them he is coming; in fact, the best such worlds are ones where he both goes to their aid and lets them know that he will do so.

However, a natural objection now emerges: what is the point of such "conditionals" if we are not allowed to detach the apparent consequents from the apparent antecedents - how do we *reason* with them? This suggests that the above line of reasoning for rejecting unqualified (FDt) is not enough, and so it was typically coupled with a *restricted* form of factual detachment, such as:

(RFDt) $(\Box p \ \& \ \mathbf{O}(q/p)) \to \mathbf{O}q$ (Restricted factual detachment)

$\Box p$ typically meant that p is now unalterable for the imagined agent.[101] The intuition is that we can conclude $\mathbf{O}q$ from $\mathbf{O}(q/p)$ only if p is not simply true but unalterably so (in the context of evaluation). This is certainly an important complement to the reasoning above for rejecting (FDt), since it does allow for a form of qualified factual detachment, and thus for reasoning from the non-deontic status of the apparent antecedent of deontic conditionals to the apparent deontic consequent. (More nuanced positions emerged, for example in [Loewer and Belzer, 1983], where the authors endorse a special form of factual detachment distinct from those above. This can perhaps be seen as further reflecting the felt need to move to more nuanced positions beyond the dilemma of having to simply choose between (FDt) and (DDt).) However, there is still the question of why this *certainly apparent* composite of a conditional and a deontic operator is actually some sort of primitive idiom and not purely derivative.

We are left with an apparent dilemma: either a) unqualified factual detachment holds and we swallow the consequence that often because someone freely will act horribly, she is obligated to do some slightly less, still horrible, thing; or b) that "if p, then ought q" contrary to appearances, is really an idiom, and the meaning of the whole is not a function of the meaning of apparent conditional and deontic parts, with all the challenges about how we learn the construction if it is not compositional. Neither option seems very satisfying.

---

[101][Greenspan, 1975] argues for this position explicitly, and many endorsed it as well.

More nuanced positions emerged, for example in [Loewer and Belzer, 1983], where the authors endorse a special form of factual detachment distinct from those above (cf. [Chisholm, 1964]). This can perhaps be seen as further reflecting the felt need to move to more nuanced positions beyond the dilemma of having to simply choose between (FDt) and (DDt). However, there is still the question of why this certainly apparent composite of a conditional and a deontic operator is actually either some sort of primitive idiom or a composite with a hidden modal antecedent.

We set the issue of (FDt) and (DDt) aside to turn briefly to two key features of the Chisholm scenario that are not represented in the standard systems and that people proposed were central to solving the puzzle.

One popular strategy for solving Chisholm's puzzle has been to carefully distinguish the times of the obligations.[102] This was reinforced by the fact that there are strong independent reasons to be concerned about differentiating the times at which things are obligatory. This was often accompanied by consideration of examples where the candidate "derived" obligations were things to be done *after* the violation (or fulfillment) of the primary obligation (a "forward" version of a CTD case), and indeed this appeared essential to many of the solutions offered. However, Chisholm's own seminal example does not fit so well here. It is naturally interpreted as either a case where at best the obligation to help and the purportedly derivable obligation to tell are *simultaneous* (a "parallel" version)[103], and at worst and more plausibly, where the telling is something to *precede* the going ("backward" versions).[104] After all, "I did help" or "I am now helping" are likely to be obvious to the neighbors, and surely letting them know you are on your way to them to provide aid when you arrive is the natural default reading. As with [Jones, 1990; Prakken and Sergot, 1994; Prakken and Sergot, 1996] stress this shortcoming with temporal solutions with a variety of examples, one being:

(8.58) The children ought not to be cycling on the street

---

[102][Thomason, 1981a; Thomason, 1981b] are classics arguing for the general importance of layering deontic logic on top of temporal logic. [van Eck, 1982; Loewer and Belzer, 1983; Åqvist and Hoepelman, 1981; Feldman, 1986] argued that attention to time is crucial in handling the Chisholm puzzle, or at least some versions thereof (among other puzzles). See also [Chellas, 1980]. For an early dissenting opinion on temporal solutions, see [Castañeda, 1977].

[103][Jones, 1990] interprets it this way and, more importantly, stresses that such a clearly possible case tells against the suggestion that distinguishing times is at the heart of the puzzle.

[104][DeCew, 1981]. [Smith, 1994] contains an illuminating discussion of the three different versions of the Chisholm puzzle (backward, parallel, and forward versions) in evaluating different approaches to solving the Chisholm paradox; in [Smith, 1993], she credits J. J. Meyer for the 'backward'-'forward' terminology.

(8.59) If the children are cycling on the street, then they ought to be cycling on the left hand side of the street

(8.60) The children are cycling on the street

They point out that the intention is surely for the first two to hold at the same time. So there are no times to separate to say the one obligation holds at t_1, but not t_2, and the other holds at t_2 not at t_1. Yet there is surely prima facia reason to think the same phenomena driving the Chisholm paradox is present above.[105]

Alternatively, some suggested that carefully separating the agential components of the example from the (non-agential) circumstancial components would solve the puzzle.[106] Again, there are plainly independently compelling reasons to pursue agency in deontic logic. However, once again, in the case of Chisholm's seminal example, it appears that the two key elements at issue are agential, for each appears to be an action, and one open to the agent as of the time of the puzzle scenario: going to the neighbors' assistance; telling the neighbors' you will help. Furthermore, there are non-agential versions of Chisholm's example such as this variant on others found in [Prakken and Sergot, 1994; Prakken and Sergot, 1996]:

(8.61) There ought to be no hurricane

(8.62) If there is no hurricane, the shutters ought not to be closed

(8.63) If there is a hurricane, the shutters ought to be closed

(8.64) There is a hurricane

Here the case seems to parallel that of Chisholm's original example rather well in broad respects, yet there is no reference to actions or agency at all; instead the reference seems to only be to different states of affairs, with the first claim telling us what is *ideally* the case, and the last telling us this ideal circumstance is not realized, with the claims in between telling us what ought to be under the respective *ideal* and *non-ideal* circumstances. (There appears to be no reference to different times here either.)

We note lastly that there have been some attempts to suggest that the problem with Chisholm's paradox might be solved by applying standard concepts of defeasibility from non-monotonic logic, such as that of excep-

---

[105] The *Forrester paradox* above quintessentially involves a parallel duties case.

[106] [Castañeda, 1981] is a salient and influential instance, arguing that we need to distinguish actions construed as circumstances and actions construed as prescribed in order to solve the problem. See also [Meyer, 1988], especially influential for its employment of dynamic logic in deontic contexts; [Meyer, 1988] also offers a solution to Chisholm's puzzle in a broadly similar vein as Castañeda (along with solutions for other puzzles in deontic logic).

tions to normative generalizations, etc.[107] However, this does not seem to jive well with the prima facia difference between violation and defeat.[108] In Chisholm's example, the natural reading is that Jones is obligated to help his neighbors unexceptionally and indefeasibly, but nonetheless, in fact he will not, so it is now true that he will (in the future) violate that obligation. This fact does not defeat that obligation, nor does the corresponding contrary to duty obligation override it or cancel it. Even if defeasibility might figure in part of the story, it seems that no discussion absent of violation concepts will suffice to cover essential features of CTD cases.[109] [Prakken and Sergot, 1996] makes the point nicely with the following example (trivially modified):

(8.65) There must be no fence
(8.66) If there is a fence, it must be a white fence
(8.67) If the cottage property includes a cliff edge, there may be a fence

Suppose (8.65) is meant defeasibly, and a cottage near a cliff edge constitutes the only defeater. Suppose now that Jones has a cottage, but not one near a cliff edge, and he has a red fence. Then he is in violation of (8.65), an undefeated (though defeasible) primary obligation for him, and he is also in violation of (8.66), since he (impermissibly) has a fence, and it is not a white one. Contrast Doe, who has a cottage near a cliff edge and a red fence. Doe is not in violation of (8.65), since it is defeated (undercut) by her exceptional circumstances. Is she in violation of (8.66)? That depends on how (8.66) is evidently meant. Imagine it comes just on the heels of (8.65) in the cottage properties manual, preceded with a "However,"; whereas (8.67) comes in the manual's appendix along with a general discussion of special exceptions to various rules. Then Doe is in full compliance with (8.65)-(8.67), since there are apparently no color restrictions for fences by a cliff edge. In contrast, if we imagine that (8.66) is intended to cover all cases,

---

[107] For example, see [McCarty, 1992; Ryu and Lee, 1991], and for an earlier work stressing defeasible principles and the Chisholm paradox, see [Loewer and Belzer, 1983; Belzer, 1986].

[108] [Smith, 1993] briefly discusses the importance of the difference for the Chisholm puzzle, and at greater length again in [Smith, 1994], stressing that the central feature of Chisholm puzzles is violation of the primary obligation, not defeat thereof. [Prakken and Sergot, 1994; Prakken and Sergot, 1996; Prakken and Sergot, 1997] also stress the difference and argue for the unresolvability of the puzzle using only defeasiblity. For a dissenting opinion however, see [Bonevac, 1998], which argues that the problem is solvable using defeasiblity, and that this also allows for the analysis of CTDs as composites of a conditional and a monadic obligation operator, pace the dyadic approach.

[109] We assume the point here stands even if the *concept* of violability itself is somehow analyzable via *defeat* concepts, for the key point is that we cannot avoid invoking the difference between a defeated or cancelled primary obligation and a violated one in standard CTD cases, nor the understanding of the CTD as conditional on said violation.

not just violations of (8.65), then Doe (along with Jones) is in violation of (8.66), but it is a CTD for Jones only.

Thus, however much temporal, agential or action-related aspects of deontic reasoning are important in their own right, it does not appear that any of them hold the key to resolving the general problem that the Chisholm puzzle indicates. Similarly, however important defeasibility is to deontic reasoning, and even if it ultimately has some role to play in a final resolution of Chisholm puzzles, it appears that the difference between defeated and violated obligations will survive that, as will the difference between a deontic conditional intended as telling us what to do if we violate an undefeated obligation, and one telling us what we are obligated to do conditional upon our performing some optional or obligatory action.

The Chisholm puzzle has been highly resistant to simple or even fully satisfying solutions, and using the term "paradox" seems less overstated than in the case of many of the other standard deontic logic puzzles. Many consider Chisholm's paradoxes to be the most important and distinctive puzzle in the development of deontic logic.[110] As noted earlier, SDL is just the normal modal logic, D, and most of the early deontic logics were extensions of SDL. This suggested deontic logic was just an interesting but simple application/interpretation of some simple normal modal logics. But the puzzles with deontic conditionals, especially Chisholm's, helped solidify deontic logic as a distinct specialization in the 1960s and 1970s, one for which normal modal logics like SDL were deemed inadequate.[111] This led to the development of alternative more complex logics for deontic conditionals, and then to a widespread (though as noted, not uncontested) perception that some sort of ordering semantics provided important and promising structures for modeling deontic conditionals, and that in all events, more elaborate expressive and semantic resources were called for.

For the reader interested in seeing what a logic for conditional obligation might look like, we provide one favoring deontic detachment that might be seen as the conditionalized counterpart to SDL, and is indeed called "SDDL" for "Standard Dyadic Deontic Logic" in [Goble, 2003]. We do not provide the semantics here, but refer the interested reader to his article and to the informal remarks above about ordering semantics for SDL.[112] We do not

---

[110] For example, [Carmo and Jones, 2002].

[111] It is also arguable that the development of conflict-tolerant deontic logics in response to puzzles like Sartre's' dilemma and Plato's dilemma has also been liberating for deontic logic (although none of the aforementioned puzzles has held deontic logicians quite as captivated as Chisholm's puzzle has).

[112] [Goble, 2003; Goble, 2004] goes through the metatheory for ordering semantic approaches to deontic conditionals in the dyadic logic tradition (as well as monadic SDL itself), but generalizing in interesting ways beyond the standard systems (e.g. to allow

present an analog in the Factual Detachment tradition–approaches in the vein of (CDC) above, since [Chellas, 1980] is widely available (deservedly), and contains a nice presentation of such a logic and its semantics.

The system below is deductively equivalent to system CD in [van Fraassen, 1972] and system VN in [Lewis, 1973], but Goble's semantics is more transparently stated via ordering relations, $\geq_u$, like those discussed above regarding the use of ordering semantics for SDL (Section 7.1) and regarding the limit assumption dilemma (Section 8.4).

The system extends a language and logic for PC as follows. A dyadic operator, $\mathbf{O}(/)$, is added and monadic $\mathbf{O}$ is defined as mentioned above: $\mathbf{O}p =_{df} \mathbf{O}(p/\top)$. Dyadic and monadic permissibility can then be defined in a typical way for dyadic approaches: $\mathbf{P}(p/q) =_{df} \neg\mathbf{O}(\neg p/q)$ and $\mathbf{P}(p/\top) =_{df} \neg\mathbf{O}(\neg p/\top)$. However these are not employed in the axiom system below, but for convenience, an ordering relation is defined for the language (not to be confused with the world-relative ordering relation in the ordering semantics, $\geq_u$, and used in the axiomatization:

(Df$\geq$) $\quad p \geq q =_{df} \neg\mathbf{O}(\neg p/p \vee q)$

(Df$\geq$) says that the proposition that p is as normatively good as the proposition that q just in case it is not obligatory that $\neg p$ on the condition that either p or q. Given the definition of the permissibility operators, this amounts to saying that p is at least as good as q iff p is permissible given $p \vee q$. (At the semantic level, assuming bests for simplicity here, this says roughly p is as good as q just in case there is some best $p \vee q$-world that is a p-world.) SDDL, the dyadic analog to SDL, is then as follows:

(A1) All PC tautologies in the language (TAUT)
(A2) $\mathbf{O}(p \to q/r) \to (\mathbf{O}(p/r) \to \mathbf{O}(q/r))$ (CKD)
(A3) $\mathbf{O}(p/q) \to \neg\mathbf{O}(\neg p/q)$ (CDD)
(A4) $\mathbf{O}(\top/\top)$ (CON)
(A5) $\mathbf{O}(q/p) \to \mathbf{O}(q\&p/p)$ (CO&)
(A6) $(p \geq q \& q \geq r) \to p \geq r$ (Trans)
(R1) If $\vdash p$ and $\vdash p \to q$ then $\vdash q$ (MP)
(R2) If $\vdash p \leftrightarrow q$ then $\vdash \mathbf{O}(r/p) \leftrightarrow \mathbf{O}(r/q)$ (CRED)
(R3) If $\vdash p \to q$ then $\vdash \mathbf{O}(p/r) \to \mathbf{O}(q/r)$ (CRMD)

A1-A4 and R1-R3 are conditional analogues of formulas we used for the standard systems so we just preface those labels (e.g. "KD") with a "C"

---

for conflicts).

(but note that a conditional analog of VW would not have axiom A4, so that is a "non-standard system" in that respect). A5 and A6 are needed to generate a complete system relative to the intended ordering semantics, and they are more unique to the dyadic conditional. The first says that if q is obligatory at all given p, then p "crosses over" from the condition side to the obligation side and joins with q. (At the semantic level, assuming bests for simplicity, it roughly reflects the idea that if there are best p-worlds and they are all q-worlds, then they all must be $p\&q$-worlds.) The second takes advantage of the definitional abbreviation for "$\geq$" to even more perspicuously reflect in the language a feature of the ordering intended in the semantics and the way "$O(\ /\ )$" is to be defined via that semantics.

BIBLIOGRAPHY

[Aikhenvald, 2010] A. Y. Aikhenvald. *Imperatives and Commands*. Oxford University Press, Oxford and New York, 2010.

[al Hibri, 1978] A. al Hibri. *Deontic Logic: A Comprehensive Appraisal and a New Proposal*. University Press of America, Washington, DC, 1978.

[Alchourrón and Bulygin, 1971] C. E. Alchourrón and E. Bulygin. *Normative Systems*. Springer-Verlag, Wien/New York, 1971.

[Alchourrón and Bulygin, 1981] C. E. Alchourrón and E. Bulygin. The expressive conception of norms. In R. Hilpinen, editor, *New Studies in Deontic Logic: Norms, Actions, and the Foundations of Ethics*, pages 95–124. Reidel, Dordrecht, 1981.

[Alchourrón and Bulygin, 1993] C. Alchourrón and E. Bulygin. On the logic of normative systems. In H. Stachowiak, editor, *Pragmatik: Handbuch pragmatischen Denkens. Band iv. Sprachphilosophie, Sprachpragmatik und formative Pragmatik*, pages 273–294. Felix Meiner Verlag, Hamburg, 1993.

[Alchourrón and Makinson, 1981] C. E. Alchourrón and D. Makinson. Hierarchies of regulations and their logic. In R. Hilpinen, editor, *New Studies in Deontic logic: Norms, Actions, and the Foundations of Ethics*, pages 125–148. Reidel, Dordrecht, 1981.

[Alchourrón, 1969] C. E. Alchourrón. Logic of norms and logic of normative propositions. *Logique et Analyse*, 12:242–268, 1969.

[Alchourrón, 1993] C. E. Alchourrón. Philosophical foundations of deontic logic and the logic of defeasible conditionals. In J.-J. Meyer and R. J. Weiringer, editors, *Deontic Logic in Computer Science: Normative System Specification*, pages 43–84. Jone Wiley and Sons, Ltd., Chichester, 1993.

[Alchourrón, 1996] C. E. Alchourrón. Detachment and defeasibility in deontic logic. *Studia Logica*, 57(1):5–18, 1996.

[Anderson, 1958] A. R. Anderson. A reduction of deontic logic to alethic modal logic. *Mind*, 67:100–103, 1958.

[Anderson, 1967] A. R. Anderson. The formal analysis of normative systems. In N. Rescher, editor, *The Logic of Decision and Action*, pages 147–213. University of Pittsburgh Press, Pittsburgh, 1967. Originally published as Technical Report No. 2, Contract No. SAR/Nonr-609 (16) New Haven: Office of Naval Research, Group Psychology Branch, 1956.

[Apostel, 1982] L. Apostel. Towards a general theory of argumentation. In E. M. Barth and J. L. Martens, editors, *Argumentation. Approaches to Theory Formation*, pages 93–122. John Benjamins B. V., Amsterdam, 1982.

[Åqvist and Hoepelman, 1981] L. Åqvist and J. Hoepelman. Some theorems about a 'tree' system of deontic tense logic. In R. Hilpinen, editor, *New Studies in Deontic Logic: Norms, Actions and the Foundations of Ethics*, pages 187–221. Reidel, Dordrecht, 1981.

[Åqvist and Mullock, 1989] L. Åqvist and P. Mullock. *Causing Harm*. Walter de Gruyter, Berlin-New York, 1989.

[Åqvist, 1963] L. Åqvist. Deontic logic based on a logic of 'Better'. In *Proceedings of a Colloquium on Modal and Many-Valued Logics; Acta Philosophica Fennica*, volume 16, pages 285–290. Societas Philosophica Fennica, 1963.

[Åqvist, 1967] L. Åqvist. Good Samaritan, contrary-to-duty imperatives, and epistemic obligations. *Noûs*, 1:361–379, 1967.

[Åqvist, 1974] L. Åqvist. A new approach to the logical theory of actions and causality. In S. Stenlund, editor, *Logical Theory and Semantic Analysis*, pages 73–79. D. Reidel, Dordrecht, 1974.

[Åqvist, 1987] L. Åqvist. *Introduction to Deontic Logic and the Theory of Normative Systems*. Biblopolis, Napoli, 1987.

[Åqvist, 2002] L. Åqvist. Deontic logic. In D. Gabbay and F. Guenthner, editors, *Handbook of Philosophical Logic*, volume 8, pages 147–264. Kluwer Academic Publishers, Dordrecht, 2nd edition, 2002.

[Armstrong, 2004] D. Armstrong. *Truth and Truthmakers*. Cambridge University Press, Cambridge, 2004.

[Artikis et al., 2012] A. Artikis, R. Craven, N. K. Cicekli, B. Sadighi, and K. Stathis, editors. *Logic Programs, Norms and Action - Essays in Honor of Marek Sergot on the Occasion of His 60th Birthday*, volume 7360 of *Lecture Notes in Computer Science*. Springer, 2012.

[Asher and Bonevac, 1996] N. Asher and D. Bonevac. "Prima facie" obligation. *Studia Logica*, 57(1):19–45, 1996.

[Austin, 1954] J. Austin. *The Province of Jurisprudence Determined and the Uses of the Study of Jurisprudence, with Introduction by L. A. Hart*. Noonday Press, New York, 1954.

[Bartha, 1999] P. Bartha. Moral preference, contrary-to-duty obligation and defeasible oughts. In P. McNamara and H. Prakken, editors, *Norms, Logics and Information Systems*, pages 93–108. IOS Press, Amsterdam, 1999.

[Bell et al., 2001] J. L. Bell, D. DeVidi, and G. Solomon. *Logical Options*. Broadview Press, Peterborough (Ontario) - Orchard Park (N.Y.), 2001.

[Belnap and Perloff, 1988] N. Belnap and M. Perloff. Seeing to it that: A canonical form for agentives. *Theoria*, 54:175–199, 1988.

[Belnap and Perloff, 1992] N. Belnap and M. Perloff. The way of the agent. *Studia Logica*, 51:463–484, 1992.

[Belnap et al., 2001] N. Belnap, M. Perloff, and M. Xu. *Facing the Future: Agents and Choices in our Indeterminist World*. Oxford University Press, New York, 2001.

[Belnap, 1991] N. Belnap. Backwards and forwards in the modal logic of agency. *Philosophy and Phenomenological Research*, 51:777–807, 1991.

[Belzer, 1986] M. Belzer. Reasoning with defeasible principles. *Synthese*, 66(1):135–158, 1986.

[Bentham, 1948] J. Bentham. *An Introduction to the Principles of Morals and Legislation*. Hafner Publishing Company, New York, 1948. First pub. date: 1789.

[Bentham, 1983] J. Bentham. Deontology; together with a table of the springs of action; and the article on utilitarianism. In *Collected Works of Jeremy Bentham*. Clarendon Press, Oxford, 1983. First pub. date: 1834.

[Boh, 1985] I. Boh. Belief, justification, and knowledge – some late-medieval epistemic concerns. *Journal of the Rocky Mountain Medieval and Renaissance Association* (current title *Quidditas*), 6:87–103, 1985.

[Boh, 1993] I. Boh. *Epistemic Logic in the Later Middle Ages*. Routledge, London and New York, 1993.

[Bonevac, 1998] D. Bonevac. Against conditional obligation. *Noûs*, 32(1):37–53, 1998.
[Bostock, 1997] D. Bostock. *Intermediate Logic*. Clarendon Press, Oxford, 1997.
[Brewka, 1989] G. Brewka. Non-monotonic logics - a brief overview. *AI Communications: The European Journal of Artificial Intelligence*, 2(2):88–97, 1989.
[Brink, 1994] D. Brink. Moral conflict and its structure. *The Philosophical Review*, 103:215–247, 1994.
[Broome, 1999] J. Broome. Normative requirements. *Ratio*, 12:398–419, 1999.
[Broome, 2004] J. Broome. Reasons. In J. Wallace, M. Smith, and et al S. Scheffler, editors, *Reason and Value: Themes from the Moral Philosophy of Joseph Raz*, pages 28–55. Oxford University Press, Oxford, 2004.
[Brown, 1992] M. A. Brown. Normal bimodal logics of ability and action. *Studia Logica*, 51:519–532, 1992.
[Brown, 1996a] M. A. Brown. Doing as we ought: Towards a logic of simply dischargeable obligations. In M. A. Brown and J. Carmo, editors, *Deontic Logic, Agency and Normative Systems*, pages 47–65. Springer Verlag, New York, 1996.
[Brown, 1996b] M. A. Brown. A logic of comparative obligation. *Studia Logica*, 57:117–137, 1996.
[Bulygin, 1982] E. Bulygin. Norms, normative propositions, and legal statements. In G. Fløistad, editor, *Contemporary Philosophy. A New survey. Vol. 3: Philosophy of action*, pages 127–152. Martinus Nijhoff, The Hague, 1982.
[Bulygin, 1992] E. Bulygin. On norms of competence. *Law and Philosophy*, 11:201–216, 1992.
[Carmo and Jones, 2002] J. Carmo and A. Jones. Deontic logic and contrary-to-duties. In D. Gabbay and F. Gueunthner, editors, *Handbook of Philosophical Logic*, volume 8, pages 265–343. Kluwer Academic Publishers, Dordrecht, 2nd edition, 2002.
[Carnap, 1956] R. Carnap. *Meaning and Necessity: A Study in Semantics and Modal Logic*. The University of Chicago Press, Chicago, enlarged edition, 1956. First pub. date: 1947.
[Castañeda, 1970] H. N. Castañeda. On the semantics of the ought-to-do. *Synthese*, 21:449–468, 1970. Reprinted in D. Davidson and G. Harman (eds), *Semantics of Natural Language*, pp. 675-694. Dordrecht: D. Reidel, 1972.
[Castañeda, 1977] H. N. Castañeda. Ought, time, and the deontic paradoxes. *Journal of Philosophy*, 74:775–788, 1977.
[Castañeda, 1981] H. N. Castañeda. The paradoxes of deontic logic: The simplest solution to all of them in one fell swoop. In R. Hilpinen, editor, *New Studies in Deontic Logic: Norms, Actions and the Foundations of Ethics*, pages 37–85. Reidel, Dordrecht, 1981.
[Chellas, 1969] B. F. Chellas. *The Logical Form of Imperatives*. Dissertation, Perry Lane Press, Stanford, 1969.
[Chellas, 1974] B. F. Chellas. Conditional obligation. In S. Stenlund, editor, *Logical Theory and Semantic Analysis*, pages 23–33. D. Reidel, Dordrecht, 1974.
[Chellas, 1980] B. F. Chellas. *Modal Logic: An Introduction*. Cambridge University Press, Cambridge, 1980.
[Chisholm, 1963] R. M. Chisholm. Contrary-to-duty imperatives and deontic logic. *Analysis*, 24:33–36, 1963.
[Chisholm, 1964] R. M. Chisholm. The ethics of requirement. *American Philosophical Quarterly*, 1:147–153, 1964.
[Chisholm, 1982] R. M. Chisholm. *Brentano and Meinong Studies (Studien Zur Öesterreichischen Philosophie iii)*, chapter Supererogation and offence: A conceptual scheme for ethics, pages 98–113. Rodopi B.V., Amsterdam, 1982. Originally published in *Ratio*, 5, pp. 1-14, 1963.
[Cresswell, 1967] M. J. Cresswell. The interpretation of some Lewis systems of modal logic. *Australasian Journal of Philosophy*, 45(2):198–206, 1967.

[Czelakowski, 1997] J. Czelakowski. Action and deontology. In S. Lindström and E. Ejerhed, editors, *Logic, Action and Cognition*, pages 47–88. Kluwer Academic Publishers, Dordrecht and Boston, 1997.
[Da Costa and Carnielli, 1986] N. C. A. Da Costa and W. A. Carnielli. On paraconsistent deontic logic. *Philosophia*, 16:293–305, 1986.
[Danielsson, 1968] S. Danielsson. *Preference and Obligation: Studies in the Logic of Ethics*. Filosofiska föreningen, Uppsala, 1968.
[Davidson, 1980a] D. Davidson. Agency. In D. Davidson, editor, *Essays on Actions and Events*, pages 43–61. Clarendon Press, Oxford, 1980.
[Davidson, 1980b] D. Davidson. The logical form of action sentences. In D. Davidson, editor, *Essays on Actions and Events*, pages 105–122. Clarendon Press, Oxford, 1980. Originally published in N. Rescher (ed), *The Logic of Decision and Action*, Pittsburgh, University of Pittsburgh Press, 1967.
[DeCew, 1981] J. W. DeCew. Conditional obligation and counterfactuals. *Journal of Philosophical Logic*, 10:55–72, 1981.
[Dignum et al., 1996] F. Dignum, J. J. C. Meyer, and R. Wieringa. Free choice and contextually permitted actions. *Studia Logica*, 57(1):193–220, 1996.
[Dubislav, 1937] W. Dubislav. Zur Unbegründbarkeit der Forderunssätze. *Theoria*, 3:330–342, 1937.
[Elgesem, 1993] D. Elgesem. *Action Theory and Modal Logic*. Ph.D. dissertation, University of Oslo, 1993.
[Elgesem, 1997] D. Elgesem. The modal logic of agency. *Nordic Journal of Philosophical Logic*, 2:1–46, 1997.
[Fehige, 1994] C. Fehige. The limit assumption in deontic (and prohairetic) logic. In G. Meggle and U. Wessels, editors, *Analyomen 1: Proceedings of the 1st Conference "Perspectives in Analytical Philosophy"*. Waiter de Gruyter, Berlin-New York, 1994.
[Feldman, 1986] F. Feldman. *Doing the Best We Can: An Essay in Informal Deontic Logic*. D. Reidel, Dordrecht, 1986.
[Føllesdal and Hilpinen, 1971] D. Føllesdal and R. Hilpinen. Deontic logic: An introduction. In R. Hilpinen, editor, *Deontic Logic: Introductory and Systematic Readings*, pages 1–35. D. Reidel, Dordrecht, 2nd edition, 1971.
[Forrester, 1984] J. W. Forrester. Gentle murder, or the adverbial samaritan. *Journal of Philosophy*, 81:193–196, 1984.
[Garson, 2001] J. Garson. Quantification in modal logic. In D. Gabbay and F. Guenthner, editors, *Handbook of Philosophical Logic*, pages 267–323. D. Reidel, Dordrecht, 2nd edition, 2001.
[Girle, 2009] R. Girle. *Modal Logics and Philosophy*. McGill-Queens University Press, Montreal and Kingston and Ithaca, 2nd edition, 2009.
[Goble, 1990a] L. Goble. A logic of good, should, and would: Part I. *Journal of Philosophical Logic*, 19(2):169–199, 1990.
[Goble, 1990b] L. Goble. A logic of good, should, and would: Part II. *Journal of Philosophical Logic*, 19:253–276, 1990.
[Goble, 1991] L. Goble. Murder most gentle: The paradox deepens. *Philosophical Studies*, 64(2):217–227, 1991.
[Goble, 1994] L. Goble. Quantified deontic logic with definite descriptions. *Logique et Analyse*, 37:239–253, 1994.
[Goble, 1996] L. Goble. 'Ought' and extensionality. *Noûs*, 30:330–355, 1996.
[Goble, 2003] L. Goble. Preference semantics for deontic logic part I- simple models. *Logique et Analyse*, 46(183-184):383–418, 2003.
[Goble, 2004] L. Goble. Preference semantics for deontic logic part II - multiplex models. *Logique et Analyse*, 47(335-363), 2004.
[Goble, 2005] L. Goble. A logic for deontic dilemmas. *Journal of Applied Logic*, 3:461–483, 2005.
[Goble, 2009] L. Goble. Normative conflicts and the logic of 'ought'. *Noûs*, 43(3):450–489, 2009.

[Goldman, 1976] H. S. Goldman. Dated rigthness and moral perfection. *The Philosophical Review*, 85(4):449–487, 1976.
[Greenspan, 1975] P. S. Greenspan. Conditional oughts and hypothetical imperatives. *Journal of Philosophy*, 72:259–276, 1975.
[Grelling, 1939] K. Grelling. Zur Logik der Sollsätze. *Unity of Science Forum*, January:44–47, 1939.
[Grue-Sörensen, 1939] K. Grue-Sörensen. Imperativsätze und Logik. Begegnung einer Kritik. *Theoria*, 5:195–202, 1939.
[Hamblin, 1987] C. L. Hamblin. *Imperatives*. Basil Blackwell, Oxford, 1987.
[Hansen, 2004] J. Hansen. Problems and results for logics about imperatives. *Journal of Applied Logic*, 2(1):39–61, 2004.
[Hansen, 2008] J. Hansen. *Imperatives and Deontic Logic: On the Semantic Foundations of Deontic Logic*. Dissertation (available from the author as a pdf), University of Leipzig, 2008.
[Hansson, 1971] B. Hansson. An analysis of some deontic logics. In R. Hilpinen, editor, *Deontic logic: Introductory and Systematic Readings*, pages 121–147. D. Reidel, Dordrecht, 1971. Reprinted from *Noûs* 3, pp. 373-98, 1969.
[Hansson, 1990] S. O. Hansson. Preference-based deontic logic (PDL). *Journal of Philosophical Logic*, 19(1):75–93, 1990.
[Hansson, 2001] S. O. Hansson. *Structures of Values and Norms*. Cambridge University Press, Cambridge, 2001.
[Hardin, 1968] G. Hardin. The tragedy of the commons. *Science*, 162(3859):1243 and 1248, 1968.
[Hart and Honoré, 1959] H. L. A. Hart and A. M. Honoré. *Causation in the Law*. Clarendon Press, Oxford, 1959.
[Hartmann, 1987] R. Hartmann. *Die Religion des Islam*. Wissenschaftliche Buchgesellschaft, Darmstadt, 1987. Originally published by Mittler and Sohn, Berlin, 1944.
[Hedenius, 1941] I. Hedenius. *Om Rätt och Moral ('On Law and Morals')*. Tidens Förlag, Stockholm, 1941.
[Henry, 1967] P. D. Henry. *The Logic of St. Anselm*. Clarendon Press, Oxford, 1967.
[Herrestad and Krogh, 1995] H. Herrestad and C. Krogh. Obligations directed from bearers to counterparts. In *Proceedings of the Fifth International Conference on Artificial Intelligence and Law (International Conference on Artificial Intelligence and Law)*, pages 210–218. ACM Press, College Park, Maryland, USA, 1995.
[Hilpinen, 1969] R. Hilpinen. An analysis of relativized modalities. In J. W. David, D.J. Hockney, and W.K. Wilson, editors, *Philosophical Logic*, Synthese library, pages 181–193. D. Reidel, Dordrecht, 1969.
[Hilpinen, 1974] R. Hilpinen. On the semantics of personal directives. In C. H. Heidrich, editor, *Semantics and Communication*, pages 162–179. North-Holland, Amsterdam, 1974.
[Hilpinen, 1993] R. Hilpinen. On deontic logic, pragmatics, and modality. In H. Stachowiak, editor, *Pragmatik: Handbuch Pragmatischen Denkens. Band iv: Sprachphilosophie, Sprachpragmatik und formative Pragmatik*, pages 295–319. Felix Meiner Verlag, Hamburg, 1993.
[Hilpinen, 1997] R. Hilpinen. On action and agency. In E. Ejerhed and S. Lindström, editors, *Logic, Action and Cognition - Essays in Philosophical Logic*, pages 3–27. Kluwer Academic Publishers, Dordrecht, 1997.
[Hilpinen, 2006] R. Hilpinen. Norms, normative utterances, and normative propositions. *Análisis Filosófico (Special Issue: Homenaje a Carlos E. Alchourrón)*, 26(2):229–241, 2006. Guest editors: E.Bulygin and G. Palau.
[Hintikka, 1957] J. Hintikka. Quantifiers in deontic logic. *Societas Scientiarum Fennica, Commentationes Humanarum Litterarum*, 23(4):1–23, 1957.
[Hintikka, 1971] J. Hintikka. Some main problems of deontic logic. In R. Hilpinen, editor, *Deontic Logic: Introductory and Systematic Readings*, pages 59–104. D. Reidel, Dordrecht, 1971.

[Hofstadter and McKinsey, 1939] A. Hofstadter and J. C. C. McKinsey. On the logic of imperatives. *Philosophy of Science*, 6:446–457, 1939.

[Hohfeld, 1919] W. N. Hohfeld. *Fundamental Legal Conceptions as Applied in Judicial Reasoning*. Yale University Press, New Haven, 1919. Edited by Walter Wheeler Cook.

[Horty, 1994] J. F. Horty. Moral dilemmas and nonmonotonic logic. *Journal of Philosophical Logic*, 23(1):35–65, 1994.

[Horty, 1996] J. F. Horty. Combining agency and obligation (preliminary version). In M. A. Brown and J. Carmo, editors, *Deontic Logic, Agency and Normative Systems*, pages 98–122. Springer Verlag, New York, 1996.

[Horty, 1997] J. F. Horty. Non-monotonic foundations for deontic logic. In D. Nute, editor, *Defeasible Deontic Logic*, pages 17–44. Kluwer Academic, Dordrecht, 1997.

[Horty, 2001] J. F. Horty. *Agency and Deontic Logic*. Oxford University Press, Oxford, 2001.

[Horty, 2003] J. F. Horty. Reasoning with moral conflicts. *Noûs*, 37(4):557–605, 2003.

[Hourani, 1975] G. F. Hourani. Ethics in medieval Islam: A conspectus. In G. Hourani, editor, *Essays on Islamic Philosophy and Science*, pages 128–135. State University of New York Press, Albany, 1975.

[Hourani, 1985] G. F. Hourani. *Reason and Tradition in Islamic Ethics*. Cambridge University Press, Cambridge, 1985.

[Hruschka, 1986] J. Hruschka. Das deontologische Sechseck bei Gottfried Achenwall im Jahre 1767. In *Joachim Jungius Gesellschaft der Wissenschaften e.V., Hamburg, 4 (1986) Heft 2*. Vandenhoeck and Ruprecht, Göttingen, 1986.

[Hursthouse, 1999] R. Hursthouse. *On Virtue Ethics*. Oxford University Press, Oxford, 1999.

[Jackson and Pargetter, 1986] F. Jackson and R. Pargetter. Oughts, options, and actualism. *Philosophical Review*, 95:233–255, 1986.

[Jackson, 1985] F. Jackson. On the semantics and logic of obligation. *Mind*, 94:177–195, 1985.

[Jackson, 1988] F. Jackson. Understanding the logic of obligation. *Aristotelian Society*, SUPP 62:255–270, 1988.

[Jennings, 1974] R. E. Jennings. A utilitarian semantics for deontic logic. *Journal of Philosophical Logic*, 3:445–456, 1974.

[Jones and Pörn, 1985] A. Jones and I. Pörn. Ideality, sub-ideality and deontic logic. *Synthese*, 65:275–290, 1985.

[Jones and Pörn, 1986] A. Jones and I. Pörn. 'Ought' and 'must'. *Synthese*, 66:89–93, 1986.

[Jones and Sergot, 1993] A. Jones and M. Sergot. On the characterization of law and computer systems: The normative systems perspective. In J.-J. C. Meyer and R. J. Wieringa, editors, *Deontic Logic in Computer Science*, pages 275–307. John Wiley and Sons, Chichester, 1993.

[Jones and Sergot, 1996] A. Jones and M. Sergot. A formal characterization of institutionalized power. *Logic Journal of the IGPL*, 4(3):429–445, 1996.

[Jones, 1990] A. Jones. Deontic logic and legal knowledge representation. *Ratio Juris*, 3(2):237–244, 1990.

[Jørgensen, 1937 and 1938] J. Jørgensen. Imperatives and logic. *Erkenntnis*, 7:288–296, 1937 and 1938.

[Jørgensen, 1938] J. Jørgensen. Imperativer og logik. *Theoria*, 4:183–90, 1938.

[Kamp, 1974] H. Kamp. Free choice permission. *Proceedings of the Aristotelian Society*, 74:57–74, 1974.

[Kamp, 1979] H. Kamp. Semantics versus pragmatics. In F. Guenthner and S. J. Schmidt, editors, *Formal Semantics and Pragmatics for Natural Languages*, pages 255–287. D. Reidel, Dordrecht, 1979.

[Kanger and Kanger, 2001] S. Kanger and H. Kanger. Rights and parliamentarism. In G. Holmström-Hintikka, S. Lindström, and R. Sliwinski, editors, *Collected Papers of Stig Kanger with Essays on his Life and Work*, volume I, pages 120–145. Kluwer Academic Publishers, Dordrecht, 2001. Originally published in 1966 in *Theoria* 32, pp. 85-115.
[Kanger, 1971] S. Kanger. New foundations for ethical theory. In R. Hilpinen, editor, *Deontic Logic: Introductory and Systematic Readings*, pages 36–58. D. Reidel, Dordrecht, 1971. Originally published in mimeographic form in 1957.
[Kanger, 1984] H. Kanger. *Human Rights in the U.N. Declaration (Acta Universitatis Upsaliensis)*. University of Uppsala, Uppsala, 1984.
[Kanger, 2001] S. Kanger. Law and logic. In G. Holmström-Hintikka, S. Lindström, and R. Sliwinski, editors, *Collected Papers of Stig Kanger with Essays on his Life and Work*, volume 1, pages 146–169. Kluwer Academic Publishers, Dordrecht, 2001. Originally published in 1972 in *Theoria* 38, pp. 105-132.
[Kelsen, 1967] H. Kelsen. *Pure Theory of Law*. University of California Press, Berkeley and Los Angeles, 1967. Translation by M. Knight of Reine Rechtslehre, 1960, 2nd ed., Verlag Franz Deuticke, Vienna.
[Kempson, 1977] R. M. Kempson. *Semantic Theory*. Cambridge texts in linguistics. Cambridge University Press, Cambridge, 1977.
[Knebel, 1991] S. K. Knebel. Necessitas moralis ad optimum. Zum historischen Hintergrund der Wahl der besten aller möglichen Welten. *Studia Leibnitiana*, 23:3–24, 1991.
[Knuuttila and Hallamaa, 1995] S. Knuuttila and O. Hallamaa. Roger Roseth and medieval deontic logic. *Logique et Analyse*, 38(149):75–87, 1995.
[Knuuttila, 1981] S. Knuuttila. The emergence of deontic logic in the fourteenth century. In R. Hilpinen, editor, *New Studies in Deontic Logic: Norms, Actions, and the Foundations of Ethics*, pages 225–248. Reidel, Dordrecht, 1981.
[Knuuttila, 1993] S. Knuuttila. *Modalities in Medieval Philosophy*. Routledge, London and New York, 1993.
[Kripke, 1963] S. Kripke. Semantical analysis of modal logic I: Normal modal propositional calculi. *Mathematical Logic Quarterly*, 9(5-6):67–96, 1963.
[Kripke, 1965] S. Kripke. Semantical analysis of modal logic II: Non-normal modal propositional calculi. In L. Hemkin, J.W. Addison, and A. Tarski, editors, *The Theory of Models (Proceedings of the 1963 International Symposium at Berkeley)*, pages 206–220. North Holland Publishing Co, Amsterdam, 1965.
[Krogh and Herrestad, 1996] C. Krogh and H. Herrestad. Getting personal: Some notes on the relationship between personal and impersonal obligation. In M. A. Brown and J. Carmo, editors, *Deontic Logic, Agency and Normative Systems*, pages 134–153. Springer Verlag, New York, 1996.
[Leibniz, 1930] G. W. Leibniz. Elementa juris naturalis. In *Sämtliche Schriften und Briefe. Sechste Reihe: Philosophische Schriften. Bd. 1*, pages 431–485. Otto Reichl Verlag, Darmstadt, 1930. First pub. date: 1671.
[Lemmon, 1962a] E. J. Lemmon. Moral dilemmas. *Philosophical Review*, 71:139–158, 1962.
[Lemmon, 1962b] E. J. Lemmon. On sentences verifiable by their use. *Analysis*, 22:86–89, 1962.
[Lewis, 1973] D. K. Lewis. *Counterfactuals*. Blackwell, Oxford, 1973.
[Lewis, 1974] D. K. Lewis. Semantic analyses for dyadic deontic logic. In S. Stenlund, editor, *Logical Theory and Semantic Analysis: Essays Dedicated to Stig Kanger on his Fiftieth Birthday*, pages 1–14. D.Reidel, Dordrecht, 1974.
[Lewis, 1978] D. K. Lewis. Reply to McMichael's "Too much of a good thing: A problem in deontic logic". *Analysis*, 38:85–86, 1978.
[Lewis, 1979] D.K. Lewis. A problem about permission. In E. Saarinen, R. Hilpinen, I. Niiniluoto, and M. Provence, editors, *Essays in Honour of Jaakko Hintikka on the Occasion of his Fiftieth Birthday*, pages 163–175. Reidel, Dordrecht, 1979.

[Lindahl and Odelstad, 2000] L. Lindahl and J. Odelstad. Normative systems represented by boolean quasi-orderings. *Nordic Journal of Philosophical Logic*, 5:161–174, 2000.
[Lindahl, 1977] L. Lindahl. *Position and Change*. D. Reidel, Dordrecht, 1977.
[Lindahl, 2001] L. Lindahl. Stig Kanger's theory of rights. In G. Holmström-Hintikka, S. Lindström, and R. Sliwinski, editors, *Collected Papers of Stig Kanger with Essays on his Life and Work*, pages 151–171. Kluwer Academic Publishers, Dordrecht, 2001. Originally published in D. Prawitz et al. (eds), *Logic, Methodology and Philosophy of Science IX*, 1994, Amsterdam: Elsevier Science B. V., pp. 889-911.
[Loewer and Belzer, 1983] B. Loewer and M. Belzer. Dyadic deontic detachment. *Synthese*, 54:295–318, 1983.
[Loewer and Belzer, 1986] B. Loewer and M. Belzer. Help for the good samaritan paradox. *Philosophical Studies*, 50:117–128, 1986.
[Loewer and Belzer, 1991] B. Loewer and M. Belzer. "Prima facie" obligation. In E. Lepore, editor, *John Searle and his Critics*, pages 359–370. Blackwell, Cambridge, 1991.
[Lokhorst and Goble, 2004] G.-J. Lokhorst and L. Goble. Mally's deontic logic. *Grazer Philosophische Studien*, 67:37-57, 2004.
[Lokhorst, 2006] G.-J. C. Lokhorst. Propositional quantifiers in deontic logic. In L. Goble and J. J. C. Meyer, editors, *Deontic Logic and Artificial Normative Systems (Deon 2006 Proceedings)*, volume 47 of *Lecture Notes in Artificial Intelligence*, pages 201–209. Springer-Verlag, Berlin and Heidelberg, 2006.
[Lokhorst, 2008] G.-J. Lokhorst. Mally's deontic logic. In E. Zalta, editor, *Stanford Encyclopedia of Philosophy*. 2008.
[Lyons, 1977] J. Lyons. *Introduction to Theoretical Linguistics*. Cambridge University Press, Cambridge, 1977.
[Makinson and van der Torre, 2000] D. Makinson and L. van der Torre. Input/output logics. *Journal of Philosophical Logic*, 29(4):383–408, 2000.
[Makinson and van der Torre, 2003] D. Makinson and L. van der Torre. What is input/output logic? In B. Löwe, W. Malzkorn, and T. Räsch, editors, *Foundations of the Formal Sciences II: Applications of Mathematical Logic in Philosophy and Linguistics*, volume 17 of *Trends in logic*, pages 163–174. Springer-Verlag, Berlin, 2003.
[Makinson, 1981] D. Makinson. Quantificational reefs in deontic waters. In R. Hilpinen, editor, *New Studies in Deontic Logic: Norms, Actions, and the Foundations of Ethics*, pages 87–91. Reidel, Dordrecht, 1981.
[Makinson, 1986] D. Makinson. On the formal representation of rights relations: Remarks on the work of Stig Kanger and Lars Lindahl. *Journal of Philosophical Logic*, 15:403–425, 1986.
[Makinson, 1993] D. Makinson. Five faces of minimality. *Studia Logica*, 52(3):339–379, 1993.
[Makinson, 1999] D. Makinson. On the fundamental problem of deontic logic. In P. McNamara and H. Prakken, editors, *Norms, Logic and Information Systems*, pages 29–53. IOS Press, Amsterdam, 1999.
[Mally, 1971] E. Mally. Elemente des Sollens. Grundgesetze der Logik des Willens. In K. Wolf and P. Weingartner, editors, *E. Mally, Logische Scriften. Grosses Logikfragment Grundgesetze des Sollens*, pages 229–324. D. Reidel, Dordrecht, 1971. Originally published in 1926, *Gurndgesetz Des Sollens. Elemente Der Logik Des Willens*. Graz, Leuschner and Lubensky.
[Marcus, 1980] R. B. Marcus. Moral dilemmas and consistency. *The Journal of Philosophy*, 77:121–136, 1980.
[Marcus, 1996] R. B. Marcus. More about moral dilemmas. In H. E. Mason, editor, *Moral Dilemmas and Moral Theory*, pages 23–35. Oxford University Press, New York, 1996.
[Mares and McNamara, 1997] E. D. Mares and P. McNamara. Supererogation in deontic logic: Metatheory for dwe and some close neighbours. *Studia Logica*, 59(3):397–415, 1997.

[Mates, 1965] B. Mates. *Elementary Logic*. Oxford University Press, Oxford, 1965.
[McCarty, 1992] L. T. McCarty. Defeasible deontic reasoning. In *Proceedings of the Fourth international Workshop on Nonmonotonic Reasoning*, pages 139–147, Plymouth, Vermont, USA, 1992.
[McConnell and Brue, 2002] C. R. McConnell and S. R. Brue. *Economics. Principles, Problems, and Policies*. McGraw-Hill, Boston, 2002.
[McMichael, 1978] A. McMichael. Too much of a good thing: A problem in deontic logic. *Analysis*, 38:83–84, 1978.
[McNamara, 1988] P. McNamara. A determination theorem for a von Wright inspired weakening of SDL. 1988.
[McNamara, 1990] P. McNamara. *The Deontic Quaddecagon*. Dissertation, University of Massachusetts, 1990.
[McNamara, 1995] P. McNamara. The confinement problem: How to terminate your mom and her trust. *Analysis*, 55(4):310–313, 1995.
[McNamara, 1996a] P. McNamara. Doing well enough: Toward a logic for commonsense morality. *Studia Logica*, 57(1):167–192, 1996.
[McNamara, 1996b] P. McNamara. Making room for going beyond the call. *Mind*, 105(419):415–450, 1996.
[McNamara, 1996c] P. McNamara. Must I do what I ought? (or will the least I can do do?). In M. A. Brown and J. Carmo, editors, *Deontic Logic, Agency and Normative Systems*, pages 154–173. Springer Verlag, New York, 1996.
[McNamara, 1999] P. McNamara. Doing well enough in an Andersonian-Kangerian framework. In P. McNamara and H. Prakken, editors, *Norms, Logics and Information Systems: New Studies in Deontic Logic and Computer Science*, pages 181–198. IOS Press, Washington, DC, 1999.
[McNamara, 2000] P. McNamara. Towards a framework for agency, inevitability, praise and blame. *Nordic Journal of Philosophical Logic*, 5:135–160, 2000.
[McNamara, 2004] P. McNamara. Agential obligation as non-agential personal obligation plus agency. *Journal of Applied Logic*, 2(1):117–152, 2004.
[McNamara, 2006] P. McNamara. Deontic logic. In D. Gabbay and J. Woods, editors, *The Handbook of the History of Logic*, volume 7, pages 197–288. Elsevier Press, Amsterdam, 2006.
[McNamara, 2010] P. McNamara. Deontic logic. In E. Zalta, editor, Stanford Encyclopedia of Philosophy. 2010. http://plato.stanford.edu.archives/fall2010/entries/logic-deontic/.
[McNamara, 2011a] P. McNamara. Praise, blame, obligation, and DWE: Toward a framework for the classical conception of supererogation and kin. *Journal of Applied Logic*, 9:153–170, 2011.
[McNamara, 2011b] P. McNamara. Supererogation, inside and out: Toward an adequate scheme for common sense morality. In M. Timmons, editor, *Oxford Studies in Normative Ethics*, volume I, pages 202–235. Oxford University Press, Oxford, 2011.
[Meinong, 1968a] A. Meinong. Ethische Bausteine (Nacgelassenes Fragment). In R. Haller and R. Kindinger, editors, *Alexius Meinong Gesamtausgabe*, vol. III, *Abhandlungen zur Werttheorie*, pages 659–724. Akademische Druck- u. Verlagsanstalt, Graz, 1968. The page numbers in the text are those of the article not the page numbers of the *Gesamtausgabe*.
[Meinong, 1968b] A. Meinong. Psychologisch-ethische Untersuchungen zur Werth-Theorie. In R. Haller and R. Kindinger, editors, *Alexius Meinong Gesamtausgabe*, vol. III, *Abhandlungen zur Werttheorie*, pages 3–94. Akademische Druck- u. Velagsanstalt, Graz, 1968. Originally published by Leuschner and Lubensky, Graz, 1894. The page numbers in the text are those of the original publication, reproduced in the *Gesamtausgabe*.
[Menger, 1939] K. Menger. A logic of the doubtful: On optative and imperative logic. In K. Menger, editor, *Reports of a Mathematical Colloquium*, 2nd series, 2nd issue, pages 53–64. Indiana University Press, 1939.

[Meyer and Weiringer, 1993] J.-J. Meyer and R. J. Weiringer. *Deontic Logic in Computer Science: Normative System Specification*. Jone Wiley and Sons, Ltd., Chichester, 1993.
[Meyer, 1988] J.-J. Meyer. A different approach to deontic logic: Deontic logic viewed as a variant of dynamic logic. *Notre Dame Journal of Formal Logic*, 29:109–136, 1988.
[Montague, 1970] R. Montague. Universal grammar. *Theoria*, 36:373–98, 1970.
[Moreau, 1996] M. Moreau. Prima facie and seeming duties. *Studia Logica*, 57:47–71, 1996.
[Mott, 1973] P. L. Mott. On Chisholm's paradox. *Journal of Philosophical Logic*, 2:197–210, 1973.
[Mulligan et al., 1984] K. Mulligan, P. Simons, and B. Smith. Truth-makers. *Philosophy and Phenomenological Research*, 44:287–321, 1984.
[Nowell Smith and Lemmon, 1960] P. H. Nowell Smith and E. J. Lemmon. Escapism: The logical basis of ethics. *Mind*, 69:289–300, 1960.
[Nute, 1997] D. Nute. *Defeasible Deontic Logic*. Synthese library, 263. Kluwer Academic, Dordrecht ; Boston, 1997.
[Parent, 2008] X. Parent. On the strong completeness of Åqvist's dyadic deontic logic G. In R. van der Meyden and L. van der Torre, editors, *Deontic Logic in Computer Science. 9th International Donference, Deon 2008*, volume 5076 of *Lecture Note in Computer Science*, pages 189–202, Berlin Heidelberg, 2008. Springer-Verlag.
[Parent, 2010] X. Parent. A complete axiom set for Hansson's deontic logic DSDL2. *Logic Journal of the IGPL*, 18(3):422–429, 2010.
[Parent, 2012] X. Parent. Why be afraid of identity? - Comments on Sergot and Prakken's views. In Artikis et al. [2012], pages 295–307.
[Peirce, 1998] C. S. Peirce. New elements. In N. Houser, J. R. Eller, and et al A. C. Lewis, editors, *The Essential Peirce. Selected Philosophical Writings, 1893-1913: Vol 2*, pages 300–324. Indiana University Press, Bloomington and Indianapolis, 1998. (Written in 1904).
[Pörn, 1977] I. Pörn. Action theory and social science: Some formal models. *Synthese Library*, 120, 1977.
[Pörn, 1989] I. Pörn. On the nature of a social order. In J. E. Fenstad, T. Frolov, and R. Hilpinen, editors, *Logic, Methodology and Philosophy of Science, VIII*, pages 553–67. Elsevier Science, New York, 1989.
[Powers, 1967] L. Powers. Some deontic logicians. *Noûs*, 1:381–400, 1967.
[Prakken and Sergot, 1994] H. Prakken and M. Sergot. Contrary-to-duty imperatives, defeasibility and violability. In A. Jones and M. Sergot, editors, *Proceedings of the Second International Workshop on Deontic Logic in Computer Science*, pages 296–318. Tano Publishers, Oslo, 1994.
[Prakken and Sergot, 1996] H. Prakken and M. Sergot. Contrary-to-duty obligations. *Studia Logica*, 57(1):91–115, 1996.
[Prakken and Sergot, 1997] H. Prakken and M. Sergot. Dyadic deontic logic and contrary-to-duty obligations. In D. Nute, editor, *Defeasible deontic logic*, Synthese library, pages 223–262. Kluwer Academic, Dordrecht-Boston, 1997.
[Prakken, 1996] H. Prakken. Two approaches to the formalisation of defeasible deontic reasoning. *Studia Logica*, 57(1):73–90, 1996.
[Priest, 2008] G. Priest. *An Introduction to Non-Classical Logics. From If to Is*. Cambridge University Press, Cambridge, second edition, 2008.
[Prior, 1954] A. N. Prior. The paradoxes of derived obligation. *Mind*, 63:64–65, 1954.
[Prior, 1955] A. N. Prior. *Formal logic*. Oxford University Press, Oxford, 1955. Second edition, 1962.
[Prior, 1958] A. N. Prior. Escapism: The logical basis of ethics. In A. I. Melden, editor, *Essays in Moral Philosophy*, pages 135–146. University of Washington Press, Seattle, 1958.
[Reach, 1939] K. Reach. Some comments on Grelling's paper 'zur logik der sollsatze'. *Unity of Science Forum*, April:72, 1939.

[Ross, 1939] W. D. Ross. *Foundations of Ethics*. Oxford University Press, Oxford, 1939.
[Ross, 1941] A. Ross. Imperatives and logic. *Theoria*, 7:53–71, 1941.
[Ross, 1944] A. Ross. Imperatives and logic. *Philosophy of Science*, 11:30–46, 1944.
[Ross, 1968] A. Ross. *Directive and Norms*. Routledge and Kegan Paul, London, 1968. B. Loar, editor.
[Ryu and Lee, 1991] Y. U. Ryu and R. M. Lee. Defeasible deontic reasoning: A logic programming model. In J.-J. Meyer and R. J. Weiringer, editors, *Deontic Logic in Computer Science: Normative System Specification*, pages 225–241. Jone Wiley and Sons, Ltd., Chichester, 1991.
[Sajama, 1988] S. Sajama. Meinong on the foundations of deontic logic. *Grazer Philosophische Studien*, 32:69–81, 1988.
[Sandu and Tuomela, 1996] G. Sandu and R. Tuomela. Joint action and group action made precise. *Synthese*, 105:319–345, 1996.
[Santos and Carmo, 1996] F. Santos and J. Carmo. Indirect action, influence and responsibility. In M. A. Brown and J. Carmos, editors, *Deontic Logic, Agency and Normative Systems*, pages 194–215. Springer Verlag, New York, 1996.
[Schotch and Jennings, 1981] P. K. Schotch and R. E. Jennings. Non-kripkean deontic logic. In R. Hilpinen, editor, *New Studies in Deontic Logic: Norms, Actions, and the Foundations of Ethics*, pages 149–162. Reidel, Dordrecht, 1981.
[Scott, 1970] D. Scott. Advice on modal logic. In K. Lambert, editor, *Philosophical Problems in Logic: Some Recent Developments*, pages 143–174. D. Reidel, Dordrecht, 1970.
[Segerberg, 1982] K. Segerberg. A deontic logic of action. *Studia Logica*, 41:269–282, 1982.
[Segerberg, 1990] K. Segerberg. Validity and satisfaction in imperative logic. *Notre Dame Journal of Formal Logic*, 31:203–221, 1990.
[Segerberg, 1992] K. Segerberg. Getting started: Beginnings in the logic of action. *Studia Logica*, 51:347–378, 1992.
[Sergot, 1999] M. Sergot. Normative positions. In P. McNamara and H. Prakken, editors, *Norms, Logic and Information Systems*, pages 289–308. IOS Press, Amsterdam, 1999.
[Sinnott-Armstrong, 1985] W. Sinnott-Armstrong. A solution to Forrester's paradox of gentle murder. *Journal of Philosophy*, 82:162–168, 1985.
[Sinnott-Armstrong, 1988] W. Sinnott-Armstrong. *Moral Dilemmas*. Basil Blackwell, Oxford, 1988.
[Slote, 1997] M. Slote. Agent-based virtue ethics. In R. Crisp and M. Slote, editors, *Virtue Ethics*. Oxford University Press, Oxford, 1997.
[Smith, 1993] T. Smith. Violation of norms. In *Proceedings of the Fourth International Conference on Artificial Intelligence and Law*, pages 60 – 65. ACM, Amsterdam, The Netherlands, 1993.
[Smith, 1994] T. Smith. *Legal Expert Systems: Discussion of Theoretical Assumptions*. Dissertation, University of Utrecht, 1994.
[Spohn, 1975] W. Spohn. An analysis of Hansson's dyadic deontic logic. *Journal of Philosophical Logic*, 4:237–252, 1975.
[Stenius, 1963] E. Stenius. The principles of the logic of normative systems. In *Proceedings of a Colloquium on Modal and Many-Valued Logics, Helsinki, 23-26 August, 1962, Acta Philosophica Fennica*, volume 16, pages 247–260, 1963.
[Stenius, 1982] E. Stenius. Ross' paradox and well-formed codices. *Theoria*, 48(2):49–77, 1982.
[Stout, 1980] R. Stout. *The League of Frightened Men*. Bantam Books, New York, 1980. First pub. date: 1935.
[Tannsjö, 2010] T. Tannsjö. *From Reasons to Norms. On the Basic Question of Ethics*. Springer, Dordrecht and New York, 2010.
[Thomason, 1981a] R. H. Thomason. Deontic logic and the role of freedom in moral deliberation. In R. Hilpinen, editor, *New Studies in Deontic Logic: Norms, Actions, and the Foundations of Ethics*, pages 177–186. Reidel, Dordrecht, 1981.

[Thomason, 1981b] R. H. Thomason. Deontic logic as founded on tense logic. In R. Hilpinen, editor, *New Studies in Deontic Logic: Norms, Actions, and the Foundations of Ethics*, pages 165–176. Reidel, Dordrecht, 1981.
[Thomson, 1971] J. Thomson. The time of a killing. *The Journal of Philosophy*, 68:115–132, 1971.
[Tomberlin, 1981] J. E. Tomberlin. Contrary-to-duty imperatives and conditional obligations. *Noûs*, 15:357–375, 1981.
[Urmson, 1958] J. O. Urmson. Saints and heroes. In A. I. Melden, editor, *Essays in Moral Philosophy*, pages 198–216. University of Washington Press, Seattle, 1958.
[van der Torre and Tan, 1997] L. van der Torre and Y. H. Tan. The many faces of defeasibility in defeasible deontic logic. In D. Nute, editor, *Defeasible Deontic Logic*, pages 79–122. Kluwer Academic, Dordrecht-Boston, 1997.
[van der Torre and Tan, 2000] L. van der Torre and Y. H. Tan. Two-phase deontic logic. *Logique et Analyse*, 171-172:411–456, 2000.
[van der Torre, 1997] L. van der Torre. *Reasoning about Obligations: Defeasibility in Preference-based Deontic Logic*. Dissertation, Tinbergen institute research series, Erasmus University Rotterdam, 1997.
[van Eck, 1982] J. van Eck. A system of temporally relative modal and deontic predicate logic and its philosophical applications. *Logique et Analyse*, 25:249–290, 339–381, 1982. Original publication, as dissertation, Groningen, University of Groningen, 1981.
[van Fraassen, 1972] B. C. van Fraassen. The logic of conditional obligation. *Journal of Philosophical Logic*, 1:417–438, 1972.
[van Fraassen, 1973] B. C. van Fraassen. Values and the heart's command. *Journal of Philosophy*, 70:5–19, 1973.
[von Wright, 1951] G. H. von Wright. Deontic logic. *Mind*, 60:1–15, 1951. Reprinted in [von Wright, 1957, pp. 58-74].
[von Wright, 1956] G. H. von Wright. A note on deontic logic and derived obligation. *Mind*, 65:507–509, 1956.
[von Wright, 1957] G. H. von Wright. *Logical Studies*. Routledge and Kegan Paul, London, 1957.
[von Wright, 1963] G. H. von Wright. *Norm and Action: A Logical Enquiry*. Humanities Press, New York, 1963.
[von Wright, 1964] G. H. von Wright. A new system of deontic logic. *Danish Yearbook of Philosophy*, 1:173–182, 1964. Reprinted in R. Hilpinen (ed), *Deontic Logic: Introductory and Systematic Readings*, D. Reidel, Dordrecht, 1971, pp. 105-115.
[von Wright, 1965] G. H. von Wright. A correction to a new system of deontic logic. *Danish Yearbook of Philosophy*, 2:103–107, 1965. Reprinted in R. Hilpinen (ed), *Deontic Logic: Introductory and Systematic Readings*, D. Reidel, Dordrecht, 1971, pp. 115-120.
[von Wright, 1968] G. H. von Wright. *An Essay in Deontic Logic and the General Theory of Action*, volume 21 of *Acta Philosophica Fennica*. North Holland Publishing Co., Amsterdam, 1968.
[von Wright, 1983] G. H. von Wright. Norms, truth, and logic. In *G. H. Von wright, Practical Reason. Philosophical Papers, vol. I*, pages 130–209. Cornell University Press, Ithaca, 1983.
[von Wright, 1996] G. H. von Wright. Is there a logic of norms? In *Six Essays in Philosophical Logic. Acta Philosophica Fennica*, volume 60, pages 35–54. Societas Philosophica Fennica, Helsinki, 1996. Originally published in *Ratio Juris*, 4, 265-283, 1991.
[Vranas, 2010] P. Vranas. In defense of imperative inference. *Journal of Philosophical Logic*, 39:59–71, 2010.
[Vranas, 2011] P. Vranas. New foundations for imperative logic: Pure imperative inference. *Mind*, 120:369–446, 2011.
[Weinberger, 1985] O. Weinberger. Freedom, range for action and the ontology of norms. *Synthese*, 65:307–324, 1985.

[Williams, 1965] B. Williams. Ethical consistency. *Proceedings of the Aristotelian Society*, Supp, vol. 39:103–124, 1965.

[Wolenski, 1990] J. Wolenski. Deontic logic and possible world semantics: A historical sketch. *Studia Logica*, 49(2):273–282, 1990.

[Xu, 1995] M. Xu. On the basic logic of Stit with a single agent. *Journal of Symbolic Logic*, 60(2):459–483, 1995.

[Zagzebski, 1996] L. Zagzebski. *Virtues of the Mind*. Cambridge University Press, New York, 1996.

Risto Hilpinen
University of Miami (Coral Gables)
Email: hilpinen@miami.edu

Paul McNamara
University of New Hampshire (Durham)
Email: paulm@unh.edu

2
Imperative Logic and Its Problems
JÖRG HANSEN

ABSTRACT. For all its history, deontic logic had to face the question whether it is a logic of descriptions or a logic of prescriptions, namely of imperatives. In this chapter, I describe how the idea that there is a 'logic of imperatives' first came about, what proposals there have been to explain it and what problems it has had difficulties to solve. I argue that the idea of a logic of imperatives rests on a mistaken parallelism between imperative and indicative language and that there is, as a matter of fact, no such logic. However, we can argue about what ought to be done or need not be done according to given imperatives, and appeal to existing imperatives to motivate new ones.

1	Introduction 137
2	Beginnings: Poincaré's proposal 138
3	Jørgensen's dilemma 140
4	Dubislav's trick and related theories 142
5	Explanations of imperative inferences 148
	5.1 Logic of satisfaction 149
	5.2 Logic of existence 152
	5.3 Logic of ideal existence 153
	5.4 Deontic logic 156
	5.5 Formalistic approaches 160
6	Ross's paradoxes 163
	6.1 Disjunctive conclusions 163
	6.2 Conjunctive premisses 169
	6.3 Summary 172
7	Ordinary language arguments 172
8	The way to go forward 183

1 Introduction

Before the arrival of modern deontic logic as a part of modal logic, there have been numerous attempts to characterize a 'logic of imperatives'. When G. H. von Wright's classical paper [von Wright, 1951] appeared in 1951, these attempts were plagued by paradoxes, most notoriously Ross's paradox which attacked the derivation, commonly accepted by imperative logicians, of an imperative 'Post the letter or burn it!' from a given imperative 'Post the letter!'. Deontic logic, which classically interprets its formulas 'OA' and 'PA' not as expressing norms, but as normative propositions, i.e. as (true or false) statements that A is obligatory or permitted *according to* some (usually unspecified) normative system, inherited these paradoxes and has since struggled to explain why they may be considered harmful in the original context of imperative logic, but not so much for a properly understood deontic logic.

During its history of now over 50 years, authors studying deontic logic and its concepts of obligation, permission and prohibition have sometimes become unsure of what their subject really is. Is it the study of prescriptively (normatively) used language, e.g. of the use of sentences in the imperative mood? Or is it the study of descriptive sentences about what norms make obligatory and permitted, and the logical relations that obtain between such descriptions? And if the second view is adopted, is not deontic logic a kind of 'ersatz theory' that only mirrors what goes on in the realm of norms, a theory that may, if properly devised, result only in dull isomorphisms of the 'real' relations that hold between the norms themselves? Troubled by this prospect, some authors have tried to answer the question in the first way and argued that deontic logic is the study of prescriptions and their logic.

This chapter describes how the idea that there is such a thing as a 'logic of imperatives' came first into being, how authors have tried to explain this idea, and what main problems such proposals have run into. I will finally argue that the idea of a 'logic of imperatives' rests on the mistaken belief that there exists a total parallelism between imperative (prescriptive) and indicative (descriptive) language, and that there is no evidence from ordinary language that there are, in fact, argument forms that resemble 'imperative inferences'. If this is true, then there is also no place for a formal theory of such a logic.

2   Beginnings: Poincaré's proposal

Can imperatives, i.e. sentences in the imperative mood, be part of logical inferences? Henri Poincaré considered this question in his 1913 essay "La Morale et la Science" [Poincaré, 1913]. He begins by observing that if the premisses are all indicatives, then so will be the conclusion, hence for an imperative conclusion at least one premiss in the imperative mood is required, and so science alone cannot establish standards of morality. However, just as steam can be put to use in different machinery, science may also be used for moral reasoning:

> "It [the moral sentiment] will give us the major premiss of our inference which, as it happens, is in the imperative mood. At

its side, science will put the minor premiss which will be in the indicative mood. From these a conclusion can be drawn that is in the imperative mood."

Poincaré seems to have in mind Aristotelian syllogisms of the following kind:

> Hang all dwellers of Sherwood Forest!
> All members of Robin's band dwell in Sherwood Forest.
> Therefore: Hang all members of Robin's band!

Poincaré then proceeds to give a second example of an inference with an imperative conclusion:

> "One can imagine inferences which are of the following type: do this, but now if one does not do that, one cannot do this, so do that. And such reasoning is not outside the field of science."

The following is an example of the suggested inference:

> Open the door!
> The door cannot be opened unless it is first unlocked.
> Therefore: Unlock the door!

So an order to do one thing includes orders to do all that is necessary to satisfy the primary command. Poincaré's proposals raise questions: by exchanging in his first example the syllogism *barbara* for *camestres* we obtain:

> Hang all dwellers of Sherwood Forest!
> No member of Robin's band is hanged.
> Therefore: No dweller of Sherwood Forest shall be a member of Robin's band!

But it did not seem as if the speaker, e.g. the Sheriff of Nottinghamshire, was creating rules for band membership. The second type of inference is problematic when there are no legal means to fulfill an imperative (cf. [Foot, 1983, p. 384]):

> Sustain your aged parents!
> I can only sustain my aged parents if I rob somebody.
> Therefore: Rob somebody!

So if commanding means also commanding all necessary acts, then even forbidden acts must be considered commanded. To improve matters, the second clause in Poincarés scheme might be changed to 'this can only *legally* be brought about by doing that'. But this introduces a normative element into a premiss that Poincaré assumed to be established by science alone.

While all this suggests that imperative inferences might require some additions and modifications, the most difficult question has turned out to be what makes them *inferences*. The problems attached to this question go under the name of 'Jørgensen's Dilemma'.

3 Jørgensen's dilemma

Logic's concern is with the soundness of arguments, or inferences. These consist of sentences that represent the 'premisses' and usually one sentence that forms the 'conclusion'. The argument is then called 'sound', 'valid' or 'logical', if it is not possible that all of its premisses are true but the conclusion is false. The premisses are then said to 'entail' the conclusion which thus 'follows' from them.[1] Expressions in the imperative mood are not, in any usual sense, true or false. Therefore they are incapable of functioning as premisses or conclusions in logical inferences, or at least not according to the textbook definitions of such inferences. However, people maintain that the opposite is true and that there are inferences that have conclusions in the imperative mood and premisses of which at least one is likewise in the imperative mood (cf. Poincaré's examples above). This is a puzzling situation, which was first pointed out by [Jørgensen, 1938]:

> "So we have the following puzzle: According to a generally accepted definition of logical inference only sentences which are capable of being true or false can function as premisses or conclusions in an inference; nevertheless it seems evident that a conclusion in the imperative mood may be drawn from two premisses one of which or both of which are in the imperative mood. How is this puzzle to be dealt with?"

To find Jørgensen's Dilemma perplexing, one must accept that imperatives cannot be meaningfully termed true or false. This seems to be the philosophical consensus, it can point to Aristotle's definition of an assertion as a grammatical entity that can be true or false, in distinction to other grammatical entities like requests that are neither true or false (*Aristotle, De interpretatione* 17 a 4). Nevertheless, a way out of the dilemma may consist in giving up just this claim. Most prominently, [Kalinowski, 1967], [Kalinowski, 1977] has argued that in the case of expressions of moral or legal norms, the attitude of the 'ordinary', non-philosophical person is to

---

[1] For such textbook definitions cf. [Mates, 1972, p. 5], [Lemmon, 1987, p. 1], [Hodges, 1977, p. 55].

treat these as true or false. E.g. people say that it is true that another person's right to live must be respected, or that slander is prohibited, and people would uphold these truths even if particular legislators did not enact such norms, or proclaimed otherwise. So Kalinowski concludes that legal or moral norms can be part of logical inferences. I think that these considerations confuse truth with the notion of a legal or moral norm's validity: the 'external' recognition of a norm as valid in a certain society. I will discuss a little later if the validity of a norm can perhaps function as a substitute for truth values. Presently, it suffices that Kalinowski himself restricts his view to legal and moral norms and does not claim that 'imperatives in the strict sense' can be said to be true or false, and in fact writes that they are not true or false.² But it is these that we are concerned with.

The fact that imperatives are traditionally not considered to be true or false finds its explanation in the different intentions in which imperatives and indicatives are used. The main use of indicatives is to convey what the speaker believes the world to be like. If it is so, then the sentence is called 'true', if not, then it is called 'false' and the recipient might point out that the speaker should perhaps change her beliefs. By use of an imperative I tell the addressee what I want to be done. If the addressee does what is demanded, the action may be qualified as 'right', or satisfactory with respect to the command, and if not, then the behavior of the addressee is in some sense 'wrong' and I will perhaps remind the agent of his or her obligation. So truth and falsity are the qualities of descriptions when things are or are not as they have been described, while 'right' or 'wrong' are the qualities of acts that are or are not in accordance with what has been prescribed. Descriptions and prescriptions have a different 'direction of fit', and true/false are the terms used to express the match/mismatch on the language side in case of a descriptive use of language, and right/wrong are the terms employed for the match/mismatch on the world side in case of a prescriptive use.³ Therefore it is a confusion of language, and indicates a misunderstanding of the intention in which the sentence has been uttered, if imperatives are termed true or false.

Accordingly, the most effort regarding Jørgensen's Dilemma has been spent on developing alternative definitions for 'imperative inferences', rather than arguing for the application of the terms of truth and falsity to imperatives – unless one is already convinced by the dilemma that such things as

---

²Cf. [Kalinowski, 1973, p. 36] and [Kalinowski, 1977, p. 107].

³This explanation of why the terms of truth and falsity are not applicable to normative uses of language originates with [Anscombe, 1957, §32]. Independent accounts can be found in [Kenny, 1966, p. 68] and [Peczenik, 1967], [Peczenik, 1968] who speaks of the norm as a 'qualifying utterance'. The dual terms right/wrong are used as corresponding qualifications e.g. by [Engliš, 1964] and [Kelsen, 1979, p. 132].

inferences with imperatives are at all impossible (e.g. [Keene, 1966]).

4 Dubislav's trick and related theories

To deal with his own 'dilemma', [Jørgensen, 1938] endorsed a proposal by Walter Dubislav [Dubislav, 1938] to transfer the 'usual definitions' of inferences between indicatives to imperatives 'by analogy'. Dubislav gives the following example:

> Though shalt not kill.
> Therefore: Cain shalt not kill Abel.

Here, he argues, the analogue of the following 'ordinary' inference is applied:

> No human being kills any other human being.
> Cain and Abel are human beings.
> Therefore: Cain does not kill Abel.

Dubislav observes that to each imperative belongs a descriptive sentence that describes the state of affairs that obtains if the subjects of the imperative realize what the commanding authority demands. The formalisms of descriptive inferences are then transferred to imperatives by what he calls a 'trick' (*Kunstgriff*): imagine the state that the commanding authority desires realized, describe it, from this description infer some other descriptive sentence, which is then again interpreted as describing a state the authority wants to see realized. He then proposes the following *convention* (DC) on the meaning of imperative inference:

> **(DC)** "An imperative F is called derivable from an imperative E if the descriptive sentence belonging to F is derivable with the usual methods from the descriptive sentence belonging to E, whereby identity of the commanding authority is assumed."

The convention is illustrated by Figure 1 (where I write $!A$ for an imperative to which the descriptive sentence A 'belongs').[4]

Dubislav's convention does not cover inferences with more than one imperative premiss, though he mentions this possibility.[5] For such inferences,

---

[4][Mally, 1926, p. 12] seems to have introduced the symbolism $!A$, which was then employed by [Hofstadter and McKinsey, 1938] for the imperative that demands that A be the case. Even though Mally introduced the symbolism, he intended $!A$ to be interpreted theoretically, as an assertion or assumption that 'A ought to be', which we now call a deontic proposition and formalize by OA.

[5]Cf. Dubislav's use of the plural when stating that "an inference from demand-sentences will now be formally facilitated by the following convention", and Dubislav's summary, in which he stresses that no demand-sentence can be derived from premisses that do not contain *at least one* demand-sentence.

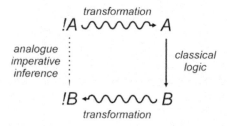

Figure 1: Dubislav's convention (DC)

(DC) can be modified as follows:

> (**DCM**) An imperative F is called derivable from the imperatives $E_1, ..., E_n$ if the descriptive sentence belonging to F is derivable with the usual methods from the descriptive sentences belonging to $E_1, ..., E_n$, identity of the commanding authority presupposed.

Dubislav then proceeds to inferences in which the imperative premiss is accompanied by another one in the indicative mood, and where the conclusion is again an imperative, for which he extends his convention:

> (**DEC**) "An imperative F is called derivable in the extended sense from an imperative E if the descriptive sentence belonging to F is at least jointly derivable from the descriptive sentence belonging to E and true descriptive sentences that are consistent with the first."

This extended convention (DEC) may again be modified to facilitate inferences with more than one imperative premiss to produce the following modified extended convention

> (**DECM**) An imperative F is called derivable in the extended sense from imperatives $E_1, ..., E_n$ if the descriptive sentence belonging to F is at least jointly derivable from the descriptive sentence belonging to $E_1, ..., E_n$ and true descriptive sentences that are consistent with these.

[Jørgensen, 1938] endorsed Dubislav's proposal as one way to deal with his dilemma and states that it seems clear to him that any imperative sentence has an indicative parallel-sentence which describes the contents of the command or wish. Jørgensen suggests that an imperative consists of an *imperative factor* and an *indicative factor*, where the first indicates *that* something is commanded, and the second *what* is commanded. The indicative factor can then be separated from the imperative and formulated in indicative sentences describing the action, change or state of affairs which is commanded. Applying the rules to these latter sentences we can thus indirectly apply the rules of logic to the imperative sentences to make their entailments explicit. Writing $\vdash A$ for the assertion that the state of affairs A holds, this slightly different analysis of imperative inferences is illustrated by the next figure:[6]

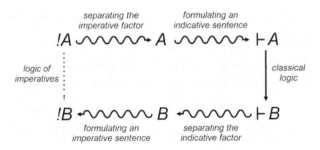

Figure 2: Jørgensen's analysis of indirect imperative entailment

Jørgensen's concept was in turn further refined by [Hare, 1949], according to whom an imperative sentence and an indicative sentence 'correspond' if they have the same 'descriptor', but different 'dictors', where what is described by the descriptor is what would be the case if the sentence is true or the command obeyed, and the dictor is what does the saying or commanding.[7] This is still not much different from Jørgensen's analysis, but Hare finds it is misleading to speak of an 'indirect', 'parallel' or 'analogous' application of logic. Instead, in Hare's view imperatives are logical in the same way as indicatives; he argues that "most inferences are inferences from descriptor to descriptor and we could add whichever set of dictors we pleased". Since most logical reasoning is done with descriptors only - this Hare calls the 'principle of the dictive indifference of logic', there is no special need for a logic of imperatives. Rather, all logic is recast as a logic of descriptors, where if the descriptors of the premises describe a state of affairs, then the descriptor of the conclusion describes, at least partially, the

---

[6]This formalization is used in [Reichenbach, 1947, §57].

[7]This distinction was anticipated by [Ledig, 1931], who wrote that norms and descriptions have an isolable imaginary content (*isolierbarer Vorstellungsinhalt*). Hare's [Hare, 1952] later terminology of a 'neustic' and 'phrastic' mirrors his earlier distinction.

same state of affairs. Hare's view of logic is pictured in the next figure:

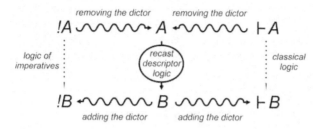

Figure 3: Hare's descriptor logic

It is immediate that neither Jørgensen's nor Hare's account make a material difference for what imperative inferences should be accepted. Hare's concept of a 'dictor' that operates on a 'descriptor' poses problems: grammatically, it is hard to see how dictors can be removed from sentences, or exchanged in them, so that the remainder or the new composite is a meaningful expression.[8] So there may be reasons not to follow Hare's analysis, but – closer to Dubislav's original concept – speak of a thematically parallel sentence in the indicative mood. However, this way or another, the idea of a descriptive sentence that parallels an imperative or of an imperative's indicative factor has become the most successful part of the Dubislav-Jørgensen-Hare analysis. [Ross, 1941] calls this element the 'theme of demand', a state the realization of which is requested by the demand, and proposed that to any imperative corresponds an ordinary indicative sentence which contains a description of the imperative's theme of demand. [Frey, 1957, p. 440] uses the term *Erfüllungsaussage* for the parallel sentence that indicates if the imperative is satisfied or violated, which [Rescher, 1966, p. 52] calls a 'command termination statement' and [Keene, 1966] the 'actualization' of the imperative. [Geach, 1958] states that for every imperative there is a future-tense statement whose 'coming true' is identical with the fulfilment of the imperative, [Sosa, 1966a], [Sosa, 1967] speaks of the 'propositional core' of an imperative and [Hanson, 1966] of a state s the commanding agent 'envisages', of which then a description S is used. Von Wright (e.g. [von Wright, 1991, p. 269]) calls the state the norm 'pronounces' as obligatory

---

[8][Opałek, 1970] points out that even if the imperative is rephrased as 'I command that ...' or 'it is obligatory that ...', the '...'-part is a Latin *ut*-expression that only due to a peculiarity of English grammar may be confused with an indicative sentence, also cf. [Opałek, 1986, ch. 2], [Kalinowski, 1974] and [Rödig, 1971] for similar criticism. On the other extreme, [Leonard, 1959] has argued that it is the descriptor that is called 'true' and 'false', and so imperatives share these properties with descriptive sentences.

or allowed the 'content of a norm'. The seemingly universal consensus is explained by the pragmatic function of imperatives, that is, the regulation of human behavior: if there were no imperative-correlated indicative sentences, it could not be understood *what* ought to be, and neither would it be possible to determine whether the norm is satisfied or violated.[9] This is why even 'anti-reductionist' authors that oppose the idea that imperatives or imperative reasoning can be reduced to indicatives or indicative logic, agree on the following principle:[10]

Principle (W)   (Weinberger's Principle).
To each imperative there corresponds a descriptive sentence that is true if the imperative is satisfied and false if it is not-satisfied (violated).

It is clear that an acceptance of (W) does not force us to also accept the Dubislav-Jørgensen-Hare account of imperative inference, which nevertheless has been accepted by a number of authors.[11] But (W) can be used to show that seemingly differing explanations of imperative inferences are in fact equivalent to this account. Thus [Rescher, 1966] defines command inferences in terms of satisfaction in the following way:

> A command inference is valid if there is no possible world in which the premisses are all satisfied and the conclusion fails to be satisfied.

which, using (W), is equivalent to

> A command inference is valid if there is no possible world in which the descriptive sentences corresponding to the premisses

---

[9] This explanation and the formulation of the principle below is most clearly expressed in [Weinberger, 1996, p. 172]. Weinberger uses the term 'coordination' instead of 'correspondence', but this suggests an onto or even one-to-one mapping.

[10] Besides Weinberger cf. [Hamblin, 1987, pp. 151-2]: "Take the exact words of the imperative, and transform them into indicative mood (...) Now the worlds which extensionally satisfy the imperative are just those of which the description is true.", and Moutafakis' theorem T3 ([Moutafakis, 1975, p. 155]), which expresses the equivalence of the statements that an imperative is satisfied and that a description of the prescribed action as performed is true.

[11] These include [Simon, 1965], who converts commands to declarative mode "by removing the imperative operators from them", obtaining a theory in which all recipients obey the commands, and then applies the 'ordinary laws of logic' to derive new relations that may be converted back into commands. According to [Niiniluoto, 1985], an "imperative !p entails imperative !q if p entails q". Very close to (DCM) is von Wright's account in [von Wright, 1963, pp. 71, 164], where he defines the content of a prescription as 'the prescribed thing', and defines that a command is entailed by a second command or by a set of commands if the content of a command is a consequence of the conjunction of the content of a command with the contents of none or one or several other commands.

are all true and the descriptive sentence corresponding to the conclusion is false.

Using the textbook definition of an argument, this is equivalent to

> A command inference is valid if the descriptive sentences corresponding to the premises entail the descriptive sentence corresponding to the conclusion.

which in turn is the modified Dubislav convention (DCM).[12]

Using the idea that norms qualify the states of affairs that satisfies or violates them as 'right' and 'wrong', one can define:[13]

> An imperative !A entails an imperative !B if and only if (iff) every state of affairs that is qualified as wrong by !B is qualified as wrong by !A.

which can then be translated into

> An imperative !A entails an imperative !B iff every state of affairs that violates !B also violates !A.

which using (W) is equivalent to

> An imperative !A entails an imperative !B iff every state of affairs in which B is false also makes A false.

which using classical logic is equivalent to

> An imperative !A entails an imperative !B iff every state of affairs in which $\neg B$ is true also makes $\neg A$ true.

which using *tertium non datur* and *modus tollens* is equivalent to

> An imperative !A entails an imperative !B iff every state of affairs in which A is true also makes B true.

---

[12]Sosa's [Sosa, 1966a], [Sosa, 1967] definition of a 'directive argument' is very similar to Rescher's, but additionally demands that the imperatives that function as premises are jointly satisfiable in order to cope with normative conflicts. Sosa adds a second condition, demanding that if the imperative conclusion is violated, at least one imperative premiss must be violated, in order to also cope with conditional imperatives. This equals Keuth's [Keuth, 1974] condition B_1 that it must be logically impossible to violate the conclusion without violating a premiss. Obviously, the second condition makes no difference for unconditional imperatives.

[13]The definition is similar to the one used by [Peczenik, 1967], [Peczenik, 1968] for forbidding norms and the quality 'forbidden'. [Kamp, 1973] uses an analogous definition for permissions, where one permission entails another if the second makes only such courses of actions permissible that were already so before.

which by definition of entailment equals

> An imperative $!A$ entails an imperative $!B$ iff A entails B.

and this is again Dubislav's convention (DC). [Lemmon, 1965] defines an entailment relation for imperatives *via* a definition of inconsistency of a set of imperatives and indicatives, where such a set is called inconsistent if it cannot be the case that all indicatives are true and all imperatives obeyed.[14] Lemmon's entailment relation is then defined as follows:

> An imperative $!A$ is entailed by a set indicatives and imperatives if this set together with $!\neg A$ is inconsistent.

where $!\neg A$ is the imperative that is satisfied if and only if A is false. Using (W) we obtain:

> An imperative $!A$ is entailed by a set indicatives and imperatives if the set of indicatives together with the set of descriptive sentences corresponding to the imperatives together with $\neg A$ is inconsistent.

which, using classical logic, is equivalent to

> An imperative $!A$ is entailed by a set indicatives and imperatives if the set of indicatives together with the set of descriptive sentences corresponding to the imperatives entails A.

and this is Dubislav's modified extended convention (DECM). This shows that, with (W), Dubislav's proposal is equivalent, or at least very similar, to a number of other proposals how imperative inferences are possible in the face of Jørgensen's dilemma.

5 Explanations of imperative inferences

Dubislav's 'trick' provides a formal method that explains *how* inferences between imperatives can be defined without having to assign them truth values. Less clear is *what* is achieved by the method, i.e. why we should think that this is what it means to 'infer' an imperative from some other imperative, or a set of imperatives, or a mixed set of imperatives and indicatives, or formally, what is means that some scheme of imperative inference

---

[14]Lemmon expresses reservations regarding his definition, but only because he thinks that it does not sufficiently restrict imperative conclusions to statements about future actions. A definition similar to Lemmon's seems to be intended by [Philipps, 1971, p. 364] who defines: 'to do p is forbidden!' is true iff the indicative 'someone does p' is incompatible with the class of valid prescriptions, where 'compatible' means that if the indicative is true, then at least one prescription is violated.

(ImpInf) $!A$
           $\therefore !B$

is valid. Dubislav's trick can easily be applied e.g. to sentences of the form 'I doubt that ...'. Then from 'I doubt that he is staying at his sister's place in San Francisco' follows 'I doubt that he is staying in San Francisco', which, though we can derive 'he is staying in San Francisco' from 'he is staying at his sister's place in San Francisco', seems wrong: I might not doubt that he is staying in San Francisco, but doubt very much that he is staying with his sister. So why should Dubislav's trick work for imperatives if it would not for other expressions?

5.1 Logic of satisfaction

The primary use of sentences in the imperative mood is to direct human activities, and so it is a basic property of imperatives that they can be satisfied (obeyed) or violated (disobeyed). Imperatives are used to make some subject do some act or achieve some state of affairs, and saying that some imperative is satisfied or violated then means that the act or state of affairs which the imperative 'envisaged' has been done or achieved, or can no longer be done or achieved by the subject(s) of the imperative (cf. [von Wright, 1963, p. 118]). E.g. the imperative "Stay at home tonight and do your homework" is satisfied if the subject stayed at home that night and did his homework and violated if the subject did not.

Some difficulties attach to the notion of satisfaction. One is whether the action or state of affairs that an imperative envisages must be possible or contingent: Can Caligula's command to a soldier to bring him the moon be meaningfully described as satisfied or violated?[15] What about the same emperor's command to an envoy to do the king of Morocco "neither good nor bad"?[16] What do we say when the state of affairs an imperative envisages is brought about not by the subject but by third parties or through some accidental turn of events? There are 'felicity conditions' for the meaningful use of imperative sentences (cf. [Fox, 2008]), and when they are not met it becomes difficult or meaningless to describe an imperative as satisfied or violated.

When there are two imperatives, and we know that one of them is satisfied, then we can sometimes infer that the other imperative is satisfied, too. E.g. if a subject has been told independently by two people at about the same time: "Go to your room!", and "Go clean your room!", and we later learn that the subject satisfied the second command then we may conclude that the first order was likewise satisfied. One solution is therefore to

---

[15]In Camus' play *Caligula*, third act, third scene.
[16]In Sueton, *Caligula*, ch. 55.

interpret the scheme

(ImpInf) $!A$
           $\therefore !B$

as stating that if the imperative $!A$ is satisfied, then it must be that the imperative $!B$ is also satisfied.

The first 'logic of imperatives' that has been devised was such a 'logic of satisfaction'. It was proposed in 1938, at almost the same time that Jørgensen's essay was published, by [Hofstadter and McKinsey, 1938]. In addition to the usual Boolean connectives between indicatives (' $\neg$ ', ' $\wedge$ ', ' $\vee$ ', ' $\rightarrow$ ', ' $\leftrightarrow$ ') the authors introduce an operator '!' that forms imperatives from indicative statements (if A is an indicative then $!A$ is the imperative 'let A be true'), three operators '$\overline{}$', '+', '×' that form an imperative out of two imperatives (where $!\overline{A}$ is the imperative that demands the negation of A to be true, $!A+!B$ is the imperative that demands A or B to be true, and $!A \times !B$ is the imperative that demands A and B to be true), one operator '$\Rightarrow$' that forms an imperative from an indicative and an imperative (where $A \Rightarrow !B$ means the conditional imperative 'if A is true then let B be true'), and finally two operators '>' and '=' that form indicatives from two imperatives ($!A > !B$ means that if A is satisfied then B is satisfied, and $!A = !B$ means that A is satisfied if and only if B is satisfied). The logic is then conducted with indicative sentences alone, and imperatives only appear within subformulas that have '>' or '=' as their main connective. E.g. the following formulas are theorems of their logic:

$!A \times !B = !(A \wedge B)$
$!(A \wedge B) > !B + !C$

The first states that the imperative which joins two imperatives $!A$ and $!B$ is satisfied if and only if the imperative $!(A \wedge B)$ is satisfied, i.e. the two are 'satisfaction-equivalent'. The second states that if an imperative that demands A-and-B is satisfied then the imperative that demands B or C is also satisfied. Up to this point, strictly speaking, Hofstadter & McKinsey's logic is not one *of* imperatives, but a logic *about* (the satisfaction of) imperatives: only statements about the satisfaction of imperatives follow from other statements about the satisfaction of imperatives. However, at the end of their paper, the authors go a bit further and define:

> "Suppose that $C_1 = !S_1$ and $C_2 = !S_2$. Then we call C_2 derivable from C_1 iff S_2 is derivable from S_1." [Hofstadter and McKinsey, 1938, p. 452]

So if an imperative is satisfaction-equivalent with an imperative $!A$ (that is satisfied if and only if A is true), and a second imperative is satisfaction-equivalent with an imperative !B (that is satisfied if and only B is true), and B is (ordinarily) derivable from A, then Hofstadter & McKinsey call the second imperative derivable from the first. It is clear that (ImpInf) is valid on this interpretation whenever the 'ordinary' argument

$$A$$
$$\therefore B$$

is valid for the descriptive sentences A and B: If A classically implies B, then it must be that if !A is satisfied then A is true, so also B is true, and hence !B is satisfied. Thus Dubislav's trick receives its semantic justification.

What remains troubling about such a logic of satisfaction is that it identifies (ImpInf) with the following inference:

!A is satisfied.
$\therefore$!B is satisfied.

But if this is what justifies (ImpInf) then it seems one should also accept these schemes

$$!A$$
$$\therefore A$$

and

$$A$$
$$\therefore !A$$

According to our (informal) convention, !A represents an imperative that is satisfied iff A is true. So it must be that if !A is satisfied, then A is true and *vice versa*, and so the above schemes are valid. But on the look of it, these schemes seem to state that from an imperative that demands A it can be inferred that A is the case, and that from a sentence that states that A is true an imperative can be inferred that demands A, which is nonsense. And this misunderstanding reveals that when we speak of the possibility of an inference in which the premisses and the conclusion are imperatives, it seems that we talk about inferring *imperatives* from other imperatives, and not about reasoning whether or not the imperatives in question are satisfied. For these reasons, [Ross, 1941, p. 61] and also [Hare, 1967] doubted that a logic of satisfaction is what one has in mind in the case of practical inferences.[17]

---

[17][Kanger, 1957, p. 49] and [Føllesdal and Hilpinen, 1971, p. 7] criticize Hofstadter & McKinsey for making !A and A 'equivalent', which is somewhat unfair since the intended interpretation of their formulas (in terms of satisfaction) is not presented.

5.2 Logic of existence

When we speak of inferring one imperative from some other imperative, this could mean that the existence of an imperative is logically deduced from the existence of some other imperative.

What is meant by saying that an imperative 'exists'? First, it could be the existence of an utterance of some sentence in the imperative mood by some commanding agent towards some commanded subject.[18] Second, one might demand that the utterance, as a performative use of the imperative sentence, was effective and established an *imperativum*, i. e. a 'command', 'demand', 'request' or the like. For this it may be required that the commanding agent had the will to command (and did not use the words for fun) as well as some authority (power to punish or reward) over the addressee.[19] Third, for an order by legal authorities in this capacity to come into 'legal existence' it may be required that the authority was competent to utter it according to the legal rules of some normative system that confers such competence, and similar for bodies that are constituted not by law but by other rules like a firm or Robin's band.[20]

Yet however much the concept of existence is thus refined, it seems to require the presence of actual facts: a (still alive?) speaker, a linguistic entity like an utterance and circumstances of speaking, a certain attitude of the speaker towards the act of speaking, a backing of the speaker by force or an authority conferred by existing and/or valid rules, etc. But it is difficult to see how logic can stipulate such an existence. This is illustrated in the following example by Aleksander Peczenik:[21]

> "The premiss 'love your neighbour' may be regarded as describing the fact that the authority – Jesus – has in fact said 'love your neighbour.' The imperative existed because it was uttered by Jesus. But the conclusion, for example, 'love Mr. X' does not describe anything which in fact has been said by Jesus."

---

[18] This existence is what [Frey, 1957], along with an additional property of 'justification', infers in imperative inferences: "If the imperatives that appear in the premisses exist and are justified, then also the imperatives derived from these exist and are justified" (p. 465). Frey's 'justification' means that what is demanded is 'good' regarding some aim of the commanding agent, called 'axiological validity' in [Ziembiński, 1976].

[19] Cf. [von Wright, 1963, pp. 120-6]. This is Ziembiński's [Ziembiński, 1976] 'thetic validity'. [Lemmon, 1965] seems to have this notion of 'validity' or 'existence' in mind when he demands that the entailment of imperatives must be defined in terms of what imperatives are *in force* at a given time.

[20] [Bulygin, 1999] uses the term 'systemic validity'. According to [Weinberger, 1975], [Weinberger, 1989, p. 259], this validity takes the place of truth as the 'hereditary trait' (*Erbeigenschaft*) that is transferred from the premisses to the normative conclusion in inferences with normative sentences (*Normsätze*).

[21] Quotation from a letter by A. Peczenik to R. Walter, in [Walter, 1997, p. 395].

Here, the intended argument from 'love your neighbour' to 'love Mr. X' is not accepted because the commanding agent 'did not actually say' what appears in the conclusion, and so unlike the premiss the conclusion did never 'exist' as a fact. An imperative sentence that has not been expressed was not received and cannot be understood by its addressee as a command or legal order. Thus the required 'existence' of the imperative, or 'validity' of the command seem to be the analogues not of truth of a descriptive sentence, but of 'stated' and 'asserted'. Yet indicative logic does not force anyone to state or assert anything, even if some other descriptive sentence was used in a way that expresses ones commitment to it. It only explains what people mean when they use an indicative sentence in order to assert some fact, by saying what other sentences must be true if the stated sentence is true.[22] Because the imperative sentence in the conclusion may not exist as an utterance, [Hamblin, 1987, p. 89] warns against speaking of inferences between imperatives. Because the 'telling part' (or attitude) of the speaker cannot be inferred, the possibility of command inferences was denied by [Sellars, 1956, pp. 239-40], and for the same reason, Lemmon's [Lemmon, 1965] attempt to define imperative inferences via the notion of an imperative's being 'in force' was rejected by [Sosa, 1967, p. 61] who argues that such notions involve attitudes by the authority or its subject that cannot be inferred. [von Kutschera, 1973] argues that a (used) imperative is an action, actions do not follow from actions, so there is no logic of imperatives. In Alchourrón & Bulygin's [Alchourrón and Bulygin, 1981] 'expressive conception of norms', the existence of a norm is seen as dependent on empirical facts and the possibility of a logic of norms is consequently denied as there are nor logical relations between facts. For similar reasons, [Philipp, 1989; Philipp, 1991] denied both the possibility of a logic of imperatives and of norms.

5.3 Logic of ideal existence

To get around this difficulty one may consider to interpret 'existence' not with respect to natural facts, but with respect to some ideal 'world of ought' or an assumed 'normative system' that is closed under consequences, where the closure operation may be understood e.g. in the sense of derivability by use of Dubislav's convention. So if the agent's use of the imperative mood has resulted in the existence of a command in the 'world of ought', or, due to the agent's legal competence as e.g. a police officer, created an order that now belongs to the normative system, then all the 'consequences' of

---

[22][Kenny, 1966] first pointed out that 'valid' and 'invalid', interpreted as meaning 'commanded' and 'not commanded', are not the analogues of 'true' and 'false', but of 'stated' and 'not stated'.

the command that can be derived by an appropriate method exist in this world or system as well.[23] What thus has ideal existence is not the imperative sentence as a spatial and temporal phenomenon or as a grammatically correct or meaningful combination of words. Rather, it is what the use of an imperative sentence expresses or accomplishes – a command, request etc. Then it must be that not only commands, requests etc. 'exist' in this sense that in fact *have been* expressed by a performative use of a sentence in the imperative mood, but also some that only *can* be expressed. For if all that 'exists' in the 'normative system' already exists as a result of a pragmatic use of language, then there would be no need to let the 'normative system' e.g. be closed under a consequence operation. That what we can express by using language (commands, requests, assertions etc.) has some existence, possibly independent from any pragmatic origin[24], in some ideal 'world of ought', is a difficult concept that possibly creates more problems than it solves.[25] But it is even difficult to see that it solves the problem of entailment between imperatives. For to say that some ideal object created by use of an imperative implies the existence of some other ideal object in some 'world of ought' or normative system, is not to say that an imperative implies another imperative. The existence of a forest might imply the existence of a tree, but to say that 'the forest implies the tree' is making a categorical mistake. The first is an indicative argument, which can be formalized in the usual way:

---

[23]The 'world of ought' terminology originates with [Walter, 1996], who is however following [Kelsen, 1979, p. 195] in that an individual norm does not exist before the general norm was applied by a judge, so orders that can 'only be deduced' do not exist in Walter's 'world of ought'. The idea to explain logical relations between 'norm sentences' (like imperative sentences) in terms of their existence in a 'system of norms' that is closed under consequences is that of [Stenius, 1963]. In [Opałek and Woleński, 1991], norms are non-linguistic entities expressed by (descriptively interpreted) deontic statements, and normative systems consist not only of norms that have been expressed by a normative authority, but also of the consequences of these 'basic obligations'. In Alchourrón & Bulygin's 'hyletic' variant of a conception of norms [Alchourrón and Bulygin, 1993], 'implicitly promulgated' norms have 'existence' in a logically closed normative system, and descriptively interpreted (deontic) norm propositions are then "propositions about the existence of norms (in that system)". [Holländer, 1993] promotes the idea of a 'deontically perfect world' where norms exist that obey logical principles, like that conflicts are excluded. [Kelsen, 1979, pp. 187-8] rejects the idea of an 'ideal existence' of norms because there is no 'ideal' act of will that creates them, and rejects the whole idea of a logic of norms.

[24]Cf. [Stenius, 1963] according to whom all normative systems include a norm that demands a tautology.

[25]Note that the topic of this discussion has not suddenly become the ontological status of notoriously difficult concepts of practical philosophy and jurisprudence, like moral obligations, natural law, human rights, laws of custom etc. Our concern are still ordinary sentences in the imperative mood, addressed e.g. to a husband, secretary, student, child or dog (cf. [Ziemba, 1976, p. 386]).

(1) $\exists x : Forest(x)$
        $\therefore \exists y : Tree(y)$

The argument is analytical when the words 'forest' and 'tree' have their usual meaning and for all that understand this meaning and thus know that there cannot be forests without trees. By starting to talk about the (ideal) existence of commands it seems that we silently changed (ImpInf) into

(2)   the command given by α to x by the use of the imperative sentence !A
        Therefore: the command given by α to x by the use of the imperative sentence !B

which appears confused. This is because the argument form is not used as it is usually used, and now we do not know what to make of it. We are used to filling in the blanks of the argument form

(3)   _____
        $\therefore$ _____

with sentences. Whether also imperative sentences can be meaningfully used to fill in the blanks is the open question. However, there is no pre-established usage of filling in the blanks with names of objects, as in

(4) a
        $\therefore b$

where a means a forest and b means a tree. At most, this is a mistaken way to try to express (1). Similarly, the scheme (2) must be corrected into (5):

(5)   There exists a command given by α to x by the use of the imperative sentence !A.
        Therefore: There exists a command given by α to x by the use of the imperative sentence !B.

Now we have arrived at something that resembles an argument – though we have not yet shown when such a scheme constitutes valid arguments. But this argument is one that uses only descriptive sentences that can be true or false. It is not a case of (ImpInf), i.e. not an argument where imperative sentences function as premisses or conclusion.[26] So by appealing

---

[26]That the 'world of ought' approach thus only provides arguments with descriptive sentences is accepted by [Walter, 1996], for he turns to identify imperatives with descriptions of the 'world-of-ought'-existence of a command that is created by the use of an imperative – consequently such sentences can be true or false and therefore part of logical inferences. Thus imperative logic is reduced to indicative logic, where the difficult part is now the verification of some descriptive sentences.

to the notion of existence we obtain at most an inference relation between sentences that describe the existence of certain commands, but not a logic *of* commands.

5.4 Deontic logic

A more recent approach tries to overcome the difficulties of a logic of imperatives by developing it in complete parallel to the logic of normative propositions, namely deontic logic. It is facilitated by three observations:

1. First, if a command like "Shut the door and close the window" has been given and is 'effective', i.e. it creates an obligation for the addressee, then the corresponding normative proposition (e.g. "I must shut the door and close the window") is, or rather: has become, true. The performative success of the act of giving a command guarantees the truth of the corresponding normative proposition.

2. The second observation is that a normative utterance like a command and the corresponding normative proposition can be expressed by the same sentence. Consider the sentence "I am commanding you to shut the door": it may be viewed as a performative, i.e. as a command that you shut the door, but also as an assertion that I am in fact commanding you to do so, and perhaps it may even be both: a command and a simultaneous assertion. It can be argued that for every imperative sentence there exists an equivalent expression that is ambiguous in this sense.

3. The third observation is that normative propositions, like all descriptive sentences, permit logical inferences. E.g. if I have been commanded to shut the door and close the window then I can conclude that I have, among other things, been commanded to shut the door.

Putting all three observations together, from some sentences that can be interpreted both as normative utterances and normative propositions it is possible to infer other sentences that can again be interpreted both as a normative utterance or as a normative proposition. The truth of the sentences used as premisses guarantees the effectiveness of the same sentences in their performative interpretation, and so likewise the truth of the sentence in the conclusion (which is guaranteed by the validity of the inference e.g. of deontic logic) should be enough to guarantee its effectiveness when interpreted again as a normative utterance. So when we infer "I am commanding you to shut the door" from "I am commanding you to shut the door and close the window", it should not matter if the sentences are interpreted descriptively, i.e. as normative propositions, or as performatives, i.e. as normative utterances. And since possibly all imperatives can be equivalently replaced by an ambiguous expression of the form "I am commanding you to...", any corresponding imperative inference must also be valid.

This solution to the problem of imperative inferences has several nice effects. First, Jørgensen's problem that imperatives cannot meaningfully be termed true or false seems to greatly diminish when instead sentences are used that can be (at least in one of their interpretations). And secondly, the way in which the double interpretation of the involved sentences facilitates imperative inferences is formally quite similar to Dubislav's trick. Only now, the indicative parallel sentence associated with a normative utterance is actually the same (or an equivalent) sentence interpreted as a normative proposition, which makes the logic that is employed on the right-hand side of his scheme a deontic logic or some other logic of normative propositions instead of "ordinary logic" (see the figure below). Thus Dubislav's trick is both refined and justified.

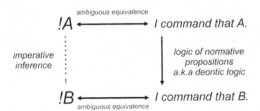

Figure 4: Parallel deontic inferences

Historically, that expressions like "I command you to ..." can be used both to command and to simultaneously assert that I do in fact command so-and-so was first recognized by [Sigwart, 1895]. Sigwart therefore claims that each imperative includes the statement that the speaker wills the act which he commands. While he adds that nevertheless the import of what is said by the imperative is not the communication of a truth but a "summons to do this, to leave that undone", [Ledig, 1928] argues that because imperatives include such assertions, one must consequently apply the terms of truth and falsity to an imperative, and that it will be true unless it is e.g. stated for fun. Later, [Opałek and Woleński, 1987] have similarly called ambiguous normative statements, i.e. statements that can be read both as performatives and descriptions (normative propositions), *qua* performative *true* if they are effective. Most influential for the above solution was, however, a passage in an article by [Kamp, 1978] who, while discussing David

Lewis' model-theoretic analysis of permissions, argues as follows:

> "It is worth noting that it is not only difficult to decide for some such cases whether the utterance was an assertion or a performative; from the point of view of the addressee it is usually also quite unnecessary. For as long as he has good reason to believe that the utterance is *appropriate*, then, whether he interprets it as a performative or as an assertion the practical consequences will be precisely the same. For either the utterance is a performative and as such creates a certain number of new options, or else it is an assertion; but then if it really is appropriate it must be true; and its truth then guarantees that these very same options already exist." [Kamp, 1978, p. 264]

This passage has been understood as claiming that both kind of utterances, a normative utterance and the corresponding normative proposition, are "informationally equivalent": the normative utterance includes its corresponding normative proposition as its content and so the information that may be gained from either of them for the purpose of subsequent normative reasoning must be the same (cf. [Hilpinen, 2006, pp. 236-7]). The logic of norms (effective normative utterances) must therefore be the same as that of normative propositions: the effectiveness (or validity) conditions of the normative utterance are also the 'truth-makers' of the corresponding normative proposition (ibid. pp. 236-8). For Hilpinen, this explicates a view earlier taken by [Alchourrón and Bulygin, 1984] that the 'logic of norms' is a 'reflection of the logic of normative propositions', and that the "logic of norm propositions yields the foundations for the logic of norms" [Alchourrón and Bulygin, 1984, p. 463]. That imperatives and normative propositions share the same semantics is also the basis of Schwager's analysis of imperatives in terms of deontic logic in [Schwager, 2006].

The proposal to use a logic of normative propositions a.k.a. deontic logic for the right hand side of Dubislav's convention is puzzling. Deontic logic is motivated by a variety of semantics, some appealing to prohairetic notions, others defining deontic truth in terms of what is necessary for a particular agent, given the situation, in order to realize everything or at least some of the things required of her. Suppose deontic logic also carries the burden of closing the set of commands with respect to which it may define its deontic operators of obligation, permission and prohibition. But this changes the adequacy conditions of deontic logic. E.g. most proposals for defining "all things considered obligations" accept the deontic axiom (D) – if A is overall obligatory then A is not also overall forbidden, and so A is permitted. Does this mean that any speaker, regardless of whether she has commanded anything (or intends to do so), commands A or permits $\neg A$? If a speaker has actually commanded A, has she also either commanded $A \wedge B$ or permitted $A \wedge \neg B$? And if this is absurd, then must not deontic logic give up (D)? Strictly speaking, the logic used is then no longer deontic logic but some other modal logic. Suppose it is of type **K**. Then if a speaker commands A she must also command $B \vee \neg B$. But some would say that this is no 'genuine' command since it fails to direct (cf. [von Wright, 1999, p. 5]), and others might object because the speaker may not have considered B. So it seems the logic must be weakened further. Perhaps a logic of relevant entailment is what is wanted. But in which sense is this then a 'logic of normative propositions'?

Normative reasoning, namely deontic logic, rarely cares about the sources of obligations: the agent might be obliged because of explicit commands given to her, and she might (also) have to consider legal norms or moral duties. Deontic logic only proposes what an agent who has accepted all of these obligations is now obliged or permitted to do. Also, since [Prior, 1955], deontic operators have been combined with those of necessity and possibility to take into account the factual situation an agent might find herself in, and such considerations are even more present in the 'tree systems' that combine deontic logic with temporal logic and a logic of agency (e.g. of [Horty, 2000]). Moreover, deontic logic has learned to deal with conflicts between norms and explain the normative reasoning of an agent who is faced with incompatible demands, demands speakers may have been illocutionary committed not to make but nevertheless did make (cf. [van Fraassen, 1973]). Even in simple cases of normative reasoning, the obligations that are recognized will no longer be ones that can be meaningfully related to any specific normative utterance a speaker has actually used, and when they are rephrased into corresponding normative utterances there may not be a single authority they can be meaningfully ascribed to. There is a distinction between norm-giving and normative advice-giving that seems lost by the above approach: not always when a statement is true that an agent must behave in a certain way there is also a specific norm or someone committed to a specific normative utterance that prescribes just this behavior.[27] The fundamental assumption of this approach, that the logic of normative propositions is the same as a logic of normative utterances, seems false.

---

[27] One could again stipulate the existence of such a norm in a 'system of norms' or 'world of ought' – this seems to be Hilpinen's proposal in [Hilpinen, 2006, p. 238] –, but, apart from giving rise to the difficulties discussed in the preceding section, this seems to lead away from an explanation of imperative inferences by the theory of speech acts.

5.5 Formalistic approaches

In ordinary life, there is often agreement that something that was not outrightly stated was nevertheless implied by a statement, even though people rarely actually carry out truth-table calculations to support their views. Rather, they rely on their understanding of the meaning of the given statement to make the inference. Someone who says that 'either a Democrat or a Republican will be the 45th President of the U.S., but it won't be a Democrat" thereby expresses the belief that the 45th President of the U.S. will be a Republican. The speaker did not say explicitly that it will be a Republican (these were not the words used), but the speaker can be said to have implicitly, or tacitly, expressed this opinion. Therefore if a person asserts something, she may be taken to 'implicitly' assert something else. Likewise, when a person commands something, she may also be 'commanding something by implication'. Consider the following example employed by [Hare, 1967]:

> Go via Coldstream or Berwick!
> Don't go via Coldstream!
> Therefore: Go via Berwick!.

Here, an officer who must go from London to Edinburgh is ordered 'go via Coldstream or Berwick,' and – a bit later – is given the order 'don't go via Coldstream'. Both commands have been taken to imply that the officer is (now) ordered to go via Berwick, and that the inference is therefore valid. For the question *what* commands should be considered to have been 'implicitly commanded' by explicitly given commands, one may then point to Dubislav's conventions or similar rules – e.g. (DCM) clearly accounts for the above inference.[28] To search for an explanation of imperative inferences in the meaning of explicitly given commands seems a much better idea than to look for a way to ascribe truth values to imperative sentences, or find truth value analogues in the notions of satisfaction or validity. Unfortunately, it is

---

[28]The term of 'commanding something by implication' was introduced by [Geach, 1963]. [Alchourrón, 1972] speaks of the consequences of what is prescribed as 'indirectly prescribed', [Alchourrón and Bulygin, 1981] write that e.g. if teacher commands that all pupils should leave the class-room, he also implicitly commands that John (who is one of the pupils) should leave the class-room, even if the teacher is not aware of the fact that John is there, and in [Alchourrón and Bulygin, 1993] they view the 'deductive consequences' of norms as 'implicitly promulgated', where the deduction process is equivalent to the modified Dubislav convention (DCM). [Hare, 1967] and [Rescher, 1966] both propose to define command inferences in terms of 'implicitly given commands' – analogously, Rescher's 'assertion logic' [Rescher, 1968] is concerned with assertions that a speaker is 'implicitly committed to' in virtue of overtly made assertions. It was shown above that Rescher's explanation of imperative inferences is equivalent to the modified extended convention (DECM).

also a circular idea: according to it one can infer a command from another command if by giving the latter command one implicitly commands the former. Or: we may use Dubislav's convention to infer imperatives from other imperatives because it provides the imperative sentences that are implicitly commanded when the first imperative sentences are used for commanding.

An interesting response to this reproach of circularity is to say that by giving rules for inferring one imperative from another, one *did all* that is required to explain the meaning of imperative inferences. It is this view that seems to lie at the root of Dubislav's proposal to "formally facilitate inferences" from demand-sentences through a 'convention' (*Übereinkunft*) or 'trick' (*Kunstgriff*). In fact, in his main work *Die Definition* [Dubislav, 1931], Dubislav gives a similarly 'formalistic' characterization of propositional logic (and later predicate logic). There, Dubislav starts with a 'pure, game-like calculus' that is played with 'pieces' and signs ('¬', '∨', brackets) by first arranging some into initial positions and then replacing pieces and re-arranging pieces into new game positions according to the rules of the game. This game then *becomes* the calculus of propositional logic by interpreting its elements as indicated: the 'pieces' as propositions, the signs as 'not', 'or', and brackets, the 'initial positions' as axiomatic basis, the game rules as the usual rules of substitution and *modus ponens*, and the achievable game positions as derivable formulas. This characterization of propositional logic is meant by Dubislav as an exposition of Boole's [Boole, 1847] idea that the "validity of the processes of analysis does not depend upon the interpretation of the symbols which are employed, but solely upon the laws of their combination". In Dubislav's view, which he calls 'the formalistic theory', this description of logic functions as a mould for all scientific theory: a theory is constituted by a pure calculus (of formulas and rules), combined with a fixed interpretation. Observational sentences are captured in formulas that can be used alongside axioms or derivable formulas of the system to derive other formulas within the calculus. Then the assignment of these derived, or better: 'calculated' formulas is reversed, i.e. they are translated back into observational sentences. If these are regularly true, then the observational sentences are 'explained' by the theory. If a calculated observational sentence turns out to be false, then the theory is erroneous. Thus it also becomes possible to decide between competing, non-isomorphic theories.

The usefulness of the Dubislav's formalistic approach for the problem of imperative logic is immediate. In fact, Dubislav's own proposal in [Dubislav, 1938] satisfies all requirements in [Dubislav, 1931] for being a theory of imperative inference: there are entities that may function as premisses and conclusions, namely imperative sentences. There is an interpretation that assigns each imperative sentence a formula, namely that of the indicative 'parallel sentence' in the calculus of 'ordinary logic'. There is a calculus, namely 'ordinary logic', that tells us what formulas can be derived from the formulas assigned to the imperative sentences that function as premisses. And finally, this assignment is reversible to provide derived imperative sentences. Other authors – taking their cues from Tarski's [Tarski, 1930] syntactical definition of consequence relations and deductive systems,[29] Tarski's [Tarski, 1935] definition of truth,[30] Gentzen's [Gentzen, 1934] idea that to define a symbol is to give rules for its introduction and elimination,[31] and Wittgenstein's dictum that the meaning of a word is its use ([Wittgenstein, 1953, §43]) – have similarly argued that instead of searching in vain for analogues of truth values, it suffices for an explanation of imperative inferences to give formal rules for obtaining imperatives from other imperatives.

If this 'formalistic' approach to the logic of imperatives is accepted, we are still not finished yet. If the assignment of formulas, calculations and back-translations of derived formulas are to be more than a game, there must be some way to judge the adequateness of the theory, and be it only to decide between competing proposals.[32] In analogy to Dubislav's general approach, where a theory is only an explanation of phenomena if its calculated observational sentences are regularly true, one should require of any proposed 'logic of imperatives' that the imperatives it 'derives' from other sentences are normally – not 'true' of course, but accepted as 'implicit' in

---

[29] Cf. [Alchourrón and Bulygin, 1993] who employ a formal consequence relation to explain what norms are 'implicitly promulgated' by a set of norms.

[30] Both [Rödig, 1972] and [Yoshino, 1978] appeal to Tarski and argue that meaningful operations with prescriptions are made possible by supposing that normative attributes like 'obligatory' or 'punishable' may be applied to actions. Rödig draws attention to the problem of objective verifiability and therefore truth of such statements. But he circumvents the problem by assuming that meta-language truth conditions *can be* given, which is sufficient to handle normative attributes as normal predicates in the object language. Rödig and Yoshino then use these predicates to formalize e.g. a norm that says that helping in an emergency situation is obligatory as $\forall acts$: $In\_emergency(act) \land Helping(act) \rightarrow Obligatory(act)$. The puzzling thing is that if this really is a prescription (norm), i.e. it makes so far unregulated acts of helping in cases of emergency obligatory, then for no such act the 'truth' of the part $Obligatory(act)$ can be established before the 'truth' of the whole is established. This at least differs from Tarski's compositional truth definition.

[31] Cf. the 'logic without truth' by [Alchourrón and Martino, 1990] who provide a calculus with an 'introduction rule' for a prescriptively interpreted O-operator, where their rule corresponds to the modified Dubislav convention (DCM) plus a requirement of joint satisfiability.

[32] It seems consensus that there must be some 'test' of adequacy. [Weinberger, 1972a] writes that one must test a rule for the logical manipulation of norm sentences for its adequacy for the area of normative thought, and [Sosa, 1966b] speaks of a 'control of commonsense' that is necessary because otherwise there would simply be no end to the possible "logics".

other sentences that are used as premises. This resembles what is called the 'soundness' of a calculus: if the calculus allows 'false' (unacceptable) conclusions to be drawn from 'true' (accepted) premises, then it must be discarded as 'unsound'.[33] I now turn to the question of adequacy in this sense.

6 Ross's paradoxes

6.1 Disjunctive conclusions

Shortly after Jørgensen's dilemma and Dubislav's workaround for a logic of imperatives had been described, Alf Ross re-considered inference schemes in 'the most simple form', where a 'new' imperative is inferred from one imperative premiss, i.e. where the scheme used is that of Dubislav's convention (DC). The following is an instance of such a scheme:

$$!A$$
$$\therefore !(A \vee B)$$

Here, $!A$ means (as now usual) an imperative sentence that is satisfied if and only if the descriptive sentence A is true, and $!(A \vee B)$ is an imperative that is satisfied if and only if either A or B are true. It is immediate that the second imperative can be inferred from the first sentence $!A$ by Dubislav's convention. Fine, said Ross, let $!A$ be interpreted as the imperative 'post the letter', so we can infer from the imperative 'post the letter' the imperative 'post the letter or burn it' $!(A \vee B)$. So

(1)   Post the letter!
        Therefore: Post the letter or burn it!

is a valid imperative inference. Ross himself points out that his paradox is not paradoxical if this 'validity' of an imperative inference is understood in the sense of a logic of satisfaction. If the letter is posted and the imperative $!A$ satisfied, then the imperative $!(A \vee B)$ will likewise be satisfied – this is no more paradoxical than that $A \vee B$ can be inferred from A. But if the meaning of 'imperative inference' refers to anything like the 'validity' or 'existence' of an imperative, then Ross claims that his inference is not only *not* immediately felt to be evident, but rather evidently false.

Why does Ross's example of an imperative inference seem paradoxical? In particular, regarding the 'formalistic theory' of imperative inference given in the last section, why should it be paradoxical to say that if one uses

---

[33] The other possibility, that the calculus does not provide *all* the inferences from premises that are acceptable (usually called 'completeness'), is less harmful and can be dealt with by e.g. refining it. For a similar definition of adequacy cf. [Chellas, 1969, p. 4], where however the terminology is vice versa.

the imperative $!A$ for commanding, then one 'implicitly' also commands $!(A \vee B)$? One explanation has been that by using a disjunctive imperative, i.e. an imperative sentence that like $!(A \vee B)$ is satisfied if some state of affairs or some other state of affairs holds, the authority has left it to the subject how to satisfy her command. Suppose Romeo hands a letter to Mercutio with the words 'Post the letter or burn it, but relieve me from deciding its fate and mine', would his friend not be free to do as he pleases? Analyzing this 'freedom', it has been argued that giving a command entails an 'imperative permission' or implicitly authorizes to carry out the actions required to satisfy the command.[34] So the imperative 'post the letter or burn it' would contain the permission 'I hereby permit you to post the letter or burn it'. Now explicit disjunctive permissions are often understood in a 'strong' sense that grants both disjuncts: when someone says 'help yourself to a cup of coffee or a cup of tea', then the guest is permitted to help herself to coffee *and* also permitted to help herself to tea (though possibly not both). So one obtains the following chain:

(2)   Post the letter!
       Therefore: Post the letter or burn it!
       Therefore: You may post the letter or burn the letter!
       Therefore: You may burn the letter!

But it seems counterintuitive to say that by ordering a letter posted one permitted it to be burned.[35] To avoid this result, one may argue that it is not the first step in (2) that is problematic, but the second, i.e. we should not be allowed to infer a strong permission from an imperative. Yet nothing seems wrong with the following piece of Mercutio's reasoning:

(3)   Romeo asked me to post the letter or burn it.
       Therefore: I may post the letter or burn the letter, as I wish.
       Therefore: I may burn the letter.

The reason why the inference from the first line of (3) to the second line seems not objectionable, while the similar inference from the second line in (2) to its third line appears somehow wrong, may lie in the fact that the imperative to 'post the letter or burn it' that is used in the reasoning is only implicit, i.e. derived, while Mercutio's reasoning was about an imperative that was explicitly used by Romeo. So one could modify one's

---

[34] Cf. [Chellas, 1969, p.19] for the term 'imperative permission' and [Keene, 1966] for the 'implicit authorization'.

[35] The idea to explain the counterintuitive nature of Ross's paradox using the also, or even more, counterintuitive inference to 'you may post the letter or burn it' was von Wright's in [von Wright, 1968, pp. 21-2], also cf. [von Wright, 1993a, pp. 121-2] and [Hintikka, 1977].

view on the second step in (2) by saying that one is only allowed to infer a strong permission to do what is commanded if this command is not itself derived. I return to such a distinction between 'explicitly given' premisses and 'implicitly given' imperatives in a moment. But consider the example from the last section, where an officer was commanded to go via Coldstream or Berwick, and (a little later) told not to go via Coldstream, where both commands were viewed as implying the command to go via Berwick. The proposed modification would still allow us to make the following inference:

(4)   I was commanded to go via Coldstream or Berwick.
       Therefore: I may go via Coldstream or Berwick, as I wish.
       Therefore: I may go via Coldstream.

So it seems the authority contradicted herself when ordering (a little later) *not* to go via Coldstream, i.e. first a choice between the two routes was granted, and later this choice was retracted, or rather: the second command modified the original command.[36] Whenever a command contradicts, cancels or modifies another command, the conflict may be absorbed e.g. by application of the rule of *lex posteriori*, which says that as a rule authorities should be considered competent to modify their own orders. But the puzzling thing is that the example was originally presented as a smooth application of imperative logic as facilitated by Dubislav's convention (DCM). Nothing made it appear as if there is some contradiction or modification involved and that more is used or required than just a flat application of the rules.[37]

An answer to these problems would be to give up the idea of strong imperative permission altogether: without it, the agent cannot reason that burning the letter is permitted.[38] But even if strong permission may be considered troublesome, or rather: troublesome to formalize, it is not clear how one could 'give it up'. It seems a dominant, if not defining feature of disjunctive permission in ordinary language. In particular, nothing seemed wrong with assuming strong permission in the case of Romeo's request (3). But whatever view is taken on strong permission, there is another point

---

[36]According to [Hare, 1967] it is just a conversational implicature that gets canceled. But it seems that by saying "go via Berwick or Coldstream" the authority *really* leaves it to the agent which route she wants to take – and later retracts this choice –, while someone who says e.g. "the tickets are upstairs or in the car", and later adds "they are not in the car" only made it *seem* as if the tickets could be in either location. If the order was only given "further orders pending", as Hare also argues, then the first order was not complete, because it left the agent unable to determine how to fulfill it. It is as if the authority had said in the middle of a sentence: "hang on, I'm not finished yet, I'll be right back."

[37]This was the point in the criticism by [Williams, 1963] of Hare's [Hare, 1967] scheme.

[38]Cf. [Stenius, 1982]: "Free choice permission is too strong a concept to be useful."

that makes Ross's paradox seem counterintuitive without appealing to some 'implied' permission: Imagine that, having been given the command 'Go via Coldstream or Berwick', the agent finds the road via Coldstream blocked. Then the following reasoning seems logical:

(5) I was commanded to go via Coldstream or Berwick.
 I cannot go via Coldstream.
 Therefore: I should go via Berwick.

It seems the kind of deliberation that one would expect of reasonable agents. Likewise, Mercutio, having been asked by Romeo to 'post the letter or burn it', might be found reasoning in the following way:

(6) Romeo asked me to post the letter or burn it.
 For fear of Tybalt's revenge, I cannot bring myself to post the letter.
 Therefore: I should burn the letter.

One might dispute whether Mercutio's fear is really on a par with a road blocked e.g. by a landslide. But if we suppose it is, then Mercutio's reasoning seems as impeccable as that of the officer. Now return to Ross's paradox: here the agent was ordered to 'post the letter'. Implicit in this imperative, so we are told by Dubislav's convention (DC), is the imperative 'post the letter or burn it'. Imagine that the agent is not able to post the letter for some cause (the postal workers are on strike and the mail bins have been locked up). So the agent could reason in the following way:

(7) I was (implicitly) ordered to post the letter or burn it.
 I cannot post the letter.
 Therefore: I should burn the letter.

But this reasoning is absurd. Just because the agent cannot fulfill her obligation to post the letter, this does not mean that she is obliged to do something that was never mentioned, and in fact could be anything: the words 'burn the letter' could be replaced e.g. by 'go to the zoo', 'kill a passer-by' or 'love your neighbor' and the inference would be just as valid – if it is valid.[39]

Now the agent, in reasoning in the above settings, used indicative statements about natural facts – like that something cannot be done – to reason about the imperatives 'go via Berwick or via Coldstream', 'post the letter' or 'post the letter or burn it'. But inferences that mix imperatives and indicatives are notoriously troublesome and should perhaps be avoided. As [MacKay, 1969] points out, both of the following 'inferences'

---

[39] This is Weinberger's [Weinberger, 1972b], [Weinberger, 1974a] explanation of why Ross's paradox poses a problem.

(8) Go fly a kite!
     You are going to drop dead.
     Therefore: Drop dead!

(8') You are going to fly a kite.
      Drop dead!
      Therefore: Go fly a kite!

are validated by Dubislav's extended convention (DEC), whereas both inferences seem plainly invalid, and so perhaps (DEC) should not be accepted. Yet consider again the case of the officer, and the two commands given to her:

(9) Go via Coldstream or Berwick!
     Don't go via Coldstream!

Imagine that, contrary to duty, and for some completely unrelated reason (maybe some superstition regarding the road through Berwick), the officer decides to go to Edinburgh via Coldstream. In her disciplinary hearing, the following remark is made:

α: At least she took one of the specified routes to Edinburgh. She did not disregard all her commands.

Given the setting of the example, α's remark seems quite reasonable: the agent was given two distinct commands, one a bit later than the other, and that the agent satisfied one of them, though not the other, should somewhat count in her favor. Now consider the two imperatives that make up Ross's paradox:

(10) Post the letter!
      Post the letter or burn it!

Imagine that the agent to whom Ross's imperative 'post the letter' was addressed, does in fact burn the letter (burning it implies not posting it, and so the imperative to post it is violated). In the discussion of her actions, the following remark is made:

β: At least she burned the letter. She did not disregard all her commands.

But β's comment appears absurd. Nothing the agent was ordered to do was achieved by burning the letter. Yet if (1) is valid, then indeed there is an imperative that was satisfied by the agent's action, namely the 'implicit' imperative to post the letter or burn it. Given the apparently completely symmetrical relation between (9) and (10) and α's and β's remarks, it seems we must agree with β that something 'right' was produced by the agent's action.

The only difference between (9) and (10) is that (9) just mentions imperatives that were explicitly given to the agent, whereas (10) contains one imperative that is only 'implicitly' contained in the other. So maybe what went wrong was that a derived imperatives was used just like an explicit one, without paying enough attention to the fact that the derived imperatives are only 'part of a system', that the 'explicitly' used imperatives have not ceased to exist, that imperatives that are only derived do not 'exist' on quite the same level as explicit imperatives, or that the agent, when reasoning about the situation, is somehow expected to make use of the logically strongest information that is available.[40] So we are back at the proposal that a difference must be made between 'explicitly used' imperatives, and imperatives that 'only derive' from explicit imperatives. But to require that reasoning with imperatives must somehow prefer the 'explicit' imperatives reveals an unusual, non-classical meaning of 'imperative inference'. For classically, logical inferences may very well be conducted by proving first that some assumptions have some desired conclusion, and then show that the assumptions follow from an accepted set of premises. This is facilitated by the transitivity of classical consequence: if $A \in Cn(B)$ and $B \in Cn(C)$ then $A \in Cn(C)$ ('consequences of the consequences are also consequences'), or the monotonicity rule: if $A \in Cn(X)$ then $A \in Cn(X \cup Y)$ (what follows from some axioms also follows from a larger set of axioms).

Ross's paradox seems to demonstrate that given the imperative inferences provided e.g. by Dubislav's convention, it becomes necessary to distinguish between the imperatives that are explicitly given and the imperatives that are only inferred: agents can use the former for their reasoning, but not always the latter, or not the latter by themselves, which makes reasoning with imperatives somehow non-classical. And so there may yet be another way to get around the difficulties: perhaps Ross's example is not really a case of an imperative inference. Perhaps (1) is simply invalid. It is obvious that the scheme is an application of Dubislav's convention, so (DC) must be modified. One way to do that is to let the logic that is used for the

---

[40][Rödig, 1972, pp. 184-5] points out that by deriving the norm to 'post the letter or burn it', the original order to 'post the letter' does not 'cease to exist', and that it is the conjunction of both norms that must be satisfied. That the entailed norms do not 'exist' in quite the same way as explicit norms is the idea of von Wright e.g. in [von Wright, 1991] and [von Wright, 1993a, p. 114,122]. According to [Stenius, 1982], the use of 'post the letter or burn it' carries the tacit information that a stronger regulation like 'post the letter' does not 'belong to the codex'. For the idea that using a weaker sentence 'post the letter or burn it' violates a conversational presupposition cf. [Hintikka, 1977]. Also cf. [Hamblin, 1987, p. 88]: " 'implicit imperatives' may be different from the real thing, and we should be wary of loading them up with the full range of imperative properties."

right hand side inference in Figure 1 be not classical logic (propositional or predicate logic), but some other logic that does not allow one to infer $A \vee B$ from A. Let Dubislav's convention therefore be reinterpreted in terms of a very strong 'relevant deduction' developed by [Weingartner and Schurz, 1986] and [Weingartner, 2003], which was tailored specifically to eliminate Ross's paradox.[41] Unlike other relevant logics, their 'R-consequence' not only blocks the inference of $A \vee B$ from A, but also that of $A \vee B$ from $(A \vee B) \wedge A$ or from $(A \vee B) \wedge \neg B$. Dubislav's scheme is then changed accordingly (cf. the next figure). Weingartner & Schurz's proposal produces

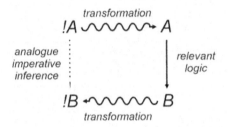

Figure 5: Dubislav's convention with relevance

a very strange consequence relation: not only monotonicity fails, but also reflexivity, i.e. $X \subseteq Cn(X)$ is not valid. But as we cannot derive $A \vee B$ from A any more, we also cannot derive $!(A \vee B)$ from $!A$, and so Ross's paradox is solved, but at a high price.

6.2 Conjunctive premisses

In [Ross, 1941] also another paradox was presented. It is a variant of the first paradox, but remains valid for Weingartner's proposal: the inference from $!(A \wedge B)$ to $!A$ ($A \wedge B$ means the conjunction of A and B). I consider the paradox in the form of 'Weinberger's paradox' or the 'window paradox':[42]

(11)   Close the window and play the piano!
        Therefore: Close the window!

(12)   Close the window and play the piano!
        Therefore: Play the piano!

---

[41] R-consequence is defined in [Weingartner and Schurz, 1986], [Weingartner, 2003] in the following way: a propositional formula A is a R-consequence of a set of formulas X iff (i) X classically implies A and (ii) it is not possible to uniformly replace a proposition letter at at least one of its occurrences in A by a random proposition letter without making the classical inference invalid.

[42] The origin of the example is unclear. The name 'Paradox of the Window' is used e.g. by [Stranzinger, 1978] and [Weinberger, 1991].

Suppose that α wants x to practice the piano, but neighbors have already complained about the disturbance and even called the police on a previous occasion. So α does not want x to play the piano while the window is open. Closing the window will reduce the noise so much that the neighbors are left with nothing to complain about. Suppose then that α sends x to play the piano, using the words 'close the window and play the piano'. A little bit later, the following discussion ensues between α and β:

α: I told her to play the piano, but I didn't hear her doing it all afternoon.
β: Well, at least she closed the window.
α: Why should she do that?

Here, the positive view on x's behavior by β is not accepted by α. Closing the window by itself is meaningless. It might even be unwanted in general – it blocks out fresh air – if it weren't for the sake of piano practice. But backed with the inference (11), β can continue in the following way:

β: You ordered her to close the window, that's what she did, so she did something right, didn't she?

Now consider the following, alternative dialogue:

α: She practised the Khachaturian with the window wide open. What shall we tell the police this time?
β: It was you that told her to play the piano.
α: But I didn't. She was also to close the window.

β's reproach for x's playing the piano is not accepted by α, because playing the piano without closing the window first was not what α had asked x to do. However, backed with the inference (12), β could reply in this way:

β: You ordered her to play the piano, that's what she did, so don't try to wiggle out of your responsibilities.

In these dialogues, α's position seems natural, while β's reaction is strange and uncomprehensible. But given the inference schemes (11) and (12), β is right: from α's command, the imperatives 'close the window' and 'play the piano' can be inferred – so we are told by Dubislav's convention (DC), and Dubislav's convention restricted by relevant implication. Moreover, these derived imperatives are used by β as imperatives are meant to be used, namely compared with reality, and reality accordingly qualified as 'right' or 'wrong'. So x did something right by closing the window, as x satisfied an (implicit) imperative, and similarly when playing the piano without closing the window. But intuitively, closing the window by itself produces nothing good, and playing the piano with the window open seems a clear violation of obligations and not satisfactory in any way.[43]

The 'window paradox' seems to arise whenever the states of affairs mentioned in the imperative are only conjunctively desired by an authority. That for this reason we cannot detach conjuncts in wishes, i.e. we cannot conclude from 'she wishes for a and b' that 'she wishes for a' was pointed out by [Menger, 1939] for the case of complementary goods, e.g. when a is 'a cigarette' and b is 'a match', for one may not wish either one of the goods by itself. [Ross, 1941] points out that the same difficulty arises for imperatives, e.g. when the imperative is to 'write a letter and post it'. Other examples have included the imperatives 'take the parachute and jump', 'pay the bill and file it' or 'fill up the boiler with water and heat it'.[44] [Goble, 1990] showed that even a seemingly innocuous obligation to 'sing and dance at Gene's party' may be planted in a setting that makes it impossible to speak of fulfilling any obligation when only one act, singing or dancing, is performed. To determine whether an imperative is 'separable' or 'inseparable', i.e. whether doing A alone produces something 'right' with respect to an imperative $!(A \wedge B)$ or not, it is necessary to examine the intentions and wishes of the authority that used the imperative, it is not a matter of logic.[45]

To solve these difficulties, [Kenny, 1966] proposes a logic of 'satisfactoriness'. This logic uses a set of propositions to represent the wishes of the authority. A *fiat* (an impersonal imperative like 'let there be light') is called satisfactory if and only if whenever the *fiat* is satisfied then every proposition in the set of wishes is true. Finally an inference of one *fiat* from another *fiat* is defined as follows:

$!B$ may be inferred from $!A$ in the logic of satisfactoriness
if and only if
if $!A$ is satisfactory then $!B$ is satisfactory.

It is clear that the troublesome inferences (1), (11) and (12) are invalidated

---

[43] It gives the paradox a further twist if we imagine that playing the piano with the window open is explicitly forbidden. For by Dubislav's convention (DC), the imperative $!\neg(A \wedge \neg B)$ ('don't play the piano while the window is open'), is derivable from the imperative $!(A \wedge B)$ ('close the window and play the piano'). But it seems that the additional prohibition is best formalized as a conditional imperative (in [Hofstadter and McKinsey, 1938] formalism: $\neg B \Rightarrow !\neg A$). Conditional imperatives pose other problems outside the current topic. In any case, one would still have to say that playing the piano with the window not closed was satisfactory with regard to some (derived) imperative.

[44] Cf. [Hare, 1967], [Weinberger, 1958] and [Weinberger, 1999]. These difficulties led Weinberger to reject the validity of an inference from $!(A \wedge B)$ to $!A$ in his publications since [Weinberger, 1958].

[45] The terminology here is that of [Hamblin, 1987, p. 184].

by this logic of satisfactoriness: when posting the letter is satisfactory for the wishes of the authority, then burning the letter need not be so. Likewise, if closing the window and playing the piano is satisfactory with respect to all wishes, then playing the piano alone does not guarantee that the wishes of the authority are also satisfied. But Kenny's approach gives rise to other paradoxes: in the logic of satisfactoriness we can e.g. derive:[46]

(13) Open the door!
       Therefore: Open the door and wear a tie today!

The inference is clearly absurd and so Kenny's logic does not help us to solve the paradoxes.

6.3 Summary

In Ross's first paradox, the imperative to 'post the letter or burn it' was 'inferred' from the imperative to 'post the letter', thus forcing one to acknowledge that some (though only inferred) imperative is satisfied by burning the letter. In the 'window paradox' we could 'infer' the imperative 'play the piano' from the imperative 'close the window and play the piano', thus forcing us to acknowledge that an (inferred) imperative is satisfied when the piano is played with the window wide open. In both cases, we would much rather say that *no* imperative was satisfied by burning the letter that was meant to be posted, and by playing the piano with the window open when it should have been closed. This, I think, is the main cause why Ross's paradox and the window paradox give rise to counterintuitive feelings, or are 'paradoxical'. So we should not be allowed to infer such imperatives. So Dubislav's convention is not an apt theory to explain how an imperative may be derived from another one. And so we are back at square one: all theories, including the formalistic approach, have so far failed to explain what it means to infer an imperative from some other imperative in spite of Jørgensen's Dilemma.

7 Ordinary language arguments

Maybe it is not really the case that all options have run out to redefine Dubislav's scheme in a way so that it avoids Ross's paradoxes. Maybe we have to replace the classical logic that appears in his scheme by yet another logic, or develop such a logic.[47] But it is hard to see what kind of logic this could be, since most logics, including other non-monotonic logics, will permit us to either infer $!(A \vee B)$ from $!A$ or $!A$ from $!(A \wedge B)$, and so at least

---

[46] A similar counterexample was given by [Gombay, 1967], also cf. [Sosa, 1966b].

[47] Cf. [Keene, 1966]: "What we wanted here is a logic of *actions*, in which a well-defined concept of *inclusion* plays a leading role."

one of the two paradoxes will arise. So I think, after all these troublesome attempts to define a 'logic of imperatives', it is worthwhile to take another look at Poincaré's proposal that originally started the controversy.

Poincaré's only explicit example of an inference with an imperative conclusion has the following form:

(1)   Do this!
        This cannot be done without that.
        Therefore: Do that!

The following is an instance of this scheme:

(2)   Drive me to the airport!
        To get to the airport, one must drive in a northerly direction.
        Therefore: Drive me in a northerly direction!

In which setting could these sentences be used? Suppose I have entered a taxi and used the above sentences. But some confusion could arise. The driver could reply: "So what do you want me to do, drive you to the airport or just drive north?" The driver needs a direction. Ordering her to go to the airport alone is sufficient for this, and the behavior expected of a passenger entering a taxi. Using two imperatives where each contains an instruction of where to go is unexpected and confusing.

So suppose I have just used the sentence 'drive me to the airport'. A little later I realize that we seem not to be going north, and I say to my partner:

> "Is she hijacking us? I ordered her to go to the airport, and the airport lies to the north. So she ought to be driving us in a northerly direction. But she is not."

This reasoning seems flawless. Yet it only involved sentences about imperatives, and did not involve sentences in the imperative mood, and so it cannot be an example of an imperative inference. But maybe this would be a good time to say to the driver:

(3)   I ordered you to go to the airport.
        To get to the airport, one must drive in a northerly direction.
        Therefore: Drive me in a northerly direction!

But here the two sentences that function as premises are both descriptive. Since Poincaré explained that an imperative cannot be derived from indicative premises alone (and there is no reason not to follow him), this cannot be an imperative inference, and there must be something more involved than the drawing of a logical conclusion. One such other function of the 'therefore' appearing at the front of the last sentence of (3) is not to reason, but to motivate, as in:

(4)   The car is broken.
       Therefore: Take the bus into town!

Here the speaker motivates the imperative to take the bus by explaining that driving into town is impossible, since the car is broken. So similarly, what seems to happen in (3) is that I motivate my (new) imperative 'drive me in a northerly direction' by an already given command and an assumed fact.

Consider again the proposed inference (2). Just like indicative inferences are explained by the fact that someone who accepts (or: assents to) the premises must also accept the conclusion, [Hare, 1952] has argued that an imperative inference is one where someone who assents to all imperative premises must also assent to the imperative conclusion:

> "A sentence p entails a sentence q if and only if the fact that a person assents to p but dissents from q is a sufficient criterion for saying that he has misunderstood one or other of the sentences. (...) A person who assented to this command ['Take all the boxes to the station'], and also to the statement 'This is one of the boxes' and yet refused to assent to the command 'Take this to the station' could only do so if he had misunderstood one of these three sentences."

But what does it mean that a person 'assents' to a command? Suppose John's mother tells him 'John, clear the table and do the washing up', and John's little sister echoes: 'John, do the washing up'. If John 'assents' to his mother's order, does he also have to 'assent' to an order by his sister, whom he might not accept as an authority? Perhaps the analysis assumes identity in the person who uses the commands. Suppose then it was not John's mother but some officer who used the imperative, and John is not obliged *qua* son, but as this officer's orderly. The second command is also used by the officer, maybe a little later. But suppose that John is only obliged to the officer if the commanding is done in a certain fashion, e.g. when the officer is standing up, or when the officer is not drunk, and that when the second imperative was used the officer was, as a matter of fact, not standing up or already had more than her fill. Or suppose that John is not an orderly, but some djinni, and the officer is the person who rubbed the lamp, but that, when the first imperative was used, this already was the last of the three wishes that had been granted. Does John, in these cases, have to 'assent' to the second command? It seems that such an interpretation of 'assent' would have to get involved into reasoning about whether the act of using an imperative 'really creates' a command. But it did not seem as if such reasoning is involved in Hare's proposal.

So let the word 'assent' be understood in its weakest possible interpretation. A person could hardly be said to assent to a command given to her if she did not satisfy or to try to satisfy it. Returning to the situation where I have asked the taxi driver to take me to the airport, when the taxi driver assents to this request, she will start driving me to where she thinks the airport lies, i.e. start to satisfy, or try to satisfy, my request. If the taxi driver agrees that the airport lies in a northerly direction, she will, in obeying my request, eventually drive in what she thinks is a northerly direction. So she might be said to additionally satisfy, or try to satisfy, a request to drive me in a northerly direction, had such a request been made. But did I request the taxi driver to drive me north? I might be absolutely sure that the airport is to the north, but I would still blame the taxi driver for not going where I requested if, opposite to what I believed, the airport is in fact to the south-west of my starting point and the driver still went north. So I did not utter such a request, and would not even imply such a request, lest I be charged by the driver for going there instead of the airport. All we can say is that the taxi driver would also be satisfying, and so seemingly assenting to, a purely hypothetical request to drive me in a northerly direction, if she satisfies the request to drive me to the airport and the airport does in fact lie in a northerly direction. But this is again not a logic that infers one imperative from some set of other imperatives and/or indicatives, but the logic of satisfaction as explained in sec. 5.1.[48]

It would be nice to have 'real life' examples, cases of 'ordinary' reasoning with imperative premises and an imperative conclusion, i.e. instances of

(ImpInf)   !A
           ∴ !B

where !A and !B are sentences in the imperative mood, and where the use of the inference – not the imperatives – is either accepted in some ordinary discourse, or opposed (and the person who uses it blamed for being 'unreasonable' or 'illogical').

Use of indicative arguments in everyday discourse often occurs in singular sentences, like

(5)   Unemployment is rising, so there are not enough jobs created.
(6)   She has got an 'A' in English, so she achieved top-marks in at least one subject area.

---

[48]This resembles the criticism by [Keene, 1966] of Hare's proposal.

(7) I have read all of Vladimir Nabokov's novels, so I have read *Pnin*.

Here two descriptive sentences are linked with the adverb 'so' (similar adverbs would be 'therefore' or 'hence'). (5) seems analytical if one understands 'enough' to be elliptical for 'enough to make up job-losses elsewhere'. (6) is analytical if one knows that 'A' is a top-grade and that English is one of several high-school subjects. (7) is made into a logical argument by the assumed background knowledge that *Pnin* is a novel by Nabokov. It is often not easy to distinguish such indicative arguments from sentences that present reasons, motives or are otherwise explanatory, for these also use the form of descriptive sentences that are concatenated by an adverb like 'so' or 'therefore', as in the following examples:

(8) I couldn't get the car started, therefore I took the bus.
(9) I wanted to make friends with her, therefore I asked her if she would go shopping with me.
(10) There were holes in the roof, so birds had come in and were roosting in the rafters.

(8) explains why today the speaker used the bus. Since the bus need not have been the only means to get into town, or the speaker may have stayed at home, the hearer cannot just conclude the second part from the first. (9) presents the psychological motive why the speaker asked the other person to go shopping with her. Other people might have been motivated differently by the desire to make friends with that person. In (10), a natural event is explained by a certain state of affairs. Again, this is not a logical argument: the birds could also have not flown in, or flown in but not nested in the ceiling. Now the adverbs 'so' and 'therefore' can also be used to meaningfully link imperatives. Consider the following examples:

(11) Stop the rise of unemployment, so see to it that more jobs are created!
(12) Make your guests comfortable, so introduce your guests to each other!
(13) Don't let vermin into you house, therefore patch up the roof!
(14) Read all of Nabokov's novels, so read *Pnin*!

(11) might be encountered in some political debate. At first it appears to be a good argument, but then doubts arise: is the speaker really appealing to logic, or is she just complementing her first imperative by a second, more specific one, as when we say: "Go there! Go there now!"? And one could also stop the rise of unemployment by e.g. prohibiting companies to dismiss their workers, or making it more difficult for them (maybe 5 was not

so analytical after all). Then (11) would seem to be rather a case of a motivating use of 'so': the imperative to see to it that more jobs are created is motivated by the primary aim to stop unemployment. Likewise, in (12), the advice to introduce guests to each other is rationalized by the more general aim to make guests comfortable. It is hard to see what could be analytical here: to ease tensions, the host may equally encourage the guests to *guess* each others names, or serve them plenty of alcohol, or maybe the guests are easygoing and do not really require any effort on the host's part to make themselves at home. Similarly, in (13) the more readily accepted advice to keep vermin out of the house is used as a rationale to make the addressee accept the drudgery of having to patch up the roof. The most promising candidate for an appeal to analyticity seems to be (14), i.e. that the imperative to read all of Nabokov's novels includes the imperative to read *Pnin*, given the background knowledge that *Pnin* is a novel by Nabokov. Note that when making the background knowledge explicit, it becomes a case of Dubislav's extended convention (DEC). Such a sentence may be used e.g. by a teacher of a literature course when addressing her students. But again we cannot rule out that this is just a case of complementing an imperative by a second, more specific one, as we sometimes do to get things done.

Adherents of Dubislav's convention (DC) must also accept the following argument:

(15) Aim for an 'A' in English, so aim for top-marks in at least one subject area!

But it seems dubious what reason the speaker could have for adding the 'so' part. Just aiming for top-marks in some subject area is clearly not what the speaker wants the addressee to do. More meaningful would be the converse,

(15a) Aim for top marks in at least one subject area, so aim for an 'A' in English!

where the advice to aim for 'A' in English is rationalized by the wish to have the student achieve top-marks somewhere. But since the student could not know from the first imperative that it was the subject of English that the speaker wanted her to achieve top marks in, this would – like (12) and (13) – rather be a 'motivating so', and not a use of 'so' that appeals to a logical capability.

Matters are further complicated by the fact that expressions of the following kind can also be meaningfully employed:

(16) The car isn't working properly, so take the bus!
(17) I forgot my keys, therefore leave your key under the mat!
(18) Gill is your best friend, so invite her to your party!

In all three sentences, the first part is descriptive and the second is in the imperative mood. We have already noted in the case of (3) that such arguments exist, but for anyone who agrees to Poincaré's thesis that imperative conclusions do not follow from an indicative premisses it is clear that (16) – (18) cannot represent valid arguments. (16) seems again a case where the 'so' is used to motivate the advice that is expressed by the imperative. The 'so' does not express a logical relation, for sometimes it is better to use a car that stutters than a coach that won't take one back. In (17) the indicative gives a reason why the speaker wants her request to be followed. According to [Hamblin, 1987], such reason-providing indicatives are often attached to advice-expressing imperatives, yet here the imperative might also be an order (e.g. of a parent). For the same reason the speaker might have ordered the agent to hand over her key, and not to leave it under the mat, and so what is expressed is again not a logical relation. (18) seems also like presenting a motive for inviting Gill to the party (she is the addressee's best friend), but here things might be a bit more complicated – the expression could be elliptical for:

(18.a) Invite your best friends to the party, Gill is your best friend, so invite her to your party!

This is very similar to what Dubislav considered a valid argument, namely his inference from 'thou shalt not kill' to 'Cain shall not kill Abel'. But then, (18) might also be elliptical for

(18.b) Gill is your best friend, one invites one's best friends to one's parties, so invite her to your party!

where the second part (which is not in the imperative mood) appeals to the existence of a rule that the speaker might consider binding, or binding for the addressee. Then this is rather a case of reason-giving, and not of a logical inference: the speaker motivates her imperative by asking the speaker to conform to some preexisting rule.

To tell the uses of 'therefore's' and 'so's' that are motivating, reason-giving or explanatory in a non-logical sense, apart from those that separate the premisses from the conclusion in an argument that is intended to be a logical one, we can use the following trick: instead of 'therefore' or 'so', use a clause like "... It follows logically from this that ..." to separate the sentences. The new phrase makes the appeal to a logical capability explicit. Where the original adverbs 'so' and 'therefore' were used to indicate a (claimed) logical inference, the new formulations

(5.a) Unemployment rates are rising. It follows logically from this that not enough jobs are created.

(6.a) She has got an 'A' in English. It follows logically from this that she achieved top-marks in at least one subject area.

(7.a) I have read all of Vladimir Nabokov's novels. It follows logically from this that yes, I have read *Pnin*.

appear only to be changes in expression. The speaker, just as before, appeals to a shared understanding of words, concepts and background knowledge, to make the second sentence seem to be expressing nothing new, but only a logical consequence from what has been said before. Note that it does not matter whether the arguments are, in fact, analytical. People sometimes think they use valid arguments when they are not. But the rephrased sentences make it clear that the speaker *intends* the sentences to be just that. And the new formulations seem not to change the meaning of the original sentences whenever a 'logical' use of the adverbs 'so' and 'therefore' was really intended. By contrast, when the first part was used to give some background information, a reason, explanation or motive, the rephrased expressions appear odd:

(8.a) I couldn't get the car started. It follows logically from this that I took the bus.

(9.a) I wanted to make friends with her. It follows logically from this that I asked her if she would go shopping with me.

(10.a) There were holes in the roof. It follows logically from this that birds had come in and were roosting in the rafters.

The phrase 'it follows logically from this' makes again an appeal to some shared understanding of used words, concepts and background. But here, this background knowledged obviously does not allow one to 'conclude' the second sentence from the first. The listener could not have known from the first sentences in these examples that the speaker took the bus, asked someone to go out shopping or has birds nesting in the roof of her house. So claiming, as the rephrased sentences do, that the second part can be concluded from the first, makes the sentences seem irritating, weird and false, while the earlier sentences appeared quite harmless.

Now consider what happens if such a method is used on imperatives. So far, (14) seemed the best candidate for a sentence that 'appeals to logic', so I will concentrate on this example. First note that

(14.a) Read all of Nabokov's novels. It follows logically from this that read *Pnin*!

is not grammatical. Now if the grammar is difficult, this may already hint at little usage of such statements, but I think that instead of the 'that' we can easily use e.g. a colon, corresponding to a pause in oral language, as in the following expression:

(14.b)   Read all of Nabokov's novels. It follows logically from this: read *Pnin*!

But here, the part that follows the colon seems strangely detached. Is this a command, i.e. is the speaker, using the expression following the colon, still giving a command? Or is the emphasis on the part before the colon, and so the purpose of the second sentence is merely to tell (truly or falsely) that some consequence relation holds? The impression that this is a strange use of words increases if we add the subject of the request:

(14.c)   John, read all of Nabokov's novels. It follows logically from this: John, read *Pnin*!

Here, the phrase 'it follows logically from this' makes it appear as if the speaker was not giving commands to John at all. It seems what the speaker really does is talking about logical relations between sentences – maybe it is a logician presenting an example of an imperative inference. So perhaps we should try out another phrase:

(14.d)   John, read all of Nabokov's novels. We can conclude from this: John, read *Pnin*!

Yet this expression also has a false ring: who is doing the commanding of the 'conclusion' – the speaker? Or the 'we' that is to do the concluding? Do the speaker and the listeners all join into giving John the command? Apparently it was wrong to use the first person plural, and so we might want to change the sentence into:

(14.e)   John, read all of Nabokov's novels. I conclude from this: John, read *Pnin*!

But this seems to be the worst alternative so far. Is the speaker concluding the last sentence? Or is the speaker commanding it? And if so, then why is she saying that she is concluding it? The performative acts of concluding and commanding seem to collide, whereas the acts of stating and concluding seemed to go hand in hand. But we have yet another phrase to try out:

(14.f)   Read all of Nabokov's novels. So you can conclude for yourself: read *Pnin*!

Though this is perhaps a less common phrase to signal logical arguments, the new sentence seems to be the most successful so far. But it appears necessary that the 'you' is the person to whom both commands are addressed. So let us make the addressees explicit. Of the following sentences

(14.g) John, read all of Nabokov's novels. So John, you can conclude for yourself: read *Pnin*!

(14.h) John, read all of Nabokov's novels. So Mary, you can conclude for yourself: read *Pnin*!

(14.i) John, read all of Nabokov's novels. So Mary, you can conclude for yourself: John, read *Pnin*!

only the first seems somehow acceptable. In (14.h) it appears as if Mary is asked to read the book, but this can hardly be 'concluded' from a command not directed at Mary. (14.i) makes it seem as if Mary is asked to give a command to John (and not just to draw a conclusion). Moreover, if the addressee is expressly included in the inferred command, then also (14.f), which seemed so promising at first, looks strange:

(14.j) John, read all of Nabokov's novels. So you can conclude for yourself: John, read *Pnin*!

It seems that in (14.f) and (14.g) the speaker has not just asked the addressee of the first command to 'draw a conclusion', but in this process to 'give himself' the command expressed by the second sentence, i.e. to 'tell himself to read *Pnin*'. When the addressee is made explicit in the 'inferred' command, it looks as if the addressee is additionally asked to use his own first name when telling himself to read *Pnin* – which is a weird thing to ask of anybody. And this points at another problem of (14.f) and (14.g): if the person who commands 'read all of Nabokov's novels' (the teacher) and the person who commands 'read *Pnin*' (John himself) are not identical, how can the second imperative be inferred from the first?

By contrast, all of the above phrases can be employed for 'deontic sentences' (non-imperative sentences that do not prescribe, but describe what ought to be done) without difficulty:

(19) You ought to read all of Nabokov's novels, therefore you ought to read *Pnin*.

(19.a) John ought to read all of Nabokov's novels, therefore John ought to read *Pnin*.

(19.b) John ought to read all of Nabokov's novels. It follows logically from this that John ought to read *Pnin*.

(19.c) John ought to read all of Nabokov's novels. We can conclude from this that John ought to read *Pnin*.

(19.d) John ought to read all of Nabokov's novels. I conclude from this that John ought to read *Pnin*.

(19.e) John ought to read all of Nabokov's novels. You can conclude for yourself that John ought to read *Pnin*.

(19.f)   John ought to read all of Nabokov's novels. Mary, you can conclude for yourself that John ought to read *Pnin*.

All these sentences seem grammatical, meaningful and not confusing. We might even view the inferences they express as sound, but this is not the question here. Yet as we have seen, all attempts to use the phrases that link these sentences, normally used to indicate logical arguments in indicative discourses, to link imperatives to indicate 'imperative inferences', result in expressions that seem somehow confused and wrong. When used to link imperatives, they mix up the roles of commanding, command-receiving, and drawing conclusions. And since the method to use such clauses to distinguish appeals to logic from e.g. motivating uses of 'therefore's and 'so's, fails to produce sentences that do not appear strange or confused in the case of imperatives, perhaps it did so because these adverbs really are not used to indicate a claimed analyticity when linking imperatives:

- A motivating use of the adverb 'so' suffices to explain why the sentence (14) seemed meaningful: the teacher, perhaps asked by John whether he also has to read *Pnin*, motivates the more specific imperative to read this book by prefixing to it the general requirement to read all of Nabokov's novels, thus making it clear that *Pnin* is in fact one of the books that John has to read.

- (14.f) appears comparatively less strange than the other reformulations because to ask John to 'give himself' the imperative to read *Pnin* may be a (roundabout) way to make sure he actually reads it.

- To understand (18) we do not need to determine whether the speaker refers to an explicit command to 'invite one's friends', or a social custom to do so, because what is in any case implicit in (18) is an appeal to a preexisting obligation to motivate the agent to do what the speaker wants her to do.

- It also explains why (15) seemed so strangely pointless: when the speaker motivates her imperative by explaining that she wants the student to achieve top marks in the subject of English, why would she only tell him to achieve top marks in *some* subject area?[49]

And so it seems that all of the imperative arguments (11)–(18) are really cases of reason-giving and motivation, and the 'so's and 'therefore's used in these expressions that like Poincaré's '*donc*', or the '*also*'s, '*daher*'s and

---

[49] Note that the same strangeness does not arise for deontic logic. One dean may say to another: "Our students are obliged to have an 'A' in English, so yes, ours – like yours – are obliged to achieve top marks in at least one subject are."

'*deshalb*'s of German language, may be used to connect both indicative and imperative sentences, provide only reasons, explanations or motives in the case of imperatives, and do not indicate claims of analyticity.

So I want to dare the hypothesis that there are no examples of imperative inferences, i.e. logical conclusions in the imperative mood, drawn from at least one premiss in the imperative mood, to be found in ordinary language arguments. They only appear in the writings of some philosophers.

8 The way to go forward

Other authors have noted before the conspicuous absence of imperative arguments from natural language. [Wedeking, 1970] argued that there are no cases in which we actually use commands in arguments, and that words like 'therefore' before imperative sentences are employed not to mark inferences, but for the purpose of reason-giving, of motivating the subject. [Harrison, 1991] argued from a point of grammar and semantics that there is no logic of imperatives; the difference between his position and my arguments above (that there are no 'imperative inferences' in ordinary language, and so a logic of imperatives has no point) seems very subtle.[50] Even more conspicuously, there have been little challenges to these arguments.[51] On the

---

[50] Harrison writes (pp. 110-1, 124-5): "The expression 'I conclude: Shut the door' does not make sense, nor does the expression 'I conclude that don't'. One can say: 'So shut the door' and 'Therefore shut the door' and 'Shut the door because ...' but the function of the words 'so', 'therefore' and 'because' is not in this context to indicate that 'Don't shut the door' is a conclusion. They have some other function, which philosophers have confused with that of indicating that what follows them is a conclusion. (...) The reason, therefore, why 'therefore' and 'so' can precede 'post the letter', but 'I conclude' can not, is that 'I conclude' can precede only propositions (and only they can be conclusions) and indicate that reasons have been given for the proposition, but 'therefore' and 'so' can precede either propositions or imperatives. When 'so' and 'therefore' precede imperatives, however, they are not reasons for the imperative, as the unwary might suppose, but for the action enjoined, advised, recommended or directed by it." Harrison concludes (p. 81): "There is no such thing as imperative logic (...) There are indeed logical relations between one imperative and another, but this simply supports a logic in which the premisses and conclusions are indicative statements *about* imperatives" (p. 81, my emphasis).

[51] [Castañeda, 1971], replying to Wedeking, grants that differences in the meaning of inferential words in indicative and imperative 'inferences' may exist, but argues that they do not prohibit a concept of imperative inferences in parallel to indicative ones. This seems to miss the point, it echoes Dubislav's convention without explaining what such formalisms are to formalize. [Vranas, 2009] has replied to the arguments in the last section that one may say: "John, watch TV if and only if you finish your homework. I conclude from this: John, if you dont finish your homework, dont watch TV", and that this constitutes a case where the second imperative is both concluded and commanded. But I think such a usage of imperative sentences seems no less weird than the examples given above. Is the speaker not sure that her first imperative was properly understood? Is she giving advice as to what John should do? (Vranas' example additionally suffers from the fact that biconditional imperatives are rather rare in ordinary language discourses –

contrary, there is a whole tradition of 'normological' or 'imperativological skepticism', of authors who have denied the existence of a logic of norms or imperatives.[52] But if there are, as a matter of fact, in ordinary language, no argument forms that resemble 'imperative inferences', then there also is no place for a formal theory for such a logic. Presenting formalizations of such a logic would be writing about what [Dubislav, 1931] called an *Unding* or *chimaera*: a non-thing that exists only as a concept, but no real object falls under the concept.

So did Poincaré commit a mistake? Did he confuse an important insight by [Hume, 1888] on the use of 'is' and 'ought' – that facts cannot be used to argue that they must be so or that other facts should be made similar to them – with a statement about grammar? Curiously, in his essay, [Poincaré, 1913] never claimed to have discovered the logic of imperatives of which he was celebrated as the pioneer. His main argument is that findings of science can influence moral reasoning. He just presumes that, like scientific arguments consist of sentences in the indicative mood, moral reasoning is conducted using sentences in the imperative mood. It is true that facts can influence the reasoning of agents about their obligations: Hare's officer, who upon being commanded to go to Edinburgh via Coldstream or Berwick finds the road via Coldstream blocked, acts quite reasonably by concluding that she now ought to go via Berwick. But this is a reasoning *about* what obligations she has, it is a deontic argument, and not a case of 'inferring' imperatives. So Poincaré's main argument is correct, but the assumed parallelism between sentences in indicative and imperative mood, that they can both feature in logical arguments, does not exist. Our language does not work that way.

There are several ways to go forward from a position of 'imperativological skepticism'. First, one might continue the 'logic of imperatives' as a logic of satisfaction. The logic of satisfaction states which imperatives must also be satisfied if some other imperatives are satisfied, and it may also be used to

---

note that John is also conditionally ordered to watch the TV.) At best, it seems a bad way in which the speaker wishes to combine a normative utterance with a normative proposition added for explanatory reasons, as in "John, don't watch the TV unless you have finished your homework, so if you haven't finished your homework yet, you are not allowed to watch TV."

[52]Such authors include G. H. von Wright who writes in [von Wright, 1993b, p. 109]: "And now I too, after a long and winding itinerary have come to the same view: logical relations, e.g. of contradiction and entailment, cannot exist between (genuine) norms.". Above we have already noted that [Hamblin, 1987, p. 89], [Sellars, 1956, p. 239-40], [von Kutschera, 1973] and [Philipp, 1989], [Philipp, 1991] have expressed scepticism or denied the possibility of a logic of imperatives altogether. For imperatives also cf. [Moritz, 1954], [Williams, 1963], [Keene, 1966], [Opałek and Woleński, 1987]. The term is coined from [Weinberger, 1986] term 'normological skepticism' which denies logical relations not only between imperatives, but any prescriptive language. The main proponent of normological scepticism is [Kelsen, 1979].

state which imperatives will be violated by satisfying other imperatives. We can use the notion of satisfaction to distinguish imperatives that might be seen as redundant in a set of imperatives in the sense that these will also be satisfied if some other, different imperatives are satisfied, or identify subsets of imperatives that cannot be all satisfied and so conflict. By providing these concepts, the logic of satisfaction, though it may appear trivial, remains a meaningful and correct way to talk about imperatives.[53]

Second, imperatives normally express the wish or desire on the part of the person or authority using the imperative that what is commanded is satisfied. But it seems unreasonable to wish for A to be realized, but also for $\neg A$ to be realized, and in this sense two wishes may exclude another. If imperatives express wishes of one particular person, we can then point out to her what wishes may be unreasonable. Likewise it might be desirable to view the norms of a particular society as if they all were the wishes of one person, the 'law giver', and logic may then give advice as to which norms must be revised so that the system is 'reasonable'. This is the position of G. H. von Wright in his late work on normative logic, cf. e.g. [von Wright, 1991; von Wright, 1993a].[54]

Finally, there is deontic logic. Deontic logic uses the modal expressions 'it is obligatory that', 'it is permitted that', 'it is prohibited that' etc. to describe what ought to be, is permitted or is prohibited *according to* given imperatives or norms.[55] Deontic logic has been disparagingly called a "kind of ersatz truth", that merely mirrors logical relations that already exist between imperatives or norms, and so we should rather look for this logic than studying a deontic logic that only reflects it and so must result in a "dull isomorphism".[56] But it has been the 'logic of imperatives' that has kept escaping us, while sentences that use deontic expressions can easily be used to form valid arguments. So maybe it is the other way round,[57] and

---

[53]Cf. C. G. Hempel's [Hempel, 1941] remark with regard to Ross's Paradox that a logic of satisfaction should not be so easily rejected.

[54]The idea that commands can be identified with wishes, which in the above sense relate to each other, goes back to [Bentham, 1970, pp. 95-7]. Note that this does not force one to acknowledge that there is a logic of commands. There is a difference between a theoretical inconsistency and a practical inconsistency, or: "what is inconsistent, and what is inconsistent to say, are two entirely different things" ([Harrison, 1991, p. 95]).

[55]I have explored the idea that deontic logic can be seen as a logic about imperatives in [Hansen, 2001], [Hansen, 2004], [Hansen, 2005], [Hansen, 2006], [Hansen, 2008].

[56]This is Hare's view in [Hare, 1967, p. 325]; also cf. [Alchourrón, 1969, pp. 264-6]; [Kalinowski, 1973, p. 134]; [Weinberger, 1986, p. 58], [Weinberger, 1991]; [Wagner and Haag, 1970, p. 102]. The idea that deontic logic reflects the logical properties of norms is that of von Wright in [von Wright, 1963, p. 134].

[57]Cf. [Alchourrón and Bulygin, 1984, p. 463]: "This logic of norms is, so to say, a reflection of the logic of normative propositions. It is because we regard as inconsistent a system in which it is true that $O_x p$ and $O_x \neg p$, that we say that the norms !p and !$\neg p$

the idea of a logic of imperatives has been a *fata morgana*, leading us to ever more futile attempts to explain inference relations between imperatives, to find analogues of truth values, or new logics to explain Dubislav's scheme, whereas any plausibility of this idea was just a reflection of the real, but distinct possibility of a logic *about* imperatives, namely of deontic logic.

But if there is no 'logic of imperatives', then it seems the task of deontic logic must be both: the study of prescriptively (normatively) used language in order to determine what it may make obligatory, permitted etc. according to the study of normative propositions. Then the question of whether it is the one or the other appears as a false dichotomy, just as the idea that there must also be a 'logic of imperatives' since there is a logic of propositions is a false parallelism. For that there is no 'logic of imperatives' does not mean that imperatives are somehow unreasonable and need not be studied. If we are addressed in the imperative mood by speakers whose authority we accept, then we will use these sentences to truly or falsely determine what we must do and what we are permitted to do. We advise norm-givers, like a club whose members reconsider their statutes, as to how their norms may be changed reasonably. We permanently reason about satisfaction: when we figure out how we can best discharge our duties, when we consider our options in a dilemma where not all commands can be fulfilled, when our actions trigger commands that were only conditional but now have to be (also) fulfilled, when we are in violation of our duties and now have to make up for it by satisfying secondary obligations. We even use commands and other norms to decide if other imperatives should be uttered or norms given, like judges who must rule according to statutes and previous rulings, even though sometimes they don't. All this can be studied regardless of the fact that sentences in the imperative mood are not parts of logical arguments.[58]

BIBLIOGRAPHY

[Alchourrón and Bulygin, 1981] C. E. Alchourrón and E. Bulygin. The expressive conception of norms. In Hilpinen [1981], pages 95–124.

[Alchourrón and Bulygin, 1984] C. E. Alchourrón and E. Bulygin. Pragmatic foundations for a logic of norms. *Rechtstheorie*, 15:453–464, 1984.

[Alchourrón and Bulygin, 1993] C. E. Alchourrón and E. Bulygin. On the logic of normative systems. In H. Stachowiak, editor, *Handbuch pragmatischen Denkens*, pages 273–293. Meiner, Hamburg, 1993.

[Alchourrón and Martino, 1990] C. E. Alchourrón and A. A. Martino. Logic without truth. *Ratio Juris*, 3:46–67, 1990.

---

are incompatible. So it is the logic of norm propositions which yields the foundations for the logic of norms.".

[58]This contribution is based on the first chapter of my thesis "Imperatives and Deontic Logic", Universität Leipzig, Inst. f. Philosophie, Leipzig, 2008. I thank L. van der Torre and L. Goble for their criticism, and X. Parent for his patience and practical advice.

[Alchourrón, 1969] C. E. Alchourrón. Logic of norms and logic of normative propositions. *Logique & Analyse*, 12:242–268, 1969.
[Alchourrón, 1972] C. E. Alchourrón. The intuitive background of normative legal discourse and its formalization. *Journal of Philosophical Logic*, 1:447–463, 1972.
[Anscombe, 1957] G. E. M. Anscombe. *Intention*. Blackwell, Oxford, 1957.
[Bentham, 1970] J. Bentham. *Of Laws in General*. Athlone Press, London, 1970. Edited by H. L. A. Hart. First appeared 1782.
[Boole, 1847] G. Boole. *The Mathematical Analysis of Logic*. Macmillan, Barclay, and Machmillan, Cambridge, 1847.
[Bulygin, 1999] E. Bulygin. Existence of norms. In Meggle [1999], pages 237–244.
[Castañeda, 1971] Hector-Neri Castañeda. There are command sh-inferences. *Analysis*, 32:13–19, 1971.
[Chellas, 1969] B. F. Chellas. *The Logical Form of Imperatives*. Perry Lane Press, Stanford, 1969.
[Dubislav, 1931] W. Dubislav. *Die Definition*. Meiner, Leipzig, 3rd edition, 1931.
[Dubislav, 1938] W. Dubislav. Zur Unbegründbarkeit der Forderungssätze. *Theoria*, 3:330–342, 1938.
[Engliš, 1964] K. Engliš. Die Norm ist kein Urteil. *Archiv für Rechts- und Sozialphilosophie*, 50:305–316, 1964.
[Føllesdal and Hilpinen, 1971] D. Føllesdal and R. Hilpinen. Deontic logic: An introduction. In Hilpinen [1971], pages 1–35.
[Foot, 1983] P. Foot. Moral realism and moral dilemma. *Journal of Philosophy*, 80:379–389, 1983.
[Fox, 2008] C. Fox. Imperatives: a logic of satisfaction. Draft paper, to be revised for publication, 2008.
[Frey, 1957] G. Frey. Idee einer Wissenschaftslogik. Grundzüge einer Logik imperativer Sätze. *Philosophia Naturalis*, 4:434–491, 1957.
[Geach, 1958] P. T. Geach. Imperative and deontic logic. *Analysis*, 18:49–56, 1958.
[Geach, 1963] P. T. Geach. Imperative inference. *Analysis*, 23:37–42, 1963.
[Gentzen, 1934] G. Gentzen. Untersuchungen über das logische Schließen. *Mathematische Zeitschrift*, 39:176–210; 405–431, 1934.
[Goble, 1990] L. Goble. A logic of good, should, and would. *Journal of Philosophical Logic*, 19:169–199, 1990.
[Gombay, 1967] A. Gombay. What is imperative inference? *Analysis*, 27:145–152, 1967.
[Hamblin, 1987] C. L. Hamblin. *Imperatives*. Blackwell, Oxford, 1987.
[Hansen, 2001] J. Hansen. Sets, sentences, and some logics about imperatives. *Fundamenta Informaticae*, 48:205–226, 2001.
[Hansen, 2004] J. Hansen. Problems and results for logics about imperatives. *Journal of Applied Logic*, 2:39–61, 2004.
[Hansen, 2005] J. Hansen. Conflicting imperatives and dyadic deontic logic. *Journal of Applied Logic*, 3:484–511, 2005.
[Hansen, 2006] J. Hansen. Deontic logics for prioritized imperatives. *Artificial Intelligence & Law*, 14:1–34, 2006.
[Hansen, 2008] J. Hansen. Prioritized conditional imperatives: Problems and a new proposal. *Autonomous Agents and Multi-Agent Systems*, 17:11–35, 2008.
[Hanson, 1966] W. H. Hanson. A logic of commands. *Logique & Analyse*, 9:329–343, 1966.
[Hare, 1949] R. M. Hare. Imperative sentences. *Mind*, 58:21–39, 1949.
[Hare, 1952] R. M. Hare. *The Language of Morals*. Oxford University Press, Oxford, 1952.
[Hare, 1967] R. M. Hare. Some alleged differences between imperatives and indicatives. *Mind*, 76:309–326, 1967.
[Harrison, 1991] J. Harrison. Deontic logic and imperative logic. In P. T. Geach, editor, *Logic and Ethics*, pages 79–129. Kluwer, Dordrecht, 1991.

[Hempel, 1941] C. G. Hempel. Review of Alf Ross: 'Imperatives and Logic'. *Journal of Symbolic Logic*, 6:105–106, 1941.
[Hilpinen, 1971] R. Hilpinen, editor. *Deontic Logic: Introductory and Systematic Readings*. Reidel, Dordrecht, 1971.
[Hilpinen, 1981] R. Hilpinen, editor. *New Studies in Deontic Logic*. Reidel, Dordrecht, 1981.
[Hilpinen, 2006] R. Hilpinen. Norms, normative utterances, and normative propositions. *Análisis Filosófico*, 26:229–241, 2006.
[Hintikka, 1977] J. Hintikka. The Ross paradox as evidence for the reality of semantical games. In E. Saarinen, editor, *Game-Theoretical Semantics*, pages 329–345. Reidel, Dordrecht, 1977.
[Hodges, 1977] W. Hodges. *Logic*. Penguin, Harmondsworth, 1977.
[Hofstadter and McKinsey, 1938] A. Hofstadter and J. C. C. McKinsey. On the logic of imperatives. *Philosophy of Science*, 6:446–457, 1938.
[Holländer, 1993] P. Holländer. *Rechtsnorm, Logik und Wahrheitswerte. Versuch einer kritischen Lösung des Jörgensenschen Dilemmas*. Nomos, Baden-Baden, 1993.
[Horty, 2000] J. F. Horty. *Agency and Deontic Logic*. Oxford University Press, Oxford, 2000.
[Hume, 1888] D. Hume. *A Treatise of Human Nature*. Oxford University Press, Oxford, 1888. Edited by L. A. Selby-Bigge. First appeared 1739.
[Jørgensen, 1938] J. Jørgensen. Imperatives and logic. *Erkenntnis*, 7:288–296, 1938.
[Kalinowski, 1967] G. Kalinowski. *Le Problème de la Vérité en Morale et en Droit*. Emmanuel Vitte, Lyon, 1967.
[Kalinowski, 1973] G. Kalinowski. *Einführung in die Normenlogik*. Athenäum, Frankfurt, 1973.
[Kalinowski, 1974] G. Kalinowski. Über die deontischen Funktoren. In H. Lenk, editor, *Normenlogik. Grundprobleme der deontischen Logik*, pages 39–63. Dokumentation, Pullach, 1974.
[Kalinowski, 1977] G. Kalinowski. Über die Bedeutung der Deontik für Ethik und Rechtsphilosophie. In A. G. Conte, R. Hilpinen, and G. H. von Wright, editors, *Deontische Logik und Semantik*, pages 101–129. Athenaion, Wiesbaden, 1977.
[Kamp, 1973] H. Kamp. Free choice permission. *Proceedings of the Aristotelian Society*, 74:57–74, 1973.
[Kamp, 1978] H. Kamp. Semantics versus pragmatics. In F. Guenthner and S. J. Schmidt, editors, *Formal Semantics and Pragmatics for Natural Languages*, pages 255–287. Reidel, Dordrecht, 1978.
[Kanger, 1957] S. Kanger. New foundations for ethical theory: Part 1. duplic., 42 p., 1957. Reprinted in [Hilpinen, 1971, pp. 36-58].
[Keene, 1966] G. B. Keene. Can commands have logical consequences? *American Philosophical Quarterly*, 3:57–63, 1966.
[Kelsen, 1979] H. Kelsen. *Allgemeine Theorie der Normen*. Manz, Wien, 1979.
[Kenny, 1966] A. J. Kenny. Practical inference. *Analysis*, 26:65–75, 1966.
[Keuth, 1974] H. Keuth. Deontische Logik und Logik der Normen. In H. Lenk, editor, *Normenlogik. Grundprobleme der deontischen Logik*, pages 64–86. Dokumentation, Pullach, 1974.
[Ledig, 1928] G. Ledig. Zur Klärung einiger Grundbegriffe. Imperativ, Rat, Bitte, Beschluß, Versprechen. *Revue Internationale de la Théorie du Droit*, 3:260–270, 1928.
[Ledig, 1931] G. Ledig. Zur Logik des Sollens. *Der Gerichtssaal*, 100:368–385, 1931.
[Lemmon, 1965] E. J. Lemmon. Deontic logic and the logic of imperatives. *Logique & Analyse*, 8:39–71, 1965.
[Lemmon, 1987] E. J. Lemmon. *Beginning Logic*. Chapman and Hall, London, 2nd edition, 1987.
[Leonard, 1959] H. S. Leonard. Interrogatives, imperatives, truth, falsity and lies. *Philosophy of Science*, 26:172–186, 1959.

[MacKay, 1969] A. F. MacKay. Inferential validity and imperative inference rules. *Analysis*, 29:145–156, 1969.
[Mally, 1926] E. Mally. *Grundgesetze des Sollens. Elemente der Logik des Willens.* Leuschner and Lubensky, Graz, 1926.
[Mates, 1972] B. Mates. *Elementary Logic.* Oxford University Press, Oxford, 2nd edition, 1972.
[Max and Raatzsch, 1998] I. Max and R. Raatzsch, editors. *Peter Philipp - Logisch-philosophische Untersuchungen.* de Gruyter, Berlin, 1998.
[Meggle, 1999] G. Meggle, editor. *Actions, Norms, Values. Discussions with Georg Henrik von Wright.* de Gruyter, Berlin, 1999.
[Menger, 1939] K. Menger. A logic of the doubtful. On optative and imperative logic. *Reports of a Mathematical Colloquium*, 2:53–64, 1939. Reprinted in [Menger, 1979] 91–102.
[Menger, 1979] K. Menger. *Selected Papers in Logic and Foundations, Didactics, Economics.* Reidel, Dordrecht, 1979.
[Moritz, 1954] M. Moritz. Der praktische Syllogismus und das juridische Denken. *Theoria*, 20:78–127, 1954.
[Moutafakis, 1975] N. J. Moutafakis. *Imperatives and their Logic.* Sterling, New Delhi, 1975.
[Niiniluoto, 1985] I. Niiniluoto. Truth and legal norms. *Archiv für Rechts- und Sozialphilosophie Beiheft*, 25:168–190, 1985.
[Opałek and Woleński, 1987] K. Opałek and J. Woleński. Is, ought, and logic. *Archiv für Rechts- und Sozialphilosophie*, 73:373–385, 1987.
[Opałek and Woleński, 1991] K. Opałek and J. Woleński. Normative systems, permission and deontic logic. *Ratio Juris*, 4:334–348, 1991.
[Opałek, 1970] K. Opałek. On the logical-semantic structure of directives. *Logique & Analyse*, 13:169–196, 1970.
[Opałek, 1986] K. Opałek. *Theorie der Direktiven und Normen.* Springer, Wien, 1986.
[Peczenik, 1967] A. Peczenik. Doctrinal study of law and science. *Österreichische Zeitschrift für öffentliches Recht*, 17:128–141, 1967.
[Peczenik, 1968] A. Peczenik. Norms and reality. *Theoria*, 34:117–133, 1968.
[Philipp, 1989] P. Philipp. Logik deskriptiver normativer Begriffe. In Max and Raatzsch [1998], pages 241–290. First published as reports of the Karl-Marx-Universität Leipzig, *Untersuchungen zur Logik und zur Methodologie* **6**:65-88, 1989, and **7**:51–82, 1990.
[Philipp, 1991] P. Philipp. Normative logic without norms. In Max and Raatzsch [1998], pages 291–301. Edited version of a presentation to the Meeting of the Central Division of the American Philosophical Association, Chicago, 1991.
[Philipps, 1971] L. Philipps. Braucht die Rechtswissenschaft eine deontische Logik? In Jahr G. and W. Maihofer, editors, *Rechtstheorie. Beiträge zur Grundlagendiskussion*, pages 352–368. Klostermann, Frankfurt, 1971.
[Poincaré, 1913] H. Poincaré. *Dernières Pensées.* Ernest Flammarion, Paris, 1913.
[Prior, 1955] A. N. Prior. *Formal Logic.* Clarendon Press, Oxford, 1955.
[Reichenbach, 1947] H. Reichenbach. *Elements of Symbolic Logic.* Collier-Macmillan, New York, 1947.
[Rescher, 1966] N. Rescher. *The Logic of Commands.* Routledge and Kegan Paul, London, 1966.
[Rescher, 1968] N. Rescher. Assertion logic. In N. Rescher, editor, *Topics in Philosophical Logic*, chapter XIV. Reidel, Dordrecht, 1968.
[Rödig, 1971] J. Rödig. Kritik des normlogischen Schließens. *Theory and Decision*, 2:79–93, 1971.
[Rödig, 1972] J. Rödig. Über die Notwendigkeit einer besonderen Logik der Normen. *Jahrbuch für Rechtssoziologie und Rechtstheorie*, 2:163–185, 1972.
[Ross, 1941] A. Ross. Imperatives and logic. *Theoria*, 7:53–71, 1941. Reprinted with only editorial changes in *Philosophy of Science*, **11**, 1944, 30–46.

[Schwager, 2006] M. Schwager. *Interpreting Imperatives*. PhD thesis, Johann-Wolfgang-von-Goethe Universität, Institut für Kognitive Linguistik, Frankfurt am Main, 2006.
[Sellars, 1956] W. Sellars. Imperatives, intentions and the logic of "ought". *Methodos*, 8:227–268, 1956.
[Sigwart, 1895] C. Sigwart. *Logic*, volume I. The Judgment, Concept, and Inference. Swan Sonnenschein & Co., London, 2nd edition, 1895.
[Simon, 1965] H. A. Simon. The logic of rational decision. *The British Journal for the Philosophy of Science*, 16:169–186, 1965.
[Sosa, 1966a] E. Sosa. The logic of imperatives. *Theoria*, 32:224–235, 1966.
[Sosa, 1966b] E. Sosa. On practical inference and the logic of imperatives. *Theoria*, 32:211–223, 1966.
[Sosa, 1967] E. Sosa. The semantics of imperatives. *American Philosophical Quarterly*, 4:57–64, 1967.
[Stenius, 1963] E. Stenius. The principles of a logic of normative systems. *Acta Philosophica Fennica*, 16:247–260, 1963.
[Stenius, 1982] E. Stenius. Ross' paradox and well-formed codices. *Theoria*, 48:49–77, 1982.
[Stranzinger, 1978] R. Stranzinger. Ein paradoxienfreies deontisches System. In I. Tammelo and H. Schreiner, editors, *Strukturierungen und Entscheidungen im Rechtsdenken*, pages 183–193. Springer, Wien, 1978.
[Tarski, 1930] A. Tarski. Über einige fundamentale Begriffe der Metamathematik. *Comptes Rendus des Séances de la Sociéte des Sciences et des Lettres de Varsovie*, 23:22–29, 1930. Published in English under the title "On Some Fundamental Concepts of Metamathematics" in [Tarski, 1956] pp. 30–37.
[Tarski, 1935] A. Tarski. Der Wahrheitsbegriff in den formalisierten Sprachen. *Studia Philosophica*, I:261–405, 1935.
[Tarski, 1956] A. Tarski. *Logic, Semantics, Metamathematics*. Oxford University Press, Oxford, 1956.
[van Fraassen, 1973] B. van Fraassen. Values and the heart's command. *Journal of Philosophy*, 70:5–19, 1973.
[von Kutschera, 1973] F. von Kutschera. *Einführung in die Logik der Normen, Werte und Entscheidungen*. Karl Alber, Freiburg-München, 1973.
[von Wright, 1951] G. H. von Wright. Deontic logic. *Mind*, 60:1–15, 1951.
[von Wright, 1963] G. H. von Wright. *Norm and Action*. Routledge & Kegan Paul, London, 1963.
[von Wright, 1968] G. H. von Wright. *An Essay in Deontic Logic and the General Theory of Action*. North Holland, Amsterdam, 1968.
[von Wright, 1991] G. H. von Wright. Is there a logic of norms? *Ratio Juris*, 4:265–283, 1991.
[von Wright, 1993a] G. H. von Wright. Gibt es eine Logik der Normen? In A. Aarnio and S. Paulson, editors, *Rechtsnorm und Rechtswirklichkeit. Festschrift für Werner Krawietz*, pages 101–123. Duncker & Humblot, Berlin, 1993.
[von Wright, 1993b] G. H. von Wright. A pilgrim's progress. In G. H. von Wright, editor, *The Tree of Knowledge and Other Essays*, pages 103–113. Brill, Leiden, 1993.
[von Wright, 1999] G. H. von Wright. Ought to be – ought to do. In Meggle [1999], pages 3–9.
[Vranas, 2009] P. Vranas. In defense of imperative inference. *Journal of Philosophical Logic*, 39:59–71, 2009.
[Wagner and Haag, 1970] H. Wagner and K. Haag. *Die moderne Logik in der Rechtswissenschaft*. Gehlen, Bad Homburg, 1970.
[Walter, 1996] R. Walter. Jörgensen's dilemma and how to face it. *Ratio Juris*, 9:168–171, 1996.
[Walter, 1997] R. Walter. Some thoughts on Peczenik's replies to 'Jörgensen's Dilemma and How to Face It' (with two letters by A. Peczenik). *Ratio Juris*, 10:392–396, 1997.

[Wedeking, 1970] G. A. Wedeking. Are there command arguments. *Analysis*, 30:161–166, 1970.
[Weinberger, 1958] O. Weinberger. *Die Sollsatzproblematik in der modernen Logik: Können Sollsätze (Imperative) als wahr bezeichnet werden?* Nakladatelství Ceskoslovenské Akademie Ved, Prague, 1958. page numbers refer to the reprint in [Weinberger, 1974b] pp. 59–186.
[Weinberger, 1972a] O. Weinberger. Bemerkungen zur Grundlegung der Theorie des juristischen Denkens. *Jahrbuch für Rechtssoziologie und Rechtstheorie*, 2:134–161, 1972.
[Weinberger, 1972b] O. Weinberger. Der Begriff der Nicht-Erfüllung und die Normenlogik. *Ratio*, 14:15–32, 1972.
[Weinberger, 1974a] O. Weinberger. Ideen zur logischen Normensemantik. In R. Haller, editor, *Jenseits von Sein und Nichtsein. Beiträge zur Meinong-Forschung*, pages 295–311, Graz, 1974. Akademische Druck- und Verlagsanstalt. Reprinted in [Weinberger, 1974b] pp. 259–277.
[Weinberger, 1974b] O. Weinberger. *Studien zur Normlogik und Rechtsinformatik.* Schweitzer, Berlin, 1974.
[Weinberger, 1975] O. Weinberger. Ex falso quodlibet in der deskriptiven und präskriptiven Sprache. *Rechtstheorie*, 6:17–32, 1975.
[Weinberger, 1986] O. Weinberger. Der normenlogische Skeptizismus. *Rechtstheorie*, 17:13–81, 1986. Reprinted in [Weinberger, 1992] pp. 431–499.
[Weinberger, 1989] O. Weinberger. *Rechtslogik.* Duncker&Humblot, Berlin, 2nd edition, 1989.
[Weinberger, 1991] O. Weinberger. The logic of norms founded on descriptive language. *Ratio Juris*, 4:284–307, 1991.
[Weinberger, 1992] O. Weinberger. *Moral und Vernunft.* Böhlau, Wien, 1992.
[Weinberger, 1996] O. Weinberger. *Alternative Handlungstheorie.* Böhlau, Wien, 1996.
[Weinberger, 1999] O. Weinberger. Logical analysis in the realm of law. In Meggle [1999], pages 291–304.
[Weingartner and Schurz, 1986] P. Weingartner and G. Schurz. Paradoxes solved by simple relevance criteria. *Logique & Analyse*, 29:3–40, 1986.
[Weingartner, 2003] P. Weingartner. Reasons from science for limiting classical logic. In P. Weingartner, editor, *Alternative Logics. Do Science need them?*, pages 233–248. Springer, Berlin, 2003.
[Williams, 1963] B. A. O. Williams. Imperative inference. *Analysis*, 23:30–36, 1963.
[Wittgenstein, 1953] L. Wittgenstein. *Philosophical Investigations.* Blackwell, Oxford, 1953.
[Yoshino, 1978] H. Yoshino. Über die Notwendigkeit einer besonderen Normenlogik als Methode der juristischen Logik. In U. Klug, editor, *Gesetzgebungstheorie, Juristische Logik, Zivil- und Prozessrecht. Gedächtnisschrift für Jürgen Rödig*, pages 141–161, Berlin, 1978. Springer.
[Ziemba, 1976] Z. Ziemba. Deontic logic, 1976. Appendix in [Ziembiński, 1976], pp. 360–430.
[Ziembiński, 1976] Z. Ziembiński. *Practical Logic.* Reidel, Dordrecht, 1976.

Jörg Hansen
Beethovenstr. 4
D-99817 Eisenach
Germany
Email: jhansen@uni-leipzig.de

PART II

CONCEPTS AND PROBLEMS

3
The Varieties of Permission
SVEN OVE HANSSON

ABSTRACT. This is an overview of the major issues in deontic logic that are specific for permission, such as: The distinction between a permission to either do or not do something (bilateral permission) and a permission to do something that does not include a permission not to do it (unilateral permission). The distinction between permissions that are explicitly stated, permissions that follow from obligations, and permissions that are inferrable from the absence of a prohibition. The interdefinability of permissions, obligations, and prohibitions. Permissions referring to disjunctions, in particular "free choice permissions" and their interpretation. Prima facie permission, conflicts between obligations and permissions, and permissions that override or are overriden by other norms. Permissions of different strengths. Different types of conditional permissions, such as counterfactual and rule-stating conditional permissions. Permissions that are conditional on someone's permitting action (grantable permission) or on the absence of someone's obligating action (revocable permission). Changes in permissions. – In conclusion it is proposed that more attention should be paid to issues of permission in deontic logic.

1	Introduction	196
2	Unilateral or bilateral permission	199
3	Explicit, implied, and tacit permission	201
4	Interdefinability with other deontic concepts	204
5	Free choice permission	206
	5.1 Definition of free choice permission	207
	5.2 Implausible derivations	207
	5.3 Analogous constructions	208
	5.4 A variety of solutions	209
	5.5 Mistranslation of "or"	210
	5.6 Conversational implicature	212
	5.7 A hidden operator	214
	5.8 Free choice operators	214
	5.9 The impossibility of single-sentence representation	218
6	Prima facie permissions and the strength of permissions	218

	6.1	Conflicts amongst prima facie permissions and obligations 219
	6.2	Overriding power 220
7	Conditional permissions 221	
	7.1	Two types of conditional norms 222
	7.2	Can prima facie norms be dispensed with? 226
8	Grantable and revocable permissions 227	
9	Changes in permission 229	
	9.1	Lewis and the problem of permission 229
	9.2	The belief change tradition 231
	9.3	Two types of normative change 232
	9.4	Bases or closed sets 232
	9.5	Change and retrieval 234
10	Conclusion 237	

1 Introduction

Deontic logic is usually defined as the logic of norms; this is also the interpretation that the etymology suggests. Permissions are arguably not norms, literally speaking. Instead permission statements indicate the absence of norms. Thus, many legislations permit marriage between cousins, but this does not make cousin marriage a norm in the sense of something recommended or commanded. Nevertheless, the logic of permissions is traditionally treated as part of deontic logic, and some of the most discussed problems in the discipline refer primarily to permissions.

Just like obligations and prohibitions, permissions are usually conceived as emanating from some source. The standard assumption is that there is some source that gives rise to deontic statements of all three types. Permissions and other deontic statements are classified as legal or moral according to whether the source is a law-maker or a codifier of ethics. It has usually been taken for granted that deontic logic should be essentially the same independently of the source. In particular it is assumed that legal and moral norms obey the same logical laws. However, this should not be taken for granted. Legal permissions and obligations usually emanate from a finite but rather extensive code, a set of rules from which all valid deontic statements can be inferred. A deontological ethicist will probably tend to ascribe a similar structure to moral permissions and obligations. Such a structure may be less adequate for other ethical theories, such as utilitarianism. Furthermore, moral theories with different views on the existence of moral conflicts can be expected to give rise to different deontic structures. Unfortunately, the relationship between moral theory and deontic principles has not been much investigated.

Another central distinction in deontic logic is that between statements that create a norm and statements that report its existence. A sentence such as "You are allowed to wear a niqab in this mall" may either be a statement that issues the permission in question, or a report that such a permission holds. Georg Henrik von Wright may have been right in saying that the Swedish philosopher Ingemar Hedenius was the first to point out this ambiguity in natural language. ([von Wright, 1999, p. 32], [Hedenius, 1941]) The distinction is usually referred to as one between "prescriptive" and "descriptive" normative sentences. Although this terminology is adequate for obligations and prohibitions, it is misleading for permissions since permissions are not prescriptions. Therefore, it would be better to distinguish between "declarative" and "descriptive" statements. A declarative norm statement can create an obligation ("You must cover your head with a kippah when you visit the synagogue"), a prohibition ("You are not allowed to wear any headgear in the cathedral") or a permission ("As a bishop you are are now allowed to wear a mitre in Church").

One of the most commonly discussed distinctions in deontic logic is that between Tun-sollen (ought to do) and Sein-sollen (ought to be, ideal ought). The first is exemplified by a sentence such as "You ought to visit your mother more often" and the latter by a sentence such as "There ought to be no earthquakes". The difference is fundamental. A Tun-sollen recommends or commands someone what to do. A Sein-sollen tells us what is desirable. A plausible analysis of this difference is that only a Tun-sollen is truly normative; a Sein-sollen expresses a value statement rather than a norm statement. This analysis is corroborated by the fact that most prescriptive predicates other than "ought" cannot be used to express a Sein-sollen, and also by the fact that neither prohibitive nor permissive statements can be used in a way that corresponds to Sein-sollen. We could not express the same idea by saying for instance: "It is obligatory that there be no earthquakes", "Earthquakes are forbidden", or "No earthquake is permitted".[1] For our present purposes it is sufficient to note that permission always seems to refer to actions, not to states that cannot be influenced by actions. This also means that the argument of permission predicates (a in Pa, "a is permitted") represents an action or at least some activity or behaviour. Traditionally, no explicit representation of action is used in deontic logic, but it should nevertheless be kept in mind that the objects of permissions, prohibitions, and (mostly) obligations are statements that refer to human action.

---

[1][Wurmbrand, 1999] offers examples such as "There may be singing but no dancing on my premises" and "There can be a party as long as it's not too loud". However, these can be regarded as examples of Tun-sollen since they can be read as abbreviated statements of permissions and prescriptions referring to actions. It would be strange to say for instance: "There may be raining but no snowing on my premises".

In what follows we will be concerned with the logical properties of a predicate P representing permission that operates on arguments that are sentences representing action or behaviour. Hence, the sentence "You may take one of these two books" is represented by a sentence $P(a \vee b)$ where a represents that you take one of the books and b that you take the other. Natural language has many different words expressing permissions, such as "may", "allowed", "permitted", and "can". ("You can use my desk when I am abroad".)

This restriction to "Tun-dürfen" (and Tun-sollen) has important implications for the use of iterated modalities in deontic logic. There is clearly a sense in which permissions may refer to other permissions. You can for instance be permitted (authorized) to allow someone else something. However, with the restriction just mentioned, this cannot be expressed with an iterated modality such as PPa. The reason for this is that Pa is not an action or behaviour, and therefore it cannot be the argument of a permission operator. Instead, representations are needed that explicitly mention the action of making a permitted. If D_iPa denotes i's act of permitting a, then a permission to perform such an act can be denoted PD_iPa. Acts of permitting will be discussed in section 8. (However, many deontic logicians have freely used iterated permissions and obligations. See [Tranøy, 1970].)

It is almost universally assumed in deontic logic that normative predicates allow for the substitution of logical equivalents. Hence, if a and a' are logically equivalent, then so are Pa and Pa' (and similarlty Oa and Oa'). This is an immensely simplifying assumption, and it is indeed difficult to develop a non-trivial deontic logic without it. It will therefore be assumed to hold although it sometimes gives rise to difficulties. (See [Hansson, 1991].)

It has been claimed that deontic logic is an oxymoron since it involves the use of (truth-valued) logic to cover subject-matter that does not refer to truth or falsehood. (Cf. [Makinson, 1999].) According to moral objectivists, statements about what is morally permitted or obligatory can be objectively true or false in the same way as statements about ordinary matters of fact. Does deontic logic require that we adhere to moral objectivism? Fortunately it does not, provided that we use the logical apparatus as a *model* of normative concepts. A model need not share all the properties of that which is modelled. An economist can use a real-valued model of monetary transactions, in which monetary value is infinitely divisible, although real monetary transfers come in discrete units. Similarly, we can use a model expressed in truth-valued logic for subject matter that does not have truth values.

For a practical example, consider a simple permissive statement such as "You may borrow my car tomorrow". In deontic logic such a statement is rendered by a sentence Pa, where P means "permitted" and a represents the permitted behaviour, in this case "You borrow my car tomorrow". In standard interpretations of sentential logic, sentences are classified as true or false. Hence it may seem natural to interpret Pa as truth-valued. However, this is not the only interpretation that is possible. We can replace the truth/false distinction in our interpretation by a distinction between statements that are sanctioned or not sanctioned by a particular moral code. Alternatively we can interpret Pa as saying that (it is true that) the moral code contains a norm to the effect that a is obligatory, instead of interpreting it as just saying that a is obligatory.

Some of the problems concerning the permission predicate P are common to the deontic notions, and apply also to prescriptive and prohibitive predicates. But there are also problems that are specific for permission. These problems will be at the focus of this chapter. Section 2 is devoted to the distinction between a permission to either do or not do something (bilateral permission) and a permission to do something that does not include a permission not to do it (unilateral permission). Section 3 has its focus on the distinction between permissions that are explicitly stated and permissions that are implicit in the system of norms. This provides a background for the discussion in section 4 of the interdefinability of permission, obligation, and prohibition. In section 5 we turn to the permission of disjunctions and to "free choice permission" that has been the subject of intense discussions. Section 6 is devoted to prima facie permission, to permissions that override or are overridden by other norms and to the overriding power of permissions. In section 7 we turn to conditional permissions and in section 8 to a variant of these, namely permissions that are conditional on someone's permitting action (grantable permission) or on the absence of someone's obligating action (revocable permission). Section 9 is devoted to changes in permissions. Some concluding remarks are offered in section 10.

2 Unilateral or bilateral permission

In ordinary language, "when saying that an action is permitted we mean that one is at liberty to perform it, that one may either perform the action or refrain from performing it". [Raz, 1975, p. 161] In formal philosophy, however, "being permitted to perform an action is compatible with having to perform it". (ibid) The former notion may be called *bilateral permission* and the latter *unilateral permission*. Unilateral permission is usually taken to be implied by obligation, i.e. if you are under an obligation to perform an action, then you are unilaterally permitted to perform it. Bilateral permission, however, is incompatible with obligation; if you have a bilateral permission concerning some action, then you cannot be under an obligation to perform it (or not to perform it).

Everyday usage of permissive terms such as "you may" or "you are permitted to" tends to conform with bilateral permission. [von Wright, 1951, p. 4] It would be strange to say of a convicted criminal that he "is allowed to spend the next ten years of his life in a penitentiary" when he is in fact under an (enforced) obligation to do so. But in spite of this, the unilateral concept dominates in formal philosophy. A major reason for this is that this practice is convenient in terms of definitional structure. If we take unilateral permission as primitive, then we can easily introduce bilateral permission as a defined concept. Let P denote unilateral permission and $\overline{P}$ bilateral permission. We can define the latter in terms of the former as follows:

$$\overline{P}a \leftrightarrow Pa \ \& \ P\neg a$$

However, we do not in general have any means to define P in terms of $\overline{P}$. (To see that, let P' be such that $P'a \leftrightarrow P\neg a$ for all a. Then $\overline{P}a \leftrightarrow Pa \ \& \ P\neg a$ if and only if $\overline{P}a \leftrightarrow P'a \ \& \ P'\neg a$, and hence we cannot distinguish between P and P' based on $\overline{P}$ alone.) Admittedly, the definition $Pa \leftrightarrow \overline{P}a \lor Oa$ makes sense, but it requires the introduction of a predicate O of obligation that is not definable in terms of $\overline{P}$. In summary, the usual convention that takes P as primitive and $\overline{P}$ as defined has important practical advantages.

The distinction between unilateral and bilateral permission has been known for long by philosophers. Quite a few terms have been used to denote what we have called here bilateral permission ($\overline{P}$). Von Wright and many others use the term "indifference" for bilateral permission. [von Wright, 1951, p. 4] However, this is a confusing terminology. In preference logic, two entities are said to be indifferent if they cannot be distinguished in value terms. It may well be the case that $\overline{P}a$, i.e. $Pa \ \& \ P\neg a$, although a is valued higher than $\neg a$. I am permitted both to give and not to give a monthly contribution to some charity, but arguably the first alternative is better so that indifference cannot be said to hold. (Note also that Knapp uses the term "indifferent" in deontic logic in a quite different way, namely to denote "neither forbidden nor permitted". [Knapp, 1981, pp. 398-99] The term "explicitly permitted" was used by Myers [1962]. It is a misleading terminology since bilateral permission need not be the result of any explicitly declared permission. The term "at liberty" used by Joseph Raz is more appropriate; if you can choose between giving and not giving a contribution to some charity then you may be said to be at liberty to do so. [Raz, 1975] Another appropriate term is "optional". If you may perform an action, and you may also refrain from performing it, then that action can be described as optional. "Of optional acts, some are morally good, others indifferent, and still others morally bad." [Forrester, 1975, p. 225]

A bilateral permission can always be expressed in two alternative ways. Since $\overline{P}a$ and $\overline{P}\neg a$ are equivalent, we can choose between saying that a is bilaterally permitted and that $\neg a$ is so. Consider the following examples:

- "The dissident is now allowed to live abroad."
- "The previously expatriated criminal is now allowed to live in his native country."

As was noted by K.E. Tranøy, "[t]he point about such cases probably is that, for reasons of an axiological or valuational nature, we emphasize and single out one of each such pair of permitted contradictories as being more desirable (or interesting or valuable or important) than the other. Declarations of human rights single out for explicit mention the rights to work and to vote rather than the rights or permissions to abstain from working and voting..." [Tranøy, 1970, p. 227] Alternatively, permission can be seen as fundamentally unilateral, acquiring bilateral meaning only through contextual implication in the Gricean sense. [Garcia, 1989]

3 Explicit, implied, and tacit permission

We tend to take for granted that what is not forbidden is permitted. But in addition to such, tacit permissions, there are also permissions that are explicitly granted. This distinction is particularly evident in legal contexts, and it has been known for long by legal scholars. It was for instance very clearly made by Gottfried Achenwall (1719-1772). [Tierney, 2007, pp. 423 and 427] Jeremy Bentham (1748-1832) was also aware of the difference between permissions inferred from "silence" and explicitly mandated permissions. [Mullock, 1979]

As was noted by von Wright, a permission can be explicitly stated in an indirect way. [von Wright, 1963, p. 86] Whatever is morally required is also (unilaterally) permitted. Strictly speaking, therefore, we have three types of permissions:

- Explicit permission: The owner of this property told me that I may use the road. Therefore, I am permitted to use the road.
- Implied permission: The judge has called me to testify in court this afternoon. Therefore I am allowed to enter this courthouse now.
- Tacit permission: There is no prohibition against diving in the harbour. Therefore I am permitted to do so.

This division into three categories was proposed by [Knapp, 1981], who used the German terms "ausdrücklich" for explicit, "implizit" for implied, and "stillschweigend" for tacit.

Both tacit and implied permissions are inferred from explicit permissions. However, an implied permission follows from the presence of some particular norm (and is monotonically inferred) whereas a tacit permission follows from the absence of any norm to the contrary (and is non-monotonically inferred).

Most writers on the subject have used a division into two categories, one of which consists of the tacit permissions and the other of the explicit and the implied ones. Several terminologies have been employed to describe this dichotomy. Myers used the terms "permitted by default" and "implicitly permitted" for that which is permitted just because no rule covers it. [Myers, 1962, p. 485] Makinson and van der Torre used the term "negative permission" for what is called here tacit permission, and "positive permission" for explicit and implied permission. [Makinson and van der Torre, 2003, pp. 391-92] The most common terminology, however, is that which distinguishes between "weak" (tacit) and "strong" (explicit or implied) permission. This terminology was justified as follows by von Wright:

> "An act will be said to be permitted in the weak sense if it is not forbidden; and it will be said to be permitted in the strong sense if it is not forbidden, but subject to norm. Acts which are strongly permitted are thus weakly permitted but not necessarily vice versa." [von Wright, 1963, p. 86]

On a later occasion he expressed himself somewhat more cautiously about the nature of the "strong" permissions:

> "I think we are well advised to distinguish between things being permitted in the weak sense of simply not being forbidden and things being permitted in some stronger sense. Exactly in what this stronger sense 'consists' may be difficult to tell. That which is in the strong sense permitted is, somehow, expressly permitted, subject to norm and not just void of deontic status altogether." [von Wright, 1981]

Tacit permissions are based on the following inference pattern:

> a is not forbidden
> *Therefore*: a is permitted

This corresponds closely to the postulate $\neg Fa \to Pa$ or equivalently $Fa \vee Pa$. This can be described as a requirement that the norm system is gapless, where a gap is something that is neither forbidden nor permitted. [von Wright, 1999, p. 32] (The term "deontic neutrality" has been used for such gaps e.g. in [Moore, 1973, p. 330]. The term "complete" has been used for systems that contain no such gap, e.g. in [Stone, 1959].) In legal contexts the absence of gaps is related to the principle "Nullum crimen sine lege" according to which what is not dealt with by the law is not a crime.

Given that the norm system is gapless, it may well be asked whether there is need for any other permissions than the tacit ones. Alf Ross claimed that all legal norms of permission have (only) the function of expressing the absence of obligation. In his view, obligation "is the fundamental directive category in which any norm may be expressed". [Ross, 1968, p. 117] Ronald Moore introduced the term "the reflex thesis" to denote the view that all legal norms of permission are merely assertions of the absence of norms of prohibition. [Moore, 1973, p. 327] (The view expressed by the reflex thesis has also been called "imperativism" since it rejects norms that cannot be expressed as imperatives; obligations differ from permissions in being expressible in terms of imperatives.) Ross defended the reflex thesis by referring to actual legislative experience:

> "I have never heard of any laws being passed with the purpose of declaring a new form of behaviour (e.g., listening to the wireless) permitted. If a legislator sees no reason to interfere by issuing an obligating prescription (a command or a prohibition) he simply keeps silent." [Ross, 1968, p. 122]

This is probably true in most cases, but it is by no means implausible that a legislator may decide to introduce a permission in a previously unregulated area in order to establish a principle (such as freedom of expression in new media). Ronald Moore argued that the reflex thesis is implausible since it can only be defended "at the expense of sacrificing powers which we do and should give to norm-authorities". [Moore, 1973, p. 331]

Opalek and Wolénski argue in favour of the reflex thesis. [Opalek and Wolénski, 1991, pp. 341-2] They claim that a normative system divides the universe of actions into the categories obligatory, prohibited, and indifferent, and that this is done by obligations and prohibitions, whereas permissions have no role in this. Therefore, they say, only obligations and prohibitions are norms. (For another modern defence of a viewpoint closely related to Ross's, see [Gert, 2003, p. 28].)

A related issue is whether explicit permissions always have the form of exceptions to pre-existing prohibitions. This is a question that Kant seems to have struggled with. In *Metaphysik der Sitten* he referred to original permissive laws for actions that were not previously prohibited or required. The permissive laws provide authorization (Befugnis) to create obligations (for instance by allowing for the creation of property rights). (Akademieausgabe VI:247; [Hruschka, 2004]) His view that such original permissive laws are possible seems to be corroborated by modern legal developments in which new permissions and rights, such as e-commerce rights, are created without there having been any previous prohibition.

4 Interdefinability with other deontic concepts

It is generally recognized that there are three major groups of normative expressions in ordinary language, namely prescriptive, prohibitive, and permissive expressions. In the formal language, they are represented by the corresponding three types of predicates. Here, prescriptive predicates will be denoted by O, permissive predicates by P, and prohibitive predicates by F. (These are abbreviations of "ought", "permitted", and "forbidden".) A close analogy can be constructed with the standard concepts of modal logic. Letting $\mathbb{N}$ stand for necessity, $\mathbb{I}$ for impossibility, and $\mathbb{P}$ for possibility, we have:

$$\mathbb{N}a \leftrightarrow \mathbb{I}\neg a$$

$$\mathbb{N}a \leftrightarrow \neg\mathbb{P}\neg a$$

$$\mathbb{I}a \leftrightarrow \neg\mathbb{P}a$$

Similarly, for the deontic concepts the following conditions seem fairly self-evident (provided of course that permission is unilateral):

$$Oa \leftrightarrow F\neg a$$

$$Oa \leftrightarrow \neg P\neg a$$

$$Fa \leftrightarrow \neg Pa$$

One way to express this is that obligation is a form of deontic necessity, prohibition a form of deontic impossibility and permission a form of deontic possibility. These analogies between modal and deontic concepts were well-known to Robert Holcot, Roger Rosetus and other fourteenth-century scholars. [Knuuttila, 1981; Tierney, 2007] These scholars were also aware of the most crucial difference between the two triplets: Whereas both $\mathbb{N}a \to a$ and $a \to \mathbb{P}a$ are valid in modal logic, neither of the corresponding principles $Oa \to a$ and $a \to Pa$ is valid.

The definition of Fa as $O\neg a$ is usually taken to be unproblematic, and no one seems to have attempted to make F the primitive operator.

There are two equivalent interdefinability formulas for P and O:

$$Oa \leftrightarrow \neg P\neg a$$

$$Pa \leftrightarrow \neg O\neg a$$

We can therefore choose to use either O or P as the primitive concept in terms of which the other two concepts (P and F, respectively O and F) are defined. The choice between these two assignments of a primitive is conventional. In his classic 1951 paper, von Wright used P as the primitive concept, and consequently defined O as $\neg P \neg$. As he later reported, this choice was more or less an accident. It was an after-effect of his recent use of possibility as the basic concept of modal logic. [von Wright, 1999, p. 37] At the same time he used the term "deontic logic" to denote the new discipline. This is a term that refers, etymologically, to duties. It had been proposed to him by C.D. Broad. [von Wright, 1963, p. v] In subsequent work by himself and others, O was very soon accepted as the primitive concept. According to Stenius, this shift must have been an advantage, since it is more difficult to get an intuitive grasp of permission than of obligation. [Stenius, 1982, pp. 66-7] It can be argued that O is a better primitive than P since it is not subject to the difficulties of interpretation that are associated with the distinction between unilateral and bilateral permission.

The logical relations between the three notions have been expressed in the so-called deontic square that also contains a fourth node: "gratuitous" (or non-obligatory), see Figure 1. If further, combined modalities such as $Oa \vee O\neg a$ and $\neg a \ \& \ Pa$ are added, then more complex geometric structures can be constructed. [Moretti, 2009; Tierney, 2007]

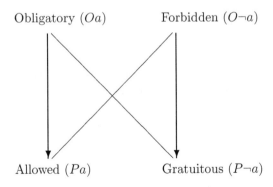

Figure 1: The deontic square. Arrows represent logical implication. The other lines represent contradiction

Under the assumption of full interdefinability between O and P (i.e. $Pa \leftrightarrow \neg O \neg a$), postulates or properties of deontic logic can be expressed either in terms of permission or obligation, as shown in Table 1.

Variant with O	Variant with P
$O(a\&b) \leftrightarrow Oa \,\&\, Ob$	$P(a \vee b) \leftrightarrow Pa \vee Pb$
$O(a \vee b) \leftrightarrow Oa \vee Ob$	$P(a\&b) \leftrightarrow Pa \,\&\, Pb$
$O(a\&b) \to Oa$	$Pa \to P(a \vee b)$
$Oa \to O(a \vee b)$	$P(a\&b) \to Pa$
$Oa \,\&\, Ob \to O(a\&b)$	$P(a \vee b) \to Pa \vee Pb$
$O(a \vee b) \to Oa \vee Ob$	$Pa \,\&\, Pb \to P(a\&b)$
$O(a\&b) \to Oa \vee Ob$	$Pa \,\&\, Pb \to P(a \vee b)$
$Oa \,\&\, Ob \to O(a \vee b)$	$P(a\&b) \to Pa \vee Pb$
$O(a \vee \neg a)$	$\neg P(a \& \neg a)$
$\neg O(a \& \neg a)$	$P(a \vee \neg a)$
$\neg(Oa \& O\neg a)$	$Pa \vee P\neg a$
$Oa \to a$	$a \to Pa$

Table 1: Equivalent versions of deontic postulates, expressed with obligation respectively permission

5 Free choice permission

Recently when a neighbour asked me if he could borrow a crowbar, I showed him my crowbars and said:

> "You may borrow either the big or the small crowbar."

The most natural interpretation of this sentence is that I offered him a choice; it was up to him which of the two he chose to borrow. In another context, the same sentence could have another meaning. Suppose that the tools belonged to someone else and that I had been authorized to lend one of them to the neighbour. However, I had forgotten which of the two he could borrow. Then I could say:

> "You may borrow either the big or the small crowbar, but I do not know which."

The first usage is the most common one. In ordinary language, a disjunctive permission usually indicates a choice. Such a notion of permission expectedly satisfies the postulate $P(a \vee b) \to Pa \,\&\, Pb$. In the second case, $P(a \vee b) \to Pa \,\&\, Pb$ does not hold, but the weaker postulate $P(a \vee b) \to Pa \vee Pb$ does. The latter postulate is valid in standard deontic logic (SDL), but the former is not.

5.1 Definition of free choice permission

As we have seen, the most common interpretation of disjunctive permission requires a logical principle that is incompatible with SDL. This seems to have been first noted by von Wright [1968, pp. 21-2]. He called the common-language form of disjunctive permission "free choice permission", which is still the most common term. However, there are several ways to define this notion more precisely. It can be defined as permission that satisfies $P(a \vee b) \to Pa \mathbin{\&} Pb$, permission that satisfies the stronger condition $P(a \vee b) \leftrightarrow Pa \mathbin{\&} Pb$ [Wolénski, 1980] or permission catching the informal notion of being allowed to choose. In the terminology introduced in section 3, a free choice permission may be either explicit, implied, or tacit. The two distinctions are independent. It is therefore a source of confusion that both free choice permission and explicit/implied permission have sometimes been called "strong permission". (See [von Wright, 1970, p. 160], [Wolénski, 1980] and [Merin, 1992, p. 105] for examples of the use of "strong permission" to denote free choice permission.)

5.2 Implausible derivations

The nature of free choice permission is probably the most discussed issue in the logic of permission. Although the postulate $P(a \vee b) \to Pa \mathbin{\&} Pb$ seems innocuous when presented in connection with a permitted choice, in combination with other deontic postulates it gives rise to a whole series of implausible results, as can be seen from the following list:

Implausible result 1: $Oa \to O(a \mathbin{\&} b)$ [Kamp, 1973, p. 61]

Requirements: Extensionality and interdefinability ($Oa \leftrightarrow \neg P \neg a$).

Derivation:

$P(\neg a \vee \neg b) \to P \neg a$

$P \neg (a \mathbin{\&} b) \to P \neg a$

$\neg P \neg a \to \neg P \neg (a \mathbin{\&} b)$

$Oa \to O(a \mathbin{\&} b)$

Implausible result 2: $Oa \to Pb$ [von Wright, 1968, p. 21]

Requirements: Extensionality, $Oa \to Pa$, and $Oa \to O(a \vee b)$.

Derivation:

$P(a \vee b) \to Pb$

$O(a \vee b) \to Pb$ (since $O(a \vee b) \to P(a \vee b)$)

$Oa \to Pb$ (since $Oa \to O(a \vee b)$)

Implausible result 3: $Pa \to Pb$ [Makinson, 1984, p. 140]

Requirements: $O(a\&b) \to Oa$ and interdefinability ($Oa \leftrightarrow \neg P\neg a$).

Derivation:

$O(\neg a \& \neg b) \to O\neg a$

$O\neg(a \vee b) \to O\neg a$

$\neg O\neg a \to \neg O\neg(a \vee b)$

$Pa \to P(a \vee b)$

$Pa \to Pb$ (since $P(a \vee b) \to Pb$))

The additional postulates used in derivations 1-3 are all valid in SDL. There could be no doubt, given these derivations, that SDL is incompatible with the postulate $P(a \vee b) \to Pa$ that is associated with free choice permission. However, this is not only a problem for SDL, as can be seen from the following derivation:

Implausible result 4: $Pa \to P(a\&b)$ [Hilpinen, 1982, pp. 176-77]

Requirements: Extensionality.

Derivation:

$P((a\&b) \vee (a\&\neg b)) \to P(a\&b) \,\&\, P(a\&\neg b)$

$Pa \to P(a\&b) \,\&\, P(a\&\neg b)$ (extensionality)

$Pa \to P(a\&b)$

Here, no SDL principles are used. All that we require of P, in addition to satisfying the "free choice postulate" $P(a \vee b) \to Pa$ is that it allows for substitution of logically equivalent sentences. Yet an utterly implausible result is obtained. It would imply for instance that if you are allowed to ask a stranger for directions to the railway station then you are allowed to ask him for directions to the railway station and then steal his wallet when he answers. This derivation has arguably not received the attention it deserves. It indicates that the free choice postulate may be faulty in itself, even if not combined with other deontic principles such as those of SDL. (Cf. section 5.9.)

5.3 Analogous constructions

Disjunctions can be used to express a choice not only in connection with permission but also in sentences that express obligations: [Aloni, 2007, p. 66]

"You must stay here or be accessible on your cellphone."

"You must turn straight left or straight right."

"Drive to your place or to my place!"

"Post this letter or burn it!"

There are also examples of "non-deontic free choice" expressed with disjunctions, such as the following three that are all taken from [Geurts and Pouscoulous, 2009b]:

"I can write a haiku or play the Moonlight Sonata."

"If things had turned out differently, I could have been a banker or a lawyer."

"Some of the guests ordered scrambled eggs or an omelet."

Implicit choice offers can be expressed not only with disjunctions but also with indefinite expressions such as:

"You may borrow one of the crowbars."

"Have a cake!"

"Take one of these!"

Only the first of these examples uses a permissive term ("may"). The others are permissions expressed with imperatives. It is not always clear whether an imperative expresses a requirement or a permission, and sometimes we do not need to know. ("Kiss me!")

Hence, free choice can be expressed either with disjunctions or with other linguistic means (indefinites), and in both cases free choice may be found either in a permissive context or some other context. However, it is the use of disjunctions embedded in permissions to express free choice that gives rise to the logical problems referred to above.

5.4 A variety of solutions

Researchers differ in their diagnosis of what is wrong with free choice permission, and consequently different types of solutions have been offered. Traditionally, the diagnoses and the proposed treatments have been divided into two major categories, semantic and pragmatic, according to whether their focus is on information that is is inherent in the language or on information that is only obtainable from the context of utterance. [Kamp, 1978] Free choice permission has often been treated as a test case for issues concerning the semantics/pragmatics boundary. [Asher and Bonevac, 2005,

p. 304] But in addition to the semantics/pragmatics boundary, many other logical and linguistic issues have been brought to the fore in discussions on free choice. In the following sections, major proposals to solve the problem of free choice permission will be discussed, namely:

- The "or" of free choice permission is not a disjunction, and the problem arises due to mistranslation from natural into logical language. (section 5.5)

- The offer to choose between the disjuncts is not inherent in the language but implied by the context of utterance. (section 5.6)

- There is a hidden, or implicit, operator of free choice, not expressed but understood by language users. (section 5.7)

- The "or" of free choice permission follows other logical laws than those of ordinary permission. (section 5.8)

- Free choice permission is not a property of the disjunction but a property of the set of disjuncts. (section 5.9)

A pragmatic approach is perhaps most clearly illustrated by the solutions discussed in section 5.6 and a semantic approach by those discussed in sections 5.5 and 5.7–5.8.

5.5 Mistranslation of "or"

In a short note published in 1973, R.Z. Parks claimed that the problem of free choice permission is simply one of mistranslation of ordinary language:

> "Translating 'you may work or relax' as '$P(p \vee q)$' rather than as 'Pp & Pq" is simply wrong in much the same way that translating 'Men and Women are welcome' as '$(x)(Mx$ & $Wx \rightarrow Cx)$' instead of '$(x)(Mx \vee Wx \rightarrow Cx)$' is wrong." [Parks, 1973]

Based on a much more detailed investigation, Stenius provided an explanation of this "mistranslation". [Stenius, 1982] It depends, in his view, on a general tendency in ordinary language to contract compound sentences. In these contractions, conjunctions (in the grammatical, not the logical sense of the word) are not always used in accordance with classical truth-functional logic. The sentence

> "Jim is not smoking and he is not drinking."

can be contracted in either of the following two ways:

"Jim is not smoking or drinking." or

"Jim is neither smoking nor drinking."

quite in accordance with de Morgan's laws. Similarly,

"Jim is forbidden to smoke and he is forbidden to drink"

is contracted to

"Jim is forbidden to smoke or drink."

in accordance with the deontic law $F(p \vee q) \leftrightarrow Fp \& Fq$. However, a sentence such as

"You may smoke and you may drink."

is not easily contracted in natural language. The most obvious alternative is probably

"You may smoke and drink."

but it will be read as $P(p\&q)$, i.e. as allowing the conjunction. Therefore the following contracted form is chosen:

"You may smoke or drink."

In this contracted sentence, "or" does not have the usual truth-functional properties of disjunction. Instead it functions as a connective for contracted sentence parts. The quoted sentence is an idiomatic contraction of "You may smoke and you may drink" and it should therefore be symbolized directly as $Pp \& Pq$. This was further clarified by David Makinson who described "or" as a "dummy connective" for contraction. "A legislator for language might have preferred to see a quite different word here, say 'uh'; but natural language has seized on the word 'or' for the purpose..." [Makinson, 1984, p. 142]

According to Makinson, this solution is "of course disappointing to the professional logician, for it suggests no new formal structure to play with". [Makinson, 1984, p. 141] He proposed a variant of the solution that provides a modicum of toys for the logician. Disjunctive permissions can be constructed as "checklist conditionals" ("enumeratively disjunctive universal conditionals") of the form:

$$(\forall x)(x=a \vee x=b \rightarrow Gx)$$

This is equivalent with Ga & Gb. The quantified notation includes a disjunction that arguably explains the choice of "or" to express the contracted sentence. However, as Makinson himself pointed out, the checklist conditional approach shares an important limitation with the dummy connective approach from which it was developed:

> "Whether one takes the 'or' of 'You may work or relax' as a dummy disjunction, as Stenius does, or as reflecting a disjunction in the antecedent of a checklist conditional, as we have suggested, one is brought to a common negative conclusion concerning the principal alternative approaches in the literature. The problem of disjunctive permission cannot be resolved by any process of fiddling with the deductive powers of formal deontic logics whilst holding the symbolic representation at $P(\alpha \vee \beta)$ ... We must either represent disjunctive permission directly and idiomatically as a conjunction of permissions or else represent it by a formula which, in addition to disjunction, makes use of quantification and identity and so transcends the limits of propositional logic." [Makinson, 1984, pp. 145 and 146]

5.6 Conversational implicature

The sentence

> "Some of the documents have disappeared from the dossier."

would normally be interpreted as "Some but not all of the documents have disappeared from the dossier". One possible explanation of this is that the word "some" really means "some but not all". However, a difficulty with that interpretation is that there are contexts where the same sentence would be interpreted differently:

> JOURNALIST: "Is it true that all the documents have disappeared from the dossier about mafia connections?"
>
> PROSECUTOR: "At this stage of the investigation I cannot answer your question."
>
> JOURNALIST: "But I have here a memorandum listing your ongoing investigations. One of the items is 'loss of documents from Dossier XR2712'. I do not see how that could be on the list unless at least some of the documents in the dossier had disappeared."
>
> PROSECUTOR: "OK. I can confirm that. Some of the documents have disappeared from the dossier."

Thus, the interpretation of "Some of the documents have disappeared from the dossier" depends on the context. If you know that all the documents have disappeared, it would be uncooperative to say (only) that some of them have done so. On the Gricean view [Grice, 1989], it is a conversational implicature that some of the documents were left. It is not part of what the sentence itself means. [Chemla, 2009; Schulz, 2005; Schulz and van Rooij, 2006; Geurts and Pouscoulous, 2009a]

Free choice permissions can be interpreted as conversationally implied. Hence, in the example referred to at the beginning of section 5,

"You may borrow either the big or the small crowbar."

does not inherently mean that you have a choice between the big and the small crowbar, but this is conversationally implied in most contexts where the sentence is used. That this information is conversationally implied (rather than inherent in the sentence) can be seen from the fact that it can be lost if the sentence is embedded in some other conversational context, such as in our example:

"You may borrow either the big or the small crowbar, but I do not know which."

A similar Gricean approach can also be used to explain why the "or" of free choice permission appears to be sometimes inclusive, sometimes exclusive. Compare the following three sentences:

"You may have either 500 US dollars or 350 British pounds."

"You may take a walk in the garden or on the meadow."

"You may have a banana or an apple."

In the first case, we hardly expect the offer to include the possibility of obtaining both the dollars and the pounds, i.e. the "or" is read as exclusive. In the second case, we would be surprised if the offer excluded the option of walking in both places. The third case is less determinate; the person to whom this is said may well wonder (and possibly have the audacity to ask) whether the offer includes the option of having both a banana and an apple. All this is fairly easily explained in terms of the (typical) situations in which these sentences are uttered.

The distinction between exclusive and inclusive interpretations of the free choice "or" has been interpreted in terms of scalar implicature (quantity implicature), i.e. the implicature that there is no reason to use a stronger term on the same scale. As we saw in the example of the lost documents, "some" often carries the contextual implication of "only some". In the same vein, an offer to have "either 500 US dollars or 350 British pounds" would expectedly not be made if the recipient could have both sums. [Chemla, 2009; Geurts and Pouscoulous, 2009a]

5.7 A hidden operator

The major competitor of the analysis of free choice permissions (and related phenomena) as conversational implicatures is to consider this information to be inherent in the language. There are two major variants of this view. According to one view, the phenomena under study depend on ambiguities in the language. This has been called the "lexical view". According to this view, "some" is ambiguous between "some and possibly all" and "some but not all". Similarly, "or" is ambiguous between plain truth-functional disjunction and a variant including "and you may choose which". (In addition, it will also have to be ambiguous between exclusive and inclusive disjunction.)

The other variant has been called "syntax-based". Just like the lexical view it holds that the language implicitly contains the information required, so that no conversational implicature is needed. However, this information is not inherent in single words, but in the syntactical structure. This is most commonly explained in terms of hidden or silent operators. On this view there is a hidden "only" in the sentence "You may have either 500 US dollars or 350 British pounds." [Geurts and Pouscoulous, 2009a, p. 3]

This approach was taken by Hans Kamp who proposed that permission statements contain a hidden one-place operator that he called the "focus operator". It puts the subformula within its scope into focus and thereby subjects it to the permission. [Kamp, 1973, pp. 69-70] Letting $\mathcal{F}$ denote the focus operator, we can insert it in two ways between the permission operator P and the disjuncts of its argument $a \vee b$, namely $P\mathcal{F}(a \vee b)$ and $P(\mathcal{F}a \vee \mathcal{F}b)$. The latter but not the former denotes free choice permission, as illustrated in two examples used by Kamp:

> $P(\mathcal{F}a \vee \mathcal{F}b)$ "You may go to the beach or to the cinema." (Said by a father to his child.)
>
> $P\mathcal{F}(a \vee b)$ "You may pillage city X or city Y. But first take counsel with my secretary." (Said by a king to his vassal.)

A similar approach was taken by Risto Hilpinen [1982]. (Cf. the discussion in [Makinson, 1984].)

5.8 Free choice operators

Several authors have tried to solve the problem of free choice permission by introducing a specific "free choice operator". It has mostly been denoted P_s, where the index letter stands for "strong". Due to the problems connected with that terminology (see section 5.1), the symbol P_c will be used instead here, with the index denoting "choice". Clearly, P_c should not be definable from obligation in the standard way, i.e. it should not hold in general that $P_c a \leftrightarrow \neg O \neg a$. It would be tempting to introduce free choice permission into SDL in the following straight-forward way [Føllesdal and Hilpinen, 1970]:

$$P_c(a \vee b) \leftrightarrow Pa \ \& \ Pb$$

However, as noted by Wolénski, this definition has implausible consequences. It gives rise to the following derivation:

Implausible result 5: $Oa \ \& \ Pb \rightarrow P_c(a \vee b)$ [Wolénski, 1980]

Requirements: $P_c(a \vee b) \leftrightarrow Pa \ \& \ Pb$ and $Oa \rightarrow Pa$.

Derivation:

$Oa \ \& \ Pb$ (assumption)

$Pa \ \& \ Pb$ (the postulate $Oa \rightarrow Pa$)

$P_c(a \vee b)$ (definition of P_c)

Hence, if it is obligatory to pay one's taxes and (unilaterally) permitted to smoke, then it is (free choice) permitted to either smoke or pay one's taxes.

The following are two additional implausible derivations that can be obtained from the same definition of P_c when inserted into SDL. (Neither of these derivations seems to have been reported previously in the literature.):

Implausible result 6: $Oa \ \& \ Ob \rightarrow P_c(a \vee b)$

Requirements: $P_c(a \vee b) \leftrightarrow Pa \ \& \ Pb$ and $Oa \rightarrow Pa$.

Derivation:

$Oa \ \& \ Ob$ (assumption)

$Pa \ \& \ Pb$ (the postulate $Oa \rightarrow Pa$)

$P_c(a \vee b)$ (definition of P_c)

Implausible result 7: $Pa \rightarrow P_c(a \vee b)$

Requirements: Extensionality, $P_c(a \vee b) \leftrightarrow Pa \ \& \ Pb$, and $Pa \rightarrow P(a \vee b)$.

Derivation:

Pa (assumption)

Pa & $P(a\vee b)$ (the postulate $Pa \to P(a\vee b)$)

$P_c(a\vee(a\vee b))$ (definition of P_c)

$P_c(a\vee b)$ (extensionality)

According to the last of these results, if you are permitted to borrow a book from the library, then you have a free choice to either borrow or steal the book. (More morbid examples are not difficult to construct.)

To avoid the problem he observed in Föllesdal's and Hilpinen's definition, Woleński proposed instead two other definitions:

$P_c(a\vee b) \leftrightarrow Pa$ & Pb & $P\neg a$ & $P\neg b$

$P_c(a\vee b) \leftrightarrow Pa$ & Pb & $P\neg a$ & $P\neg b$ & $P(a\&b)$ & $P\neg(a\&b)$

Unfortunately, the addition of either of these definitions to SDL gives rise to inconsistencies. To begin with, consider the first of them. Let a and b be sentences such that Pa, $P(\neg a\&\neg b)$, and $\neg Pb$. (This will presumably be the case for instance if a denotes that you give your cousin a birthday present and b that you embezzle all her money.) It follows from SDL postulates that $P(a\vee b)$, $P\neg a$, and $P\neg b$. Due to the definition of P_c it follows directly from $\neg Pb$ that $\neg P_c(a\vee b)$. However, we also have:

Pa & $P\neg a$ & $P(a\vee b)$ & $P\neg(a\vee b)$

and thus, by the same definition, $P_c(a\vee(a\vee b))$ which is equivalent with $P_c(a\vee b)$. This contradiction shows that the definition is inconsistent.

The same example can be used to prove the inconsistency of Woleński's second definition. Again, it follows directly from $\neg Pb$ that $\neg P_c(a\vee b)$. We then have:

Pa & $P\neg a$ & $P(a\vee b)$ & $P\neg(a\vee b)$ & $P(a\&(a\vee b))$ & $P\neg(a\&(a\vee b))$

from which it follows that $P_c(a\vee(a\vee b))$ and thus $P_c(a\vee b)$ just as for the other definition.

In 1970 von Wright proposed a system in which $P_c(a\vee b) \leftrightarrow P_ca$ & P_cb holds whenever both a and b are contingent. [von Wright, 1970, pp. 164-65] But unfortunately, any system with this property leads to an absurd conclusion, as follows:

Implausible result 8: $P_c(a\vee b)$ & $P_c(c\vee d) \to P_c(a\vee c)$ if a, b, c, and d are contingent.

Requirements: $P_c(a\vee b) \leftrightarrow P_ca$ & P_cb when a and b are contingent.

Derivation: Let a, b, c, and d be contingent.
$P_c(a \vee b)$ & $P_c(c \vee d)$ (assumption)
$P_c a$ & $P_c c$
$P_c(a \vee c)$

Suppose that you are unmarried and Kim wants to marry you. Then you presumably have a free choice permission to marry Kim or not to marry Kim. You also have, we may assume, a free choice permission to buy or not to buy a new garbage can. It would then follow, according to this derivation, that you have a free choice permission to marry Kim or buy a new garbage can. However, these are (hopefully) not two options to choose between. We suppose free choice permissions to concern options that we have a choice between.

The following derivation contributes to showing that $P_c(a \vee b) \leftrightarrow P_c a$ & $P_c b$ is implausible, even when a and b are contingent.

Implausible result 9: $P_c(a \vee b) \rightarrow P_c((a\&c) \vee (b\&c))$ if a, b, $a\&c$, $a\&\neg c$, $b\&c$, and $b\&\neg c$ are contingent.

Requirements: Extensionality, $P_c(a \vee b) \leftrightarrow P_c a \& P_c b$ when a and b are contingent.

Derivation:
$P_c(a \vee b)$ (assumption)
$P_c a$ & $P_c b$
$P_c((a\&c) \vee (a\&\neg c))$ & $P_c((b\&c) \vee (b\&\neg c))$
$P_c(a\&c)$ & $P_c(b\&c)$
$P_c((a\&c) \vee (b\&c))$

In a course I gave some years ago, students were free to choose between writing an essay on Kant (a) and writing an essay on Hume (b). They did not, however, have a choice between writing an essay on Kant and do it by plagiarizing ($a\&c$) and writing an essay on Hume and do it by plagiarizing ($b\&c$).

These results are absurd enough to block any further consideration of a P_c operator such that $P_c(a \vee b) \leftrightarrow P_c a$ & $P_c b$ holds for contingent a and b. (Note that $\vee$ is a truth-functional connective, so that the extensionality used in this and previous derivations can hardly be questioned.)

The negative results reported in this section probably give the impression that there is some common, underlying reason why it is difficult to reconstruct free choice permission in terms of an operator to which disjunctions are attached as arguments. Indeed, there is such a reason.

5.9 The impossibility of single-sentence representation

The attempts at symbolizing free choice permission reported in the previous section are all based on a common assumption that has usually been taken for granted. We can call it the "single sentence assumption":

> *The single sentence assumption:* Free choice between a and b can be represented as a property of a single sentence, namely $a \vee b$.

The single sentence assumption has an obvious consequence:

> If $a \vee b$ is equivalent with $c \vee d$, then there is a free choice permission between a and b if and only if there is a free choice permission between c and d.

It is not difficult to find examples showing that this leads to absurd conclusions:

> *The vegetarian's free lunch*
>
> You may have a meal with meat or a meal without meat. Therefore you may either have a meal and pay for it or have a meal and not pay for it.
>
> Proof: Let m denote that you have a meal with meat, v that you have a meal without meat, and p that you pay. $P(m \vee v)$ is equivalent with $P(((m \vee v)\&p) \vee ((m \vee v)\&\neg p))$.

To sum up, (free choice) permission to perform either a or b is not a function of a single sentence $a \vee b$ but a function of the two sentences a and b. It is a function of two variables, not one. Similarly, (free choice) permission to perform either a, b, or c is a function of three variables, etc. Therefore, free choice permission should be represented as a property of the set of action-describing sentences ($\{a, b\}$ respectively $\{a, b, c\}$) rather than a property of the disjunction of these sentences ($a \vee b$ respectively $a \vee b \vee c$). [Hansson, 2001, pp. 130-31]

6 Prima facie permissions and the strength of permissions

It was W. D. Ross who introduced the notion of a prima facie duty. [Ross, 1930; Ross, 1939] One has a prima facie duty to act in a particular way if and only if there is a valid moral reason that one should do so. However, there may be valid reasons pointing in different directions. If the valid reasons against the action in question are stronger than the valid reasons for it, then the duty is overridden, in other words it is not a duty all things considered.

Just as there are prima facie duties there are prima facie permissions. A prima facie permission may or may not be overridden by some competing norm.

6.1 Conflicts amongst prima facie permissions and obligations

If two prima facie obligations refer to contradictory actions, then they are in conflict. In other words, if $a\&b$ is inconsistent, then Oa and Ob are in conflict. This applies irrespectively of whether a and b refer to actions by the same or different persons. If you are under an obligation to keep the window open, and I am under an obligation to keep it closed, then we have a situation of conflict. The same applies if one and the same person has to keep the window open and keep it closed.

Conflicts between prima facie obligations and prima facie permissions are also common. Such conflicts may concern an obligation and a permission for the same person, or they may concern an obligation for one person and a permission for another. The former type is exemplified by conflicts between freedom of the press and prohibitions against libel (i.e. permissions to make statements and obligations not to make statements), between our moral freedom to live our lives as we choose and our obligations to help others in need, etc. The latter type is exemplified by a conflict in Swedish environmental legislation. Owners of ecologically valuable land areas are required to protect them against damage. Sweden also has a far-reaching right of public access to the wilderness (right to roam). Since visitors to sensitive areas can threaten ecological values, there is a conflict between the property owner's duty to preserve these areas and everyone's permission to visit them.

In conflicts between an obligation and a permission, either the obligation or the permission may prevail, i.e. become an overall obligation respectively permission. Declarations of human rights contain permissions such as those to speak and to organize that have a strong standing and can often override obligations.

Conflicts between permissions only arise if the permissions pertain to different persons. If a and b are actions by the same person, then Pa and Pb can coexist even if $a\&b$ is inconsistent. Indeed, without such coexistence, moral choice would be impossible. A permission to either perform or not perform some action (Pa and $P\neg a$) involves an inconsistent combination of sentences (a and $\neg a$), yet such permissions are essential parts of moral life. However, if a and b are actions by different agents, then conflicts can arise. No conflict arises if I am allowed to keep the window open and at the same

Constellation of norms	Conflict potential
Obligation for one person vs. obligation for another person	conflicts may arise
Obligation for one person vs. obligation for the same person	conflicts may arise
Obligation for one person vs. permission for another person	conflicts may arise
Obligation for one person vs. permission for the same person	conflicts may arise
Permission for one person vs. permission for another person	conflicts may arise
Permission for one person vs. permission for the same person	no conflicts

Table 2: The potential of different constellations of norms to give rise to conflicts

time allowed to keep it shut. However, if I am allowed to keep the window open and you are allowed to keep it shut, then there is a normative conflict (that may or may not lead to a conflict in practice depending on whether we use our respective permissions).

In summary, we have the pattern of possible conflicts shown in Table 2.

6.2 Overriding power

In order to clarify how conflicts can be resolved, it is useful to think of permissions and obligations as differing in their degrees of overriding power (strength). This can be done by the introduction of some measure μ of overriding power, such that $\mu(Pa)$ is a number representing the overriding power of Pa. Then a conflict between Pa and $O\neg b$ will be resolved in the favour of Pa if $\mu(Pa) > \mu(O\neg a)$. (Ways to deal with ties, e.g. the case when $\mu(Pa) = \mu(O\neg a)$, will have to be introduced in such a framework.) Alternatively we can use a linear order $\gg$ to represent overriding power. Then a conflict between Pa and $O\neg b$ will be resolved in favour of Pa if $Pa \gg O\neg a$.

For obligations, the introduction of degrees of strength or overriding power is supported by linguistic evidence. Different expressions for moral requirement have different strengths. "Must" is more stringent than "ought", and "ought" is more stringent than "should". ([Guendling, 1974]. On differences in strength between moral requirements, see also [Ladd, 1957, p. 125], [Sloman, 1970, p. 391], [Harman, 1977, pp. 117 and 118], [Jones and Pörn, 1985], [Meyer, 1987, p. 87], [Garcia, 1989], [Brown, 1996], and [Fintel and Iatridou, 2008].) Presumably, a prima facie obligation that is expressed with "must" will prevail over a conflicting obligation that can (only) be expressed with "should". For permission, there is not much support in the English language for the introduction of degrees. There is no obvious difference in stringency between saying that an action is permitted, that it is allowed, or that the agent may perform it. However, although the difference is not expressed lexically, it is obvious from the way in which we treat permissions that some permissions (such as those conferred by basic human rights) have a higher status than others.

A useful way to deal with degrees of strength is to assume that there is an operator corresponding to each level of strength. Consider two permission operators P_α and P_β. We can say that P_α is stronger than P_β (i.e. has more overriding power) if and only if it holds for all a that if $P_\alpha a$ then $P_\beta a$, but it does not hold for all a that if $P_\beta a$ then $P_\alpha a$. Under the assumption that strength is linearly ordered it should be the case for all permission operators P_α and P_β that:

$$\text{either } \{x \mid P_\alpha x\} \subseteq \{x \mid P_\beta x\} \text{ or } \{x \mid P_\beta x\} \subseteq \{x \mid P_\alpha x\}.$$

The strength of a permission to a can then be characterized by the strongest permission operator that allows it.[2] Degrees of obligation can be introduced with a similar structure. [Hansson, 2001, pp. 132-33]

An obvious way to combine degrees of permissions and obligations is to have, for each obligation operator O_α, a permission operator P_α satisfying $P_\alpha a \leftrightarrow \neg O_\alpha \neg a$ for all a, and vice versa. Under the assumption that obligation and permission operators are connected in this way, it follows directly that:

P_α is stronger than P_β if and only if O_β is stronger than O_α.

7 Conditional permissions

Many permissive statements are conditional. Traditionally they have been treated in deontic logic by the introduction of a symbol for conditional permission:

$$P(a \mid b),$$

meaning that a is permitted if b holds. However, there are several types of conditional permission statements, and it should not be taken for granted that one and the same symbolic representation can cover all of them.

---

[2]If the number of distinct permission operators is infinite, then there need not be any such strongest operator. The set of permission operators that allow a can then be used instead.

7.1 Two types of conditional norms

A particularly important distinction concerns the scope of the deontic statements, i.e. what situations they refer to. It is essential to distinguish between those permissions (and obligations) that refer to situations in general and those that refer to a particular situation. Unfortunately, this distinction is not obvious, since the English language (like many others) employs the same linguistic forms for both purposes (and this both for obligations and permissions). It is common practice in deontic logic to follow natural language in this respect, and use the same symbolic form (Pa respectively Oa) to express that something is permitted or obligatory in a particular situation and to express a permissive or obligative rule for situations in general. In a logical analysis, however, this distinction has to be made. Consider the following statements:

(1) "You may now take a short break."

(2) "You must leave this room immediately."

These are statements about what is permitted respectively obligatory at the time of utterance. No conclusion about what holds in other situations can be drawn from these statements. They report veritable norms, i.e. (over-all or only prima facie) norms that obtain in a particular situation, in these examples in the present state of the world. (It would be tempting to refer to these as "actual" permissions and obligations, but that is bound to lead to misunderstandings since Ross and others have used that term to denote over-all obligations. See [Ross, 1930, p. 20].) Veritable norms may also refer to situations other than the present one:

(3) "You have always been permitted to use my car."

(4) "You will be allowed to use a dictionary in next year's exam tests."

(5) "Two years ago my son was not allowed to be out after nine o'clock."

Furthermore, veritable statements about norms may refer to hypothetical situations:

(6) "If you had paid your previous bills, then you would have been allowed to buy a computer on credit."

(7) "If we had a more competent management, then we would have been allowed to use the work methods we prefer."

(8) "If the recession had not hit the company as hard as it did, then we would still have been permitted to take Friday afternoons off."

These are conditional statements, saying that *if* the situation satisfied a certain characteristic, *then* certain actions would have been permitted (obligatory, etc.). These statements do not report any normative rules, they only tell us what would have been the case (normatively) under certain conditions.

In order to capture such, veritable, conditional permissions (and obligations) we need a good account of counterfactual statements. There are important similarities between (8) and a statement such as the following:

(8') "If the recession had not hit the company as hard as it did, then we would have had a decent wage increase."

In both cases, "if ... then" induces us to consider what the world would have been like under a certain condition (that the recession had been more lenient to the company), with focus on some particular aspect (wage increase respectively permission to go home early in Fridays). In both cases, we need a good account of counterfactuality. Letting $P_v(\ |\ )$ and $O_v(\ |\ )$ represent veritable conditional permission respectively obligation, we should therefore expect that the following conditions:

$$P_v(a\mid b) \text{ if and only if } b \Rightarrow Pa$$
$$O_v(a\mid b) \text{ if and only if } b \Rightarrow Oa$$

hold for some relation $\Rightarrow$ that represents counterfactual conditionality. Furthermore, provided that $\Rightarrow$ satisfies Conditional Excluded Middle (i.e. for all a and b it is either the case that $a \Rightarrow b$ or that $a \Rightarrow \neg b$) we should expect that exactly one of Pa and $\neg Pa$ holds under the conditions specified by b, i.e.:

$$\text{Either } b \Rightarrow Pa \text{ or } b \Rightarrow \neg Pa \text{ but not both}$$

This gives rise to the following derivation:

$P_v(a\mid b)$
if and only if $b \Rightarrow Pa$
if and only if $\neg(b \Rightarrow \neg Pa)$
if and only if $\neg(b \Rightarrow O\neg a)$
if and only if $\neg O_v(\neg a\mid b))$,

thus confirming for this interpretation the interdefinability $P(a\mid b) \leftrightarrow \neg O(\neg a\mid b))$ that has a long tradition in deontic logic.[3] [Føllesdal and

---

[3]It is much less plausible for normative rules.

Hilpinen, 1970, p. 27] (It has also often been criticized. See [Chellas, 1974, p. 27].)

However, not all permissions are veritable, and neither are all obligations. Consider the following two examples:

(9) "You must release Molly from that terribly undersized cage."

(10) "You are not allowed to be cruel to animals."

Whereas (9) is a veritable obligation, saying what must be done in the present situation, (10) expresses a rule that holds (presumably) in all situations. Many such rules are conditional:

(11) "If you borrow money, then you must pay it back."

(12) "If you pay the exam fee at least one week in advance then you will be permitted to take part in the exam."

(11) is an obligative and (12) a permissive rule. As already indicated, such rules cannot be distinguished from veritable conditional obligations respectively permissions based on the linguistic form. We use the same lexical elements to express normative rules and veritable conditional norms. It is from the context that we can tell the difference. For instance, compare (12) to the following:

(13) "If you bribe the headmaster then you will be permitted to take part in the exam."

It is our knowledge of what legal and administrative rules usually look like that makes us infer that (12) reports a permissive rule and (13) a conditional veritable permission, i.e. a statement about what will be permitted under certain conditions. The distinction is not always clear, but it is nevertheless important since we cannot expect the two types of statements to follow the same logical laws. Let $P_r(\ |\ )$ denote permissive rules and $O_r(\ |\ )$ obligative rules. Intuitively we can read $P_r(a\ |\ b)$ as saying that the circumstance b triggers or activates the permission Pa.

In order to see why the two types of "conditional permission" and "conditional obligation" should be distinguished between, consider the following two postulates for conditional veritable permission respectively obligation:

If b is true and Pa holds, then so does $P_v(a\ |\ b)$.

If b is true and Oa holds, then so does $O_v(a\ |\ b)$.

It is reasonable to expect both of these postulates to hold, and this even if a and b are completely unrelated. This is due to the counterfactual interpretation of the relationship. In a non-normative context, we would admit the following inference as valid (albeit somewhat awkward):

> Xiu-xiu has a blue shirt.
> Xiu-xiu knows the ancient Greek language.
> ———
> If Xiu-xiu has a blue shirt then she knows the ancient Greek language.

For the same reason we should accept the following inference:

> Xiu-xiu has a blue shirt.
> Xiu-xiu is permitted to read classified government documents.
> ———
> If Xiu-xiu has a blue shirt then she is permitted to read classified government documents.

However, the corresponding principles for permissive and obligative rules:

> If b is true and Pa holds, then so does $P_r(a \mid b)$.
>
> If b is true and Oa holds, then so does $O_r(a \mid b)$.

are patently absurd. We can use the same example to show this. From the facts that Xiu-xiu has a blue shirt and that she is permitted to read classified government documents we cannot conclude that there is a rule to the effect that if she has a blue shirt then she is permitted to read classified government documents. [Hansson, 201x]

We can conclude from this example that conditional veritable norms and conditional normative rules obey different logical laws. This is a strong reason to treat them as two conceptually distinct categories, and to make this distinction also in informal philosophical contexts, in spite of the fact that ordinary language does not distinguish between them.

Generally speaking, a much weaker logic can be expected for normative rules than for conditional veritable normative statements. The interesting logical issues seem to arise, not on the level of determining what (prima facie) norms are in the system but on the level of drawing overall normative conclusions from the combination of a normative system and a situation. This will require the weighing of normative rules against each other. Even if a (prima facie) permission $P_r(a \mid b)$ holds and b obtains, it does not follow that Pa holds all things considered. In order to determine whether that is the case we need to know (i) whether there are c and d such that some $O_r(c \mid d)$ is a norm in the system, d is true in the situation in question, and a and c are in conflict, and (ii) in that case, which of the two norms $P_r(a \mid b)$ and $O_r(c \mid d)$ will prevail over the other. (See Hansson and Makinson 1997 for a model of such competition among normative rules.)

7.2 Can prima facie norms be dispensed with?

Conditional norms can to some extent fill the same function in a norm system as prima facie norms. Consider for instance the freedom of movement, the permission to travel to where one wants. This is a prima facie permission that has one major exception: it does not apply to persons who have been sentenced by a court of law to imprisonment, home detainment, or other types of restraining orders. We can deal with this by including in the norm system a prima facie obligation for people sentenced to such punishments to follow the restrictions in question. If this obligation has more overriding power than the freedom to travel, then it will serve to represent the intended exception to that freedom. However, we can also account for this exception in a seemingly somewhat simpler way. We can incorporate the exception into the permission itself by making it a conditional rule: "If a person is not subject to (specified) legal restrictions, then that person is permitted to travel where she wants."

How far can this method be used? Can we get rid of all conflicts between norms, and therefore also all exceptions to norms, if we include the appropriate exception rules in the antecedents of conditional rules?

Some authors have believed this to be possible:

> "[L]et us consider the following situation: $S_n = \{C_1, C_2, ...C_n\}$, where $C_1...C_n$ are morally significant circumstances. Suppose that upon deliberation (or inspection), we assert that $O(A \mid C_1)$. It follows by the above discussion that $O(A \mid C_1)$ is a statement of prima facie obligation because it rests on C_1 and not on all the other morally significant circumstances in S_n as well. Suppose now that we proceed with our deliberations to conclude that $O(B \mid C_1\&...C_n)$. Well clearly this statement is one of actual obligation because it rests on the totality of the morally significant circumstances in S_n. But now, just as clearly, this actual obligation is conditional too, resting as it does on $C_1\&...C_n$.
>
> It should be immediately clear from this interpretation that if prima facie obligations are conditional upon one aspect of the situation, then actual conditions are in turn conditional upon all aspects of the situation." [Al-Hibri, 1980, p. 80]

Promising though this approach may seem, it cannot be used in practice to represent human norm systems in their full complexity. A major reason for this is that it is not possible in practice to enumerate all the conditions that could invalidate a certain rule. As an example, suppose that I have undertaken to meet my six year old nephew, who will be arriving alone at the railway station to live in my home for a few days. Then I have an obligation to turn up in time as promised. But on my way to the railway station, I see people gathering around an accident victim. If I can help him with mouth-to-mouth resuscitation, I seem to be excused from my obligation to be at the station in time. However, I recognize the accident victim as a person suffering from a deadly disease transmitted by blood. He is bleeding from his mouth, and I myself have a sore lip. In this case I am permitted to refrain from applying the mouth-to-mouth method that is the method at my disposal. If there is nothing else that I can do to help him, I am – again – obliged to meet my nephew in time. The example could be carried still further, but it should suffice to show that conditional rules about over-all duties would have to be limitlessly complex. In particular, "weak" moral rules would have to contain implicitly all the stronger moral rules that could influence their validity in different circumstances. Thus there could not, for instance, be any moral rule about the keeping of promises that did not contain stipulations about murder, rape, and the prevention of wars. Similarly, a rule codifying the requirement to speak the truth would have to contain an extraordinarily complex set of exceptions, counterexceptions, and counter-counterexceptions. In summary, it is not possible to foresee and enumerate all the conditions that could invalidate a permission or obligation. A model with prima facie permissions and obligations to be weighed against each other is much more suitable to deal with this complexity than a model based on complete listings of exceptions.

8 Grantable and revocable permissions

One of the most interesting applications of deontic logic is the characterization and classification of legal relations such as rights, claims, and powers, some of which may be seen as extended versions of permissions.

Several authors, notably Stig Kanger, Frederick Fitch, and Lars Lindahl have used deontic logic to characterize and classify legal relations. ([Kanger, 1957; Kanger and Kanger, 1966; Fitch, 1967; Lindahl, 1977; Lindahl, 1994]. For an overview, see [Herrestad, 1996].) The basic framework, first developed by Stig Kanger, makes use of two operators: An ought operator O and an action predicate D, such that $D_i p$ means "i sees to it that p". The atomic sentences on which the classification is built have the form $[\neg]O[\neg]D_i[\neg]p$, where $[\neg]$ is a placeholder that can either be deleted or replaced by a negation sign. (Hence, $OD_i\neg p$ and $\neg O\neg D_i p$ are atomic sentence-types in this language, and there are six others.) Obviously, the choice of O rather than P here is arbitrary.

These systems are clarifying in important respects, but there are prominent features of normative systems that they do not cover. Perhaps foremost among these is that many norms are activated or deactivated by symbolic actions of different kinds, typically (but not exclusively) by speech-acts such as those of permitting and commanding. I am for instance permitted to enter your home (only) if you permit me to do so. The symbolic actions in question all have in common that, in the given legal system, they count as declarations by a person and that they have normative effects in the particular situation. To express such symbolic actions in the formal language we need to specify in each case (i) a normative sentence ("you may enter this house") and (ii) the agent by whom it is asserted. The declaration operator Dc has been introduced for that purpose. [Hansson, 1990] The sentence $Dc_i\delta$ denotes that an action has been performed that counts as a declaration by i that δ. A sentence of the form $Dc_i\delta$ is an atomic declarative sentence, and any non-tautologous truth-functional combination of such sentences is also a declarative sentence.

Declaration operators are building-blocks in normative rules that have the form $\alpha \mapsto \delta$, where α is a declarative sentence, $\mapsto$ denotes the "if ... then" of normative rules, and δ is a normative statement. This is a highly versatile framework; it allows for instance for rules such as:

$Dc_i Pa$ & $Dc_j Pa \mapsto Pa$ (a is permitted if both i and j permit it.)

$\neg Dc_i \neg Oa \mapsto Oa$ (a is obligatory unless i revokes the obligation to perform it)

$Dc_i Oa \vee Dc_j Oa \mapsto Oa$ (a is obligatory if either i or j demands that it be performed)

etc.

Probably the most important norm categories that this framework adds to the repertoire of deontic logic are the following four:

$Dc_i Pa \mapsto Pa$ (grantable permission[4])

$\neg Dc_i \neg Pa \mapsto Pa$ (revocable permission[5])

$Dc_i Oa \mapsto Oa$ (claimable obligation)

$\neg Dc_i \neg Oa \mapsto Oa$ (revocable obligation)

A simple example of a *grantable permisson* is the following:

"If Alan allows Betty to enter this garden, then she may do so."

---

[4] Close in meaning to the German "Verbot mit Erlaubnisvorbehalt".
[5] Close in meaning to the German "Erlaubnis mit Verbotsvorbehalt".

Grantable permissions can be seen as a type of rights. Although seldom referred to, they have important roles both in our legal and our moral systems. It is an important part of our sovereignty as individuals that there is a large set of actions affecting us (including various infringements into our private sphere) that others are allowed to perform only if we allow them to do so.

The following is a revocable permission:

> "Alan is allowed to walk along King's Street unless the police forbids him to walk there."

Revocable permissions differ from grantable ones in that the permission holds in the default case, so that a declaration is needed to revoke it, not to activate it. Revocable permissions can be found in many contracts. As a tenant I am permitted to live in the flat I rent unless the landlord terminates the contract with the stipulated notice period.

The following are examples of claimable respectively revocable obligations:

> "If Alan demands that Betty shall repay this loan, then she has to do so."

> "Alan is forbidden to enter Betty's house unless she lets him do so."

The revocation of a revocable obligation is a symbolic action very close to (and in practice essentially identical to) the granting of a permission.

9 Changes in permission

An account of permissions that treats them as static can only give a very rudimentary understanding of how permissions function in human interactions. The act of permitting is central in social life, and it differs fundamentally from stating or reporting that a permission obtains. [Raz, 1975, p. 163] An act of permitting can cancel or override an earlier obligation, or establish exceptions that limit its applicability. Similarly, an act that creates an obligation or prohibition can cancel or override an earlier permission. Some such operations can be expressed with notions such as grantable and revocable permissions, as shown in the previous section. However, much more versatility can be achieved in a framework for changes in permissions.

9.1 Lewis and the problem of permission

Much of the discussion on the dynamics of norms has referred to an article by David Lewis in which he asked the question: What exactly is the effect of permitting something that was not previously permitted? [Lewis, 1979] To exemplify this question, consider a child who was previously not allowed to walk home alone from school. Then one day her parents decide that from now on she may walk home alone. This does not mean that she can walk home in any way she pleases. She is not allowed to walk on the roadway or in private gardens, and neither is she allowed to smash windows or do other mischief on her way home. Her permission to walk home on her own is (implicitly) only a permission to walk home alone in certain ways. But which ways?

To deal with problems like this Lewis introduced the idea of a sphere of permissibility, containing all the permitted courses of action. The introduction of a new permission leads to an extension of that sphere, whereas a new prohibition (or obligation) will reduce it. However, the introduction of a permission is much more complicated than that of an obligation. To begin with the latter case, let Φ be the permissible sphere at a particular point in time, i.e. it is the set of all permitted courses of action. If a prohibition against the action class a is introduced (i.e. $O\neg a$ is made valid), then the new permissible sphere will simply consist in those courses of action in Φ that do not include a. If you were previously allowed to eat your own food in the canteen, then a prohibition against doing so will simply remove all courses of action in which you eat your own food in the canteen from your sphere of permissibility. In contrast, if the permissible sphere Φ is extended to make some action class a permitted (i.e. Pa is made valid), then it is not extended by all courses of action in which a takes place. If you were previously not allowed to eat your own food in the canteen, then a permission to do so will add some but not all of the courses of action in which you eat your own food in the canteen to the permissible sphere. Probably, you will not be allowed to use the canteen's tableware, to eat in the management's dining room, etc. Hence, whereas the effects of a new prohibition (or obligation) can be determined on strictly logical grounds, the effects of a new permission depend on a choice beween different realizations that cannot be based on logic alone.

Lewis described this in terms of a fictive game in which a Master controls the actions of a Slave by uttering commands and permissions. The Slave's actions depend on the present extent of the sphere of possibility, which in its turn depends on the series of previous commands and permissions given by the Master. The problem of permission appears very clearly in this game:

> "When the Master permits something, he does not thereby permit that thing to come about in whatever way the Slave pleases – not if the game is to be realistic. Suppose the Slave has been commanded to carry rocks every day of the week, but on Thurs-

day the Master relents and says to the slave '¡ *the Slave does no work tomorrow*'. That is all he says. He has thereby permitted a holiday, but not just any possible sort of holiday... *Some* of the accessible worlds where the Slave does no work on Friday have been brought into permissibility, but not all of them. The Master has not said which ones. He did not need to; somehow, that is understood." [Lewis, 1979, p. 169]

One way to solve the problem is to assume that there is some ordering that ranks the possible worlds from a moral point of view. When x is permitted, then this means that the best x-worlds (not all the x-worlds) are added to the sphere of possibility (set of permissible worlds). However, as van Rooij pointed out, with this solution we have no guarantee that the introduction of a permission $P(a \vee b)$ will result in a being permitted. [Rooij, 2006, p. 386] If b is better than a, then none of the best $a \vee b$-worlds will be a-worlds. Rooij proposed a way to include free choice permission in this framework. He did this by treating disjunctions as existential quantifications in a way not dissimilar to Makinson's checklist construction that was referred to in section 5.5. (Cf. also [Rooy, 2000; Merin, 1992].)

9.2 The belief change tradition

Beginning in the 1970's a focused discussion on rational belief change has taken place in the philosophical community. Early studies by Isaac Levi set the stage for much of these developments. [Levi, 1977; Levi, 1980] Other important studies were performed jointly by Carlos Alchourrón and David Makinson, whose early work focused on changes in legal codes. [Alchourrón and Makinson, 1981] They joined forces with Peter Gärdenfors, whose early work was concerned with the connections between belief change and conditional sentences. [Gärdenfors, 1978] The three wrote a paper that provided a new, much more general and versatile formal framework for studies of belief change, the AGM model. [Alchourrón *et al.*, 1985] This model has been varied and extended in what is now a very large literature in which the AGM model has a status similar to that of SDL in deontic logic. [Fermé and Hansson, 2011] The early topic of normative change has largely been lost in this development. However, conceptual and technical developments in AGM have provided us with improved tools that we can now use for the analysis of normative change.

In the AGM model, changes are performed on a set of sentences that is closed under logical consequence. This set is called the belief set, and represents everything that the person is committed to believe. It is usually denoted K. There are three types of operations on belief sets, namely contraction, expansion, and revision. They all take a sentence as input and produce a new belief set as output. Contraction is the operation that removes a sentence from the belief set. Its specific form in the AGM framework is called partial meet contraction. A partial meet contraction on K is an operation $\div$ such that for any non-tautology a, the contraction outcome $K \div a$ is equal to $\bigcap \gamma(K \perp a)$, where $K \perp a$ is the set of maximal subsets of K not implying a and γ a selection function such that $\emptyset \neq \gamma(K \perp a) \subseteq K \perp a$. If a is a tautology, then the contraction leaves K unchanged.

Expansion is an operation that adds the input sentence to the belief state. The defining equation is $K + a = \text{Cn}(K \cup \{a\})$, i.e. the expansion outcome consists of the logical consequences of $K \cup \{a\}$. Revision is a consistency-preserving operation that adds the input sentence a to the belief set, but removes enough from the original belief set to ensure that the resulting outcome is consistent. It is based on contraction, and the defining equation is $K * a = (K \div \neg a) + a$.

9.3 Two types of normative change

In developing a theory of normative change, a distinction must be made between changes in veritable norms (in the sense explained in section 7.1) and changes in the norm system. Changes in veritable norms, i.e. in what is actually permitted, forbidden or obligatory in some specific situation, can originate either in (i) changes in the system of normative rules or (ii) changes in the factual situation. For example, as a half-time employee Amira is allowed to use the company gym. There are two ways in which this permission can be rescinded: (i) the company may change the rules so that half-time employees are no longer allowed to use the gym, and (ii) she may cease to be a half-time employee.

Changes in the norm system are, of course, what give rise to changes of type (ii) of veritable norms. However, changes in the norm system need not always give rise to changes in veritable norms. Legislation sometimes contains stipulations for unusual cases, for instance criminal acts that have not yet been committed. Such stipulations can be changed without any direct effects on veritable norms.

In what follows, the focus will be on changes in the norm system. Furthermore, the focus will be on legal systems since the structure of a legal system is more well-defined than that of a moral system of norms.

9.4 Bases or closed sets

Due to their logical closure belief sets contain many strange elements. Hence, suppose that the belief set contains the sentence a, "Verdi wrote *Otello*". Due to logical closure it then also contains the sentence $a \vee b$, "Either Verdi wrote *Otello* or Palestrina wrote *Otello*". The latter sentence is a "mere logical consequence" and intuitively speaking it should have no standing of its own in the belief state represented by this belief set.

Belief bases have been introduced to capture this feature of the structure of human beliefs. [Hansson, 1994; Hansson, 1999] A belief base is a set of sentences that is not closed under logical consequence. Its elements represent beliefs that are justified independently of any other belief or set of beliefs. Those elements of the belief set that are not in the belief base are merely derived, i.e., they have no independent standing. Although the original AGM model was applied to logically closed belief sets, a considerable part of the developments of the framework have been devoted to models in which changes are performed on a belief base. The underlying intuition is that the merely derived beliefs are not worth retaining for their own sake. If one of them loses the support that it had in basic beliefs, then it will be automatically discarded.

For every belief base B, there is a belief set $Cn(B)$ consisting of the logical consequences of B. It represents the beliefs held according to B. On the other hand, one and the same belief set can be represented by different belief bases. In this sense, belief bases have more expressive power than belief sets. As an example, the two belief bases $\{a,b\}$ and $\{a, a \leftrightarrow b\}$ have the same logical closure. They are statically equivalent, in the sense of representing the same beliefs. On the other hand, the following example shows that they are not dynamically equivalent in the sense of behaving in the same way under operations of change. They can be taken to represent different ways of holding the same beliefs.

> "Let a denote that the Liberal Party will support the proposal to subsidize the steel industry, and let b denote that Ms. Smith, who is a liberal MP, will vote in favour of that proposal.
>
> Abe has the basic beliefs a and b, whereas Bob has the basic beliefs a and $a \leftrightarrow b$. Thus, their beliefs (on the belief set level) with respect to a and b are the same.
>
> Both Abe and Bob receive and accept the information that a is wrong, and they both revise their belief states to include the new belief that $\neg a$. After that, Abe has the basic beliefs $\neg a$ and b, whereas Bob has the basic beliefs $\neg a$ and $a \leftrightarrow b$. Now, their belief sets are no longer the same. Abe believes that b whereas Bob believes that $\neg b$." [Hansson, 1999, p. 20]

(In belief set models, cases like these are taken care of by assuming that although Abe's and Bob's belief states are represented by the same belief set, this belief set is associated with different selection mechanisms in the two cases. Abe has a selection mechanism that gives priority to b over $a \leftrightarrow b$, whereas Bob's selection mechanism has the opposite priorities.)

The distinction between static and dynamic equivalence is directly applicable to norms. Consider the following example:

> The school rules say: "All pupils must do at least four hours of schoolwork each day."
>
> Case (1): The headmaster, who is authorized to make exceptions to the school rules, declares: "On Sundays, all pupils should be outdoors the whole day instead of doing schoolwork."
>
> Case (2): The school board decides to change the rulebook, and replaces the old rule by two new ones: "All pupils must do at least four hours of schoolwork each day except Sundays", and "On Sundays, all pupils should be outdoors the whole day instead of doing schoolwork."

The systems of rules that emerge in cases 1 and 2 are statically equivalent, i.e. exactly the same behaviour is permitted respectively prohibited in the two cases. But now suppose that a member of the school board manages to convince his colleagues that spending a whole day out is unnecessary and a waste of time. Then:

> The school board decides to revoke the rule that requires pupils to spend the whole Sunday outdoors.

In case (1), the resulting rule system after the revocation will require that students do at least four hours of schoolwork each Sunday. In case (2), there is no such stipulation. Hence, although the two sets of rules are statically equivalent (support the same norms) they are dynamically non-equivalent (can support different norms after receipt of the same input). This is the same distinction that was described above for belief bases. It can be concluded from this example that the dynamics of legal (or otherwise explicitly stipulated) norms follows the pattern of belief bases, not that of belief sets.

9.5 Change and retrieval

As was noted above, belief revision theory is based on an input-assimilating framework. It describes how the epistemic agent transforms her state of belief upon receipt of an input. Between the inputs, the state of belief is assumed to be constant. [Hansson, 1999, pp. 3-11] This intermediate constancy is an idealization. Actual subjects change their minds as a result of deliberations that are not induced by external inputs.

In belief revision, conflicting information makes epistemic choices necessary. When constructing a belief revision model, we have a choice between

(1) making these selections as part of the operations of change when new information is received, and (2) letting operations of change leave conflicts unresolved, and instead make the necessary epistemic choices when information is retrieved from the system. [Rott, 2001; Hansson, 2010] There is a trade-off in simplicity between retrieval and change. In the AGM model, the retrieval operation is as simple as possible – it is just the identity operation. The change operations of AGM are much more complex. In belief base models we have a somewhat more complex retrieval mechanism, namely a consequence operator. (If the belief base is B, then the set of beliefs to which the agent is committed is $Cn(B)$.) On the other hand, operations of change tend to be somewhat less complex in belief base models than in belief set models. (The expansion of a belief base B by a sentence a is equal to the set-theoretical union $B \cup \{a\}$. The expansion of a belief set K by a sentence a is equal to $Cn(K \cup \{a\})$.) We can go further than this, and transfer much more from the operations triggered by the receipt of new information to the operations triggered by a need for information retrieval. The crucial step would be to move the selection mechanism from the receipt to the retrieval part of the model. However, this does not seem to be a good account of most processes of belief change.

> I previously believed that La Paz is the capital of Bolivia. Then
> I received information that Sucre is the capital of Bolivia.

Typically, such a conflict is dealt with immediately. If the new information is accepted, old beliefs are given up to the extent needed to retain consistency. We do not retain the conflicting pieces of information to adjudicate between them only when information is retrieved (i.e. we do not wait until we need to know whether La Paz or Sucre is the capital of Bolivia). Doing so except in a small number of cases would rapidly make the belief representation forbiddingly complex. In order to avoid that, the selection has to take place already when the new information is received.

However, this is not how we deal with conflicts in norm systems. Consider the following example:

> The legal code of a country contains the following rule: "Employers are free to hire and fire as they please."
>
> The country's parliament decides to add the following rule to the legal code: "All employees have a right not to be discriminated against because of their gender or ethnic origin."

The introduction of the second norm gives rise to a normative conflict; situations when the two norms point in different directions are in practice sure to arise. This is analogous to situations in which a new belief contradicts an old one. However, there is an important difference. As we have seen, conflicts in beliefs are usually dealt with immediately so that the two beliefs do not coexist in the same belief state. Conflicting norms are typically allowed to coexist in the same norm system. The conflicts that arise are dealt with separately in each situation where they arise. Hence, the conflict between the legal stipulations in our example is not dealt with once and for all by removing or modifying one of the conflicting norms. Instead, this conflict is dealt with repeatedly in the various situations in which it arises. It may well be the case that one of the two norms prevails in certain situations and the other norm in other situations, depending on the particular circumstances. In this way the actual conflict resolution, i.e. the selection among competing normative rules, takes place in a process of retrieval, not in the assimilation of inputs.

In legal norm systems, operations of change are very simple. New normative rules can be added through an operation very similar to expansion in belief revision. Furthermore, old normative rules can be removed. However, the removal of a normative rule is quite different from belief contraction. When contracting by a sentence a, we do not only remove a but also other sentences that imply a. In contrast, legal stipulations are removed by specific decisions that mention exactly what is going to be annulled. There is no propagation leading to further removals as in belief change.

In summary, the dynamics of legal (and other explictly stated) norms differs from the standard AGM model in at least three ways:

- The set of norms that hold in a legal system is not closed under logical consequence. Therefore, the dynamics of a norm system is closer to a belief change model employing belief bases than to the standard AGM theory in which (logically closed) belief sets are the primary objects of change.
- Removals of norms take place through a simple process more similar to set subtraction than to belief contraction. In belief change, when a specified belief is removed, a selection of other beliefs is removed with it in order to ensure that the sentence is not implied by the resulting new belief set. In contrast, decisions to remove normative rules from a normative system specify what is to be removed, and the process of removal does not propagate to other rules than these.
- In normative change, the selection among competing norms takes place when the normative system is used to determine a particular issue (i.e. in a process of retrieval from the system), rather than in the process of assimilating new norms. Defeated norms are not removed from the norm system but are retained and may be activated

in other situations.

It should again be emphasized that these conclusions refer to changes in a system of legal norms. It is less clear how changes in a system of moral norms should be modelled. If moral norms are taken to have about the same structure as legal norms, then the conclusions offered above are relevant for moral norms as well. However, the structure of a system of moral norms is much more difficult to determine than that of a system of legal norms, and it may also differ between moral standpoints or theories.

10 Conclusion

Attempts have been made to treat permissions as trivially derivable from obligations, but such reductions only seem to be possible at the price of leaving out important structural features of norm systems. The need for a separate treatment of permissions becomes particularly evident when changes in norm systems are taken into account. In a static perspective, it may seem as if there is no important difference between a norm system containing no stipulation about postings on notice boards and one containing an explicit permission to express one's opinion in such postings. However, the difference will be obvious if a proposal is made to forbid the posting of antigovernment statements on notice boards. The second case differs from the first in that a legal defence is available against such a new rule, namely the permission to make postings. Permissions can have a high degree of overriding power, and thereby override obligations and prohibitions with which they run into conflict. Permissions with a strong overriding power are a central feature of any legal or moral system that puts value on freedom. This should be reflected in (formal and informal) accounts of such systems. Therefore, the specific issues related to permission should have a central place in deontic logic.

Acknowledgements

I would like to thank Kai von Fintel and an anonymous referee for unusually helpful comments on an earlier version of this chapter.

BIBLIOGRAPHY

[Al-Hibri, 1980] A. Al-Hibri. Conditionality and Ross's deontic distinction. *Southwestern Journal of Philosophy*, 11:79–87, 1980.

[Alchourrón and Makinson, 1981] C. Alchourrón and D. Makinson. Hierarchies of regulation and their logic. In R. Hilpinen, editor, *New Studies in Deontic Logic*, pages 125–148. Reidel, Dordrecht, 1981.

[Alchourrón et al., 1985] C. Alchourrón, P. Gärdenfors, and D. Makinson. On the logic of theory change: Partial meet contraction and revision functions. *Journal of Symbolic Logic*, 50:510–530, 1985.

[Aloni, 2007] M. Aloni. Free choice, modals, and imperatives. *Natural Language Semantics*, 15:65–94, 2007.
[Asher and Bonevac, 2005] N. Asher and D. Bonevac. Free choice permission is strong permission. *Synthese*, 145:303–323, 2005.
[Brown, 1996] M. A. Brown. A logic of comparative obligation. *Studia Logica*, 57:117–137, 1996.
[Chellas, 1974] B. F. Chellas. Conditional obligation. In S. Stenlund, editor, *Logical Theory and Semantic Analysis*, pages 23–33. Reidel, Dordrecht, 1974.
[Chemla, 2009] E. Chemla. Universal implicatures and free choice effects: experimental data. *Semantics & Pragmatics*, 2(2):1–33, 2009.
[Fermé and Hansson, 2011] E. Fermé and S. O. Hansson. AGM 25 years: Twenty-five years of research in belief change. *Journal of Philosophical Logic*, 40:295–331, 2011.
[Fintel and Iatridou, 2008] K. von Fintel and S. Iatridou. How to say 'ought' in foreign: The composition of weak necessity modals. In J. Guéron and J. Lecarme, editors, *Time and Modality*, volume 75 of *Studies in Natural Language and Linguistic Theory*, pages 115–141. Springer, 2008.
[Fitch, 1967] F. B. Fitch. A revision of Hohfeld's theory of legal concepts. *Logique et Analyse*, 10:269–276, 1967.
[Føllesdal and Hilpinen, 1970] D. Føllesdal and R. Hilpinen. Deontic logic: An introduction. In R. Hilpinen, editor, *Deontic Logic: Introductory and Systematic Readings*, pages 1–35. Reidel, Dordrecht, 1970.
[Forrester, 1975] M. Forrester. Some remarks on obligation, permission, and supererogation. *Ethics*, 85:219–226, 1975.
[Garcia, 1989] J. L. A. Garcia. The problem of comparative value. *Mind*, 98:277–283, 1989.
[Gärdenfors, 1978] P. Gärdenfors. Conditionals and changes of belief. *Acta Philosophica Fennica*, 30:381–404, 1978.
[Gert, 2003] J. Gert. Requiring and justifying: two dimensions of normative strength. *Erkenntnis*, 59:5–36, 2003.
[Geurts and Pouscoulous, 2009a] B. Geurts and N. Pouscoulous. Embedded implicatures. *Semantics & Pragmatics*, 2(4):1–34, 2009.
[Geurts and Pouscoulous, 2009b] B. Geurts and N. Pouscoulous. Free choice for all: a response to Emmanuel Chemla. *Semantics & Pragmatics*, 2(5):1–10, 2009.
[Grice, 1989] P. Grice. *Studies in the Way of Words*. Harvard University Press, Cambridge, MA, 1989.
[Guendling, 1974] J. E. Guendling. Modal verbs and the grading of obligations. *Modern Schoolman*, 51:117–138, 1974.
[Hansson, 1990] S. O. Hansson. A formal representation of declaration-related legal relations. *Law and Philosophy*, 9:399–416, 1990.
[Hansson, 1991] S. O. Hansson. The revenger's paradox. *Philosophical Studies*, 61:301–305, 1991.
[Hansson, 1994] S. O. Hansson. Taking belief bases seriously. In D. Prawitz and D. Westerståhl, editors, *Logic and Philosophy of Science in Uppsala*, pages 13–28. Kluwer, 1994.
[Hansson, 1999] S. O. Hansson. *A Textbook of Belief Dynamics*. Kluwer, Dordrecht, 1999.
[Hansson, 2001] S. O. Hansson. *The Structure of Values and Norms*. Cambridge University Press, Cambridge, 2001.
[Hansson, 2010] S. O. Hansson. Changing the scientific corpus. In E. J. Olsson and S. Enqvist, editors, *Belief Revision Meets Philosophy of Science*, pages 43–58. Springer, Dordrecht, 2010.
[Hansson, 201x] S. O. Hansson. Beyond language. *Frontiers of Philosophy in China*. In press, 201x.
[Harman, 1977] G. Harman. *The Nature of Morality*. Oxford University Press, New York, 1977.

[Hedenius, 1941] I. Hedenius. *Om Rätt och Moral.* Tidens Förlag, Oslo, 1941.
[Herrestad, 1996] H. Herrestad. *Formal Theories of Rights.* Juristforbundets Forlag, Oslo, 1996.
[Hilpinen, 1982] R. Hilpinen. Disjunctive permissions and conditionals with disjunctive antecedent. *Acta Philosophica Fennica,* 35:175–194, 1982.
[Hruschka, 2004] J. Hruschka. The permissive law of practical reason in Kant's Metaphysics of Morals. *Law and Philosophy,* 23:45–72, 2004.
[Jones and Pörn, 1985] A. Jones and I. Pörn. Ideality, sub-ideality and deontic logic. *Synthese,* 65:275–290, 1985.
[Kamp, 1973] H. Kamp. Free choice permission. In *Proceedings of the Aristotelian Society,* pages 57–74, 1973.
[Kamp, 1978] H. Kamp. Semantics vs. pragmatics. In F. Guenthner and S. J. Schmidt, editors, *Formal Semantics and Pragmatics for Natural Language,* pages 255–287. Reidel, Dordrecht, 1978.
[Kanger and Kanger, 1966] S. Kanger and H. Kanger. Rights and parliamentarism. *Theoria,* 32:85–115, 1966.
[Kanger, 1957] S. Kanger. *New Foundations for Ethical Theory.* Almqvist and Wiksell, Stockholm, 1957. Reprinted in R. Hilpinen (ed.), *Deontic Logic: Introductory and Systematic Readings,* Synthese Library, Dordrecht, 1970, pp. 36-58.
[Knapp, 1981] V. Knapp. Einige Probleme der deontischen Modalität 'erlaubt'. *Archiv für Rechts- und Sozialphilosophie,* 67:397–406, 1981.
[Knuuttila, 1981] S. Knuuttila. The emergence of deontic logic in the fourteenth century. In R. Hilpinen, editor, *New Studies in Deontic Logic,* pages 225–248. Reidel, Dordrecht, 1981.
[Ladd, 1957] J. Ladd. *The Structure of a Moral Code.* Harvard University Press, Cambridge Mass., 1957.
[Levi, 1977] I. Levi. Subjunctives, dispositions and chances. *Synthese,* 34:423–455, 1977.
[Levi, 1980] I. Levi. *The Enterprise of Knowledge.* MIT Press, Cambridge, Mass., 1980.
[Lewis, 1979] D. Lewis. A problem about permission. In E. Saarinen, R. Hilpinen, I. Niiniluoto, and M. Provence, editors, *Essays in Honour of Jaakko Hintikka,* pages 163–175. Reidel, Dordrecht, 1979.
[Lindahl, 1977] L. Lindahl. *Position and Change. A Study in Law and Logic.* Reidel, Dordrecht, 1977.
[Lindahl, 1994] L. Lindahl. Stig Kanger's theory of rights. In D. Prawitz, B. Skyrms, and D. Westerståhl, editors, *Logic, Methodology and Philosophy of Science,* pages 889–912. Kluwer, Dordrecht, 1994.
[Makinson and van der Torre, 2003] D. Makinson and L. van der Torre. Permission from an input/output perspective. *Journal of Philosophical Logic,* 32:391–416, 2003.
[Makinson, 1984] D. Makinson. Stenius' approach to disjunctive permission. *Theoria,* 50:138–147, 1984.
[Makinson, 1999] D. Makinson. On a fundamental problem of deontic logic. In P. McNamara and H. Prakken, editors, *Norms, Logics and Information Systems. New Studies in Deontic Logic and Computer Science,* volume 49 of *Frontiers in Artificial Intelligence and Applications,* pages 29–53. IOS Press, Amsterdam, 1999.
[Merin, 1992] A. Merin. Permission sentences stand in the way of Boolean and other lattice-theoretic semantics. *Journal of Semantics,* 9:95–162, 1992.
[Meyer, 1987] J.-J. Ch. Meyer. A simple solution to the 'deepest' paradox in deontic logic. *Logique et Analyse,* pages 81–90, 1987.
[Moore, 1973] R. Moore. Legal permission. *Archiv für Rechts- und Sozialphilosophie,* 59:327–346, 1973.
[Moretti, 2009] A. Moretti. The geometry of standard deontic logic. *Logica Universalis,* 3:19–57, 2009.
[Mullock, 1979] P. Mullock. Logic and liberty. *Philosophical Studies,* 35:217–238, 1979.
[Myers, 1962] G. E. Myers. Prescriptions, permission, and obligations. *Philosophy and Phenomenological Research,* 22:481–489, 1962.

[Opalek and Woleński, 1991] K. Opalek and J. Woleński. Normative systems, permission and deontic logic. *Ratio Juris*, 4:334–348, 1991.
[Parks, 1973] R. Z. Parks. Note on an argument of von Wright's. *Philosophical Studies*, 24(1):64, 1973.
[Raz, 1975] J. Raz. Permissions and supererogation. *American Philosophical Quarterly*, 12:161–168, 1975.
[Rooij, 2006] R. van Rooij. Free choice counterfactual donkeys. *Journal of Semantics*, 23:383–402, 2006.
[Rooy, 2000] R. van Rooy. Permission to change. *Journal of Semantics*, 17:119–145, 2000.
[Ross, 1930] D. Ross. *The Right and the Good*. Clarendon Press, Oxford, 1930.
[Ross, 1939] D. Ross. *Foundations of Ethics*. Clarendon Press, Oxford, 1939.
[Ross, 1968] A. Ross. *Directives and Norms*. Routledge and Kegan Paul, London, 1968.
[Rott, 2001] H. Rott. *Change, Choice and Inference. A Study of Belief Revision and Nonmonotonic Reasoning*. Clarendon, Oxford, 2001.
[Schulz and van Rooij, 2006] K. Schulz and R. van Rooij. Pragmatic meaning and nonmonotonic reasoning: the case of exhaustive interpretation. *Linguistics and Philosophy*, 29:205–250, 2006.
[Schulz, 2005] K. Schulz. A pragmatic solution for the paradox of free choice permission. *Synthese*, 147:343–377, 2005.
[Sloman, 1970] A. Sloman. 'Ought' and 'better'. *Mind*, 79:385–394, 1970.
[Stenius, 1982] E. Stenius. Ross' paradox and well-formed codices. *Theoria*, 48:49–77, 1982.
[Stone, 1959] J. Stone. Non liquet and the function of law in the international community. *British Year Book of International Law*, 35:124–161, 1959.
[Tierney, 2007] B. Tierney. Obligation and permission: On a 'deontic hexagon' in Marsilius of Padua. *History of Political Thought*, 28:419–432, 2007.
[Tranøy, 1970] K. E. Tranøy. Deontic logic and deontically perfect worlds. *Theoria*, 36:221–231, 1970.
[von Wright, 1951] G. H. von Wright. Deontic logic. *Mind*, 60:1–15, 1951.
[von Wright, 1963] G. H. von Wright. *Norm and Action. A Logical Enquiry*. Routledge and Kegan Paul, London, 1963.
[von Wright, 1968] G.H. von Wright. An essay in deontic logic and the general theory of action. *Acta Philosophica Fennica*, 21:1–110, 1968.
[von Wright, 1970] G. H. von Wright. Deontic logic and the theory of conditions. In R. Hilpinen, editor, *Deontic Logic: Introductory and Systematic Readings*, pages 159–177. Reidel, Dordrecht, 1970.
[von Wright, 1981] G. H. von Wright. On the logic of norms and actions. In R. Hilpinen, editor, *New Studies in Deontic Logic*, pages 3–35. Reidel, Dordrecht, 1981.
[von Wright, 1999] G. H. von Wright. Deontic logic: A personal view. *Ratio Juris*, 12:26–38, 1999.
[Woleński, 1980] J. Woleński. A note on free choice permissions. *Archiv für Rechts- und Sozialphilosophie*, 66:507–510, 1980.
[Wurmbrand, 1999] S. Wurmbrand. Modal verbs must be raising verbs. In S. Bird, A. Carnie, J. D. Haugen, and P. Norquest, editors, *Proceedings of the 18th West Coast Conference on Formal Linguistics (WCCFL 18)*, pages 599–612, Somerville, MA, 1999. Cascadilla Press.

Sven Ove Hansson
Division of Philosophy
Royal Institute of Technology, Stockholm
Email: soh@kth.se

4
Prima Facie Norms, Normative Conflicts, and Dilemmas

Lou Goble

ABSTRACT. There is a normative conflict when one ought to do each of a number of things but cannot do them all. These pose a dilemma for logics of normative propositions. On one hand, they seem common; on the other, core principles of deontic logic entail they are impossible. This chapter examines several strategies to resolve this dilemma. Some declare normative conflicts indeed to be impossible, and then try to explain their widespread appearance. Here especially we study the distinction between *prima facie* and all-things-considered obligations. While the former might be in conflict, the latter are supposed not to be. We look at ways that could be. Other strategies consider normative conflicts to be possible. Their task is to devise appropriate principles to govern normative propositions that do not render such conflicts inconsistent, while still doing the work one expects. We examine several proposals to that end, including some that would significantly revise the foundations of deontic logic.

1	The dilemma of normative conflicts	242
2	Preliminaries	245
	2.1 Target concepts	245
	2.2 Formalities	248
3	No Conflict 1: Multiple operators	251
4	No Conflict 2: *Prima facie* and all-things-considered oughts	256
	4.1 A traditional view	257
	4.2 Reasons and *prima facie* oughts	266
	4.3 *Prima facie* oughts and *ceteris paribus* principles	276
	4.4 Basic *prima facie* oughts revisited	286
	4.5 A hybrid account	294
	4.6 Reprise	295
5	Conflict 1: Revisionist strategies	296
	5.1 Non-Kantian systems	297
	5.2 Non-aggregative systems	299
	5.3 Non-distributive systems	304
	5.4 Systems with limited replacement	315

	5.5	Reprise 318
6	Conflict 2: More radical strategies 321	
	6.1	Paraconsistent deontic logics 321
	6.2	Two-phase deontic logic 326
	6.3	An imperatival approach 330
	6.4	Adaptive deontic logics 337
	6.5	Reprise 344
7	Review 345	

Ada promised to take her son to the circus Friday afternoon, and so presumably she ought to spend that afternoon with him at the circus. It also happens, however, that there is an important meeting of her committee that same afternoon, and she ought to be present for that. She cannot do both. Ada seems stuck; whatever she does, it seems she will not do something she ought to do.

1 The dilemma of normative conflicts

Ada faces a normative conflict. Her situation is typical of what we will be discussing in this chapter. Generally speaking, there is a normative conflict when an agent ought to do a number of things, each of which is possible for the agent, but it is impossible for the agent to do them all. The prospect of such conflicts poses a dilemma for a logic of normative concepts. On one hand, they appear to be commonplace. We have all, no doubt, found ourselves in positions like Ada's from time to time. The literature on normative conflicts and moral dilemmas is replete with examples, from the pedestrian to the poignant; we will see some classic ones below. On the other hand, common principles of deontic logic entail that normative conflicts are literally impossible. Any system of logic that is supposed to apply to a broad range of normative discourse must somehow reconcile these two positions, but there is little consensus how that should be done. In this chapter we will examine several strategies to resolve the dilemma.

Common examples, like Ada's case above, illustrate the first horn of the dilemma. Of course, there is more to say about them, as we will see below. The second horn, that principles of deontic logic entail that normative conflicts are impossible, is easily demonstrated by two arguments.[1]

---

[1] These are both widely found in the literature. See, for example, [McConnell, 2010] for a summary; and [Brink, 1994] for a more detailed development. There is a third argument often presented along with these two (e.g., in the sources mentioned), but since it applies an equivalence between PA and $\neg O \neg A$, it seems merely a variation on the second argument given here.

Argument I draws on the principle that 'ought' implies 'can', that one ought to do something only if one can do it, or, in the usual notation,[2]

P) $OA \to \Diamond A$

together with the principle of aggregation, often called agglomeration and sometimes factoring and sometimes simply (AND), or as we now say (C), viz. that if one ought to do each of A and B, then one ought to do both,

C) $(OA \land OB) \to O(A \land B)$

Both principles are widely considered logically valid.

The argument then proceeds: Suppose there were a normative conflict in which all of OA_1, ..., OA_n hold ($n \geq 2$), but $\neg \Diamond (A_1 \land \cdots \land A_n)$ also holds. By principle (C), $O(A_1 \land \cdots \land A_n)$ must hold, and then by (P), $\Diamond(A_1 \land \cdots \land A_n)$ must hold, a contradiction. Hence there could be no such normative conflict.

Argument II applies the rule of distribution, or inheritance, or sometimes called necessitation, that if A entails B then if one ought to do A then one ought to do B,

RM) If $\vdash A \to B$, then $\vdash OA \to OB$

or, more generally, its modal counterpart, that if A necessitates B, then if one ought to do A then one ought to do B,

NM) $\Box(A \to B) \to (OA \to OB)$

together with the principle that if one ought to do something, then it is not the case that one ought not to do it,

D) $OA \to \neg O \neg A$

Argument II then proceeds: Suppose a case, like Ada's, in which both OA and OB but $\neg \Diamond(A \land B)$. By ordinary modal logic, A necessitates $\neg B$, $\Box(A \to \neg B)$. Given OA, then $O \neg B$ by (NM). By (D), however, since OB holds, $\neg O \neg B$ must also hold. Hence, both $O \neg B$ and $\neg O \neg B$, a contradiction. Thus there could be no such normative conflict.[3]

These four principles, (P), (C), (RM) or (NM), and (D), comprise what I will call the core principles of deontic logic. Historically they have figured that way. Along with (N) below, they form the basis of normal modal

---

[2]We say more about notation, both informally and more formally, in Section 2 below.

[3]Unlike Argument I, Argument II only excludes what we will call 'binary' normative conflicts, in which there are just two competing oughts. To generalize the argument to conflicts of arbitrarily many competitors requires also principle (C) above.

logics of type **KD** for the operator O, which is often considered the benchmark deontic logic. Also central, perhaps even more so, is the principle of replacement for logical equivalents within deontic contexts,

RE)   If $\vdash A \leftrightarrow B$, then $\vdash OA \leftrightarrow OB$

which, of course, follows from (RM). (RM) also entails the principle of simplification, which is in effect the converse of (C),

M) $O(A \wedge B) \rightarrow OA$

This, together with (RE), in turn implies (RM), and so all three belong together as core principles.

In the absence of the alethic modality, $\Diamond$, just as (RM) would apply rather than (NM), so (P) would be replaced by

P)'   If $\vdash \neg A$ then $\vdash \neg OA$

While it does not figure in Arguments I and II, another principle worth mentioning for later reference is the rule of necessitation for O,

N) $\Box A \rightarrow OA$, or, in the absence of alethic modalities,
       if $\vdash A$ then $\vdash OA$

Although philosophers often question this principle, it is contained in most familiar systems. Usually it is included merely for technical reasons, and considered innocuous. As we will see in §5.1 and elsewhere, however, there may be substantial formal reason for adopting it, and in some cases reason to reject it, §5.4, and even to require its contrary, §5.3.

Arguments I and II demonstrate that if there are, or could be, normative conflicts for a given sense of ought, then at least one member of each pair [(P), (C)] and [(NM), (D)] must be rejected or revised for that sense of ought. In Sections 5 and 6 we examine several suggestions along those lines. Before that, however, in Sections 3 and 4, we follow the other way to resolve the dilemma, by denying the possibility of genuine normative conflicts.

Within both perspectives, we draw on numerous proposals found in the literature. We aim at broad generality, however, and do not try to catalog every approach that has been taken, or even to be entirely true to the ones we do consider. We may scant details in order to present the general purport of an account. Some of these strategies are designed primarily to address the question of normative conflicts; others might be part of a larger vision and apply to other issues concerning normative concepts. Here we concentrate just on the aspects that pertain to normative conflicts.

The purpose of this discussion is not to decide among the several proposals put forward to resolve the basic dilemma, so much as to lay out the logical landscape of the issue, so that one can see the commitments and consequences of the various alternatives. Along the way we will, however, present some false starts, proposals that seem attractive at first as ways to solve one problem or another, but which fall short, by the standards of the perspective itself. The purpose in doing that is to reveal the depth and difficulty of the problems we are dealing with. Many natural suggestions do not work as one might expect them to; it is important to become aware of that, and learn from it, and seek more sophisticated solutions. Inevitably, as with all issues that go to the foundations of a form of logic, the question of how to treat normative conflicts raises controversies, and many questions remain open. Much fruitful research remains to be done in this area.

2 Preliminaries

Before entering our full explorations, it is useful to draw some boundaries, both informal and formal, around our topic. First, we distinguish normative conflicts as we will discuss them from some related concepts that we will not. After that, we describe quite briefly the formal languages employed here.

2.1 Target concepts

In this chapter we concentrate on normative conflicts of the sort illustrated above with Ada's situation. These are conflicts of obligations or of 'oughts', in which an agent ought to do several things but cannot do them all.[4] We may generalize that to include cases in which several states of affairs ought to be but where they are not all conjointly possible. In this chapter, we do not distinguish propositions of what an agent ought-to-do from propositions of what ought-to-be. Notationally, both are represented by expressions 'OA', which might be read indiscriminately to say that agent a ought to do α, or that it ought to be that a does α, or that it ought to be that A.[5]

---

[4]In this chapter, we apply the terms 'ought' and 'obligation', even 'duty', 'requirement', 'should', 'must', and others, more or less interchangeably, subject to constraints of grammar and style. Some philosophers, especially those impressed by the claims of ordinary language, take these different terms to mark significant conceptual distinctions. Perhaps they do, though there seems little consensus as to which word expresses which concept. Moreover, other vernacular languages will have other resources that may not match the English distinctions at all. If the ambiguities implicit in our present usage risk a loss of clarity, any of these locutions can be paraphrased uniformly for whatever sense is appropriate in context.

[5]While the distinction between expressions of the ought-to-do vs. the ought-to-be might be significant, we gloss over it since the issues it raises are generally orthogonal to present concerns. When rigor demands it, the default position here is the ought-to-be, treating the 'O' of 'OA' as a modal operator applied to the propositional formula given by 'A'. Even if propositions of what an agent ought-to-do differ from such forms, nevertheless most of the considerations discussed here will apply, *mutatis mutandis*, to them as well. We also abstract from issues of time, so that when we speak of a conflict

In all cases we take OA to be a statement descriptive of a state of affairs, and thus true or false, rather than an expression of a norm *per se*, e.g., a command, if that is something different.

Many discussions of normative conflicts and moral dilemmas concentrate on what we might call 'simple' or 'strict' conflicts, those of the form OA, $O\neg A$. Conflicts like Ada's are 'binary', OA, OB and $\neg\Diamond(A \wedge B)$, when B is not logically equivalent to $\neg A$. These too are often discussed. It is important to remember, however, that there might be n-ary conflicts, $OA_1, \ldots, OA_n$, with $A_1, \ldots A_n$ each individually possible but not jointly possible, for arbitrary $n \geq 2$. We will see some examples below.

The idea of normative conflict depends on a notion of possibility, on what an agent can or cannot do, or on what states of affairs are possible. Possibility can be construed in different ways, from logical possibility or consistency to physical possibility to whatever is practicable in a situation, including the agent's own abilities. Here, in the interest of generality, we allow for any of these readings, and take the notation $\Diamond$ to express any sort of non-normative possibility that seems appropriate to the context. We assume it satisfies the usual principles of alethic modal logic, but otherwise say little about it. We take necessity, $\Box$, to be the dual of possibility, in whatever sense is being applied, so that $\Box A$ is logically equivalent to $\neg\Diamond\neg A$.

The initial characterization of normative conflicts above concerned a single agent, such as Ada, who ought to do several things but cannot do them all. Such single-agent conflicts will continue to be the focus of this chapter. There might also be multi-agent conflicts in which a number of agents ought each to do something but where it is not possible for them all to succeed. Games provide natural examples, e.g., each player ought to win but not all can. Such multi-agent conflicts add dimensions of complexity that we will not take up here, though they deserve further research.[6] For present purposes, multi-agent conflicts may be included as conflicts of what ought-to-be: it ought to be that player a wins, OA, and it ought to be that player b wins, OB, but it is not possible for both to win.

Some philosophers[7] distinguish conflicts of obligations from conflicts of prohibitions, or of prohibitions and obligations or prohibitions and permissions. Insofar as prohibition and permission are construed as equivalent to functions of obligation and negation, these are merely variants on the kinds of conflicts we shall discuss. For other senses of prohibition or permission, it is plausible that many of the same issues will arise, *mutatis mutandis*,

---

between OA and OB we presume the times of the oughts, if not their executions, to be the same.

[6]See [Kooi and Taminga, 2008] for an investigation into such multi-agent conflicts; for some earlier informal background see [McConnell, 1988].

[7]E.g., Valentine [1987; 1989]; [Almeida, 1989] is a reply.

for their conflicts, but we shall not pursue that here. In what follows, we always construe permission as the dual of obligation, with PA definitionally equivalent to $\neg O \neg A$. Prohibition, as such, will not enter in.

Most of the philosophical discussion about normative conflicts addresses whether moral dilemmas are possible, whether it is possible that an agent morally ought to do something, A, and also morally ought to do something, B, that is incompatible with A.[8] Our interest here is broader than that, with conflicts of oughts of any normative kind, and not merely the moral. Conflicts could appear within what the law prescribes, or within what the rules of a game require, not to mention prudence, and courtesy, in short, any domain of normative discourse. There could also be conflicts between what is required in one domain and what is required in another. Here we will be primarily interested in the formal properties of the ought-operator, our O, and not with any specific conceptual interpretation of it.

Our focus is on normative conflicts like Ada's situation. We do not raise questions of the general consistency of normative systems, morality, law, etc. Marcus [1980] argues that the possibility of normative conflicts does not entail that a normative system is necessarily inconsistent or irrational. That is not our concern here, but only the possibility of the conflict itself.

We shall consider only conflicts among simple, monadic oughts, not of conditional obligations. In many respects those echo the concerns we address; in some respects they open further challenging problems.

A normative conflict does not imply that the agent must face a difficult choice about what to do. One of the oughts could have priority over the others. Even Ada might have a clear course of action. Her presence at the meeting might not be very important, while her promise to her son is. Then she might naturally choose to take her son to the circus. Or it could be the reverse. The meeting might be crucial to her career and the success of the committee's work, while her son has little desire to go to the circus, and is even a bit afraid of it. Then Ada would be well advised to attend the meeting, and perhaps do something else with her son some other time. If one of the oughts does have priority over the others in the conflict, the conflict is 'resolvable', resolvable in favor of that with highest priority. If none has higher priority over all the others, then the conflict is 'irresolvable'.

This distinction between resolvable and irresolvable conflicts presupposes a notion of priority upon what the agent ought to do or what ought to be. Such a notion may be applicable in certain domains, less so in others. In ethical theory, since at least [Ross, 1930], it is common to draw a distinc-

---

[8]For a sample of that literature, see the papers in the collections, [Gowans, 1987b] and [Mason, 1996], or, for a quick summary, see [McConnell, 2010], which also contains a useful bibliography.

tion between *prima facie* obligations and all-things-considered obligations. That distinction rests on a notion of the relative weight of one's *prima facie* obligations. In legal theory laws are often ranked according to various principles, e.g., the superiority of the law-making authority, or when statutes are enacted, etc. Other sorts of regulations or commands may likewise be applied in order of importance. In some normative domains priority may have less application, e.g., the rules of games. For those areas where priority relations are appropriate, it is often argued that all conflicts are resolvable. In Section 4 below, we examine in detail how such resolution might operate, and defer until then further discussion of priority relations.

Even if a conflict is resolvable, the agent still ought, in the relevant sense, to do each of the several competing actions. Resolvability should be distinguished from forms of escaping from the conflict, which could also apply to irresolvable conflicts. For example, the circumstances that produce the conflict might be changed. Ada could resign from her committee, and then would be under no obligation to attend the meeting, and so would be free to take her son to the circus without conflict. Or the body of laws or regulations might be amended according to various procedures to eliminate its conflicts. And so on. In this chapter we shall not be concerned with that sort of revision or escape that changes the obligations one is under. We adopt instead a static perspective, to consider the logical consequences of there being a normative conflict in the moment.

In a similar vein, we do not discuss strategies, if any, for deciding what to do when one faces an irresolvable conflict. Here we are concerned only with the fact, if it is a fact, of there possibly being normative conflicts, and what that entails for the concept of ought and its logic.

2.2 Formalities

In this chapter we assume a language, $\mathcal{L}_D$, typical for monadic propositional deontic logic. It contains atomic formulas $p, q, r \ldots$, etc., and the familiar connectives $\land, \lor, \neg, \rightarrow$, as well as the alethic operator $\Diamond$ and the deontic operator O with the usual formation rules, except that formulas $\Diamond A$ and OA are limited to A containing neither $\Diamond$ nor O.[9] Such formulas containing

---

[9] Iterated alethic modalities are excluded merely for convenience. They could easily be included without affecting the results to come. Alethic modalities could also be excluded from the language altogether; one often sees discussion of normative conflicts without them. Then formal consistency and entailment could be fair surrogates for possibility and necessity. In that case, one would address the formal distribution rule (RM) rather than (NM) and (P)′ instead of (P). Because our discussion here is mainly informal, and should apply both to languages containing alethic modalities and to languages without them, we will often give results in both terms, and will frequently speak of a logic containing (NM)/(RM) or (P)/(P)′ to refer ambiguously to the principle appropriate to the language. Iterated deontic modalities are also excluded chiefly for convenience, but some of the

neither modal operator comprise the base language $\mathcal{L}_B$. $A \leftrightarrow B$ is defined as $(A \rightarrow B) \wedge (B \rightarrow A)$; $\Box A$ as $\neg \Diamond \neg A$; and PA as $\neg O \neg A$. We use $\bot$ to refer to an arbitrary contradiction, and $\top$ for $\neg \bot$.

From time to time in what follows we will introduce other operators into the language, including multiple deontic operators O that will be distinguished by various devices such as subscripts, superscripts, etc. These will be explained when they occur.

Unless specified otherwise, the usual logical connectives will be considered classical, and the alethic modalities assumed to follow standard principles of modal logic, such as the principles of **T–S5** for first-degree formulas.

To interpret formulas of $\mathcal{L}_D$, we will apply models of the general form

- $M = \langle w_0, W, v, \mathcal{S} \rangle$

W is to be a set of so-called 'possible worlds' and $w_0 \in W$ is to be the 'actual world' within that set. These are for the interpretation of alethic modalities. v is a function assigning each elementary formula p an extension, a set of worlds at which it holds, so that $v(p) \subseteq W$.[10] For formulas A and B of $\mathcal{L}_D$, unless specified otherwise, as usual

$M, w \models p$ if and only if $w \in v(p)$
$M, w \models \neg A$ iff $M, w \not\models A$
$M, w \models A \wedge B$ iff $M, w \models A$ and $M, w \models B$
$M, w \models A \vee B$ iff $M, w \models A$ or $M, w \models B$
$M, w \models A \rightarrow B$ iff $M, w \not\models A$ or $M, w \models B$
$M, w \models \Diamond A$ iff $M, w' \models A$ for some $w' \in W$
$M, w \models \Box A$ iff $M, w' \models A$ for every $w' \in W$

The component $\mathcal{S}$ of M represents a structure of some kind for the interpretation of deontic formulas, OA. Just what sort of structure it is, and how it interprets such formulas, will depend on the particular accounts to be considered. On some of these accounts $\mathcal{S}$ will contain linguistic, or quasi-linguistic, objects. That invites developing finer grained models without such elements, but we will not be concerned with that here. The present schema should be compatible with any such development; we use it now primarily for convenience to establish some uniformity across our discussion.

---

accounts to be discussed are difficult to adapt to iterated contexts. $\mathcal{L}_D$ does include mixed formulas, expressions containing both deontic and non-deontic components, such as $p \wedge Op$. Principle (P), $OA \rightarrow \Diamond A$, requires this, but even without alethic modalities it seems a useful part of any language for normative discourse, although one often finds deontic logics without such expressions. Most of our discussion would be unaffected by excluding such mixed formulas. (An exception is in Section 4.3.)

[10]If one would eschew talk of possible worlds altogether, W may be taken as a set of valuations and $w_0 = v \in W$.

Indeed, the precise model theory of these accounts will generally lie in the background. We are more interested in what arguments and what principles should be considered valid, or otherwise worthy, when normative conflicts are taken into consideration, and expect both model theory and proof theory to be adapted accordingly, at least as much as plausible.

For a model $M = \langle w_0, W, v, \mathcal{S}\rangle$, we say that A holds on M *simpliciter* just in case it holds on M for the actual world.

- $M \models A$ iff $M, w_0 \models A$

Then logical validity and logical consequence are defined as usual. For a class of models $\mathcal{M}$, and A in $\mathcal{L}_D$ and a set of formulas Γ from $\mathcal{L}_D$,

- A is logically valid for $\mathcal{M}$ — $\Vdash_\mathcal{M} A$ — iff $M \models A$, for every $M \in \mathcal{M}$

- A is a logical consequence of Γ for $\mathcal{M}$ — $\Gamma \Vdash_\mathcal{M} A$ — iff for all $M \in \mathcal{M}$, if $M \models C$ for every $C \in \Gamma$ then $M \models A$.

(The subscript $\mathcal{M}$ may be omitted when clear in context.)

In what follows, it will often be convenient to speak of one state of affairs, A, necessitating another, B, or of a set Γ necessitating B, according to a model M, and similarly to say that B is incompatible with A or with Γ. For $A, B \in \mathcal{L}_B$ and $\Gamma \subseteq \mathcal{L}_B$, for a given model, $M = \langle w_0, W, v, \mathcal{S}\rangle$,

- A necessitates B (on M) — $A \vdash_\Box^M B$ — iff for all $w \in W$, if $M, w \models A$ then $M, w \models B$,

- Γ necessitates B (on M) — $\Gamma \vdash_\Box^M B$ — iff, for all $w \in W$, if $M, w \models C$ for every $C \in \Gamma$, then $M, w \models B$

- B is incompatible with A (Γ) (on M) iff A (Γ) necessitates $\neg B$ (on M),

- B is compatible (or co-possible) with A (Γ) (on M) iff B is not incompatible with A (Γ) (on M),

- Γ is co-possible (on M) iff for all $B \in \Gamma$, B is compatible with $\Gamma - \{B\}$.

These relations are designed to hold whether or not the language actually contains the alethic modalities, $\Diamond$ and $\Box$. (Here too the annotation 'M' on $\vdash_\Box^M$ may be omitted.)

3 No Conflict 1: Multiple operators

We saw in Section 1 how the prospect of normative conflicts poses a dilemma for deontic logic. In this section we begin to look at strategies to resolve this dilemma by denying that such conflicts are ever real, at least not for the sorts of oughts to which the principles of deontic logic are supposed to apply. This view is further refined in the next section. The denial of true normative conflicts might be based on a particular conceptual analysis of the meaning of ought, or it might be based on Arguments I and II themselves, or on something else. For present purposes it does not matter. In this section and the next we simply assume without question that the core principles, which generate Arguments I and II, are to be accepted.

The challenge for this perspective is to explain, or explain away, the widespread appearance of normative conflicts, manifest in the plethora of examples from the literature, and our experience. If one rejects the possibility of normative conflicts because of a particular conceptual analysis of ought, then most likely one would respond to offered examples in light of that analysis. We do not consider such accounts here. Likewise we say nothing about epistemological claims that the appearance of normative conflict is simply that, appearance, or perhaps uncertainty or ignorance about what the agent ought to do.[11] Without delving into the epistemology of normative judgements it is impossible to evaluate such claims. If they are correct, then the appearance of conflicts presents no particular problem for the logic of normative concepts.

A more interesting account of the appearance of conflicts draws on semantical considerations. This maintains that there are hidden ambiguities in the descriptions of the typical examples. It might be, for example, that an agent morally ought to do something, A, but legally ought to do something else, B, incompatible with A. If the first, the moral, 'ought' has a distinct sense from the second, the legal, 'ought', then there is no real conflict. In a perspicuous language, these would be represented by different deontic operators. O might be decorated with appropriate subscripts, $O_m A$, $O_\ell B$, to indicate which 'ought' is being expressed. Each operator might then follow the core principles given above, but no contradiction ensue. Thus, this sort of 'conflict' is fully compatible with the core principles.[12]

In a similar vein, if one considers the basis of normative requirements to be a set of imperatives issued by some authority, one might want to relativize

---

[11]This seems to be McConnell's position on moral dilemmas in [1978; 2010].

[12]Castañeda, in numerous papers, e.g., [1981; 1982], indexed his operators in this way to indicate distinct norm-making institutions, though his was a very fine-grained view in which individual laws or regulations determine distinct institutions and each calls for its own indexed operator.

the ought-operators to distinct authorities. It could be that i commands the agent to do A and j commands the agent to do B, when the agent cannot do both. From this it might follow that O_iA and O_jB, but these too would be consistent when $i \neq j$. At a more abstract level, there is significant current research into contextualized deontic logic in which multiple deontic operators are relativised to contexts. So long as each indexed operator obeys the core principles, conflicts are excluded for it, but it remains possible for there to be conflicts between what is required within one context and what is required within another.[13]

While this appeal to ambiguity might account for some appearances of normative conflict, it seems unlikely to account for all. It requires that each of the allegedly different senses of 'ought' be conflict-free, which seems unwarranted in the general case. Some examples of apparent conflict, like that introduced by Marcus [1980] discussed below, are specifically designed to reflect a conflict within a single normative domain, indeed stemming from a single normative principle, which might even derive from the command of a single issuing authority. Moreover, merely to distinguish multiple ought-operators by source, authority, context, etc., without accounting for their interaction imposes a severe limitation on the application of the logic.

To see that, consider the following case, involving no conflict, that was introduced by Horty in several sources, e.g., [1994; 1997; 2003; 2012].[14] We call it the Smith Argument. Suppose,

i) Smith ought to fight in the army or perform alternative national service.
ii) Smith ought not to fight in the army.

From these it seems correct to infer,

iii) Smith ought to perform alternative national service.

As Horty presents the case, he imagines that the two premises stem from different sources, (i) from the laws of Smith's country and (ii) from his religious convictions or conscience. While that is accidental to the full significance of the example, here it is useful. If obligations stemming from separate sources should be represented by separate operators, we should represent the premises of the argument by, for example,

---

[13]Yamada [2008] indexes the deontic operator to authorities, as well as to agents, to allow for normative conflicts without untoward consequences. On contextual deontic logic, see, e.g., [Krabbendam and Meyer, 1999]. For a more abstract, more general account, see [Gabbay and Governatori, 1999].

[14]Horty's point with the example was rather different from the present; we will discuss his point further in Section 5.2 below. (Horty himself never names the agent; I do.)

i)' $O_\ell(f \vee s)$
ii)' $O_c \neg f$

But now there is no way to represent the conclusion. That Smith perform alternative service is not required by the laws of his country, nor by his religious convictions. Even if what one ought to do is often determined by different sources or authorities, insofar as propositions of what one ought to do serve as guides to action or as standards of evaluation of an agent's overall actions, there must be a common ought derived from those separate sources. We want to be able to represent the inference of the Smith Argument by

i)* $O(f \vee s)$
ii)* $O \neg f$
∴ iii)* Os

where the O represents that common ought. If there are cases like $O_m A$ and $O_\ell B$, when $\neg \Diamond (A \wedge B)$, then there still seems to be a normative conflict, OA and OB, for that common sense of ought, for which the appeal to separate domains or separate authorities or, more generally, separate contexts has no purchase. Hence the challenge remains to explain the examples of conflict when they apply either to the ought of a single domain or to the common ought derived from the oughts of different domains.

In all that follows we shall assume that the operator O represents a univocal sense of 'ought' in its setting. When distinctions are appropriate, they will be indicated by subscripts, superscripts or other devices.

There is another way the oughts in the examples might be considered ambiguous. At the end of Section 2.1, I mentioned the distinction, familiar from ethical theory, between *prima facie* and all-things-considered obligations. When Ross [1930] introduced this distinction, or this terminology for it, he illustrated it (p. 18) with the well-known example of a person who has promised to meet a friend for some trivial purpose, but who could also help the victims of a serious accident, though only by breaking the promise. Because of the promise, presumably the person ought to meet the friend. Because of the need of the accident's victims, presumably the person ought to help them. Each is possible, but they are not jointly possible. Because this example is so common in the literature, and because we will refer to it frequently, we henceforth call it simply the Ross Example.

As just presented, the person seems to face a normative conflict; the person ought to meet the friend, the person ought to help the victims, and the person cannot do both. If this is an example of a normative conflict, however, it is what we have called a resolvable conflict. In typical cases like this, there is no question of what the person really ought to do. The person really ought to help the victims, and because of that, the person would be fully justified in breaking the promise. The person has a *prima facie* obligation to keep the promise, simply because it's a promise; the person also has a *prima facie* obligation to help the victims, because of their need and humane requirements of beneficence. The latter *prima facie* obligation, however, outweighs the former, so that, all-things-considered, the person ought to help the victims. The person's actual duty, duty *sans phrase* in Ross's parlance, is to help the victims.

There are thus two separate readings for statements of what an agent ought to do, one for what is the agent's *prima facie* obligation and another for the agent's all-things-considered, or actual, obligation. Let us represent these with two distinct operators, $O_{pf}A$ and $O_{atc}A$, within $\mathcal{L}_D$. The Ross Example could then be represented by $O_{pf}m$, $O_{pf}h$, $\neg\Diamond(m \wedge h)$, and $O_{atc}h$ but $\neg O_{atc}m$, and even $O_{atc}\neg m$. There is thus a conflict between the *prima facie* oughts, $O_{pf}m$, $O_{pf}h$, but not between the all-things considered propositions.

To the challenge that apparent normative conflicts present to deontic logic, one could argue in general that while the typical examples cited to illustrate such conflicts, like the Ross Example itself, reveal conflict between *prima facie* oughts, they do not present conflicts between all-things-considered oughts. One could maintain, further, that the core principles of deontic logic apply only to the latter, to all-things-considered oughts, and not to *prima facie* oughts. Hence, there is no contradiction between the examples and the principles of deontic logic.[15]

While this reply may apply to resolvable conflicts, it leaves the question of irresolvable conflicts, cases in which an agent ought to do a number of things, cannot do them all, and there is no unique one that outweighs or overrides all the others. Marcus [1980, p.125] offers this kind of example, which is widely cited in the literature. We call it the Marcus Example: The lives of two twins are in jeopardy. An agent is in a position to save one of them, either one, but not both. Because of the need, the agent ought to save the one twin, and the agent ought to save the other, but the agent cannot do both, and so faces a normative conflict. Here we may have two *prima facie* obligations, $O_{pf}s_1$, $O_{pf}s_2$, but because the needs of both are so much the same, and, we suppose, all other factors are in balance, the weight of each *prima facie* obligation is the same as the other. Since neither overrides the other, the conflict seems genuine.

Another example, too common in the literature, comes from J.-P. Sartre, [1946, pp. 295-6]. It tells of a student during WWII who felt the need to

---

[15] This is a common position. It is perhaps first articulated by Al-Hibri [1978, p. 49], but see also, for example, [Brink, 1994], and the summaries of [McConnell, 2010] and [Gowans, 1987a].

join the Free French forces to fight the Nazis, and equally felt the need to stay at home to care for his mother, but he could not do both. As Sartre presents the case, it seems apt to say both that the student ought to join the Free French forces and also that he ought to remain at home, even though the two exclude each other. The student thus faced a normative conflict. Call this the Sartre Example.

The Ross Example illustrates competing *prima facie* oughts where one outweighs another, and so becomes the all-things-considered ought. The Marcus Example presents competing *prima facie* oughts that have equal weight. While it is impossible to say what Sartre's own purposes were, his example is usually taken to illustrate competing *prima facie* oughts neither of which outweighs the other, not because they have equal weight, but because they are too disparate to be compared. Both the Marcus and the Sartre Examples are irresolvable conflicts in our sense.

Cases like these remain a challenge for the view that admits conflicts for *prima facie* oughts but denies them for all-things-considered oughts. For the Marcus Example, for example, it looks as though both $O_{atc}s_1$ and $O_{atc}s_2$ fail. So the oughts of the situation must be merely *prima facie*. If the agent has no actual duty to save twin-1 and no actual duty to save twin-2, then it seems the agent might as well just go home and watch reality shows on tv, leaving the twins to their fate. Perhaps one of the programs will even feature their plight.

No one subscribes to that conclusion, of course. To avoid it, one might maintain that there are no irresolvable conflicts, that the relation of greater weight over *prima facie* oughts is always total, and of a finite set of competing *prima facie* oughts one must always outweigh the others, though we might not know which it is.[16] This, however, simply reiterates the original position opposed to the possibility of genuine normative conflicts, and raises the same questions. Another common reply maintains that, for an irresolvable conflict like the Marcus Example, while it is true that neither $O_{atc}s_1$ nor $O_{atc}s_2$ holds in the situation, the agent is not absolved of all responsibility to act. The agent does have an all-things-considered obligation at least to save one twin or the other; $O_{atc}(s_1 \vee s_2)$ remains in force, and for the agent to do nothing but go home and watch tv would be an egregious violation of it. More generally, if there are a number of *prima facie* oughts, $O_{pf}A_1, \ldots, O_{pf}A_n$, that cannot all be fulfilled but where none overrides or outweighs any of the others, then the agent's all-things-considered obligation is to fulfill the disjunction of those requirements, $O_{atc}(A_1 \vee \cdots \vee A_n)$. Call this the Disjunctive Response. We will examine it in more detail in the next section where we consider the duality of *prima facie* and all-things-

---

[16]McConnell [1978; 2010] seems to hold this view.

considered oughts more closely, and the relations between the two, and how the distinction applies to the question of normative conflicts.[17]

4 No Conflict 2: *Prima facie* and all-things-considered oughts

The concept of *prima facie* obligation is a vexed notion in philosophy. Many would eschew it altogether, and more would shun the term.[18] Even Ross was not happy with this phrase, because of its epistemic connotations. Ross's notion was explicitly not epistemic. To say that an agent has a *prima facie* duty to do something is not merely to say that, *prima facie*, the agent has a duty to do that, or that it *seems as if* the agent has that duty. It is to say that the agent stands in a real normative position with respect to the act, even though in light of all relevant factors that is not the agent's final duty. In what follows here, we will not be concerned to explicate Ross's concept *per se*, nor even to be faithful to all of his intentions. Instead we take this notion more or less for granted in an informal, preliminary way, as we see it appear in examples like the Ross Example above. We are more interested to track the two formal notions, which we represent by the operators O_{pf} and O_{atc}, and their interplay as they are supposed to explain the alleged examples of normative conflicts.

The distinction between *prima facie* and all-things-considered oughts comes from ethical theory and concerns moral obligations. It seems less appropriate for other domains of normative discourse, such as the law, or bodies of regulations, etc. Nevertheless, within those other domains there may well be a sense of priority regarding competing requirements that is analogous to the notion of one *prima facie* obligation being more significant or outweighing another. To the extent that is so, then most of the discussion to follow should apply as well to the oughts of those domains, whether they are called *prima facie* or all-things-considered or not.

We assume little from the start about these two notions, except as they are supposed to explain the appearance of normative conflict within a context that accepts the core principles introduced in Section 1. This, however, is sufficient to frame three criteria of adequacy that any account of *prima*

---

[17]For the Disjunctive Response see, e.g., [Al-Hibri, 1978, p. 71], [Brink, 1994, p. 238 *et seq.*], [Donagan, 1987, pp. 286-7], [Hansson, 2001, pp. 174-5], [Sinnott-Armstrong, 1996, p. 51], [Zimmerman, 1996, p. 209], amongst others. Horty [2003; 2012] discusses this response at some length. Our commentary below will be somewhat different.

[18]Nowadays one is more likely to hear of *pro tanto* obligations or *pro tanto* reasons, or something like that. If one prefers to avoid the terms '*prima facie*' and 'all-things-considered', one could substitute 'Type 1' and 'Type 2', or any other similarly neutral phrase, so long as the fundamental presuppositions of the distinction are preserved, such as we see in the criteria below.

facie and all-things-considered oughts should meet.

a) It accepts that *prima facie* oughts can conflict, and that all-things-considered oughts cannot;
b) it maintains the core principles for all-things-considered oughts;
c) it supports the Disjunctive Response for irresolvable conflicts.

For the present perspective, we simply assume that the core principles are to hold for all-things-considered oughts, criterion (b). A concept for which that is not so belongs more naturally to the discussion of Section 5 or 6 below. Insofar as *prima facie* oughts might conflict, the question of what logic, if any, is appropriate for them also falls under the purview of those sections, although we will occasionally nod in that direction here.

In this section we introduce five different ways of looking at *prima facie* and all-things-considered oughts and the relations between them. We approach these with an eye toward generality, rather than focussing on the details of any specific author's account, expecting that what we say here will apply, with suitable adaptation, to many particular proposals that have been made. We are more interested now with how, in a general way, these accounts meet, or try to meet, the criteria above, than with any precise interpretation, much less any specific model theory, for the operators.[19]

4.1 A traditional view

We call this first view 'traditional' because it is widely found in the philosophical literature and discussion as an expression of Ross's original idea, or at least part of that idea. It begins with a body of *prima facie* oughts applicable in a situation that are ordered by a relation of relative weight or significance. The applicable *prima facie* oughts may conflict. From among those that conflict, the all-things-considered ought is the *prima facie* ought that has the greatest significance. (Any unconflicted *prima facie* ought is automatically an all-things-considered ought.)[20]

Following the pattern suggested in Section 2.2, we suppose a model $M = \langle w_0, W, v, \mathcal{S} \rangle$, in which the first three elements determine the non-normative

---

[19] The first four patterns illustrate main trends found in the literature; the fifth is a hybrid of two. As usual here, we do not try to be entirely comprehensive and consider every sort of account that has been offered to describe *prima facie* and all-things-considered oughts. In particular, we do not consider certain views that do not seem motivated by the criteria given above. For example, [Alchourrón, 1996] and [van Eck, 1982] each propose accounts of *prima facie* oughts that are governed by all the core principles, and conflicts among them thus excluded, contrary to criterion (a).

[20] Cf. [Ross, 1930, p. 19], [Brink, 1994, pp. 216-7], who also calls this a 'traditional' view, [Harman, 1975, p. 115], amongst other sources. We find this kind of language in almost every informal discussion of *prima facie* and all-things-considered obligation. As far as I know, however, it has not received formal attention or articulation before; the present construction is meant to be faithful to the underlying commitments of the view.

facts, including what is possible and impossible. The component $\mathcal{S}$ will consist in a pair $(\Delta, \geq)$. Δ represents the *prima facie* oughts that hold in a situation. For present purposes we may take Δ literally to be a set of formulas $O_{pf}A$; what further constraints should apply remains to be seen. $\geq$ represents the relation of relative weight or significance that is defined over Δ. '$O_{pf}A \geq O_{pf}B$' says that the *prima facie* obligation for A is at least as significant as the *prima facie* obligation for B. We suppose this relation to be reflexive and transitive. $>$ is its strict counterpart, so that $O_{pf}A > O_{pf}B$ if and only if $O_{pf}A \geq O_{pf}B$ and not-$(O_{pf}B \geq O_{pf}A)$. The *prima facie* obligation for A *outweighs* or *overrides* that for B when A and B are incompatible and $O_{pf}A > O_{pf}B$. To say that two *prima facie* oughts have equal weight, we write $O_{pf}A \approx O_{pf}B$, defined as $O_{pf}A \geq O_{pf}B$ and $O_{pf}B \geq O_{pf}A$. This would be the case in the Marcus Example, for example. $\geq$ might not be total; there could be *prima facie* oughts that cannot be compared, as in the Sartre Example. Further conditions on $\geq$ will be proposed below. Let $\mathcal{M}_t$ be the class of models with such an $\mathcal{S}$.

Rather trivially, we take a *prima facie* ought statement to hold for a model just in case it is within Δ.

- $M \models O_{pf}A$ if and only if $O_{pf}A \in \Delta$

The key question for this account is what determines when an all-things-considered ought statement holds for the model. By this present, traditional viewpoint, the all-things-considered ought is that *prima facie* ought that outweighs all competitors.

    Def 1) $M \models O_{atc}A$ if and only if $O_{pf}A \in \Delta$ and for all B, if $O_{pf}B \in \Delta$ and A is incompatible with B then $O_{pf}A > O_{pf}B$.

This must be distinguished from saying an all-things-considered ought is a *prima facie* ought that is not overridden by any other. That would allow for conflicts between all-things-considered oughts, contrary to the view now being considered. Nevertheless, it is convenient to be able to represent *prima facie* oughts that are not overridden. Let us speak of 'strong *prima facie* ought', with the notation $O_{spf}A$ in $\mathcal{L}_D$, such that

    Def 2) $M \models O_{spf}A$ if and only if $O_{pf}A \in \Delta$ and there is no B such that $O_{pf}B \in \Delta$ and $O_{pf}B$ outweighs $O_{pf}A$.

An irresolvable normative conflict (of *prima facie* obligations) is then a case in which there are a number of things an agent strongly *prima facie* ought to do but where the agent cannot do all of them. The Marcus Example and the Sartre Example illustrate this.

(Def 1) provides a natural account of how all-things-considered oughts are determined by *prima facie* oughts and their relative weights. As we will see, however, it fails to meet the three criteria above.

4.1.1 Principles for O_{atc}

By criterion (b), formulas $O_{atc}A$ should follow all the core principles. That imposes constraints on the membership of Δ and the properties of $\geq$.

Principle (D), $O_{atc}A \to \neg O_{atc}\neg A$, requires no special conditions. Its validity (for $\mathcal{M}_t$) is immediate from (Def 1) and the transitivity and asymmetry, hence irreflexivity, of $>$. Similarly for (P), $O_{atc}A \to \Diamond A$.[21]

On the other hand, since (Def 1) calls for two conditions to be met for $O_{atc}A$ to hold, that $O_{pf}A$ and that $O_{pf}A$ outweigh all competing *prima facie* oughts, each of the other core principles (NM)/(RM) and (C) for all-things considered oughts calls for two conditions to be imposed on models, one governing membership in Δ, the other constraining $\geq$. For (RM) and (NM) these conditions are, for any $M \in \mathcal{M}_t$,

For (RM): (rm) If $O_{pf}A \in \Delta$ and $\Vdash A \to B$ then $O_{pf}B \in \Delta$
(ent) If $O_{pf}A, O_{pf}B \in \Delta$ and $\Vdash A \to B$, then
$O_{pf}B \geq O_{pf}A$
For (NM): (nm) If $O_{pf}A \in \Delta$ and A necessitates B, then
$O_{pf}B \in \Delta$
(nent) If $O_{pf}A, O_{pf}B \in \Delta$ and A necessitates B,
then $O_{pf}B \geq O_{pf}A$

It is easy to verify that these conditions validate the principles (RM) and (NM) for formulas $O_{atc}A$. Without them, those principles would not be valid. It is also easy to see that (rm) and (nm) validate the corresponding rule for formulas $O_{pf}A$.

RM)$_{pf}$ If $\Vdash A \to B$ then $\Vdash O_{pf}A \to O_{pf}B$
NM)$_{pf}$ $\Vdash \Box(A \to B) \to (O_{pf}A \to O_{pf}B)$

which seem as plausible as the principle for all-things-considered oughts.[22]

It is harder to say whether (ent) and (nent) are correct. The literature on *prima facie* oughts does not say enough about how relative weight is supposed to work to be confident one way or the other. It is not our concern,

---

[21] For (D), consider any $M \in \mathcal{M}_t$, and suppose $M \models O_{atc}A$ and $M \models O_{atc}\neg A$. From the first, $O_{pf}A > O_{pf}B$ for every B such that $O_{pf}B \in \Delta$ and B is incompatible with A. That includes $\neg A$ since $O_{pf}\neg A \in \Delta$ by virtue of the second assumption. So $O_{pf}A > O_{pf}\neg A$. By the same reasoning, $O_{pf}\neg A > O_{pf}A$. By transitivity, $O_{pf}A > O_{pf}A$, contrary to the irreflexivity of $>$. Hence, if $M \models O_{atc}A$ then $M \not\models O_{atc}\neg A$, or $M \models \neg O_{atc}\neg A$. A similar argument applies for (P).

[22] [Al-Hibri, 1978, p. 82] accepts this or (M)$_{pf}$, which with (RE)$_{pf}$ yields (RM)$_{pf}$. [Brink, 1994, p.244] too recommends the principle analogous to (NM)$_{pf}$.

however, to argue for or against these conditions, but only to point out that they are required if (RM) and (NM) are to be validated for all-things-considered oughts, as called for by criterion (b).

The situation with regard to aggregation (C) for all-things-considered oughts is similar, but a bit more complicated. On the face of it, this too calls for corresponding rules for Δ and $\geq$ that will be the converse of the conditions for (RM), as (C) is the converse of (M), i.e.,

For (C): (c)   If $O_{pf}A, O_{pf}B \in \Delta$ then $O_{pf}(A \wedge B) \in \Delta$
          (cent) If $O_{pf}A, O_{pf}B \in \Delta$ and $\Vdash A \to B$ then $O_{pf}A \geq O_{pf}B$

With both of these (C) is easily validated. Condition (c) also validates aggregation for *prima facie* oughts, $(C)_{pf}$, $(O_{pf}A \wedge O_{pf}B) \to O_{pf}(A \wedge B)$, however, and one might not want to be committed to that since such oughts may well conflict.[23]

Interestingly, validation of (C) does not demand the full power of condition (c). It would suffice to close Δ under a weaker condition of 'consistent (or co-possible) aggregation'.

For (C): (cc) If $O_{pf}A \in \Delta$ and $O_{pf}B \in \Delta$ and A is compatible with B then $O_{pf}(A \wedge B) \in \Delta$

For *prima facie* oughts this validates only the weaker aggregation principle, $(CC)_{pf}$, $(O_{pf}A \wedge O_{pf}B \wedge \Diamond(A \wedge B)) \to O_{pf}(A \wedge B)$, which might seem more plausible than the full-throated $(C)_{pf}$.

Condition (cc) ensures that if $O_{atc}A$ and $O_{atc}B$ both hold, then so does $O_{pf}(A \wedge B)$, the first step to establishing $O_{atc}(A \wedge B)$.[24] The second step, that the *prima facie* obligation for $A \wedge B$ outweighs all competing *prima facie* obligations in Δ, i.e., that if $O_{pf}C \in \Delta$ and $A \wedge B$ necessitates $\neg C$ then $O_{pf}(A \wedge B) > O_{pf}C$, is ensured by condition (cent).[25]

---

[23] Al-Hibri [1978, p. 82] does accept this. To accept (C) for *prima facie* oughts requires that (P) be rejected for them. That is an option. On the other hand, to have both (C) and (NM)/(RM) for oughts that might conflict has the further untoward consequence that everything would be required under that sense of ought. We call this 'deontic explosion', and discuss it further in Section 5.

[24] Suppose $M \models O_{atc}A$ and $M \models O_{atc}B$ then immediately $O_{pf}A \in \Delta$ and $O_{pf}B \in \Delta$. Moreover, A must be compatible with B, for otherwise, since $O_{atc}A$ and $O_{atc}B$ both hold, $O_{pf}A > O_{pf}B$, and $O_{pf}B > O_{pf}A$. But then $O_{pf}A > O_{pf}A$ by transitivity, and contrary to the irreflexivity of $>$. With A and B compatible, and $O_{pf}A, O_{pf}B \in \Delta$, by (cc) then $O_{pf}(A \wedge B) \in \Delta$ as desired.

[25] Consider a C such that $O_{pf}C \in \Delta$ and $A \wedge B$ is incompatible with C. This C is compatible with A, or it is not. Suppose the first, $M \models \Diamond(C \wedge A)$ Then we have $O_{pf}(A \wedge C) \in \Delta$ by closure under (cc). Moreover, since $M \models O_{atc}B$ and $M \models \neg\Diamond(B \wedge (A \wedge C))$, $O_{pf}B > O_{pf}(A \wedge C)$. Since $\Vdash (A \wedge B) \to B$, by (cent) $O_{pf}(A \wedge B) \geq O_{pf}B$, whence $O_{pf}(A \wedge B) > O_{pf}(A \wedge C)$ by transitivity. Further, since $\Vdash (A \wedge C) \to C$,

That is all very well, but these provisions face serious problems. For one thing, consistent aggregation, $(CC)_{pf}$, for *prima facie* oughts may be as dubious as the richer principle $(C)_{pf}$, and so the condition (cc) that validates it must be suspect.[26] Moreover, if (cent) is adopted along with (ent) and its companion condition (rm) then any two *prima facie* oughts in Δ must have equal weight, in which case the whole point of distinguishing *prima facie* oughts from all-things-considered oughts collapses.[27]

Since (ent) and (cent) cannot both be plausibly required, either (RM) or (C) must fail to be valid for all-things-considered oughts under (Def 1). These results do not decide between (RM) and (C), but one must be rejected, which means that this account fails to satisfy criterion (b), that all core principles of deontic logic apply to all-things-considered oughts.

4.1.2 The Disjunctive Response

To articulate the Disjunctive Response, it is not enough simply to say that, for incompatible A and B, if $O_{pf}A$ and $O_{pf}B$ and neither outweighs the other then $O_{atc}(A \vee B)$, for the conflict might arise between three or more alternatives. Perhaps there are triplets in jeopardy. We need a principle or a schema that allows for indefinitely many disjuncts.

Given $(\Delta, \geq)$, let Θ be the set of all members of Δ such that each one is a strong *prima facie* ought, i.e., not outweighed by any competitor, and also that the enjoined content of any two members of Θ are pairwise incompatible. I.e., given $M = \langle w_0, W, v, (\Delta, \geq) \rangle$

- $\Theta = \{O_{pf}A \in \Delta : M \models O_{spf}A$ and for all B, if $\not\Vdash A \leftrightarrow B$ and $M, \models O_{spf}B$ then A is incompatible with $B\}$

The Disjunctive Response then maintains, for non-empty, finite Θ,

DR) Given Θ as specified, $M \models O_{atc} \bigvee \Theta^*$.

where $\bigvee \Theta^*$ is the disjunction of the enjoined contents of all the members of Θ, i.e., $\bigvee \Theta^* = A_1 \vee \cdots \vee A_n$ for each A_i such that $O_{pf}A_i \in \Theta$.

$O_{pf}(A \wedge C) \geq O_{pf}C$ by (cent), and then $O_{pf}(A \wedge B) > O_{pf}C$ by transitivity, as desired. On the other hand, if $M \models \neg \Diamond (A \wedge C)$, then since $M \models O_{atc}A$, $O_{pf}A > O_{pf}C$, and since $O_{pf}(A \wedge B) \geq O_{pf}A$ by (cent), then $O_{pf}(A \wedge B) > O_{pf}C$, again by transitivity. Hence, in either case we have what is needed to show that $O_{pf}(A \wedge B)$ outweighs any incompatible alternative.

[26] Section 5.2.1 below discusses problems with consistent aggregation; the fatal problem is that $(CC)_{pf}$ faces much the same sort of deontic explosion described in Note 23.

[27] Suppose $O_{pf}A, O_{pf}B \in \Delta$. Since $\Vdash A \rightarrow (A \vee B)$, by (rm) we must have $O_{pf}(A \vee B) \in \Delta$. With (cent), $O_{pf}A \geq O_{pf}(A \vee B)$), and with (ent) $O_{pf}(A \vee B) \geq O_{pf}A$. Hence $O_{pf}A \approx O_{pf}(A \vee B)$. In a similar way, $O_{pf}B \approx O_{pf}(A \vee B)$. Then, since $\approx$ is an equivalence relation, $O_{pf}A \approx O_{pf}B$. Hence our arbitrary *prima facie* oughts in Δ must bear equal weight.

According to criterion (c), (DR) should fall out from (Def 1) on this account. Unfortunately, it does not.

Consider the following case. We call it the Mission Example 1. An agent, Brown, is under a directive to travel to Amsterdam at a certain time, A, so $O_{pf}A$. Brown is also under another equally compelling directive to travel to Barcelona at the same time, B, so $O_{pf}B$. Brown cannot do both, $\neg\Diamond(A \wedge B)$. If that were all, then under the Disjunctive Response we would conclude $O_{atc}(A \vee B)$, while $\neg O_{atc}A$ and $\neg O_{atc}B$. This much parallels the Marcus and Sartre examples. But now there is more. Brown is under another directive to travel to Cairo at that time, so $O_{pf}C$, and Brown cannot go to Cairo and Amsterdam both, nor to Cairo and Barcelona both, $\neg\Diamond(A \wedge C)$ and $\neg\Diamond(B \wedge C)$. Moreover, suppose these three missions are all of equal weight, $O_{pf}A \approx O_{pf}B \approx O_{pf}C$. If that were all, then under the Disjunctive Response all-things-considered Brown ought to go either to Amsterdam or to Barcelona or to Cairo, $O_{atc}(A \vee B \vee C)$, even while $\neg O_{atc}A$, $\neg O_{atc}B$ and $\neg O_{atc}C$. But there is still more to fill in the example. Brown has a fourth assignment, to travel to Dublin, that is definitely more important than his mission to Cairo, so $O_{pf}D$ and $O_{pf}D > O_{pf}C$, and he can't do both, $\neg\Diamond(C \wedge D)$. Since Brown's obligation to go to Dublin overrides his obligation to go to Cairo, the *prima facie* obligation to go to Cairo is not a strong *prima facie* obligation. It remains possible, however, for Brown to complete his mission in Dublin and then travel on to complete his mission in Amsterdam, $\Diamond(A \wedge D)$, and likewise possible that he do what is required in Dublin and carry on to Barcelona, $\Diamond(B \wedge D)$.

Assuming nothing else is relevant, it seems natural to conclude that all-things-considered Brown ought to go to Dublin, and, under the Disjunctive Response, that he ought to go to Amsterdam or to Barcelona, though nothing decides between those. For $M = \langle w_0, W, v, (\Delta, \geq)\rangle$, with Δ the least set containing $O_{pf}A$, $O_{pf}B$, $O_{pf}C$, $O_{pf}D$ and closed appropriately and $\geq$ as specified, along with the given possibilities, then $M \models O_{atc}D$ well enough, since $O_{pf}D$ outweighs all incompatible alternatives, viz. $O_{pf}C$. But even with (rm) and (ent), $O_{pf}(A \vee B)$ does not outweigh $O_{pf}C$; hence $M \not\models O_{atc}(A \vee B)$, contrary to the desired response.

To this the proponent of that response might propose that the restriction of Θ to pairwise incompatible strong *prima facie* obligations is too strong. Let Θ then simply the set of all strong *prima facie* obligations from Δ. Call the modified principle (DR)$'$. This does follow from (Def 1) and (Def 2), with other plausible assumptions.[28]

---

[28]Consider any $M \in \mathcal{M}_t$, and suppose Θ as now specified to be finite, so that $\Theta = \{O_{pf}A_1, \ldots, O_{pf}A_n\}$, and $\Theta^* = \{A_1, \ldots, A_n\}$ and $\bigvee \Theta^* = A_1 \vee \cdots \vee A_n$. To show that $O_{atc}(A_1 \vee \cdots \vee A_n)$ requires first that $M \models O_{pf}(A_1 \vee \cdots \vee A_n)$. So long as Δ is

Nevertheless, (DR)′ still falls short. For the Mission Example 1, it supports only $M \models O_{atc}(A \vee B \vee D)$ for the disjunctive all-things-considered obligation, not $M \models O_{atc}(A \vee B)$, which is what is desired. Moreover, this $O_{atc}(A \vee B \vee D)$ is merely a trivial consequence of $O_{atc}D$ by (RM), and does not require (DR)′ at all.

Here is another problematic case for the Disjunctive Response under (Def 1); call it the Mission Example 2. Agent Brown now has a mission to Amsterdam, A, and one to Berlin, B, and another to Copenhagen, C. (Dublin is out of the picture.) Any two of these are possible, but all three would be too much; $\Diamond(A \wedge B), \Diamond(A \wedge C), \Diamond(B \wedge C)$, but $\neg\Diamond(A \wedge B \wedge C)$. The missions are all of equal importance, $O_{pf}A \approx O_{pf}B \approx O_{pf}C$. Given no other pertinent *prima facie* obligations in the case, what ought Brown to do, all-things-considered?

In the spirit of the Disjunctive Response, all-things-considered Brown ought at least to go to Amsterdam, or to Berlin, or to Copenhagen, but he has no such obligation to go to any one of them. For such a situation, $M \models O_{atc}(A \vee B \vee C)$, while $M \not\models O_{atc}A$, $M \not\models O_{atc}B$ and $M \not\models O_{atc}C$. And this will hold under the present definitions. The situation seems to call for more, however. Since these missions are pairwise co-possible, the Disjunctive Response would seem to call for saying that all-things-considered Brown ought to do any two of these, though it might be that Brown does not have an obligation, all-things-considered, to do any particular pair. I.e., $M \models O_{atc}((A \wedge B) \vee (A \wedge C) \vee (B \wedge C))$, though $M \not\models O_{atc}(A \wedge B)$, $M \not\models O_{atc}(A \wedge C)$ and $M \not\models O_{atc}(B \wedge C)$.

Under (Def 1), however, there is little prospect for deriving $O_{atc}((A \wedge B) \vee (A \wedge C) \vee (B \wedge C))$ given the situation as described. That would require first establishing $O_{pf}((A \wedge B) \vee (A \wedge C) \vee (B \wedge C)) \in \Delta$, and that is not forthcoming. Though $O_{pf}A \in \Delta$ and $O_{pf}B \in \Delta$, that does not suffice for $O_{pf}(A \wedge B) \in \Delta$, even though A and B are jointly possible, not without the closure condition (cc), which, as we saw, is unacceptable in combination

---

closed under (rm) that will follow since $O_{pf}A_i \in \Delta$ for each $A_i \in \Theta^*$. Next it must be shown that for any C such that $O_{pf}C \in \Delta$ with C incompatible with $(A_1 \vee \cdots \vee A_n)$, $O_{pf}(A_1 \vee \cdots \vee A_n) > O_{pf}C$. Suppose such an incompatible C. Then, by basic modal logic, C is incompatible with each A_i. Suppose for *reductio* $\neg(O_{pf}(A_1 \vee \cdots \vee A_n) > O_{pf}C)$. Since $O_{pf}C$, then either $M \models O_{spf}C$ or there is a D such that $M \models O_{spf}D$ and $O_{pf}D > O_{pf}C$. If the first, then $C = A_i$ for some $A_i \in \Theta^*$, by the revised definition of Θ. But that contradicts that C is incompatible with A_i. On the other hand, if there is such a D, then $D = A_i$ for some $A_i \in \Theta^*$, and so $O_{pf}(A_1 \vee \cdots \vee A_n) \geq O_{pf}D$ by (ent). In that case, $O_{pf}(A_1 \vee \cdots \vee A_n) > O_{pf}C$ by transitivity to contradict the assumption. Hence, in either case there is a contradiction and the assumption must be denied. So, $O_{pf}(A_1 \vee \cdots \vee A_n) > O_{pf}C$, and so $M \models O_{atc}(A_1 \vee \cdots \vee A_n)$, as desired. This presumes that Δ satisfies (rm) and that $\geq$ respects (ent). Otherwise, even the revised version of (DR) will fail.

with the conditions (rm)/(nm) and (ent)/(nent), which are also required in the derivation. The same goes for any of the other conjunctions of the *prima facie* oughts. Without such conjunctive *prima facie* oughts, it seems impossible to establish the disjunctive combination of them, and without that $O_{atc}((A \wedge B) \vee (A \wedge C) \vee (B \wedge C))$ is unwarranted.

These examples show that the Disjunctive Response, however plausible it might seem at first, cannot be justified on the basis of (Def 1). Perhaps there are other variations on a principle like (DR) that can do better; perhaps there are more epicycles to introduce. It is doubtful, however, that that can be done in a way that rules out further counterexamples, remains sufficiently intuitive, and is firmly grounded in (Def 1), which is the target here.

4.1.3 The final collapse

So far we have seen that this account of how all-things-considered oughts are derived from *prima facie* oughts by (Def 1) fails criterion (b) since at least one of the core principles, (NM)/(RM) or (C), must be abandoned. We have also seen that it fails criterion (c) by not supporting the Disjunctive Response. Now we discover further shortcomings. The account misdescribes the all-things-considered oughts one would expect to apply in many situations. It both undergenerates and overgenerates. For some situations there are all-things-considered oughts that hold that it fails to produce, and for others it produces all-things-considered oughts it should not, including full-fledged normative conflicts for the all-things considered ought, contrary to the avowed perspective of this account. Hence, it fails criterion (a) as well.

Undergeneration appears in the following example; call it Mission 3. Brown again has three assignments, to go to Amsterdam, A, to Barcelona, B, and to Copenhagen, C, $O_{pf}A$, $O_{pf}B$ and $O_{pf}C$. It is not possible for him to go both to Amsterdam and to Barcelona, nor to Barcelona and Copenhagen; $\neg \Diamond (A \wedge B)$, $\neg \Diamond (B \wedge C)$, but it is possible for Brown to go to both Amsterdam and Copenhagen, $\Diamond (A \wedge C)$. In this case the weights of the missions are such that the obligation to go Amsterdam outweighs that for Barcelona, which outweighs the obligation to go to Copenhagen, $O_{pf}A > O_{pf}B > O_{pf}C$. Under (Def 1) it now follows that all-things-consisdered Brown ought to go to Amsterdam, $O_{atc}A$, and not that he go to Barcelona, $\neg O_{atc}B$. It also follows, however, that it is not the case that Brown ought, all-things-considered, to go to Copenhagen, $\neg O_{atc}C$, since the *prima facie* obligation to go there is outweighed. This seems wrong. It overlooks the fact that what outweighs going to Copenhagen has itself been overruled. That would seem to reinstate the obligation for Copenhagen. $O_{atc}C$ should hold, but nothing in the present account warrants that.

For later reference we call the general pattern at work here Pattern A, when $O_{pf}A > O_{pf}B > O_{pf}C$ and $\neg \Diamond (A \wedge B)$, $\neg \Diamond (B \wedge C)$, $\Diamond (A \wedge C)$.

The next three Patterns illustrate overgeneration, where more all-things-considered oughts are generated by (Def 1) than seems right. Perhaps it is not necessary to tell a story to go with these. All three Patterns apply variations of a situation in which $O_{pf}\neg A$, $O_{pf}\neg B$ and $O_{pf}(A \vee B)$ hold, and they are pairwise compatible.

For Pattern B, these *prima facie* oughts are ordered so that $O_{pf}\neg A > O_{pf}\neg B > O_{pf}(A \vee B)$. Given M with $(\Delta, \geq)$ so constituted, by (Def 1), $M \models O_{atc}\neg A$ and $M \models O_{atc}\neg B$ just as one expects. Unfortunately, also $M \models O_{atc}(A \vee B)$, contrary to expectation. This is because there is no *prima facie* ought in Δ incompatible with $O_{pf}(A \vee B)$. If $O_{pf}(\neg A \wedge \neg B)$ were in Δ and had greater weight than $O_{pf}(A \vee B)$, that would take care of things. But that is not given explicitly, and otherwise it requires conditions (cc) for Δ and (cent) for $\geq$, which are problematic.

This Pattern thus reveals the need to allow some way for a combination of *prima facie* oughts to outweigh another, and not just one on one. It is the pair $O_{pf}\neg A$, $O_{pf}\neg B$ taken together that in some way cancels $O_{pf}(A \vee B)$. On the present account, however, there is no mechanism for that kind of outweighing.

It could go the other way too; there could be cases in which a single *prima facie* ought outweighs a combination of other *prima facie* oughts. This is Pattern C, where $O_{pf}\neg A > O_{pf}\neg B$ and $O_{pf}\neg B \approx O_{pf}(A \vee B)$. Here one expects $O_{atc}\neg A$ to hold, and it does under (Def 1). But one also expects neither $O_{atc}\neg B$ nor $O_{atc}(A \vee B)$. Under (Def 1), however, both are supported. In some sense the *prima facie* obligation $O_{pf}\neg A$ overrules the pair of $O_{pf}\neg B$ and $O_{pf}(A \vee B)$, though neither one individually. The present account has no mechanism to handle this either.

Finally, Pattern D represents cases in which none of the *prima facie* oughts outweighs any of the others, e.g., $O_{pf}\neg A \approx O_{pf}\neg B \approx O_{pf}(A \vee B)$. For such a situation one might expect none of the *prima facie* oughts to rise to an all-things-considered ought. Under (Def 1), however, all three do since none is incompatible with any other.

This Pattern, like the other two, reveals that (Def 1) not only generates more all-things-considered oughts than seems warranted, it also allows for cases in which an agent all-things-considered ought to do a number of things which cannot all be done together. In short, it allows for full-fledged normative conflicts for all-things-considered oughts. Thus it fails criterion (a) as well as (b) and (c). Accounts like (Def 1) are often thought to exclude normative conflicts for all-things-considered oughts easily by definition. While they do indeed exclude binary conflicts, nevertheless, as these Patterns illustrate, conflicts of greater arity remain possible. That is surely anathema to those who would deny the possibility of genuine normative conflicts.

Since all three criteria fail, accounts like this with (Def 1) must appear seriously inadequate for the purposes of those who would promote them, however plausible they might seem at first.

4.2 Reasons and *prima facie* oughts

Here we begin to look more deeply into the nature of *prima facie* oughts. The previous account ignored that there is something inherently conditional or relational about such oughts. That was always part of Ross's own characterization. We say, for example, that Ada ought to take her son to the circus because she so promised, or that, given her promise, she ought to take her son to the circus, or that she ought because her promising provides a reason for taking her son to the circus. It is easy to see these locutions as conditionals. At a minimum, it seems as if Ada ought to take her son to the circus because of a particular relation between the reason, her promise, and what she is to do, what she has promised.

To capture this conditional or relational element, I introduce a new notation, $\leftrightarrowtail$, that is supposed to be unencumbered by commitments from established accounts. This lets us to explore in a general way what properties are appropriate to it when dealing with *prima facie* oughts and their conflicts. '$A \leftrightarrowtail B$' says informally 'The state of affairs A would provide a reason for B', or 'A would require B', where typically B expresses an agent's doing something, but might also represent other sorts of states of affairs. We take $\leftrightarrowtail$ to carry normative force, so that $A \leftrightarrowtail B$ says that A would provide a normative reason for, or would normatively require, B. $A \leftrightarrowtail B$ could even be read to say that, if A, then it ought to be that B, though that phrasing might suggest other conditional oughts. We say that A *is* a reason for B when both A and $A \leftrightarrowtail B$ obtain. Formally speaking, $A \leftrightarrowtail B$ is a well-formed member of $\mathcal{L}_D$ whenever A and B are formulas of $\mathcal{L}_B$.[29]

While $A \leftrightarrowtail B$ may be a sort of conditional, it does not satisfy *modus ponens*, for it may be that A would provide reason for doing B, and A is true, but one does not do B. Ada's promising to take her son to the circus

---

[29]The present account stems primarily from Chisholm, who writes 'pRq' for 'p requires q' [1964] or 'p would require q' [1974]. [Loewer and Belzer, 1991, p. 365] presents the approach more formally, using standard notation, '$O(q/p)$', which they read 'p is a moral reason to do q', or again [Belzer and Loewer, 1997, p. 46] '$\bigcirc(A/B)$' to mean that B defeasibly requires A. [Asher and Bonevac, 1996] write '$A >_O B$' to say that the truth of A is a reason for doing B. (Asher and Bonevac do not, however, follow (Def 4) given below.) Our notation $A \leftrightarrowtail B$ is meant to do duty for all of these, and also for the concepts of other authors who write in a similar vein, in various ways, of reasons or requirements, especially in relation to *prima facie* oughts, but who do not introduce a formal notation, e.g. Broome [1999; 2004; 2007], Harman [1975], Raz [1978], Sinnott-Armstrong [1988]. Horty [2007; 2012] also speaks of reasons and *prima facie* oughts, but his account corresponds more to that of Section 4.4 below.

would give reason for her to do that, $p \leadsto c$, she does so promise, p, but maybe she does not take him to the circus. That she promised implies she ought to take him, but that cannot mean that all-things-considered she ought to, for something of greater moment might override this requirement, such as her need to attend the meeting. Detachment to a *prima facie* ought-statement does, however, seem appropriate. Generally speaking, one ought *prima facie* to do that which one has a (normative) reason to do. That is the basis of the present account.

Models in $\mathcal{M}_r$ are structures $M = \langle w_0, W, v, \Re \rangle$ in which $\Re$ represents whatever it takes to interpret expressions $A \leadsto B$. For present purposes it will suffice to let it be a set of such expressions themselves. Then,

- $M \models A \leadsto B$ if and only if $A \leadsto B \in \Re$

much as in the previous subsection for expressions $O_{pf}A$. The key question then is, What conditions should $\Re$ meet in order that it provide an appropriate account of *prima facie* and all-things-considered oughts?

In keeping with the description above for *prima facie* oughts, for this account, with such models $M \in \mathcal{M}_r$,

Def 3) $M \models O_{pf}A$ if and only if there is a B such that
$M \models B$ and $B \leadsto A \in \Re$.

We continue to use Δ to represent the set of *prima facie* oughts that hold in a situation. I.e., given $M \in \mathcal{M}_r$, $\Delta_M = \{O_{pf}A : M \models O_{pf}A\}$.

(Def 3) plainly allows the possibility of *prima facie* oughts in conflict, as in the Ross Example. There could be reason to do something, as the promise would provide reason to meet the friend, $p \leadsto m$, and reason to do something else, as the accident and the victims' need would provide reason to help them, $a \leadsto h$, where both reasons obtain, $p \wedge a$, so for M to represent this situation, $M \models O_{pf}m$ and $M \models O_{pf}h$, but $M \models \neg \Diamond(m \wedge h)$.

Given that, and given that there is no primitive relation of relative weight in play, how should the fact that one *prima facie* ought takes precedence over another be captured in this picture? With the Ross Example as usually understood, where the obligation to help the victims is more incumbent than the obligation to meet the friend, it is natural to suppose that the combined occurrence, $p \wedge a$, itself provides abiding reason to help the victims. Hence, $(p \wedge a) \leadsto h$ obtains, while $(p \wedge a) \leadsto m$ does not. Indeed, there is even reason not to meet the friend, $(p \wedge a) \leadsto \neg m$. This is the key to how one *prima facie* ought defeats another.

Defeat comes in two flavors, one stronger than the other. On the one hand, it could be that a reason A for something C is defeated by something else B that makes it the case that one has reason not to do that C. In that case, let us say that the further reason B *overrides* the former A for C. On the other hand, it could be that B simply undercuts the reason for C, making it no longer sufficient for C.[30] Given a model $M \in \mathcal{M}_r$,

- B overrides$_M$ a reason A for C if and only if A is a reason for C (i.e., $M \models A$ and $M \models A \leadsto C$) and $M \models B$ and $M \models (A \wedge B) \leadsto \neg C$;

- B undercuts$_M$ a reason A for C iff A is a reason for C, and $M \models B$ and $M \not\models (A \wedge B) \leadsto C$;

- B defeats$_M$ a reason A for C iff B overrides$_M$ or undercuts$_M$ A for C;

- a reason A for C is overridden$_M$ (undercut$_M$, defeated$_M$) *simpliciter* iff there is a B that overrides$_M$ (undercuts$_M$, defeats$_M$) A for C.

- $O_{pf}C \in \Delta_M$ is overridden$_M$ (undercut$_M$, defeated$_M$) iff every reason A for C is overridden$_M$ (undercut$_M$, defeated$_M$).

(When M is clear in context, the subscript may be omitted.)

This characterizes overriding, undercutting and defeat in terms of a situation or circumstance undoing a *prima facie* ought. These notions can be extended to describe one *prima facie* ought overriding, undercutting or defeating another in the obvious way. Given $M \in \mathcal{M}_r$,

- $O_{pf}D$ overrides $O_{pf}C$ iff $O_{pf}C, O_{pf}D \in \Delta_M$, C and D are incompatible and for every reason A for C there is a reason B for D such that $M \models (A \wedge B) \leadsto (D \wedge \neg C)$;

- $O_{pf}D$ undercuts $O_{pf}C$ iff $O_{pf}C, O_{pf}D \in \Delta_M$, C and D are incompatible and for every reason A for C there is a reason B for D such that B undercuts A for C;

- $O_{pf}D$ defeats $O_{pf}C$ iff $O_{pf}D$ overrides or undercuts $O_{pf}C$.

Note that $O_{pf}A$ is overridden if and only if there is some $O_{pf}B$ that overrides $O_{pf}A$. $O_{pf}A$ could, however, be undercut but not by any $O_{pf}B$.

These notions of defeat, being overridden, being undercut, do not rely on any relation of relative weight for the *prima facie* oughts or their supporting reasons, but simply on the (lack of) persistence of reasons in the face of other reasons. Because of that, now, instead of saying that one's all-things-considered obligation is to do the weighiest of one's *prima facie* duties, we

---

[30] What we now call undercutting, Chisholm [1964, p. 148], [1974, p. 8], called overriding. Given principles for $\leadsto$ to which Chisholm subscribes, our sense of overriding entails his.

say merely that one ought, all-things-considered, to do those *prima facie* oughts that are undefeated in the circumstances. They are those whose reasons persist against all comers. Given $M \in \mathcal{M}_r$

Def 4) $M \models O_{atc}A$ iff $O_{pf}A \in \Delta_M$ and $O_{pf}A$ is undefeated$_M$, i.e., neither overridden nor undercut.

With (Def 3) and (Def 4) to characterize *prima facie* and all-things-considered oughts, respectively, what properties the two oughts possess and whether this account satisfies the criteria (a)–(c) above, depends in great part on the properties of how reasons or requirements are related to their consequences, on the nature of $A \leftrightsquigarrow B$ itself. We turn to that next.

4.2.1 Principles for $\leftrightsquigarrow$

What principles are appropriate for formulas $A \leftrightsquigarrow B$ depends on precisely how that connection is supposed to be understood. Rather than engage that question, however, as if to provide a full theory of $\leftrightsquigarrow$, we now follow the method of the preceding section, and consider what minimally must be the case for the account to meet the primary criteria (a)–(c), and especially (b) that O_{atc}-oughts obey deontic logic's core principles.

We assume without remark replacement for logical equivalents, though only a very limited form is actually required for the present results.[31]

$\mathfrak{R}$e)   If $\Vdash A \leftrightarrow B$ and $\Vdash C \leftrightarrow D$, then $A \leftrightsquigarrow C \in \mathfrak{R}$ iff $B \leftrightsquigarrow D \in \mathfrak{R}$

which validates

$\leftrightsquigarrow$E)   If $\Vdash A \leftrightarrow B$ and $\Vdash C \leftrightarrow D$, then $\Vdash (A \leftrightsquigarrow C) \leftrightarrow (B \leftrightsquigarrow D)$

Beyond that, two fundamental conditions on $\mathfrak{R}$ are required corresponding to the principles of distribution and aggregation. For $M = \langle w_0, W, v, \mathfrak{R} \rangle$,

$\mathfrak{R}$rm)   If $A \leftrightsquigarrow B \in \mathfrak{R}$ and $\Vdash B \to C$ then $A \leftrightsquigarrow C \in \mathfrak{R}$, or
$\mathfrak{R}$nm)   If $A \leftrightsquigarrow B \in \mathfrak{R}$ and B necessitates C on M then $A \leftrightsquigarrow C \in \mathfrak{R}$,
$\mathfrak{R}$c)   If $A \leftrightsquigarrow B \in \mathfrak{R}$ and $A \leftrightsquigarrow C \in \mathfrak{R}$ then $A \leftrightsquigarrow (B \wedge C) \in \mathfrak{R}$

These obviously validate these principles for $\leftrightsquigarrow$,[32]

---

[31] This sort of full replacement is assumed by [Asher and Bonevac, 1996] and [Belzer and Loewer, 1997].

[32] ($\leftrightsquigarrow$C) is Chisholm's [1974] axiom A6. Although he does not mention ($\leftrightsquigarrow$RM) or ($\leftrightsquigarrow$M), nothing in his discussion suggests he would deny them. Other authors do provide for these, e.g., Asher and Bonevac [1996] and Belzer and Loewer [1997]. (Belzer and Loewer present this as an option. Their preferred system in [1997], M3D, is very weak and does not include (M), for all-things-considered oughts, nor does it include (C), though they acknowledge that both could be added.)

↝RM)	If $\Vdash B \to C$ then $\Vdash (A \↝ B) \to (A \↝ C)$
↝NM)	$\Vdash \Box(B \to C) \to ((A \↝ B) \to (A \↝ C))$
↝M)	$\Vdash (A \↝ (B \land C)) \to (A \↝ B)$
↝C)	$\Vdash ((A \↝ B) \land (A \↝ C)) \to (A \↝ (B \land C))$

While one might question both distribution and aggregation for ↝, their validity is a prerequisite for the all-things-considered ought $O_{atc}A$ to satisfy the corresponding principles. Without (ℜrm)/(ℜnm) there is no guarantee that $O_{pf}B$ will hold when $O_{atc}A$ and $\Vdash A \to B$ or $\Box(A \to B)$, and without $O_{pf}B$, $O_{atc}B$ will not follow. Similarly for the consequent of (C).[33] Hence we now take them for granted.

These conditions suffice for all the core principles for all-things-considered oughts. That (ℜnm)/(ℜrm) yields the validity of (NM)/(RM) under (Def 4) is to be expected. Interestingly, (C) calls for both (ℜc) and (ℜrm).[34]

Condition (ℜrm) also suffices for (P),[35] and (C) and (P) yield (D) for all-things considered oughts. Thus, with just the two conditions (ℜnm)/(ℜrm) and (ℜc) the present account meets criterion (b).

Some authors present other principles to govern their versions of ↝. In addition to (↝ C), Chisholm [1964, p. 148], [1974, pp. 6ff.] postulates analogues of (P) and (D). In our notation,

---

[33] For a countermodel to both, without (ℜrm) and (ℜc), let $M = \langle w_0, W, v, \Re \rangle$ with $w_0 \in v(p)$, $\Re = \{A \↝ r : M \models A\} \cup \{B \↝ s : M \models B\}$, and the rest of M may be anything. Then $p \↝ r \in \Re$ and $p \↝ s \in \Re$, so $M \vdash O_{pf}r$ and $M \vdash O_{pf}s$. Neither of those is defeated in M, as is easily verified. Hence $M \models O_{atc}r$ and $M \models O_{atc}s$. Yet $M \not\models O_{pf}(r \lor t)$ and $M \not\models O_{pf}(r \land s)$, as is easily verified. So $M \not\models O_{atc}(r \lor t)$, contrary to (RM), and $M \not\models O_{atc}(r \land s)$, contrary to (C).

[34] Verification for (NM)/(RM) is easy, and left to the reader. For (C), suppose $M \models O_{atc}A$ and $M \models O_{atc}B$. By the non-defeat clause of (Def 4) that entails that there are C, D, E and F such that $M \models C \land (C \↝ A)$ and for all G if $M \models G$ then $M \models (C \land G) \↝ A$ and $M \models D \land (D \↝ B)$ and for all H if $M \models H$ then $M \models (D \land H) \↝ B$ (non-undercut), and $M \models E \land (E \↝ A)$ and for all I if $M \models I$ then $M \not\models (C \land I) \↝ \neg A$ and $M \models F \land (F \↝ B)$ and for all J if $M \models J$ then $M \not\models (F \land J) \↝ \neg B$ (non-overridden). Let $\Phi = C \land D \land E \land F$. Then $M \models \Phi \land (\Phi \↝ A)$ and $M \models \Phi \land (\Phi \↝ B)$. Hence, by (ℜc), $M \models \Phi \land (\Phi \↝ (A \land B))$, so by (Def 3) $M \models O_{pf}(A \land B)$. To show that that is not overridden in M, suppose it were. Then there would be a K such that K overrides Φ for $A \land B$, i.e., $M \models K$ and $M \models (\Phi \land K) \↝ \neg (A \land B)$. By the original non-undercut provision, $M \models (K \land \Phi \land C) \↝ A$ and $M \models (K \land \Phi \land D) \↝ B$, but the clauses C and D are redundant, so $M \models (K \land \Phi) \↝ (A \land B)$. By (ℜc), $M \models (K \land \Phi) \↝ (A \land B \land \neg (A \land B))$, whence $M \models (K \land \Phi) \↝ \neg A$ by (ℜrm). By the original non-overridden clause, however, we know that $M \not\models (K \land \Phi \land E) \↝ \neg A$, or, since E is redundant $M \not\models (K \land \Phi) \↝ \neg A$, a contradiction. Hence $O_{pf}(A \land B)$ must not be overridden. That it is not undercut is argued similarly, but (ℜc) alone suffices.

[35] Suppose $M \models O_{atc}A$. Then $M \models O_{pf}A$. Since that is not overridden, there is a B such that $M \models B$ and $M \models B \↝ A$ and for all C, if $M \models C$ then $M \not\models (C \land B) \↝ \neg A$. Hence $M \not\models B \↝ \neg A$. Suppose, for *reductio*, A were impossible, $M \models \neg \Diamond A$. By basic modal logic, $M \models \Box(A \to \neg A)$. Since $M \models B \↝ A$, $M \models B \↝ \neg A$, by (↝nm). Contradiction.

⇝P) $(A \leadsto B) \to \Diamond(A \wedge B)$
⇝D) $(A \leadsto B) \to \neg(A \leadsto \neg B)$

(With (⇝C) the former entails the latter.)[36]

Asher and Bonevac [1996, p. 43] argue against (⇝P). Since one can promise the impossible, and the fact of a promise provides a reason for doing what one has promised, there can be a reason, the promise, and that for which it is a reason, the impossible, that are not jointly possible. This would not jeopardize (P) for O_{atc} since any such reason would be self-defeating. On the other hand, if there could be reasons for the impossible, then, by (⇝NM), everything would be *prima facie* obligatory.[37] This suggest one might want at least this weaker principle.

⇝P)′ $(A \leadsto B) \to \Diamond B$

Chisholm [1974] also gives, his A5 and A7,

⇝DISJ 1) $((A \leadsto C) \wedge (B \leadsto C)) \to ((A \vee B) \leadsto C)$
⇝DISJ 2) $((A \vee B) \leadsto C) \to ((A \leadsto C) \vee (B \leadsto C))$

Asher and Bonevac [1996] subscribe to (⇝DISJ 1), though perhaps not (⇝DISJ 2). Belzer and Loewer [1997] discuss these principles and also a principle of rational monotony they ascribe to von Wright

vW) $((A \leadsto B) \wedge \neg(A \leadsto \neg C)) \to ((A \wedge C) \leadsto B)$

We leave these as options since they are not required to derive the core principles for all-things-considered oughts within the present picture, which is all that is needed to apply the distinction between *prima facie* and all-things-considered oughts to the issue of normative conflicts.

Not generally discussed in this framework is the question of what basic principles govern *prima facie* oughts themselves. Conditions ($\Re$nm)/($\Re$rm) and ($\Re$c) validate distribution $(NM)_{pf}/(RM)_{pf}$, and that's about all. One expects aggregation $(C)_{pf}$ to fail, and it does. There could be a reason C for A and a reason D for B but no reason E for $A \wedge B$. Even $(P)_{pf}$ does not hold unless (⇝P) or (⇝P)′ obtains. $(P)_{pf}$ and $(RM)_{pf}$ together yield a logic for *prima facie* oughts that is close to the system **P** discussed below in Section 5.2 as a logic for ought that allows normative conflicts.

---

[36] Given (⇝P) and (⇝C), or the conditions on $\Re$ to validate them, overriding entails undercutting, in which case the all-things-considered ought can be defined as the *prima facie* ought that is not undercut, which is Chisholm's definition, though he says not 'overridden'. Also, with (⇝P), aggregation (C) for all-things-considered oughts is validated by ($\Re$c) without appeal to ($\Re$rm).

[37] Suppose $\neg \Diamond B$ and $A \leadsto B$. Since B is impossible, $\Box(B \to C)$, so $A \leadsto C$ for any C. Cf. the discussion of (P) and (DEX_{oi}) in Section 5.2 below.

4.2.2 The Disjunctive Response

As we have seen, in this picture conflicts among all-things-considered oughts are impossible. That is guaranteed by (P), (C) and (NM)/(RM) for O_{atc}. Since all-things-considered oughts are now identified with *prima facie* oughts that are undefeated, conflicts among undefeated *prima facie* oughts are impossible. How then shall situations like the Marcus Example be represented where there seem to be two obligations neither of which overrides the other?

Neither overrides, but they may yet undercut each other, and so both be defeated. In the Marcus Example, twin-1 is in need, n_1, which provides a reason to save twin-1, $n_1 \hookrightarrow s_1$, and also twin-2 is in need, n_2, which provides a reason to save twin-2, $n_2 \hookrightarrow s_2$. Hence, for the potential rescuer, $O_{pf}s_1$ and $O_{pf}s_2$ both hold, while $\neg \Diamond(s_1 \wedge s_2)$. It is also given that neither *prima facie* ought overrides the other. That entails that neither $(n_1 \wedge n_2) \hookrightarrow \neg s_1$ nor $(n_1 \wedge n_2) \hookrightarrow \neg s_2$ obtains. By ($\Re$nm), that entails further that neither $(n_1 \wedge n_2) \hookrightarrow s_1$ nor $(n_1 \wedge n_2) \hookrightarrow s_2$ holds. Hence both *prima facie* oughts are undercut, and neither rises to an all-things-considered obligation. This seems a fair representation of the scenario.

According to the Disjunctive Response, this scenario should also make $O_{atc}(s_1 \vee s_2)$ true. Unfortunately, this does not seem to follow from anything given so far. While $O_{pf}(s_1 \vee s_2)$ follows from each of $O_{pf}s_1$ and $O_{pf}s_2$, to establish that it is not defeated requires $(n_1 \wedge n_2) \hookrightarrow (s_1 \vee s_2)$. While that may seem correct in context, it does not follow from any of the provisions given with the example. If there were a general principle,

$\hookrightarrow$CSA) $\Vdash ((A \hookrightarrow C) \wedge (B \hookrightarrow C)) \to ((A \wedge B) \hookrightarrow C)$

a sort of very cautious strengthening of the antecedent, that would provide what is needed. As far as I know, however, no one has recommended such a principle, and we do not recommend it now. Yet without it, or something similar, the Disjunctive Response is not supported for this example. The first Mission Example suffers much the same.

Mission-2 is even more problematic since it calls for conjunctive disjuncts. While it might be that r_a provides a reason for Brown to go to Amsterdam, $r_a \hookrightarrow A$ and r_b a reason to go to Berlin, $r_b \hookrightarrow B$, and r_c reason to go to Copenhagen, $r_c \hookrightarrow C$, there is little prospect of providing reason to go to Amsterdam and Berlin, or Amsterdam and Copenhagen, or Berlin and Copenhagen even with ($\hookrightarrow$CSA). A stronger principle seems to be required, one that allows conjunction in both antecedent and consequent, e.g.,

$\hookrightarrow$CAC) $\Vdash ((A \hookrightarrow B) \wedge (C \hookrightarrow D) \wedge \Diamond(A \wedge B \wedge C \wedge D)) \to$
$((A \wedge C) \hookrightarrow (B \wedge D))$

This would provide $(r_a \wedge r_b) \hookrightarrow (A \wedge B)$, etc. Without that, or the others, $O_{pf}((A \wedge B) \vee (A \wedge C) \vee (B \wedge C))$ will not be forthcoming, much less $O_{atc}((A \wedge$

$B) \vee (A \wedge C) \vee (B \wedge C))$. Principle ($\looparrowright$CAC) is, however, clearly unacceptable in the present context. It would defeat the very idea of defeasibilty.[38]

Without general rules like ($\looparrowright$CSA) or ($\looparrowright$CAC) one can, of course, claim that the requisite *prima facie* obligations hold, e.g., for the Marcus Example, that $(n_1 \wedge n_2) \looparrowright (s_1 \vee s_2)$, and similarly for the others. This seems entirely *ad hoc*, however. There should be a principle by which to derive the all-things-considered oughts the Disjunctive Response declares. Moreover, it should be a principle that grows out of the natural properties of $\looparrowright$, however that is taken. None seems in the offing. Hence, we must conclude this account of *prima facie* and all-things-considered oughts with (Def 3) and (Def 4) fails to support of the Disjunctive Response, and so fails criterion (c).

4.2.3 Other issues: Under- and overgeneration

In Section 4.1.3 we saw the initial picture of all-things-considered oughts derived from *prima facie* oughts finally collapse under the weight of two sorts of problems. On one hand, it undergenerated certain all-things-considered oughts. There were cases, such as the Mission Example 3 under Pattern A, where a *prima facie* ought was determined not to be an all-things-considered ought even though it looked as though it should be. And on the other hand the picture overgenerated other all-things-considered oughts, determining certain *prima facie* oughts to be all-things-considered oughts even though they should not be, including triadic normative conflicts for all-things-considered oughts. That was seen in Patterns (B)–(D). Here we look at how the present picture copes with those problems, and we find it fares better, though some questions remain.

In Pattern A, $O_{pf}C$, is overridden by $O_{pf}B$ which is overridden by $O_{pf}A$ compatible with the first. Under (Def 1), because $O_{pf}C$ is overridden, it is excluded as a candidate for an all-things-considered ought, $O_{atc}C$, even though its claim should be reinstated by $O_{pf}B$ being overridden. The present account with (Def 4) faces much the same problem.

Simply put, $O_{pf}C$ cannot be both overridden by $O_{pf}B$, as in the Pattern, and not overridden at all, as required for $O_{atc}C$ by (Def 4). Nevertheless, this framework offers a more nuanced account of the kind of situation that gives rise to Pattern A, for it can articulate that a reason, r_c for C can be overridden by a reason, r_b for B, and that reason overridden in turn by a reason r_a, for A, and the combination of all three still provide a reason for C. This means, at least, $M \models r_c$ and $M \models r_c \looparrowright C$ and $M \models r_b$ and $M \models r_b \looparrowright B$, and $M \models (r_b \wedge r_c) \looparrowright \neg C)$, and thus r_b overrides r_c for C, while also $M \models r_a$ and $M \models r_a \looparrowright A$ and $M \models (r_a \wedge r_b) \looparrowright \neg B)$, so

---

[38] Consider the Ross Example, where it is assumed $p \looparrowright m$, $a \looparrowright h$ and $\neg((p \wedge a) \looparrowright m)$, and $\Diamond(p \wedge a \wedge m)$. By ($\looparrowright$RM), $a \looparrowright \top$, hence $(p \wedge a) \looparrowright (m \wedge \top)$, by ($\looparrowright$CAC), since $\Diamond(p \wedge a \wedge m \wedge \top)$. That is equivalent to $(p \wedge a) \looparrowright m$, a contradiction.

that r_a overrides r_b for B. Then, when all three reasons are considered, it is plausible that $M \models (r_a \wedge r_b \wedge r_c) \leadsto A$, and $M \models (r_a \wedge r_b \wedge r_c) \leadsto \neg B$, and also $M \models (r_a \wedge r_b \wedge r_c) \leadsto C$. If these are all the relevant conditions, $M \models O_{atc}A$, $M \models O_{atc}\neg B$ and $M \models O_{atc}C$, as expected. In this way, the present picture can represent the reinstatement of the obligation for C consistently, unlike the first account with (Def 1). This is so at least so long as those assumptions about the compound reasons are granted.

We could elevate that to a principle to govern the connection $\leadsto$. Van der Torre [1997, p. 139] and van der Torre and Tan [1997, p. 115] propose this rule of obligation reinstatement for *prima facie* conditional obligations that are subject to overriding. In our notation, for any model $M \in \mathcal{M}_r$

RIO) If $M \models B_1 \leadsto A_1$ and $M \models (B_1 \wedge B_2) \leadsto (\neg A_1 \wedge A_2)$ and $M \models (B_1 \wedge B_2 \wedge B_3) \leadsto \neg A_2$, then $M \models (B_1 \wedge B_2 \wedge B_3) \leadsto A_1$.

This would apply to cases following Pattern A as illustrated above.[39]

The other problem that beset the first, traditional picture with (Def 1) concerned the overgeneration of unwanted all-things-considered oughts. These arose because there could be cases where it seems a combination of *prima facie* oughts, rather than a single one, precludes another *prima facie* ought, or that a *prima facie* ought supercedes a combination even though it does not defeat any individual one in that combination. Under the original picture with (Def 1) there is no way to represent that sort of cancelling, and so the intuitively defeated *prima facie* oughts emerge as all-things-considered oughts when they should not.

Patterns B–D concerned the combination of three *prima facie* oughts, $O_{pf}\neg A$, $O_{pf}\neg B$ and $O_{pf}(A \vee B)$ with different priorities in the different Patterns. In the present framework we suppose models that provide reasons for each *prima facie* ought, i.e., r_1, r_2, r_3 such that $M \models r_1 \wedge r_2 \wedge r_3$ and $M \models r_1 \leadsto \neg A$ and $M \models r_2 \leadsto \neg B$ and $M \models r_3 \leadsto (A \vee B)$, and then consider different arrangements of overriding.

For Pattern B, the pair of $O_{pf}\neg A$ and $O_{pf}\neg B$ should defeat $O_{pf}(A \vee B)$, the first two counting as all-things-considered oughts and the third not. Here that amounts to assuming when all three reasons are considered, they provide reason for $\neg A$ and $\neg B$ but not $A \vee B$. Thus, $M \models (r_1 \wedge r_2 \wedge r_3) \leadsto \neg A$ and $M \models (r_1 \wedge r_2 \wedge r_3) \leadsto \neg B$, which combine by $(\leadsto C)$ to $M \models (r_1 \wedge r_2 \wedge r_3) \leadsto (\neg A \wedge \neg B)$, hence $M \models (r_1 \wedge r_2 \wedge r_3) \leadsto \neg(A \vee B)$. Assuming no other relevant factors, this provides for $M \models O_{atc}\neg A$, $M \models O_{atc}\neg B$ and $M \not\models O_{atc}(A \vee B)$, just as we should have. Thus, this picture can consistently

---

[39]These authors propose two other principles, their (RI) and (FC), to govern $\leadsto$ but we bypass them now since they are not directly applicable to our concerns with contexts of normative conflict.

represent the appropriate all-things-considered oughts, and non-oughts, as the previous, traditional account could not.

Pattern C is much the same, except that while the *prima facie* ought for $\neg A$ becomes an all-things-considered ought, those for $\neg B$ and $A \vee B$ have the same status and should not both be considered all-things-considered oughts. This admits $M \models (r_1 \wedge r_2 \wedge r_3) \leadsto \neg A$, but now $M \not\models (r_1 \wedge r_2 \wedge r_3) \leadsto \neg B$ and $M \not\models (r_1 \wedge r_2 \wedge r_3) \leadsto (A \vee B)$. Hence, assuming nothing else pertinent, $M \models O_{atc} \neg A$, as before, but now $M \not\models O_{atc} \neg B$ and $M \not\models O_{atc}(A \vee B)$.

With Pattern D, we have no priority rankings, no overriding, of the *prima facie* oughts, and so no all-things-considered obligation to fulfill any one of them. Here $M \not\models (r_1 \wedge r_2 \wedge r_3) \leadsto \neg A$, $M \not\models (r_1 \wedge r_2 \wedge r_3) \leadsto \neg B$ and $M \not\models (r_1 \wedge r_2 \wedge r_3) \leadsto (A \vee B)$. Hence, by (Def 4), $M \not\models O_{atc} \neg A$, $M \not\models O_{atc} \neg B$ and $M \not\models O_{atc}(A \vee B)$, as should be.

Pattern D generates no all-things-considered oughts, unlike the approach of Section 4.1 where all three $O_{atc} \neg A$, $O_{atc} \neg B$ and $O_{atc}(A \vee B)$ obtain, and with them a full fledged normative conflict for all-things-considered oughts. The present framework is spared that.

Even so, this account may suffer another shortcoming, another form of undergeneration; this affects the first picture as well. Patterns B–D reveal that *prima facie* oughts $O_{pf}(A \vee B)$ and $O_{pf} \neg A$ do not imply $O_{atc}B$. Nevertheless, it seems plausible that if one has reason to do A-or-B and also reason not to do A, then by those facts one has reason to do B, and so $O_{pf}(A \vee B)$ and $O_{pf} \neg A$ should imply $O_{pf}B$, i.e., for any model if $M \models O_{pf}(A \vee B)$ and $M \models O_{pf} \neg A$ then $M \models O_{pf}B$, or for $\mathcal{M}_r$, $O_{pf}(A \vee B), O_{pf} \neg A \Vdash O_{pf}B$.

Recall the Smith Argument from Section 3. In the present framework one might say there is a reason in the law, r_ℓ, for Smith to fight with the army or perform alternative service, $r_\ell \leadsto (f \vee s)$, and so Smith has a *prima facie* obligation to do that, $O_{pf}(f \vee s)$. Smith also has reason, r_c, from his religious convictions not to fight with the army, $r_c \leadsto \neg f$, and so Smith *prima facie* ought not to do that, $O_{pf} \neg f$. There is no conflict here. One might think that in some way these reasons, r_ℓ and r_c, combine, so that Smith *prima facie* ought to perform alternative service, $O_{pf}s$. Perhaps this is not what Smith ought, all-things-considered, to do; there could be something more compelling that Smith ought to do that is incompatible with his performing alternative service. At the level of *prima facie* oughts themselves, however, the conclusion $O_{pf}s$ seems plausible. Yet it does not follow within the present account. In general, there could be $M \in \mathcal{M}_r$ such that, for some C, $M \models C$ and $M \models C \leadsto (A \vee B)$ and some D, with $M \models D$ and $M \models D \leadsto \neg A$, but no E such that $M \models E$ and $M \models E \leadsto B$. As with the Smith Example, one might think C and D together provide reason for B. That would be so, by ($\leadsto$RM), if $M \models (C \wedge D) \leadsto ((A \vee B) \wedge \neg A)$, but this does not follow by any of the principles so far assumed. Nor should $C \leftrightarrowtail A$ and $D \leftrightarrowtail B$ in general entail $(C \wedge D) \leftrightarrowtail (A \wedge B)$. That is the rule ($\leftrightarrowtail$CAC) rejected above.

4.3 *Prima facie* **oughts and** *ceteris paribus* **principles**

The previous two accounts of the relation between *prima facie* and all-things-considered oughts took the former to be conceptually prior and derived the latter from it. Here that picture is reversed. The notion of all-things-considered ought is taken to be fundamental and *prima facie* ought construed in terms of it.[40]

Section 4.2 presented one sort of conditionality inherent in *prima facie* oughts; this was the normative connection, represented by $\leftrightarrowtail$, between a reason and what it calls for. Here we look at the conditionality in a rather different way. For the Ross Example, for example, instead of saying that the making of the promise provides a reason for the person to meet the friend, $p \leftrightarrowtail m$, we now say that the fact of there being that promise, p, provides grounds to conclude that the person ought to meet the friend. The promise *explains*, at least in part, why the person ought to meet. Now, however, we take this 'ought' in a full-fledged, unqualified sense, Om. This is 'ought *sans phrase*' to use Ross's term; this is actual or even all-things-considered ought, although now to say 'all-things-considered' is redundant.

Another way to look at it, is to take a principle, such as that one ought to keep one's promises, to express a hedged generalization: Other things being equal, one ought to keep one's promises, or, *ceteris paribus* one ought to keep one's promises. So long as it is presumed that things are normal, or that other things are equal, then one concludes one ought to do what one promised. As in the Ross Example, if given just p and a principle that if p then Om, then one is warranted to conclude Om. But if it turns out that there are extraordinary factors, such as the accident a, then one does not conclude one ought to do what one promised, Om. One might even conclude that $O\neg m$. Here 'ought', O, itself is unqualified; the hedge affects how one reasons with such principles, not the nature of the conclusion.

That one would withdraw an inference on being given more information, e.g., that other things are not equal, indicates that reasoning with such principles is defeasible, or nonmonotonic. If general *ceteris paribus* principles are a sort of conditional, then that conditional too should be nonmono-

---

[40]We see this kind of picture in Prakken [1996], Asher and Bonevac [1996], and Morreau [1996], amongst others, though there are significant differences among the specific proposals of all these authors, as well as the present account. Pietroski [1993] provides philosophical support for this approach, but does not give any formal explication. Brink [1994] refers to Pietroski's view favorably, though his own discussion of moral dilemmas seems more in line with the picture we suggested in Section 4.1.

tonic. It should not be subject to Strengthening of the Antecedent, and *modus ponens* itself should hold at most as a defeasible inference. Let us use the notation $A \gg B$ as a surrogate for the different notations various authors use for this sort of conditional.[41] $A \gg B$ is read to say if A, then normally B, or if A then *ceteris paribus* B, or in other ways to that effect. Call the language extending $\mathcal{L}_D$ of Section 2.2 with $\gg$, $\mathcal{L}_C$. Expressions $A \gg B$ are well-formed when $A, B \in \mathcal{L}_D$. For present purposes we do not consider more complex formulas containing $\gg$. Unlike $\leadsto$ of the previous account, $\gg$ carries no normative force. All aspects of normativity are to be manifest in the standard deontic operator O itself. Otherwise we assume very little about $\gg$, although we will say a bit below about some properties that might be expected of it, and we will say more about how conditionals $A \gg B$ figure in defeasible reasoning.

Given such *ceteris paribus* principles, a particular *prima facie* ought, $O_{pf}A$, obtains when the condition of the principle obtains; i.e., $O_{pf}A$ holds just when there is a B such that both B and $B \gg OA$ obtain.[42] Following the general pattern of the previous sections, we take $\mathcal{M}_c$ to be the class of models $M = \langle w_0, W, v, (\mathcal{C}, \mathcal{O}) \rangle$, where $\mathcal{C}$ is whatever is required to model conditionals $A \gg B$ and $\mathcal{O}$ is whatever models formulas OA in such a way that they conform to all the core principles of ordinary deontic logic, as called for by criterion (b). In other words, we take $M \models A \gg B$ and $M \models OA$ to be well defined, and are now interested in how M models *prima facie* oughts, $O_{pf}A$, and what that requires for $\mathcal{C}$ and $\mathcal{O}$.

By the present picture, we should assume

Def 5) $M \models O_{pf}A$ if and only if there is a B such that $M \models B$ and $M \models B \gg OA$

As before, let Δ_M be the set of *prima facie* oughts that hold on M; $\Delta_M = \{O_{pf}A : M \models O_{pf}A\}$

With (Def 5) it is easy to see that *prima facie* oughts can conflict, in keeping with criterion (a) above. There might be models on which C and

---

[41] Asher and Bonevac [1996] and Morreau [1996] use '>' for this purpose, reading it as 'common sense entailment'; Nute [1999] uses '⇒' for a different concept of defeasible rule; '⤳' is common. Here we want to be neutral between specific formalizations of the connection. Asher and Bonevac also apply another binary concept, represented $A >_O B$, whose account is closer to the descriptions of Section 4.2, though not with (Def 4).

[42] Cf. Asher and Bonevac [1996, p. 42]. They actually distinguish two sorts of unconditional defeasible oughts, $O_{cp}A$ and O_gA, the first being given by there being a B such that B and, in our notation, $B \leadsto A$, the second by there being a B such that B and $B \gg OA$. The first corresponds to $O_{pf}A$ as given in Section 4.2, the second corresponds to the present account. When Asher and Bonevac come to define *prima facie* ought *per se* they specify $O_{pf}A$ if and only if $O_{cp}A \lor O_gA$. Morreau [1996] gives an account of 'seeming' oughts that corresponds closely with the present picture.

$C \gg OA$ both hold and also D and $D \gg OB$ even while $\neg\Diamond(A \wedge B)$ also obtains, so that $O_{pf}A \in \Delta_M$ and $O_{pf}B \in \Delta_M$ despite the incompatibility. That all-things-considered oughts do not conflict is given by the assumption that O obeys all the core principles, which directly guarantees criterion (b).

This leaves the question of what logical principles, if any, govern *prima facie* oughts. That question is less important now than before since there is no expectation that the core principles for all-things-considered oughts should be derived from the behavior of *prima facie* oughts. Nevertheless, it may have some interest of its own.

4.3.1 Logic for *prima facie* oughts

What principles govern *prima facie* oughts depends in large measure on the principles that govern $\gg$ itself. If $\gg$ satisfies a rule of weakening of the consequent, or its modal counterpart,

$\gg$WC)   If $\Vdash B \to C$ then $A \gg B \Vdash A \gg C$, or
$\gg$M) $A \gg (B \wedge C) \Vdash A \gg B$, or
$\gg$NWC) $A \gg B, \Box(B \to C) \Vdash A \gg C$

then, O_{pf} will satisfy the corresponding rules $(RM)_{pf}$, $(M)_{pf}$, or $(NM)_{pf}$. These rules, at least ($\gg$WC) and ($\gg$M), are widely assumed for the defeasible conditional in these kinds of account.

Further, since O satisfies (P), $OA \to \Diamond A$, then if $\gg$ satisfies ($\gg$WC), then O_{pf} must also satisfy $(P)_{pf}$ for *prima facie* oughts, i.e., $O_{pf}A \Vdash \Diamond A$.[43]

Since *prima facie* oughts can conflict, principle (D) must fail for them. (C), and even (CC), also fail for *prima facie* oughts, since there might be models M on which C and $C \gg OA$ both hold and also D and $D \gg OB$, but no E such that E and $E \gg O(A \wedge B)$ hold, even when $\Diamond(A \wedge B)$ obtains.

This leaves a core logic of *prima facie* oughts that has just the rules $(NM)_{pf}/(RM)_{pf}$ and $(P)_{pf}$. This logic, **P**, was briefly mentioned in Section 4.2.1; we meet it again in Section 5.2 for oughts that may conflict.

For the accounts of Sections 4.1 and 4.2, the hard question was how are the core principles for all-things-considered oughts to be justified given the way they are generated from *prima facie* oughts. Here the core principles are assumed to hold for formulas OA. The hard question now is how are such formulas to be detached from given *prima facie* oughts. Such detachment must be defeasible. As we have seen, one might have a *prima facie* obligation to meet someone, $O_{pf}m$, because one has promised and normally one ought

---

[43] As remarked in Note 36, Asher and Bonevac [1996, p. 43] reject (P) for *prima facie* oughts, but this they only say for the sense of *prima facie* ought as $A >_O B$, not $A > OB$. For the latter, since they are committed to ($\gg$WC) for their $>$ and presumably to (P) for their O, since it is supposed to be standard, they must accept (P) for this other sense of *prima facie* ought.

to do what one has promised, hence p and $p \gg Om$. If one takes things to be normal, one might detach Om. One might also have a *prima facie* obligation to help the victims of an accident, a and $a \gg Oh$, and other things being equal, one might detach Oh. But given the combination $p \wedge a$, things are not normal, other things are not equal, and then both detachments might fail. Thus, to answer the question of how to conclude what one really ought to do given background *prima facie* oughts, we must sail briefly through the rocky waters of defeasible reasoning.

4.3.2 Defeasible inference

The task of a theory of defeasible, or nonmonotonic, reasoning is to explain how some inferences seem acceptable even while other inferences including the very same premises and conclusion do not.[44]

There is little consensus for how that explanation should go. Different researchers offer different accounts. Some are syntactical, others semantical or model theoretic. Rather than privilege any one of them here, and rather than get caught in unnecessary detail and complexity, I will instead describe a fairly simple toy model of a theory of defeasible inference as it might apply to the kinds of cases we have been discussing.[45] I do not claim this is an adequate theory in general, or indeed for very much beyond the present discussion. My purpose is simply to illustrate a way of looking at certain issues, though I do intend it to be within the spirit of better, more fully developed accounts, especially those that would apply to problems concerning reasoning about normative conflicts. It should agree with such accounts about the examples we have discussed here.

In fact, I will present two models of a theory of defeasible reasoning, both syntactical. The first works well enough for simple cases, but not for all the examples we have been considering. It sets the stage for the second model, which in turn lays groundwork for methods applied in the next subsection

---

[44]For a general introduction to theories of nonmonotonic reasoning, their motivations and issues concerning them, see [Horty, 2001] and the further readings suggested there. See also [Makinson, 1994]. The literature on defeasible reasoning is vast. For particular application to issues in deontic logic, Nute's collection [1997] is a valuable source. See also the papers [Asher and Bonevac, 1996], [Morreau, 1996] and [Prakken, 1996] already mentioned, also [Nute, 1999]. Asher and Bonevac's and Morreau's accounts of defeasible reasoning are semantical, Prakken's and Nute's more syntactical, though different. Horty's [1994; 1997; 2003] also provide useful introductions to certain kinds of defeasible reasoning and their application to deontic logic, although his account of how all-things-considered oughts are derived from *prima facie* principles belongs more to the next proposal than to this one. For more on that see also [Horty, 2012]. [Prakken, 1996] criticizes Horty's approach, in its early forms, to favor something more like what we see here.

[45]This model stands to real accounts much as a toy piano stands to a real piano. It will play some music, but not much.

for a different account of all-things-considered oughts.

The language $\mathcal{L}_C$ contains two sorts of formulas, those of the original $\mathcal{L}_D$ and those of the form $B \gg C$, for $B, C \in \mathcal{L}_D$. We assume the former follow the principles of a classically based modal logic and at least the core principles of deontic logic. $\vdash_d$ represents such a classical consequence relation over that original language. We are now interested in defining a relation, $\Gamma \vdash\!\sim A$, of defeasible validity that holds between a set of formulas Γ from $\mathcal{L}_C$ and a formula A of $\mathcal{L}_D$. That is a limitation of the present model. We ask, Under what conditions does some information Γ support, or warrant, the conclusion A, when A expresses simple information about a situation? We do not ask when Γ supports a conclusion that is itself a conditional, $\Gamma \vdash\!\sim B \gg C$. For present purposes we are most interested in drawing conclusions about what one ought to do, something of the form OB, given various facts and *prima facie* oughts operative in the situation.

The set Γ is thus divisible into exclusive subsets, $\Gamma = \Gamma^d \cup \Gamma^c$, where $\Gamma^d \subseteq \mathcal{L}_D$ and Γ^c is the set of conditionals in Γ. We assume that Γ^d is consistent, as defined by $\vdash_d$. That is another limitation. For a conditional $B \gg C$, we naturally call B the antecedent and C the consequent of that conditional. For a set of conditionals, Δ, let its consequent set be the set of the consequents of its members; $\mathcal{C}(\Delta) = \{C : \exists B(B \gg C \in \Delta)\}$. For convenience, $\mathcal{C}(\gamma) = \mathcal{C}(\{\gamma\})$ for $\gamma \in \Gamma^c$.

Given Γ, we suppose that all of the conditionals in Γ^c are 'triggered', that their antecedents are entailed by Γ^d. This is a third limitation, but all of the examples under review here fall within it.[46] Given $\Gamma = \Gamma^d \cup \Gamma^c$, let

- $\Gamma^{d+c} = \Gamma^d \cup \mathcal{C}(\Gamma^c)$

All the members of Γ^{d+c} are formulas of $\mathcal{L}_D$.

This set Γ^{d+c} gives all the useful information that might support a conclusion A from Γ. On the other hand, given that conditionals in Γ may have inconsistent consequents, or consequents that are inconsistent with other information in Γ^d, this set Γ^{d+c} may well be inconsistent. As a result, we cannot say simply that A is supported by Γ just in case it is a logical consequence of Γ^{d+c}. Instead we look at maximal consistent subsets of Γ^{d+c}, and the information on which they all agree, and we expect they all agree about the information in Γ^d, that they are faithful to the basic information given.

---

[46]To include untriggered conditionals raises interesting and challenging problems for a theory of defeasible reasoning, especially in the context of a theory of prioritized *prima facie* principles, problems we simply bypass here. For a useful discussion of some of those issues, see [Hansen, 2008].

- A set Γ^* is a faithful maximal consistent subset of Γ^{d+c} if and only if (i) $\Gamma^d \subseteq \Gamma^*$ and (ii) $\Gamma^* \subseteq \Gamma^{d+c}$, and (iii) Γ^* is consistent (as defined by $\vdash_d$) and (iv) there is no $B \in \Gamma^{d+c}$ such that $B \notin \Gamma^*$ and $\Gamma^* \cup \{B\}$ is consistent.

(i) provides faithfulness, (ii) inclusion in Γ^{d+c}, (iii) consistency, and (iv) maximality within Γ^{d+c}. This gives our first definition for $\Gamma \mathrel{\vert\!\sim} A$ that represents the defeasible inference of A from Γ.

Def 6) $\Gamma \mathrel{\vert\!\sim} A$ if and only if $\Gamma^* \vdash_d A$, for each Γ^* that is a faithful maximal consistent subset of Γ^{d+c}.

We can apply this definition to the Ross Example to see its virtues, and its limitations. Given just the fact of the promise to meet and the principle that if one has promised to meet then *ceteris paribus* one ought to meet, one concludes one ought to meet. $\Gamma = \{p, p \gg Om\}$. Hence $\Gamma^d = \{p\}, \Gamma^c = \{p \gg Om\}, \mathcal{C}(\Gamma^c) = \{Om\}, \Gamma^{d+c} = \{p, Om\}$, and that is the only faithful maximal consistent extension of itself. Since $\{p, Om\} \vdash_d Om$, $\Gamma \mathrel{\vert\!\sim} Om$. The same applies with respect to the accident and helping the victims. If $\Gamma = \{a, a \gg Oh\}$ then $\Gamma \mathrel{\vert\!\sim} Oh$.

On the other hand, given both facts of the promise and the accident and the two principles, one expects not to infer Om. Here $\Gamma = \{p, a, p \gg Om, a \gg Oh, \neg\Diamond(m \wedge h)\}$, and $\Gamma^d = \{p, a, \neg\Diamond(m \wedge h)\}$ and $\Gamma^c = \{p \gg Om, a \gg Oh\}$. $\mathcal{C}(\Gamma^c) = \{Om, Oh\}$. Since O follows all the core principles, $Om \wedge Oh \vdash_d \Diamond(m \wedge h)$. Hence $\Gamma^d \vdash_d \neg(Om \wedge Oh)$. As a result, Om and Oh cannot both be in any faithful maximal consistent subset of Γ^{d+c}. Instead there are two such subsets, $\Gamma^1 = \{p, a, \neg\Diamond(m \wedge h), Om\}$ and $\Gamma^2 = \{p, a, \neg\Diamond(m \wedge h), Oh\}$. Since Om is not a consequence of both of those, $\Gamma \mathrel{\not\vert\!\sim} Om$, as it should be.

Less satisfying, (Def 6) also yields $\Gamma \mathrel{\not\vert\!\sim} Oh$. The problem is that it does not take relations of priority of *prima facie* oughts into consideration. All the consequent oughts are treated equally when they should not be. That leads to our second proposal that will take such relations into account.

So far, however, we have no account of when one *certeris paribus* principle takes priority over another. Following common practice, we now take that to be determined by specificity, that more specific principles take precedence over less specific ones. A proposition A is more specific than another, B, if A necessitates B but B does not necessitate A. In the context of a given Γ we take that A necessitates B with respect to Γ to mean that the necessities within Γ together with A entail B, i.e., if $\Gamma^n = \{\Box C : \Box C \in \Gamma^d\}$, then A necessitates B for Γ just in case $\Gamma^n, A \vdash_d B$.

The relative strength of conditionals is determined by the specificity of their antecedents.

- $A \gg B$ is stronger than $C \gg D$ for Γ — $(A \gg B) \sqsupset_\Gamma (C \gg D)$ — just in case $\Gamma^n, A \vdash_d C$ and $\Gamma^n, C \nvdash_d A$.

This relation, $\sqsupset_\Gamma$, is a strict partial order, irreflexive and transitive.

This relation orders the conditionals in Γ^c somewhat, but it could easily be that neither $(A \gg B) \sqsupset_\Gamma (C \gg D)$ nor $(C \gg D) \sqsupset_\Gamma (A \gg B)$. Let a complete strict order, $\succ$, over Γ^c be said to 'respect' $\sqsupset_\Gamma$ just in case $\sqsupset_\Gamma \subseteq \succ$. Such relations merely fill in the gaps left by $\sqsupset_\Gamma$. There can be many such respectful total orders for a given Γ. These determine the 'preferred' maximal consistent subsets of Γ^{d+c} that define $\Gamma \mathrel{\vdash\mkern-9mu\sim} A$, preferred because they do take priorities into account.

For a complete strict order, $\succ$, over Γ^c that respects $\sqsupset_\Gamma$, let the ordering of Γ^c by $\succ$ be $\langle c_1, \ldots, c_i, \ldots \rangle$, so that $c_i < c_j$ in the sequence just in case $c_i \succ c_j$.[47] Then define $\Phi_\succ$ from a sequence of subsets, $\Phi_\succ^i$, of Γ^{d+c}, thus:

- $\Phi_\succ^0 = \Gamma^d$;

- for $0 \leq i < j = i+1$, $\Phi_\succ^j = \Phi_\succ^i \cup \{\mathcal{C}(c_j)\}$ if that is consistent, otherwise $\Phi_\succ^j = \Phi_\succ^i$;

- $\Phi_\succ = \bigcup \Phi_\succ^i$.

Each such set, $\Phi_\succ$, is a faithful maximal consistent subset of Γ^{d+c}, and so is appropriate to apply to define $\Gamma \mathrel{\vdash\mkern-9mu\sim} A$.

Def 7) $\Gamma \mathrel{\vdash\mkern-9mu\sim} A$ iff $\Phi_\succ \vdash_d A$, for each complete strict order, $\succ$, that respects $\sqsupset_\Gamma$.

(Def 7) differs from (Def 6) only when Γ^c contains more than one member. Hence the little examples from the start of the Ross Example are treated the same as before. For the full Ross Example, however, in which we suppose also $(p \wedge a) \gg Oh \in \Gamma$ in order to catch the priority of the obligation to help over the obligation to meet the friend, then we see, $\Gamma^d = \{p, a, \neg\Diamond(m \wedge h)\}$ and $\Gamma^c = \{p \gg Om, a \gg Oh, (p \wedge a) \gg Oh\}$. Call those c_1, c_2, c_3 respectively. Then $c_3 \sqsupset_\Gamma c_1$ and $c_3 \sqsupset_\Gamma c_2$, and no priority obtains between c_1 and c_2. That offers two respectful complete orders, $c_3 \succ_1 c_1 \succ_1 c_2$ and $c_3 \succ_2 c_2 \succ_2 c_1$. Then $Oh \in \Phi_{\succ_1}^1$ because it is the consequent of the most favored conditional by $\succ_1$. $Om \notin \Phi_{\succ_1}^2$ because it is incompatible with what is already in, and $\Phi_{\succ_1}^2 = \Phi_{\succ_1}^1$. $Oh \in \Phi_{\succ_1}^3$ because it is compatible with what is already in, to wit itself. Hence $\Phi_{\succ_1} = \{Oh\}$. Similarly, for $\succ_2$, $\Phi_{\succ_2} = \{Oh\}$. Hence, by (Def 7), $\Gamma \mathrel{\vdash\mkern-9mu\sim} Oh$, as it should be, and, of course, still $\Gamma \mathrel{\not\vdash\mkern-9mu\sim} Om$ as well.

---

[47] We ignore for convenience the possibility that Γ^c might be empty; that case is easily incorporated into the account to follow.

We will see below how (Def 7) also supports the Disjunctive Response, defeasibly, and how it applies to Patterns A–D from Section 4.1.3.[48]

4.3.3 Overriding *prima facie* oughts

The Marcus and Mission Examples and the other hard cases from Section 4.1.3 rely on a sense of some *prima facie* oughts outweighing or overriding or defeating others. In Section 4.2 these concepts were defined in terms of relative specificity. Something similar applies here.

Asher and Bonevac [1996, p. 20] state a rule of 'deontic specificity', that more specific *prima facie* oughts take precedence over less specific, which in turn yields a defeasible form of detachment for all-things-considered oughts, OA. In our notation that becomes

> Spec) If $A \gg OC$ and $B \gg O\neg C$, and if A is more specific than B, and both A and B, then (defeasibly) OC.

This is supported, as a defeasible inference, by (Def 7). For suppose $\Gamma = \{A, B, \square(A \to B), A \gg OC, B \gg O\neg C\}$. Since $(A \gg OC) \sqsupset_\Gamma (B \gg O\neg C)$, and that relation is now total, it is easy to establish $\Gamma \hspace{2pt}\vdash\hspace{-7pt}\sim\hspace{2pt} OC$. For the interesting cases, the trick will be to ensure the appropriate specificity conditions are met.

For simple cases, like the Marcus Example, where nothing overrides anything else, the Disjunctive Response is easily verified. For the twins and their needs to be saved, we have $\Gamma = \{n_1, n_2, \neg\Diamond(s_1 \land s_2), n_1 \gg Os_1, n_2 \gg Os_2\}$. Here $\sqsupset_\Gamma$ is empty, and there are two complete strict orders that respect it. $\Phi_{\succ_1} = \{Os_1\}$; $\Phi_{\succ_2} = \{Os_2\}$. Since neither ought is entailed by both Φs, $\Gamma \hspace{2pt}\not\vdash\hspace{-7pt}\sim\hspace{2pt} Os_1$ and $\Gamma \hspace{2pt}\not\vdash\hspace{-7pt}\sim\hspace{2pt} Os_2$. Nevertheless, since the rules for O include (RM), $\Phi_{\succ_1} \vdash_d O(s_1 \lor s_2)$ and $\Phi_{\succ_2} \vdash_d O(s_1 \lor s_2)$. Hence $\Gamma \hspace{2pt}\vdash\hspace{-7pt}\sim\hspace{2pt} O(s_1 \lor s_2)$, just as called for by the Disjunctive Response.

More complicated cases like the Mission Examples, though more complicated, work in much the same way. For Mission 1, we have $O_{pf}A$, $O_{pf}B$, $O_{pf}C$ and $O_{pf}D$, with A, B, and C pairwise incompatible, and D incompatible with C and compatible with each of A and B. Moreover, the *prima facie* oughts for A, B and C are of equal weight, but $O_{pf}D$ overrides $O_{pf}C$. In the present framework this means $\Gamma = \{E, F, G, H, \neg\Diamond(A \land B), \neg\Diamond(A \land C), \neg\Diamond(B \land C), \neg\Diamond(D \land C), \Diamond(A \land D), \Diamond(B \land D), E \gg OA, F \gg OB, G \gg OC, H \gg OD, (G \land H) \gg OD\}$. E, F, G, H represent the reasons why Brown ought to go to each of the several cities. The final conditional is there to

---

[48]If the word 'each' in (Def 7), or (Def 6), were changed to 'some', the definitions would allow the defeasible appearance of normative conflicts even while O itself follows the core principles strictly. It could be that $\Gamma \hspace{2pt}\vdash\hspace{-7pt}\sim\hspace{2pt} OA$ and $\Gamma \hspace{2pt}\vdash\hspace{-7pt}\sim\hspace{2pt} OB$ when $\neg\Diamond(A \land B) \in \Gamma$, even as (NM) and (D) both hold for formulas OA. This change from 'each' to 'some' corresponds to the contrast, familiar in discussions of defeasible reasoning, between 'skeptical' and 'credulous' strategies respectively. This contrast appears again in Sections 4.4 and 6.3.

make the obligation for D defeat that for C, as by (Spec). If we call the conditionals in Γ c_1, c_2, c_3, c_4, c_5 respectively, then we have $c_5 \sqsupset_\Gamma c_4$ and $c_5 \sqsupset_\Gamma c_3$, but otherwise there are no priorities. This leaves room for several complete strict orders, $\succ$, respectful to $\sqsupset_\Gamma$, which the reader may fill in. It is then easy to see that for each generated $\Phi_\succ$, $\Phi_\succ \vdash_d OD$; hence $\Gamma \mid\!\sim OD$. It is also easy to see that for each $\Phi_\succ \not\vdash_d OC$ since OD would always be put into $\Phi_\succ$ before OC was considered. Hence $\Gamma \mid\!\not\sim OC$. Also, for some $\Phi_\succ$, $\Phi_\succ \vdash_d OA$ and for the rest $\Phi_\succ \vdash_d OB$ but none would entail both. Hence $\Gamma \mid\!\not\sim OA$ and $\Gamma \mid\!\not\sim OB$. But every $\Phi_\succ$ that entails OA also entails $O(A \vee B)$, as does every $\Phi_\succ$ that entails OB, and that is all the $\Phi_\succ$s. Hence $\Gamma \mid\!\sim O(A \vee B)$, which accords with the Disjunctive Response. The Mission Example 2 is similar, but with no priorities, and conjunctive disjuncts. I leave that to the reader.

In this way, this account supports the Disjunctive Response, and satisfies criterion (c). Next let us consider how it handles the problems of under- and overgeneration with Patterns A–D from Section 4.1.3.

For undergeneration and Pattern A, suppose $\Gamma^d = \{D, E, F, \neg\Diamond(A \wedge B), \neg\Diamond(B \wedge C), \Diamond(A \wedge C)\}$ and $\Gamma^c = \{c_1, c_2, c_3, c_4, c_5, c_6\}$, with $c_1 = D \gg OA$, $c_2 = E \gg OB$, $c_3 = F \gg OC$, $c_4 = (D \wedge E) \gg O(\neg C \wedge B)$, $c_5 = (E \wedge F) \gg O(\neg B \wedge A)$, and $c_6 = (D \wedge E \wedge F) \gg O(\neg B \wedge C)$, where the c_4–c_6 express the crucial relations of overriding presumed by the Pattern. This yields quite a number of relationships by $\sqsupset_\Gamma$, and even more possible total orderings that respect those. I leave those to the reader to determine. Likewise the various sets $\Phi_\succ$ that correspond to each such total ordering. Working out the details reveals, however, that each such set contains OA, $O\neg B$, and OC, as well as the members of Γ^d, and so $\Gamma \mid\!\sim OA$, $\Gamma \mid\!\sim O\neg B$ (and $\Gamma \mid\!\not\sim OB$), and significantly $\Gamma \mid\!\sim OC$, all as should be for this Pattern.

For overgeneration and Pattern B, since $O_{pf}\neg A$, $O_{pf}\neg B$ and $O_{pf}(A \vee B)$ are all to hold with the first weightier than the second and the second weightier than the third, assume some D, E, F to hold such that $D \gg O\neg A$, $E \gg O\neg B$, and $F \gg O(A \vee B)$ also hold, and for the relative weights in terms of specificity, $(E \wedge F) \gg O\neg B$, $(D \wedge E \wedge F) \gg O\neg(A \vee B)$. Thus $\Gamma^d = \{D, E, F\}$ and $\Gamma^c = \{D \gg O\neg A, E \gg O\neg B, F \gg O(A \vee B), (E \wedge F) \gg O\neg B, (D \wedge E \wedge F) \gg O\neg(A \vee B)\}$, which, for convenience we label, c_1, c_2, c_3, c_4, c_5, in the order given. Then c_5 takes precedence by $\sqsupset_\Gamma$ over c_4, which takes precedence over each of c_1, c_2, c_3, and no particular ordering applies to those.

This means six possible orderings $\succ$ that respect $\sqsupset_\Gamma$, and accordingly six sets $\Phi_\succ$, except that they will all be the same. All will include all the members of Γ^d, and all will include the consequent of c_5, $O\neg A$, since that is put in first, and all will include the consequent of c_4, $O\neg B$, which is put in second since it is consistent with what is already in. None will contain the consequent of c_3, $O(A \vee B)$, since that is incompatible with the two members already in, and the consequents of c_2 and c_1 are already in. Thus for every $\succ$, $\Phi_\succ = \Gamma^d \cup \{O\neg A, O\neg B\}$. Hence, $\Gamma \mathrel{\vert\!\sim} O\neg A$, $\Gamma \mathrel{\vert\!\sim} O\neg B$ and $\Gamma \mathrel{\vert\!\not\sim} O(A \vee B)$, just as should be for this case.

Pattern C is similar, except that the *prima facie* oughts for $\neg B$ and $(A \vee B)$ are put on a par. Hence Γ^d remains the same, but Γ^c no longer contains c_4. With the elimination of c_4, there will now be two distinct preferred sets $\Phi_\succ$ generated by the several completions.

$\Phi_{\succ,1} = \Gamma^d \cup \{O\neg A, O(A \vee B)\}$
$\Phi_{\succ,2} = \Gamma^d \cup \{O\neg A, O\neg B\}$

Since $O\neg A$ is in both, $\Gamma \mathrel{\vert\!\sim} O\neg A$, as we expect. Since neither $O\neg B$ nor $O(A \vee B)$ is in both, $\Gamma \mathrel{\vert\!\not\sim} O\neg B$ and $\Gamma \mathrel{\vert\!\not\sim} O(A \vee B)$, as we also expect.

Pattern D, where all the *prima facie* oughts have equal status, works similarly. None of $O\neg A$, $O\neg B$ or $O(A \vee B)$ can be detached. On the other hand, any preferred set will contain two of these, whence it follows by (C) and (RM) that $\Gamma \mathrel{\vert\!\sim} O((\neg A \wedge \neg B) \vee (\neg A \wedge (A \vee B)) \vee (\neg B \wedge (A \vee B)))$, which fits the Disjunctive Response.

The present account thus treats the hard cases as they should be, at least as defeasible inferences. Nevertheless, there remains a significant respect in which it might fall short of expectations. Like the approach of Section 4.2, here the only sense of priority among *prima facie* oughts, the only account of overriding or defeat, is given in terms of specificity, as manifest in the relation $\sqsupset_\Gamma$. This required some *ad hoc* postulation of how things stood with respect to multiple conditionals and their antecedents as called for by Patterns A–D, as well as those needed to ensure the Disjunctive Response for the Marcus and Mission Examples.

Even so, confining overriding and defeat to specificity is a severe constraint on the present account. Often a *prima facie* principle takes precedence over another for quite different reasons that are difficult to capture in this way, as when one law overrides another. Even simple cases, like the Ross Example, in which the need to help the accident victims overrides the keeping of a promise, do not seem much like a matter of specificity, even if it can be gerrymandered to fit that model. One might, perhaps, consider introducing separate, primitive relations of priority, $>$, to hold for principles of the form $B \gg OA$, and then apply those in place of $\sqsupset_\Gamma$ in the definitions for (Def 7). That could cover some cases where specificity *per se* does not seem to be the relevant factor. It might not cover all, however. With the Ross Example, for example, it is not so much that the principle $a \gg Oh$ has greater weight than $p \gg Om$, as that, in the situation, the *prima facie* ought for helping itself outweighs the *prima facie* ought for meeting. This suggests that relations $>$ should be defined for *prima facie* oughts themselves, not the general principles or conditionals. That is how it works on the next proposal, where we return to the viewpoint of Section 4.1, taking *prima facie* oughts to be basic and specifying relations of priority $>$ on them, though these will now be treated quite differently, more in line with the account of this section.

4.4 Basic *prima facie* oughts revisited

This next proposal within the perspective that would distinguish *prima facie* oughts from all-things-considered oughts and consider the latter to be immune from conflicts, returns to something like the view of the first, more traditional account of Section 4.1 to derive all-things-considered oughts from unanalyzed *prima facie* oughts that have a priority relation, or relation of relative weight, $>$, defined over them. This proposal, however, abandons the idea that all-things-considered oughts are simply those *prima facie* oughts that override all competitors. Instead, the way all-things-considered oughts are derived will be quite different from (Def 1) or even (Def 4); it will be more like the nonmonotonic generation described in Section 4.3.2, (Def 7).[49]

Basic *prima facie* oughts are represented by formulas $O^b_{pf}A$ with the superscript now to indicate their particular role within this proposal. As in Section 4.1, very little is assumed about these, except that they are ranked by relative weight. Much as in Section 4.1, we now consider the class $\mathcal{M}_d$ of models $M = \langle w_0, W, v, (\Delta, >) \rangle$, where Δ is a set of basic *prima facie* oughts, $O^b_{pf}A$, which may be thought to include all, and only, those *prima facie* oughts relevant in context. Unlike the models of $\mathcal{M}_t$ of Section 4.1, however, no further conditions on Δ are imposed, e.g., that it be closed under various rules. $>$ is a well-founded strict partial order defined over Δ,

---

[49] The account given here, with (Def 8) below, is something of an amalgam of work of Hansen [2006] and Horty [2007; 2012], but limited to unconditional basic *prima facie* oughts and to unconditional all-things-considered oughts as well. Interesting, and hard, questions arise when conditionals are included in the mix. See, e.g., [Hansen, 2008] for more on that. Hansen's work draws much from the work of Brewka and Nebel, among others, while Horty's is based more on Reiter's default logic. Horty's [2003] provides useful introduction to the kind of approach sketched here, though with significant differences. As I speak here of basic *prima facie* oughts, Hansen speaks of imperatives or imperative norms, which are located entirely within the semantics for a deontic language, with no suggestion that they bear truth-values, or have a logic of their own. This seems apt. Horty [1994] also speaks of imperatives, writing !A as here we write $O^b_{pf}A$. In [1997] he speaks only of oughts, writing $\bigcirc A$, without distinguishing *prima facie* from all-things-considered oughts. In [2003] he adopts the language of *prima facie* oughts, much as here, though taking them to be imperatives, and again writing !A (or the conditional counterpart). In [2007; 2012] he speaks more of 'reasons'.

with $O^b_{pf}A > O^b_{pf}B$ read to say $O^b_{pf}A$ is (strictly) weightier than $O^b_{pf}B$.[50]
For any $\Theta \subseteq \Delta$, we take $\Theta^* = \{A : O^b_{pf}A \in \Theta\}$, the enjoined contents of Θ.
As in Section 2.2, we say a set of formulas Γ of $\mathcal{L}_B$ necessitates a formula B (on a model M) — $\Gamma \vdash^M_\square B$ — iff, for all $w \in W$ if $M, w \models C$ for every $C \in \Gamma$ then $M, w \models B$. And we often omit the annotation M.

For such $M = \langle w_0, W, (\Delta, >)\rangle$, immediately

- $M \models O^b_{pf}A$ iff $O^b_{pf}A \in \Delta$

The crucial question, however, is now, How are all-things-considered oughts, $O_{atc}A$ determined, or generated, by such a declared set of basic *prima facie* oughts, Δ? As before, the answer should satisfy criteria (a)–(c), that (a) while conflicts are possible for *prima facie* oughts, they are not possible for all-things-considered oughts, (b) all-things-considered oughts obey all the core principles of deontic logic, and (c) the Disjunctive Response applies in cases of irresolvable conflict among *prima facie* oughts.

Given Δ, one might say that, all-things-considered, one ought to do what is necessitated by the combined contents of the *prima facie* oughts in Δ.

- $M \models O_{atc}A$ if and only if $\Delta^* \vdash_\square A$

This, however, ignores that there could be conflicts among the *prima facie* oughts in Δ, in which case there would be conflicts among the generated all-things-considered oughts, contrary to criterion (a).

That could be corrected by taking $O_{atc}A$ to hold just in case A is necessitated by the maximal consistent subsets of Δ^*.[51]

- $M \models O_{atc}A$ if and only if $\Sigma \vdash_\square A$, for each Σ that is a maximal consistent (co-possible) subset of Δ^*.

While this supports the core principles for all-things-considered oughts, and so eliminates conflicts among them, it fails to take the relative weight or significance of *prima facie* oughts into account, though that is a primary point of distinguishing *prima facie* from all-things-considered oughts. In the Ross Example, for example, where $\Delta = \{O^b_{pf}m, O^b_{pf}h\}$, for the *prima facie* obligations to meet the friend and to help the accident victims, where $\neg\Diamond(m \wedge h)$ and $O^b_{pf}h > O^b_{pf}m$, we expect to conclude $O_{atc}h$ but not $O_{atc}m$. In this case, $\Delta^* = \{m, h\}$ and there are two maximal co-possible subsets $\{m\}$ and $\{h\}$. Since m is not necessitated by both, $M \not\models O_{atc}m$, which is

---

[50] Here, unlike Section 4.1, it is not necessary to have a weak ordering $\geq$, though that is an option, with $>$ its strict counterpart.

[51] This is essentially the proposal, Definition 4, of [Horty, 1997, p. 32]; cf. also [Hansen, 2004a, p. 149], and [Hansen, 2005, p. 487].

as it should be. But likewise h is not entailed by both, and so $M \not\models O_{atc}h$, which is not as should be. This is the very point of Ross's example, to bring out the role of the relation $>$ over *prima facie* oughts.

This relation $>$ selects a preferred set, Π, of 'premium' basic *prima facie* oughts from Δ, which then determines all-things-considered oughts. It is tempting to think of Π as the result of purging from Δ all those *prima facie* oughts that conflict with others of greater weight, $\Pi_M = \{O_{pf}^b A \in \Delta : \neg \exists B(O_{pf}^b B \in \Delta \,\&\, O_{pf}^b B > O_{pf}^b A \,\&\, M \models \neg \Diamond (A \wedge B))\}$. Such a Π_M might still be inconsistent, as in the Marcus or Sartre Examples, and so we would take the all-things-considered oughts generated by Δ to be those whose content is necessitated by the contents of the maximal consistent subsets of the premium basic *prima facie* oughts in Π_M.[52]

- $M \models O_{atc} A$ if and only if $\Sigma \vdash_\Box A$, for each Σ that is a maximal consistent (co-possible) subset of Π_M^*

For the Ross Example, with Δ as above, $\Pi_M = \{O_{pf}^b h\}$, and $\Pi_M^* = \{h\}$. Since that is the only maximal consistent subset of itself, $M \models O_{atc}h$, as it should be, and $M \not\models O_{atc}m$, also as should be.

While this works for simple cases like the Ross Example, it is not suited for more complex situations, like those of Pattern A–D, with more than two *prima facie* oughts in play. For Pattern A, with $\Delta = \{O_{pf}^b A, O_{pf}^b B, O_{pf}^b C\}$ and $\neg \Diamond(A \wedge B), \neg \Diamond(B \wedge C), \Diamond(A \wedge C)$, and $O_{pf}^b A > O_{pf}^b B > O_{pf}^b C$, we expect $M \models O_{atc}A$, $M \not\models O_{atc}B$ and $M \models O_{atc}C$. Here, $\Pi_M = \{O_{pf}A\}$. That will yield the first $M \models O_{atc}A$, and not $M \models O_{atc}B$, as should be, but also not $M \models O_{atc}C$, contrary to expectation. $O_{pf}^b C$ is excluded from Π_M by its conflict with $O_{pf}^b B$, and there is no provision for its reinstatement by virtue of that being itself overridden.

Patterns B and C also fail since, in these cases, the role of $>$ is lost in the specification of Π_M because there are no pairwise conflicts among the basic *prima facie* oughts. Only Pattern D survives since there there are no (non-trivial) all-things-considered oughts to be derived.

To correct this, we now take premium sets Π to be developed stepwise, much like the sets Φ in Section 4.3.2. Before really defining Π, however, let us briefly illustrate how the procedure works with Patterns A–D.

With Pattern A, as above, the set of premium *prima facie* oughts begins by including $O_{pf}^b A$ since that is the weightiest of all in Δ. $O_{pf}^b B$ is the second weightiest, but it will not be included since it conflicts with $O_{pf}^b A$. Then

---

[52]This is essentially the 'disjunctive' account of [Horty, 2003, p. 569], though that was for conditional oughts and hence more complicated; cf. also the preliminary disjunctive account of [Horty, 2012, ch. 1-4], though not the final, refined version of Ch. 8.

$O^b_{pf}C$, the least significant in Δ, will be included because it is compatible with all the members so far in the premium set, viz. $O^b_{pf}A$, regardless of its incompatibility with $O^b_{pf}B$ since that is not included. Thus, for this case, the desired $\Pi = \{O^b_{pf}A, O^b_{pf}C\}$, and $\Pi^* = \{A, C\}$. Since its only maximal consistent subset is itself and it entails both A and C we will have $M \models O_{atc}A$ and $M \models O_{atc}C$, as should be.

With Pattern B, the outline is similar. Here $\Delta = \{O^b_{pf}\neg A, O^b_{pf}\neg B, O^b_{pf}(A \vee B)\}$ with $O^b_{pf}\neg A > O^b_{pf}\neg B > O^b_{pf}(A \vee B)$, and $\Diamond(\neg A \wedge \neg B)$, $\Diamond(\neg A \wedge (A \vee B))$, $\Diamond(\neg B \wedge (A \vee B))$. To form Π the weightiest, $O^b_{pf}\neg A$ is included first, and then the second, $O^b_{pf}\neg B$, since it is compatible with the first member in. The least weighty, $O^b_{pf}(A \vee B)$, however, cannot be included since it is incompatible with the members already in, taken in combination. Thus $\Pi = \{O^b_{pf}\neg A, O^b_{pf}\neg B\}$. $\Pi^* = \{\neg A, \neg B\}$, and $M \models O_{atc}\neg A$, $M \models O_{atc}\neg B$, and $M \not\models O_{atc}(A \vee B)$, as we expect.

If the ordering relation $>$ is total over Δ, this will deliver a set Π appropriate to generate the requisite all-things-considered oughts. The procedure fails, however, if there are irresolvable conflicts among the members of Δ, e.g., cases where $\neg\Diamond(A \wedge B)$, and neither $O^b_{pf}A > O^b_{pf}B$ nor $O^b_{pf}B > O^b_{pf}A$. Then these steps will not produce a definite class Π at all.

Consider, for example, Pattern C. Take Δ to be the same as above, but now $O^b_{pf}\neg A > O^b_{pf}\neg B$ and $O^b_{pf}\neg A > O^b_{pf}(A \vee B)$, but neither $O^b_{pf}\neg B > O^b_{pf}(A \vee B)$ nor $O^b_{pf}(A \vee B) > O^b_{pf}\neg B$. If Π were formed as described above, taking the weightiest member first, $O^b_{pf}\neg A$, and then $O^b_{pf}\neg B$ second, since it is compatible with the membership already in, then $O^b_{pf}(A \vee B)$ will not be included since it is not compatible with the membership that precedes it. But there is no reason to prefer $O^b_{pf}\neg B$ to $O^b_{pf}(A \vee B)$. It was considered second only because it was listed second in the membership of Δ, but that is artificial. $O^b_{pf}(A \vee B)$ could just as well have been included second, since it is compatible with the first member in, and then $O^b_{pf}\neg B$ would not be taken. There is nothing to privilege one formation over the other as 'the' premium class. There is no unique such class.

To work around that, we proceed hypthetically or provisionally, as if the relation $>$ were total. Given a set of basic *prima facie* oughts Δ and a relation, $>$, not necessarily total, defined over Δ, let $\succ$ be a 'completion' of $>$ over Δ if $\succ$ is a strict partial order that is total over Δ and $> \subseteq \succ$. Given that $>$ is well-founded, there will always be at least one such completion.[53] In general, there may be more than one. We define sets Π_M now with respect to completions, rather than $>$ itself.

---

[53][Hansen, 2006, theorem 1], from Brewka.

So for Pattern C with Δ and $>$ as described, there are two completions of $>$, $O^b_{pf}\neg A \succ_1 O^b_{pf}\neg B \succ_1 O^b_{pf}(A\vee B)$, and also $O^b_{pf}\neg A \succ_2 O^b_{pf}(A\vee B) \succ_2 O^b_{pf}\neg B$. For each relation $\succ_1$ and $\succ_2$, the procedure described above now generates $\Pi_{\succ_1} = \{O^b_{pf}\neg A, O^b_{pf}\neg B\}$ and $\Pi_{\succ_2} = \{O^b_{pf}\neg A, O^b_{pf}(A\vee B)\}$. Thus there are two classes of premium *prima facie* oughts.

For Pattern D, there will be more relations and more premium sets. Δ is the same, but none of the *prima facie* oughts outweighs any others. So there are six completions of (the empty) $>$ representing the various possible orders of the three members of Δ. For each of them, $\Pi_{\succ_i}$ will contain just the first two in the ordering, to yield three distinct premium classes.

In general, if the relation $>$ is total over Δ then, trivially, there will be only one relation $\succ$, and $\Pi_\succ$ will be unique. If there are gaps in $>$, there will be multiple completions $\succ$ of $>$. Nevertheless, if every conflict of basic *prima facie* oughts in Δ is resolvable, so that for every $O^b_{pf}A, O^b_{pf}B \in \Delta$ if $\neg\Diamond(A\wedge B)$ then $O^b_{pf}A > O^b_{pf}B$ or $O^b_{pf}B > O^b_{pf}A$, then even though there will be multiple completions of $>$, each one will deliver the same set of premium *prima facie* oughts from Δ. I.e., if $\succ_1$ and $\succ_2$ are both completions of $>$ over Δ, then $\Pi_{\succ_1} = \Pi_{\succ_2}$. In case of irresolvable conflicts the sets $\Pi_{\succ_1}$ and $\Pi_{\succ_2}$ would be distinct.

Let us now define this procedure more precisely. Suppose Δ has two or more members, since otherwise $>$ would have no non-empty completions, and let $\succ$ be a completion of $>$. Take $\sigma_\succ$ for the sequence of the members of Δ as ordered by $\succ$. That is, if $\Delta = \{O^b_{pf}A_1, \ldots, O^b_{pf}A_i, O^b_{pf}A_j, \ldots\}$, then $\sigma_\succ = \langle O^b_{pf}A_1, \ldots, O^b_{pf}A_i, O^b_{pf}A_j, \ldots \rangle$ such that $O^b_{pf}A_i < O^b_{pf}A_j$ if and only if $O^b_{pf}A_i \succ O^b_{pf}A_j$. Then given $\sigma_\succ$ let

- $\Pi^1_\succ = O^b_{pf}A_1$ if $\not\vdash_\Box \neg A_1$, otherwise $\Pi^1_\succ = \emptyset$;

- For $1 \leq i < j = i+1$, if $\not\vdash_\Box \neg(\bigwedge \Pi^i_\succ \wedge A_j)$, then $\Pi^j_\succ = \Pi^i_\succ \cup \{O^b_{pf}A_j\}$, otherwise $\Pi^j_\succ = \Pi^i_\succ$.

- $\Pi_\succ = \bigcup \Pi^i_\succ$ for all i that mark a member of $\sigma_\succ$.

For each $\Pi_\succ$, $\Pi^*_\succ$, the set of enjoined contents of the members of $\Pi_\succ$, is a maximal co-possible subset of Δ^*. It could be that $\Pi_\succ$, and hence $\Pi^*_\succ$, are empty, as when all the members of Δ^* are individually impossible. This will not make a difference.

We now specify how all-things-considered oughts are generated from a set of basic *prima facie* oughts, Δ, with a relation $>$ defined over it.[54]

---

[54] (Def 8) is close to Hansen's (td-6) of [2006], though that provides for the generation of conditional-oughts, $O(A/C)$. See that paper's §4 for comparison of this approach

Def 8)   i)   If $\Delta = \emptyset$ or if $\Delta = \{O^b_{pf}B\}$, for some B, and $\vdash^M_\square \neg B$, then $M \models O_{atc}A$ if and only if $\emptyset \vdash^M_\square A$;
         ii)   if $\Delta = \{O^b_{pf}B\}$, for some B, and $\nvdash^M_\square \neg B$ then $M \models O_{atc}A$ if and only if $B \vdash^M_\square A$;
         iii)   otherwise, $M \models O_{atc}A$ if and only if $\Pi^*_\succ \vdash^M_\square A$, for each $\succ$ that is a completion of $>$ over Δ.

The first two clauses are for bookkeeping. The interesting cases, when there are normative conflicts, fall under clause (iii). We speak here of necessitation in order to allow for generality. If one did not want to say that that all necessities are *ipso facto* all-things-considered obligatory, one could replace necessitation with logical entailment and logical validity.

4.4.1   The criteria

In contrast to the accounts in Section 4.1 with (Def 1) and Section 4.2 with (Def 4), the present approach satisfies all three criteria (a)–(c). (Def 8) provides for all the core principles of deontic logic, (D), (P), (C) and (NM)/(RM), as well as (N), for formulas $O_{atc}A$ with no special assumptions regarding the basic *prima facie* oughts, $O^b_{pf}B$, in Δ.[55]

Because it validates all the core principles, this account also satisfies criterion (a), conflicts among all-things-considered oughts are excluded, even while there might be conflicts among basic *prima facie* oughts.

It is not hard to show how (Def 8) also satisfies criterion (c) by providing for the Disjunctive Response in cases of irresolvable conflicts among the basic *prima facie* oughts in Δ. For example, for any $M \in \mathcal{M}_d$, if $O^b_{pf}A_1, \ldots, O^b_{pf}A_n \in \Delta$, and individually $M \models \Diamond A_i$, but pairwise, for any

---

with several others that are similar. (Def 8) is designed to exclude normative conflicts for all-things-considered oughts, as is fitting for the present perspective. Changing the word 'each' to 'some' in clause (iii) gives a definition that would allow such conflicts. See Horty's [2003; 2012] for a comparison of the two approaches. We discuss that other account that allows conflicts for all-things-considered oughts in Section 6.3 below.

[55] For convenience, consider only the case where Δ has at least two members, and all are fulfillable; the other cases are similar. For (D), suppose $M \models O_{atc}A$, but $M \models O_{atc}\neg A$. Then, each $\Pi^*_\succ$ necessitates A and also $\neg A$. Moreover, there must be at least one such $\Pi^*_\succ$. By ordinary modal logic, it necessitates $A \wedge \neg A$, contrary to the fact that $\Pi^*_\succ$ is a mutually co-possible subset of Δ^*. Hence, if $M \models O_{atc}A$ then $M \not\models O_{atc}\neg A$ and $M \models \neg O_{atc}\neg A$. The argument for (P) is similar.

For (NM), suppose $M \models \square(A \to B)$ and $M \models O_{atc}A$. Consider any appropriate $\Pi^*_\succ$. Since $M \models O_{atc}A$, $\Pi^*_\succ \vdash_\square A$. Since $M \models \square(A \to B)$, $\Pi^*_\succ \vdash_\square B$ by ordinary modal logic. Hence, $M \models O_{atc}B$. (Without necessity, $\square$, the argument for (RM), hence (M), is similar.)

For (C), suppose $M \models O_{atc}A$ and $M \models O_{atc}B$, and consider any appropriate $\Pi^*_\succ$. By the first assumptions, $\Pi^*_\succ$ necessitates both A and B, hence, by ordinary modal logic, it necessitates $A \wedge B$. That suffices for $M \models O_{atc}(A \wedge B)$, as desired.

Necessitation, (N), should be obvious.

$i \neq j$, $M \models \neg\Diamond(A_i \wedge A_j)$ and neither $O^b_{pf}A_i > O^b_{pf}A_j$ nor $O^b_{pf}A_j > O^b_{pf}A_i$, then $M \not\models O_{atc}A_i$, for each $1 \leq i \leq n$, but $M \models O_{atc}(A_1 \vee \cdots \vee A_n)$.[56]

This is most easy to appreciate in the very simple cases like the Marcus and Sartre Examples. For the first, suppose $\Delta = \{O^b_{pf}s_1, O^b_{pf}s_2\}$, where $\Diamond s_1$, $\Diamond s_2$, $\neg\Diamond(s_1 \wedge s_2)$, and neither $O^b_{pf}s_1 > O^b_{pf}s_2$ nor $O^b_{pf}s_2 > O^b_{pf}s_1$. There are two completions of $>$, $O^b_{pf}s_1 \succ_1 O^b_{pf}s_2$ and $O^b_{pf}s_2 \succ_2 O^b_{pf}s_1$, and accordingly, two premium sets $\Pi_{\succ_1} = \{O^b_{pf}s_1\}$ and $\Pi_{\succ_2} = \{O^b_{pf}s_2\}$. Obviously, $\Pi^*_{\succ_1} = \{s_1\}$ does not necessitate s_2 nor $\Pi^*_{\succ_2} = \{s_2\}$ necessitate s_1, so neither $M \models O_{atc}s_1$ nor $M \models O_{atc}s_2$. Nevertheless, $\{s_1\}$ does necessitate $s_1 \vee s_2$ and $\{s_2\}$ necessitates $s_1 \vee s_2$. So $\Pi^*_{\succ_1}$ necessitates $s_1 \vee s_2$ and likewise $\Pi^*_{\succ_2}$ necessitates $s_1 \vee s_2$. Since those are the only sets $\Pi_\succ$, $M \models O_{atc}(s_1 \vee s_2)$, as should be under this perspective.

The same works for more complex cases like the Mission Examples. For the first, with the agent's missions to Amsterdam, Barcelona, Cairo and Dublin, where $\Delta = \{O^b_{pf}A, O^b_{pf}B, O^b_{pf}C, O^b_{pf}D\}$ and $\neg\Diamond(A \wedge B)$, $\neg\Diamond(A \wedge C)$, $\neg\Diamond(B \wedge C)$, $\Diamond(A \wedge D)$, $\Diamond(B \wedge D)$, $\neg\Diamond(C \wedge D)$, while $O^b_{pf}A \approx O^b_{pf}B \approx O_{pf}C$, but $O^b_{pf}D > O^b_{pf}C$, whence $O^b_{pf}D > O^b_{pf}A$ and $O^b_{pf}D > O^b_{pf}B$. From that, there are six completions of $>$ with $O^b_{pf}D$ in first place for all, and differing in the priorities for $O^b_{pf}A, O^b_{pf}B, O^b_{pf}C$. Of the six corresponding sets of premium *prima facie* oughts, $O^b_{pf}D$ will be in all, $O^b_{pf}C$ in none, and $O^b_{pf}A$ and $O^b_{pf}B$ each in some but not all.

This means that $M \models O_{atc}D$, as expected, but not $M \models O_{atc}A$, nor $M \models O_{atc}B$, nor $M \models O_{atc}C$. Nevertheless, $M \models O_{atc}(A \vee B)$ in keeping with the Disjunctive Response since each $\Pi_{\succ_i}$ that necessitates A also necessitates $A \vee B$ and likewise each that necessitates B. We even have $M \models O_{atc}(D \wedge (A \vee B))$, and equivalently $M \models O_{atc}((D \wedge A) \vee (D \wedge B))$, as it should be.

The second Mission Example, with the missions to Amsterdam, Berlin and Copenhagen, works similarly. Details are left to the reader.

We have already seen how this approach applies to Patterns A–D that were destructive for the traditional account of Section 4.1 with (Def 1).

4.4.2 Derived *prima facie* oughts

While the account so far presented does satisfy all three criteria (a)–(c), here too there remains a respect in which it might still fall short of expectations

---

[56] For any pair A_i, A_j, each will be in some $\Pi^*_\succ$, but because $\neg\Diamond(A_i \wedge A_j)$, they will not both be in the same such set $\Pi^*_\succ$. Hence, not every $\Pi^*_\succ$ defined for M necessitates A_i, and not every such $\Pi^*_\succ$ necessitates A_j. Hence neither $M \models O_{atc}A_i$ nor $M \models O_{atc}A_j$. On the other hand, each $\Pi^*_\succ$ that necessitates A_i does necessitate $A_1 \vee \cdots \vee A_i \vee \cdots \vee A_n$, and similarly for those that necessitate A_j, and for all the other A's from 1 to n. Hence, every $\Pi^*_\succ$ defined for M necessitates $A_1 \vee \cdots \vee A_n$, and so $M \models O_{atc}(A_1 \vee \cdots \vee A_n)$, as desired.

for an understanding of *prima facie* oughts and their relations to all-things-considered oughts.

The basic *prima facie* oughts in Δ are treated as unanalyzed elements, with nothing much assumed about them, especially not that they follow any particular logic of their own. Perhaps it is a misnomer to call them *prima facie* oughts at all.[57] Considering how that notion has developed in philosophical contexts, one might expect something more. If an agent has a *prima facie* obligation to do A and B, $O_{pf}(A \wedge B)$, one expects to infer the agent has a *prima facie* obligation to do A, $O_{pf}A$. Perhaps Ada is secretary of her committee; she *prima facie* ought to attend the meeting and take notes, $O_{pf}(m \wedge n)$. One might expect to infer that she *prima facie* ought to attend the meeting, $O_{pf}m$. That is so even if, all-things-considered, she ought to take her son to the circus. Smith *prima facie* ought to fight in the army or perform alternate service, $O_{pf}(f \vee s)$, and also Smith *prima facie* ought not to fight in the army, $O_{pf}\neg f$. But perhaps there is a stronger injunction against alternative service, $O_{pf}\neg s$. In that case, all-things-considered Smith ought not perform alternative service, $O_{atc}\neg s$. Even so, one might expect, on the basis of all the information, that Smith has a *prima facie* obligation to perform alternative service, $O_{pf}s$. That is an active normative position for Smith, though it has been overridden by the more compelling injunction. Nothing in the account so far covers these inferences among *prima facie* oughts.

Accordingly, let us distinguish the basic *prima facie* oughts, $O_{pf}^{b}A$, given by the membership of Δ, from 'derived' *prima facie* oughts that are themselves generated by Δ. For this we write $O_{pf}^{d}A$. These are oughts whose content is necessitated by the collected contents of consistent basic *prima facie* oughts.

Def 9) $M \models O_{pf}^{d}A$ if and only if $\Sigma \vdash_{\Box}^{M} A$, for some consistent (co-possible) set $\Sigma \subseteq \Delta^*$

Here the relation $>$ has no role to play. It is only significant for determining all-things-considered oughts from the basic *prima facie* oughts in Δ.

(Def 9) is essentially the same as the account to be offered in Section 6.3 below for oughts that admit normative conflicts, and so we defer to that section further discussion of the properties of operators defined in this way.

One question, however, that remains open for this as a specification of distinctively *prima facie* oughts is whether they can stand in relations of relative weight and can override each other. (Def 9) offers no provision for that. The relation $>$ applies only to the membership of Δ, and it is

---

[57][Horty, 2012, ch. 2 and 3] speaks instead of an 'austere theory of reasons', with emphasis on 'austere'.

unclear how it could be extended to derived *prima facie* oughts, or whether it should be. So long as that is an open question, it is uncertain how well this picture of *prima facie* oughts corresponds to the notion philosophers have developed.[58]

4.5 A hybrid account

Finally, we indicate briefly a hybrid between the account of Section 4.2, analyzing oughts in terms of reasons, $\looparrowright$, and that of Section 4.4 with its distinctive treatment of priority relations.

Any model $M \in \mathcal{M}_r$ from Section 4.2 determines a model $M' \in \mathcal{M}_d$. Given $M = \langle w_0, W, v, \Re \rangle$, (Def 3) determines a set of *prima facie* oughts for M, Δ_M. These are (partially) ordered by the overriding relation, so that for $O_{pf}A, O_{pf}B \in \Delta_M$, $O_{pf}A >_o O_{pf}B$ just in case $O_{pf}A$ overrides$_M$ $O_{pf}B$. Take $M' = \langle w_0, W, v, (\Delta_M, >_o) \rangle$. All of the machinery of Section 4.4 applies to this model, and by extension to M. Accordingly, corresponding to (Def 8), in combination with (Def 3), for $M = \langle w_0, W, v, \Re \rangle \in \mathcal{M}_r$,[59]

Def 10) i) If $\Delta_M = \emptyset$ or if $\Delta_M = \{O_{pf}B\}$, for some B, and $\vdash_{\Box}^{M} \neg B$, then $M \models O_{atc}A$ if and only if $\vdash_{\Box}^{M} A$;
ii) if $\Delta_M = \{O_{pf}B\}$, for some B, and $\nvdash_{\Box}^{M} \neg B$ then $M \models O_{atc}A$ if and only if $B \vdash_{\Box}^{M} A$;
iii) otherwise, $M \models O_{atc}A$ if and only if $\Pi_{\succ}^{*} \vdash_{\Box}^{M} A$, for each $\succ$ that is a completion of $>_o$ over Δ_M.

This has all the virtues of the preceding account, including satisfaction of all three basic criteria of adequacy. It also provides the correct responses to Patterns A–D since it applies the very same apparatus as the preceding.

The primary advantage of this hybrid lies in the more robust theory of *prima facie* oughts that it offers by way of (Def 3). It does not distinguish basic from derivative *prima facie* oughts, and it includes a natural sense of how *prima facie* oughts might override others. Like the account of Section 4.2, however, this approach still does not provide for the Smith Argument when that pertains to *prima facie* oughts.

---

[58] Notice that the treatment of Patterns A–D above and the examples for the Disjunctive Response presupposes that the respective *prima facie* oughts are basic and included in Δ. If they were derived, as described here, the story would be different. Since these examples rely on relative weights for the various *prima facie* oughts it is an open question what that story would be.

[59] By virtue of its membership in $\mathcal{M}_r$, M should satisfy ($\Re$nm)/($\Re$rm) and ($\Re$c). These conditions would not now be required to establish the core principles for all-things-considered oughts; that operates as under (Def 8). They are pertinent to the theory of *prima facie* oughts.

4.6 Reprise

In this section we have looked at some proposed accounts of the relation between *prima facie* and all-things-considered oughts. Section 4.1 described a traditional approach that reflects much philosophical language. This assumes that *prima facie* oughts are ordered by a relation of relative weight or significance and takes all-things-considered oughts to be those *prima facie* oughts that outweigh all competing *prima facie* oughts (Def 1). Section 4.2 presented an account that analyzes *prima facie* oughts in terms of a more fundamental relation of a fact being a (normative) reason for an action or state of affairs, here represented by $A \hookrightarrow B$. *Prima facie* oughts are then those acts or states of affairs for which there is some reason that obtains (Def 3). All-things-considered oughts are then specified to be those *prima facie* that are undefeated (Def 4).

While these first two accounts determine all-things-considered oughts from *prima facie* oughts, the approach of Section 4.3 takes the concept of all-things-considered ought to be fundamental, and considers *prima facie* oughts to be expressions of *ceteris paribus* principles regarding such oughts, so that an agent *prima facie* ought to do something just in case, in the circumstances, *ceteris paribus* the agent ought (all-things-considered) to do that (Def 5). To express such *ceteris paribus* principles, a defeasible, non-normative conditional $A \gg B$ was introduced. Under the first two accounts, the central question was, How are all-things-considered oughts determined by *prima facie* oughts? Under this third account, the key question is, How are full-fledged (all-things-considered) oughts to be detached from *prima facie* oughts? Given A and $A \gg OB$, what does it take to infer OB? Such inferences are taken to be defeasible, or nonmonotonic. We described a simple model for how that might work (Def 7). The fourth approach of Section 4.4 returned to the idea of basic *prima facie* oughts, or as we might call them, directives, with an ordering of relative weight or priority defined over them. This, however, derived all-things-considered oughts from such basic *prima facie* oughts quite differently from the first, traditional account, see (Def 8). In Section 4.5 we pointed to a hybrid, (Def 10), of the second and fourth approaches.

Each of these accounts is measured against three criteria of adequacy that reflect the general perspective of this section, that (a) while *prima facie* oughts might conflict, genuine all-things-considered oughts cannot, (b) indeed, all-things-considered oughts follow the familiar core principles of deontic logic, which exclude the possibility of conflicts for such oughts, and further (c) for irresolvable normative conflicts the Disjunctive Response should apply, i.e., when *prima facie* oughts for A and for B are in conflict but neither outweighs the other (and no other oughts are pertinent), then neither A nor B is by itself all-things-considered obligatory, but there is an all-things-considered obligation for the disjunction A-or-B, and similarly for conflicts of greater arity.

As we saw, the different accounts have varying success in meeting these criteria. The first, traditional account failed on all three. The second 'reasons' account succeeded with (a) and (b) but failed to given an account of the Disjunctive Response (c). The third, 'ceteris paribus' approach does meet all three, at least so long as one is satisfied with the Disjunctive Response being taken as a defeasible inference. The fourth, directives, account also satisfies all three criteria handily, as does the hybrid.

We also saw how these accounts fared on the problems of under- and overgeneration, Patterns A–D, introduced in Section 4.1. For this we distinguish the simple reasons account of Section 4.2, with just $(\Re nm)/(\Re rm)$ and $(\Re c)$ from the version with the further rule (RIO) since that is required for Pattern A.

Table 1 summarizes these findings, where a '✓' indicates the account meets the criterion, or that it provides the proper responses to the Patterns, and '×' that it does not. With all of the accounts there are, of course, other questions that may be pressed, and which invite further results.

| | Criteria | | | Patterns | |
Account	(a)	(b)	(c)	A	B – D
traditional	×	×	×	×	×
reasons	✓	✓	×	×	✓
reasons + (RIO)	✓	✓	×	✓	✓
ceteris paribus	✓	✓	✓	✓	✓
basic directives	✓	✓	✓	✓	✓
hybrid	✓	✓	✓	✓	✓

Table 1

5 Conflict 1: Revisionist strategies

In this section and the next we follow the other direction taken to reconcile a logic of normative statements with apparent conflicts. This course accepts the possibility of genuine normative conflicts for the sense of 'ought' to which the principles of a deontic logic should apply, but, to escape Arguments I and II from Section 1, it rejects or revises some of the core principles used in those arguments. This section presents some fairly straight-forward proposals to do that. The next describes some other approaches that call for a more radical rethinking of the foundations of a logic of normative concepts.

For the discussion of these two sections the operator O represents any sort of normative ought for which conflicts are admitted. We do not now distinguish *prima facie* oughts from all-things-considered oughts, as in the preceding section. If one would draw that distinction, then the present O could express either of them, if it is allowed that all-things-considered oughts can conflict. Otherwise, it could represent just the *prima facie* ought. By the same token, it could be used to express the ought of any specific normative domain, such as morality, law, etc., or the common ought of practical reasoning. We do, of course, suppose O to be univocal in context.

The primary concern of this section is to determine what logical principles are appropriate for a logic of O that admits the possibility of normative conflicts. Clearly, a central desideratum for such a logic is that it accept such conflicts as consistent. For a range of individually possible A's,

Desideratum 1) $\neg \Diamond (A_1 \wedge \cdots \wedge A_n), OA_1, \ldots, OA_n \nvdash \bot$, or
even if $\vdash \neg (A_1 \wedge \cdots \wedge A_n)$, yet $OA_1, \ldots, OA_n \nvdash \bot$

where $\vdash$ represents logical consequence for the proposed system, including principles for the alethic modality $\Diamond$, if present, and the recommended principles for the deontic O, and $\bot$ represents a fundamental absurdity expressible in the system, e.g., a logical contradiction. Other desiderata will emerge as we look at attempts to meet this one, and their shortcomings.

Arguments I and II in Section 1 reveal that satisfying Desideratum 1 requires rejecting or revising at least one from each pair of core principles [(P), (C)] and [(NM)/(RM), (D)]. In this section we examine the effects of each option in turn. As we will see, the resulting systems struggle between being too weak, not being able to do all that is expected of them, and being too strong, generating unacceptable consequences. All the systems discussed in this section are extensions of classical propositional logic, CL. (In Section 6.1 we will consider modifying even that.)

5.1 Non-Kantian systems

Early on, in order to accommodate normative conflicts Lemmon [1962] proposed denying the so-called Kantian principle that 'ought' implies 'can' (P) while retaining both aggregation (C) and distribution (NM)/(RM). Since (C) and (NM)/(RM) entail that (P) is equivalent to (D) this has the effect of denying the 'no-conflicts' provision (D) as well. After Lemmon, this is a fairly common suggestion in the literature on normative conflicts. The system that results is a normal modal logic of type **K** for O,[60] and such

---

[60]The real **K** also contains the rule (N) of necessitation for O. Other non-core postulates could also be added to form other systems, e.g., of type **K45**. The same issues

logics are known to be very stable and robust. Because it undermines both Argument I and Argument II, this proposal easily satisfies Desideratum 1; normative conflicts are logically consistent.

Logics including **K** for O are, however, too strong to accommodate normative conflicts well. Any system that contains all of classical logic, including that a contradiction entails everything, *ex contradictione quodlibet*,

ECQ) $\quad A, \neg A \vdash B$

as well as (C) and (NM)/(RM) will produce 'deontic explosion' from any normative conflict. If there is any strict normative conflict, $OA, O\neg A$, then everything is obligatory, i.e., OB, for every B.[61]

DEX$_s$) $\quad OA, O\neg A \vdash OB$

This generalizes to apply to all conflicts, given both (C) and (NM)/(RM) on a classical base.[62]

DEX) $\quad \neg\Diamond(A_1 \wedge \cdots \wedge A_n), OA_1, \ldots, OA_n \vdash OB$, or
if $\vdash \neg(A_1 \wedge \cdots \wedge A_n)$, then $OA_1, \ldots, OA_n \vdash OB$

Presumably this is unacceptable under the present view. It is one thing to admit the possibility of normative conflicts, and quite another to agree that everything is obligatory. Hence a second desideratum for a logic that would allow normative conflicts is that it not blow up in this way.

Desideratum 2)   The logic should not contain (DEX), or anything like it.

The reason for the last rider will appear below. The proposal to eliminate (P) and (D) while keeping the other core principles, (C) and (NM)/(RM), of deontic logic fails to meet this desideratum.[63]

It is worth remarking at this point that any logic that contains (ECQ), or its modal counterpart $\Box\neg A \to \Box(A \to B)$, and unrestricted (NM)/(RM) will also contain 'permissive' explosion. For every B,

DEX$_{pi}$) $\quad \Box A, \neg OA \vdash PB$, or
if $\vdash A$ then $\neg OA \vdash PB$

---

discussed here apply to them as well.

[61] Suppose OA and $O\neg A$. By (C), $O(A \wedge \neg A)$. From (ECQ) by (RM) $\vdash O(A \wedge \neg A) \to OB$. Hence OB.

[62] If $\neg\Diamond(A_1 \wedge \cdots \wedge A_n)$ then $\Box((A_1 \wedge \cdots \wedge A_n) \to B)$, by basic (classical) modal logic. By repeated (C) $\vdash (OA_1 \wedge \cdots \wedge OA_n) \to O(A_1 \wedge \cdots \wedge A_n)$, whence OB by (NM). In the absence of the alethic modalities, the same result applies for normative conflicts in which $\{A_1, \ldots, A_n\}$ is inconsistent through the use of (RM).

[63] To hold there is something not obligatory, some A such that $\neg OA$, is a substantive position. Accordingly, this is a desideratum for those who take that position.

which says that if any necessity is not obligatory, or equivalently, if there is any permitted impossibility (pi), then everything is permitted.[64] Presumably, unless one is a complete nihilist, one does not want to assert that everything is permitted, or equivalently, that nothing is obligatory. Hence, if one accepts (NM)/(RM), one is committed to every necessity, especially every validity, being obligatory. In other words, one is committed to the principle of necessitation, (N), from Section 1. So long as one accepts that commitment, (DEX$_{pi}$) is not as disasterous as (DEX) itself, which has no such escape under the present proposal.

With (DEX$_{pi}$) we see that (N) is not easily avoided if (NM)/(RM) is maintained. This is so regardless of what other principles, such as (C), are assumed. Indeed, this is so whether or not one admits the possibility of normative conflicts in the first place, and so applies to the positions discussed in Sections 3 and 4 above. We will mention (N) from time to time in what follows, but for the most part will henceforth take it for granted.[65]

5.2 Non-aggregative systems

After Lemmon's proposal to reject (P) in the face of normative conflicts, the leading alternative is to reject (C) instead.[66] That too defeats Argument I. In the absence of (C), (P) ceases to be equivalent to (D); so it can be retained even while (D) is denied to defeat Argument II. With Arguments I and II out, Desideratum 1 is now met. So too is Desideratum 2, although this proposal is usually made only with an eye to consistency. Deontic explosion (DEX) is not derivable in the absence of (C).

Under this proposal (P) is generally preserved in deference to the Kantian intuitions that 'ought' does imply 'can'. Within this system, this is more than just a philosophical nicety. Just as any logic that assumes (NM)/(RM) without restriction must contain (N) to defuse (DEX$_{pi}$), so it must also contain (P) to disarm

---

[64]Suppose $\Box A$ and $\neg OA$. By classical modal logic $\Box(\neg B \to A)$. Suppose $\neg PB$, i.e., $O\neg B$. By (NM), OA, a contradiction. Hence if $\Box A$ and $\neg OA$ then PB. The argument is similar without alethic modalities.

[65]That there is something not permitted, like the claim that there is something not obligatory, is a substantive position. Someone disagreeing would not find (DEX$_{pi}$) troubling, and so might not find (N) so necessary. The need for (N) can also be obviated by restricting (NM)/(RM) to cases where B is contingent. Some philosophers would do that. Since the question of whether necessities can properly be considered obligatory is a separate issue from our concerns with normative conflicts we do not pursue it here.

[66]The proposal to deny (C) in order to admit moral dilemmas, while keeping principle (P), perhaps first appeared in philosophical discussion in [Williams, 1965] and was renewed by Marcus [1980]. It is given more formal treatment, in different ways, in [van Fraassen, 1973], [Chellas, 1980], [Schotch and Jennings, 1981], and my [Goble, 2000; Goble, 2003; Goble, 2004a]. It now reappears frequently.

DEX$_{oi}$) ¬◇$A, OA \vdash OB$, or
if $\vdash \neg A$ then $OA \vdash OB$

that if there is any obligatory impossibility (oi), then everything is obligatory.[67] Hence, any counterexample to (P)/(P)′ entails the trivialization or collapse of the normative system. Excluding such counterexamples renders (DEX$_{oi}$) harmless. Under this proposal then, (P)/(P)′ seems to be mandated if Desideratum 2 is to be met.

The logic consisting of (NM)/(RM) and (P)/(P)′ together with the rule (N) above, and a base of classical propositional logic, is a non-normal, but classical modal logic of type **EMN** for the operator O, in Chellas's nomenclature. It is often called 'minimal deontic logic' (MDL); elsewhere, e.g., [Goble, 2000; Goble, 2003; Goble, 2004a], I have called it **P**, and will so call it here. Though weak in output, this logic too is very stable and robust. It is amenable to a variety of semantic treatments.

Adopting the general schema of Section 2.2 for models $M = \langle w_0, W, v, \mathcal{S} \rangle$, each of the following characterizations of $\mathcal{S}$ is sound and complete for **P**, with allowance for not including iterated modalities.[68]

(i), multi-optimal: $\mathcal{S} = \mathcal{R}$, a non-empty set of (non-empty) sets $R \subseteq W$ (or, much the same, a non-empty set of serial binary relations over W), and

- $M \models OA$ if and only if there is an $R \in \mathcal{R}$ such that $R \subseteq |A|_M$

where $|A|_M = \{w \in W : M, w \models A\}$; this is the proposition expressed by A, according to M.

On this account normative conflicts are understood to occur when the individual oughts arise under different standards of optimality R. This might correspond to the view that such conflicts result from disagreements among normative systems or other sources of obligations.

(ii), simple preference: $\mathcal{S} = P$, where P is a binary preference relation over W, with wPw' understood to say that w is at least as good, preferable, desirable, etc., as w', and

---

[67]Suppose ¬◇A and OA. By the former, □¬A. Since $\vdash \neg A \rightarrow (A \rightarrow B)$, $\vdash \Box \neg A \rightarrow \Box(A \rightarrow B)$ by basic modal logic. So $\Box(A \rightarrow B)$. By (NM), $\vdash \Box(A \rightarrow B) \rightarrow (OA \rightarrow OB)$. So $OA \rightarrow OB$, and OB follows by *modus ponens*. A similar argument applies via (RM) in the absence of alethic modalities.

[68]For (i), cf. [Schotch and Jennings, 1981] and [Goble, 2000]. For (ii) and (iii), cf. [Goble, 2000; Goble, 2003; Goble, 2004a]. For (iv), cf. [Chellas, 1980, p. 202]. For (v), cf. the first account of van Fraassen's [1973]; see also [Horty, 1994, p. 38] and [Hansen, 2004b, pp. 40-2], esp. the definition for O^3. (**P** is also sound and (weakly) complete for the more sophisticated imperatival account we describe in Section 6.3.)

- $M \models OA$ if and only if there is a $w \in W$ such that $w \in |A|_M$ and for any $w' \in W$ such that $w'Pw$, $w' \in |A|_M$.

The relation P can be as you like it; it might be reflexive and transitive, but that is not required. If it is required to be not only reflexive and transitive, but also connected, so that for any $w, w' \in W$, wPw' or $w'Pw$, then this account validates (C) as well as (RM) and (P)', hence (D), and so models all the core principles. Situations of normative conflict can be understood as counterexamples to connectedness in this framework. This applies to conflicts that might arise within a single normative system as well as among systems.

(iii), multi-preference: $\mathcal{S} = \mathcal{P}$ with $\mathcal{P}$ a non-empty set of preference relations, P, much as above; these could be required to be reflexive, transitive and connected on their fields, but need not be. Then

- $M \models OA$ if and only if there is a $P \in \mathcal{P}$ and $w \in W$ such that $w \in |A|_M$ and for any $w' \in W$ if $w'Pw$ then $w' \in |A|_M$.

This account understands normative conflicts much as (i) but in terms of relations of preference rather than optimality, and also much as (ii) since the relations P in $\mathcal{P}$ need not be connected.

(iv), neighborhoods or minimal models: $\mathcal{S} = \mathcal{O}$, in which $\mathcal{O}$ represents the extension of O, the set of propositions that are obligatory (according to M), where a proposition is taken to be a set of worlds, so that $\mathcal{O} \subseteq \wp W$. Then

- $M \models OA$ if and only if $|A|_M \in \mathcal{O}$.

To validate (M) and (RM) it is required that $\mathcal{O}$ be closed under supersets, that for $X, Y \subseteq W$, if $X \in \mathcal{O}$ and $X \subseteq Y$ then $Y \in \mathcal{O}$. To validate (P), assume each $X \in \mathcal{O}$ is non-empty; to validate (N) assume $W \in \mathcal{O}$. On this approach normative conflicts are simply cases in which incompatible propositions fall within the extension of O.

(v), simple imperatival: similar to the account of Section 4.4, except that priority relations do not come into play, and conflicts are allowed. Here, $\mathcal{S} = \Delta$, a nonempty class of directives, or imperatives, d_i, such that each has propositional content, that which must be so for the directive to be fulfilled. Writing !B for such directives, $B \in \mathcal{L}_B$ being the propositional content,

- $M \models OA$ if and only if there is a $d \in \Delta$ such that $d = !B$ and $M \models \Diamond B$ (or $\nvdash \neg B$), and $B \vdash A$.

Thus one ought to do whatever is necessary to fulfill a (doable) directive one is under.[69] Normative conflicts arise from incompatible directives.

Each of these accounts not only offers a rigorous theory by which to interpret formulas OA, they also, each in its own way, provide a rationale for understanding normative conflicts to be possible.

That is all to the good. Nevertheless, by rejecting (C), these accounts all fall short in a significant respect. In Section 3, and again in Section 4.2.3 for *prima facie* oughts, we introduced what we call the Smith Argument:

  i) Smith ought to fight in the army or perform alternative national service. — $O(f \vee s)$
  ii) Smith ought not to fight in the army. — $O \neg f$
∴ iii) Smith ought to perform alternative national service. — Os

On the face of it, this argument looks valid. It looks valid for any normative reading of the oughts, assuming they are univocal throughout. It looks valid for moral oughts, and legal oughts, and oughts stemming from other sorts of sources. It looks valid for the common ought of practical reasoning. It looks valid if the oughts are taken to be all-things-considered oughts, and also if they are merely *prima facie* oughts. Any system that purports to be a logic of 'ought' for any of these senses should thus be able to account for that sense of validity. The present system fails in this regard.

Indeed, when Horty, [1994; 1997; 2003; 2012] presents this example,[70] his purpose is exactly to challenge logics, like the present **P**, that abandon the aggregation principle (C). With (C) and (RM), the validity of the Smith Argument is assured. By (C), (i) and (ii) entail $O((f \vee s) \wedge \neg f)$, which entails (iii) Os by (RM). Without (C), however, there is no prospect for drawing the conclusion.

Logics with (DEX) that produce deontic explosion from normative conflicts are too strong. Logics that cannot account for the Smith Argument, or others like it, are too weak. An adequate logic of normative concepts should therefore satisfy this third desideratum of adequate strength.

 Desideratum 3) The logic should explain in a plausible way the apparent validity of several paradigm arguments, including the Smith Argument.

Non-aggregative logics like **P** fail this standard. This is also a stumbling block for many of the proposals we will be considering below.

This desideratum may call for more explanation than the previous two. In the ensuing discussion we will meet a few more paradigm arguments.

---

 [69]That B be possible validates (P); that Δ be non-empty validates (N).
 [70]Following a concern raised by van Fraassen [1973, p. 18], who cites Stalnaker as his source.

To include these within the desideratum presumes that they, like the Smith Argument, do appear to be valid. For some, or even all of them, this could be controversial. If one thinks that any is not valid for certain sorts of ought, then, of course, one would not desire a logic for such oughts to proclaim that it is, and it would seem no shortcoming of an approach that it fails to account for such arguments. It is not our purpose here to argue for, or against, the virtues of any of these examples. They are offered rather as standards of comparison by which to weigh one approach against another.

Furthermore, we should allow some latitude in what counts as a plausible account of the apparent validity of the paradigm arguments. With the Smith Argument, for example, it must be conceded that it is not strictly valid in the sense of being an instance of a logically valid pattern of inference, Deontic Disjunctive Syllogism,

DDS) $O(A \vee B), O\neg A \vdash OB$

that is valid for all instances of A and B. In logics like **P** that have the full power of (RM), this rule is deductively equivalent to (C). Hence, to demand (DDS) along with (RM) returns deontic explosion (DEX) and a violation of Desideratum 2. Instead, we should anticipate a looser sense of validity or an account by which the Smith Argument, or something quite close to it, would rate as a piece of good reasoning, even while another superficially similar argument might not. In what follows we will see several such accounts.

5.2.1 Consistent aggregation

The Smith Argument reveals the need for some aggregation, but perhaps it does not call for the full power of (C). The most immediate suggestion to revise this rule would limit its application to cases where there is no conflict between the oughts to be conjoined. Call this 'consistent aggregation'.[71]

CC) $\Diamond(A \wedge B) \to ((OA \wedge OB) \to O(A \wedge B))$, or
if $\nvdash A \to \neg B$ then $\vdash (OA \wedge OB) \to O(A \wedge B)$

where the second form would be used if alethic modalities are not included. Call the result of adding this principle to **P**, **Pcc**.

Weakening aggregation in this way blocks Argument I for the inconsistency of normative conflicts, and without (D) Argument II cannot be

---

[71] Cf. Section 4.1.1. While this move does seem a natural proposal, it rarely appears in print. [van der Torre, 1997, p. 94] and [van der Torre and Tan, 2000, p. 411] seem to endorse it, citing [van Fraassen, 1973], but only within a specialized setting that avoids the problems presented here. (See Section 6.2 below.) The same may be said for [Prakken, 1996, p. 74 and p. 84]. [McNamara, 2004, p. 137] proposes a form of this principle, but then (p. 147) retreats from it. Brink [1994, p. 229] and Zimmerman [1996, p. 214] both suggest this as a way for advocates of moral dilemmas to preserve (P) while retaining a measure of aggregation, but neither Brink nor Zimmerman endorses the suggestion; neither accepts moral dilemmas either.

completed. Thus, a logic like **Pcc**, containing (CC) along with (NM)/(RM) and (P)/(P)' but not (D), meets Desideratum 1. It also accounts nicely for the Smith Argument since $f \vee s$ and $\neg f$ are presumed co-possible, which allows $O(f \vee s)$ and $O\neg f$ to be combined by (CC) to $O((f \vee s) \wedge \neg f)$, which entails Os by (RM). In effect, this treats the original Smith Argument as an enthymeme with the tacit premise $\Diamond((f \vee s) \wedge \neg s)$. In this way, this logic satisfies Desideratum 3; it is not too weak.

Unfortunately, this logic is much too strong, and fails Desideratum 2.[72] It validates this form of deontic explosion, that any normative conflict entails that anything possible is obligatory, which is surely unacceptable.[73]

DEX-1) $\neg \Diamond(A_1 \wedge \cdots \wedge A_n), OA_1, \ldots, OA_n \vdash \Diamond B \to OB$, or
if $\vdash \neg(A_1 \wedge \cdots \wedge A_n)$ and $\nvdash \neg B$ then $OA_1, \ldots, OA_n \vdash OB$

Similar results are likely to afflict other attempts to restrict aggregation with a condition that is amenable to replacement of logical equivalents. It is too easy to dodge such constraints.[74]

5.3 Non-distributive systems

Although the distribution principle (NM)/(RM) and its consequence (M) have been questioned for many reasons, especially their complicity in the familiar deontic paradoxes, they are rarely discussed in connection with normative conflicts. Nevertheless, to reject them could be a useful strategy for admitting conflicts.

Without (NM)/(RM), Argument II of Section 1 is blocked, and Argument I can be blocked by denying (P) while preserving (C). Strictly speaking, (D) can be kept if one would exclude strict normative conflicts but allow other non-strict conflicts, but since that seems an unlikely position we suppose

---

[72] (CC), along with (RM), overgenerates oughts in a number of ways; for examples see, e.g., [Horty, 2003, p. 581], [McNamara, 2004, p. 147], with credit to Hansen, and [Goble, 2004b, pp. 81-2], [Goble, 2005, p. 468], [Goble, 2009, p. 462], though these examples may be controversial. By contrast, the explosion of (DEX-1) seems fatal.

[73] Suppose $\neg \Diamond(A_1 \wedge \cdots \wedge A_n)$ and $OA_1, \ldots, OA_n$. Since we presume principle (P), $\Diamond A_1$ and ... and $\Diamond A_n$. Let $\Phi = \{A_{j_1}, \ldots, A_{j_m}\}$ be a maximal co-possible or consistent subset of $\{A_1, \ldots, A_n\}$, so that $\Diamond(A_{j_1} \wedge \cdots \wedge A_{j_m})$ but $\neg \Diamond(A_{j_1} \wedge \cdots \wedge A_{j_m} \wedge A_k)$ for any $A_k \notin \Phi$. There must be such an A_k. By (CC), $O(A_{j_1} \wedge \cdots \wedge A_{j_m})$. By basic modal logic, $\Box((A_{j_1} \wedge \cdots \wedge A_{j_m}) \to \neg A_k)$. Hence, given (NM), $O\neg A_k$. (This corresponds to the first part of Argument II.) OA_k and $O\neg A_k$ entail $O(A_k \vee B)$ and $O(\neg A_k \vee B)$ respectively, by (RM). If $\Diamond B$, then $\Diamond((A_k \vee B) \wedge (\neg A_k \vee B))$ by basic modal logic. Hence we have the conditions for (CC), which yields $O((A_k \vee B) \wedge (\neg A_k \vee B))$. $(A_k \vee B) \wedge (\neg A_k \vee B)$ is (classically) equivalent to B. Hence OB by (RE). The argument for the second version, in the absence of alethic modalities, is much the same.

[74] E.g., in [Goble, 2004b, §2.4.3], [Goble, 2005, §2.4.4], [Goble, 2009, §3.2], I mention restricting (C) to 'permitted aggregation', (PC) $P(A \wedge B) \to ((OA \wedge OB) \to O(A \wedge B))$. While this resolves some of the issues in the sources mentioned in note 72, it suffers much the same explosion as (CC); given a conflict, anything permitted is obligatory.

(D) is rejected too. That leaves a logic with just (C), and, we assume, (RE), replacement for logical equivalents, of the original core principles.

Such a logic is a classical modal logic of type **EC**. As such, it is amenable to a neighborhood semantics, or minimal models, along the lines of method (iv) for **P** above, except that instead of being closed under supersets, which validates (RM), the extension of O is to be closed under intersection, if $X \in \mathcal{O}$ and $Y \in \mathcal{O}$ then $X \cap Y \in \mathcal{O}$, which validates (C).

If full aggregation (C) is replaced with consistent aggregation (CC) and the principle (P), we have the logic **ECc**.

Logics like these satisfy Desideratum 1; normative conflicts are consistent. Moreover, the arguments for deontic explosion, (DEX) and (DEX-1), are blocked, and so they meet Desideratum 2. Nevertheless, such logics fail Desideratum 3; they do not account for the Smith Argument. With (C), (i) and (ii) entail $O((f \vee s) \wedge \neg f)$, whence $O(\neg f \wedge s)$ by (RE), well enough, but without (RM) one cannot proceed to (iii) Os, as desired.

Hansson [1990, §3] presents a non-distributive logic for prescriptive oughts that extends **EC** with the principle of Disjunctive Closure, which might be seen as a very weak distribution principle.

DC) $(OA \wedge OB) \to O(A \vee B)$.

We call this **ECdc**.[75] Like **EC**, this too admits normative conflicts and avoids deontic explosion. But, like **EC**, it too fails to provide an account of the Smith Argument, and so falls short of Desideratum 3.

EC can also be extended in another way that does directly support the Smith Argument. Add the analog of the principle (K) of alethic modal logic, or its necessitative counterpart,

K) $O(A \to B) \to (OA \to OB)$, or
NK) $\Box(C \to A) \to (O(A \to B) \to (OC \to OB))$

With (NM)/(RM) these are deductively equivalent to (C); without those rules these are independent and so must be introduced separately. Call this extension **ECK**.[76]

In **ECK** the Smith Argument is immediately validated as an instance of (K) itself, and so Desideratum 3 is so far fulfilled. We must also consider, however, whether the other desiderata have now been jeopardized. Desideratum 1 is safe; normative conflicts remain consistent. It is more difficult to

---

[75]As a portion of Hansson's full 'preference-based deontic logic' (PDL), it could also be called **PDL⁻**. Full (PDL), for obeyable oughts, adds (D) and (P), and so excludes normative conflicts. For later misgivings about (DC), see [Hansson, 2001, pp. 154-5].

[76]I discuss this system in [Goble, 2009, §4.1], under the name **S**. It is based on a proposal of Stranzinger [1978], his logic PF, though he was more concerned with the familar deontic paradoxes, not normative conflicts. Lacking alethic modalities, the original PF contains (C) and (K), and also (D), and presumably (RE).

say about Desideratum 2. The arguments for (DEX) and (DEX-1) cannot proceed; so in that sense Desideratum 2 is met. On the other hand, the logic does contain this variant of deontic explosion,

DEX-2) $\quad \neg \Diamond (A_1 \wedge \cdots \wedge A_n), OA_1, \ldots, OA_n \vdash O(A_i \wedge B)$, or
if $\vdash \neg(A_1 \wedge \cdots \wedge A_n)$ then $OA_1, \ldots, OA_n \vdash O(A_i \wedge B)$

for any B.[77] E.g, $O(A_1 \wedge B)$ follows from a binary conflict OA_1, OA_2 with $\neg \Diamond (A_1 \wedge A_2)$. Since without (M) one cannot conclude OB, this fares better than the proposal of Section 5.1 above, but perhaps $O(A_1 \wedge B)$ is unintuitive enough to think Desideratum 2 is not met.

(DEX-2) can be avoided by dropping aggregation (C) since this is no longer needed to validate the Smith Argument; that comes by way of (NK)/(K) alone. Call that system **EK**. Nevertheless, since (K) suffices for this qualified aggegation[78]

$C_\top$) $\quad O\top \to ((OA \wedge OB) \to O(A \wedge B))$

(DEX-2) will hold in the qualified form that if any tautology is obligatory, then any normative conflict involving OA implies $O(A \wedge B)$, for any B.

More than that, because they contain (K) and (RE) both **EK** and **ECK** contain the similar conditional form of (M),

M_1) $\quad O\top \to (O(A \wedge B) \to OA)$

This means that a conditional form of (DEX) will be forthcoming by arguments similar to those of Section 5.1.

DEX-3) $\quad \neg \Diamond (A_1 \wedge \cdots \wedge A_n), OA_1, \ldots, OA_n \vdash O\top \to OB$, or
if $\vdash \neg(A_1 \wedge \cdots \wedge A_n)$ then $OA_1, \ldots, OA_n \vdash O\top \to OB$

As a result, any logic with (K) must not contain the rule of necessitation (N), discussed above in Section 5.1, lest full (DEX) be returned. So too logics like **ECdc** with (DC) must not contain (NK)/(K), for that returns (DEX$_s$).[79]

This illustrates that a logic with (NK)/(K) must not only not contain the rule (N), it should force that no tautology be obligatory if explosion is to be avoided. It should contain a rule of 'anti-necessitation',

---

[77]Suppose $\neg \Diamond (A_1 \wedge \cdots \wedge A_n)$ and $OA_1, \ldots, OA_n$. By the first $\Box ((A_1 \wedge \cdots \wedge A_n) \to \neg A_i)$, by basic modal logic. By the second $O(A_1 \wedge \cdots \wedge A_n)$ by (C). Classically, $\vdash A_i \leftrightarrow (\neg A_i \to (A_i \wedge B))$; hence $O(\neg A_i \to (A_i \wedge B))$ by (RE), and then $O(A_i \wedge B)$ by (NK). The argument for the second form is similar using (RM) and (K).

[78]Suppose $O\top$. Since $\vdash \top \leftrightarrow (A \to (B \to (A \wedge B)))$, $O(A \to (B \to (A \wedge B)))$ by (RE), whence $(OA \wedge OB) \to O(A \wedge B)$ by (K) twice.

[79]Suppose a strict conflict $OA, O\neg A$; then by (DC) $O(A \vee \neg A)$, or $O\top$ by (RE), whence OB, for any B by (DEX-3).

AN) $\Box A \to \neg OA$, or
      if $\vdash A$ then $\vdash \neg OA$,

This would effectively defuse (DEX-3). (AN) is essentially a principle that 'ought' implies 'can not', $OA \to \Diamond \neg A$, which some philosophers have recommended, though it is not usually counted among the core principles. Call the systems with this principle **ECK**(AN) and **EK**(AN).

Whether or not (DEX-2) or (DEX-3) fall within the bounds of Desideratum 2, these logics face another concern. The logics **EC** and **ECdc** with (C) alone or with (C) and (DC) are too weak; they fail to account for the validity of the Smith Argument, and so fail Desideratum 3. That is rectified with (K) or (NK). But the resulting logics **ECK**(AN) or **EK**(AN) still appear too weak. They fail to account for the validity of another wide range of arguments that seem quite acceptable and unproblematic. Suppose, at a party, Jones ought to tell a joke and sing a song. It seems reasonable to infer that Jones ought to tell a joke.[80] Call this the Jones Argument.

i)   Jones ought to tell a joke and sing a song. — $O(j \wedge s)$
∴ ii) Jones ought to tell a joke. — Oj

This looks valid. It looks as valid as the Smith Argument, as do others like it, and they look valid for as wide a range of oughts. Accordingly, we include the Jones Argument and other simple arguments like it among the paradigms of Desideratum 3. Any logic containing the full power of (M) will obviously serve for this. The present proposal, however, does not.

5.3.1 Consistent distribution

The call to account for the Jones Argument reveals the need for some measure of distribution, much as the validity of the Smith Argument points to the need for some aggregration as well as something like distribution, e.g., (K). Some aggregation, and some distribution. But not too much.

Just as it was suggested that (C) could be replaced by (CC), one might consider replacing (M) and (NM)/(RM) with a rule of consistent simplification or distribution, that if one ought to see to it that A and B, then one ought to see to it that A, provided that A and B are consistent (or mutually possible), and similarly for the other forms.[81]

---

[80]Cf. [Al-Hibri, 1978, p. 19].

[81]Sinnott-Armstrong [1988, p. 165] proposed a slightly more complicated version of this rule, (NCM)' $(\Diamond A \wedge \Diamond \neg B \wedge \Box(A \to B)) \to (OA \to OB)$, with much the same purpose, to avoid deontic explosion (DEX) while keeping unrestricted aggregation (C) and also a significant amount of distribution. For Sinnott-Armstrong, $\Diamond$ is understood as logical possibility. Moreover, his preferred strategy to allow for moral dilemmas was not only to restrict distribution in this way but also to reject or restrict aggregation, though he did suggest that just the restriction on distribution would suffice (p. 167). If (C) is rejected altogether, then the resulting system will not account for the Smith Argument, as we

(CM) If $\nvdash \neg(A \wedge B)$, then $\vdash O(A \wedge B) \to OA$,
(RCM) If $\nvdash \neg A$ and $\vdash A \to B$, then $\vdash OA \to OB$,
(NCM) $(\Diamond A \wedge \Box(A \to B)) \to (OA \to OB)$

Given (RE), (CM) and (RCM) are deductively equivalent; (NCM) is their modal extension. It implies the other two.

Such a restriction will defeat Argument II for the inconsistency of (binary) normative conflicts. Argument I is defeated if (P) is rejected or if (C) is replaced by (CC). Given either (C) or (CC), the Smith Argument is validated since (i) $O(f \vee s)$ and (ii) $O \neg f$ can be combined to $O((f \vee s) \wedge \neg f)$ (presuming that $(f \vee s)$ and $\neg f$ are consistent or jointly possible to suit the constraints of (CC)). Then, since $\vdash ((f \vee s) \wedge \neg f) \to s$, and hence $\vdash \Box(((f \vee s) \wedge \neg f) \to s)$, it follows by (RCM) or (NCM) that $\vdash O((f \vee s) \wedge \neg f) \to Os$, whereupon (iii) Os follows by *modus ponens*. Furthermore, the Jones Argument is immediately validated by (CM) so long as it is presumed that $j \wedge s$ is possible. Thus, Desideratum 3 is met as so far stipulated.

That is all to the good, but these principles too are still too strong. Given either (C) or (CC), as required for Desideratum 3, deontic explosion for binary conflicts in the form of (DEX) or (DEX-1) returns to be derivable.[82] So once again Desideratum 2 is not met.

5.3.2 Unconflicted distribution

The distribution rule can also be restricted to cases where the antecedent A is normatively possible, i.e., unconflicted or permitted.[83] This has the effect of allowing ought to distribute across necessitation for those oughts A that satisfy the (D) postulate themselves. Not all satisfy this, since conflicts are allowed, but presumably many, even most, do. For them, these rules are tantamount to the original (NM)/(RM).

---

saw in Section 5.2. If (C) is retained or restricted to (CC), the system is vulnerable to the problems presented here.

[82] For example, for (DEX-1) with (CC) and (NCM), suppose $OA_1, \ldots, OA_n$ and $\neg \Diamond (A_1 \wedge \cdots \wedge A_n)$. Given (P), $\Diamond A_1$ and ... and $\Diamond A_n$. Let Φ be a maximal co-possible subset of $\{A_1, \ldots, A_n\}$, much as in Note 73 for the original (DEX-1), let $\bigwedge \Phi$ be the conjunction of members of Φ and $A_k \notin \Phi$. Consider any B such that $\Diamond B$. By elementary modal logic, $\Diamond(\bigwedge \Phi \vee B)$, $\Diamond(A_k \vee B)$, $\Diamond((\bigwedge \Phi \vee B) \wedge (A_k \vee B))$, $\Diamond((\bigwedge \Phi \wedge A_k) \vee B)$. Given $\Diamond \bigwedge \Phi$ and $\Diamond A_k$, $O((\bigwedge \Phi \vee B)$ and $O(A_k \vee B)$ both by (RCM), whence $O((\bigwedge \Phi \vee B) \wedge (A_k \vee B))$ by (CC) and $O((\bigwedge \Phi \wedge A_k) \vee B)$ by (RE). Since $\neg \Diamond (\bigwedge \Phi \wedge A_k)$, $\Box(((\bigwedge \Phi \wedge A_k) \vee B) \to B)$ by modal logic. Hence, OB by (NCM). Thus, given any conflict anything possible is obligatory. The arguments for the others are similar. Even without principle (P), we nevertheless can draw on the presumption that in a normative conflict each individual ought is possible, although collectively they are not.

[83] I introduced these systems in [Goble, 2004b; Goble, 2005; Goble, 2009], calling them 'deontic logics of permitted distribution' or **DPM**. I now prefer to emphasize the unconflicted aspect rather than the permitted. See those sources for more complete development of the systems. I was more optimistic then than now about their adequacy.

We use the notation UA to say that A is (strictly) unconflicted, i.e., UA abbreviates $\neg(OA \land O\neg A)$. Then the rule of unconflicted distribution is

NUM) $\Box(A \to B) \to (UA \to (OA \to OB))$, or without alethic modality,
RUM)   if $\vdash A \to B$ then $\vdash UA \to (OA \to OB)$

The logics of unconflicted distribution, **LUM**, add this rule to a base of classical logic together with (RE) and either (C), (CC), or perhaps this version of 'permitted aggregation',[84]

PC)' $(PA \land PB) \to ((OA \land OB) \to O(A \land B))$

in order to provide enough aggregation to cover the Smith Argument. If the system contains full (C), then it must not contain (P)/(P)' if it is to preserve consistency in the face of normative conflicts. With (CC) or (PC)', (P)/(P)' is acceptable, though not required. The necessitation rule (N) is, however, necessary, as shown below.

The rules (NUM) and (RUM) are deductively equivalent to these 'permitted' forms, which are often easier to work with.

NPM) $\Box(A \to B) \to (PA \to (OA \to OB))$
RPM)   if $\vdash A \to B$ then $\vdash PA \to (OA \to OB)$

and given (RE), (RUM) and (RPM) are deductively equivalent to

UM) $U(A \land B) \to (O(A \land B) \to OA)$
PM) $P(A \land B) \to (O(A \land B) \to OA)$

The first part of Argument II of Section 1 showed that, given the unrestricted rule (NM)/(RM), any binary normative conflict of the sort OA, OB and $\neg\Diamond(A \land B)$ entails the strict normative conflicts, $OA \land O\neg A$ and $OB \land O\neg B$. With the restricted rules (NUM)/(RUM), that is not the case, but something similar is. Given a normative conflict OA, OB, $\neg\Diamond(A \land B)$ then either $OA \land O\neg A$ or $OB \land O\neg B$.[85]

SCon) $OA, OB, \neg\Diamond(A \land B) \vdash (OA \land O\neg A) \lor (OB \land O\neg B)$

Thus, any binary conflict entails a conflict of the form OC and $O\neg C$, although it might be the case that only one of the conflicting oughts, A and B, is so self-conflicted.[86] Hence these logics must not contain principle (D).

---

[84] Cf. Note 74. This alternative form is due to Straßer [2010a; 2012]

[85] Suppose OA, OB, $\neg\Diamond(A \land B)$ and also $\neg(OA \land O\neg A)$. Then $\neg O\neg A$, or PA. Since $\Box(A \to \neg B)$, $O\neg B$ follows by (NPM). Hence if not $OA \land O\neg A$, then $OB \land O\neg B$.

[86] (SCon) can be generalized for n-ary conflicts with $n > 2$; the precise form of the generalization depends on which form of aggregation applies.

Just as logics with the full distribution rule (NM)/(RM) require (N) in order to defuse (DEX$_{pi}$), as in Section 5.1, so **LUM** call for (N) in order to deactivate the similar

DEX$_{pic}$) $\Box A, \neg OA \vdash OB \rightarrow O\neg B$, or
         if $\vdash A$ then $\neg OA \vdash OB \rightarrow O\neg B$

that if any necessity is not obligatory then every ought is strictly conflicted.[87] (N) also serves to disarm the counterpart of (DEX$_{oi}$),

DEX$_{opi}$) $\neg \Diamond A, OA, PA \vdash OB$, or
         if $\vdash \neg A$ then $OA, PA \vdash OB$

that if any impossibility were obligatory without conflict, then everything is obligatory. With (N) this ceases to be catastrophic since any obligatory impossibility would necessarily be conflicted. Thus (P)/(P)′ is not required, as it was for logics with full (NM)/(RM). Of course, (P)/(P)′ may still be posited for other reasons, provided (C) is replaced with (CC) or (PC)′.

Since it contains (N), **LUM** must not be extended to include (NK)/(K), which would return the full power of (NM)/(RM) and with it (DEX), etc.

The **LUM** logics are non-normal but classical modal logics for the operator O. As such, they are amenable to the sort of neighborhood semantics or minimal models of the kind described for the logic **P** in Section 5.2, item (iv) with models $M = \langle w_0, W, v, \mathcal{O} \rangle$ and $M \models OA$ if and only if $|A|_M \in \mathcal{O}$.[88] Now, however, since (RM) is not to be validated we do not presume that $\mathcal{O}$ is closed under supersets, but instead this more restrictive condition.

- For $X, Y \subseteq W$, if $-X \notin \mathcal{O}$ and $X \subseteq Y$, then if $X \in \mathcal{O}$ then $Y \in \mathcal{O}$.

This validates (RUM). To validate (C), let $\mathcal{O}$ be closed under intersections, as for **EC**. Logics with (P)′ and (CC) or (PC)′ require

- For (P)′: $\emptyset \notin \mathcal{O}$

- For (CC): If $X \cap Y \neq \emptyset$, $X \in \mathcal{O}$ and $Y \in \mathcal{O}$, then $X \cap Y \in \mathcal{O}$

- For (PC)′: If $X \in \mathcal{O}, Y \in \mathcal{O}, -X \notin \mathcal{O}$ and $-Y \notin \mathcal{O}$, then $X \cap Y \in \mathcal{O}$

The **LUM** systems defeat Argument I of Section 1 either by not containing (P)/(P)′, or by restricting aggregation to consistent or permitted

---

[87]Suppose $\Box A$ and $\neg OA$. By basic modal logic $\Box(\neg B \rightarrow A)$. Then, by (NPM), $P\neg B \rightarrow (O\neg B \rightarrow OA)$, whence $\neg OA \rightarrow (OB \rightarrow O\neg B)$ by contraposition, etc., and so $OB \rightarrow O\neg B$ from the suppositon.

[88]In [Goble, 2004b] I present the logics and their semantics in formal detail, and demonstrate their axiomatizations to be sound and complete.

aggregation, (CC) or (PC)′, and they defeat Argument II by not containing principle (D). So these systems allow for the possibility of normative conflicts, and satisfy Desideratum 1. They also do not contain any of the disasterous forms of deontic explosion (DEX$_s$), (DEX), (DEX-1), or (DEX-3), or even the peculiar (DEX-2).[89] Nor is there any other form of deontic explosion in the offing that has not been defused, as (DEX$_{pic}$) and (DEX$_{opi}$) are by including (N). Hence, these logics satisfy Desideratum 2.

The crucial remaining question is whether these logics are too weak. Do they meet Desideratum 3? Because it lacks (K), or (DDS), **LUM** cannot validate the Smith Argument directly. Instead, it treats that Argument as an enthymeme containing the tacit premise that it is permitted that Smith not fight in the army but perform alternative service, $P(\neg f \wedge s)$ or, equally well, the premise that Smith's not fighting in the army is unconflicted, $U\neg f$, both of which seem implicit in the setting of the original example. With either extra premise, the argument is valid in **LUM**.[90]

The Jones Argument is treated similarly. Given (i) that Jones ought to tell a joke and sing a song, $O(j \wedge s)$, and that this is a no-conflict situation, so that (ii) Jones is permitted, i.e., not forbidden, to tell a joke and sing a song, $P(j \wedge s)$, i.e., $\neg O\neg(j \wedge s)$, then the conclusion (iii) that Jones ought to tell a joke, Oj, follows immediately by (PM). Thus the original Jones Argument too is given account, as an enthymeme for this more explicit form, which is fully valid. So long as these representations as enthymemes are accepted then the present approach seems to satisfy all three desiderata. It is the first we have seen that does so.

While that might seem satisfying, there is nevertheless a strong air of the *ad hoc* to the **LUM** logics. There is little to recommend their restriction on the rule (NM)/(RM) except its success. This air might be dissipated somewhat if there were other, more interesting semantics for these logics, model theories that would reflect more deeply the natural understanding of ought. The kind of neighborhood semantics described above, while valuable for establishing results about the logics, such as determining what is derivable from what within the systems, do not yield much illumination into the concepts being formalized. The conditions on the neighborhoods that validate the various principles merely mimic, at the level of propositions, the principles being validated. In the present setting, they do not explain why the restrictions on the rule (NM)/(RM) are appropriate. It is a significant

---

[89]This is demonstrated by models in the semantics as described. See, e.g., [Goble, 2005, p. 482], or [Goble, 2009, p. 478].

[90]For the first version, aggregate $O(f \vee s)$ and $O\neg f$ and then apply (RPM). For the second version, $U\neg f$ and $O\neg f$ entail $O(\neg f \vee s)$, which can be aggregated with $O(f \vee s)$ for $O((f \vee s) \wedge (\neg f \vee s))$, which is equivalent, by (RE) to Os. (With the restricted forms of aggregation, we assume also the appropriate conditions, e.g., that $\Diamond((f \vee s) \wedge \neg f)$, etc.)

open problem to develop other, more illuminating interpretations for these systems.

These **LUM** logics may also be less successful at meeting Desideratum 3 than first advertized. While they may account for the Smith and Jones Arguments, as plausible enthymemes, they cannot account for other, more complex arguments. Consider this variant on the Jones Argument.

    i)   Roberts ought to pay federal taxes and register for national service. — $O(t \wedge r)$

    ii)   Roberts ought not to pay federal taxes but volunteer to help the homeless in his community. — $O(\neg t \wedge v)$

As with the Smith Argument, (i) might stem from the laws of the land and (ii) from Roberts' religious or political convictions, or those of an organization to which he belongs.

There is a conflict between (i) and (ii). Even so, it seems reasonable to infer from (i) that Roberts ought to register for national service, Or, and similarly to infer from (ii) that he ought to volunteer to help the homeless, Ov. In short, the argument from (i) and (ii) to

    iii)   Roberts ought to register for national service and ought to volunteer to help the homeless. — $Or \wedge Ov$

appears valid, and so wants an account. We call this the Roberts Argument.

This argument is not, however, valid in the **LUM** logics, even as a plausible enthymeme. By virtue of (SCon) above, the implicit permission $P(t \wedge r)$ required to infer Or from (i) and the permission $P(\neg t \wedge v)$ required to infer Ov from (ii) cannot both be true. At least one of the oughts in the premises must be strictly conflicted. Hence at least one of the conjuncts of (iii) cannot be inferred; perhaps both.

If the Roberts Argument is included among the paradigm examples of Desideratum 3, then the **LUM** logics fall short of full adequacy.

Going beyond this first form of the Roberts Argument, one might consider too whether this stronger conclusion

    iii)′   Roberts ought to register for national service and volunteer to help the homeless. — $O(r \wedge v)$

should also be considered a reasonable inference from (i) and (ii). Call this argument Roberts-2. Not surprisingly, it has no more warrant in the **LUM** logics than the original (iii), though given (iii), (iii)′ would follow by the appropriate form of aggregation.

The inference to (iii)′ might be more controversial than to (iii).[91] One could maintain that (iii)′ is otiose. Given (iii), $Or \wedge Ov$, if Roberts does as he

---

[91]Some of the examples deployed against consistent aggregation might challenge the

ought by (iii) he will *ipso facto* satisfy the alleged ought of (iii)', and so there is no need for a separate conclusion. Compare this situation to the Smith Argument. Horty originally introduced this example to challenge logics, like **P** from Section 5.2, that simply reject aggregation (C). Their advocates too could claim that conclusions by (C) are otiose, since if an agent ought to do A and ought to do B, and the agent does as it ought, it will necessarily do A and B. (Cf., e.g., [Brink, 1994, p. 229].) Horty [2003, pp. 278f.] replies, however, that this view of the role of deontic logic is too narrow. It limits its scope merely to be action-guiding, without regard for broader descriptions of the normative situation. In Smith's case, that situation includes not only $O(f \vee s)$ and $O\neg f$, which might suffice to guide Smith in such a way that he does perform alternative service, but also the very fact that Smith ought to perform such service, Os. For example, it might be part of the general situation that if Smith ought to perform alternative service then he ought to enroll in a training program, $Os \to Oe$. Given (i) and (ii) one might expect to infer Oe. Without the inference to Os this will not be forthcoming.

Similarly, a full description of Roberts's normative situation may call for some form of aggregation. To bring this out, consider this argument, which combines aspects of Roberts with the Smith Argument. Call this the Thomas Argument.

i) Thomas ought to pay federal taxes and either fight in the army or perform alternative national service. — $O(t \wedge (f \vee s))$

ii) Thomas ought neither to pay federal taxes nor fight in the army. — $O(\neg t \wedge \neg f)$

Plainly, (i) and (ii) put Thomas in a normative conflict. Nevertheless,

iii) Thomas ought to perform alternative national service. — Os

seems to follow from (i) and (ii) since that part of Thomas's duty is unconflicted. That inference seems much like the inference of the Smith Argument, and if that seems valid, then so does this. If a logic has the full power of distribution, (RM) or (M), then the Smith Argument can be extracted from the Thomas Argument, and so the conclusion Os should follow apace. It seems a genuine part of the full description of the normative situation.

If that is so, there seems a need both to detach $O(f \vee s)$ and $O\neg f$ from (i) and (ii) and then to aggregate those to $O((f \vee s) \wedge \neg f)$ in order to conclude (iii). So similarly for Roberts-2.

None of this works in **LUM**, which cannot even draw the first inferences. If Roberts-2 and the Thomas Argument are included among the paradigms,

---

inference here; cf. Note 72, and the references there. These examples might, however, be disputed.

then once again the **LUM** logics fall short. So, for that matter, do some of the more sophisticated proposals discussed in the next section.

If one does concede a certain appearance of validity to the Thomas Argument, one might, nevertheless, explain that by arguing that it has been misrepresented in the formalism. One could hold that, if the inference from (i) and (ii) to (iii) seems correct, that is because one takes the premises to have the forms (i)' $Ot \land O(f \lor s)$ and (ii)' $O\neg t \land O\neg f$, in which case the Thomas Argument adds nothing to the Smith Argument, and so it would be treated in just the same way as that argument, likewise Roberts-2.

On the other hand, if one can be so free to restructure the (vernacular) argument in this way, then one could do the same with the two Roberts arguments and restore **LUM** to favor. Even the Jones Argument would no longer count against **ECK**(AN) and **EK**(AN). For that matter, one could say, if one takes the Smith Argument to be valid, then one must consider the premises to be properly construed as $O((f \lor s) \land \neg f)$, in which case it is valid in the basic non-aggregative system **P**, though the example was introduced precisely to challenge that system.

If these latter moves seem unconvincing, then so should the response to the Thomas Argument. There seems to be a presumption to take the vernacular at face value, and to take the Thomas Argument to have the form given first, and to take it to seem valid in that very form.[92] In that case, it does pose a significant challenge to the **LUM** logics, and others that we will look at later. No doubt there is more that could be said on this question. For present purposes, we may leave it open whether this Argument, and Roberts-2, properly belong within Desideratum 3.

More generally, examples like these raise the question of how much should one ask of a system of logic when confronted with normative conflicts. Should the Smith and Jones Arguments be accepted but the first form of Roberts not, so that the system should simply shut down in case of a conflict, as the **LUM** logics do? Should the first form of Roberts be accepted but not Roberts-2 and Thomas; is there a principled difference between applications of aggregation and of distribution in case of conflict? For that matter, should the two Roberts Arguments be accepted but the Smith and Thomas Arguments not? Or should all of them be accepted as good reasoning, so that the system continues once it extricates itself from the conflicted premises? Each proposal considered in this section and especially those in the next provide their own answers to these questions. Before turning to

---

[92]There might be even less inclination to restructure the premises if, for example, (ii) stemmed from Thomas being under an instruction: "Don't pay taxes or fight in the army", or (i) from: "You must either pay taxes and fight in the army or pay taxes and perform alternative service", the immediate renderings of which would be equivalent to (i) and (ii) as given, and the result still deemed valid.

that next section, however, let us consider one more modification of the core principles.

5.4 Systems with limited replacement

All of the logics discussed so far in this section have preserved the rule of replacement (RE) for classical logical equivalents, either as a spin-off from the distribution rule (RM) or by explicit postulation. Here we consider weakening even this rule, while keeping most of the other core principles.[93]

Such rules as deontic disjunctive syllogism (DDS), simplification (M) and aggregation (C) do have strong initial plausibility in their full generality. Moreover, these principles,

DDS) $\quad (O(A \vee B) \wedge O\neg A) \rightarrow OB$
M) $\qquad O(A \wedge B) \rightarrow OA$
C) $\qquad (OA \wedge OB) \rightarrow O(A \wedge B)$

suffice to support all of the paradigm arguments introduced above. (DDS) validates Smith, (M) yields Jones, Roberts-1, and, with (DDS), Thomas, while (C) with (M) provides Roberts-2. Thus a system with all of these would naturally meet Desideratum 3.

To deny the rule of replacement for logical equivalents (RE) requires denying also the distribution rule (RM), but without (RE), (M) is not equivalent to (RM), and so they can be separated, (M) preserved, (RM) denied.

Without (NM)/(RM), Argument II of Section 1 for the inconsistency of conflicts is defeated. And if (P)/(P)' is not considered valid, as in Section 5.1, then Argument I is also defeated. (D) itself must naturally be denied to allow for normative conflicts. That would satisfy Desideratum 1. Alternatively, (P)/(P)' could be preserved if (C) is modified to consistent aggregation (CC).

Without (RE), and hence without (NM)/(RM), the derivations of deontic explosion are likewise blocked, and so such a system would meet Desideratum 2 as well. Thus a system without (RE) and with just the core principles (DDS), (M) and (C) easily satisfies all three desiderata of this section.

Nevertheless, there seems to be something seriously wrong with a system that does not allow for replacement. Without that, or something like that, the sense of ought-statements becomes highly sensitive to their syntactical structure. $O(\neg A \wedge \neg B)$ would not necessarily be equivalent to $O\neg(A \vee B)$. If, for example, one were informed that Thomas ought neither to pay taxes nor fight with the army, then, one would have to consider whether that should be formalized as $O(\neg t \wedge \neg f)$ or as $O\neg(t \vee f)$ before one could know

---

[93]This proposal has not been previously published, though the idea of limiting replacement along these lines derives from a similar suggestion in [Straßer and Beirlaen, 2012] for a quite different context. The present development is my own.

whether to infer that Thomas ought not to fight in the army, $O\neg f$. This risks the account of the Thomas Argument. For that matter, $O(A \land B)$ would not be equivalent to $O(B \land A)$, $O(A \lor B)$ to $O(B \lor A)$, OA to $O\neg\neg A$, and so on, for countless other forms that seem so trivially equivalent they do not even require comment.

With that in mind, we allow a certain amount of replacement for basic 'analytic' equivalents. Just what counts as a basic analytic equivalence, we leave open, but for the usual propositional connectives we expect it to include the DeMorgan equivalences and such aspects of conjunction and disjunction as their idempotence, commutativity, associativity, etc. To capture this idea we proceed syntactically, and introduce a narrow equivalence relation, $\Leftrightarrow_A$, over non-deontic formulas, A, B, C, etc., with the postulates:[94]

i) $A \Leftrightarrow_A A$
ii) If $A \Leftrightarrow_A B$, then $B \Leftrightarrow_A A$
iii) If $A \Leftrightarrow_A B$ and $B \Leftrightarrow_A C$, then $A \Leftrightarrow_A C$
iv) $A \Leftrightarrow_A (A \land A)$; $A \Leftrightarrow_A (A \lor A)$
v) $(A \land B) \Leftrightarrow_A (B \land A)$; $(A \lor B) \Leftrightarrow_A (B \lor A)$
vi) $(A \land (B \land C)) \Leftrightarrow_A ((A \land B) \land C)$;
     $(A \lor (B \lor C)) \Leftrightarrow_A ((A \lor B) \lor C)$
vii) $(A \land (B \lor C)) \Leftrightarrow_A ((A \land B) \lor (A \land C))$;
     $(A \lor (B \land C)) \Leftrightarrow_A ((A \lor B) \land (A \lor C))$
viii) $A \Leftrightarrow_A \neg\neg A$
ix) $(\neg A \land \neg B) \Leftrightarrow_A \neg(A \lor B)$; $(\neg A \lor \neg B) \Leftrightarrow_A \neg(A \land B)$
x) $(A \rightarrow B) \Leftrightarrow_A (\neg A \lor B)$
xi) If $A \Leftrightarrow_A B$ then $(A \land C) \Leftrightarrow_A (B \land C)$, $(A \lor C) \Leftrightarrow_A (B \lor C)$, $\neg A \Leftrightarrow_A \neg B$, and $\Diamond A \Leftrightarrow_A \Diamond B$

(i)–(iii) make $\Leftrightarrow_A$ an equivalence relation; (iv)–(ix) catch the basic equivalences mentioned above. (x) is there because the material conditional, $\rightarrow$, is now being treated entirely classically. (xi) yields monotonicity with respect to this relation.

These postulates are meant to provide a natural minimal platform for a theory of $\Leftrightarrow_A$. We leave it open whether other principles would also be appropriate. For our application of this relation, however, these should *not* hold, lest deontic explosion be derivable.

X-a) $A \Leftrightarrow_A (A \land (A \lor B))$; $A \Leftrightarrow_A (A \lor (A \land B))$
X-b) $A \Leftrightarrow_A (A \lor (B \land \neg B))$; $A \Leftrightarrow_A (A \land (B \lor \neg B))$

---

[94]We acknowledge the classical bias of these postulates. If one objects to that for $\Leftrightarrow_A$, one would probably want to base the deontic logic itself on a non-classical logic, in which case the postulates for $\Leftrightarrow_A$ should be adapted accordingly.

With $\Leftrightarrow_A$ we can also introduce a limited sort of analytic implication over non-deontic formulas by the definition $A \Rightarrow_A B =_{df} A \Leftrightarrow_A (A \wedge B)$. These familiar properties are then derivable:

- xii) $A \Leftrightarrow_A B$ iff $(A \Rightarrow_A B$ and $B \Rightarrow_A A)$
- xiii) If $A \Rightarrow_A B$ and $B \Rightarrow_A C$ then $A \Rightarrow_A C$
- xiv) $(A \wedge B) \Rightarrow_A A$; $(A \wedge B) \Rightarrow_A B$
- xv) If $A \Rightarrow_A B$ and $A \Rightarrow_A C$ then $A \Rightarrow_A (B \wedge C)$
- xvi) If $A \Rightarrow_A C$ and $B \Rightarrow_A C$, then $(A \vee B) \Rightarrow_A C$

as well as the implications immediately derivable from (i)–(xi) via (xii). Without (X-a), (X-b), this relation also does not satisfy

- X-c) $A \Rightarrow_A (A \vee B)$; $B \Rightarrow_A (A \vee B)$
- X-d) If $A \Rightarrow_A B$ then $\neg B \Rightarrow_A \neg A$
- X-e) $(A \vee (B \wedge \neg B)) \Rightarrow_A A$; $A \Rightarrow_A (A \wedge (B \vee \neg B))$

Given (i)–(xi) it should be obvious that

Fact 5.1 *If $A \Leftrightarrow_A B$, then $\vdash A \leftrightarrow B$; if $A \Rightarrow_A B$ then $\vdash A \to B$*

(with $\vdash$ for validity in the underlying classical, and perhaps alethic modal, logic.) Of course, their converses do not hold. Less obvious, perhaps, but easily shown, is

Fact 5.2 *If $A \Leftrightarrow_A B$, then every atom in A is an atom in B.*

This would be proved by induction on a derivation of $A \Leftrightarrow_A B$ from (i)–(xi). It is also easy to establish

Fact 5.3 *If $A \Leftrightarrow_A B$, then $C \Leftrightarrow_A C[A/B]$, and hence $C \Rightarrow_A C[A/B]$*

where $C[A/B]$ is the result of replacing one or more occurrences of A in C by B. This is proved like any replacement theorem.

With $\Leftrightarrow_A$ in place, we can now posit the limited rule of replacement for basic equivalents in deontic contexts

RBE) If $A \Leftrightarrow_A B$, then $\vdash OA \leftrightarrow OB$.

The system of 'basic deontic logic', **BDL**, is then defined by the postulates (DDS), (M), (C) and (RBE), over a basis of classical logic CL and, if desired, standard alethic modal logic for $\Diamond$.

By Fact 5.3 (RBE) yields the more general replacement rule.

Fact 5.4 *If $A \Leftrightarrow_A B$ then $\vdash OC \leftrightarrow OC[A/B]$.*

It is also easy to see that **BDL** satisfies this limited distribution rule

RBM) If $A \Rightarrow_A B$ then $\vdash OA \to OB$.

BDL meets all three desiderata for a logic for normative conflicts as described above. Normative conflicts are consistent in the system; deontic explosion is not derivable,[95] and the several paradigm arguments are all valid in this logic.

Let **BDLcc** be **BDL** with (P), $\vdash OA \to \Diamond A$ or (P)′, If $\vdash \neg A$ then $\vdash \neg OA$, and (CC) in place of (C); it too meets all three desiderata.

Neither system contains the principle

OR) $OA \to O(A \vee B)$

and it must not be added lest (DEX) be derivable. **BDL** does, however, already contain the narrower principle of disjunctive closure (DC), $(OA \wedge OB) \to O(A \vee B)$, from Section 5.3. (**BDLcc** has this for co-possible A, B.)

Just as neither **BDL** nor **BDLcc** should be extended with (OR), neither should be extended with the general rule of deontic necessitation (N), $\Box A \to OA$, or If $\vdash A$ then $\vdash OA$, lest that too return (DEX), as well as the full (RM). On the other hand, it is not now necessary to include the 'anti-necessitation' rule (AN) that we saw in Section 5.3 for the systems **EK** and **ECK** in order to defuse (DEX-3) from (K), i.e., (DDS). In the **BDL** systems (C_T) and (M_T) that produced (DEX-3), are not derivable without the full replacement rule (RE) that was present in those systems.

Of all the systems discussed in this section, **BDL** and **BDLcc** offer the only approach that meets all three of the desiderata for a logic for normative conflicts. Nevertheless, much work remains to be done. Philosophically, there must be a serious study to see if limiting replacement in this way really is appropriate. This means taking a hard look at why replacement rules seem requisite for a logic of normative discourse, and what the proper limits to those rules might be. It also means developing a robust understanding of the relation $\Leftrightarrow_A$ itself, to determine if its application here is adequate. In the absence of that, these systems seem *ad hoc*. On the formal front, **BDL** and **BDLcc** so far lack any semantics or model theory, and it is difficult to see how that might be developed, while respecting the limits necessary to protect their treatment of normative conflicts.

5.5 Reprise

In this section we have looked at a number of proposals for logics to accommodate normative conflicts. In a sense, they all begin with the logic **KD** that contains all the core principles, and so excludes normative conflicts.

---

[95]This is proved by showing that if $OA_1, \ldots OA_n \vdash OB$ then every atom of B is an atom of $A_1 \wedge \cdots \wedge A_n$ by induction on the derivation of OB from $OA_1, \ldots OA_n \vdash OB$ and Fact 5.2.

These other, weaker systems then reject or revise various combinations of those principles. In so doing, however, they remain within the general framework of modal logic for the operator O. The several logics are summarized in Table 2, with reference to where they are introduced. All extend classical propositional logic, CL, and perhaps standard alethic modal logic, and all but the **BDL** systems contain (RE) for classical equivalence.

System	
KD (§1)	(C) + (NM)/(RM) + (N), and (P) and (D)
K (§5.1)	(C) + (NM)/(RM) + (N), and not (P) and not (D)
P (§5.2)	(P) + (NM)/(RM) + (N), and not (D)
Pcc (§5.2.1)	**P** + (CC)
EC (§5.3)	(C), and not (P) and not (D)
ECc (§5.3)	(CC) + (P), and not (D)
ECdc (§5.3)	**EC** + (DC)
EK(AN) (§5.3)	(NK)/(K) + (AN), and not (P) and not (D)
ECK(AN) (§5.3)	**EK(AN)** + (C)
ECcm (§5.3.1)	(C) (or (CC)) + (NCM)/(RCM), and not (D)
LUM (§5.3.2)	(C) (or (CC) or (PC)′) + (NUM)/(RUM) + (N), and not (D) (but perhaps (P) if (CC) or (PC)′)
BDL (§5.4)	(DDS), (M), (C) + (RBE), and not (D), not (N), and not (P)
BDLcc (§5.4)	(DDS), (M), (CC), (P) + (RBE), and not (D) and not (N)

Table 2

We have weighed these logics against three desiderata: Whether (1) they allow normative conflicts to be consistent; (2) they are not too strong, especially that they do not contain destructive deontic explosion (DEX) or (DEX-1); and (3) they are not too weak, but provide a plausible account of the apparent validity of several paradigm arguments, notably the Smith (S), Jones (J), Roberts (R), Roberts-2 (R-2) and Thomas (T) Arguments. In that regard, we take a plausible account to include when the argument as stated is fully valid within the logic, as the Smith Argument is in the logic **K** or in **BDL**, and also when the argument may reasonably be considered an enthymeme for a fully valid argument, as the Smith Argument is in **Pcc** and **LUM** with the tacit premises $\Diamond((f \vee s) \wedge \neg f)$ and $U(\neg f)$, respectively. With each paradigm, there may be question whether or not it should be considered acceptable. Hence, failure to account for one or another need not be grounds to consider a logic inadequate. That would depend on the case made for, or against, the argument, which is a discussion we have not engaged here. That an adequate logic for normative conflicts satisfy the first two desiderata seems more clear-cut.

How the several logics measure on these standards is summarized in Table 3, where '✓' signifies that the logic does meet the desideratum, and '×' that it does not, or, for Desideratum 3, that it accounts for the argument, or that it does not. We put a '?' under Desideratum 2 for **ECK**(AN) out of uncertainty whether (DEX-2) counts as pernicious as the others. We do not consider (DEX-3) or (DEX$_{pi}$), (DEX$_{oi}$), (DEX$_{pic}$), or (DEX$_{opi}$) so objectionable for the logics that defuse them through inclusion of (AN) or (N) or (P), as appropriate.

	Desiderata						
	1	2	3				
System			S	J	R	R-2	T
KD	×	×	✓	✓	✓	✓	✓
K	✓	×	✓	✓	✓	✓	✓
P	✓	✓	×	✓	✓	×	×
Pcc	✓	×	✓	✓	✓	✓	✓
EC	✓	✓	×	×	×	×	×
ECc	✓	✓	×	×	×	×	×
ECdc	✓	✓	×	×	×	×	×
EK(AN)	✓	✓	✓	×	×	×	×
ECK(AN)	✓	?	✓	×	×	×	×
ECcm	✓	×	✓	✓	✓	✓	✓
LUM	✓	✓	✓	✓	×	×	×
BDL	✓	✓	✓	✓	✓	✓	✓
BDLcc	✓	✓	✓	✓	✓	✓	✓

Table 3

Of all these systems only **BDL** and **BDLcc** meet all the desiderata in all their aspects. For the rest, there seems to be a tension between Desideratum 2 and Desideratum 3 insofar as all those logics strong enough to account for all of the paradigm arguments also engender deontic explosion. Of the logics that avoid explosion, how adequate one considers them depends in part on what one thinks of the different paradigms. There is also much more to say about each of these systems, as we saw when they were introduced. They may have other virtues, and other vices, not reflected in this table.

6 Conflict 2: More radical strategies

The previous section presented several logical systems that try to accommodate normative conflicts within the framework of typical modal logics by rejecting or restricting various combinations of the core principles that give rise to Arguments I and II. Here we look at some diverse strategies that call for more radical revisions of the foundations of a logic of normative concepts. The same desiderata of the preceding section continue to apply, however. We look for a logic that is not too strong, one that (1) allows normative conflicts to be consistent and (2) does not entail deontic explosion in any of its destructive forms, and also not too weak, one that (3) provides a plausible account for the seeming validity of the paradigm arguments, Smith, Jones, Roberts, and perhaps Roberts-2 and Thomas. As we will see, all of these accounts meet the first two standards. How well they fare on the third is a more complex question.

6.1 Paraconsistent deontic logics

In some respects this first approach is quite conservative, and could have been treated in the preceding section, for it preserves the basic framework of modal logics and merely denies the no-conflict principles (P) and (D), much like Lemmon's proposal discussed in Section 5.1. In another respect, however, it is quite radical, for it rejects the underlying platform of classical logic itself, which has so far been taken for granted.

Lemmon's proposal runs into trouble because it contains deontic explosion (DEX). In the preceding section we took that to call for rejecting or revising the aggregation principle (C) or the distribution principle (NM)/(RM) that were key to the derivation of (DEX). There is, however, a third principle at work in that derivation. That is the classical *ex contradictione quodlibet*, (ECQ), that a contradiction entails everything. One might argue that this is the real culprit in the derivation of (DEX).[96] In that case, one might consider preserving both (C) and (NM)/(RM) in their full strength, but apply them in a logic that lacks (ECQ).

Logics that lack (ECQ) are called 'paraconsistent' logics. Such nonclassical logics can be extended to include deontic operators very much as classical logic is extended, and those operators treated in very much the same way, e.g., semantically in terms of Kripke-accessibility relations or in terms of neighborhoods or minimal models, etc. If the logic is **K**-like for O, containing (C) and (K) as well as (NM)/(RM), but not (P) or (D), the result will indeed be a deontic logic that admits normative conflicts as con-

---

[96]There are other arguments for (DEX) that do not use (ECQ), but since they apply principles, like replacement for classical equivalents, that imply (ECQ) they are just as suspect from the present point of view.

sistent, Desideratum 1, and also does not derive (DEX) or anything like it, Desideratum 2. We will return to Desideratum 3.[97]

Paraconsistent logics can take various forms, and their deontic extensions will differ accordingly. Here we describe briefly two that are typical and that have been applied to the question of normative conflicts.[98]

One natural proposal is to add deontic principles to a relevant (or relevance) logic.[99] In these logics (ECQ) fails because the conclusion, B, may be 'irrelevant' to the premises, A and $\neg A$, something everyone feels on first meeting the principle. Classical equivalence likewise fails, as for example, $A \wedge \neg A$ is classically equivalent to $B \wedge \neg B$, but the two might have nothing in common, and so they are not relevantly equivalent. The deontic extensions of these kinds of systems are quite straight-forward, though in addition to (C) and (NM)/(RM), the (K) principle $O(A \to B) \to (OA \to OB)$ must be assumed separately, when $\to$ is relevant implication.

Another approach, advocated, e.g., by Priest [2006], would base deontic logic on a logic that allows propositions to be both true and false, as well as true only or false only, in effect a three-valued logic, such as Priest's **LP**, a logic for paradox. If, for example, A is both true and false, then $\neg A$ will also be both true and false, but B might be only false. In these logics, logical consequence means truth-preservation, so that if all the premises of a valid argument are at least true (and possibly both), then the conclusion must likewise be at least true. Since A and $\neg A$ can be true (as well as false) and B not at all true, (ECQ) is not valid. In [2006, §13.3], Priest proposes what amounts to a neighborhood semantics for formulas OA, much like the proposal (iv) of Section 5.2, except that O is taken to have both a positive

---

[97]Some, though a minority of, proponents of paraconsistent logic, most notably Priest, promote a position, dialetheism, that maintains that some contradictions are true. For such a view, paraconsistent logic is almost required, lest, by (ECQ), everything be considered true. Within this position, one could maintain that not only are there normative conflicts of the form OA and $O\neg A$, say, but also of the form OA and $\neg OA$. Priest does hold that, [2006, §13.2]. Though Priest does not recommend it, one could even maintain principle (D), that every true (strict) normative conflict entails a dialetheia. Bohse [2005] advocates this position. This would violate Desideratum 1, but such inconsistency would not be troubling to the dialetheist. Neither deontic explosion (DEX), nor full-throated explosion (ECQ) would result, if the underlying logic were suitably paraconsistent.

[98]For some other paraconsistent deontic logics, especially some based on da Costa logics, see [da Costa, 1996], [da Costa and Carnielli, 1986], [Ausín and Peña, 2000], [Grana, 1990a; Grana, 1990b], [Loparić and Puga, 1986]; see [Kouznetsov, 2004] for a simple matrix approach. The remarks below concerning the Smith Argument apply to these proposals as well.

[99]Routley and Plumwood [1989] argued for this, and in [Goble, 1999; Goble, 2001] I suggested using the logic **R** of relevant implication for this purpose. Generally speaking, although relevant logics are paraconsistent, that is not their only motivation. For an overview of such logics, see [Mares and Meyer, 2001] or [Dunn and Restall, 2002], and the references they provide.

and a negative extension. If $|A|$ is in the positive extension, then OA is true; if $|A|$ is in the negative extension, then OA is false. Since the positive and negative extensions need not be exclusive, OA might be both true and false. So might $O\neg A$. Both might be true, but OB not at all true. Hence (DEX) as well as (ECQ) fails. That is one way to interpret formulas OA in this framework. A more typical Kripke-semantics could also be used, though with the proviso that propositions can be both true and false at worlds; cf. [Priest, 2006, §19.15].[100]

Whether it is a relevant logic like **R** or a logic that allows truth-value 'gluts' like **LP**, or another form of paraconsistent logic, the base logic must lack the principle of Disjunctive Syllogism.

DS) $\quad A \vee B, \neg A \vdash B$

for otherwise (ECQ) would be derivable. As a result, the deontic extension of the logic will lack Deontic Disjunctive Syllogism (DDS), $O(A \vee B), O\neg A \vdash OB$, mentioned in Section 5.2. For example, in an **LP**-sort of system, A could be both true and false at all permissible worlds and B uniquely false at some, which would make $O(A \vee B)$ and $O\neg A$ true but OB not true. Thus (DDS) fails. Indeed, (DDS) must fail, for if it were valid deontic explosion (DEX) would be derivable even if (ECQ) is not.

Without (DDS), however, these systems will not validate the Smith Argument even if they contain (C), (RM) and (K) without restriction. As a result, these paraconsistent deontic logics seem to fail Desideratum 3, not being strong enough to do all that is expected of them. (These systems do support the Jones Argument and both forms of the Roberts Argument, but not the Thomas Argument since that too relies on (DDS).)

The Smith argument appears valid. So, for that matter, do arguments applying (DS) itself. One might have non-deontic premises, (i) Smith either fights in the army or performs alternate service, $f \vee s$, (ii) Smith does not fight in the army, $\neg f$, and expect to infer, (iii) Smith performs alternate service, s. In paraconsistent logics this inference is not valid.

There are two main lines of defense for paraconsistent logic to explain the appearance of validity for arguments by (DS). These lines might also explain the appearance of validity for arguments with normative statements by (DDS). If they succeed then Desideratum 3 would be met.

---

[100]For those less inclined to abandon classical logic, McGinnis [2007a; 2007b], offers a 'semi-paraconsistent' deontic logic. This preserves all of classical propositional logic, including (ECQ) and classical equivalence, for non-deontic propositions, but treats propositions in deontic contexts as if they were paraconsistent, much as on Priest's proposal. Because normative propositions behave much as in the fully paraconsistent systems, the remarks to come apply in much the same way to McGinnis's account.

The first defense, usually applied for relevant logics, is to maintain that 'or' is ambiguous. It has both an extensional, truth-functional meaning, typically represented by $\vee$, and also an intensional, non-truth-functional meaning, perhaps represented by $+$, under which a statement $A + B$ is (relevantly) equivalent to $\neg A \to B$ (where the $\to$ represents relevant implication). Under the first reading of 'or', (DS) is not valid; under the second reading it is. When arguments using (DS) seem valid, that is because it is the second interpretation at work.[101]

In a similar vein, one could maintain that with the Smith Argument when one takes it to be valid, when one says Smith ought to fight in the army or perform alternative service, one is really saying that it ought to be that if Smith does not fight in the army then he performs alternative service, i.e., $O(f + s)$ or $O(\neg f \to s)$. Then with the premise $O\neg f$, the desired conclusion, Os, will follow by the (K) principle. How successful this reply is, depends on how plausible it is to find that ambiguity in our language, and if it is there, how plausible it is that the major premise of the Smith Argument has this intensional reading.

The second defense, more applicable when propositions are allowed to be both true and false, would maintain that if an argument by (DS) seems valid, that is because one presumes the minor premise, or perhaps both premises, to be univalent, not to be both true and false but to be true only. While the rule (DS) is not strictly valid, it might be considered enthymematic for

DS)' $A \vee B, \neg A, \star A \vdash B$

which is properly valid, where $\star$ indicates univalency.[102] That is, given a 3-valued table, with **t**, **f**, **b** for being true-only, false-only, and both-true-and-false, respectively,

A	$\star A$
t	t
f	t
b	f

So too for arguments by (DDS). One could hold that if such an argument appears valid that is because one presumes that in alternative worlds each content (or at least the content $\neg A$) is univalent. After all, this is supposed to be a situation in which there is no conflict. Thus the correct representation of the Smith Argument would be

---

[101] Arguments that purport to derive (ECQ) then equivocate between these readings. See [Read, 1988] for extended defense along these lines. See [Burgess, 1981] for (earlier) sharp criticism of this kind of proposal.

[102] Paraconsistent logics containing operators like $\star$ are often called Logics of Formal Inconsistency (LFI); cf. [Carnielli et al., 2007].

$$O(f \vee s), O\neg f, O \star \neg f \vdash Os$$

This would be valid in a **LP**-sort of system extended to include the operator $\star$ for univalency. Since the original Smith Argument can be seen to be enthymematic for this, its appearance of validity is explained.

How successful this reply is depends on how plausible one finds the additional premise. Is it really present in the Smith Argument as we understand it? Do we understand what it means to say that it ought to be that it is univalent that Smith not fight in the army, $O \star \neg f$?

I leave those questions open. The introduction of the operator $\star$, required by the reply, raises another significant issue, however. Given $\star$, we can define another, Boolean, negation $\sim A$ that represents pure falsehood, so that $\sim A$ holds whenever A is not at all true, i.e., not true only, and not both true and false. Let $\sim A =_{df} \neg A \wedge \star A$.[103] Then

A	$\sim A$
t	f
f	t
b	f

With this the Smith Argument might be construed as

$$O(f \vee s), O \sim f \vdash Os$$

Since this is valid in **LP**-like systems with $\star$, the appearance of validity for the Smith Argument is explained by way of an alleged ambiguity in the notion of negation. This is another way to meet Desideratum 3.

The introduction of $\sim$ raises the spectre of normative conflicts of the sort $OA, O \sim A$. If there is that sort of negation in the language, then the arguments Priest and others put forward to support the widespread occurrence of normative conflicts, even normative dialetheias, would seem to support the possibility of this sort of normative conflict too. Such a conflict, however, generates once again deontic explosion.

DEX$_s^\sim$) $OA, O \sim A \vdash OB$

That should be anathema to advocates of this sort of view.

These advocates face a dilemma. Either $\sim$ is admissible in the language or it is not. If it is, then (DEX$_s^\sim$) is valid, and the proposal fails to meet Desideratum 2. If it is not admissible, then the Smith Argument is without an explanation, and the proposal fails to meet Desideratum 3. To escape the dilemma, seemingly one would have to argue that normative conflicts of the sort $OA, O \sim A$ are not possible, in which case (DEX$_s^\sim$) could be harmless. I leave such questions open.

---

[103] Alternatively, $\sim$ could be primitive and $\star$ defined in terms of it, $\star A =_{df} \sim(A \wedge \neg A)$.

Inevitably, the chief concern to put to a paraconsistent deontic logic is the adequacy of its underlying, non-classical, propositional base. This is not the place, however, to discuss the virtues and vices of paraconsistent logic in general. It is enough to see the strains that arise even here with respect to Desiderata 2 and 3.

In what follows, we revert to the more traditional stance that takes the logic of normative concepts to be founded on a platform of classical propositional logic, to ask if there is an adequate set of principles there to accommodate normative conflicts.[104]

6.2 Two-phase deontic logic

Van der Torre [1997] and van der Torre and Tan [2000] introduced a 'two-phase' deontic logic expressly designed to address the problem of validating the Smith Argument while admitting normative conflicts and avoiding deontic explosion. They call this van Fraassen's paradox.[105] This proposal works quite differently from those of Section 5. Instead of putting constraints on the formulas to which the core principles might apply, it controls the order of application of the rules.

In the reasoning to derive a conclusion from a set of premises, inferences by aggregation (or consistent aggregation) are allowed, as are inferences based on distribution, but the latter must always come after the former. One clusters all (normative) information together into consistent or co-possible packages, and only then draws out implications. In this way, for example, the premises of the Smith Argument, (i) $O(f \vee s)$ and (ii) $O\neg f$, may first be aggregated to $O((f \vee s) \wedge \neg f)$, which is equivalent to $O(\neg f \wedge s)$, and then the conclusion (iii) Os drawn by (M). That is legitimate. By contrast, arguments that would support (DEX) or (DEX-1) are not. Since aggregation is limited to (CC), given a conflict, say, OA, OB, when $\vdash A \rightarrow \neg B$, the two oughts cannot be combined to exploit $\vdash O(A \wedge B) \rightarrow OC$, based on (ECQ). While one could infer by (RM) $O(A \vee C)$ and $O(B \vee C)$, whose contents are consistent if C is, nevertheless, these cannot be combined to $O((A \vee C) \wedge (B \vee C))$, which would yield OC by (RE), since that combination could only occur after the inferences by (RM) were drawn, and that is inadmissible. In this way (DEX) and (DEX-1) are avoided.

---

[104] The advocate of paraconsistent logic might wonder why a proponent of normative conflicts in a classical setting would be content to swallow the camel of explosion in general (ECQ) yet strain at the gnat of deontic explosion (DEX). The point is well taken. The proponent of normative conflicts might, however, reply that the difference is that there are, or seem to be, genuine normative conflicts whereas there are not, and do not seem to be, true contradictions. There is more to say on this question, but it would go well beyond the present discussion.

[105] The full two-phase approach extends to conditional-oughts and applies also to other vexing problems in deontic logic, but that is beyond the scope of this chapter.

These controls on the application of the rules are embodied in the language through the use of two deontic operators, ①A and ②A, for the two phases. Premises of arguments, like the Smith Argument, or even the arguments for deontic explosion, are taken to be of the first sort, while conclusions are of the second. Formulas in the first phase, ①A, are amenable to consistent aggregation, but not distribution, while formulas of the second phase, ②A, allow distribution, but not any sort of aggregation. The two phases are sequenced by virtue of ①A implying ②A, but not conversely. Thus these are taken to be valid, along with (RE) for both forms,

CC-1) $(\Diamond(A \wedge B) \wedge ①A \wedge ①B) \to ①(A \wedge B)$, or
if $\not\vdash \neg(A \wedge B)$ then $\vdash (①A \wedge ①B) \to ①(A \wedge B)$
NM-2) $\Box(A \to B) \to (②A \to ②B)$, or
RM-2) if $\vdash A \to B$ then $\vdash ②A \to ②B$, including
M-2) $②(A \wedge B) \to ②A$, and
OR-2) $②A \to ②(A \vee B)$
REL) $①A \to ②A$

None, however, is valid if the modalities are interchanged. Also valid are

P-1) $①A \to \Diamond A$, or
P-1)' if $\vdash \neg A$ then $\vdash \neg ①A$
P-2) $②A \to \Diamond A$, or
P-2)' if $\vdash \neg A$ then $\vdash \neg ②A$
N-1) $\Box A \to ①A$, or
if $\vdash A$ then $\vdash ①A$
N-2) $\Box A \to ②A$, or
if $\vdash A$ then $\vdash ②A$

but not

D-1) $①A \to \neg①\neg A$
D-2) $②A \to \neg②\neg A$

and not

K-1) $①(A \to B) \to (①A \to ①B)$
K-2) $②(A \to B) \to (②A \to ②B)$

(And *a fortiori* (NK) is not valid for either.)

This duality of oughts is not at all the same as the duality of *prima facie* and all-things-considered oughts discussed in Section 4. It should be understood instead in terms of the roles of the statements in a process of reasoning, the collection of information and the extraction of conclusions from that collection.

The logics of each operator, ①, ②, are non-normal classical modal logics. The first is essentially **ECc** mentioned in Section 5.3 plus (N-1), while the second is **P** from Section 5.2. As such they can readily be given a neighborhood semantics of the usual sort, as in Section 5.2 item (iv).[106]

More interesting, these modalities can be interpreted in terms of preference relations on worlds, along the lines of Section 5.2, item (ii), ① representing ordering and ② minimizing according to that relation. Let $M = \langle w_0, W, v, P \rangle$, with P a binary reflexive and transitive relation on W, with wPw' read to say that w is at least as preferable as w'. Then

- $M \models ①A$ if and only $|A|_M \neq \emptyset$ and for all $w, w' \in W$, if wPw' and $w' \in |A|_M$ then $w \in |A|_M$.

- $M \models ②A$ if and only if there is a $w \in W$ such that $w \in |A|_M$ and for any $w' \in W$, if $w'Pw$ then $w' \in |A|_M$.

where, as before, $|A|_M = \{w : M, w \models A\}$. The first rule says, in effect, that it ought to be that A just in case A is possible and all worlds that are at least as preferable as *any* A-world are themselves A-worlds, where the second rule says that it ought to be that A just in case there is an A-world that marks a threshold, as it were, such that any other world at least as preferable as it is also an A-world. This is the difference between ordering, which takes the whole configuration of W into account, and minimizing, which looks only to the class of worlds than which none are better, or in the case of infinite chains, worlds that offer such a threshold.

These rules also validate what should be valid and provide counterexamples to what should not be valid. (Notice that the rule for ②A is essentially the same as the rule (ii) for the logic **P** of Section 5.2.)

By virtue of the restriction to consistent aggregation for formulas ①A, Argument I applied to such formulas is blocked, and Argument II fails by the lack of both (NM-1)/(RM-1) and (D-1). Hence, normative conflicts among such oughts are possible. For formulas ②A, the absence of any sort of aggregation defeats Argument I, and without (D-2) Argument II cannot be completed, much as for the logic **P** of Section 5.2. Hence, for both sorts of ought, normative conflicts are consistent; for both Desideratum 1 is met.

By virtue of (RM-1) not being valid, arguments for (DEX-1) cannot get started for ①-formulas, likewise the other destructive forms of deontic explosion. By virtue of (C) and (CC) both failing for ②-formulas, arguments for (DEX) and (DEX-1) cannot be completed. By the failure of both, mixed forms of deontic explosion, e.g., ①A, ①B, $\neg\Diamond(A \wedge B) \vdash ②C$, also fail. Hence, both oughts, and their combination, satisfy Desideratum 2.

---

[106]There will be two extensions, $\mathcal{O}_1$ and $\mathcal{O}_2$, where both are non-empty and $\mathcal{O}_1$ is closed under the consistent intersection condition, mentioned in Section 5.3.2, while $\mathcal{O}_2$ is closed under the superset condition from Section 5.2, Also, to validate (REL), $\mathcal{O}_1 \subseteq \mathcal{O}_2$. These validate the principles listed above, and enable the non-validities to be falsified.

With regard to Desideratum 3, the Jones Argument is immediately validated. If given ①$(j \wedge s)$, then ②$(j \wedge s)$ follows by (REL), whence ②j by (M-2), as desired. The first Roberts Argument is validated similarly.

For the Smith Argument, if it is represented as

SA$_{1,2}$)   (i) ①$(f \vee s)$, (ii) ①$\neg f$ ∴ (iii) ②s

then it too is easily validated (with $\Diamond((f \vee s) \wedge \neg f)$ a tacit premise). From (i) and (ii), we have ①$((f \vee s) \wedge \neg f)$ by (CC), whence ②$((f \vee s) \wedge \neg f)$ by (REL), whence (iii) ②s, by (RM-2).

The key question then is whether that is an adequate representation of the argument. When the Smith Argument is judged, informally, to be valid, the oughts of the premises and the conclusion are naturally supposed to be univocal. There is no evidence for the sort of ambiguity apparent in (SA$_{1,2}$). So one would expect to find perhaps either of

SA$_1$)   (i) ①$(f \vee s)$, (ii) ①$\neg f$ ∴ (iii) ①s
SA$_2$)   (i) ②$(f \vee s)$, (ii) ②$\neg f$ ∴ (iii) ②s

to be valid. But neither is, on the present approach. Thus, the Smith Argument is only given an account in a very limited way.

Neither the Roberts-2 Argument nor the Thomas Argument is validated on this approach, even in the basic forms

R-2$_{1,2}$)   ①$(t \wedge r)$, ①$(\neg t \wedge v)$ ∴ ②$(r \wedge v)$
TA$_{1,2}$)   ①$(t \wedge (f \vee s))$, ①$(\neg t \wedge \neg f)$ ∴ ②s

As discussed at the end of Section 5.3.2, it may be debatable how desirable that is. We leave that question open.

In general, this two-phase approach assumes that the premises of arguments, like the Smith Argument, always apply the ordering operator ① while the conclusion applies the minimizing operator ②. That is the reason for the first representation (SA$_{1,2}$). In the absence of linguistic evidence for this sort of ambiguity in vernacular language, this assumption seems suspiciously *ad hoc*. Perhaps, however, the use of two operators here is not supposed to mark a semantic difference of the oughts in question, so much as to serve as indicators or guides to how the ought-statements are to be used at various stages, or phases, of a reasoning process. They are reminders that the process should begin with aggregation and finish with distribution, and not be mixed in between.

Nevertheless, one might wonder what motivates this kind of constraint on the process of reasoning, except for its accounting for the Smith Argument in this way, while rendering normative conflicts consistent and avoiding the pitfalls of deontic explosion. Given the Smith Argument with a semantically univocal reading of the oughts, so that the premises are really just (i) $O(f \vee$

s) and (ii) $O\neg f$, one might reason to the conclusion (iii) Os as above. Or one might reason this way: From (ii) $O(\neg f \vee s)$ by (RM). By (CC) that and (i) yield $O((f \vee s) \wedge (\neg f \vee s))$, which is logically equivalent to (iii) Os by (RE). Under the two-phase approach this latter pattern of reasoning is not legitimate, since it applies distribution before aggregation. One might wonder, however, what is the rationale whereby the former is considered good reasoning while the latter is not.

To speak of 'phases', as this approach does, seems more appropriate to describing procedures of reasoning than the relation of logical consequence itself. With logical consequence one is interested to know what must be true if certain other propositions are true. It should not matter how one demonstrates that. If one is looking for an account of the validity of the Smith Argument as something like logical consequence, rather than something procedural, then this approach seems to fall short of Desideratum 3, even while it may have many virtues as an account the dynamics of inference-making.

6.3 An imperatival approach

Imperatival accounts maintain that ought-propositions are derived from more fundamental norms, e.g., commands or directives.[107] We met one such account in Section 4.4 in which the basic directives were construed as non-derivative *prima facie* oughts and all-things-considered oughts were determined from them. That account was designed to preclude normative conflicts. Here we describe a similar approach that does admit the possibility of such conflicts. Since we do not now draw the distinction between *prima facie* and all-things-considered oughts we shall take directives simply to be particular imperatives, which we write as $!A$, where A represents the propositional content of the command, i.e., that which must be so in order that the command be satisfied. Also, we do not suppose that directives are ordered by a relation of priority or relative weight, though this account could be readily adapted to include that.[108]

Accordingly, we now take imperatival models $\mathcal{M}_i$ to be structures $M = \langle w_0, W, v, \Delta \rangle$, where Δ is a set of directives $!A$. As in Section 4.4, and unlike Section 4.1, this is taken as given, with no closure conditions imposed. As before too, given $\Theta \subseteq \Delta$, $\Theta^* = \{B : !B \in \Theta\}$.

Given such a model, the key question is, How are ought-statements, OA, determined? One ought to do what the directives tell one to do. But since

---

[107]The present account is derived in great part from the work of Horty [1994; 1997; 2003; 2012] and Hansen, esp. [2004a; 2005]. Both Horty and Hansen trace their work to van Fraassen [1973].

[108]As in Section 4.4 we simplify discussion by not considering conditional imperatives, though there might be nonconditional imperatives of the form $!(A \rightarrow B)$. As there, too, we do not count iterated deontic modalities in the language.

directives might conflict, one might not be able to do what all of them tell one to do. In that case, on the present view, one ought to do as much as one can, though that might engender conflicting oughts. One ought to do what any maximal nonconflicting set of directives tells one to do, or since maximality turns out not to be significant in this setting,

- $M \models OA$ if and only if $\Sigma \vdash_{\Box}^{M} A$, for some co-possible $\Sigma \subseteq \Delta^*$.

This resembles the second account of Section 4.4 except as it calls for A to be necessitated merely by some consistent subset of Δ^*, rather than by every maximal consistent subset. It resembles the simple imperatival proposal (v) of Section 5.2 except that it takes into account when A is necessitated by multiple (consistent) directives together, rather than merely one.[109]

The former aspect allows for normative conflicts to be generated; there might well be, e.g., $\Delta = \{!p, !\neg p\}$, whence $M \models Op$ and $M \models O\neg p$ for any M with that Δ. The latter aspect enables more oughts to be generated than under the earlier approach, such as called for by the situation of the Smith Argument. One could have a set of directives, $\Delta = \{!(f \lor s), !\neg f\}$. If M contains that Δ, it will follow that $M \models Os$, whereas that would not hold under the proposal (v) of Section 5.2 since there is no single directive in Δ whose content necessitates s.

It is easily shown that this account supports the core principles (P)/(P)', (NM)/(RM), and (N), but not (D), and not (C), or even (CC).

P) $\Vdash OA \to \Diamond A$, or (P)' if $\Vdash \neg A$ then $\Vdash \neg OA$
NM) $\Vdash \Box(A \to B) \to (OA \to OB)$
RM) If $\Vdash A \to B$ then $\Vdash OA \to OB$
N) $\Vdash \Box A \to OA$, or if $\Vdash A$ then $\Vdash OA$
XD) $\nVdash OA \to \neg O\neg A$
XC) $\nVdash (OA \land OB) \to O(A \land B)$

Because (D) and (C) are not supported, normative conflicts are consistent within this account and none of the destructive forms of deontic explosion discussed in Section 5 are derivable. This approach satisfies Desiderata 1 and 2 quite elegantly.

The question whether it provides all that Desideratum 3 calls for is harder. Since (RM), and hence (M), is valid without restriction, the Jones Argument and the first Roberts Argument are immediately validated. Of

---

[109] It also corresponds closely to the account of derived *prima facie* oughts of Section 4.4, (Def-9), except that now we do not draw the distinction between *prima facie* and all-things-considered oughts.

greater concern, however, is the Smith Argument. This is not valid on the present account. I.e., $O(f \vee s), O\neg f \not\Vdash Os$.[110]

It is ironic that the Smith Argument fails on this account since this approach is advertized, at least by Horty, as making up for the deficiency of non-aggregative systems like **P** that could not account for this inference.[111] As we saw above, the present approach does have the advantage over the simpler imperatival approach of Section 5.2, which underwrites **P**, of enabling Os to be generated at least when given the direct commands $!(f \vee s)$ and $!\neg f$. That is a different matter, however, from establishing the validity of the Smith Argument itself, which seemed to be the original purpose.

Indeed, when it comes to matters of consequence, of valid arguments, this account is equivalent to, or very close to, that of **P** itself. That is, with $\Gamma \vdash_\mathbf{P} A$ for derivability within **P**, then

Observation 6.1 *For finite* Γ, $\Gamma \Vdash A$ *if and only if* $\Gamma \vdash_\mathbf{P} A$.[112]

We have said that Desideratum 3 does not require the full-fledged validity of the Smith Argument. It could be enough to provide for something reasonably close to validity, another relation between the premises and the conclusion that represents an appropriate form of good reasoning. For this, we might look to what the present account does offer that the earlier, simpler account did not, namely an explanation for why Smith ought to perform alternative service when given the direct commands to fight in the army or perform such service and not to fight in the army.

Accordingly, one might propose a weaker quasi-consequence relation, $\Gamma \mathrel{\vert\!\sim} A$, that imposes just such a condition of direct command on Γ.[113]

For present purposes, let us restrict Γ, A to propositions of the form OB. This is a severe limitation, but it still allows all the paradigms we have been tracking. Then,

---

[110] For a counterexample, consider a model $M = \langle w_0, W, v, \Delta \rangle$ with $\Delta = \{!f, !\neg f\}$. Then $M \models O(f \vee s)$, $M \models O\neg f$ but $M \not\models Os$. Models corresponding to the Thomas Argument of Section 5.3.2, with $\Delta = \{!(t \wedge (f \vee s)), !(\neg t \wedge \neg f)\}$, also invalidate this inference.

[111] [Horty, 2003, p. 603, n. 11], "It was the desire to allow for some measure of agglomeration that led van Fraassen to move from the simple deontic logic presented in Section 6 of his [1973] to the more complicated system presented in Section 7. It was this desire also that lies behind my criticism in [1997] of proposals, such as that set out in Chapter 6 of Brian Chellas's [1980], to formalize deontic reasoning with conflicting oughts within the framework of weak, nonnormal modal logics."

[112] This result is derived from [Hansen, 2004a, p. 49, theorem 5.1], which shows that the axiomatizaion for **P** is sound and weakly complete for the present semantics. That it is only weakly complete calls for Γ to be finite here.

[113] This $\mathrel{\vert\!\sim}$ is similar to Horty's $\vdash_F$ in [1997]. It is suggested by his discussion of the rule he calls 'consistent consequent agglomeration' in [Horty, 2003, p. 580], which I describe as 'weak consistent aggregation' in [Goble, 2009, p. 465]. Nair [201x] develops and further motivates a concept very like $\mathrel{\vert\!\sim}$.

- $\Gamma \mathrel{\vdash\!\!\!\sim} A$ iff for all $M = \langle w_0, W, v, \Delta \rangle$, if $M \models C$ for every $C \in \Gamma$ and for all $OD \in \Gamma$, $!D \in \Delta$, then $M \models A$.

In case $\Gamma = \{A_1, \ldots, A_n\}$ is co-possible, then $\mathrel{\vdash\!\!\!\sim}$ is no different from the standard consequence relation of **KD**.[114]

Observation 6.2 *If $\{A_1, \ldots, A_n\}$ is co-possible, then $OA_1, \ldots, OA_n \mathrel{\vdash\!\!\!\sim} OB$ if and only if $OA_1, \ldots, OA_n \vdash_{\mathbf{KD}} OB$.*

Hence it is only in case of reasoning with normative conflicts that the special character of $\mathrel{\vdash\!\!\!\sim}$ comes into play.

Since $\{f \vee s, \neg f\}$ is presumed consistent, $O(f \vee s), O\neg f \mathrel{\vdash\!\!\!\sim} Os$. Hence the Smith Argument now has an account under this relation. Moreover, if $\Gamma \Vdash A$, then obviously $\Gamma \mathrel{\vdash\!\!\!\sim} A$. So the Jones and first Roberts Arguments are likewise covered. The Roberts-2 and Thomas Arguments are not, but we leave them aside for the moment.

Given that technically $O(f \vee s), O\neg f \mathrel{\vdash\!\!\!\sim} Os$, one might still ask whether this really does provide a satisfactory account for the Smith Argument. Does this describe the argument appropriately as a piece of good reasoning, even if not truly valid in classical terms?

The relation $\mathrel{\vdash\!\!\!\sim}$ has many properties associated with a proper relation of good reasoning. E.g., adapted to the limitations of $\mathrel{\vdash\!\!\!\sim}$,[115]

Reflex) $\Gamma \mathrel{\vdash\!\!\!\sim} OA$ when $OA \in \Gamma$
SupraCn) $\Gamma \Vdash A$, then $\Gamma \mathrel{\vdash\!\!\!\sim} A$
LLE)   If $\Gamma, OA \mathrel{\vdash\!\!\!\sim} OC$ and $OA \dashv\vdash OB$ then $\Gamma, OB \mathrel{\vdash\!\!\!\sim} OC$
RW)   If $\Gamma \mathrel{\vdash\!\!\!\sim} OA$ and $OA \Vdash OB$, then $\Gamma \mathrel{\vdash\!\!\!\sim} OB$
Mono)   If $\Gamma \mathrel{\vdash\!\!\!\sim} OA$, then $\Gamma, OB \mathrel{\vdash\!\!\!\sim} OA$

On the other hand, $\mathrel{\vdash\!\!\!\sim}$ fails to satisfy cumulative transitivity, or CUT.[116]

CUT)   If $\Gamma \mathrel{\vdash\!\!\!\sim} OA$ and $\Gamma, OA \mathrel{\vdash\!\!\!\sim} OB$, then $\Gamma \mathrel{\vdash\!\!\!\sim} OB$

Indeed, it must not satisfy (CUT), lest deontic explosion (DEX) be justified in these terms. By definition of $\mathrel{\vdash\!\!\!\sim}$, both (i) $Op, O\neg p \mathrel{\vdash\!\!\!\sim} O(p \vee q)$ and (ii) $Op, O\neg p, O(p \vee q) \mathrel{\vdash\!\!\!\sim} Oq$ hold. If (CUT) were sound for $\mathrel{\vdash\!\!\!\sim}$, then (iii) $Op, O\neg p \mathrel{\vdash\!\!\!\sim} Oq$ would follow.

This rule, (CUT), holds classically, and in theories of nonmonotonic reasoning it is widely, though not universally, considered essential for any rela-

---

[114] Cf. [Horty, 1997, p. 31, theorem 5].

[115] Other familiar properties that call on disjunction or conjunction on the left or right cannot be given in terms of $\mathrel{\vdash\!\!\!\sim}$.

[116] Cf. [Horty, 1997, p. 29]. Horty's relation $\vdash_F$ there also fails to be reflexive; $OA \vdash_F OA$ fails when A is inconsistent.

tion purporting to be a consequence relation.[117] To have it means that one can reuse conclusions that have been drawn from some information as further premises in conjunction with that information, confident that the final conclusion drawn is implied by the original information. This is ubiquitous in mathematical and scientific reasoning. It is a sign of the stability of the procedures of inference. Not to have (CUT) means a line is drawn between what can serve as a premise in a process of reasoning and what has been derived as a conclusion.

We see this in the attempt to derive (DEX) described above. By (i) $O(p \vee q)$ is legitimately inferable from $\{Op, O\neg p\}$, but then with (ii) the same proposition $O(p \vee q)$, is applied as a premise along with that original set to derive Oq. While either role may be all right by itself, the double use of $O(p \vee q)$ is not, and so (iii) should be rejected. Much the same occurs in attempts to support the Thomas Argument, or Roberts-2. Whether one thinks reasoning in normative contexts, especially in the face of normative conflicts, should be limited in this way, in contrast to reasoning in mathematical or scientific practice, may depend a lot on what one expects a reasoning agent to do, and how extensive its reasoning is expected to be. That question is beyond the scope of this chapter, as is the broader question of whether a relation reflecting good reasoning in general should satisfy (CUT).

In the present setting, however, the Smith Argument raises a concern for the particular relation $\vdash$ and its use to represent good reasoning for normative discourse. By that representation, since $O(f \vee s), O\neg f \vdash Os$, it is supposed to be a piece of good reasoning to conclude that Smith ought to perform alternative service from the information that Smith ought either to fight in the army or perform such service and that he ought not to fight in the army. It would seem, though, that before that conclusion is warranted, it must be known that the set of imperatives binding Smith includes precisely that he fight in the army or perform alternative service and that he not fight in the army. That information, however, is not conveyed by the Smith Argument itself. As the Argument is presented, one is informed of the two oughts incumbent on Smith, but not whether those oughts are, in effect, basic and underived, and hence capable of serving as premises, or whether they have in fact already been drawn as conclusions from some more basic information and thus not in a position to be reused as input to the inference to Os. If, for example, the original imperatives were as for the Thomas Argument, $\{!(t \wedge (f \vee s)), !(\neg t \wedge \neg f)\}$, then, under the present account, even though both $O(f \vee s)$ and $O\neg f$, it would not be appropriate to conclude Os.

---

[117]Cf. [Makinson, 1994, p. 43 et seq.], also [Kraus et al., 1990, p. 169 and §3]. Both Makinson and Kraus et al. refer to Gabbay.

Very often, as one reasons from normative premises, like those of the Smith Argument, one might be quite unaware of what are the original underlying imperatives that give rise to them. In that case, the relation ⊢ gives little or no guidance for what one should conclude.

This can be rectified to some extent by enriching the deontic language, to express exactly the cues that are necessary. Let O^ι be a monadic operator such that $O^\iota A$ holds just in case A is logically equivalent to the content of a member of Δ.[118] Given $M = \langle w_0, W, v, \Delta \rangle$,

- $M \models O^\iota A$ if and only if there is a B such that $!B \in \Delta$ and $\Vdash A \leftrightarrow B$.

By itself, this operator O^ι has a very weak logic. It satisfies replacement for logical equivalents (RE), but that is all among the core principles. It does not agree to any form of aggregation or any form of distribution. Even (P) need not hold for it, unless Δ is limited to directives that can be fulfilled. Nor does (N) hold, nor obviously (D).

With O^ι, we can return to genuine logical consequence, $\Vdash$, from the relation ⊢, at least when each A_i is itself possible. This is easy to verify.

Observation 6.3 *If each A_i is possible, then $OA_1, \ldots, OA_n$ ⊢ OB if and only if $O^\iota A_1, \ldots, O^\iota A_n \Vdash OB$.*

In light of that we could represent the validity of the Smith Argument by

SA$_a$ $\quad O^\iota(f \vee s), O^\iota \neg f \Vdash Os$

Observation 6.3 suggests that perhaps that is what the previous representation $O(f \vee s), O\neg f$ ⊢ Os comes to.

The validity of the argument in this form, however, relies on the double oughts, O^ι and O, the first for the premises, the second for the conclusion, much as we saw in the two-phase account of Section 6.2, this despite the fact that nothing in the original presentation of the Argument suggests such an ambiguity for 'ought'. If the 'ought's are treated univocally, with either operator, validity is broken.

SA$_b$ $\quad O^\iota(f \vee s), O^\iota \neg f \not\Vdash O^\iota s$
SA$_c$ $\quad O(f \vee s), O\neg f \not\Vdash Os$

(The first should be obvious; the second we have already seen.)

---

[118]This O^ι corresponds to Hansen's O^2 of [2004a; 2005], while our O corresponds to his O^F, 'F' for van Fraassen. One could think of O^ι as expressing basic *prima facie* oughts, in the strict sense of those enjoined by directly given directives or imperatives, not the sort of derived *prima facie* oughts described in Section 4.4.2, and O could be thought of as expressing an all-things-considered ought, though now in a sense that allows for normative conflicts.

Thus, for all its merits, the present proposal seems to fall short of Desideratum 3. It accounts for the Smith Argument only in the limited form of (SA$_a$), which requires the oughts of the premises to have the very weak sense of O^ι and the ought of the conclusion the sense of the full-blooded O.

The Smith Argument as originally stated, $O(f \vee s), O\neg f \therefore Os$, might also be construed as an enthymeme for

SA$_d$ $O(f \vee s), O\neg f, O^\iota(f \vee s), O^\iota\neg f \Vdash Os$

which is valid. Indeed, (SA$_d$) is a rather redundant extension of (SA$_a$), and, in light of Observation 6.3, it is equivalent to $O(f \vee s), O\neg f \mathrel{\vdash\!\!\!\!\!\!\!-} Os$. Perhaps that is the intended interpretation of $\mathrel{\vdash\!\!\!\!\!\!\!-}$ in this kind of setting. Whether it is plausible to think of the original Smith Argument as this sort of enthymeme may be left open. It adds little to (SA$_a$).

As the validity of this argument thus seems to rely on a duality of oughts, the present account fares much like the two-phase account of Section 6.2 although the two pairs of ought-operators are rather different. In Section 6.2 it was suggested that the two operators might be thought to represent not so much two different meanings of 'ought', but rather to mark two different roles that ought-statements might play in argumentation. One might think of the two 'oughts' of (SA$_a$) in a similar way. Indeed, in general,

Observation 6.4 $O^\iota A_1, \ldots, O^\iota A_n \Vdash OB$ iff $①A_1, \ldots ①A_n \vdash ②B$.[119]

By Observation 6.3, then

Observation 6.5 *If each A_i is possible, then $OA_1, \ldots OA_n \mathrel{\vdash\!\!\!\!\!\!\!-} OB$ if and only if $①A_1, \ldots ①A_n \vdash ②B$.*

This may reflect that reading of the two-phase operators as indicating disparate roles in inference-making since $\mathrel{\vdash\!\!\!\!\!\!\!-}$ is committed to that by virtue of not satisfying (CUT).

Thus, despite the differences between the present imperatival approach and the two-phase approach of Section 6.2, they come out in much the

---

[119]Proof sketch: L-to-R, suppose $O^\iota A_1, \ldots, O^\iota A_n \Vdash OB$, and consider $M = \langle w_0, W, v, \Delta \rangle$ with $\Delta = \{!A_1, \ldots, !A_n\}$. Then $M \models OB$; so there is a consistent $\Sigma \subseteq \{A_1, \ldots, A_n\}$ such that $\Sigma \vdash B$. Let $\Sigma = \{C_1, \ldots, C_m\}$. Suppose all of $①A_1, \ldots, ①A_n$; amongst them are $①C_1, \ldots, ①C_m$. They can be aggregated to $①(C_1 \wedge \cdots \wedge C_m)$ since $\{C_1, \ldots C_m\}$ is consistent. By (Rel), $②(C_1 \wedge \cdots \wedge C_m)$, whence $②B$ by (RM-2) since $\vdash (C_1 \wedge \cdots \wedge C_m) \to B$.

R-to-L, suppose $①A_1, \ldots, ①A_n \vdash ②B$, and consider an arbitrary model $M = \langle w_0, W, v, \Delta \rangle$ such that $M \models O^\iota A_1, \ldots, M \models O^\iota A_n$. Hence $\{!C_1, \ldots, !C_n\} \subseteq \Delta$ with each C_i logically equivalent to A_i. Let $D_1, \ldots, D_m$ be a derivation of $②B$ from $①A_1, \ldots, ①A_n$. For each D_i, let D_i^* be the result of replacing each occurrence of $①$ and of $②$ by O. Show by induction on the derivation that for each D_i^*, $1 \leq i \leq m$, $M \models D_i^*$. Since $D_m = ②B$, $D_m^* = OB$. Hence $M \models OB$, as desired.

same way in their treatment of the Smith Argument. Either they rely on an underlying ambiguity in the oughts of the premises and the ought of the conclusion, or they impose a bar to using a drawn conclusion again as a premise in combination with previously given information. I leave it open whether either attitude offers an account of the Smith Argument sufficient to satisfy Desideratum 3.

The present account fails to provide for the Roberts-2 and Thomas Arguments, even in the sense of $O(t \wedge r), O(\neg t \wedge v) \vdash O(r \wedge v)$ and $O(t \wedge (f \vee s)), O(\neg t \wedge \neg f) \vdash Os$. If one would have these, there is a variation, suggested by Horty, that would apply.[120] This calls for articulating the meanings of the premises more fully than perhaps initially stated, and drawing conclusions from that articulation.

More precisely, for an ought statement, OA, with A formed solely in terms of $\neg$, $\wedge$ and $\vee$, we say that an occurrence of a subformula of A is positive or negative in A according as it lies within the scope of an even or an odd number of negations. Then for a set, Γ, of such ought statements, its *articulation*, Γ^a, is the smallest superset of Γ that contains both $O(\ldots B \ldots)$ and $O(\ldots C \ldots)$ whenever it contains either a formula of the form $O(\ldots(B \wedge C)\ldots)$ when the occurrence of the conjunction is positive or one of the form $O(\ldots(B \vee C)\ldots)$ when the occurrence of the disjunction is negative. Then we may say OB is inferable$_a$ from Γ just in case OB is inferable in the previous sense from the articulation of that set; i.e.,

- $\Gamma \vdash_a OB$ iff $\Gamma^a \vdash OB$.

Applied to the Roberts-2 Argument with the premise set $\Gamma = \{O(t \wedge r), O(\neg t \wedge v)\}$, its articulation $\Gamma^a = \{O(t \wedge r), O(\neg t \wedge v), Ot, Or, O\neg t, Ov\}$. Since r and v are co-possible, any model whose Δ contains imperatives for all of that articulated set will verify $O(r \wedge v)$. Hence $\{O(t \wedge r), O(\neg t \wedge v), Ot, Or, O\neg t, Ov\} \vdash O(r \wedge v)$, so that $O(t \wedge r), O(\neg t \wedge v) \vdash_a O(r \wedge v)$. The Thomas Argument is similar. Whether this provides a satisfactory way to meet Desideratum 3 in all its parts is much the same question as whether the account of the Smith Argument in terms of the original $\vdash$ sufficed.

6.4 Adaptive deontic logics

In Section 5.1 we saw how the combination of aggregation (C) and distribution (NM)/(RM), if left unchecked, leads to deontic explosion (DEX). In the remainder of Section 5 we saw how various proposals to constrain these rules still lead to unsatisfactory results. Here we consider a rather different method for putting reins on problematic principles. These are the adaptive

---

[120][Horty, 1997, pp. 33f.], though his purpose was somewhat different.

deontic logics. They should avoid all destructive forms of deontic explosion while still accounting for the paradigm arguments of Desideratum 3.[121]

In general, adaptive logics are a type of dynamic, nonmonotonic system of reasoning designed to apply problematic rules, such as aggregation or distribution, *provisionally*. A use of the rule is accepted until it makes trouble, as gauged against a specified class of abnormalities, at which point, but only at that point in context, it is rejected. The motivation is to view situations as normal as possible given an initial premise set. This is supposed to explicate actual processes of reasoning. In this setting a conflict-free argument like the Smith Argument should pass, even while a similar argument that would generate explosion from conflicting oughts would not.

A particular adaptive logic, **AL**, is defined in standard format as a triple $\langle \mathbf{LLL}, \Omega, \text{Strategy}\rangle$. **LLL** is the 'lower limit logic', a logic whose consequence relation is reflexive, transitive, monotonic and compact, has a characteristic semantics, and contains all of classical logic. Ω is the specified class of abnormalities, a set of formulas characterized by a logical form that is **LLL**-contingent and contains at least one logical symbol. The third element is the 'adaptive strategy' that determines how abnormalities are to be treated. There are two primary strategies developed in the literature of adaptive logics that could be applied here, 'reliablility' and 'minimal abnormality'. Here we work just with the first, reliablility, since it is easier to describe; minimal abnormality is somewhat bolder in the inferences it supports.

Given **LLL** characterized by a class of models $\mathcal{M}$, its adaptive extension, **AL**, is characterized by those models in $\mathcal{M}$ that are as normal or, for now, as reliable as possible. Given $M \in \mathcal{M}$, its abnormal part, $Ab(M)$, is the set of abnormalities from Ω that it validates, $Ab(M) = \{A \in \Omega : M \models A\}$. A *Dab*-formula is a disjunction of members of Ω. For a finite $\Theta \subset \Omega$, $Dab(\Theta) = \bigvee \Theta$. (If Θ is a singleton $\{A_1\}$, then $Dab(\Theta) = A_1$.) $Dab(\Theta)$ is a 'minimal *Dab*-consequence of Γ' if and only if $\Gamma \models_{\mathbf{LLL}} Dab(\Theta)$ and there is no set $\Theta' \subset \Theta$ such that $\Gamma \Vdash_{\mathbf{LLL}} Dab(\Theta')$. All the disjuncts of such minimal *Dab*-consequences of Γ are considered 'unreliable'. If $Dab(\Theta_1), \ldots, Dab(\Theta_n), \ldots$ are all the minimal *Dab*-consequences of Γ, the unreliabilities for Γ are $U(\Gamma) = \Theta_1 \cup \cdots \cup \Theta_n \cup \ldots$. An **LLL** model M is 'reliable for Γ' if and only if $M \models \Gamma$ and $Ab(M) \subseteq U(\Gamma)$. Then A is an **AL**-consequence of Γ, $\Gamma \Vdash_{\mathbf{AL}} A$ if and only if $M \models A$ for all $M \in \mathcal{M}$ that are reliable for Γ.

It is, however, in the proof-theory that the dynamics of adaptive logics stands out. This is given by three generic rules, or proof procedures, where

---

[121][Goble, 201x] presents an expanded account of the results of this section. The inspiration for applying the framework of adaptive logics to the problem of normative conflicts in deontic logic comes, however, from Joke Meheus, her work and the work of some of her students, see, e.g., the references in Note 123 below. For a general account of adaptive logic, see Batens, [2007], who has done the most to develop this method.

the first item on the right represents the formula entered on a line and the second represents an annotation, defined in the description of the rules:

PREM If $A \in \Gamma$,

$$\frac{\cdots \qquad \cdots}{A \qquad \emptyset}$$

RU If $A_1, \ldots, A_n \vdash_{\mathbf{LLL}} B$,

$$\begin{array}{cc} A_1 & \Delta_1 \\ \vdots & \vdots \\ A_n & \Delta_n \\ \hline B & \Delta_1 \cup \cdots \cup \Delta_n \end{array}$$

RC If $A_1, \ldots, A_n \vdash_{\mathbf{LLL}} B \vee Dab(\Theta)$,

$$\begin{array}{cc} A_1 & \Delta_1 \\ \vdots & \vdots \\ A_n & \Delta_n \\ \hline B & \Delta_1 \cup \cdots \cup \Delta_n \cup \Theta \end{array}$$

Rule (RC) expresses the use of the potentially problematic rule: From $A_1, \ldots, A_n$ infer B, under the constraint that none of the abnormalities in Θ be derivable from Γ. If some were, then B would be marked as 'unreliable'. Marked formulas are not considered derived from a given Γ. The adaptive strategy determines the rules for marking at each stage or section of a proof. For the strategy Reliability, say that $Dab(\Theta)$ is a minimal Dab-formula at a stage s of a derivation just in case $Dab(\Theta)$ is derived at stage s on condition $\emptyset$ and there is no $Dab(\Theta')$ derived at s on $\emptyset$ with $\Theta' \subset \Theta$. Where $Dab(\Theta_1), \ldots, Dab(\Theta_n)$ are the minimal Dab-formulas derived from Γ on condition $\emptyset$ at stage s, the 'unreliable' formulas at stage s are $U_s(\Gamma) = \Theta_1 \cup \cdots \cup \Theta_n$. If Δ is the condition on a line i, then line i is marked at stage s just in case $\Delta \cap U_s(\Gamma) \neq \emptyset$. Marking is dynamic; a line can be unmarked at one stage, marked at a later stage, and unmarked again still later.

A formula A is considered finally derivable from a set Γ at line i of a stage s just in case A is derived at i of s and every extension of the proof in which i is marked has an extension in which it is unmarked. $\Gamma \mathrel{\mspace{-1mu}\sim\mspace{-4mu}}_{\mathbf{AL}^r} A$ if and only if A is finally derivable from Γ at a line i in a proof from Γ. As we will see, the relation $\mathrel{\mspace{-1mu}\sim\mspace{-4mu}}_{\mathbf{AL}^r}$ is nonmonotonic. In general, for $\mathbf{AL}^r$ in standard format, $\Gamma \mathrel{\mspace{-1mu}\sim\mspace{-4mu}}_{\mathbf{AL}^r} A$ if and only if $\Gamma \Vdash_{\mathbf{AL}^r} A$.[122]

This pattern can now be applied in the framework of logics for normative conflicts. We consider briefly some systems that apply adaptive versions of distribution and of aggregation, or both. For convenience, we now ignore alethic modalities, though they could easily be incorporated.[123]

---

[122] [Batens, 2007, p. 233, theorem 7, corollary 2].
[123] I present these logics, and others, in [Goble, 201x]. The first is an adaptation of

We begin with an adaptive extension of the **LUM** logics of Section 5.3.2 because that is the most direct. Although those logics lack the full distribution rule (RM), with (RUM) they do contain

If $\vdash A \to B$, then $\vdash OA \to (OB \vee (OA \wedge O\neg A))$

which looks just like the rule (RM) with the adaptive constraints of (RC) as above, when Ω contains all formulas of the form $OC \wedge O\neg C$.

Accordingly, take **ALUM**^{r} to be the triple $\langle$**LUM**, Ω, Reliablity$\rangle$, where **LUM** is formulated as in Section 5.3.2 with either the full rule (C) and not (P)' or else with (P)' and either (CC) or (PC)'. $\Omega = \{A : \exists C(A = OC \wedge O\neg C)\}$, and Reliability is the adaptive strategy that determines how lines in a proof are marked, as described above.

To see how this works, consider how **ALUM**^{r} treats the Smith Argument. To show $O(f \vee s), O\neg f \vdash_{\textbf{ALUM}^r} Os$, the reasoning could go this way:

i) $O(f \vee s)$ PREM $\emptyset$
ii) $O\neg f$ PREM $\emptyset$
iii) $O((f \vee s) \wedge \neg f)$ i, ii RU $\emptyset$
iv) $O(\neg f \wedge s)$ iii RU $\emptyset$
v) Os iii RC $\{O(\neg f \wedge s) \wedge O\neg(\neg f \wedge s)\}$

Since there is no way to continue the proof that would derive from $\{O(f \vee s), O\neg f\}$ alone a minimal Dab-formula that contains a disjunct matching the condition on the right of (v), line (v) will stand unmarked, and it will be right to say that $O(f \vee s), O\neg f \vdash_{\textbf{ALUM}^r} Os$.

The proof thus operates as if it were given that $\neg O\neg(\neg f \wedge s)$ even though that was not stated among the premises. It is not even necessary to consider it a tacit premise, as in the treatment of the Smith Argument under the **LUM** logics themselves.

Furthermore, if it were given that $O\neg s$, so that this would be a conflict situation, then the proof might be extended:

✓v) Os iii RC $\{O(\neg f \wedge s) \wedge O\neg(\neg f \wedge s)\}$
vi) $O\neg s$ PREM $\emptyset$
vii) $(O(\neg f \wedge s) \wedge O\neg(\neg f \wedge s)) \vee (Os \wedge O\neg s)$ iv, vi RU $\emptyset$

Line (vii) follows in **LUM** from (iv) and (vi) by (SCon). Since its first disjunct is a member of the condition on line (v), that line must be marked.

---

a proposal of Straßer's [2010a; 2010b; 2012], though he applied the alternative strategy of minimal abnormality. [Meheus et al., 2010; Meheus et al., 201x] present a rather different adaptive deontic logic that contextually restricts aggregation; that system plays off the bi-modal logic **SDL**$_a$**P**$_e$ of [Goble, 2000; Goble, 2004a], which adds an additional complication to the picture. One could also have adaptive paraconsistent deontic logics; e.g., [Beirlaen et al., 2013; Goble, 201x].

Then Os is no longer considered derived from the (extended) premise set, and so $O(f \vee s), O\neg f, O\neg s \not\vdash_{\mathbf{ALUM}^r} Os$. This reveals the nonmonotonic character of adaptive logics. (The same holds for the other adaptive systems presented below.)

Withdrawing the conclusion Os contrasts with **LUM** itself. There, if Os is derived from $O(f \vee s)$ and $O\neg f$ because of the assumption of the tacit premise $U\neg f$, say, then Os will still be derived even when $O\neg s$ is included as an additional premise, so long as $U\neg f$ is preserved. While the inclusion of $O\neg s$ might incline one to withdraw that tacit assumption, it does not force it to be withdrawn. This also stands in contrast to the two prior approaches. In the two-phase logic of Section 6.2, ①$(f \vee s)$, ①$\neg f$, ①$\neg s \vdash$ ②s will be true, and on the imperatival approach of Section 6.3, $O(f \vee s), O\neg f, O\neg s \vdash Os$ will hold, just as for the original Smith Argument.

Contrast the derivation (i)–(v) for the original Smith Argument with an attempt to demonstrate (DEX$_s$) deriving an arbitrary Oq from a strict normative conflict, $\Gamma = \{Op, O\neg p\}$,

i)	Op	PREM	$\emptyset$
ii)	$O\neg p$	PREM	$\emptyset$
✓ iii)	$O(p \vee q)$	i RC	$\{Op \wedge O\neg p\}$
✓ iv)	$O((p \vee q) \wedge \neg p)$	iii, iv RU	$\{Op \wedge O\neg p\}$
✓ v)	$O(\neg p \wedge q)$	iv RU	$\{Op \wedge O\neg p\}$
✓ vi)	Oq	v RC	$\{Op \wedge O\neg p, O(\neg p \wedge q)$ $\wedge\, O\neg(\neg p \wedge q)\}$
vii)	$Op \wedge O\neg p$	i, ii RU	$\emptyset$

The proof is blocked because of (vii), which marks lines (iii), (iv), (v) and especially (vi). Other attempted derivations would fail similarly, as would attempts to derive more complicated versions of deontic explosion by more roundabout arguments.

ALUM^r thus avoids all the destructive forms of explosion, even while it gives an account of the Smith Argument, and the Jones Argument, in a way that may be smoother than **LUM** itself since the adaptive version does not rely on introducing extra premises supposing them to be tacit in context.

On the other hand, just as the **LUM** logics themselves are unable to account for either the Roberts or the Roberts-2 Argument, or the Thomas Argument, so also their adaptive extensions **ALUM**^{r} fail to account for them. Because of the conflicts that occur within the premises, (SCon) will generate a minimal Dab-formula from those premises that will force the conclusions of the Arguments to be marked, much as in the extended (conflicted) Smith Argument above. Consequently, **ALUM**^{r} falls short of Desideratum 3.

The next systems do meet this desideratum. The first of these offers a contextually restricted form of aggregation. For this we would like to draw on the logic **P** from Section 5.2, but at first blush that will be difficult. By analogy to **ALUM**^{r} above, we would like to have the effect of aggregation with the condition against absurdity, i.e.,

$$\vdash_{\mathbf{P}} (OA \wedge OB) \to (O(A \wedge B) \vee (OA \wedge O\neg A) \vee (OB \wedge O\neg B))$$

or something like that. But **P** does not contain that. Hence the difficulty.

This can be overcome by enriching what counts as absurd or abnormal. The paradigms of abnormality are, of course, strict normative conflicts, $OC \wedge O\neg C$, as above, but when applying adaptive restrictions on the aggregation rule, we should also consider as abnormal those cases where aggregation is explicitly contradicted, i.e., cases of the form $OA \wedge OB \wedge \neg O(A \wedge B)$. While close, that isn't quite right, however. Deontic explosion teaches that some failures of aggregation are quite appropriate, even desirable. Instead we might think of the abnormal as those cases where aggregation is explicitly contradicted *even though* all the components of the aggregate are innocent of conflict.

To express this, suppose $B_1, \ldots, B_n$ are all the subformulas of A; let '$\mho(A)$' abbreviate $UB_1 \wedge \cdots \wedge UB_n$, where UB abbreviates $\neg(OB \wedge O\neg B)$. Notice that $\neg \mho(A \wedge B)$ is equivalent to a formula that is a disjunction of strict conflicts, $(OC_1 \wedge O\neg C_1) \vee \cdots \vee (OC_m \wedge O\neg C_m)$.[124]

Though **P** does not contain the above, it does have

$$\vdash_{\mathbf{P}} (OA \wedge OB) \to (O(A \wedge B) \vee (OA \wedge OB \wedge \neg O(A \wedge B) \wedge \mho(A \wedge B)) \vee \neg \mho(A \wedge B))$$

Thus, in **P** either aggregation is all right, or else it fails when none of the components of the aggregate are tainted by (strict) conflict, or one of those components is so tainted. This underwrites use of aggregation in restricted contexts (RC) provided that Ω takes all the latter disjuncts into account.

Accordingly, let

$\Omega = \{C : \exists A \exists B(C = (OA \wedge OB \wedge \neg O(A \wedge B) \wedge \mho(A \wedge B)) \vee \neg \mho(A \wedge B))\}$, or equivalently,

$\Omega = \{C : \exists A \exists B(C = (OA \wedge OB \wedge \neg O(A \wedge B)) \vee \neg \mho(A \wedge B))\}$

Then **AP**^{r} is $\langle \mathbf{P}, \Omega, \text{Reliability} \rangle$.

---

[124] Note too that $\mho$ is not a sort of operator on formulas the way U is; it might fail replacement for logical equivalents. Thus A might be equivalent to B but $\mho(A)$ not equivalent to $\mho(B)$ since A and B might have quite differnt subformulas. $\mho$ is just a notational device for presenting general schemas.

This system treats the Smith Argument very much as **ALUM**^{r} does, but with appropriately different conditions for the application of (RC). It defuses deontic explosion in much the same way too. The reader may work out the details.

AP^r also provides for both Roberts Arguments and the Thomas Argument since **P** contains the full power of (RM), which allows detaching the unconflicted parts of the premises and then treating the results as for the original Smith and Jones Arguments. Hence these systems satisfy Desideratum 3 in full, along with Desiderata 1 and 2.

This method of extending **P** can also be applied for a different version of adaptive distribution. Recall the logic **EC** from Section 5.3 with (RE) and (C) but not (P) or (D) and not any form of distribution (RM) or (M). Extend that with the rule (N) to form **ECN**.[125] Though it lacks (M) $O(A \wedge B) \to OA$, **ECN** does contain $\vdash_{\mathbf{ECN}} O(A \wedge B) \to (OA \vee [(O(A \wedge B) \wedge \neg OA) \wedge \mho(A \wedge B)] \vee \neg\mho(A \wedge B))$. That sets up **AECN**^{r} = $\langle \mathbf{ECN}, \Omega, \text{Reliability} \rangle$ much like **AP**^{r}, but with

$$\Omega = \{C : \exists A \exists B (C = (O(A \wedge B) \wedge \neg OA) \vee \neg\mho(A \wedge B))\}$$

This too blocks deontic explosion much as **ALUM**^{r} and **AP**^{r} do, though with different applications of (RC). It also provides for the Smith and Jones Arguments a lot like **ALUM**^{r}. Moreover, unlike that system, now the two Roberts Arguments and the Thomas Argument are also supported. Again all three desiderata are satisfied. (These details too are left to the reader.)

Finally, we note the methods of **AP**^{r} and **AEC**^{r} can be combined to provide for an adaptive extension, **AEN**^{r}, of the fundamental logic **EN** that contains neither (C) nor (RM). Then both those rules would function under adaptive restrictions. For this let Ω be the union of the two Ωs for those systems. This too meets all the desiderata.

Interestingly, for all of the adaptive deontic logics described here, although normative conflicts are consistent and do not generate deontic explosion, nevertheless, the (D) principle also holds for them. That is to say, it holds *defeasibly*. For all these **AL**^{r}, if A is possible, then $OA \mathrel{\vdash\!\!\sim}_{\mathbf{AL}^r} \neg O \neg A$. More generally, if a premise set Γ is entirely normal, i.e., consistent with the standard **KD** principles, then the adaptive consequences of Γ are exactly its **KD** consequences; i.e., for normal Γ, $\Gamma \mathrel{\vdash\!\!\sim}_{\mathbf{AL}^r} A$ iff $\Gamma \vdash_{\mathbf{KD}} A$.[126]

This raises questions for this approach to the problem of normative conflicts. (D) holds defeasibly because strict normative conflicts, $OC \wedge O\neg C$, are included as (disjunctive parts of) abnormalities, the members of Ω. This also

---

[125](N) is necessary to block (DEX) in the present framework. Another way would apply **ECc** containing (P) and (CC) in place of (C). Then the rule (N) would be optional. The account of **AECN**^{r} to follow applies equally to the variants **AECc**^{r} and **AECcN**^{r}.

[126]This results as **KD** is the upper limit logic of **AL**^{r}; cf., [Batens, 2007, theorem 12].

accounts for the nonmonotonicity remarked earlier with regard to the Smith Argument, that $O(f \vee s), O\neg f \mathrel{\smash{\vrule height 1.2ex}\sim}_{\mathbf{AL}^r} Os$, but $O(f \vee s), O\neg f, O\neg s \not\mathrel{\smash{\vrule height 1.2ex}\sim}_{\mathbf{AL}^r} Os$. Nevertheless, normative conflicts do not seem at all abnormal or unreliable; they seem a simple fact of life. There seems no warrant to conclude (defeasibly) $\neg O\neg A$ from OA. For the Smith Argument, there seems no evidence that one should withdraw the conclusion Os when given the additional information $O\neg s$. One need merely remark that in this conflict situation, the conflict extends to Smith's performing alternative service.

Strict normative conflicts must be included as (disjunctive parts of) members of Ω to avoid deontic explosion. Yet, under the interpretation of adaptive logics generally, such inclusion presumes that normative conflicts are never really genuine, though one might often have to reason from information expressing such conflicts. That presumption is contrary to the point of view we have followed in this section, and the preceding.

6.5 Reprise

The systems discussed in this section make more radical departures from the standard picture of deontic logic than those of the preceding section. The paraconsistent deontic logics of §6.1 deny parts of classical propositional logic, especially the principle *ex contradictione quodlibet* (ECQ) and its relatives. Beyond that, however, their treatment of the deontic modalities is quite orthodox. They are modelled in familiar modal logical ways; principles (C) and (NM)/(RM) are accepted without qualification, etc. The other three approaches considered here preserve classical logic. The two-phase deontic logic of §6.2 applies two deontic operators, ① and ②, where the first follows the logic **ECc** of §5.3 (augmented by (N-1)), and so admits consistent aggregation but not distribution, and the second follows the logic **P** of §5.2 to apply distribution but not any form of aggregation. The two together provide a mechanism to control the shape of derivations or procedures of reasoning. The third, imperatival approach of §6.3, like the account of all-things-considered oughts in §4.4, offers a distinctive way to interpret formulas OA, though here these allow for normative conflicts. It too provides a new relation, $\vdash$, to characterize acceptable reasoning for normative contexts. An alternative relation, $\vdash_a$, was also described that was based on a more articulated form of the contents of given normative premises. Finally, the four adaptive deontic logics of §6.4 offer a different mechanism to explicate the procedures of proper reasoning. This allows potentially problematic principles, like aggregation or distribution, to be applied provisionally, under contextually determined restrictions.

All of these accounts allow for normative conflicts and all avoid deontic explosion, and so they all meet the first two desiderata that guided the previous section and this one. Their differences stand out with respect to the third desideratum, the paradigm arguments they account for. Those results are summarized in Table 4. For this we allow some latitude for what counts as an account of an argument's apparent validity. We accept, e.g., that the two-phase approach accounts for the Smith Argument by virtue of its validating, $①(f \vee s), ①\neg f \vdash ②s$, and we accept that the imperatival approach accounts for that argument by virtue of its yielding $O(f \vee s), O\neg f \vdash Os$, even though we raised questions in the respective sections about how adequate those accounts really are. (The row 'Imperatival(a)' refers to the imperatival account with the alternate relation $\vdash_a$.)

| | \multicolumn{7}{c}{Desiderata} |
| System | 1 | 2 | \multicolumn{5}{c}{3} |
			S	J	R	R-2	T
Paraconsistent	✓	✓	✗	✓	✓	✓	✗
Two-phase	✓	✓	✓	✓	✓	✗	✗
Imperatival	✓	✓	✓	✓	✓	✗	✗
Imperatival(a)	✓	✓	✓	✓	✓	✓	✓
Adaptive **ALUM**^{r}	✓	✓	✓	✓	✗	✗	✗
Adaptive **AP**^{r}	✓	✓	✓	✓	✓	✓	✓
Adaptive **AECN**^{r}	✓	✓	✓	✓	✓	✓	✓
Adaptive **AEN**^{r}	✓	✓	✓	✓	✓	✓	✓

Table 4

Because these accounts do take large steps away from traditional deontic logic, naturally the simple facts of their meeting, or failing to meet, any of these desiderata will not decide their true success in providing a logic for normative conflicts. That rests with their ability to provide for reasoning with basic normative concepts in general.

7 Review

In this chapter we have taken a long excursion through numerous responses to the central challenge that normative conflicts pose for basic deontic logic, namely, to explain how it can seem plausible there are such conflicts even while seemingly valid principles of deontic logic entail they are impossible. If one would accept those principles, and thus deny the reality of normative conflicts, then one must somehow explain the examples of apparent conflicts. If, on the other hand, one accepts that such conflicts can be genuine, then one must provide alternatives to the standard, core principles in a way that will still provide for proper reasoning in normative settings, as illustrated by the several paradigms, while yet avoiding deontic explosion.

Following the first course, one might try to explain the appearance of conflict in the examples by appealing to hidden ambiguities in the normative terms that describe the cases. Perhaps the agent ought to do one thing as directed by one normative system and ought to do something else incompatible with the first as directed by a different normative system. Insofar as that makes the 'ought's semantically distinct, there would be no real conflict. We looked at such a response briefly in Section 3. While this sort of analysis might apply to some examples, it leaves serious questions open. One might ask about the agent's overall normative position. That still looks like a conflict. One might also ask what if a single normative system directs an agent to a conflict. The Marcus Example would be a case in point.

Another way to explain the examples, including the Marcus Example, applies the distinction between *prima facie* oughts and all-things-considered oughts. When there appear to be conflicts, it might be said, they obtain between oughts of the first kind, the *prima facie* oughts, but when the agent's responsibilities are sorted out in light of all relevant factors, the agent's all-things-considered oughts will be found to be conflict-free. The core principles of deontic logic are supposed to govern only the latter, all-things-considered ought. In Section 4 we examined several proposals to explain just how *prima facie* oughts are to be sorted out in light of their priority relations to yield all-things-considered oughts. These had different degrees of success when measured against the standards of this point of view. Those results are summarized in §4.6.

Following the other course, and allowing the possibility of genuine normative conflicts, one must provide a logic that undermines Arguments I and II from Section 1, while at the same time finding a safe course between the reefs of excessive strength, entailing a form of deontic explosion, and the shoals of undue weakness, not being able to account for the several paradigms that seem to be acceptable inferences. Section 5 presented several logical systems to this end that keep within the general view of deontic logic as a type of simple modal logic. These had varying success, summarized in §5.5.

Section 6 offered more radical means to navigate these waters. These all meet the two desiderata of taking normative conflicts to be consistent and of avoiding destructive forms of deontic explosion. Whether or not they are strong enough, and give proper accounts of the various paradigm arguments, depends in great part on what one expects of a logic for normative discourse. The paraconsistent deontic logics of §6.1 deny parts of classical logic itself, but in so doing they seem unable to support some of the paradigm arguments that seem valid, not to mention many arguments in the non-deontic base.

The two-phase deontic logic of §6.2, the imperatival approach of §6.3 and the adaptive deontic logics of §6.4 all introduce different concepts of what counts as good reasoning, and so what counts as an account of the paradigms. How well they fare is summarized in §6.5.

After this long journey, it should be apparent that all of the proposals considered invite further exploration. Many questions remain open, and much remains to be done, both formally and philosophically, before the challenge of normative conflicts for deontic logic is fully resolved.

BIBLIOGRAPHY

[Al-Hibri, 1978] A. Al-Hibri. *Deontic Logic: A Comprehensive Appraisal and a New Proposal.* University Press of America, Washington, D. C., 1978.

[Alchourrón, 1996] C. E. Alchourrón. Detachment and defeasibility in deontic logic. *Studia Logica*, 57:5–18, 1996.

[Almeida, 1989] M. J. Almeida. Deontic problems with prohibition dilemmas. *Logique et Analyse*, 32(127–128):163–175, 1989.

[Asher and Bonevac, 1996] N. Asher and D. Bonevac. 'Prima facie' obligation. *Studia Logica*, 57:19–45, 1996.

[Ausín and Peña, 2000] F. J. Ausín and L. Peña. Deontic logic with enforceable rights. In D. Batens, editor, *Frontiers of Paraconsistent Logic*, pages 29–47. Research Studies Press, Philadelphia, 2000.

[Batens, 2007] D. Batens. A universal logic approach to adaptive logics. *Logica Universalis*, 1:221–242, 2007.

[Beirlaen et al., 2013] M. Beirlaen, C. Straßer, and J. Meheus. An inconsistency-adaptive deontic logic for normative conflicts. *Journal of Philosophical Logic*, 42:285–315, 2013.

[Belzer and Loewer, 1997] M. Belzer and B. Loewer. Deontic logics of defeasibility. In D. Nute, editor, *Defeasible Deontic Logic*, pages 45–57. Kluwer, Dordrecht, 1997.

[Bohse, 2005] H. Bohse. A paraconsistent solution to the problem of moral dilemmas. *South African Journal of Philosophy*, 24:77–86, 2005.

[Brink, 1994] D. Brink. Moral conflict and its structure. *The Philosophical Review*, 103:215–247, 1994. Reprinted in [Mason, 1996], pp. 102–126.

[Broome, 1999] J. Broome. Normative requirements. *Ratio*, 12:398–419, 1999.

[Broome, 2004] J. Broome. Reasons. In R. J. Wallace, M. Smith, S. Scheffer, and P. Pettit, editors, *Reason and Value: Essays on the Moral Philosophy of Joseph Raz*, pages 28–55. Oxford University Press, Oxford, 2004.

[Broome, 2007] J. Broome. Requirements. In T. Rønnow-Rasmussen, B. Petersson, J. Josefsson, and D. Egonsson, editors, *Hommage à Wlodek. Philosophical Papers Dedicated to Wlodek Rabinowicz.* www.fil.lu.se/hommageawlodek, 2007.

[Burgess, 1981] J. Burgess. Relevance: a fallacy? *Notre Dame Journal of Formal Logic*, 22:76–84, 1981.

[Carnielli et al., 2007] W. Carnielli, M. Coniglio, and J. Marcos. Logics of formal inconsistency. In D. Gabbay and F. Guenthner, editors, *Handbook of Philosophical Logic*, volume 14, pages 15–107. Springer, Dordrecht, 2nd edition, 2007.

[Castañeda, 1981] H.-N. Castañeda. The paradoxes of deontic logic: The simplest solution to all of them in one fell swoop. In R. Hilpinen, editor, *New Studies in Deontic Logic*, pages 37–85. D. Reidel, Dordrecht, 1981.

[Castañeda, 1982] H.-N. Castañeda. The logical structure of legal systems: A new perspective. In A. A. Martino, editor, *Deontic Logic, Computational Linguistics and Legal Information Systems*, volume II, pages 21–37. North Holland Publishing, Dordrecht, 1982.

[Chellas, 1980] B. Chellas. *Modal Logic: an introduction*. Cambridge University Press, Cambridge, 1980.

[Chisholm, 1964] R. Chisholm. The ethics of requirement. *American Philosophical Quarterly*, 1:147–153, 1964.

[Chisholm, 1974] R. Chisholm. Practical reason and the logic of requirement. In S. Körner, editor, *Practical Reason*, pages 1–17. Yale University Press, New Haven, 1974. Reprinted in J. Raz, editor, *Practical Reasoning*, Oxford University Press, Oxford, 1978, pp. 118–127.

[da Costa and Carnielli, 1986] N. C. A. da Costa and W. A. Carnielli. On paraconsistent deontic logic. *Philosophia*, 16:293–305, 1986.

[da Costa, 1996] N. C. A. da Costa. New systems of predicate deontic logic. *Journal of Non-Classical Logic*, 5:75–80, 1996.

[Donagan, 1987] A. Donagan. Consistency in rationalist moral systems. In C. Gowans, editor, *Moral Dilemmas*, pages 271–290. Oxford University Press, Oxford, 1987.

[Dunn and Restall, 2002] J. M. Dunn and G. Restall. Relevance logic. In D. Gabbay and F. Guenthner, editors, *Handbook of Philosophical Logic*, volume 6, pages 1–128. Kluwer, Dordrecht, 2nd edition, 2002.

[Gabbay and Governatori, 1999] D. Gabbay and G. Governatori. Dealing with label dependent deontic modalities. In P. McNamara and H. Prakken, editors, *Norms, Logics and Information Systems: New Studies on Deontic Logic and Computer Science*, pages 311–330. IOS Press, Amsterdam, 1999.

[Goble, 1999] L. Goble. Deontic logic with relevance. In P. McNamara and H. Prakken, editors, *Norms, Logics and Information Systems: New Studies on Deontic Logic and Computer Science*, pages 331–345. IOS Press, Amsterdam, 1999.

[Goble, 2000] L. Goble. Multiplex semantics for deontic logic. *Nordic Journal of Philosophical Logic*, 5:113–134, 2000.

[Goble, 2001] L. Goble. The Andersonian reduction and relevant deontic logic. In B. Brown and J. Woods, editors, *New Studies in Exact Philosophy: Logic, Mathematics and Science*, pages 213–246. Hermes Scientific Publishers, Oxford, 2001.

[Goble, 2003] L. Goble. Preference semantics for deontic logic, Part I – simple models. *Logique et Analyse*, 46(183–184):383–418, 2003.

[Goble, 2004a] L. Goble. Preference semantics for deontic logic, Part II – multiplex models. *Logique et Analyse*, 47(185–188):335–363, 2004.

[Goble, 2004b] L. Goble. A proposal for dealing with deontic dilemmas. In A. Lomuscio and D. Nute, editors, *Deontic Logic in Computer Science, 7th International Workshop on Deontic Logic in Computer Science, DEON 2004*, pages 74–113. Springer, Berlin, 2004.

[Goble, 2005] L. Goble. A logic for deontic dilemmas. *Journal of Applied Logic*, 3:461–483, 2005.

[Goble, 2009] L. Goble. Normative conflicts and the logic of 'ought'. *Noûs*, 43:450–489, 2009.

[Goble, 201x] L. Goble. Deontic logic (adapted) for normative conflicts. *Logic Journal of IGPL*, 201x. forthcoming; DOI: 10.1093/jigpal/jzt022.

[Gowans, 1987a] C. W. Gowans. Introduction: The debate on moral dilemmas. In C. Gowans, editor, *Moral Dilemmas*, pages 3–33. Oxford University Press, Oxford, 1987.

[Gowans, 1987b] C. W. Gowans. *Moral Dilemmas*. Oxford University Press, Oxford, 1987.

[Grana, 1990a] N. Grana. *Logica Deontica Paraconsistente*. Liguori Editore, Napoli, 1990.

[Grana, 1990b] N. Grana. On a minimal non-alethic logic. *Bulletin of the Section of Logic*, 19:25–29, 1990.

[Hansen, 2004a] J. Hansen. Conflicting imperatives and dyadic deontic logic. In A. Lomuscio and D. Nute, editors, *Deontic Logic in Computer Science, 7th International Workshop on Deontic Logic in Computer Science, DEON 2004*, pages 146–164. Springer, Berlin, 2004.

[Hansen, 2004b] J. Hansen. Problems and results for logics about imperatives. *Journal of Applied Logic*, 2:39–61, 2004.

[Hansen, 2005] J. Hansen. Conflicting imperatives and dyadic deontic logic. *Journal of Applied Logic*, 3:484–511, 2005.

[Hansen, 2006] J. Hansen. Deontic logics for prioritized imperatives. *Artificial Intelligence and the Law*, 14:1–34, 2006.

[Hansen, 2008] J. Hansen. Prioritized conditional imperatives: Problems and a new proposal. *Autonomous Agents and Multi-Agent Systems*, 17:11–35, 2008.

[Hansson, 1990] S. O. Hansson. Preference-based deontic logic (PDL). *Journal of Philosophical Logic*, 19:75–93, 1990.

[Hansson, 2001] S. O. Hansson. *The Structure of Values and Norms*. Cambridge University Press, Cambridge, 2001.

[Harman, 1975] G. Harman. Reasons. *Critica*, 7:3–18, 1975. Reprinted in J. Raz, editor, *Practical Reasoning*, Oxford University Press, Oxford, 1978, pp. 110–117.

[Horty, 1994] J. F. Horty. Moral dilemmas and nonmonotonic logic. *Journal of Philosophical Logic*, 23:35–65, 1994.

[Horty, 1997] J. F. Horty. Nonmonotonic foundations for deontic logic. In D. Nute, editor, *Defeasible Deontic Logic*, pages 17–44. Kluwer, Dordrecht, 1997.

[Horty, 2001] J. F. Horty. Nonmonotonic logic. In L. Goble, editor, *The Blackwell Guide to Philosophical Logic*, pages 336–361. Blackwell, Oxford, 2001.

[Horty, 2003] J. F. Horty. Reasoning with moral conflicts. *Noûs*, 37:557–605, 2003.

[Horty, 2007] J. F. Horty. Defaults with priorities. *Journal of Philosophical Logic*, 36:367–413, 2007.

[Horty, 2012] J. F. Horty. *Reasons and Defaults*. Oxford University Press, Oxford, 2012.

[Kooi and Taminga, 2008] B. Kooi and A. Taminga. Moral conflicts between groups of agents. *Journal of Philosophical Logic*, 37:1–21, 2008.

[Kouznetsov, 2004] A. Kouznetsov. Quasi-matrix deontic logic. In A. Lomuscio and D. Nute, editors, *Deontic Logic in Computer Science, 7th International Workshop on Deontic Logic in Computer Science, DEON 2004*, pages 191–208. Springer, Berlin, 2004.

[Krabbendam and Meyer, 1999] J. Krabbendam and J.-J. Meyer. Contextual deontic logics. In P. McNamara and H. Prakken, editors, *Norms, Logics and Information Systems: New Studies on Deontic Logic and Computer Science*, pages 347–362. IOS Press, Amsterdam, 1999.

[Kraus et al., 1990] S. Kraus, D. Lehmann, and M. Magidor. Nonmonotonic reasoning, preferential models and cumulative logics. *Artificial Intelligence*, 44:167–207, 1990.

[Lemmon, 1962] E. J. Lemmon. Moral dilemmas. *The Philosophical Review*, 70:139–158, 1962. Reprinted in part in [Gowans, 1987b] pp. 101–114.

[Loewer and Belzer, 1991] B. Loewer and M. Belzer. Prima facie obligation: Its deconstruction and reconstruction'. In E. Lepore and R. van Gulick, editors, *John Searle and his Critics*, pages 359–370. Blackwell, Oxford, 1991.

[Loparić and Puga, 1986] A. Loparić and L. Puga. Two systems of deontic logic. *Bulletin of the Section of Logic*, 15:137–144, 1986.

[Makinson, 1994] D. Makinson. General patterns in nonmonotonic reasoning. In D. Gabbay, C. Hogger, and J. Robinson, editors, *Handbook of Logic in Artificial Intelligence*, pages 35–110. Oxford University Press, Oxford, 1994.

[Marcus, 1980] R. B. Marcus. Moral dilemmas and consistency. *Journal of Philosophy*, 77:121–136, 1980. Reprinted in [Gowans, 1987b] pp. 115–137, page references to the original.

[Mares and Meyer, 2001] E. D. Mares and R. K. Meyer. Relevant logics. In L. Goble, editor, *The Blackwell Guide to Philosophical Logic*, pages 280–308. Blackwell, Oxford, 2001.

[Mason, 1996] H. E. Mason. *Moral Dilemmas and Moral Theory*. Oxford University Press, Oxford, 1996.

[McConnell, 1978] T. McConnell. Moral dilemmas and consistency in ethics. *Canadian Journal of Philosophy*, 8:269–287, 1978. Reprinted in [Gowans, 1987b], pp. 115–137.

[McConnell, 1988] T. McConnell. Interpersonal moral conflicts. *American Philosophical Quarterly*, 25:25–35, 1988.

[McConnell, 2010] T. McConnell. Moral dilemmas. In E. Zalta, editor, *Stanford Encyclopedia of Philosophy*. Stanford University, summer 2010 edition, 2010. URL = <http://plato.stanford.edu/archives/sum2010/entries/moral-dilemmas>.

[McGinnis, 2007a] C. McGinnis. *Paraconsistency and Deontic Logic: Formal Systems for Reasoning with Normative Conflicts*. PhD thesis, University of Minnesota, 2007. University of Michigan Microforms, UMI Number: 3292966.

[McGinnis, 2007b] C. McGinnis. Semi-paraconsistent deontic logic. In J.-Y. Beziau, W. Carnielli, and D. Gabbay, editors, *Handbook of Paraconsistency*, pages 103–125. College Publications, London, 2007.

[McNamara, 2004] P. McNamara. Agential obligation as non-agential personal obligation plus agency. *Journal of Applied Logic*, 2:117–152, 2004.

[Meheus et al., 2010] J. Meheus, M. Beirlaen, and F. Van de Putte. Avoiding deontic explosion by contextually restricting aggregation. In G. Governatori and G. Sartor, editors, *Deontic Logic in Computer Science, DEON2010*, pages 148–165. Springer, Berlin, 2010.

[Meheus et al., 201x] J. Meheus, M. Beirlaen, C. Straßer, and F. Van de Putte. Non-adjunctive deontic logics that validate aggregation as much as possible. *Journal of Applied Logic*, 201x. forthcoming.

[Morreau, 1996] M. Morreau. 'Prima facie' and seeming duties. *Studia Logica*, 57:47–71, 1996.

[Nair, 201x] S. Nair. Consequences of reasoning with conflicting obligations. *Mind*, 201x. forthcoming.

[Nute, 1997] D. Nute. *Defeasible Deontic Logic*. Kluwer, Dordrecht, 1997.

[Nute, 1999] D. Nute. Norms, priorities, and defeasibility. In P. McNamara and H. Prakken, editors, *Norms, Logics and Information Systems: New Studies on Deontic Logic and Computer Science*, pages 201–218. IOS Press, Amsterdam, 1999.

[Pietroski, 1993] P. M. Pietroski. Prima facie obligations, ceteris paribus laws in moral theory. *Ethics*, 103:489–515, 1993.

[Prakken, 1996] H. Prakken. Two approaches to the formalisation of defeasible deontic reasoning. *Studia Logica*, 57:73–90, 1996.

[Priest, 2006] G. Priest. *In Contradiction: A Study of the Transconsistent*. Oxford University Press, Oxford, expanded edition, 2006.

[Raz, 1978] J. Raz. Reasons for action, decisions and norms. In J. Raz, editor, *Practical Reasoning*, pages 128–143. Oxford University Press, Oxford, 1978. Excerpts reprinted from *Mind*, (1975), 481–499.

[Read, 1988] S. Read. *Relevant Logic*. Blackwell, Oxford, 1988.

[Ross, 1930] W. D. Ross. *The Right and the Good*. Clarendon Press, Oxford, 1930.

[Routley and Plumwood, 1989] R. Routley and V. Plumwood. Moral dilemmas and the logic of deontic notions. In G. Priest, R. Routley, and J. Norman, editors, *Paraconsistent Logic: Essays on the Inconsistent*, pages 653–690. Philosophia Verlag, Munich, 1989.

[Sartre, 1946] J.-P. Sartre. *L'Existentialisme est un Humanisme*. Nagel, Paris, 1946. Translated as "Existentialism is a Humanism" in W. Kaufmann, editor, *Existentialism from Dostoevsky to Sartre*, Meridian Press, 1975, pp. 287–311; page references to this translation.

[Schotch and Jennings, 1981] P. Schotch and R. Jennings. Non-kripkean deontic logic. In R. Hilpinen, editor, *New Studies in Deontic Logic*, pages 149–162. D. Reidel, Dordrecht, 1981.

[Sinnott-Armstrong, 1988] W. Sinnott-Armstrong. *Moral Dilemmas*. Blackwell, Oxford, 1988.

[Sinnott-Armstrong, 1996] W. Sinnott-Armstrong. Moral dilemmas and rights. In H. E. Mason, editor, *Moral Dilemmas and Moral Theory*, pages 48–65. Oxford University Press, Oxford, 1996.

[Stranzinger, 1978] R. Stranzinger. Ein paradoxienfreies deontisches system. In I. Tammelo and H. Schreiner, editors, *Strukturierungen und Entscheidungen im Rechtsdenken*, pages 183–192. Springer-Verlag, Vienna and New York, 1978.

[Straßer and Beirlaen, 2012] C. Straßer and M. Beirlaen. An Andersonian deontic logic with contextualized sanctions. In T. Agotnes, J. Broersen, and D. Elgesem, editors, *Deontic Logic in Computer Science, 11th International Conference on Deontic Logic in Computer Science, DEON2012*, pages 151–169. Springer, Berlin, 2012.

[Straßer et al., 2012] C. Straßer, J. Meheus, and M. Beirlaen. Tolerating deontic conflicts by adaptively restricting inheritance. *Logique et Analyse*, 52(219):477–506, 2012.

[Straßer, 2010a] C. Straßer. *Adaptive Logic Characterizations of Defeasible Reasoning with Applications in Argumentation, Normative Reasoning and Default Reasoning*. PhD thesis, Ghent University, 2010.

[Straßer, 2010b] C. Straßer. An adaptive logic framework for conditional obligations and deontic dilemmas. *Logic and Logical Philosophy*, 19:95–128, 2010.

[Vallentyne, 1987] P. Vallentyne. Prohibition dilemmas and deontic logic. *Logique et Analyse*, 30(117–118):113–122, 1987.

[Vallentyne, 1989] P. Vallentyne. Two types of moral dilemmas. *Erkenntnis*, 30:301–318, 1989.

[van der Torre and Tan, 1997] L. van der Torre and Y.-H. Tan. The many faces of defeasibility in defeasible deontic logic. In D. Nute, editor, *Defeasible Deontic Logic*, pages 79–121. Kluwer, Dordrecht, 1997.

[van der Torre and Tan, 2000] L. van der Torre and Y.-H. Tan. Two-phase deontic logic. *Logique et Analyse*, 43(171-172):411–456, 2000.

[van der Torre, 1997] L. van der Torre. *Reasoning about Obligations: Defeasibility in Preference-Based Deontic Logic*. Thesis Publishers, Amsterdam, 1997.

[van Eck, 1982] J. A. van Eck. A system of temporally relative modal and deontic predicate logic and its philosophical applications. *Logique et Analyse*, 25:249–290 and 339–381, 1982.

[van Fraassen, 1973] B. van Fraassen. Values and the heart's command. *Journal of Philosophy*, 70:5–19, 1973. Reprinted in [Gowans, 1987b] pp. 138–153; page references to the original.

[Williams, 1965] B. Williams. Ethical consistency. *Proceedings of the Aristotelian Society*, 39 (supplemental):103–124, 1965. A revised version appears in *Problems of the Self: Philosophical Papers 1956–1972*, Cambridge University Press, 1973, pages 166–186; reprinted in [Gowans, 1987b] pp. 115–137.

[Yamada, 2008] T. Yamada. Logical dynamics of some speech acts that affect obligations and preferences. *Synthese*, 165:295–315, 2008.

[Zimmerman, 1996] M. J. Zimmerman. *The Concept of Moral Obligation*. Cambridge University Press, Cambridge, 1996.

Lou Goble
Eugene, Oregon, USA
Email: lgoble@willamette.edu

5
Normative Positions
MAREK SERGOT

ABSTRACT. The Kanger-Lindahl theory of normative positions is an attempt to apply the tools of modal logic to the formalisation of Hohfeld's 'fundamental legal conceptions', to the construction of a formal theory of duties and rights, and to the formal characterisation generally of complex normative relations that can hold between (pairs of) agents with regard to an action by one or other of them. The theory employs a standard deontic logic, a logic of action/agency of the 'brings it about' or 'sees to it' kind, and a method of mapping out in a systematic and exhaustive fashion the complete space of all logically possible normative relations—or 'positions'—of some given type. The article presents a generalised version of the methods and a brief dicussion of its limitations as a comprehensive theory of duty and right.

1	Introduction	 353
2	Preliminary discussion	 356
3	Motivating examples	 360
4	The Kanger-Lindahl theory	 364
5	Partitions	 375
6	Normative positions	 382
	6.1 Maxi-conjunctions for logics of type $EMCP$	 383
	6.2 Refinement structures	 385
7	Example	 388
8	Discussion	 393
	8.1 Alchourrón-Bulygin's normative systems, and conditional positions	 393
	8.2 Extended forms of act expression	 395
	8.3 Limitations	 400
9	Conclusion	 403

1 Introduction

The theory of normative positions is an attempt to apply the tools of modal logic to the formalisation of the 'fundamental legal conceptions' (duty, right, privilege, power, immunity, etc.) most closely associated with the American jurist W.N. Hohfeld [1913], to the construction of a formal theory of duties and rights, and to the formal characterisation generally of complex normative relations that can hold between (pairs of) agents with regard to an action by one or other of them. The development was initiated by Stig Kanger and subsequently extended and refined, most notably by Lars Lindahl. Ingmar Pörn applied similar techniques to the study of 'control and influence' relations in social interactions.

The theory employs a standard deontic logic, a logic of action/agency of the 'brings it about' or 'sees to it' kind, and a method for mapping out in a systematic and exhaustive fashion the complete space of all logically possible normative relations between two agents with respect to some given act type. Kanger called these relations the 'atomic types of rights relation'; we will follow later usage and refer to them generally as normative 'positions'. The methods are presented in [Kanger, 1971; Kanger, 1985; Kanger and Kanger, 1966] with a more general account of related issues in [Kanger, 1972]. As described later in the article, Lars Lindahl [1977] developed Kanger's account in several important respects, providing also a commentary on the relationships to Hohfeld's work and the jurisprudential tradition, of Jeremy Bentham and John Austin, within which it falls. Ingmar Pörn [1977] applied similar techniques to the study of what he called 'control' and 'influence' relations in which there are iterations of the action/agency modalities in place of the deontic logic component. For further discussion of the theory and some of its features and possible applications, see e.g. [Talja, 1980; Makinson, 1986; Lindahl, 1994; Jones and Sergot, 1992; Jones and Sergot, 1993; Herrestad and Krogh, 1995; Herrestad, 1996; Krogh, 1997; Sergot and Richards, 2000; Jones and Parent, 2008]. The technical account presented in this article is extracted from [Sergot, 2001].

The concepts treated by the theory of normative positions are usually discussed within the context of law and legal relations. Hohfeld himself referred to them as the 'fundamental legal conceptions'. These are not exclusively *legal* concepts, however, but characteristic of all forms of regulated and organised agent interaction. Although the theory does address fundamental issues in the formal representation of laws and regulations and legal contracts—Allen and Saxon [1986; 1993] for example long argued that proper attention to the Hohfeldian concepts is essential for legal knowledge representation—it also finds applications in other areas, such as the specification of aspects of computer systems (see e.g. [Jones and Sergot, 1993; Krogh, 1997; Jones and Parent, 2008]), as a contribution to the formal theory of organisations in the analysis of notions such as responsibility, entitlement, authorisation and delegation, and in the field of multi-agent systems, where the notion of commitment in particular, in the sense of a directed obligation of an agent a to another agent b, features prominently in the literature on co-ordinated action, joint planning, and agent communication languages. (See e.g. [Jennings, 1993; Shoham, 1991; Shoham, 1993; Singh, 1998; Singh, 1999; Colombetti, 1999; Colombetti, 2000] for some early references.)

The theory of normative positions has a number of important and well-documented limitations. As a theory of rights, it lacks a treatment of the role of *counterparty*, the agent who is the beneficiary of a right relation or to whom a duty is owed. As a formalisation of the Hohfeldian framework, it does not deal with the feature Hohfeld called '(legal) power', also referred to sometimes as 'legal capacity' or 'competence'. See e.g. [Makinson, 1986; Lindahl, 1994] for some of these points, and the discussion that follows in Section 8 below. The theory of normative positions is therefore best seen as a *component* of a formal theory of duty and right, and not as a complete theory of all aspects of these complex concepts. Its methods need to be augmented: with a treatment of 'power', with temporal constructs, and with a richer set of action concepts, at the very least.

Nevertheless, the Kanger-Lindahl theory is generally regarded as the most comprehensive and best developed attempt to formalize distinctions such as Hohfeld's. For example, Hohfeld identified four distinct legal/normative relations that could hold between any two agents with respect to some given act type. Some examples are given later in Section 3. Kanger's systematic, formal analysis yielded 26 distinct 'atomic types of rights relations' or 'normative positions' as a refinement of Hohfeld's four. Lindahl's subsequent analysis produced 35 of the same basic kind as Kanger's and 127 if a more precise set of possible relationships is considered instead. Section 4 discusses the methods in more detail. It also explains why there are more possibilities still than are accounted for in Lindahl's version: employing the same logics, 255 distinct relationships can be generated refining Kanger's 26 and Lindahl's 127, and many more if we include more complex act types and more agents than two.

This article follows the formal treatment presented in [Sergot, 2001] which generalised the Kanger-Lindahl accounts in the following respects. (1) The generalised theory deals with interaction between any number of agents, not just two, including 'ought-to-be' statements where no agent is specified. (2) The Kanger-Lindahl-Pörn theories deal with act expressions of the form 'agent x brings it about that F'. The generalised theory allows any number of such act expressions in any combination, and allows compound acts, that is to say, boolean compounds of propositions in the scope of the 'brings

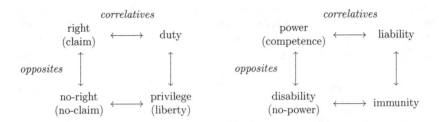

Figure 1: Hohfeld's 'fundamental legal conceptions'

it about' operator. (3) Building on a suggestion by David Makinson it is possible to give an abstract characterisation of classes of 'positions' and relationships between them, and a complete separation of the method of generating the space of 'positions' from properties of the underlying modal logics. The generalised theory does not rely on any in-built assumptions about the specific deontic or action logics employed. It also means that, in principle at least, a richer combination of modalities could be used to represent more complex notions.

[Sergot, 2001] also shows how the methods for generating 'positions' can be automated without the need for theorem provers for the modal logics, and presents an automated computer system intended to facilitate application of the theory to the analysis of practical problems. Those methods will not be covered in this article.

2 Preliminary discussion

Hohfeld's seminal work [1913] is still often taken as the starting point for much that is written in this field. It identified two groups of four concepts with various relationships between them, as summarized in Figure 1. Right and duty are 'correlatives' in the sense that when x has a right (a 'claim-right') against y that F (be done by y) then y owes a duty to x that F (be done by y); and conversely. The relationships may be summarised semi-formally by the following scheme, adapted from [Lindahl, 1977]:

$$Right(x, y, F) \leftrightarrow Duty(y, x, F)$$
$$Right(x, y, \text{not-}F) \leftrightarrow Duty(y, x, \text{not-}F)$$

Here not-F is intended to stand for y's refraining from doing F. Of course it remains to explain how this notion of refraining is to be represented formally; this is one of the features of Kanger's framework.

Duty and privilege (some authors prefer 'liberty') are 'opposites' in the Hohfeldian scheme in the sense that x has a privilege/liberty from y with respect to F when x does not owe a duty to y to refrain from F; x has a privilege/liberty from y to refrain from F when x does not owe a duty to y that F (be done by x). In the semi-formal notation these relationships may be summarised as follows:

$$Privilege(x, y, F) \leftrightarrow \neg Duty(x, y, \text{not-}F)$$
$$Privilege(x, y, \text{not-}F) \leftrightarrow \neg Duty(x, y, F)$$

Similarly, right/no-right and no-right/privilege are also opposite and correlative pairs in the Hohfeldian scheme, in the following sense:

$$Right(x, y, F) \leftrightarrow \neg No\text{-}right(x, y, F)$$
$$Right(x, y, \text{not-}F) \leftrightarrow \neg No\text{-}right(x, y, \text{not-}F)$$
$$No\text{-}right(x, y, F) \leftrightarrow Privilege(y, x, \text{not-}F)$$
$$No\text{-}right(x, y, \text{not-}F) \leftrightarrow Privilege(y, x, F)$$

One can see already, however, as pointed out in [Lindahl, 1977, pp. 26–7] and in [Kanger and Kanger, 1966], that there are discrepancies in Hohfeld's account: the right/duty and no-right/privilege correlative pairs are not exactly of the same form, and nor are the right/no-right and duty/privilege opposites.

There is further inexactitude in Hohfeld's scheme for his second group of concepts, those on the right of the diagram in Figure 1. This second group is concerned with *changes* of legal/normative relations, as when it is said, for example, that x has power (competence) to impose a duty on y that such-and-such or to grant a privilege or right to z that such-and-such. Discussion of this second set of concepts raises a new set of questions however and is beyond the scope of this article. The second part of [Lindahl, 1977] is concerned with this group of concepts. See [Jones and Sergot, 1996] for an alternative account of power/competence.

For present purposes, the point is that Hohfeld's writings, and much else that has been written on these topics in legal theory, provide a wealth of examples and the beginnings of a systematic account, but are not precise enough to give a formal theory. Kanger attempted to provide such a theory by applying the formal tools of modal logic to this task.

The Kanger-Lindahl theory has a deontic logic component, an action logic component, and a method for generating the space of all logically possible positions. The language is that of propositional logic augmented with modal operators O (for 'obligation') and its dual P (for 'permission'), and relativised modal operators $E_a, E_b, \ldots$ for act expressions, where $a, b, \ldots$ are the names of individual agents. (This notation is slightly different from Kanger and Lindahl's, who use *Shall* and *May* for O and P, and *Do* for act expressions. The alternative notation is chosen simply because it is more concise and reduces the size of the formal expressions to be manipulated.)

An expression of the form OA may be read as 'it is obligatory that A' or 'it ought to be the case that A'. P is the dual of O: $PA =_{\text{def}} \neg O \neg A$. The expression PA may be read as 'it is permissible that A'. We will also say 'permitted'. The deontic logic employed by Kanger and Lindahl is—for all intents and purposes—the system usually referred to as Standard Deontic Logic (SDL). Specifically, the deontic logic employed is the smallest system containing propositional logic (*PL*) and the following axiom schemas and rules:

$$\text{O.RE} \quad \frac{A \leftrightarrow B}{OA \leftrightarrow OB}$$

$$\text{O.M} \quad O(A \wedge B) \to (OA \wedge OB)$$

$$\text{O.C} \quad (OA \wedge OB) \to O(A \wedge B)$$

$$\text{O.P} \quad \neg O \bot$$

The names of axiom schemas and rules in this article are based on those of [Chellas, 1980]: the logic of O is a classical modal logic of type *EMCP*. For comparison, Standard Deontic Logic (SDL) is a normal modal logic of type *KD*, which is type *EMCP* together with the additional rule of necessitation

$$\text{O.RN} \quad \frac{A}{OA}$$

or, equivalently, the axiom schema $O\top$ ($\top$ any tautology). The absence or presence of rule O.RN plays no role in the generation of normative positions: this is why we say that Kanger's choice of deontic logic is to all intents and purposes Standard Deontic Logic. The 'deontic axiom' of Standard Deontic Logic

$$\text{O.D} \quad OA \to PA$$

follows from O.C and O.P.

Of course Standard Deontic Logic (of type *KD* or *EMCP*) has many well-known limitations and its inadequacies are taken as the starting point for many of the developments in the field. Both axioms O.M and O.C can be criticised as simplistic, for example. However, in combination with the logic of action, and in the restricted ways it is employed in the generation of normative positions, these inadequacies are relatively benign. In any case, the extended theory of normative positions to be presented in later sections is not dependent on specific choices for the deontic and action logics employed. These can be changed, as explained below.

As regards the action component, expressions of the form $\mathrm{E}_x A$ stand for 'agent x sees to it that, or brings it about that, A'. This approach to the logic of action has been extensively studied in analytical philosophy and philosophical logic though is perhaps not so familiar in Computer Science. The *stit* operator of [Belnap and Perloff, 1988; Belnap and Perloff, 1992] and *dstit* of [Horty and Belnap, 1995] are instances of the general approach that have had some exposure in the AI literature. The focus of attention is not on transitions and state changes as in most treatments of action in AI and Computer Science, but rather on the end result A and the agent x whose actions are responsible, in some appropriate sense, for this end result; the specific means or actions employed by agent x to bring about A are not expressed.

The logic of each E_x is that of a (relativised) classical modal system of type ET in the Chellas classification, i.e. the smallest system containing PL, closed under the rule E.RE:

$$\text{E.RE} \qquad \frac{A \leftrightarrow B}{\mathrm{E}_x A \leftrightarrow \mathrm{E}_x B}$$

and containing the axiom schema

$$\text{E.T} \qquad \mathrm{E}_x A \to A$$

The schema E.T indicates that this is a notion of *successful* action. It does not matter, for the purposes of this article, whether x brings about A intentionally or unintentionally, knowingly or unknowingly.

The E_x notation is from [Pörn, 1977]. For present purposes, however, the (relativised) operators E_x should be regarded as standing for one of a range of possible action modalities rather than any one of them specifically. For a discussion of some candidates and their relative merits see e.g. [Chellas, 1969; Pörn, 1970; Pörn, 1974; Pörn, 1977; Pörn, 1989; Åqvist, 1974; Segerberg, 1985; Segerberg, 1989; Segerberg, 1992; Belnap and Perloff, 1988; Belnap and Perloff, 1992; Perloff, 1991; Horty and Belnap, 1995; Elgesem, 1992; Hilpinen, 1997; Horty, 2001] as well as more recent works on 'stit' logics in particular. It is likely that a comprehensive theory of rights and/or organisations would require several different notions of action and agency. In [Santos and Carmo, 1996; Santos *et al.*, 1997], for instance, it is suggested that distinguishing between direct and indirect action may be important for describing certain organisational structures. Nothing in the present account depends on such detailed choices. As in the Kanger-Lindahl framework, the only properties assumed for the action modalities E_x are the schema E.T and closure under logical equivalence, E.RE.

3 Motivating examples

We conclude this introductory discussion with some brief examples to illustrate the expressive power of the language and to motivate the formal development to be undertaken in the remainder of the article. These examples are intended to be simple and familiar. They are the same as those used in [Sergot, 2001].

Example 3.1 (Library book) *Let b name a borrower in a library who has some book out on loan. Let R represent that this book is returned to the library by the date due. b has an obligation to return the book by date due. In the Kanger framework this obligation on b can be represented by the following expression.*

(1) $$\text{OE}_b R$$

Expression (1) is not the only, nor perhaps even an adequate, representation of what we mean by saying that b has an obligation to return the book. It employs what some authors refer to as the Meinong-Chisholm analysis, whereby 'x ought to bring it about that F' is taken to mean 'it ought to be that x brings it about that F'. It is possible to question whether these expressions are in fact equivalent. See e.g. the discussions in [Horty, 1996; Horty, 2001; Sergot and Richards, 2000; Brown, 2000] among others. There are also some senses of 'obligation'—as when we say e.g. 'x is responsible for, or held accountable for, ensuring that F is the case'—which are not adequately represented by this construction. Possible formalisations of these other senses will not be discussed in this article.

Studies of duty and right, such as Hohfeld's, adopt a relational perspective: the focus is on relationships between pairs of agents. So, given the truth of e.g. $\text{OE}_b R$, one is led to ask about the obligations and permissions of other agents, a say, with respect to the returning of the book. One can see that, according to the logics employed, the following three possibilities are all consistent with $\text{OE}_b R$:

1. a is obliged to return the book: $\text{OE}_a R$;

2. a is permitted but not obliged to return the book:
$$(\text{PE}_a R \wedge \neg \text{OE}_a R) = (\text{PE}_a R \wedge \text{P}\neg \text{E}_a R);$$

3. a is not permitted to return the book: $\neg \text{PE}_a R$.

Note that the first of these is logically possible given (1): the expression $OE_a R \land OE_b R$ is not inconsistent. In the logics employed, it is equivalent to $O(E_a R \land E_b R)$, but there is no principle in the logic of action to say that a and b could not both act in such a way that they both see to it that R.

Are there any other possibilities besides the three listed above? It is the systematic exploration of all such possible relations that motivates in large part the construction of the Kanger-Lindahl theories.

Notice that the three possibilities above may be distinguished by asking in turn whether $PE_a R$ is true, and if so, whether $P\neg E_a R$ is true. This is the kind of analysis that the automated system described in [Sergot, 2001] is designed to support.

Example 3.2 (Fence) *The following example is adapted from [Lindahl, 1977]. Again, no claim is made here for the completeness or adequacy of the representation. The aim is merely to illustrate some of the distinctions and nuances that can be expressed with the resources available.*

Suppose a and b are neighbours, and let F represent that there is a fence on the boundary between their adjoining properties. We want to say that a has a 'right' to erect such a fence, or more generally, that a has a 'right' to see to it that there is such a fence.

We build up a (partial) representation in stages. In the first instance it seems reasonable to assert that the following is true:

(2) $$PE_a F \land \neg PE_b \neg F$$

The second conjunct of expression (2) captures something of the idea that the neighbour b is not permitted to *prevent* a from seeing to it that F. One could also add a conjunct $\neg PE_b \neg E_a F$ to cover a different sense in which b is forbidden to prevent a from seeing to it that F. The ability to iterate action operators in this fashion has been seen as one of the main advantages of using the E_x device in the treatment of action. 'x refrains from seeing to it that F' can be represented as $E_x \neg E_x F$, for example. We shall not study iterated act expressions in any detail in this article, however. Some examples and some possible lines of development are discussed briefly in Section 8. Iterated act expressions are the basis of the 'control' and 'influence' positions examined in [Pörn, 1977].

Of course a is not *obliged* to see to it that F, so also $\neg OE_a F$ is true in the example. Furthermore, a's permission to see to it that F does not depend on b's actions, in the sense that the following is also true: $P(E_a F \land \neg E_b F)$. Putting these together:

(3) $$PE_a F \land \neg PE_b \neg F \land \neg OE_a F \land P(E_a F \land \neg E_b F)$$

Expression (3) is an approximation to the concept of a 'vested right'. It is only an approximation because as already observed there are other possible ways in which b can be said to 'prevent' a's seeing to it that F, e.g. as expressed by $E_b \neg E_a F$. It also fails to capture the idea that a's rights may already be infringed by *unsuccessful* attempts by b to *interfere* with a's actions [Makinson, 1986]. Moreover (3) does not say what further rights and obligations are created if b should so interfere.

In this example b's normative status in relation to F is clearly symmetrical to a's and so we may add also:

(4) $\qquad PE_b F \land \neg PE_a \neg F \land \neg OE_b F \land P(E_b F \land \neg E_a F)$

Still there are a number of unresolved questions. Is it the case that $P \neg F$, or is it obligatory, $\neg P \neg F$, that there is a fence? Is it the case that $P(\neg F \land \neg E_a \neg F \land \neg E_b \neg F)$: is it permitted that there is no fence when neither a nor b brought this about? As a matter of fact, in the logics employed (3) and (4) together imply

(5) $\qquad P \neg F \leftrightarrow P(\neg F \land \neg E_a \neg F \land \neg E_b \neg F)$

i.e., it is obligatory that there is a fence iff $O(\neg F \rightarrow (E_a \neg F \lor E_b \neg F))$ is true. On the other hand, (3) and (4) together do not imply $P(F \land \neg E_a F \land \neg E_b F)$. That question remains unresolved. Perhaps some other agent, besides a and b, is permitted to see to it that there is a fence between their adjoining properties, perhaps not.

The example is intended in part to demonstrate why there is a need for automated support even for the analysis of simple examples. Questions such as those above can can be explored systematically by means of the automated inference methods described in [Sergot, 2001] and summarised in Section 7 below.

The fence example also demonstrates that there may be an obligation on a and b together, without there being an obligation on either of them individually: it is possible that $O(E_a F \lor E_b F)$ is true while both $OE_a F$ and $OE_b F$ are false.

Example 3.3 (Car park) *Ronald Lee [1988] presents a rule-based language intended for specifying permitted, obligatory and forbidden actions. The example used for illustration concerns the rules governing a University car park. For simplicity "assume that administrators have unrestricted parking privileges. Faculty, however, must obtain a parking permit to park on campus. Students must park off campus." Lee represents such rules in the form of if/then rules whose antecedent ('body') is a conjunction of* *factual conditions ('is an administrator', 'has a parking permit', etc.) and whose consequent specifies an action (here, 'park') that can be permitted, obligatory, or prohibited.*

Leaving aside the details of the language, one might ask whether these primitives 'permitted', 'obligatory', 'prohibited' are enough, whether they cover all imaginable cases. Notice first that they are not mutually exclusive: an obligatory action is also (presumably) permitted. It may be that the primitive 'permitted' in Lee's language was intended to be understood in the sense of permissible but not obligatory, or what is sometimes referred to as 'facultative'. In the logic we are using, A is facultative when $PA \land \neg OA$ is true, or equivalently when $PA \land P\neg A$ is true.

One can see a very close connection between this rule based representation and the conception of a normative system as introduced and developed in Alchourrón and Bulygin's classical work [1971]. There a normative system $\mathcal{N}$ is defined in terms of a 'universe of cases'—these are all the possible fact combinations that can be expressed using some fixed set $\mathcal{P}rops$ of propositional atoms—and a set of actions. For each action and case (set of factual circumstances) a normative system assigns a 'solution' which specifies whether that action is obligatory, prohibited, or facultative in that factual circumstance. The normative system is consistent when no case is assigned different solutions for any given action, and complete when every case is assigned a solution for every action.

But again, taking a relational perspective, one is led to think in terms of interactions between the administrator who is permitted to park and other agents: other users of the car park, passers by, the gatekeepers who control access to the car park, the University who owns the car park and to whom the gatekeepers are responsible, and so on. An analysis based on the Hohfeldian scheme, for example, would ask not whether there is a permission to park *simpliciter* but whether the administrator has a 'privilege' to park or whether this is in fact a 'claim-right' (vis-à-vis, in turn, other users of the car park, the gatekeepers, the University). And likewise for other pairs of agents.

If in place of the informal Hohfeldian scheme, we employ the formal machinery offered by the Kanger-Lindahl theories or the extended scheme of [Sergot, 2001], the if/then rules of the representation language would take the form

$$\text{if } conditions \text{ then } normative\text{-}position$$

where *normative-position* is one of some appropriately chosen class of normative positions. Lee's rule-based language, and solutions in Alchourrón and Bulygin's formalisation of a normative system, can be regarded as a special case where the class of candidate normative positions is a particularly simple one. For more precision, more complex classes of normative positions should be considered.

We will return to this point in Section 8 after the formal machinery has been introduced, and we will look again at the car park example in more detail in Section 7.

One might ask why anyone would be interested in representing the rules of a library or the rules of a car park at these levels of precision. One answer is that a precise specification may be essential if we were assigned the task of constructing a system that advises the employees and users of a library about their duties and rights, or if we were given the task of designing a system for controlling access to a car park. Or instead of controlling who may put cars in a car park, imagine for instance that the car park is a computer file of some kind, and that $p(x)$ represents not that car x is parked in the car park but that data entry x is stored in the file. The task is then to specify with precision which agents (computer agents or human) are to be permitted to insert and delete data entries in this file, in which circumstances and in which combinations. A gatekeeper agent g who controls access to a car park is not so different from a 'file monitor' (human or computer agent) which controls access to a computer file. And likewise for many of the other forms of interactions that take place in regulated human and electronic societies.

4 The Kanger-Lindahl theory

The focus in the Kanger-Lindahl theory is on mapping out the space of logically possible legal/normative relations of given forms that can hold between pairs of agents. In order to examine the possibilities systematically, Kanger considers first what he called the 'simple types of rights relations' of two agents a and b with respect to some state of affairs F. They are represented by the expressions falling under the scheme:

$$(6) \qquad \pm O \pm \begin{pmatrix} E_a \\ E_b \end{pmatrix} \pm F$$

The notation was suggested by David Makinson [1986]. $\pm$ stands for the two possibilities of affirmation and negation; the *choice-scheme* $\begin{pmatrix} E_a \\ E_b \end{pmatrix}$ indicates the (here, two) alternatives E_a and E_b. There are thus sixteen expressions falling under the scheme (6), ranging from $OE_a F$ to $\neg O \neg E_b \neg F$. The choice-scheme notation can be seen as shorthand for a *set* of expressions and so will be mixed freely with standard set notation.

The 'simple types' were given names by Kanger in addition to their symbolic explication. Following Lindahl's summary [1994], from the perspective of a's rights versus b, those in the scheme $O \pm E_b \pm F$ are called *Claim, Counter-claim, Immunity, Counter-immunity*; those in the scheme $\neg O \pm E_a \pm F$ (equivalently, $P \pm E_a \pm F$) are called *Power, Counter-power, Freedom, Counter-freedom*. The Appendix of Henning Herrestad's doctoral dissertation [1996] lists out the correspondence between names of the 'simple types' and their symbolic expression. We will not reproduce the details here since the naming scheme is of less importance than the symbolic scheme. We note only that the choice of some of these names is unfortunate, since they do not all correspond to Hohfeld's terminology. 'Power' in particular means something quite different in the Hohfeldian scheme (it is to do with the capacity or competence to effect *changes* in rights relations).

Of more interest than the 'simple types' are the various compounds that may be formed from them, or what Kanger called the 'atomic types of rights relation'. Makinson's observation [1986] was that Kanger's 'atomic types', for two agents a, b with respect to the bringing about of some state of affairs F, can be characterised as the expressions belonging to the set:

(7) $$\left[\!\!\left[\pm O \pm \begin{pmatrix} E_a \\ E_b \end{pmatrix} \pm F \right]\!\!\right]$$

The brackets denote *maxi-conjunctions*: where Φ is a choice-scheme (or set of sentences) $[\![\Phi]\!]$ stands for the set of *maxi-conjunctions* of Φ —the maximal consistent conjunctions of expressions belonging to Φ. 'Consistent' refers to some underlying logic, here the specific logics for O and E_x employed by Kanger and Lindahl. 'Conjunction' means a conjunction without repetitions, and with some standard order and association of conjuncts. A conjunction is 'maximal consistent' when addition of any other conjunct from Φ yields an inconsistent conjunction: in other words, a conjunction Γ is a maxi-conjunction of Φ if and only if Γ is consistent, and every expression of Φ either appears as a conjunct in Γ or is inconsistent with Γ. Note that maxi-conjunctions may contain logical redundancies (one or more conjuncts may be logically implied by the others). We shall occasionally abuse the notation and write also $[\![\Phi]\!]$ for the set of conjunctions obtained by removing all logical redundancies from the maxi-conjunctions of Φ. A justification for this practice will be provided in later sections.

As can readily be checked, and will be shown more generally later (Theorem 4.1), Kanger's 'atomic types' (7) can be written as conjunctions of two simpler expressions:

(8) $$\left[\!\!\left[\pm O \pm \begin{pmatrix} E_a \\ E_b \end{pmatrix} \pm F \right]\!\!\right] = \left[\!\!\left[\pm O \pm E_a \pm F \right]\!\!\right] \cdot \left[\!\!\left[\pm O \pm E_b \pm F \right]\!\!\right]$$

Here the notation is as follows: when **P** and **Q** represent sets of expressions, **P**·**Q** stands for the set of all the *consistent* conjunctions that can be formed by conjoining an expression from set **P** with an expression from set **Q**. (For technical reasons, it is convenient to take $\mathbf{P} \cdot \emptyset =_{\text{def}} \emptyset \cdot \mathbf{P} =_{\text{def}} \mathbf{P}$.) In order to reduce the need for parentheses, we adopt the convention that the · binds more tightly than other operators. So, for example, the choice-scheme expression $(\pm O \pm \Phi_1 \cdot \Phi_2)$ is to be read as $(\pm O \pm (\Phi_1 \cdot \Phi_2))$.

The maxi-conjunctions in

(9) $$[\pm O \pm E_a \pm F]$$

are, in the terminology of [Jones and Sergot, 1993], Kanger's *normative one-agent act positions*. According to the logic employed by Kanger, there are six elements in (9). Following the numbering at [Lindahl, 1977, p. 100] and eliminating logical redundancies, they are:

(K$_1$) $PE_a F \wedge PE_a \neg F$
(K$_2$) $O \neg E_a F \wedge O \neg E_a \neg F$
(K$_3$) $OE_a F$
(K$_4$) $PE_a F \wedge P \neg E_a F \wedge O \neg E_a \neg F$
(K$_5$) $OE_a \neg F$
(K$_6$) $O \neg E_a F \wedge PE_a \neg F \wedge P \neg E_a \neg F$

These six expressions, by construction, are consistent, mutually exclusive, and their disjunction is a tautology. In any given situation precisely one of them must be true, according to the logical principles employed.

One can see that (K$_1$)–(K$_6$) are symmetric in F and $\neg F$ (as is obvious from the form of the expression (9)). (K$_3$) expresses an obligation on a, in the Meinong-Chisholm sense, to bring it about that F. In (K$_1$) a is permitted to bring it about that F and permitted to bring it about that $\neg F$. (K$_2$) can be written equivalently in a number of different ways.

(K$_2'$) $\neg PE_a F \wedge \neg PE_a \neg F$

says that a is neither permitted to bring it about that F nor permitted to bring it about that $\neg F$. Following Lindahl, it is convenient to define the following abbreviation:

(10) $$\text{Pass}_a F =_{\text{def}} \neg E_a F \wedge \neg E_a \neg F$$

$\text{Pass}_a F$ represents a kind of 'passivity' of agent a with respect to state of affairs F. (K$_2$) can be written equivalently as:

(K$_2''$) $O(\neg E_a F \wedge \neg E_a \neg F) = O\text{Pass}_a F$

and so expresses an obligation on a to remain 'passive' with respect to F. (K_4) is equivalent to

$$(K_4') \quad PE_a F \wedge P \neg E_a F \wedge \neg PE_a \neg F$$

According to (K_4'), a is permitted to bring it about that F and permitted to refrain from bringing it about that F, but a is not permitted to bring it about that $\neg F$.

For Kanger's 'atomic types' for two agents, expression (8), there are $6 \times 6 = 36$ conjunctions to consider. Of these, 10 turn out to be logically inconsistent. On Kanger's analysis, therefore, there are 26 atomic types of right (for two agents with respect to the bringing about of some given state of affairs). Again, by construction these 26 'atomic types' are internally consistent, mutually exclusive, and their disjunction is a tautology. In any given situation precisely one of them must be true, according to the logics employed. It is in this sense that Kanger can be said to provide a complete and exhaustive analysis of all the logically possible normative positions.

Kanger's 26 'atomic types' are listed in full in [Kanger and Kanger, 1966, pp. 93–4] and [Lindahl, 1977, p. 56] and in several other works. In these works however each position (atomic type) is described by listing the names (i.e., claim, freedom, power, etc) of the constituent single-agent types rather than the symbolic expressions.

For example, the first of the 26 atomic types in the standard table is listed as 'Power, not Immunity, Counter-power, not Counter-immunity' which corresponds to the conjunction of one-agent act positions (K_1) for a and (K_1) for b and thus the symbolic expression:

$$PE_a F \wedge PE_a \neg F \wedge PE_b F \wedge PE_b \neg F$$

The 15th atomic type in the table, to pick just one other example, is listed as 'Liberty, not Power, Immunity, Counter-power, Counter-immunity'. This corresponds to the conjunction of one-agent act positions (K_6) for a and (K_2) for b and thus the symbolic expression:

$$O \neg E_a F \wedge PE_a \neg F \wedge P \neg E_a \neg F \wedge O \neg E_b F \wedge O \neg E_b \neg F$$

The complete listing and numbering used by the previous authors together with the corresponding symbolic expressions in each case can be found in [Herrestad, 1996, Appendix].

Each of Kanger's 26 atomic types can be expressed as a conjunction of two of the 6 single-agent types (K_1)–(K_6) by virtue of equation (8) (and Theorem 4.1 below).

Kanger gives a complete and exhaustive analysis of all the logically possible atomic types. In general, all maxi-conjunctions of the form $[\![\pm\,\Phi]\!]$ have this property of exhaustiveness. Moreover, all (consistent) boolean compounds of expressions in Φ are logically equivalent to a (non-empty) disjunction of elements from $[\![\pm\,\Phi]\!]$. As observed by Makinson [1986], the maxi-conjunctions can be given an algebraic interpretation (as atoms of a Boolean algebra). For certain logics (those of type *EMCP*, though not for weaker ones), they give the constituents of a distributive normal form in the underlying modal logics. (They are not quite yet a normal form: for that we would need to consider not just the sentences of Φ but also all of their subsentences.)

The value of Makinson's suggestion, besides the conciseness of the notation, is that the characterisation of positions in terms of maxi-conjunctions emphasises their character rather than the specific procedures by which they happen to be generated. There are many different ways of generating the same set of maxi-conjunctions. The following elementary property of maxi-conjunctions is particularly useful, and is the basis for a whole family of such procedures.

Theorem 4.1 *For any choice scheme* $\Phi = \Phi_1 \cup \Phi_2$ *(Φ_1 and Φ_2 not necessarily distinct):*

1. $[\![\Phi_1]\!] \cdot [\![\Phi_2]\!] \subseteq [\![\Phi]\!]$
2. $[\![\pm\,\Phi]\!] = [\![\pm\,\Phi_1]\!] \cdot [\![\pm\,\Phi_2]\!]$

Proof. Straightforward. See [Sergot, 2001]. ∎

Computationally: to generate the set of maxi-conjunctions $[\![\pm\Phi]\!]$, decompose the scheme (or set of sentences) Φ into smaller, not necessarily disjoint, subsets Φ_1 and Φ_2 (there are many different strategies for this step); (recursively) compute the sets of maxi-conjunctions $[\![\pm\,\Phi_1]\!]$ and $[\![\pm\,\Phi_2]\!]$, possibly in parallel; form all conjunctions of expressions from these sets of maxi-conjunctions; discard those conjunctions that are logically inconsistent. The steps, especially the last two steps, may be co-routined for efficiency. It is straightforward to code any such procedure as a computer program, requiring only an implementation of the inconsistency check for the generated conjunctions. Although this is not difficult—it is only fragments of the underlying modal logics that are required—it is not particularly useful either. In Section 6 we show how a little additional manipulation eliminates the need for theorem-proving techniques altogether, at least for the most common types of modal logic.

As an example, the method used to generate classes of normative positions in [Jones and Sergot, 1993] (and in [Jones and Parent, 2008]) is a special case of Theorem 4.1. For illustration, in [Jones and Sergot, 1993] the generation of what are there called the 'normative fact positions'

(11) $$[\![\pm O \pm F]\!]$$

proceeds as follows. Form two tautologies $OF \vee \neg OF$ and $O\neg F \vee \neg O\neg F$. Their conjunction is another tautology. Re-write it as a disjunction of conjunctions by picking one disjunct from each in all combinations, to obtain $(OF \wedge O\neg F) \vee (OF \wedge \neg O\neg F) \vee (\neg OF \wedge O\neg F) \vee (\neg OF \wedge \neg O\neg F)$. The first disjunct of this expression is logically inconsistent and so can be deleted; the others can be simplified. That procedure can be presented as a special case of Theorem 4.1 as follows:

$$\begin{aligned}
[\![\pm O \pm F]\!] &= [\![\pm OF]\!] \cdot [\![\pm O\neg F]\!] \quad \text{(by Theorem 4.1)} \\
&= \begin{pmatrix} OF \\ \neg OF \end{pmatrix} \cdot \begin{pmatrix} O\neg F \\ \neg O\neg F \end{pmatrix} \\
&= \begin{pmatrix} OF \\ O\neg F \\ PF \wedge P\neg F \end{pmatrix} \quad \text{(with logical redundancies removed)}
\end{aligned}$$

Equation (8) expressing Kanger's two-agent atomic types as conjunctions of one-agent types is also a special case of Theorem 4.1. This follows immediately from:

$$\pm O \pm \begin{pmatrix} E_a \\ E_b \end{pmatrix} \pm F \;=\; \pm O \pm E_a \pm F \;\cup\; \pm O \pm E_a \pm F$$

There will be other examples presently.

Lars Lindahl [1977] presents a refinement and further development of Kanger's analysis. The second part of his book deals also with aspects of 'change' of normative positions. That part of Lindahl's account will not be pursued here.

Lindahl constructs his analysis on the following set of normative one-agent act positions:

(12) $$[\![\pm P [\![\pm E_a \pm F]\!]\,]\!]$$

where now there is a maxi-conjunction expression within the scope of the P operator. In words, (12) is the set of maxi-conjunction expressions of the form $\pm PA$, where each A is itself a maxi-conjunction of sentences of the form $\pm \mathrm{E}_a \pm F$. The iterated bracket notation is again from [Makinson, 1986].

There are three *act positions* in the set

(13) $$[\![\pm \mathrm{E}_a \pm F]\!]$$

They are:

(A_1) $\mathrm{E}_a F$
(A_2) $\mathrm{E}_a \neg F$
(A_3) $\neg \mathrm{E}_a F \wedge \neg \mathrm{E}_a \neg F$

The third of these (A_3) is the 'passivity' of agent a with respect to state of affairs F, which following Lindahl we also write using the abbrevation $\mathrm{Pass}_a F$.

There are $2^3 - 1 = 7$ expressions in the set (12). They are, numbered as in [Lindahl, 1977] and with logical redundancies removed:

(T_1) $\mathrm{PE}_a F \wedge \mathrm{PE}_a \neg F \wedge \mathrm{PPass}_a F$
(T_2) $\mathrm{PE}_a F \wedge \mathrm{O} \neg \mathrm{E}_a \neg F \wedge \mathrm{PPass}_a F$
(T_3) $\mathrm{PE}_a F \wedge \mathrm{PE}_a \neg F \wedge \neg \mathrm{PPass}_a F$
(T_4) $\mathrm{O} \neg \mathrm{E}_a F \wedge \mathrm{PE}_a \neg F \wedge \mathrm{PPass}_a F$
(T_5) $\mathrm{OE}_a F$
(T_6) $\mathrm{OPass}_a F$
(T_7) $\mathrm{OE}_a \neg F$

(T_2) and (T_4) can be written equivalently as:

(T_2') $\mathrm{PE}_a F \wedge \neg \mathrm{PE}_a \neg F \wedge \mathrm{PPass}_a F$
(T_4') $\neg \mathrm{PE}_a F \wedge \mathrm{PE}_a \neg F \wedge \mathrm{PPass}_a F$

Lindahl's construction gives a finer-grained analysis than Kanger's. For the one-agent types, five of the six in Kanger's (9) are logically equivalent to five of the seven in Lindahl's (12), as summarized in Table 1.

Normative Positions

K_1	is logically equivalent to	$(T_1 \vee T_3)$
K_2	... ...	T_6
K_3	... ...	T_5
K_4	... ...	T_2
K_5	... ...	T_7
K_6	... ...	T_4

Table 1: Normative one-agent act positions

On Lindahl's analysis, therefore, Kanger's type (K_1) can be decomposed:

(K_1) $\quad \mathrm{PE}_a F \wedge \mathrm{PE}_a \neg F$

is logically equivalent to a disjunction of two of Lindahl's types, viz.

(T_1) $\quad \mathrm{PE}_a F \wedge \mathrm{PE}_a \neg F \wedge \mathrm{PPass}_a F$
(T_3) $\quad \mathrm{PE}_a F \wedge \mathrm{PE}_a \neg F \wedge \neg \mathrm{PPass}_a F$

For an example of (T_3), consider a judge (a) who is permitted to see to it that the prisoner is imprisoned (F) and permitted to see to it that the prisoner is not imprisoned ($\neg F$); but a is not permitted to do neither of these: $\neg \mathrm{PPass}_a F$.

In place of Kanger's two-agent types (8), Lindahl has the following set of positions:

(14) $\qquad \left[\!\left[\pm \mathrm{P} \left[\!\left[\pm \mathrm{E}_a \pm F \right]\!\right] \right]\!\right] \cdot \left[\!\left[\pm \mathrm{P} \left[\!\left[\pm \mathrm{E}_b \pm F \right]\!\right] \right]\!\right]$

There are $7 \times 7 = 49$ conjunctions to consider, of which 35 are internally consistent. These are Lindahl's 'individualistic' normative two-agent act positions. The significance of 'individualistic' will be explained in a moment. Lindahl's construction again gives a finer-grained analysis than Kanger's: some of Kanger's 26 two-agent 'atomic types' (7) are logically equivalent to disjunctions of Lindahl's corresponding 35 types (14). We omit the details: the next section presents a general result and a computational method to perform this kind of calculation.

Notice that, since P is the dual of O, Kanger's one-agent positions (9) may be written equivalently as $\left[\!\left[\pm \mathrm{P} \pm \mathrm{E}_a \pm F \right]\!\right]$. The expression within the maxi-conjunction brackets may be seen in two ways: either as a scheme of *four* (not mutually exclusive) act positions $\pm \mathrm{E}_a \pm F$ prefixed by $\pm \mathrm{P}$, or as *two* mutually exclusive act positions $\mathrm{E}_a \pm F$ prefixed by $\pm \mathrm{P} \pm$. What is obtained by combining the second view, $\pm \mathrm{P} \pm$, with the three mutually exclusive act positions $[\![\pm E_a \pm F]\!]$ used by Lindahl? In other words, consider the following:

(15) $\quad [\![\pm P \pm [\![\pm E_a \pm F]\!]]\!] = [\![\pm O \pm [\![\pm E_a \pm F]\!]]\!]$

(The equality here is because P and O are duals.) This is the construction used in Jones and Sergot's account of normative positions [1992; 1993]. It turns out that for the logics employed by Kanger and Lindahl the positions in set (15) are exactly the same seven as those in Lindahl's simpler form (12). By Theorem 4.1 the following holds irrespective of the logic of O:

(16) $\quad [\![\pm O \pm [\![\pm E_a \pm F]\!]]\!] = [\![\pm P [\![\pm E_a \pm F]\!]]\!] \cdot [\![\pm O [\![\pm E_a \pm F]\!]]\!]$

But when the logic of O is of type *EMCP* (or stronger), then also (as shown later in Section 6, Theorem 6.1):

(17) $\quad [\![\pm O \pm [\![\pm E_a \pm F]\!]]\!] = [\![\pm P [\![\pm E_a \pm F]\!]]\!]$

For weaker logics the equality (17) does not hold. In that case the Jones-Sergot form (15) gives a more refined analysis than Lindahl's (12).

There is another important respect in which Lindahl extends Kanger's analysis of two-agent 'atomic types'. In [Lindahl, 1977, Ch. 5] the account is extended to what are called 'collectivistic two-agent types', to cover the case where, for instance, there is an obligation on two agents which does not apply to either of them individually:

$$O(E_a F \vee E_b F) \wedge \neg O E_a F \wedge \neg O E_b F$$

Lindahl is there addressing the *co-ordination* of a and b's actions, which introduces distinctions that cannot be expressed by conjunctions of the 'individualistic' types (14). The reason is simply that, in the logics employed, P does not distribute over conjunction (nor O over disjunction): $(PA \wedge PB) \to P(A \wedge B)$ is *not* a theorem for arbitrary A and B. For instance, $PE_a F \wedge PE_b F$ is consistent with both $P(E_a F \wedge E_b F)$ and $\neg P(E_a F \wedge E_b F)$.

Lindahl's 'collectivistic' two-agent positions are obtained by the following construction:

(18) $\quad \left[\!\!\left[\pm P \left[\!\!\left[\pm \binom{E_a}{E_b} \pm F\right]\!\!\right]\right]\!\!\right] = \left[\!\!\left[\pm P \left([\![\pm E_a \pm F]\!] \cdot [\![\pm E_b \pm F]\!]\right)\right]\!\!\right]$

In the *EMCP*-equivalent Jones-Sergot form these positions are:

(19) $\quad \left[\!\!\left[\pm O \pm \left[\!\!\left[\pm \binom{E_a}{E_b} \pm F\right]\!\!\right]\right]\!\!\right] = \left[\!\!\left[\pm O \pm \left([\![\pm E_a \pm F]\!] \cdot [\![\pm E_b \pm F]\!]\right)\right]\!\!\right]$

For the logics employed by Kanger and Lindahl, there are $2^7 - 1 = 127$ 'collectivistic normative two-agent act positions' in the sets (18) and (19). Each collectivistic type implies one of the 'individualistic' types (14); each of the 'individualistic' types is logically equivalent to a disjunction of one or more of the collectivistic types. This can be seen by reference to the table compiled by [Lindahl, 1977, p. 180], or, as shown in later sections, from a general property of maxi-conjunctions which holds when the logic of O is of type *EMCP*.

[Sergot, 2001] presents a generalised theory of normative positions that builds upon Makinson's maxi-conjunction characterisation. It is summarised in the next two sections, and addresses the following questions in particular:

(1) How can the account be generalised to the case of n agents? This is a possibility mentioned by Lindahl but not developed by him, presumably because of the size and number of the symbolic expressions to be manipulated.

(2) How can the account be generalised to deal with related states of affairs, in the same kind of way that the 'collectivistic' positions generalise the 'individualistic'? Consider two neighbours, a and b. Let F represent that there is a fence at the front of their adjoining properties, and G that there is a fence at the back of their properties. Suppose both neighbours are permitted to see to it that there is a fence at the front, $\mathrm{PE}_a F \wedge \mathrm{PE}_b F$, and permitted to see to it that there is a fence at the back, $\mathrm{PE}_a G \wedge \mathrm{PE}_b G$. We might nevertheless want to distinguish between the case represented by $\mathrm{P}(\mathrm{E}_a F \wedge \mathrm{E}_a G) \wedge \mathrm{P}(\mathrm{E}_b F \wedge \mathrm{E}_b G)$ and the case represented by $\neg \mathrm{P}(\mathrm{E}_a F \wedge \mathrm{E}_a G) \wedge \neg \mathrm{P}(\mathrm{E}_b F \wedge \mathrm{E}_b G)$. It is conceivable that there could be other constraints, such as that represented by $\mathrm{O}(\mathrm{E}_a F \leftrightarrow \mathrm{E}_a G)$, i.e. $\neg \mathrm{P}(\mathrm{E}_a F \wedge \neg \mathrm{E}_a G) \wedge \neg \mathrm{P}(\mathrm{E}_a G \wedge \neg \mathrm{E}_a F)$. These distinctions cannot be expressed in the Kanger-Lindahl framework.

(3) To what extent can these various constructions be generalised to other, weaker logics than those employed by Kanger and Lindahl? Which features of the theory are properties of the specific logics employed, and which of maxi-conjunctions in general?

(4) Lindahl's construction yields a finer-grained analysis than Kanger's. Is there similarly a finer-grained analysis than Lindahl's? Is there a finest analysis?

The last question can be answered as follows. For one agent a and one state of affairs F, Lindahl bases his analysis on the set of three act positions $[\![\pm \mathrm{E}_a \pm F]\!]$. But a finer analysis can be obtained by taking instead the act positions from the following scheme:

(20) $$[\pm E_a \pm F] \cdot [\pm F]$$

We might call these 'cumulative fact/act positions'. There are four such positions:

(A$_1$) $E_a F$
(A$_2$) $E_a \neg F$
(A$_{3a}$) $F \wedge \neg E_a F$ (which is equivalent to $\text{Pass}_a F \wedge F$)
(A$_{3b}$) $\neg F \wedge \neg E_a \neg F$ (which is equivalent to $\text{Pass}_a F \wedge \neg F$)

Lindahl's 'passive' act position (A$_3$) does not distinguish between (A$_{3a}$) and (A$_{3b}$).

The corresponding single-agent 'normative act positions' are:

(21) $$\big[\pm O \pm [\pm E_a \pm F] \cdot [\pm F]\big]$$

There are $2^4 - 1 = 15$ conjunctions in the set (21), as compared with the seven (T$_1$)–(T$_7$) constructed in Lindahl's analysis. They are listed in Table 2. Three are identical to Lindahl's (T$_3$), (T$_5$) and (T$_7$); the other four of Lindahl's types are each logically equivalent to a disjunction of three conjunctions from (21). Just as Lindahl is able to give examples to illustrate the ambiguity in Kanger's type (K$_1$), so it is easy to find examples to illustrate the ambiguities in Lindahl's types (T$_1$), (T$_2$), (T$_4$), (T$_6$). Consider (T$_1$) for example, and suppose that a neighbour a is permitted to see to it that there is a fence (F), permitted to see to it that there is no fence, and permitted to remain passive with respect to there being a fence. It may be, however, that if there is a fence then a must see to it, in other words that $O(F \rightarrow E_a F)$, equivalently $\neg P(F \wedge E_a F)$, is true. That possibility is covered by the second of the (T$_1$) refinements in Table 2 but not by the other two.

For two-agent positions, the corresponding expressions for 'individualistic' and 'collectivistic' positions are, respectively:

(22) $$\big[\pm O \pm [\pm E_a \pm F] \cdot [\pm F]\big] \cdot \big[\pm O \pm [\pm E_b \pm F] \cdot [\pm F]\big]$$

(23) $$\Big[\pm O \pm \big[\pm \binom{E_a}{E_b} \pm F\big] \cdot [\pm F]\Big] = \big[\pm O \pm [\pm E_a \pm F] \cdot [\pm E_b \pm F] \cdot [\pm F]\big]$$

T_1	$\begin{cases} \mathrm{PE}_a F \wedge \mathrm{PE}_a \neg F \wedge \mathrm{P}(F \wedge \neg \mathrm{E}_a F) \wedge \mathrm{P}(\neg F \wedge \neg \mathrm{E}_a \neg F) \\ \mathrm{PE}_a F \wedge \mathrm{PE}_a \neg F \wedge \neg \mathrm{P}(F \wedge \neg \mathrm{E}_a F) \wedge \mathrm{P}(\neg F \wedge \neg \mathrm{E}_a \neg F) \\ \mathrm{PE}_a F \wedge \mathrm{PE}_a \neg F \wedge \mathrm{P}(F \wedge \neg \mathrm{E}_a F) \wedge \neg \mathrm{P}(\neg F \wedge \neg \mathrm{E}_a \neg F) \end{cases}$
T_2	$\begin{cases} \mathrm{PE}_a F \wedge \neg \mathrm{PE}_a \neg F \wedge \mathrm{P}(F \wedge \neg \mathrm{E}_a F) \wedge \mathrm{P}(\neg F \wedge \neg \mathrm{E}_a \neg F) \\ \mathrm{PE}_a F \wedge \neg \mathrm{PE}_a \neg F \wedge \neg \mathrm{P}(F \wedge \neg \mathrm{E}_a F) \wedge \mathrm{P}(\neg F \wedge \neg \mathrm{E}_a \neg F) \\ \mathrm{PE}_a F \wedge \neg \mathrm{PE}_a \neg F \wedge \mathrm{P}(F \wedge \neg \mathrm{E}_a F) \wedge \neg \mathrm{P}(\neg F \wedge \neg \mathrm{E}_a \neg F) \end{cases}$
T_3	$\{\mathrm{PE}_a F \wedge \mathrm{PE}_a \neg F \wedge \neg \mathrm{PPass}_a F$
T_4	$\begin{cases} \neg \mathrm{PE}_a F \wedge \mathrm{PE}_a \neg F \wedge \mathrm{P}(F \wedge \neg \mathrm{E}_a F) \wedge \mathrm{P}(\neg F \wedge \neg \mathrm{E}_a \neg F) \\ \neg \mathrm{PE}_a F \wedge \mathrm{PE}_a \neg F \wedge \neg \mathrm{P}(F \wedge \neg \mathrm{E}_a F) \wedge \mathrm{P}(\neg F \wedge \neg \mathrm{E}_a \neg F) \\ \neg \mathrm{PE}_a F \wedge \mathrm{PE}_a \neg F \wedge \mathrm{P}(F \wedge \neg \mathrm{E}_a F) \wedge \neg \mathrm{P}(\neg F \wedge \neg \mathrm{E}_a \neg F) \end{cases}$
T_5	$\{\mathrm{OE}_a F$
T_6	$\begin{cases} \mathrm{OPass}_a F \wedge \mathrm{O} F \\ \mathrm{OPass}_a F \wedge \mathrm{O} \neg F \\ \mathrm{OPass}_a F \wedge \mathrm{P} F \wedge \mathrm{P} \neg F \end{cases}$
T_7	$\{\mathrm{OE}_a \neg F$

Table 2: Normative one-agent cumulative fact/act positions

When the logic of O is of type *EMCP* or stronger, constructions (21) for one agent and (23) for any pair of agents are—effectively—the finest-grained set of normative positions that can be constructed for a given state of affairs, respectively. The next section explains what is meant by 'finest-grained'.

The account can be generalised, to any (finite) number of agents $\{a, b, \ldots\}$ not just two, and any (finite) number of separate states of affairs $\{F, G, \ldots\}$ not just one. Consider for instance the following construction:

$$(24) \quad \left[\pm \mathrm{O} \pm \left[\pm \begin{pmatrix} \mathrm{E}_a \\ \mathrm{E}_b \\ \vdots \end{pmatrix} \pm \begin{pmatrix} F \\ G \\ \vdots \end{pmatrix}\right] \cdot \left[\pm \begin{pmatrix} F \\ G \\ \vdots \end{pmatrix}\right]\right]$$

There are still more complex classes of normative positions if we allow also iterations of the action modalities. We will give some examples in Section 7 below.

5 Partitions

Lindahl's construction yields a finer-grained analysis than Kanger's. But Kanger's analysis is also exhaustive, in the sense that his 'atomic types' are logically consistent, mutually exclusive, and their disjunction is a tautology. Kanger's analysis and Lindahl's analysis are both exhaustive, but Lindahl's is finer than Kanger's. We now formalise these notions.

We begin by defining a syntactic version of the standard notion of a *partition* of a set whereby a set is partitioned into non-empty disjoint subsets. All definitions are given with respect to some underlying logic Λ. Since Λ is usually obvious from context we write $\vdash A$ for $A \in \Lambda$. The only assumption we make in this section is that Λ includes classical propositional logic, i.e. contains all tautologies PL and is closed under modus ponens.

Definition 5.1 *Let* $\mathbf{P} = \{P_1, P_2, \ldots\}$ *be a set of sentences and* Q *a sentence of the language of* Λ. *Then* $\mathbf{P} = \{P_1, P_2, \ldots\}$ *is a Λ-partition of* Q *iff it satisfies the following conditions:*

1. *every element P_i of $\mathbf{P}$ is logically consistent:* $\not\vdash \neg P_i$;

2. *every element P_i of $\mathbf{P}$ logically implies Q:* $\vdash P_i \to Q$;

3. *distinct elements of $\mathbf{P}$ are mutually exclusive:* $\vdash \neg(P_i \wedge P_j) \quad (i \neq j)$;

4. *the set $\mathbf{P}$ 'exhausts' Q:* $\vdash Q \to \bigvee_{P \in \mathbf{P}} P$.

Conditions (2) and (4) together are: $\vdash Q \leftrightarrow \bigvee_{P \in \mathbf{P}} P$.

When Q is a tautology we shall say that $\mathbf{P}$ is a complete Λ-partition, or simply a Λ-partition. Where context permits we omit the Λ-prefix and simply say 'partition'. In what follows partitions will be finite sets.

Example 5.2 *All of the following (the terminology is from [Jones and Sergot, 1993]) are (complete) partitions:*

- *fact positions:* $[\![\pm F]\!] = \{F, \neg F\}$;

- *Lindahl's one-agent act positions:*

$$[\![\pm \mathrm{E}_a \pm F]\!] = \{\mathrm{E}_a F, \mathrm{E}_a \neg F, \mathrm{Pass}_a F\};$$

- *normative fact positions:* $[\![\pm \mathrm{O} \pm F]\!] = \{\mathrm{O}F, \mathrm{O}\neg F, \mathrm{P}F \wedge \mathrm{P}\neg F\}$;

- *Lindahl's normative one-agent act positions (T_1)–(T_7):*

$$\left[\!\!\left[\pm \mathrm{P} [\![\pm \mathrm{E}_a \pm F]\!]\right]\!\!\right]$$

In general, any maxi-conjunction of the form $[\![\pm \Phi]\!]$ is a (complete) partition. In contrast:

- The act positions used by Kanger, $\pm\mathrm{E}_a \pm F$, are not mutually exclusive, whereas $\mathrm{E}_a \pm F = \{\mathrm{E}_a F, \mathrm{E}_a \neg F\}$ are mutually exclusive but do not form a complete partition.

Naturally, if $\{P_1, \ldots, P_n\}$ is a set of consistent, mutually exclusive sentences, then $\{P_1, \ldots, P_n\}$ is a partition of $P_1 \vee \ldots \vee P_n$.

Λ-partitions are just syntactic analogues of the standard notion of a partition of a set. The two are easily related. For any model $\mathcal{M}$ of Λ, let $\|Q\|^{\mathcal{M}}$ denote the 'truth set' of Q, i.e. the set of possible worlds of $\mathcal{M}$ at which Q is true. The exact structure of $\mathcal{M}$ does not matter. Then the set of sentences $\mathbf{P} = \{P_1, P_2, \ldots\}$ is a Λ-partition of Q when, for all models $\mathcal{M}$ of Λ, the sets $\|P_1\|^{\mathcal{M}}, \|P_2\|^{\mathcal{M}}, \ldots$ partition the set $\|Q\|^{\mathcal{M}}$.

In view of this observation, it would be possible to eliminate the need for Definition 5.1 altogether and use instead the set-theoretic language indicated above, identifying each sentence with the set of all maximal consistent sets that contain it, and taking the notion of partition in its ordinary set-theoretic sense. We will stick to the syntactic version of Definition 5.1, however, because its application is more immediate in the present context. Furthermore, given a set of sentences, it is still necessary to check whether they constitute a partition, and for this purpose Definition 5.1 is more useful. We record in this section a number of properties of (syntactic) partitions that will be used later. All of them are easy to check, either directly from Definition 5.1 or by translating first to the set-theoretic analogue.

Proposition 5.3 *Let $\mathbf{P}$ and $\mathbf{Q}$ be partitions of some sentence R. Then the set of conjunctions $\mathbf{P} \cdot \mathbf{Q}$ is non-empty and is also a partition of R.*

In the above, $\mathbf{P} \cdot \mathbf{Q}$ must be non-empty, else R is logically inconsistent and $\mathbf{P}$ and $\mathbf{Q}$ could not be partitions. We now define some relations between partitions.

Definition 5.4 *Let $\mathbf{P}$ and $\mathbf{Q}$ be partitions of some sentence R. $\mathbf{P}$ and $\mathbf{Q}$ are equivalent ($\mathbf{P} \equiv \mathbf{Q}$) iff their elements are pairwise logically equivalent, i.e. iff there is a bijection $f: \mathbf{P} \to \mathbf{Q}$ such that $\vdash P \leftrightarrow f(P)$ for all elements P of $\mathbf{P}$.*

Definition 5.5 *Let $\mathbf{P}$ and $\mathbf{Q}$ be partitions of some sentence R. $\mathbf{P}$ is a refinement of $\mathbf{Q}$ ($\mathbf{P} \geq \mathbf{Q}$) iff every element of $\mathbf{P}$ logically implies some element of $\mathbf{Q}$:*

$$\mathbf{P} \geq \mathbf{Q} \quad \textit{iff} \quad \forall P \in \mathbf{P} \; \exists Q \in \mathbf{Q} \;\; \textit{such that} \;\; \vdash P \to Q.$$

When $\mathbf{P} \geq \mathbf{Q}$ we shall also say that partition $\mathbf{P}$ *refines* partition $\mathbf{Q}$.

Proposition 5.6 *Let* **P**, **Q**, **R** *be partitions of some sentence S.*

1. $\mathbf{P} \equiv \mathbf{Q}$ *iff* $\mathbf{P} \geq \mathbf{Q}$ *and* $\mathbf{Q} \geq \mathbf{P}$;

2. $\mathbf{P} \cdot \mathbf{Q} \geq \mathbf{P}$ *and* $\mathbf{P} \cdot \mathbf{Q} \geq \mathbf{Q}$;

3. $\mathbf{P} \cdot \mathbf{Q} \equiv \mathbf{P}$ *iff* $\mathbf{P} \geq \mathbf{Q}$;

4. *Moreover, the conjunction operator* · *is the 'meet' operator (glb) for partitions: if* $\mathbf{R} \geq \mathbf{P}$ *and* $\mathbf{R} \geq \mathbf{Q}$ *then* $\mathbf{R} \geq \mathbf{P} \cdot \mathbf{Q}$.

Example 5.7

- *Here is an instance of a general property to be established in a moment:*

$$[\![\pm P]\!] \cdot [\![\pm Q]\!] \geq [\![\pm P]\!]$$

- *[Lindahl, 1977, p. 100] provides a table comparing his atomic (one-agent) types with those of Kanger, reproduced as Table 1 above. From the table it is clear that Lindahl's types (which are a (complete) partition) are a refinement of Kanger's:*

$$[\![\pm P [\![\pm E_a \pm F]\!]]\!] \geq [\![\pm O \pm E_a \pm F]\!]$$

In later sections we shall be able to establish this relationship without having to compute these sets explicitly. It holds when the logic of O *is of type EMCP. See Example 5.9 and Theorems 6.1 and 6.3 below.*

- *The procedure used in [Jones and Sergot, 1993] constructs a set of maxi-conjunctions that is a refinement of Lindahl's normative one-agent act positions:*

$$[\![\pm O \pm [\![\pm E_a \pm F]\!]]\!] \geq [\![\pm P [\![\pm E_a \pm F]\!]]\!]$$

This is just a corollary of Theorem 4.1 and does not depend on the logic of O. *See Example 5.9 below. When the logic of* O *is of type EMCP we have also*

$$[\![\pm P [\![\pm E_a \pm F]\!]]\!] \geq [\![\pm O \pm [\![\pm E_a \pm F]\!]]\!]$$

i.e., an equivalence. See Theorem 6.1.

- Lindahl's 'collectivistic' two-agent types are a refinement of the 'individualistic' types:

$$\left[\pm P\left[\pm \begin{pmatrix} E_a \\ E_b \end{pmatrix} \pm F\right]\right] \geq \left[\pm P\left[\pm E_a \pm F\right]\right] \cdot \left[\pm P\left[\pm E_b \pm F\right]\right]$$

This can be seen by examination of the table compiled by [Lindahl, 1977, p. 180] but again it can be established, without evaluating the two expressions in full, by means of general properties of maxi-conjunctions. It holds when the logic of O is of type EMCP. See Theorems 6.1 and 6.3 below.

- Normative positions based on cumulative fact/act postions (21) are a refinement of Lindahl's normative one-agent act positions:

$$\left[\pm O \pm \left[\pm E_a \pm F\right] \cdot \left[\pm F\right]\right] \geq \left[\pm O \pm \left[\pm E_a \pm F\right]\right]$$
$$\geq \left[\pm P\left[\pm E_a \pm F\right]\right]$$

This can be seen by inspection of Table 2 above. It holds because $\left[\pm E_a \pm F\right] \cdot \left[\pm F\right] \geq \left[\pm E_a \pm F\right]$. In general when O is of type EMCP, $\mathbf{A} \geq \mathbf{B}$ implies $\left[\pm O \pm \mathbf{A}\right] \geq \left[\pm O \pm \mathbf{B}\right]$. See Theorem 6.3 below.

- There is a similar relationshp between the corresponding two agent 'collectivistic' positions:

$$\left[\pm O \pm \left[\pm \begin{pmatrix} E_a \\ E_b \end{pmatrix} \pm F\right] \cdot \left[\pm F\right]\right] \geq \left[\pm O \pm \left[\pm \begin{pmatrix} E_a \\ E_b \end{pmatrix} \pm F\right]\right]$$

The following property is very useful. It follows from Theorem 4.1 and Proposition 5.6, part (2).

Proposition 5.8 *For sets of sentences* $\Phi_1 \subseteq \Phi_2$: $\left[\pm \Phi_2\right] \geq \left[\pm \Phi_1\right]$.

Example 5.9 *Since* P *is the dual of* O, $\pm P\left[\pm E_a \pm F\right] \subseteq \pm O \pm \left[\pm E_a \pm F\right]$, *and hence*

$$\left[\pm O \pm \left[\pm E_a \pm F\right]\right] \geq \left[\pm P\left[\pm E_a \pm F\right]\right]$$

as observed in Example 5.7 above. Similarly, $E_a \pm F \subseteq \left[\pm E_a \pm F\right]$ *so*

$$\left[\pm O \pm \left[\pm E_a \pm F\right]\right] \geq \left[\pm O \pm E_a \pm F\right]$$

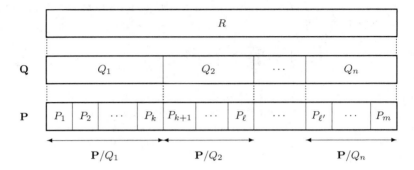

Figure 2: Partitions **P** and **Q** of R with $\mathbf{P} \geq \mathbf{Q}$

Definition 5.10 *For* **P** *a set of sentences and Q any expression:*

$$\mathbf{P}/Q =_{\text{def}} \{P \in \mathbf{P} \mid P \wedge Q \text{ consistent}\}.$$

For example: suppose that in the analysis of some scenario or set of regulations, it is determined that $\text{OE}_a F$ is true. The library example of Section 3 is of this form. Then

$$\left[\!\left[\bot\, \text{O} \pm [\!\pm \text{E}_b \pm F]\!\right]\!\right] \Big/ \text{OE}_a F$$

represents the (Jones-Sergot) normative one-agent act positions consistent with $\text{OE}_a F$. The 'collectivistic' two-agent act positions consistent with $\text{OE}_a F$ are given by the expression:

$$\left[\!\left[\pm\, \text{O} \pm \left[\!\pm \binom{\text{E}_a}{\text{E}_b} \pm F\right]\!\right]\!\right] \Big/ \text{OE}_a F$$

We can say much more about the structure of partitions **P** and **Q** in the case that **P** is a refinement of **Q**. When $\mathbf{P} \geq \mathbf{Q}$ and Q is an element of **Q** then $\mathbf{P}/Q$ is also the set of elements of **P** that logically imply Q. Indeed, when $\mathbf{P} \geq \mathbf{Q}$ and Q is an element of **Q** then $\mathbf{P}/Q$ is a Λ-partition of Q. And further: the set **P** itself is partitioned (standard set notion) into the collection of disjoint subsets $\mathbf{P}/Q_i$ where the Q_i are the elements of **Q**. The relationships are summarised in Figure 2. (The rectangles can be seen as Venn diagrams of the corresponding truth sets, moved apart to show the structure of the two partitions.)

We are now in a position to summarise the relationship between Kanger's (one-agent) 'atomic types', Lindahl's more refined version, the more complicated construction used in [Jones and Sergot, 1993], and the maxi-conjunctions identified at the end of Section 4 as a further refinement still. We include for completeness the set of 'normative fact positions' $[\![\pm O \pm F]\!]$. The Kanger and Lindahl forms are not refinements of this last one. They have a weaker relationship which we term an *elaboration*.

Definition 5.11 *Let* **P** *and* **Q** *be partitions of some sentence R.* **P** *is an elaboration of* **Q** *(***P** $\succeq$ **Q***) iff for every* $Q \in \mathbf{Q}$ *there is a* $P \in \mathbf{P}$ *such that* $\vdash P \to Q$.

Example 5.12 *Consider the 'one-agent act positions' used by Lindahl:*

$$[\![\pm E_a \pm F]\!] = \{E_a F,\ E_a \neg F,\ \mathrm{Pass}_a F\}$$

Since E_a *is a 'success' operator,* $[\![\pm E_a \pm F]\!]$ *is an elaboration of* $[\![\pm F]\!]$. *But* $[\![\pm E_a \pm F]\!]$ *is not a refinement of* $[\![\pm F]\!]$ *because* $\mathrm{Pass}_a F = \neg E_a F \wedge \neg E_a \neg F$ *does not imply any element of* $[\![\pm F]\!]$.

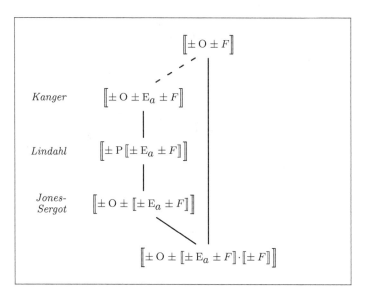

Figure 3: Normative one-agent act positions

It is possible to establish various relationships between refinements, elaborations and equivalences of partitions, but we shall not do so here. The relationships between the various forms of one-agent positions are summarised in Figure 3. The broken line represents an elaboration. The solid lines are

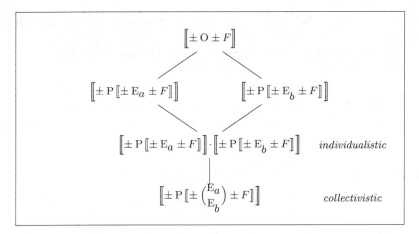

Figure 4: Lindahl's individualistic and collectivistic positions

refinements. The partitions at the bottom of the diagram are refinements (elaborations) of those higher up.

The relationships between Lindahl's individualistic and collectivistic normative positions are summarised in Figure 4.

Finally, the following properties are useful for performing (hand) computations.

Proposition 5.13 *Let* $\mathbf{P}$, $\mathbf{Q}$, $\mathbf{R}$ *be partitions of some sentence* S *such that* $\mathbf{P} \geq \mathbf{R}$. *Then for any* $R \in \mathbf{R}$: $\mathbf{P} \cdot \mathbf{Q}/R = (\mathbf{P}/R) \cdot (\mathbf{Q}/R)$.

As a special case, for any choice schemes (or sets of sentences) Φ_1 *and* Φ_2, *and any sentence* $A \in (\pm \Phi_1 \cup \pm \Phi_2)$:

$$([\![\pm \Phi_1]\!] \cdot [\![\pm \Phi_2]\!])/A \;=\; ([\![\pm \Phi_1]\!]/A) \cdot ([\![\pm \Phi_2]\!]/A).$$

We will refer to these properties when looking at some small examples later.

6 Normative positions

There are two main questions to consider:

(a) Given logic Λ and scheme (set of sentences) Φ, what is the set of maxi-conjunctions $[\![\pm \Phi]\!]$?

(b) For given logic Λ, which schemes (sets of sentences) Φ yield the most meaningful, or useful, sets of maxi-conjunctions $[\![\pm \Phi]\!]$?

6.1 Maxi-conjunctions for logics of type *EMCP*

We begin by looking at a special case of question (a), focussing on maxi-conjunctions of the form:

(25) $\quad [\![\pm O \pm \mathbf{A}]\!] = [\![\pm P \pm \mathbf{A}]\!] \quad$ (**A** a complete partition)

The equality is because O and P are duals.

We assume only that **A** is a complete partition. We shall not take into account the structure of sentences in **A** and the possibility of rules and axiom schemas in Λ that would allow reductions of certain iterated modalities. In this article we restrict attention to the logics employed by Kanger and Lindahl: type *EMCP* for the logic of O and type *ET* for the action modalities E_x. Elsewhere [Sergot, 1996] we set out the structure of maxi-conjunctions of the form (25) for a range of logics from type *EP* to type *EMCP*, and beyond.

For O of type *EMCP* and **A** a complete partition, the maxi-conjunctions in $[\![\pm O \pm \mathbf{A}]\!]$ have a particularly simple form.

Theorem 6.1 *Let* $\mathbf{A} = \{A_1, \ldots, A_n\}$ *be a complete partition. When the logic of O is of type EMCP the set of maxi-conjunctions:*

$$[\![\pm O \pm \mathbf{A}]\!] = [\![\pm P \pm \mathbf{A}]\!]$$

is equivalent (Definition 5.4) to the set of conjunctions of the form

(26) $\quad \pm P A_1 \wedge \ldots \wedge P A_j \wedge \ldots \wedge \pm P A_n$

that is, conjunctions such that, for each $A_i \in \mathbf{A}$, there is a conjunct of the form $P A_i$ or $\neg P A_i$, and at least one conjunct is of the form $P A_j$.

We write $\pi \mathbf{A}$ to stand for any conjunction of the form (26). $\pi^+ \mathbf{A}$ is the set of the permissible A_i in $\pi \mathbf{A}$, i.e.

$$\pi^+ \mathbf{A} =_{\text{def}} \{A_i \in \mathbf{A} \mid \pi \mathbf{A} \vdash P A_i\}$$

$\pi^- \mathbf{A}$ is the set of the 'prohibited' A_i, i.e.

$$\pi^- \mathbf{A} =_{\text{def}} \{A_i \in \mathbf{A} \mid \pi \mathbf{A} \vdash \neg P A_i\} = \mathbf{A} - \pi^+ \mathbf{A}.$$

Proof. See [Sergot, 2001]. In outline: every conjunction $\pi \mathbf{A}$ of the form (26) is consistent, and maximal for expressions falling under the scheme $\pm \mathrm{PA}$. The conjunction $\neg P A_1 \wedge \ldots \wedge \neg P A_n$, where there is no conjunct of the form $P A_j$, is inconsistent. The remaining expressions to consider are those falling under the scheme $\pm P \neg \mathbf{A}$, i.e. those of the form $\pm P \neg A_j$, $A_j \in \mathbf{A}$. It can be readily checked that every such expression is either inconsistent with or implied by every conjunction of form (26). ∎

Corollary 6.2 *When the logic of* O *is of type EMCP, and* **A** *is a complete partition:*

$$[\![\pm\mathrm{O}\pm\mathbf{A}]\!] \equiv [\![\pm\mathrm{P}\mathbf{A}]\!].$$

The corollary generalises the remarks in Section 4 on the equivalence, when O is of type *EMCP*, between Lindahl's form for normative one-agent and two-agent act positions, (12) and (18) respectively, and the forms (15) and (19) employed in [Jones and Sergot, 1993] for the same purpose.

Notice that in order to specify any element $\pi\mathbf{A}$ of $[\![\pm\mathrm{O}\pm\mathbf{A}]\!]$ it is sufficient to specify the permissible elements $\pi^+\mathbf{A}$. For O of type *EMCP* and **A** a complete partition, $[\![\pm\mathrm{O}\pm\mathbf{A}]\!]$ can thus be represented by the set of non-empty subsets of **A**. [Talja, 1980] takes a special case of this observation as the starting point for an algebraic treatment of the [Lindahl, 1977] account of 'change' of normative positions. Notice also that when $\pi^+\mathbf{A}$ is a singleton, and O is of type *EMCP* (or stronger), the conjunction $\pi\mathbf{A}$ can be written equivalently in a simpler form: when $\pi^+\mathbf{A} = \{A_j\}$, $\pi\mathbf{A}$ is logically equivalent to $\mathrm{O}A_j$.

For example, Lindahl's normative one-agent act positions (12) are given by the expression $[\![\pm\mathrm{P}[\![\pm\mathrm{E}_a\pm F]\!]]\!]$. There are three act positions in the partition $[\![\pm\mathrm{E}_a\pm F]\!]$, viz. $\{\mathrm{E}_a F, \mathrm{E}_a\neg F, \mathrm{Pass}_a F\}$. There are $2^3 - 1 = 7$ non-empty subsets of $[\![\pm\mathrm{E}_a\pm F]\!]$, and hence 7 elements in (12). They were listed earlier using Lindahl's numbering (T_1)–(T_7). The application of Theorem 6.1 is clearer when (T_2) and (T_4) are re-written in the logically equivalent forms (T'_2) and (T'_4).

As one more example, Jones and Parent [2008] study what they call *normative-informational* positions as a contribution to the investigation of such rights as the right to silence, the right to know and the right to conceal information.

Let $\mathrm{I}_j A$ represent that 'agent j is informed/told that A'. Let $\mathrm{O}_k A$ represent that 'it is obligatory for agent k that A'. P_k is the dual. The logic of each O_k is a normal logic of type *KD*, which is type *EMCP* together with a rule of necessitation $A/\mathrm{O}_k A$. As observed earlier, the rule of necessitation plays no role in the generation of normative positions for logics of this type.

The Jones-Parent normative-informational positions are given by the expression:

(27) $$[\![\pm\mathrm{O}_k\pm[\![\pm\mathrm{I}_j\pm A]\!]]\!]$$

The logic of I_j is taken to be a classical logic of type *K*. There are four *informational positions* in the set $[\![\pm\mathrm{I}_j\pm A]\!]$:

$$(I_1) \quad I_j A \wedge \neg I_j \neg A$$
$$(I_2) \quad I_j \neg A \wedge \neg I_j A$$
$$(I_3) \quad \neg I_j A \wedge \neg I_j \neg A$$
$$(I_4) \quad I_j A \wedge I_j \neg A$$

(I_1) and (I_2) are called the *straight truth/straight lie* positions, depending on whether A is or is not the case. (I_3) represents the *silence* position. (I_4) represents the *conflicting information* position.

There are $2^4 - 1 = 15$ non-empty subsets of $[\![\pm I_j \pm A]\!]$ and so 15 normative-informational positions of type (27). They are symmetric in A and $\neg A$. Jones and Parent re-write some of them in more readable equivalent form but our purpose here is merely to illustrate the application of Theorem 6.1.

Note that the difference between a 'straight truth' and a 'straight lie' is the difference between $A \wedge I_j A \wedge \neg I_j \neg A$ on the one hand and $\neg A \wedge I_j A \wedge \neg I_j \neg A$ (or $A \wedge I_j \neg A \wedge \neg I_j A$) on the other. Suppose then we consider the following more refined class of normative-informational positions:

$$(28) \qquad [\![\pm O_k \pm [\![\pm I_j \pm A]\!] \cdot [\![\pm A]\!]]\!]$$

There are 8 informational positions in $[\![\pm I_j \pm A]\!] \cdot [\![\pm A]\!]$ and so $2^8 - 1 = 255$ normative-informational positions of type (28), symmetric in A and $\neg A$.

6.2 Refinement structures

As will be established presently, when O is of type *EMCP*, then $\mathbf{A} \geq \mathbf{B}$ implies $[\![\pm O \pm \mathbf{A}]\!] \geq [\![\pm O \pm \mathbf{B}]\!]$. There is much more that can be said about the structure of such maxi-conjunctions, however.

We now summarise the structure of conjunctions of the form

$$[\![\pm O \pm \mathbf{A}]\!] / \pi \mathbf{B} \qquad (\mathbf{A} \geq \mathbf{B})$$

The question is also of considerable practical significance. (It is the basis of the automated inference methods presented in [Sergot, 2001].)

Suppose $B_j \in \pi^+ \mathbf{B}$, i.e. $\pi \mathbf{B}$ is an element of $[\![\pm O \pm \mathbf{B}]\!]$ containing a conjunct $P B_j$. Since $\mathbf{A} \geq \mathbf{B}$ there is some set of elements $\mathbf{A}/B_j = \{A_1^j, \ldots, A_{m_j}^j\}$ such that $\vdash B_j \leftrightarrow (A_1^j \vee \ldots \vee A_{m_j}^j)$. By O.RE, $\vdash P B_j \leftrightarrow P(A_1^j \vee \ldots \vee A_{m_j}^j)$, and when O is of type *EMCP*, then also $\vdash P B_j \leftrightarrow (P A_1^j \vee \ldots \vee P A_{m_j}^j)$. It follows that every element $\pi \mathbf{A}$ of $[\![\pm O \pm \mathbf{A}]\!]/\pi \mathbf{B}$ must have at least one conjunct $P A_i^j$, i.e. every $\pi^+ \mathbf{A}$ contains at least one element of $\mathbf{A}/B_j$.

Conversely, suppose $B_j \in \pi^- \mathbf{B}$. Then, since $\vdash A_i^j \to B_j$ for every A_i^j in $\mathbf{A}/B_j$, it follows when O is of type *EMCP* that $\vdash \neg P B_j \to \neg P A_i^j$.

Theorem 6.3 *Let $\mathbf{A}$ and $\mathbf{B}$ be complete partitions such that $\mathbf{A} \geq \mathbf{B}$. Suppose $\pi\mathbf{B}$ is an element of $[\![\pm \mathrm{O} \pm \mathbf{B}]\!]$; $\pi\mathbf{B}$ is logically equivalent to a conjunction of the form:*

$$\neg PB_1 \wedge \ldots \wedge \neg PB_k \wedge PB_{k+1} \wedge \ldots \wedge PB_n \qquad (k \geq 1)$$

i.e. $\pi^-\mathbf{B} = \{B_1, \ldots, B_k\}$ and $\pi^+\mathbf{B} = \{B_{k+1}, \ldots, B_n\}$. When O is of type EMCP, every element of $[\![\pm \mathrm{O} \pm \mathbf{A}]\!]/\pi\mathbf{B}$ is logically equivalent to a conjunction of the form:

$$\neg PB_1 \wedge \ldots \wedge \neg PB_k \wedge \pi(\mathbf{A}/B_{k+1}) \wedge \ldots \wedge \pi(\mathbf{A}/B_n).$$

Proof. In the previous discussion. ∎

It follows that when O is of type *EMCP*, $\mathbf{A} \geq \mathbf{B}$ implies $[\![\pm \mathrm{O} \pm \mathbf{A}]\!] \geq [\![\pm \mathrm{O} \pm \mathbf{B}]\!]$.

Example 6.4 *Suppose we are given the truth of $\mathrm{O}F$ (F represents, let us suppose, that there is a fence between two adjoining properties) and we wish to investigate what this implies about obligations of some agent a. We wish to determine the normative positions of form (21) that are consistent with $\mathrm{O}F$, i.e.*

$$(29) \qquad [\![\pm \mathrm{O} \pm [\![\pm \mathrm{E}_a \pm F]\!] \cdot [\![\pm F]\!]]\!]/\mathrm{O}F$$

Proceed as follows. $\mathrm{O}F$ can be written equivalently as $\mathrm{P}F \wedge \neg \mathrm{P}\neg F$. All conjunctions (29) will thus be equivalent to conjunctions $\neg \mathrm{P}\neg F \wedge C$ where C is a conjunction of the form $\pi([\![\pm \mathrm{E}_a \pm F]\!] \cdot [\![\pm F]\!]/F)$. Consider now $[\![\pm \mathrm{E}_a \pm F]\!] \cdot [\![\pm F]\!]/F$. By Proposition 5.13 this is $\{F \wedge \mathrm{E}_a F, F \wedge \neg \mathrm{E}_a F\} \equiv \{\mathrm{E}_a F, F \wedge \neg \mathrm{E}_a F\}$. There are three non-empty subsets of this set, and so, by Theorem 6.3, three normative positions in set (29). They are (equivalent to):

$$\begin{pmatrix} \neg \mathrm{P}\neg F \wedge \mathrm{PE}_a F \wedge \neg \mathrm{P}(F \wedge \neg \mathrm{E}_a F) \\ \neg \mathrm{P}\neg F \wedge \neg \mathrm{PE}_a F \wedge \mathrm{P}(F \wedge \neg \mathrm{E}_a F) \\ \neg \mathrm{P}\neg F \wedge \mathrm{PE}_a F \wedge \mathrm{P}(F \wedge \neg \mathrm{E}_a F) \end{pmatrix} \equiv \begin{pmatrix} \mathrm{OE}_a F \\ \mathrm{O}F \wedge \neg \mathrm{PE}_a F \\ \mathrm{O}F \wedge \mathrm{PE}_a F \wedge \mathrm{P}\neg \mathrm{E}_a F \end{pmatrix}$$

In similar fashion we may calculate which of the 'collectivistic' normative positions of form (23) for two agents a and b are consistent with, say $\mathrm{OE}_a F$:

$$\left[\!\!\left[\pm \mathrm{O} \pm \left[\!\!\left[\pm \binom{\mathrm{E}_a}{\mathrm{E}_b}\right) \pm F\right]\!\!\right] \cdot [\![\pm F]\!]\right]\!\!\right]/\mathrm{OE}_a F =$$

$$\bigl[\!\!\bigl[\pm \mathrm{O} \pm \bigl[\!\!\bigl[\pm \mathrm{E}_a \pm F\bigr]\!\!\bigr]\cdot\bigl[\!\!\bigl[\pm \mathrm{E}_b \pm F\bigr]\!\!\bigr]\cdot\bigl[\!\!\bigl[\pm F\bigr]\!\!\bigr]\bigr]\!\!\bigr]\big/\mathrm{OE}_a F$$

These positions will be (equivalent to) conjunctions of the form $\mathrm{OE}_a F \wedge C$: *to determine* C *we need to consider*

$$\bigl[\!\!\bigl[\pm \mathrm{E}_a \pm F\bigr]\!\!\bigr]\cdot\bigl[\!\!\bigl[\pm \mathrm{E}_b \pm F\bigr]\!\!\bigr]\cdot\bigl[\!\!\bigl[\pm F\bigr]\!\!\bigr]\big/\mathrm{E}_a F$$
$$= (\bigl[\!\!\bigl[\pm \mathrm{E}_a \pm F\bigr]\!\!\bigr]/\mathrm{E}_a F)\cdot(\bigl[\!\!\bigl[\pm \mathrm{E}_b \pm F\bigr]\!\!\bigr]/\mathrm{E}_a F)$$
$$= \{\mathrm{E}_a F \wedge \mathrm{E}_b F,\ \mathrm{E}_a F \wedge \neg \mathrm{E}_b F\}$$

There are three non-empty subsets, and so again three normative positions of the form we seek. They are (equivalent to):

$$\begin{pmatrix} \mathrm{OE}_a F \wedge \mathrm{PE}_b F \wedge \neg \mathrm{P}\neg \mathrm{E}_b F \\ \mathrm{OE}_a F \wedge \neg \mathrm{PE}_b F \wedge \mathrm{P}\neg \mathrm{E}_b F \\ \mathrm{OE}_a F \wedge \mathrm{PE}_b F \wedge \mathrm{P}\neg \mathrm{E}_b F \end{pmatrix} \equiv \begin{pmatrix} \mathrm{OE}_a F \wedge \mathrm{OE}_b F \\ \mathrm{OE}_a F \wedge \mathrm{O}\neg \mathrm{E}_b F \\ \mathrm{OE}_a F \wedge \mathrm{PE}_b F \wedge \mathrm{P}\neg \mathrm{E}_b F \end{pmatrix}$$

The procedure illustrated in the previous example is quite mechanical, and is readily automated. It is the basis of the automated inference methods presented in [Sergot, 2001].

The example also illustrates an important advantage of basing the generation of normative positions on cumulative fact/act positions of the form $\bigl[\!\!\bigl[\pm \mathrm{E}_a \pm F\bigr]\!\!\bigr]\cdot\bigl[\!\!\bigl[\pm F\bigr]\!\!\bigr]$ in preference to the simpler act positions $\bigl[\!\!\bigl[\pm \mathrm{E}_a \pm F\bigr]\!\!\bigr]$ employed by Lindahl. Not only is the resulting analysis more precise, but Lindahl's act positions are not a refinement of $\bigl[\!\!\bigl[\pm F\bigr]\!\!\bigr]$ and so the computational methods just described cannot be exploited, except in a messy and rather indirect way.

For O of type *EMCP*, $\bigl[\!\!\bigl[\pm\mathrm{O}\pm\mathbf{A}\bigr]\!\!\bigr]$, and hence $\bigl[\!\!\bigl[\pm\mathrm{P}\mathbf{A}\bigr]\!\!\bigr]$, is the most refined set of normative positions that can be constructed from a partition $\mathbf{A}$. It is essentially the basis of a disjunctive normal form for the fragment of the logic consisting of sentences falling under the schemes $\pm\mathrm{O}\pm\mathbf{A}$ and $\mathbf{A}$ and their subsentences [Sergot, 2001].

What of the act positions? Which act positions $\mathbf{A}$ yield the most refined set of normative positions $\bigl[\!\!\bigl[\pm\mathrm{O}\pm\mathbf{A}\bigr]\!\!\bigr]$? Here the answer is more complicated because it depends on the specific properties of the action modalities employed besides E.RE and E.T. Full discussion of the possibilities is far beyond the scope of this article. For practical purposes it seems reasonable to restrict attention to act expressions containing propositional atoms or their negations within the scope of an action operator. That rules out of consideration act expressions such as $\mathrm{E}_a(p \wedge q)$, $\mathrm{E}_a(p \wedge \neg q)$, $\mathrm{E}_a(p \vee q)$, and so on. In principle there is nothing problematic about allowing these more general forms of act expressions; in practice, it is not clear that the added level of precision is worth the extra trouble.

So, as a practical compromise, for a (finite) set of agents $\mathcal{A}g = \{a, b, \ldots\}$ and a (finite) set of propositional atoms $\mathcal{P}rops = \{p, q, \ldots\}$ it seems reasonable to focus on act positions of the following form:

$$(30) \quad \left\langle \pm \begin{pmatrix} E_a \\ E_b \\ \vdots \end{pmatrix} \pm \begin{pmatrix} p \\ q \\ \vdots \end{pmatrix} \right\rangle =_{\text{def}} \left[\!\left[\pm \begin{pmatrix} E_a \\ E_b \\ \vdots \end{pmatrix} \pm \begin{pmatrix} p \\ q \\ \vdots \end{pmatrix} \right]\!\right] \cdot \left[\!\left[\pm \begin{pmatrix} p \\ q \\ \vdots \end{pmatrix} \right]\!\right]$$

This is the form of act expression supported by the automated analysis program described in [Sergot, 2001].

The number of positions in $[\![\pm O \pm \mathbf{A}]\!]$ when O is of type *EMCP* is $2^{|\mathbf{A}|} - 1$. When $\mathbf{A}$ is of the form (30) and there are m agents in $\mathcal{A}g$ and n propositional variables in $\mathcal{P}rops$, the number of act positions is $2^{(m+1)n}$. The number of normative positions is then $2^{2^{(m+1)n}} - 1$. Although it is easy to write a computer program to generate all these expressions, that is a *very* large number of positions to examine even when m and n are small. It can nevertheless be practical to examine positions of this complex form because the analysis can be broken down into simple stages using the refinement results outlined in this section.

7 Example

The previous sections presented an extended and generalised version of the Kanger-Lindahl theory of normative positions. This framework is an important but still incomplete component of a full formal theory of duties, rights and other complex normative relations. We comment on some of the missing ingredients in Section 8 below.

[Sergot, 2001] describes how the procedures described in previous sections can be implemented in a computer program that is intended to facilitate application of the theory to the analysis of practical examples, either for the purpose of interpretation and disambiguation of legal texts, rules, and regulations, or in the design and specification of a new set of norms. A typical example is the case discussed in [Jones and Sergot, 1992; Jones and Sergot, 1993] concerning access 'rights' to sensitive medical information in a hospital database [Ting, 1990]. The problem here is to clarify and expand an incomplete and very imprecise statement of requirements into a precise specification at some desired level of detail.

In order to conduct such an analysis, the general strategy is to pick some scheme $[\![\pm O \pm \mathbf{A}]\!]$ which represents the problem under consideration at the appropriate level of detail. The objective of the analysis is to identify

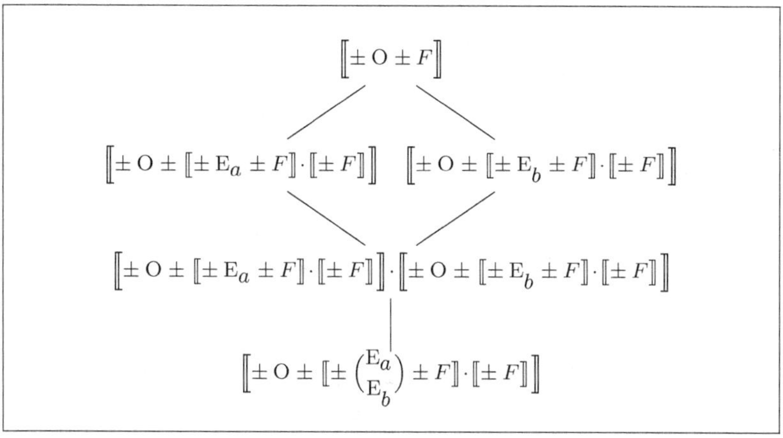

$$\left[\pm O \pm F\right]$$

$$\left[\pm O \pm \left[\pm E_a \pm F\right]\cdot\left[\pm F\right]\right] \qquad \left[\pm O \pm \left[\pm E_b \pm F\right]\cdot\left[\pm F\right]\right]$$

$$\left[\pm O \pm \left[\pm E_a \pm F\right]\cdot\left[\pm F\right]\right]\cdot\left[\pm O \pm \left[\pm E_b \pm F\right]\cdot\left[\pm F\right]\right]$$

$$\left[\pm O \pm \left[\pm \binom{E_a}{E_b} \pm F\right]\cdot\left[\pm F\right]\right]$$

Figure 5: Positions for two agents a and b and one state of affairs F

which position in this *target partition* holds in the (real or hypothetical) circumstances under consideration. In practice, there will often be points of detail on which we will be unable or unwilling to decide. In that case the result of the analysis will be a disjunction of positions.

As suggested in previous sections such an analysis can be conducted by a process of progressive refinement. At each stage the analysis completed so far is used to constrain the choice of possible positions at the next level of detail. Given a target partition $[\pm O \pm \mathbf{A}]$, find a sequence of refinements $\mathbf{A}_0 \leq \mathbf{A}_1 \leq \ldots \leq \mathbf{A}_N \leq \mathbf{A}$ and proceed as follows. First determine which position $\pi_0 \mathbf{A}_0$ of $[\pm O \pm \mathbf{A}_0]$ holds in the given circumstance. Then consider the candidate positions at the next level of detail: determine position $\pi_1 \mathbf{A}_1$ from the candidate set $[\pm O \pm \mathbf{A}_1]/\pi_0\mathbf{A}_0$. Now consider $[\pm O \pm \mathbf{A}_2]/\pi_1\mathbf{A}_1$, and so on, until left with the task of identifying a position from the target partition, which will be an element of $[\pm O \pm \mathbf{A}]/\pi_N\mathbf{A}_N$. As described in the previous section, the calculation of the candidate positions at each individual step is simple (especially when O is of type *EMCP*) and quite mechanical.

In practice the procedure is more complicated because usually it will not be a *sequence* of refinements that has to be considered but a more elaborate structure. Figure 5 shows the refinement structure for the case of two agents a and b and one state of affairs F. Figure 6 shows the structure for the case of one agent a and $\mathcal{P}rops = \{F, G\}$. In each case, the analysis would begin with the partitions at the top of the figure and work its way down to the more refined partitions shown lower down.

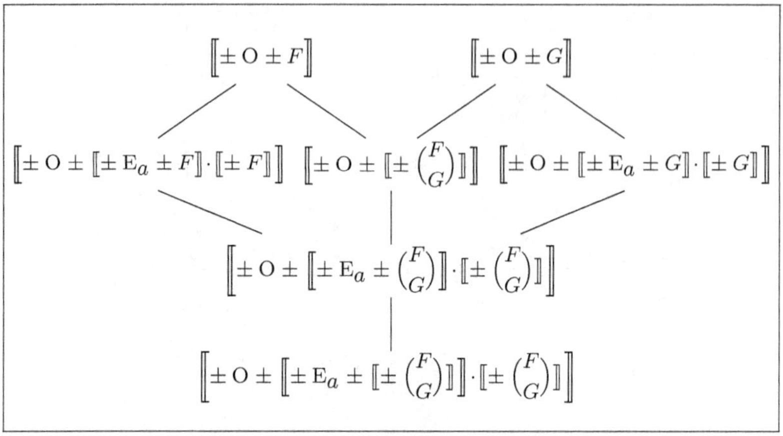

$$[\pm O \pm F]\qquad [\pm O \pm G]$$

$$\left[\pm O \pm [\pm E_a \pm F]\cdot[\pm F]\right]\quad \left[\pm O \pm [\pm \binom{F}{G}]\right]\quad \left[\pm O \pm [\pm E_a \pm G]\cdot[\pm G]\right]$$

$$\left[\pm O \pm [\pm E_a \pm \binom{F}{G}]\cdot[\pm \binom{F}{G}]\right]$$

$$\left[\pm O \pm [\pm E_a \pm [\pm \binom{F}{G}]]\cdot[\pm \binom{F}{G}]\right]$$

Figure 6: Positions for one agent a and two states of affairs F and G

We present here a small example of how this can work. The example is for illustration only; longer accounts with detailed transcripts from the automated system and supplementary comments are provided in [Sergot, 2001] and in [Sergot and Richards, 2000].

The example is a modified version of Ronald Lee's [1988] car park example discussed briefly in Section 3. It concerns the specification of which categories of staff are permitted and not permitted to park in a car park. We will use it to make a number of different points to Lee's. We choose it because it is familiar and requires no further explanation. In Lee's example, administrators are permitted to park in the car park. We will ignore other categories of staff here.

Consider the following scenario:

> a is an administrator, permitted to park in the car park. a has two cars, car-a_1 and car-a_2. b is a disgraced administrator, banned from the car park. b has one car, car-b. c is a passer-by. g is the gatekeeper, charged with controlling access to the car park and ensuring the rules are obeyed.

We will not attempt to cover every feature of the example. In particular the representation of what it means to say that the gatekeeper g is responsible for ensuring that the rules of the car park are obeyed raises a number of difficult points which are outside the scope of this article.

Let $p(a_1)$, $p(a_2)$, $p(b)$ represent that cars car-a_1, car-a_2, car-b are parked in the car park, respectively. We take it that the following at least is implicit and obvious from the scenario description as given above: that it is not permitted that *car-b* is parked in the car park, $\neg \mathrm{P}p(b)$; that it is permitted but not obligatory that *car-a_1* is parked in the car park, $\mathrm{P}p(a_1) \wedge \mathrm{P}\neg p(a_1)$; and that it is permitted but not obligatory that *car-a_2* is parked in the car park, $\mathrm{P}p(a_2) \wedge \mathrm{P}\neg p(a_2)$.

What else holds according to the rules of the car park (as we imagine them to be from the scenario and previous experience of typical car parks)? In order to investigate the possibilities in a systematic fashion, and to identify any points requiring further clarification, the task is to pick out one or, in the case of some residual uncertainty, several of the positions from the following *target partition*:

(31) $$\left[\!\left[\pm \mathrm{O} \pm \left\langle \pm \begin{pmatrix} \mathrm{E}_a \\ \mathrm{E}_b \\ \mathrm{E}_c \\ \mathrm{E}_g \end{pmatrix} \pm \begin{pmatrix} p(a_1) \\ p(a_2) \\ p(b) \end{pmatrix} \right\rangle \right]\!\right]$$

We want to restrict attention to those positions in the target partition that are consistent with the initial assertions and thus to compute

(32) $$\left[\!\left[\pm \mathrm{O} \pm \left\langle \pm \begin{pmatrix} \mathrm{E}_a \\ \mathrm{E}_b \\ \mathrm{E}_c \\ \mathrm{E}_g \end{pmatrix} \pm \begin{pmatrix} p(a_1) \\ p(a_2) \\ p(b) \end{pmatrix} \right\rangle \right]\!\right] / $$
$$\neg \mathrm{P}p(b) \wedge (\mathrm{P}p(a_1) \wedge \mathrm{P}\neg p(a_1)) \wedge (\mathrm{P}p(a_2) \wedge \mathrm{P}\neg p(a_2))$$

The problem can be simplified by focussing first on, say, the two car owners a and b, and analyzing

(33) $$\left[\!\left[\pm \mathrm{O} \pm \left\langle \pm \begin{pmatrix} \mathrm{E}_a \\ \mathrm{E}_b \end{pmatrix} \pm \begin{pmatrix} p(a_1) \\ p(a_2) \\ p(b) \end{pmatrix} \right\rangle \right]\!\right] / $$
$$\neg \mathrm{P}p(b) \wedge (\mathrm{P}p(a_1) \wedge \mathrm{P}\neg p(a_1)) \wedge (\mathrm{P}p(a_2) \wedge \mathrm{P}\neg p(a_2))$$

This in turn can be simplified to sub-problems

(34) $$\left[\!\left[\pm \mathrm{O} \pm \left\langle \pm \begin{pmatrix} \mathrm{E}_a \\ \mathrm{E}_b \end{pmatrix} \pm \begin{pmatrix} p(a_1) \\ p(a_2) \end{pmatrix} \right\rangle \right]\!\right] / $$
$$\neg \mathrm{P}p(b) \wedge (\mathrm{P}p(a_1) \wedge \mathrm{P}\neg p(a_1)) \wedge (\mathrm{P}p(a_2) \wedge \mathrm{P}\neg p(a_2))$$

and

(35) $\left[\pm O \pm \left\langle \pm \begin{pmatrix} E_a \\ E_b \end{pmatrix} \pm p(b) \right\rangle \right] /$

$\neg P p(b) \land (P p(a_1) \land P \neg p(a_1)) \land (P p(a_2) \land P \neg p(a_2))$

The automated analysis program described in [Sergot, 2001] provides a graphical interface to help visualize the structure of these sub-problems, and to keep track of the analysis as it proceeds.

Consider (34). Some questions are immediate. Presumably $P(\neg p(a_1) \land \neg p(a_2))$ is true in the car park. But is it the case that $P(p(a_1) \land p(a_2))$? Is it permitted for both of administrator a's cars to be parked at the same time? In a practical setting, this would need to be checked with the car park authorities, or left undetermined if it were not regarded as important. One purpose of the analysis to identify points of detail that may have remained undetected otherwise.

Similarly $PE_a p(a_1)$ and $PE_a \neg p(a_1)$ seem straightforward. But what of $P(p(a_1) \land \neg E_a p(a_1))$ and $P(\neg p(a_1) \land \neg E_a \neg p(a_1))$, equivalently, $O(p(a_1) \to E_a p(a_1))$ and $O(\neg p(a_1) \to E_a \neg p(a_1))$? It might be tempting to read the first as saying that if car-a_1 is parked then it must have been the administrator a who parked it. But note that expression $E_a p(a_1)$ does not necessarily signify 'a parks car-a_1'; a may bring about $p(a_1)$ in some different way, perhaps even unintentionally. The correct reading of $E_a p(a_1)$ depends on which version of the logic of action is employed and its semantics. There are many variations. We will make a few further remarks in Section 8 below. And similarly for the question $O(\neg p(a_1) \to E_a \neg p(a_1))$.

What of $PE_b p(a_1)$ and $PE_b \neg p(a_1)$? Again, we might be tempted to read the first as asking whether the banned administrator b is permitted to park a's car, though again that really depends on how precisely the action modality is to be read. And similarly for the second question. Note that in general $E_a F$ does not imply $\neg E_b F$ for other agents $b \neq a$. a and b could act jointly to bring about F, or could even act unintentionally in such a way that each brings about F.

Switching now to the sub-problem (35): $\neg P p(b)$ implies both $\neg PE_a p(b)$ and $\neg PE_b p(b)$ in the logics we are employing. Presumably $PE_b \neg p(b)$ is true in the car park. But is it the case that $PE_a \neg p(b)$? Is a permitted to see to it that b's car is not parked? That is far from clear. It will depend on what precisely the act expression $E_a \neg p(b)$ represents. We will return briefly to some of these points in Section 8.

[Sergot, 2001] and [Sergot and Richards, 2000] present detailed transcripts of a full exploration of partition (33) in the example. Depending on the answers given to earlier questions, about a dozen questions are required to determine a unique position in the partitions (34) and (35); from that about a dozen more pick out a unique position from the partition (33). An exploration of the original target position (31) where there are other agents c and g to consider in addition can be undertaken in similar fashion.

8 Discussion

8.1 Alchourrón-Bulygin's normative systems, and conditional positions

We will comment briefly on the connnection between the mapping out of classes of normative positions and Alchourrón and Bulygin's [1971] formalisation of a normative system. A normative system $\mathcal{N}$ maps a *universe of cases* to *solutions*. The universe of cases is the set of all possible fact combinations that can be constructed from a given set $\mathcal{P}rops$ of propositional variables. In the maxi-conjunction notation, it is $[\![\pm \mathcal{P}rops]\!]$. Where there is one action F for which solutions are specified, a (consistent and complete) normative system $\mathcal{N}$ is a mapping of the form:

$$(36) \qquad \mathcal{N} : [\![\pm \mathcal{P}rops]\!] \mapsto [\![\pm O \pm F]\!]$$

As observed earlier, when the logic of O is type *EMCP* (or Standard Deontic Logic, type *KD*) $[\![\pm O \pm F]\!]$ is (with logical redundancies removed) the set of mutually exclusive normative 'fact positions' $\{OF, O\neg F, PF \wedge P\neg F\}$, or in words, 'obligatory', 'prohibited/forbidden', 'facultative'.

More generally, for a set of propositional variables $\mathcal{P}rops$ and actions $\{F_1, \ldots, F_n\}$, a consistent and complete normative system $\mathcal{N}$ maps the universe of cases to solutions as follows:

$$(37) \qquad \mathcal{N} : [\![\pm \mathcal{P}rops]\!] \mapsto [\![\pm O \pm F_1]\!] \cdot \ldots \cdot [\![\pm O \pm F_n]\!]$$

which is

$$\mathcal{N} : [\![\pm \mathcal{P}rops]\!] \mapsto \begin{pmatrix} OF_1 \\ O\neg F_1 \\ PF_1 \wedge P\neg F_1 \end{pmatrix} \cdot \ldots \cdot \begin{pmatrix} OF_n \\ O\neg F_n \\ PF_n \wedge P\neg F_n \end{pmatrix}$$

Note that a mapping $\mathcal{N}'$ of this alternative form:

$$(38) \qquad \mathcal{N}' : [\![\pm \mathcal{P}rops]\!] \mapsto [\![\pm O \pm \begin{pmatrix} F_1 \\ \vdots \\ F_n \end{pmatrix}]\!]$$

defines normative system $\mathcal{N}'$ as a refinement (in the sense used by Alchourrón and Bulygin) of the normative system $\mathcal{N}$: the set of solutions in $\mathcal{N}'$ is a refinement (in the sense of this article) of the set of solutions in $\mathcal{N}$.

Viewed in this way, the solutions in expressions (36)–(38) are classes of normative positions of a rather simple kind, where no agent is specified. More generally then, one could define a normative system as mapping a universe of cases to sets of *normative positions*, of arbitrary degrees of precision, as exemplified by the following possible forms (among many others):

$$\mathcal{N} : [\![\pm \mathcal{P}rops]\!] \mapsto [\![\pm O \pm [\![\pm E_a \pm F]\!]]\!]$$

$$\mathcal{N} : [\![\pm \mathcal{P}rops]\!] \mapsto [\![\pm O \pm [\![\pm E_a \pm F]\!]]\!] \cdot [\![\pm O \pm [\![\pm E_b \pm F]\!]]\!]$$

$$\mathcal{N} : [\![\pm \mathcal{P}rops]\!] \mapsto \left[\!\!\left[\pm O \pm \left[\!\!\left[\pm \binom{E_a}{E_b} \pm F\right]\!\!\right]\right]\!\!\right]$$

$$\mathcal{N} : [\![\pm \mathcal{P}rops]\!] \mapsto \left[\!\!\left[\pm O \pm \left[\!\!\left[\pm \begin{pmatrix} E_a \\ \vdots \\ E_b \end{pmatrix} \pm \begin{pmatrix} F_1 \\ \vdots \\ F_n \end{pmatrix}\right]\!\!\right] \cdot \left[\!\!\left[\pm \begin{pmatrix} F_1 \\ \vdots \\ F_n \end{pmatrix}\right]\!\!\right]\right]\!\!\right]$$

Alchourrón and Bulygin's formalisation can thus be seen as a special case of a much more general account.

Similarly, a rule-based representation language such as that employed in [Lee, 1988] (Example 3.3, Section 3) can be seen as a set of if/then rules whose consequents are agent-free normative positions of a very simple kind. A more general representation language would have if/then rules of the form

if *conditions* then *normative-position*

where *normative-position* is one of some class of normative positions, of arbitrary complexity and precision depending on the needs of the application.

The representation of *conditional* (normative) positions is far from straightforward. It is not just the additional combinatorial complexity that would have to be addressed; there are also strong interactions between conditional structures and deontic logic, and between conditional structures and the treatment of action adopted. For example, unless all actions can be assumed to be instantaneous (an assumption which is made in some of the works cited above) there is a great deal to sort out. If we say that Alice is permitted to park her car if, and only if, it is raining, and if the action of parking takes some significant length of time, do we check that it is raining when she begins to park, or when she completes the job? Do we require it to be raining throughout the entire process? The first of these seems the most natural but that would require quite far-reaching adjustments to the logic of action that has been employed.

8.2 Extended forms of act expression

For certain purposes we might consider extending the initial class of act expressions from which the normative positions are constructed. Some regulations pertain not to *individual* agent positions of the form $E_x F$, but to what have been termed interpersonal *control* positions, e.g. of type $E_x E_y F$ or $E_x \neg E_y F$. Indeed, the ability to iterate action operators in this way is one of the generally perceived benefits of employing this approach to the treatment of action.

Consider the car park example. The banned administrator b's car may not be parked, $\neg P\, p(b)$. It follows in the logic that the banned administrator b may not see to it that his car is parked, $\neg P\, E_b\, p(b)$. Consider now the responsibilities of the gatekeeper. It seems reasonable to say that the gatekeeper g is permitted to see to it that the banned administrator does not park his car, or more generally that $P\, E_g \neg E_b\, p(b)$ holds according to the rules of the car park. (One might even be tempted to say that there is an *obligation* on the gatekeeper g to see to it that b does not park his car. However, as discussed in the introductory sections, an expression of the form $O\, E_g \neg E_b\, p(b)$ does not represent such an obligation adequately. We will not discuss it further.) One would surely not *insist*, however, that g sees to it that $E_b \neg p(b)$—surely we would expect that $P \neg E_g\, E_b \neg p(b)$ holds in the car park. Are there any other possibilities?

Ingmar Pörn [1977] has applied similar position-generating techniques to the systematic study of what he called 'control' and 'influence' positions, and in particular to classes of positions of the following forms:

(39) $\quad [\pm E_b \pm [\pm E_a \pm F]]$

(40) $\quad [\pm E_b \pm Can \pm [\pm E_a \pm F]]$

Here *Can* is a modality for a notion of (practical) possibility.

[Sergot and Richards, 2000] have considered normative positions of the following general form:

(41) $\quad \left[\pm O \pm \left[\pm \begin{pmatrix} E_x \\ \vdots \\ E_y \end{pmatrix} \pm \begin{pmatrix} E_x \\ \vdots \\ E_y \end{pmatrix} \pm \begin{pmatrix} F \\ \vdots \\ G \end{pmatrix}\right] \cdot [\pm \begin{pmatrix} F \\ \vdots \\ G \end{pmatrix}]\right]$

The general principles and methods of construction are exactly as presented in previous sections, though much more complicated in application. For simplicity [Sergot and Richards, 2000] consider in detail only the simpler case of normative positions of the following form:

(42) $$\left[\pm O \pm [\pm E_x \pm E_y F] \cdot [\pm E_x \pm E_y \neg F] \cdot [\pm E_y \pm F] \cdot [\pm F]\right]$$

The act-expressions are

(43) $$\left[\pm E_x \pm E_y F\right] \cdot [\pm E_x \pm E_y \neg F] \cdot [\pm E_y \pm F] \cdot [\pm F]$$

There are 16 act-expressions in this set, symmetric in F and $\neg F$, and hence $2^{16} - 1$ normative positions of type (42).

Note that in some versions of action/agency, notably the 'stit' logics, it is not meaningful to say x 'sees to it' that y 'sees to it' that F for $x \neq y$ (see e.g. the discussion in [Belnap and Perloff, 1988]). In those logics, $\neg E_x E_y F$ is a theorem for all $x \neq y$. If a 'stit' version is adopted for E_x, then the list of act positions (43) can be simplified. There are 12 act-expressions in that case, and $2^{12} - 1$ corresponding normative positions.

If we look now at the car park and the gatekeeper's control over the banned administrator then we need to consider which of the following act expressions can be permitted given P ($E_g \neg E_b\, p(b) \wedge \neg p(b)$) (supposing, as we do, that this is true in the car park):

(44) $$\left[\pm E_g \pm E_a\, p(b)\right] \cdot [\pm E_g \pm E_b\, \neg p(b)] \cdot [\pm E_b \pm p(b)] \cdot [\perp p(b)]$$

Applying the methods of the previous sections, we obtain the following set of mutually exclusive act expressions. At least one of them must be permitted, but there may be more than one.

(a) $E_g\, E_b \neg p(b) \wedge E_g \neg E_b\, p(b)$
(b) $E_b \neg p(b) \wedge \neg E_g\, E_b \neg p(b) \wedge E_g \neg E_b\, p(b)$
(c) $\neg p(b) \wedge E_g \neg E_b \neg p(b) \wedge E_g \neg E_b\, p(b)$
(d) $\neg p(b) \wedge \neg E_g \neg E_b \neg p(b) \wedge E_g \neg E_b\, p(b)$

In the case of a 'stit' logic for the action modalities, the first of these can be eliminated as it is logically inconsistent, and the second can be simplified by removing the second conjunct.

In each of these expressions b's car is not parked and g's actions are such as to ensure that b does not see to it that b's car is parked. In each case however the interaction between g and b is subtly different. Which of these acts are permitted in the car park (as we imagine it to be)?

It is not easy to give a concise reading to these expressions. A careful reading of each would be quite involved, and more importantly, would again depend critically on what precisely the action modalities are taken to represent. Apart from the huge number of new positions that are created, even with a relatively small number of agents and states of affairs, it is very far from clear whether there is any real value in providing this level of analysis. As the example illustrates, deciphering these complex expressions is far from straightforward. One may be offering a level of precision that is simply unusable in practice.

One of the main difficulties in deciphering the control positions is in interpreting negatives. It is hard to decide what 'x does not see to it that F is not the case' actually means. This is made all the harder because it is unclear what 'not being parked' means exactly: do we mean that the car was never in the car park, or that it was in the car park and was then removed? This can make a big difference. We turn to that next.

It is very easy to imagine a car park in which the gatekeeper g is permitted to prevent a banned car from parking but not permitted to remove a car even if it is illegally parked. With the presently available resources all we can say is that $PE_g \neg p(b)$ —the gatekeeper is permitted to see to it that b's car is not parked. Clearly some kind of termporal extension is required.

One possible approach is to follow a suggestion made by von Wright [1968; 1983], Segerberg [1992], and Hilpinen [1997]. We will follow the terminology of Hilpinen's version; the others are essentially the same. There are two components: first, the idea that actions are associated with transitions between states; and second, a distinction between transitions corresponding to the agent's activity and transitions corresponding to the agent's inactivity. The latter are transitions where the agent lets 'nature take its own course'. There are then eight possible modes of agency, and because of the symmetry between F and $\neg F$, four basic forms to consider:

- x brings it about that F ($\neg F$ to F, x active);

- x lets it become the case that F ($\neg F$ to F, x inactive);

- x sustains the case that F (F to F, x active);

- x lets it remain the case that F (F to F, x inactive).

As discussed by Segerberg and Hilpinen there remain a number of fundamental problems to resolve in this account. Moreover, not discussed by those authors, the picture is considerably more complicated when there are the actions of other agents to take into account and not just the effect of nature's taking its course.

To illustrate one possible line of development, Sergot [2008a; 2008b] presents a formalism which combines a logic of action of the 'brings it about' kind with a transition-based treatment of action. Leaving aside the details, an expression $0{:}F$ is true at a transition when F is true at its initial state; $1{:}F$ is true when F is true at the final state of a transition. The distinctions above can then be expressed as follows. The first ('brings it about that') and third ('sustains the case that') are:

(45) $\quad E_x(0{:}\neg F \wedge 1{:}F)$, equivalently (as it turns out) $\quad 0{:}\neg F \wedge E_x 1{:}F$
(46) $\quad E_x(0{:}F \wedge 1{:}F)$, equivalently $\quad 0{:}F \wedge E_x 1{:}F$

The second and fourth cases, where x is inactive, can be expressed as follows

(47) $\quad (0{:}\neg F \wedge 1{:}F) \wedge \neg E_x(0{:}\neg F \wedge 1{:}F)$
(48) $\quad (0{:}F \wedge 1{:}F) \wedge \neg E_x(0{:}F \wedge 1{:}F)$

These four cases are mutually exclusive.

With these additional resources we are able to distinguish between seeing to it that a car not parked in the car park remains not parked (approximately, preventing a car from entering), and seeing to it that a car which was parked is no longer parked (approximately, removing it). In the (imaginary) car park, the first is permitted for the gatekeeper g, the second is not:

(49) $\quad PE_g(0{:}\neg p(b) \wedge 1{:}\neg p(b)) \quad \text{and} \quad \neg PE_g(0{:}p(b) \wedge 1{:}\neg p(b))$

In the logic, these expressions are equivalent to, respectively

(50) $\quad P(0{:}\neg p(b) \wedge E_g 1{:}\neg p(b)) \quad \text{and} \quad \neg P(0{:}p(b) \wedge E_g 1{:}\neg p(b))$

One could make a case that in the car park we have in mind, $P(0{:}\neg p(b) \wedge E_g 1{:}\neg p(b))$ could be strengthened to

(51) $\quad O(0{:}\neg p(b) \to E_g 1{:}\neg p(b))$

These brief examples are offered as suggestions for further lines of development. We will not discuss them further here.

More generally, the various examples in this article are intended in part to illustrate some of the difficulties of employing the 'brings it about' or 'sees to it that' treatment of action in the representation of practical problems. These are very abstract treatments of action. There is often a temptation in particular to read expressions containing E_x with emphasis on the 'end result' feature and insufficient attention to the agency component. Where $p(x)$ stands for 'x's car is parked', for example, it can be tempting to read the expression $E_x p(x)$ as 'x parks his car', and further, $OE_x p(x)$ as a representation of an ought-to-do statement that 'x ought to park his car'. But this is not what these expressions say. What they do say depends on the semantics of the action logic adopted. One problem is that in most versions the semantics of the action operators is very abstract indeed, making it very difficult to see how to interpret some expressions in a practical setting.

For example, in the car park it seems intuitively right to say that the banned administrator b is not permitted to park the administrator a's car, or rather, not permitted to see to it that the administrator a's car is parked. But is this correctly represented by $\neg P\, E_b\, p(a_1)$? In the logics employed, $P\, E_b\, p(a_1)$ is consistent with the following

(52) $$P\,(E_a\, p(a_1) \wedge E_b\, p(a_1))$$

Are the administrator a and the banned administrator b, perhaps when acting together, permitted to park the administrator a's car? Perhaps they act in such a way that both bring it about that the car is parked (or remains parked). One can imagine circumstances where that would seem to be reasonable, and we could certainly create other similar examples where it would be so. We have $P\,(E_a\, p(a_1) \wedge E_b\, p(a_1))$, and since the logic contains all instances of $P\,(A \wedge B) \to P\,B$, we have also:

$$P\,(E_a\, p(a_1) \wedge E_b\, p(a_1)) \to PE_b\, p(a_1)$$

It seems that $P\, E_b\, p(a_1)$ is likely to be true in the car park after all, if we consider all possible imaginable combinations of actions by a and b.

The erroneous reading of such expressions seems very easy to slip into. For instance, Lindahl uses the example of two adjoining properties, one of which is owned by an agent called John. When discussing the possible normative relations between John and his neighbour in regard to various kinds of acts, including the painting of the neighbour's house white, Lindahl suggests: "...a case in which John is completely unauthorized to influence the situation (since it is no business of his): John may neither bring about nor prevent the main building on his neighbour's property being painted white."[Lindahl, 1977, pp. 93–4]. In the light of the previous discussion, this is unlikely to be correct. More likely, there are permitted circumstances in which John and his neighbour *between them* act in such a way that they both bring about that the neighbour's property is painted white. It would then follow that John *is* permitted to influence the situation, even though the colour of his neighbour's house is no business of his. The conjunction $E_a\, p(a_1) \wedge E_b\, p(a_1)$ may but does not necessarily signify (intentional) joint action by a and b. It could be that both a and b choose independently to see to it that $p(a_1)$. It could be that both bring it about that $p(a_1)$ by chance. It could be that one does it intentionally and the other by chance. Nor does the conjunction represent a composite agent a-and-b-together.

Whether or not these general observations apply for a particular choice of action logic will depend on the details of that choice and on the semantics. The point is that the theory of normative positions makes only minimal assumptions about the properties of the action modalities. For practical applications, it will be necessary to look at some of the detailed choices.

8.3 Limitations

The Kanger-Lindahl theories have several well-documented limitations. Lindahl [1994] himself argues that Kanger's attempted classification of types of *rights* is better seen as a typology of *duties*.

There are two main shortcomings. As a formalisation of the Hohfeldian scheme, the theory of normative positions does not address the feature Hohfeld called '(legal) power'. It has long been understood that 'power' in the sense of (legal) capacity or 'competence' cannot be reduced to permission, and must also be distinguished from the 'can' of practical possibility. An agent can have 'power', to effect a marriage say, without necessarily having the permission nor the practical possibility of exercising that power. The example is from [Makinson, 1986]. Jones and Sergot [1996] argue that 'power' in this Hohfeldian sense is to be understood as a special case of a more general phenomenon, whereby in the context of a given normative system or institution, designated kinds of acts, when performed by designated agents in specific circumstances, *count as* acts that create or modify specific kinds of institutional relations and states of affairs. This switches attention from the formalisation of permission to the formalisation of the *count as* relation more generally.

The second shortcoming of the theory of normative positions, when viewed as a theory of duties and rights or as a formalisation of the Hohfeldian framework, is that it fails to deal with the notion of *counterparty*—the idea that when a party x owes an obligation or duty to party y that such-and-such, or when y has a claim-right against x that such-and-such, then the *counterparty* y has a special relationship in the normative relation between x and y that is not shared by other agents.

There are two main views of how to treat the *counterparty*: as *claimant* or as *beneficiary*. Discussions of the relative merits are sometimes framed as if they were competing accounts for the same notion. It is more helpful to see them not as competitors but as meaningful and distinct notions in their own right. In some cases claimant and beneficiary coincide, in other cases they do not. Both views however present severe challenges to an adequate formal characterisation.

The counterparty as claimant notion is associated with 'power'. Thus a commonly expressed view of what it means to be a counterparty is in terms of a conditional power: 'A relative duty in the law is owed to the party who has the legal power to initiate proceedings to enforce that duty.' [Wellman, 1989]

Makinson [1986] puts it like this:

> "The informal account that suggests itself is that x bears an obligation to y that F under the system N of norms iff in the case that F is not true then y has the power under the code N to initiate legal action against x for non-fulfillment of F (or in the case of a moral rather than a legal code, iff in such a case y is 'entitled to complain' of x for non-fulfillment of F)." [Makinson, 1986, p. 423]

There is nevertheless a fundamental difficulty. Generally speaking, a party y has a power to *initiate* legal action against x even when x has no obligation to y, even when the legal action is initiated on what will turn out to be completely unsubstantiated grounds, or perhaps even frivolously. What is missing is the idea that when x does bear an obligation to y, y has the power to initiate legal action with some expectation of *success*. One could not say there is a *guarantee* of success because legal action by its nature is never that certain. But some extra ingredient is essential to eliminate speculative, unsubstantiated or frivolous legal actions. It is very far from clear how one might approach a characterisation of that idea.

Some authors have preferred to take the view that what it means to be a counterparty is to be the *beneficiary* of another's duty or obligation. That notion also remains a serious challenge to formal characterisation. Herrestad and Krogh [1995] for instance, along with others, have proposed adding an index to the obligation operator to designate the beneficiary. Let $\underset{x \to y}{O} F$ represent that there is directed obligation that F on the bearer x that is for the benefit of the counterparty y. (It is the obligation that is of benefit to y, not necessarily the content F of the obligation itself.) This device allows useful distinctions to be expressed though adding an index in itself obviously does not provide any insight into what the beneficiary is.

One simple suggestion, which nevertheless shows much promise, has been made by Lars Lindahl [1994] as a variation of the Andersonian reduction.

Where x and y are (names of) agents, let propositional constants $W(x, y)$ be read as x 'is wronged by' y. Let $\underset{x \to y}{O} F$ represent that x is the bearer of a directed obligation (relative duty) to y that F, or on Lindahl's suggested reading, that 'y has a *right-proper* versus x to the effect that F'. Define $\underset{x \to y}{O}$ in terms of $W(y, x)$ as follows:

(53) $\qquad \underset{x \to y}{O} F =_{\text{def}} \Box(\neg F \to W(y, x))$

In words, x owes an obligation to y that F (y has a right-proper versus x that F) when, if it is not the case that F, then y is wronged by x. Lindahl takes $\Box$ to be a normal (alethic) modality of type KT; one could consider other options.

Let S be the Andersonian propositional constant representing that a violation or Something Bad has occurred. It is natural to add the axiom schema:

(54) $\qquad W(x, y) \to S \qquad$ (for all x and y)

The usual Andersonian reduction

(55) $\qquad OF \leftrightarrow \Box(\neg F \to S)$
(56) $\qquad \neg \Box S$

then makes the logic of each $\underset{x \to y}{O}$ Standard Deontic Logic (a normal logic of type KD). We also get, for all x, y, w and z:

(57) $\qquad O \neg W(x, y)$
(58) $\qquad \underset{x \to y}{O} F \to \neg \underset{z \to w}{O} \neg F$
(59) $\qquad \underset{x \to y}{O} F \to OF$

The idea is simple but it can be refined in several interesting respects. For instance, as Lindahl points out, one can make a case for the following additional schema:

(60) $\qquad \Box(E_x W(x, y) \to W(y, x))$

If x himself sees to it that x is wronged by y, then y is wronged by x.

With the addition of some rather simple general properties of E_x and $\Box$, which we omit here in the interests of space, it is possible to derive the following:

(61) $\qquad \underset{x \to y}{O} E_x F \to \underset{y \to x}{O} \neg E_y \neg F$

If x owes a duty to y to see to it that F then y owes a duty to x not to see to it that $\neg F$. This seems entirely plausible. One can investigate several variations along these lines.

9 Conclusion

We have presented an account of the theory of normative positions, as originally developed by Kanger and Lindahl, and in the generalised and extended form developed in [Sergot, 2001] building on David Makinson's maxi-conjunction characterisation. The methods for mapping out and investigating classes of 'positions' are quite general and are independent of the choice of specific deontic and action logics, though specific results can be obtained for the special case where the underlying logics are those employed by Kanger and Lindahl. The deontic logic component is (a very slightly weakened version of) Standard Deontic Logic. The action logic component makes minimal assumptions: the action logic could be strengthened and refined in many ways.

A secondary aim of this article has been to illustrate the inherent complexity of normative concepts such as duty, right, authorisation, responsibility, commitment, which are encountered not just in legal discourse, but in any description of regulated and organised agent interaction. The theory of normative positions as presented here is an important but limited component of a formal treatment of this complex network of concepts. It is already clear even from this limited theory that there is no point in searching for some, possibly large but nevertheless identifiable, set of basic types—'lowest common denominators' in Hohfeld's words—in terms of which all normative relations between any (two) agents could be articulated. The representation of such relations can be taken to arbitrary levels of detail and complexity. There are nevertheless grounds to believe that a more comprehensive formal account could be developed, together with the automated support tools necessary for its practical use.

Acknowledgements

I am indebted to Andrew Jones for introducing me to the Kanger-Lindahl theories and for numerous valuable discussions on topics related to this article. I am grateful to David Makinson for a number of detailed comments and suggestions on the technical development.

BIBLIOGRAPHY

[Alchourrón and Bulygin, 1971] C. E. Alchourrón and E. Bulygin. *Normative Systems.* Springer-Verlag, Wien-New York, 1971.

[Allen and Saxon, 1986] L. E. Allen and C. S. Saxon. Analysis of the logical structure of legal rules by a modernized and formalized version of Hohfeld fundamental legal conceptions. In A. A. Martino and F. Socci, editors, *Automated Analysis of Legal Texts*, pages 385–451. North-Holland, Amsterdam, 1986.

[Allen and Saxon, 1993] L. E. Allen and C. S. Saxon. A-Hohfeld: A language for robust structural representation of knowledge in the legal domain to build interpretation-assistance expert systems. In J.-J. Ch. Meyer and R. J. Wieringa, editors, *Deontic*

Logic in Computer Science: Normative System Specification, chapter 8, pages 205–224. John Wiley & Sons, Chichester, England, 1993.

[Åqvist, 1974] L. Åqvist. A new approach to the logical theory of actions and causality. In S. Stenlund, editor, *Logical Theory and Semantic Analysis*, number 63 in Synthese Library, pages 73–91. D. Reidel, Dordrecht, 1974.

[Belnap and Perloff, 1988] N. Belnap and M. Perloff. Seeing to it that: a canonical form for agentives. *Theoria*, 54:175–199, 1988. Corrected version in [Belnap and Perloff, 1990].

[Belnap and Perloff, 1990] N. Belnap and M. Perloff. Seeing to it that: a canonical form for agentives. In H. E. Kyburg, Jr., R. P. Loui, and G. N. Carlson, editors, *Knowledge Representation and Defeasible Reasoning*, volume 5 of *Studies in Cognitive Systems*, pages 167–190. Kluwer, Dordrecht, Boston, London, 1990.

[Belnap and Perloff, 1992] N. Belnap and M. Perloff. The way of the agent. *Studia Logica*, 51:463–484, 1992.

[Brown, 2000] M. A. Brown. Conditional obligation and positive permission for agents in time. *Nordic Journal of Philosophical Logic*, 5(2):83–112, December 2000.

[Chellas, 1969] B. F. Chellas. *The Logical Form of Imperatives*. Dissertation, Stanford University, 1969.

[Chellas, 1980] B. F. Chellas. *Modal Logic—An Introduction*. Cambridge University Press, 1980.

[Colombetti, 1999] M. Colombetti. Semantic, normative and practical aspects of agent communication. In *Preprints of the IJCAI'99 Workshop on Agent Communication Languages, Stockholm*, pages 51–62, 1999.

[Colombetti, 2000] M. Colombetti. A commitment-based approach to agent speech acts and conversations. In *Proc. Workshop on Agent Languages and Conversation Policies, Autonomous Agents 2000, Barcelona*, June 2000.

[Elgesem, 1992] D. Elgesem. *Action Theory and Modal Logic*. Doctoral thesis, Department of Philosophy, University of Oslo, 1992.

[Herrestad and Krogh, 1995] H. Herrestad and C. Krogh. Obligations directed from bearers to counterparties. In *Proc. 5th International Conf. on Artificial Intelligence and Law, Univ. of Maryland*, pages 210–218. ACM Press, 1995.

[Herrestad, 1996] H. Herrestad. *Formal Theories of Rights*. Doctoral thesis, Department of Philosophy, University of Oslo, 1996.

[Hilpinen, 1997] R. Hilpinen. On action and agency. In E. Ejerhed and S. Lindström, editors, *Logic, Action and Cognition—Essays in Philosophical Logic*, volume 2 of *Trends in Logic, Studia Logica Library*, pages 3–27. Kluwer Academic Publishers, Dordrecht, 1997.

[Hohfeld, 1913] W. N. Hohfeld. Some fundamental legal conceptions as applied in judicial reasoning. *Yale Law Journal*, 23, 1913. Reprinted with revisions as [Hohfeld, 1919 1923 1964] and [Hohfeld, 1978].

[Hohfeld, 1919 1923 1964] W. N. Hohfeld. (Revised version). In W. W. Cook, editor, *Some Fundamental Legal Conceptions as Applied in Judicial Reasoning, and Other Legal Essays*. Yale University Press, 1919, 1923, 1964.

[Hohfeld, 1978] W. N. Hohfeld. (Revised version). In W. C. Wheeler, editor, *Some Fundamental Legal Conceptions as Applied in Judicial Reasoning, and Other Legal Essays*. Greenwood Press, 1978.

[Horty and Belnap, 1995] J. F. Horty and N. Belnap. The deliberative stit: a study of action, omission, ability, and obligation. *Journal of Philosophical Logic*, 24(6):583–644, 1995.

[Horty, 1996] J. F. Horty. Agency and obligation. *Synthese*, 108:269–307, 1996.

[Horty, 2001] J. F. Horty. *Agency and Deontic Logic*. Oxford University Press, 2001.

[Jennings, 1993] N. R. Jennings. Commitments and conventions: the foundation of coordination in multi-agent systems. *Knowledge Engineering Review*, 8(3):223–250, 1993.

[Jones and Parent, 2008] A. J. I. Jones and X. Parent. Normative-informational positions: a modal-logical approach. *Artifical Intelligence and Law*, 16:7–23, 2008.
[Jones and Sergot, 1992] A. J. I. Jones and M. J. Sergot. Formal specification of security requirements using the theory of normative positions. In Y. Deswarte, G. Eizenberg, and J.-J. Quisquater, editors, *Computer Security—ESORICS 92*, LNCS 648, pages 103–121. Springer-Verlag, Berlin Heidelberg, 1992.
[Jones and Sergot, 1993] A. J. I. Jones and M. J. Sergot. On the characterisation of law and computer systems: The normative systems perspective. In J.-J. Ch. Meyer and R. J. Wieringa, editors, *Deontic Logic in Computer Science: Normative System Specification*, chapter 12, pages 275–307. John Wiley & Sons, Chichester, England, 1993.
[Jones and Sergot, 1996] A. J. I. Jones and M. J. Sergot. A formal characterisation of institutionalised power. *Journal of the IGPL*, 4(3):429–445, 1996.
[Kanger and Kanger, 1966] S. Kanger and H. Kanger. Rights and Parliamentarism. *Theoria*, 32:85–115, 1966.
[Kanger, 1971] S. Kanger. New foundations for ethical theory. In R. Hilpinen, editor, *Deontic Logic: Introductory and Systematic Readings*, pages 36–58. D. Reidel, Dordrecht, 1971. Originally published as Technical Report, Stockholm University, 1957.
[Kanger, 1972] S. Kanger. Law and Logic. *Theoria*, 38:105–132, 1972.
[Kanger, 1985] S. Kanger. On realization of human rights. In G. Holmström and A. J. I. Jones, editors, *Action, Logic and Social Theory*. Acta Philosophica Fennica, Vol. 38, 1985.
[Krogh, 1997] C. Krogh. *Normative Structures in Natural and Artificial Systems*. Doctoral thesis, University of Oslo, 1997.
[Lee, 1988] R. M. Lee. Bureaucracies as deontic systems. *ACM Transactions on Information Systems*, 6(2):87–108, 1988.
[Lindahl, 1977] L. Lindahl. *Position and Change—A Study in Law and Logic*. Number 112 in Synthese Library. D. Reidel, Dordrecht, 1977.
[Lindahl, 1994] L. Lindahl. Stig Kanger's theory of rights. In D. Prawitz, B. Skyrms, and D. Westerståhl, editors, *Logic, Methodology and Philosophy of Science IX*, pages 889–911, New York, 1994. Elsevier Science Publishers B.V.
[Makinson, 1986] D. Makinson. On the formal representation of rights relations. *Journal of Philosophical Logic*, 15:403–425, 1986.
[Perloff, 1991] M. Perloff. 'Stit' and the language of agency. *Synthese*, 86:379–408, 1991.
[Pörn, 1970] I. Pörn. *The Logic of Power*. Blackwells, Oxford, 1970.
[Pörn, 1974] I. Pörn. Some basic concepts of action. In S. Stenlund, editor, *Logical Theory and Semantic Analysis*, number 63 in Synthese Library, pages 93–101. D. Reidel, Dordrecht, 1974.
[Pörn, 1977] I. Pörn. *Action Theory and Social Science: Some Formal Models*. Number 120 in Synthese Library. D. Reidel, Dordrecht, 1977.
[Pörn, 1989] I. Pörn. On the nature of a social order. In J. E. Fenstad et al., editors, *Logic, Methodology and Philosophy of Science VIII*, pages 553–567. Elsevier Science Publishers, 1989.
[Santos and Carmo, 1996] F. Santos and J. Carmo. Indirect action, influence and responsibility. In M. A. Brown and J. Carmo, editors, *Deontic Logic, Agency and Normative Systems—Proc. DEON'96: 3rd International Workshop on Deontic Logic in Computer Science, Sesimbra (Portugal)*, Workshops in Computing Series, pages 194–215. Springer-Verlag, Berlin-Heidelberg, 1996.
[Santos et al., 1997] F. Santos, A. J. I. Jones, and J. Carmo. Action concepts for describing organised interaction. In *Proc. 13th Annual Hawaii International Conf. on System Sciences*, volume V. IEEE Computer Society Press, Los Alamitos, California, 1997.
[Segerberg, 1985] K. Segerberg. Routines. *Synthese*, 65:185–210, 1985.
[Segerberg, 1989] K. Segerberg. Bringing it about. *Journal of Philosophical Logic*, 18:327–347, 1989.

[Segerberg, 1992] K. Segerberg. Getting started: Beginnings in the logic of action. *Studia Logica*, 51(3-4):347-378, 1992.

[Sergot and Richards, 2000] M. J. Sergot and F. C. M. Richards. On the representation of action and agency in the theory of normative positions. In *Proc. Fifth International Workshop on Deontic Logic in Computer Science (DEON'00), Toulouse*, January 2000.

[Sergot, 1996] M. J. Sergot. A computational theory of normative positions. II Non-regular logics. Technical report, Department of Computing, Imperial College, January 1996.

[Sergot, 2001] M. J. Sergot. A computational theory of normative positions. *ACM Transactions on Computational Logic*, 2(4):581-622, October 2001.

[Sergot, 2008a] M. J. Sergot. Action and agency in norm-governed multi-agent systems. In A. Artikis, G.M.P. O'Hare, K. Stathis, and G. Vouros, editors, *Engineering Societies in the Agents World VIII. 8th Annual International Workshop, ESAW 2007, Athens, October 2007, Revised Selected Papers*, LNCS 4995, pages 1-54. Springer, 2008.

[Sergot, 2008b] M. J. Sergot. The logic of unwitting collective agency. Technical Report 2008/6, Department of Computing, Imperial College London, 2008.

[Shoham, 1991] Y. Shoham. Implementing the intentional stance. In R. Cummins and J. Pollock, editors, *Philosophy and AI: Essays at the Interface*, pages 261-277. MIT Press, Cambridge, Mass., 1991.

[Shoham, 1993] Y. Shoham. Agent-oriented programming. *Artificial Intelligence*, 60:51-92, 1993.

[Singh, 1998] M. P. Singh. Agent communication languages: Rethinking the principles. *IEEE Computer*, 31:40-47, 1998.

[Singh, 1999] M. P. Singh. A social semantics for agent communication languages. In *Preprints of the IJCAI'99 Workshop on Agent Communication Languages, Stockholm*, pages 75-88, 1999.

[Talja, 1980] J. Talja. A technical note on Lars Lindahl's *Position and Change*. *Journal of Philosophical Logic*, 9:167-183, 1980.

[Ting, 1990] T. C. Ting. Application information security semantics: A case of mental health delivery. In D. L. Spooner and C. E. Landwehr, editors, *Database Security: Status and Prospects III*. North-Holland, Amsterdam, 1990.

[von Wright, 1968] G. H. von Wright. *An Essay in Deontic Logic and the General Theory of Action*. Number 21 in Acta Philosophica Fennica. 1968.

[von Wright, 1983] G. H. von Wright. *Practical Reason*. Blackwell, Oxford, 1983.

[Wellman, 1989] C. Wellman. Relative duties in the law. *Philosophical Topics*, 18(1):183-202, 1989.

Marek Sergot
Department of Computing
Imperial College
London SW7 2AZ, UK
Email: m.sergot@imperial.ac.uk

6
Constitutive Norms and Counts-as Conditionals

DAVIDE GROSSI AND ANDREW J. I. JONES

ABSTRACT. The chapter introduces the theory of constitutive rules and counts-as statements from a philosophical/informal point of view and addresses existing attempts to provide a formalization of it. These attempts are concisely described and compared along three main lines: one pertaining to their contribution to the clarification of a set of selected benchmark problems (e.g., institutional power, classificatory rules, conventions etc.); the second pertaining to their methodology (axiomatic/syntactic vs. model-theoretic/semantic approaches); the third pertaining to their strictly formal properties. On the grounds of such systematic comparisons the chapter also identifies open questions and points to future research directions that the authors consider essential in order to shed further light on constitutive rules and counts-as.

1 Introduction 408
2 Theory of counts-as—informal contributions 408
  2.1 Constitutive vs. regulative norms 409
  2.2 Brute vs. institutional facts, and contextual nature of constitutive norms 411
  2.3 Constitutive rules as a "technique of presentation" . . 412
3 Theory of counts-as—formal contributions 415
  3.1 Jones *et al.* 416
  3.2 Gelati *et al.* 419
  3.3 Grossi *et al.* 420
  3.4 Lorini *et al.* 423
4 Alternative formalisms for counts-as 424
  4.1 Governatori *et al.* 424
  4.2 Boella *et al.* 425
  4.3 Lindahl *et al.* 427
5 Classifying the formal approaches to counts-as 428
  5.1 Thematic classification 428
  5.2 Methodological classification 429
  5.3 Logical classification 432

6   Some open problems 435
7   Conclusions 437

1 Introduction

Constitutive norms—or rules[1]—are a commonplace of social reality as we know it. They make possible basic 'institutional' actions such as the making of contracts, the issuing of fines, the decreeing of divorces. With the work of Searle (in particular [Searle, 1969] and [Searle, 1995]), to which we will often return in the chapter, these norms have acquired a somewhat canonical form, the one of *counts-as conditionals*:

$$X \text{ counts as } Y \text{ in context } C.$$

This canonical presentation of constitutive norms paved the way for the natural question of what the logic of these rules is, in terms of the logic of counts-as conditionals. The present chapter reviews the attempts that have been made at understanding this logic since the first paper on the issue was published in 1996 [Jones and Sergot, 1996]. As we will see, these investigations have given rise to a lively inter-disciplinary research field which has produced a rich and varied landscape of logical systems.

Outline of the chapter.
The chapter will develop along the following lines. Section 2 provides an overview of that philosophical work on constitutive norms and counts-as conditionals from which later formal work has developed, and which has inspired all the different formal approaches which will constitute the core of the chapter. Those approaches are dealt with in Section 3. That section provides a bird's eye view of the landscape of formal accounts of constitutive norms and counts-as conditionals and sets the ground for their analysis and comparison, which is developed in Section 5. We consider this latter section as the main contribution of this chapter, where the various approaches to counts-as are classified and compared with respect to three key criteria: one, the aspects of constitutive rules they aim at accounting for formally; two, the methods they use in their formal analyses; three, the formal properties of the different formalizations. Finally, Section 6 deals with some of the most challenging—in the authors' view—open problems in the field. Section 7 briefly recapitulates and concludes the chapter.

2 Theory of counts-as—informal contributions

The concepts of counts-as conditional and constitutive rule have been informally discussed, under different names, in several philosophical sub-

---
[1] We use the two terms interchangeably in the chapter.

disciplines such as: the theory of institutions [Searle, 1969; Cherry, 1973; Searle, 1995], the theory of action [Goldman, 1976], the theory of norms [Von Wright, 1963; Alchourrón and Bulygin, 1971], the theory of law [Rawls, 1955; Ross, 1957; Peczenik, 1989; Bulygin, 1992], the theory of communication [Searle, 1969; Jones and Parent, 2004; Fornara *et al.*, 2007].

The present section is devoted to a brief discussion of the main features of counts-as and constitutive rules as they emerge from some of the philosophical literature just mentioned. This has to be intended as a non-exhaustive overview, emphasizing some of the aspects which have been given particular attention by the formal approaches to counts-as that will be discussed later.[2] These aspects are: the opposition between regulative and constitutive norms; the opposition between brute and institutional facts and the contextual nature of the latter; the classificatory and definitional role played by constitutive norms; finally, their use as a basic technique of presentation of the law.

2.1 Constitutive vs. regulative norms

Regulative norms are what most commonly go simply under the name 'norm'. They have deontic content and they indicate what is obligatory, permitted, forbidden. A very much emphasised feature of constitutive rules is that they do not regulate actions or states-of-affairs, but rather they define new possible actions or states of affairs.

The distinction is very explicitly stated in Searle and Bulygin, as the following quotes illustrate:

> "As a start, we might say that regulative rules regulate antecedently or independently existing forms of behavior [...]. But constitutive rules do not merely regulate, they create or define new forms of behavior." [Searle, 1969, p. 33]

> "Where the rule is purely regulative, behaviour which is in accordance with the rule could be given the same description or specification (the same answer to the question 'What did he do?') whether or not the rule existed, provided the description or specification makes no explicit reference to the rule. But where the rule (or system of rules) is constitutive, behaviour which is in accordance with the rule can receive specifications or descriptions which it could not receive if the rule did not exist." [Searle, 1969, p. 35]

---

[2]There are of course also other philosophical works that address these topics, but we have chosen to focus on those that have perhaps most influenced the development of formal-logical theories of counts-as conditionals.

> "If we do not comply with such rules [constitutive rules], the result is not a sanction or a punishment, for it is not breach or violation of any obligation, nor an offence, but nullity." [Bulygin, 1992, p. 208]

Although the difference between regulation and constitution might be clear, it is much less clear what the notion of constitution precisely amounts to. In the philosophical literature, a common way to describe the notion of constitution is by interpreting it as the fact that the very existence of constitutive norms is a necessary condition for the existence of certain social practices like games, such as baseball or chess:

> "In the case of actions specified by practices it is logically impossible to perform them outside the stage-setting provided by those practices, for unless there is the practice, and unless the requisite proprieties are fulfilled, whatever one does, whatever movements one makes, will fail to count as a form of action which the practice specifies. What one does will be described in some other way.
>
> One may illustrate this point from the game of baseball. Many of the actions one performs in a game of baseball one can do by oneself or with others whether there is the game or not. For example, one can throw a ball, run, or swing a peculiarly shaped piece of wood. But one cannot steal base, or strike out, or draw a walk, or make an error, or balk; although one can do certain things which appear to resemble these actions such as sliding into a bag, missing a grounder and so on. Striking out, stealing a base, balking, etc., are all actions which can only happen in a game. No matter what a person did, what he did would not be described as stealing a base or striking out or drawing a walk unless he could also be described as playing baseball, and for him to be doing this presupposes the rule-like practice which constitutes the game". [Rawls, 1955, p. 25].

Or, similarly, as Searle puts it about the game of chess:

> "[W]hat the 'rule' seems to offer is part of a definition of 'checkmate' [...] That, for example, a checkmate in chess is achieved in such and such way can appear now as a rule, now as an analytic truth based on the meaning of 'checkmate in chess'. That such statements can be construed as analytic is a clue to the fact that the rule in question is a constitutive one. The rules for checkmate [...] must 'define' *checkmate in chess* [...] in

the same way that [...] the rules of chess define 'chess' [...]."
[Searle, 1969, p. 34]

Let us elaborate these observations by means of a concrete example of the simple type of constitutive norms consisting of the rules of chess.

Example 2.1 (Checkmate) *The following are constitutive rules of the game of chess: a checkmate occurs if a king is under direct attack and all of its moves lead to a position which is also under direct attack; a piece is under direct attack if an opponent's piece has an available move to its square; an available move of a piece is a move according to the piece's codified style of moving. These rules describe a class of situations on the chessboard, all those situations in which a checkmate occurs. Figure 1 depicts a simple instance of that class.*

Figure 1: A configuration on a chess board instantiating a situation, viz. the fact that the black king cannot move, which *counts as* a checkmate (of black)

2.2 Brute vs. institutional facts, and contextual nature of constitutive norms

To continue with Example 2.1, the situation illustrated in Figure 1 can be crudely described by giving all the pieces' coordinates on the chessboard. In virtue of the rules of chess, that configuration of pieces is such that the black king cannot move according to its style of moving, and hence it is checkmated. Searle would call the description given in terms of coordinates a 'brute fact', and checkmate an 'institutional fact'. The link between the two is granted by the rules of chess.

In other words, the way constitutive norms define new forms of actions or new states-of-affairs is by relating them to something already existing or established. So, constitutive norms may relate "brute facts" [Anscombe, 1958] to "institutional facts".[3] A precursor of the distinction brute/institutional is the German legal philosopher Pufendorf who, already in the 17th century, made a similar distinction between physical and moral entities:

> "Now, as the original manner of producing physical entities is creation, there is hardly a better way to describe the production of moral entities than by the word 'imposition' [*impositio*]. For moral entities [*entia moralia*[4]] do not arise from the intrinsic substantial principles of things but are superadded to things already existent and physically complete [read brute facts]." [Pufendorf, 1688, pp. 100-1]

This distinction, however, plays a central role in Searle's theory who also stresses a further aspect of it, namely the contextual nature of the 'constitution'. Context has been incorporated by Searle as an explicit component of what we called above counts-as conditionals:

> "[...] 'institutions' are systems of constitutive rules. Every institutional fact is underlain by a (system of) rule(s) of the form 'X counts as Y in context C'." [Searle, 1969, pp. 51-2]

The contextual nature of constitutive norms is not obvious in examples inspired by game-playing such as Example 2.1. It becomes instead evident when talking about social practices such as, for instance, marriage.

Example 2.2 (The bringing about of a marriage) *"By the power vested in me by the State c, I now pronounce you husband and wife". The declaration makes explicit the context c in which the new state-of-affairs occurs, and that this state occurs as result of the declaration itself. The power to which the declaration refers is rooted in a rule of State c, stating that such a declaration in the wedding ritual,* counts as *the creation of a state-of-affairs in which the couple is married.*

2.3 Constitutive rules as a "technique of presentation"

We conclude the section by briefly reporting on work by Ross [Ross, 1957]. This work offers a quite illuminating view of constitutive rules which focuses on the *'raison d'être'* of such rules within legal systems. Constitutive rules seem to be a pervasive feature of legal systems, but why is it so? Or, said otherwise, what are constitutive rules actually good for?

Ross provides an answer to the question by reporting a lively story, of which we quote an excerpt:

---

[3] As we shall see later, they may also relate institutional facts to other institutional facts.

[4] Cf. [Ricciardi, 1997].

> "On the Nosulli Islands in the South Pacific lives the Noît-cif tribe, generally regarded as one of the more primitive peoples to be found in the world today [...]. This tribe [...] holds the belief that in the case of an infringement of certain taboos—for example, if a man encounters his mother-in-law, or if a totem animal is killed, or if someone has eaten of the food prepared for the chief—there arises what is called tû-tû. The members of the tribe also say that the person who committed the infringement has become tû-tû. It is very difficult to explain what is meant by this. [...] tû-tû is conceived as a kind of dangerous force [...] a person who has become tû-tû must be subjected to a special ceremony of purification." [Ross, 1957, p. 812]

In Ross's view a term such as tû-tû is a word devoid of any meaning, it is a term without reference.[5] Nonetheless, terms of this type do play a key role in the specification of norms. They are the bridge—in logical terms the interpolant—which enables inferences connecting concrete facts, to normative consequences. To use Ross's example:

(i) If a person has eaten of the chief's food she is *tû-tû*.

(ii) If a person is *tû-tû* she has to be subjected to a ceremony of purification.

(iii) If a person has eaten of the chief's food she has to be subjected to a ceremony of purification.

To say it in the manner of Searle, tû-tû is an institutional fact and Statement (i) connects it to a specific brute fact. In the counts-as terminology: the fact that a person has eaten of the chief's food counts, in the context of Noît-cif tribal laws, as the fact that she is tû-tû. Statement (ii) then introduces a normative consequence linked to the institutional fact 'tû-tû', stating what the effects of being tû-tû are (in this case normative effects). Taken together, they allow the inference of Statement (iii), where tû-tû does not occur any more, and which makes the connection between the fact at issue and its normative consequences explicit.

Ross stresses the analogy of the above inference pattern to the sort of rulings we are bound to encounter in modern legal codes.

> "We find the following phrases, for example, in legal language:

---

[5] We do not endorse this view here, as it is actually a matter of controversy in philosophy (cf. [Hindriks, 2009]), nor do we intend to dig deeper into the issue of reference and denotation of tû-tû-like terms. What matters for our purposes here is Ross's analysis of the function of such terms within a normative system.

(1) If a loan is granted, there comes into being a claim;

(2) If a claim exists, then payment shall be made on the day it falls due;

This is only a roundabout way of saying:

(3) If a loan is granted, then payment shall be made on the day it falls due.

The claim mentioned in (1) and (2), but not in (3), is obviously, like tû-tû, not a real thing; it is nothing at all, merely a word, an empty word devoid of any semantic reference." [Ross, 1957, p. 817-8]

The point is thus made that "our legal rules are in a wide measure couched in a 'tû-tû' terminology" [Ross, 1957, p. 817]. Terms such as claim, right, duty, ownership work exactly like tû-tû allowing us to connect a set of concrete circumstances to a set of legal or, more generally, normative consequences. This detour via tû-tû-terms might be dispensed with, but the price to pay is a rather cumbersome formulation. To realize this, suppose you were asked to connect, by means of rules, each of n (brute) facts $F_1, \ldots, F_n$ to m (normative) consequences $C_1, \ldots, C_m$. The naive way to do that would consist in connecting each fact to each consequence, thereby producing $n \cdot m$ different rules of the form $F_i - C_j$, with $1 \leq i \leq n$ and $1 \leq j \leq m$. This solution is displayed in Figure 2.

$$
\begin{array}{cccc}
F_1 - C_1 & F_2 - C_1 & \ldots & F_n - C_1 \\
F_1 - C_2 & F_2 - C_2 & \ldots & F_n - C_2 \\
\vdots & \vdots & & \vdots \\
& & & F_{n-1} - C_{m-1} \\
F_1 - C_n & F_2 - C_n & \ldots & F_n - C_m
\end{array}
$$

Figure 2: $n \cdot m$ rules connecting n (brute) facts to m (normative) consequences

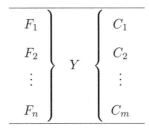

Figure 3: $n + m$ rules connecting n (brute) facts to m (normative) consequences

But now observe what the addition of a tû-tû-like term—let us call it Y—would allow you to do. The same situation would be expressible via $n + m$ rules: n rules of the form $F_i - Y$ connecting each brute fact to the term Y, and m rules of the form $Y - C_j$, with $1 \leq j \leq m$, connecting Y to each normative consequence. This solution is displayed in Figure 3.[6] Each of the $F_i - Y$ rules can be consistently thought of as a counts-as statement of the form F_i counts as Y, in the context of the thereby defined normative system or institution.

Ross's point is precisely that tû-tû-like terms or, in Searlean terminology, institutional facts, enable a very manageable and effective "technique of presentation" [Ross, 1957, p. 821] for systems of norms. And within this picture constitutive rules play therefore a central role and have to be considered as a basic building block for the construction of normative systems.

3 Theory of counts-as—formal contributions

The thrust to the development of a formal analysis of constitutive rules could be traced back to Searle's work itself where, in both [Searle, 1969] and [Searle, 1995], constitutive rules are constantly related to a specific syntactic form: "Constitutive rules have the form: 'X counts as Y in context C'".[7] Once a special syntactic form is in focus, the natural question arises as to what the logic of that form is. Spanning across several techniques and methods, this section summarizes the findings of those authors that took up the quest for a logic of statements of the form 'X counts as Y in context

---

[6] The interested reader might already glance over the formalization of Figure 3 provided later on within Section 3 in Formula 6.

[7] However, Searle was certainly not the only—nor indeed the first—philosopher in the modern period to describe constitutive rules in terms of the ordinary English verb 'count as'. See, e.g., [Rawls, 1955, p. 25].

C'.

We can identify five main groups of contributions to the formal analysis of counts-as and constitutive norms: by Jones et al. [Jones and Sergot, 1996; Jones and Parent, 2004; Jones and Parent, 2007]; by Gelati et al. [Gelati et al., 2002; Artosi et al., 2004; Gelati et al., 2004]; by Boella et al. [Boella and van der Torre, 2003; Boella and Van der Torre, 2004; Boella and van der Torre, 2006]; by Lindahl et al. [Lindahl and Odelstad, 2006; Lindahl and Odelstad, 2008a; Lindahl and Odelstad, 2008b]; by Grossi et al. [Grossi et al., 2005; Grossi, 2007; Grossi et al., 2006a; Grossi et al., 2006b; Grossi et al., 2008; Grossi, 2008; Grossi, 2011]; by Lorini et al. [Gaudou et al., 2008; Lorini and Longin, 2008; Lorini et al., 2009]; and by Governatori et al. [Governatori and Rotolo, 2008].

The present section describes the key features of each of these approaches and offers a concise overview of the different techniques used up till now in the formal analysis of counts-as. We will refrain from providing full details about each of the approaches, and will rather focus on their key definitions, trying to emphasize the characteristic features of each of them. This will give us a solid basis on which to ground the systematic comparison of these approaches to be articulated in Section 5. The section divides the formalisms into two groups: the ones based on modal logic, and the ones based on alternative formalisms.

Modal logics of counts-as

To this group belong the works by Jones et al., Gelati et al., Grossi et al. and Lorini et al.

3.1 Jones et al.

The formal analysis of constitutive norms and counts-as conditionals starts with [Jones and Sergot, 1996]. This work is concerned with isolating a core of logical principles governing the use of statements of the form "X counts as Y in context (or, institution) c", which are given a representation in a logical language by means of a connective $\Rightarrow_c$. Counts-as conditionals are thus represented by formulae of the type $\varphi_1 \Rightarrow_c \varphi_2$, and are studied within the framework of conditional logics in the Chellas tradition [Chellas, 1980].

The resulting conditional logic of counts-as is taken to validate, on top of propositional logic, the following principles:

(1) $\quad \varphi_2 \leftrightarrow \varphi_3 \ / \ (\varphi_1 \Rightarrow_c \varphi_2) \leftrightarrow (\varphi_1 \Rightarrow_c \varphi_3)$

(2) $\quad \varphi_1 \leftrightarrow \varphi_3 \ / \ (\varphi_1 \Rightarrow_c \varphi_2) \leftrightarrow (\varphi_3 \Rightarrow_c \varphi_2)$

(3) $\quad ((\varphi_1 \Rightarrow_c \varphi_2) \wedge (\varphi_1 \Rightarrow_c \varphi_3)) \rightarrow (\varphi_1 \Rightarrow_c (\varphi_2 \wedge \varphi_3))$

(4) $\quad ((\varphi_1 \Rightarrow_c \varphi_2) \wedge (\varphi_3 \Rightarrow_c \varphi_2)) \rightarrow ((\varphi_1 \vee \varphi_3) \Rightarrow_c \varphi_2)$

(5) $\quad (\varphi_1 \Rightarrow_c \varphi_2 \wedge \varphi_2 \Rightarrow_c \varphi_3) \to (\varphi_1 \Rightarrow_c \varphi_3)$

The inference rules in Formulae 1 (right logical equivalence) and 2 (left logical equivalence) simply state the rather uncontroversial property that counts-as conditionals are closed under substitution of provable equivalents in the antecedent as well as the consequent.

Formulae 3 and 4 express the properties of conjunction of the consequent and, respectively, disjunction of the antecedent, both to be considered quite natural for statements of the type "X counts as Y in c". In fact they allow for a natural rendering of the logical content of the "technique of presentation" enabled by counts-as statements, which we briefly discussed in Section 2.3. We could recast the content of Figure 3 by means of operator $\Rightarrow_c$ in the following way:

(6) $\quad \displaystyle\bigvee_{1 \leq i \leq n} F_i \Rightarrow_c Y \wedge Y \Rightarrow_c \bigwedge_{1 \leq j \leq m} C_j$

assuming consequences C_j to be expressed as institutional facts too. In other words, Formulae 3 and 4 allow one to cluster n counts-as statements of the form $F_i \Rightarrow_c Y$ connecting n brute facts F_i to the institutional fact Y, and m similar statements of the form $Y \Rightarrow_c C_j$ from the institutional fact Y to m institutional facts C_j.

Notice that not quite as intuitive and natural are the converses of Formulae 3 and 4, and in particular the converse of Formula 4 (antecedent strengthening[8]), for the reason perspicuously illustrated by the following quote:

> "The point is essentially this: suppose that x is empowered to marry couple a and b by performing ritual R. Now suppose that some other agent y brings it about that x performs ritual R—y, let us imagine, successfully exercises influence over x by some means or other. So x performs the ritual and the couple a and b are married. Despite his successful exercise of influence, we would not here want to say that y too was empowered, by institution s, to marry the couple. Institutional power is not *transferable* in that way." [Jones and Sergot, 1996, p. 434]

Finally, Formula 5 expresses the transitivity of counts-as statements, which [Jones and Sergot, 1996] accepts on the grounds of the property of *conventional generation* as studied in [Goldman, 1976].

---

[8]In the current notation such a property is formulated as follows:

$$\varphi_1 \Rightarrow \varphi_2 \to \varphi_1 \wedge \varphi_3 \Rightarrow \varphi_2.$$

As we will see later in Section 5, Formulae 1-5 constitute a 'minimal core' of logical principles for the logic of counts-as and have hardly been criticized (except for the more controversial property of transitivity, Formula 5) by later proposals for the formal analysis of counts-as.

In addition to the above principles, [Jones and Sergot, 1996] links the logic of counts-as conditionals to the logic of a modality D_c. The intuitive reading of a formula $D_c\varphi$ is "it is a constraint of institution c that φ". The logic of D_c is taken to be that of the normal modal system $\mathbf{KD}$[9] and it is linked to the logic of counts-as conditionals by the following schemata:

(7) $\qquad (\varphi_1 \Rightarrow_c \varphi_2) \to D_c(\varphi_1 \to \varphi_2)$
(8) $\qquad (\varphi_1 \Rightarrow_c \varphi_2) \to (\varphi_1 \to D_c\varphi_1)$.

Formula 7 states, intuitively, that counts-as conditionals of a given institution c are a subset of the constraints operative in institution c. The second one, Formula 8, states that, if a state-of-affairs occurs as an antecedent in a counts-as conditional, then, if that state-of-affairs is the case it is also "institutionally" the case, that is, it is recognized by the institution concerned.

The rationale for the choice of schemata such as Formulae 7 and 8, is motivated in [Jones and Sergot, 1996] by the attempt to provide a systematization of inference patterns that arise in the common use of counts-as statements. The sort of logic arising from the interaction of Formulae 1-5 with the logic of the D_c modality allows for specific reasoning patterns concerning counts-as to be given a formal systematization. In particular, the analysis proposed in [Jones and Sergot, 1996], besides being philosophically informed by contributions such as those of Bulygin, Goldman, Rawls and Searle, among others, aims precisely at accounting for the following sort of inference.

Example 3.1 (Institutional detachment) *Let us provide a first formalization of Example 2.2. The counts-as rule at issue can be expressed as* $p \Rightarrow_c m$. *Roughly, the state-of-affairs in which the officer pronounces the couple husband and wife (p in symbols), in the context of institution c, counts as the couple being married (m in symbols). Now, by assuming that p is the case, we would like to be able to infer that m actually holds in the context*

---

[9]That is, the modal logic with modality D_c containing the axioms of propositional logic, axioms

$\qquad \mathbf{K} \qquad D_c(\varphi_1 \to \varphi_2) \to (D_c\varphi_1 \to D_c\varphi_2)$
$\qquad \mathbf{D} \qquad \neg D_c \bot$

and the rules of modus ponens (RM) and necessitation (RN).

of institution c. This reasoning pattern is sound in the Jones et al.'s logic:

(9) $\quad \{p \Rightarrow_c m, p\} \vdash D_c m$

To prove this rule, from $p \Rightarrow_c m$, Formula 8 and modus ponens we can infer $(p \to D_c p)$, from which, by p, modal principles and modus ponens we can conclude $D_c m$.

Besides triggering essentially all the further proposals on the formalization of counts-as statements, the framework developed in [Jones and Sergot, 1996] has been applied to the theory of signaling conventions in [Jones and Parent, 2004; Jones and Parent, 2007].

3.2 Gelati *et al.*

Gelati *et al.* [Gelati *et al.*, 2002; Gelati *et al.*, 2004] define a counts-as operator $\Rightarrow_c$ in terms of a more basic form of conditional $\Rightarrow$, which they call *normative*, and, following in the footsteps of [Jones and Sergot, 1996], in terms of a D_c modality:

(10) $\quad \varphi_1 \Rightarrow_c \varphi_2 \;:=\; (\varphi_1 \Rightarrow D_c \varphi_2) \land (D_c \varphi_1 \Rightarrow D_c \varphi_2)$

The rationale of this definition, the authors claim, resides in interpreting statements of the type "X counts as Y in context c" as statements asserting that "from X follows that Y is institutionally the case in c *and* that from the fact that X is institutionally the case in c it also follows that Y is institutionally the case in c". To capture that intuitive reading, and thereby substantiate the definition in Formula 10, the authors choose non-normal logics for both the $\Rightarrow_c$ and the D_c operators:

- The $\Rightarrow$ operator is taken to correspond to cumulative reasoning[10] together with the rather non-standard inference rule inspired by non-monotonic reasoning:

$$\frac{\Phi \vdash \varphi_1 \quad \Phi \vdash \varphi_1 \Rightarrow_c \varphi_2}{\Phi \vdash \varphi_2}$$

if for any φ_1' such that $\Phi \vdash \varphi_1' \to \varphi_1$, $\Phi \nvdash \varphi_1' \Rightarrow_c \neg \varphi_2$, and where Φ is a set of formulae.

- The D_c operator is taken to be a non-normal modality obeying only the two principles: $D_c \varphi \to \neg D_c \neg \varphi$ and $(D_c \varphi_1 \land D_c \varphi_2) \to D_c(\varphi_1 \land D_c \varphi_2)$.

---

[10]To be precise, in [Gelati *et al.*, 2002] it is argued that the logic of counts-as corresponds to preferential reasoning, while in [Gelati *et al.*, 2004] it is considered to correspond *at least* to cumulative reasoning, i.e., preferential reasoning without the property of disjunction of the antecedents (see [Kraus *et al.*, 1990]).

The structural properties of $\Rightarrow_c$ arising from Formula 10 are not explored and no metalogical results (e.g., soundness, completeness) are investigated for the proposed logic of counts-as.

3.3 Grossi *et al.*

In a series of papers starting with [Grossi *et al.*, 2005] Grossi *et al.* develop a theory of counts-as conditionals within the framework of normal modal logic (i.e., of extensions of the modal system **K**). The upshot of their analysis consists in isolating a family of different operators, in ascending logical strength, all capturing different semantic components which seem to be involved in statements of the type "X counts as Y in context c". Four notions of counts-as are studied, which are informally presented in the following list:

Classificatory counts-as: situations of the type X are all of the type Y in context c. This interpretation considers counts-as conditionals simply as contextual classifications and is rooted in a simple observation made in [Jones and Sergot, 1996]:

> "There are usually constraints within any institution according to which certain states of affairs of a given type count as, or *are to be classified as*, states of affairs of another given type." [Jones and Sergot, 1996, p. 139]

Proper classificatory counts-as: situations of the type X are all of the type Y in context c, and this does not hold in general. This interpretation builds on the previous one requiring that the statement "situations of the type X are all of the type Y" be not a universal truth or, in other words, that it be something proper of context c.

Ascriptive counts-as: situations of the type X are all of the type Y in context c, and type Y is a newly introduced concept. This interpretation also builds on the first one, and makes explicit that what happens in a counts-as statement, besides the classificatory content, is also the 'creation' of a new concept, without which the classification would not be possible.

Constitutive counts-as: situations of the type X are all of the type Y in context c, and this does not hold in general and, in addition, context c is explicitly defined by a set Γ of statements including that X implies Y. Finally, this last interpretation takes into consideration that:

> "Rules are constitutive if and only if they are part of a set of rules. Strictly speaking, there is no such thing as *a* rule that is constitutive in isolation." [Ricciardi, 1997, p. 5]

This means that a counts-as statement, besides its classificatory content, is also always part of a set of rules which, together, define the context c of the statement.[11]

The fundamental component of the formalization of all these different notions is, essentially, a simple modal logic of contexts where contexts are represented as modal operators $[c]$ with the following interpretation:

(11) $$\mathcal{M}, s \models [c]\varphi \quad \leftrightarrow \quad S_c \subseteq ||\varphi||_\mathcal{M}$$

where $||\varphi||_\mathcal{M}$ denotes the truth-set of φ in $\mathcal{M}$ and S_c is the set of states corresponding to context c.[12] That is, φ holds in context c if and only if all states in c are φ states. [13] This is all that is needed to express the classificatory core meaning of counts-as statements, the so-called *classificatory counts-as*:

(12) $$\varphi_1 \Rightarrow_c^{cl} \varphi_2 \; := \; [c](\varphi_1 \rightarrow \varphi_2)$$

On the top of this simple definition, Grossi et al. then investigate a number of normal modal operators which, in interaction with the context modality, allow one to capture all the aforementioned different interpretations of counts-as statements. Here we will only report the semantics of such operators, and show how they are used to obtain the desired formalizations. Such operators are the well-known universal modality $[U]$ (cf. [Blackburn et al., 2001]), a linguistic release operator $\Delta_{\sigma(\varphi)}$ borrowed from [Krabbendam and Meyer, 2000; Krabbendam and Meyer, 2003], and a contextual complement operator $[-c]$ introduced in [Grossi et al., 2006b; Grossi et al., 2008]:

(13) $\quad \mathcal{M}, s \models [U]\varphi \quad \leftrightarrow \quad S \subseteq ||\varphi||_\mathcal{M}$
(14) $\quad \mathcal{M}, s \models \Delta_{\sigma(\varphi_2)}\varphi_1 \quad \leftrightarrow \quad \forall s' \text{ S.T. } s' \sim_{\sigma(\varphi_2)} s : \mathcal{M}, s \models \varphi_1$
(15) $\quad \mathcal{M}, s \models [-c]\varphi \quad \leftrightarrow \quad S \backslash S_c \subseteq ||\varphi||_\mathcal{M}$

where $\sigma(\varphi_2) \subseteq \mathbf{P}$ denotes the vocabulary of formula φ_2 and the relation $\sim_{\sigma(\varphi_2)}$ is an indistinguishability relation holding between states which are equivalent with respect to the atoms in the vocabulary $\sigma(\varphi_2)$. These operators support the following extensions of the definition of classificatory counts-as given in Formula 12:

(16) $\quad \varphi_1 \Rightarrow_c^{cl+} \varphi_2 \; := \; [c](\varphi_1 \rightarrow \varphi_2) \wedge \neg[U](\varphi_1 \rightarrow \varphi_2)$

---

[11] Note that this is in line with the warning raised in [Makinson, 1999]: "no logic of norms without attention to a system of which they form part".
[12] For a thorough exposition we refer the reader to, e.g., [Grossi, 2007].
[13] "A set of constitutive rules defines a logical space" [Ricciardi, 1997, p. 6].

(17) $\quad \varphi_1 \Rightarrow_c^{As} \varphi_2 \; := \; [c](\varphi_1 \to \varphi_2) \wedge \neg[c]\Delta_{\sigma(\varphi_2)}(\varphi_1 \to \varphi_2)$

(18) $\quad \varphi_1 \Rightarrow_{c,\Phi}^{co} \varphi_2 \; := \; [c] \bigwedge \Phi \wedge [-c]\neg \bigwedge \Phi \quad \text{with} \quad \varphi_1 \to \varphi_2 \in \Phi$

Formula 16 defines counts-as as a contextual implication which is not valid in the model—$\neg[U](\varphi_1 \to \varphi_2)$. Formula 17, along a similar line, defines counts-as as a contextual implication which would not be valid in the context any more if we 'release' or 'forget' the vocabulary of the consequent. Finally, Formula 18 defines counts-as as a contextual implication belonging to a set Φ which defines the context of reference c. The definition of the context is rendered by stating that all formulae in Φ are valid in c—$[c] \bigwedge \Phi$—and that some are false in the complement of c—$[-c]\neg \bigwedge \Phi$.

The work of Grossi et al. has provided a detailed study of the structural properties of all the introduced operators[14] and, also, of their relative logical strength. In particular, the following logical relations hold between the different forms of counts-as:

(19) $\qquad\qquad\qquad\qquad \Rightarrow_{c,\Phi}^{co} \; \subset \; \Rightarrow_c^{As} \; \subset \; \Rightarrow_c^{cl+} \; \subset \; \Rightarrow_c^{cl} .$

Furthermore, all the logics used to define the above operators have been proven sound and strongly complete.

So, the approach of Grossi et al. has focused on the identification of different meanings of the counts-as locution, and has formalized those meanings within the framework of normal modal logic. This perspective determined significant differences with respect to the formalization proposed by Jones et al. and it is worth illustrating these differences by showing how the formalization of Grossi et al. handles Example 2.2 in contrast to the formalization provided in Example 3.1.

Example 3.2 (Institutional detachment in Grossi et al.) *Let us now denote with Φ the rules of the institution at issue. This set Φ contains the implication $p \to m$, and Φ defines a context c. We have then a constitutive counts-as: $p \Rightarrow_{c,\Phi}^{co} m$. Now, suppose we are in a situation, let us call it s, in which the officer pronounces the couple husband and wife (p). We have two possibilities. If s belongs to the context defined by Φ we can conclude that in s the couple is married (m) by the classificatory counts-as: $p \Rightarrow_c^{cl} m$ which follows from the constitutive one (recall Formula 19). In other words, if we are indeed in a situation where context c applies, then we can conclude that the couple is married. Note that this is an alternative version of the*

---

[14] We will come back to this aspect in Section 5.

institutional detachment of Example 3.1, albeit semantical:[15]

(20) $\qquad \{p \Rightarrow_{c,\Phi}^{co} m, \bigwedge \Phi, p\} \models m$

On the other hand, if the context does not apply, i.e., if we drop assumption $\bigwedge \Phi$, the conclusion can no longer be drawn.

Examples 3.1 and 3.2 nicely illustrate one of the most striking differences between the approaches of Jones *et al.* and Grossi *et al.*, which consists precisely in the rendering of institutional detachment: Formula 9 vs. Formula 20. As Examples 3.1 and 3.2 nicely show, the key difference between the two approaches consists in the consequences that can be drawn from the assumption of the 'factual' truth p. In Jones *et al.* the truth of p always determines, via counts-as, the institutional truth of m or, more precisely, the truth of $D_c m$ (Formula 9). In Grossi *et al.*, instead, p does not determine the truth of a modalized occurrence of m, but of m itself, although this entailment is conditional under the assumption that the rules defining the context of the counts-as are actually in force (Formula 20). This is rather interesting and reveals a radical difference in the basic set up of the formalization of counts-as conditionals ultimately concerning the rendering of a notion of institutional truth. In Jones *et al.* institutional truth is represented by the modality D_c. So, m is institutionally true in a state s if $\mathcal{M}, s \models D_c m$. On the other hand, in Grossi *et al.* institutional truth can be viewed just as standard truth (i.e., satisfaction in a pointed model) where the evaluation state belongs to the context defined by the set of formulae Φ, that is, $\mathcal{M}, s \models \bigwedge \Phi \wedge m$ and m belongs to the vocabulary of Φ.

3.4 Lorini *et al.*

Lorini *et al.* follow a research line similar to the one pursued by Grossi *et al.*, but focus on an aspect of constitutive rules which was neglected in the latter analysis, namely the fact that constitutive rules, in order to be in force, need to be *accepted* by the members of the society concerned. To pursue this aim, they propose a study of counts-as conditionals where the basic building block is not a normal logic of context, as in Grossi *et al.*, but a logic of acceptance by groups of agents in a set N who are members of institutions in a set C. The primitive modal operator in this logic is $\mathcal{A}_{X:c}$, whose intuitive reading is: "all agents in group X acting as members of institution C accept that ...". The operator is interpreted according to the standard satisfaction relations in Kripke semantics:

(21) $\qquad \mathcal{M}, s \models \mathcal{A}_{X:c} \varphi \;\leftrightarrow\; \forall s' \text{ S.T. } s' \in ACC_{X:c}(s) : \mathcal{M}, s' \models \varphi$

---

[15] A fully syntactical version can be provided by introducing nominals (see [Grossi, 2007; Grossi *et al.*, 2008]).

where $c \in C$, $\emptyset \subset X \subseteq N$, $ACC_{X:c}$ is a function assigning to each state the set of states 'accepted' by X as members of c. We will not provide the formal properties of that function, but for our purposes it suffices to say that a set $ACC_{X:c}(s)$ behaves in a slightly weaker way than a context in Grossi et al., as it represents some kind of collective mental attitude of a group of agents, rather than an external objective institutional reality.

This said, Lorini et al. refine the pattern of Grossi et al. proper classificatory rules to define a counts-as conditional based on acceptance:

$$(22) \quad \varphi_1 \triangleright_c \varphi_2 := \left(\bigwedge_{\emptyset \subset X \subseteq N} \mathcal{A}_{X:c}(\varphi_1 \to \varphi_2) \right) \\ \wedge \neg \left(\bigwedge_{c' \in C} \left(\bigwedge_{\emptyset \subset Y \subseteq N} \mathcal{A}_{Y:c'}(\varphi_1 \to \varphi_2) \right) \right)$$

In other words, φ_1 counts as φ_2 in c if and only if all groups of agents X acting as members of c accept that φ_1 implies φ_2, and there exists at least one institution c' and group of agents Y which does not accept the implication.

Lorini et al. provide a thorough analysis of the definition in Formula 22 as well as soundness and completeness results for the logic of acceptance.

4 Alternative formalisms for counts-as

To this group belong the works by Governatori et al., based on defeasible logic, by Boella et al., based on Input/Output logic and by Lindahl et al., which makes use of an algebraic formalism.

4.1 Governatori et al.

Governatori et al. tackle the analysis of constitutive rules and counts-as conditionals from a legally-informed point of view. In particular, they stress the importance of incorporating in the analysis the typical feature of legal reasoning known as defeasibility which, in the case of counts-as, roughly amounts to the following observation: the inference from X to Y via a statement "X counts as Y in context C" need not be logically valid, and it can be retracted in the presence of further information (e.g., if X is of a somehow exceptional kind). In short, they propose an analysis of counts-as allowing for defeasible institutional detachment (see Example 4.1 below).

This aim is achieved within the framework of defeasible logic [Nute, 1987] whose key structure is the so-called defeasible theory, that is, a triple $(F, R, >)$ where:

- F is a finite set of literals representing basic facts;

- R is a set of rules including *strict rules*, whose conclusions are indisputable, *defeasible rules*—denoted $\Rightarrow$—whose conclusions can be defeated, and *defeaters* which are essentially statements to the effect that a given defeasible rule is not applicable;
- $>$ is a priority relation among rules whose function is to resolve conflicts between rules.

The core intuition of Governatori *et al.* consists in representing institutional scenarios such as Example 2.2 as defeasible theories where F represents the set of facts and where counts-as conditionals are rendered as defeasible rules.[16] To illustrate this idea, we resort to our running example, Example 2.2, by adapting examples to be found in [Governatori and Rotolo, 2008].

Example 4.1 (Defeasible institutional detachment) *Consider the scenario given in Example 2.2, and assume also that "the officer performing the wedding is under threat of death by the couple" (in symbols, d). We have the following defeasible theory $(F, R, >)$ with:*

- $F = \{p, d\}$
- $R = \{r_1 : p \Rightarrow m;\ r_2 : p, d \Rightarrow \neg m\}$
- $\gt = \{(r_2, r_1)\}$

In the terminology of defeasible logic, from this defeasible theory we can defeasibly prove that $\neg m$, while conclusion m obtainable via rule r_1 is overridden by rule r_2.

A further important characteristic of this approach, with respect to those presented above, consists in its tractability from a computational point of view.

4.2 Boella *et al.*

Like the work by Governatori *et al.* the analysis of counts-as proposed in the series of papers [Boella and van der Torre, 2003; Boella and Van der Torre, 2004; Boella and van der Torre, 2006] is also an attempt to formalize constitutive rules by means of techniques coming from the area of defeasible reasoning. The technique is, in this case, the so-called input/output logic (IOL) [Makinson and van der Torre, 2000].

The key idea behind the application of IOL to the analysis of counts-as, consists in representing constitutive norms simply as ordered pairs (a, b)

---

[16] Governatori *et al.* propose actually an articulated extension of defeasible theories, but here we want to expose just the basic idea underlying their approach leaving the details to the interested reader.

where a represents the antecedent of the rule, and b its consequent: "a counts as b". Typically, both a and b are taken to be formulae from propositional logic. Each set of such ordered pairs can be seen as an inferential mechanism which, given an input, determines an output based on such rules.

Various definitions can be given of how to produce the output on the basis of a set of pairs, and all consist in ways of closing the given set of pairs by adding new pairs in accordance to some principles, of which we give two examples:

$$(23) \qquad SI : \frac{(a,b)}{(a \wedge c, b)} \qquad CT : \frac{(a,b),(a \wedge b, c)}{(a,c)}$$

where SI stands for strengthening of the input—essentially an antecedent strengthening property—and CT stands for cumulative transitivity. Formally, given a set $CONS$ of pairs, a closure operation C defined in terms of some of the above principles, and a set of facts A, the output of $CONS$ given C and a set of input formulae I is:

$$(24) \qquad out_C(CONS, A) = \{b \mid (a,b) \in C(CONS) \text{ and } s \in A\}$$

The freedom available in defining the output operation makes IOL an extremely versatile framework. As to the analysis of counts-as, Boella et al. usually employ the output operation which uses a closure based only on the two above principles SI and CT. The technical name in the IOL literature for such an output operation is *simple-minded reusable output*.

The work of Boella et al. does not focus further on the study of structural properties of counts-as statements as such, but is rather interested in the application of such statements to the formal specification of multi-agent systems. In particular, the authors focus on the interaction between the IOL representation of constitutive norms and the representation of regulative norms in the same logic.[17] The representation of regulative norms follows the very same formulation: regulative norms are pairs (d, e) of conditions and normative consequences, and a set of such norms, under a given closure operation, can be used to yield the set of normative consequences of a given set of propositional formulae.

This simple approach naturally lends itself to a formal representation of the sort of nesting of constitutive norms (from brute to institutional facts) and regulative norms (from institutional facts to normative consequences). So, given a set of pairs $CONS$ representing constitutive norms, and a set of pairs REG representing regulative norms, the set of normative consequences

---

[17]The idea that regulative and constitutive rules should be formalized in terms of the same logic is also argued for in [Gelati et al., 2002].

of a given set of facts A is determined by the nested application of an output operation:

(25) $$out_{C'}(REG, out_{C''}(CONS, A))$$

where C' is the set of principles for the output operation on regulative norms, and C'' the set of principles for the output operation on constitutive norms (like in Formula 23). Boella *et al.* develop this intuition further to more complex forms of interaction between the two output operations, but Formula 25 provides the basic idea.

Finally, it is worth mentioning that the authors assume an original conceptualization of normative systems as the set of beliefs and desires of the system itself viewed as one agent, where constitutive rules represent the system's beliefs, and the normative rules its desires.

4.3 Lindahl *et al.*

Based on work by the authors from the late 90s [Lindahl and Odelstad, 2000] in which they advocated an algebraic analysis of normative systems, the series of papers [Lindahl and Odelstad, 2006; Lindahl and Odelstad, 2008a; Lindahl and Odelstad, 2008b] focuses on 'intermediate concepts', that is, the kind of concepts such as "ownership", or "marriage", that are most typically introduced via counts-as statements. In focusing on this issue, the authors provide a formal account of counts-as that highlights its character of 'technique of presentation' which has been discussed earlier in this chapter in Section 2.3.

The approach chosen by Lindahl *et al.* is very close in spirit to the one, discussed above, of IOL. However, the formal machinery deployed is considerably more complex as it hinges on several algebraic and order-theoretical notions. In this section we provide just a brief sketch of the basic technical ideas underlying the framework, referring the interested reader to the authors' chapter in this same volume for more details. Furthermore, as the framework has undergone several modifications, it is appropriate to mention that our presentation will be based, specifically, on the paper [Lindahl and Odelstad, 2008a].

The key notion in Lindahl *et al.*'s work is the one of *Boolean joining system*. The idea behind it is that norms can be seen—exactly as in IOL—as simple pairs (a, b) connecting (factual) conditions to (normative) consequences. Both conditions and consequences are viewed as structured according to a supplemented Boolean algebra, that is, a structure $B = \langle X, \sqcap, -, \bot, \rho \rangle$ where $\langle X, \sqcap, -, \bot \rangle$ is a Boolean algebra, and $\rho \subseteq X^2$ a binary relation including the order $\preceq$ associated to the Boolean algebra.[18] Relation ρ captures, intuitively, a relation of entailment between the elements of the algebra, which strengthens the classical entailment relation given by $\preceq$. Now, once conditions and consequences are represented via such structures, a *Boolean joining system* is a structure $\langle B_1, B_2, J\rangle$ where B_1 and B_2 are supplemented Boolean algebras for conditions and, respectively, consequences, and J is a set of pairs (b_1, b_2) *joining* elements of B_1 with elements of B_2. That set, which is required to satisfy further conditions which we refrain from mentioning here, represents the stipulation of the (regulative) norms of a given normative system.

So how are constitutive norms represented in this framework? The idea is that such norms introduce *intervenients* of existing joinings (b_1, b_2), where an intervenient is an element b of a supplemented subalgebra B of B_1 and B_2 such that b_1 entails b—in symbols, $b_1 \rho b$—and b entails b_2—in symbols, $b \rho b_2$—and in addition b_1 is, roughly, the weakest ground for b and b_2 the strongest consequence of it. According to Lindahl *et al.* the notion of interpolant characterizes the structural properties of constituted concepts such as "ownership" within a given normative system.[19]

5 Classifying the formal approaches to counts-as

The previous sections have provided a somewhat historical overview of the development of the theory of counts-as and constitutive rules. The present section proposes a systematization of these different contributions according to a few different criteria. The various approaches are compared with one another according to the aspects of constitutive rules they deal with, the method they follow in pursuing their analysis, and the formal properties of the resulting models.

Before starting this section, we want to stress that our aim here is not to argue in favour of or against any of the approaches presented, but rather to provide some useful guidelines for the reader to navigate the existing literature.

5.1 Thematic classification

Each of the formal approaches to counts-as we have presented focus their analysis on some features of constitutive rules, abstracting from others. In particular, for each of them it is easy to recognize one main focus of attention in the development of the analysis.

---

[18] A partial linear order can always be associated to a given Boolean algebra as follows:
$$(26) \qquad a \preceq b \quad \leftrightarrow \quad a \sqcap b = a.$$

[19] Lindahl *et al.*'s approach is treated in detail in a dedicated chapter in this Handbook.

- Contextual aspects of counts-as. These have been stressed since the Jones *et al.* work and the reference to a context has been recognized by almost all approaches as essential for the syntax of counts-as conditionals.

- Classificatory aspects of counts-as. These have been highlighted in particular in [Grossi *et al.*, 2006a; Grossi *et al.*, 2008].

- Counts-as and actions (counts-as as the basis of institutional power). The grounding of institutional power on counts-as statements was one of the key issues stressed in [Jones and Sergot, 1996].

- Counts-as and conventions. The relation between counts-as and convention, in particular within communication theory, has received attention in [Jones and Parent, 2004; Jones and Parent, 2007].

- Counts-as as grounded on dedicated agents' mental attitudes. This topic has been systematically investigated, within modal logic, in [Gaudou *et al.*, 2008; Lorini and Longin, 2008; Lorini *et al.*, 2009].

- Counts-as as related to regulative norms. The topic of how regulative norms (e.g., obligations, permissions, etc.) are related to constitutive ones—a topic much discussed in Searle's work [Searle, 1995]—has been studied, in particular, in [Boella and Van der Torre, 2004; Boella and van der Torre, 2006].

- Counts-as as related to the definition of legal terms (e.g., contract) in legal systems. This aspect is highlighted in [Lindahl and Odelstad, 2006; Lindahl and Odelstad, 2008a; Lindahl and Odelstad, 2008b].

Table 1 on page 430 compactly records the thematic focuses of each approach to counts-as considered in this overview.

5.2 Methodological classification

The formal analysis of counts-as and constitutive rules is an exercise in applied logic or, more broadly, in applied mathematics.[20] We recognize

---

[20]It might be instructive to recall an excerpt from [Tarski, 1944] neatly describing how logic is applied to the analysis of concepts:

> "[...] it seems to me obvious that the only rational approach to such problems [of concept analysis] would be the following: [1] We should reconcile ourselves with the fact that we are confronted, not with one concept, but with several different concepts which are denoted by one word; [2] we should try to make these concepts as clear as possible (by means of definition, or of an axiomatic procedure, or in some other way); [3] to avoid further confusions, we should agree to use different terms for different concepts; and then

	Grossi et al.	Lorini et al.	Jones et al.	Gelati et al.
Contexts	✓	✓	✓	✓
Classification	✓	✓		
Power			✓	✓
Conventions			✓	
Mental attitudes		✓		
Const. vs. reg.	✓	✓	✓	✓
Legal concepts				

	Governatori et al.	Lindahl et al.	Boella et al.
Contexts			
Classification			
Power			
Conventions			
Mental attitudes			✓
Const. vs. reg.	✓	✓	✓
Legal concepts	✓	✓	

Table 1: An inventory of the themes addressed in the literature on counts-as conditionals

three salient methodological features of the formal approaches to counts-as discussed in this chapter. Such features are not properties of the formal analysis of counts-as alone, but are common to any logico-mathematical analysis of informal philosophical notions.

Object vs. meta-language

The analysis of counts-as is either carried out within the object language, by means of dedicated operators, or within the meta-language, by characterizing dedicated notions of non-classical logical consequence.

At the object level, the analysis moves from a given logic whose language is expanded with a suitable operator—in the case of counts-as conditionals, typically, a ternary operator with places for antecedent, consequent and context—which denotes the to-be-analyzed notion. The new operator is then studied, axiomatically or semantically, within the framework given by the background logic—e.g. normal modal logic [Blackburn et al., 2001] or classical modal logic [Chellas, 1980]. In the meta-language case, instead, a background logic for the analysis is selected—in the case of counts-as, typically, propositional logic—but its language is not expanded. Instead, the notion of logical consequence or, equivalently, of derivation of the original logic is strengthened or weakened in order to capture certain features which are considered characteristic of the reasoning involved with the to-be-analyzed notion. Roughly, while approaches of the first type are interested in the logic of statements of the sort "φ counts as ψ in context c", approaches of the second type are interested in which ψs can be inferred by assuming φ and a given set c of constitutive rules.

Defined vs. primitive

Be it studied as an operator or as a relation of logical consequence (or derivability), counts-as is formally characterized either by defining it in terms of simpler components (that is, simpler logical operators, respectively, simpler logical relations) whose logic is already available, or assuming it as a logical primitive and studying it in its own right. Approaches of the first type are reductionistic in the sense that they consider statements "φ counts as ψ in context c" to be synonymous with other statements (e.g., "φ implies ψ in context c" [Grossi et al., 2006a]), thereby reducing the logic of counts-as to the interaction of logics of simpler components (in this case the logics of "implies" and "in context").

---

we may proceed to a quiet and systematic study of all concepts involved, which will exhibit their main properties and mutual relations." [Tarski, 1944, p. 355]

Syntactic (axiomatic) vs. semantic (model-theoretic)

All the formal frameworks considered in the chapter, with few exceptions, provide both a proof-theory and a semantics for the logic of counts-as. However, differences arise regarding which one of the two perspectives is privileged during the initial set up of the formal theory. If a semantics is fixed, then a sound and complete axiomatization is looked for and, vice versa, if an axiomatics is established, then a semantics is looked for with respect to which the axiomatics is sound and complete.

In the first case the formal analysis moves initially from insights concerning the set up of the formal models on which counts-as statements can be interpreted. For example, as in the case of Grossi *et al.*, the assumption that counts-as statements are essentially of a classificatory type, leads the authors to use models on which counts-as statements are interpreted, essentially, as concept subsumptions. So, these approaches develop their formal analysis by first establishing under what conditions the to-be-formalized statements are true in the dedicated models (e.g., if and only if "all φ-states in context c are ψ-states" [Grossi *et al.*, 2006a]). The proof-theoretic properties of the various statements are studied as a consequence of the semantic set up.

In the second case, the analysis is driven instead by considering plausible-looking candidates for logical truths as axioms, by looking at the natural language counterparts of the to-be-formalized statements. Driving questions are, in this case, whether, for instance, counts-as statements are reflexive, or transitive, or symmetric, etc. Once the set of axioms is fixed, then suitable models are developed, on which the axiomatized statements can be interpreted.

The distinction between syntactic (axiomatic) and semantic (model-theoretic) is a very typical methodological dichotomy to be found in work in applied logic, and the analysis of counts-as is no exception.[21]

Table 2 shows how the various approaches discussed in this chapter can be placed with respect to the methodological standpoints described above.

5.3 Logical classification

The present section compares the various approaches introduced in Section 3 from the point of view of the formal frameworks specifically used in the analysis.

First we look at the structural properties of counts-as conditionals in those approaches that deal with them at an object-language level (see Ta-

---

[21] We will not discuss in this chapter what the pros and cons are of each of these approaches. Interesting considerations on this issue can be found in [Tarski, 1944; Tarski, 1983].

	Level of analysis	Characterization	Intuition
Boella et al.	metalanguage	primitive	syntactic
Jones et al.	object language	primitive	syntactic
Gelati et al.	object language	primitive	syntactic
Governatori et al.	metalanguage	primitive	syntactic
Grossi et al.	object language	defined	semantic
Lindahl et al.	metalanguage	primitive	syntactic
Lorini et al.	object language	defined	semantic

Table 2: Methodological classification of formal approaches to counts-as

ble 2). From a technical point of view, this is arguably the most informative comparison and it expands the comparison that was first elaborated in [Grossi, 2007; Grossi et al., 2008]. As Section 3 has made clear, the formal analysis of counts-as has been pursued in rather different—and thus difficult to compare—logical paradigms. The two technically closest approaches are the ones of Grossi et al. and Lorini et al., which work in normal modal logic. This section compares the relative strength of the logics presented in those works, and also relates them to the non-normal modal logic of, respectively, Gelati et al., and Jones et al..

Structural properties of counts-as conditionals

Conditional $\Rightarrow_c^{cl}$ enjoys strong properties (in particular reflexivity, antecedent strengthening, and transitivity) and displays, therefore, a very classical behavior. Instead, conditional $\Rightarrow_c^{cl+}$ behaves much less classically, rejecting reflexivity, strengthening of the antecedent, even the weaker version of cautious monotonicity, and transitivity. On the other hand, it still retains a weaker form of transitivity, namely cumulative transitivity.

As we have seen in Section 3, Jones et al. have developed a logic for counts-as conditionals (denoted by the operator $\Rightarrow_c$) obeying the following principles: left logical equivalence, right logical equivalence, disjunction of antecedents, conjunction of the consequents and transitivity. Recall, though, that it does not enjoy cumulative transitivity and cautious monotonicity.

Gelati *et al.* argue, instead, that the logic of counts-as conditionals, which they denote via the operator $\Rightarrow$, amounts to the logic of preferential reasoning [Kraus *et al.*, 1990], preferential reasoning being characterized by the following properties: reflexivity, left logical equivalence, weakening of the consequent, conjunction of the consequents, cut, cautious monotonicity and disjunction of the antecedents.

An overview of the main properties enjoyed by each object-level formal characterization of counts-as is provided in Table 3.

		$\Rightarrow_c^{cl}$	$\Rightarrow_c^{cl+}$	$\Rightarrow_c$	$\Rightarrow$	$\triangleright_c$
A	Reflexivity	✓			✓	
B	Antecedent Strengthening	✓				
C	Transitivity	✓		✓		
D	Disjunction of the Antecedents	✓	✓	✓	✓	✓
E	Conjunction of the Consequents	✓	✓	✓	✓	✓
F	Left Logical Equivalence	✓	✓	✓	✓	✓
G	Right Logical Equivalence	✓	✓	✓	✓	✓
H	Consequent Weakening	✓			✓	
I	Cumulative Transitivity	✓	✓		✓	✓
L	Cautious Monotonicity	✓			✓	

Table 3: Properties of counts-as operators, where: $\Rightarrow_c^{cl}$ is the contextual classification of Grossi *et al.* ; $\Rightarrow_c^{cl+}$ the proper contextual classification of Grossi *et al.*; $\Rightarrow_c$ is the counts-as conditional of Jones *et al.*; $\Rightarrow$ is the counts-as conditional of Gelati *et al.* ; and $\triangleright_c$ is the acceptance-based conditional of Lorini *et al.*

This overview provides grounds for a number of interesting observations. First of all, notice that there seems to be a structural hard core of all characterizations of counts-as including Grossi *et al.*, which corresponds to properties (D) to (G) inclusive in Table 3. These properties are exactly the ones recognized as a sort of minimal characterization of counts-as in [Jones and Sergot, 1996]. There are then two remarkable facts to be noticed, which concern the relation between our notions of contextual and proper contextual classification and the notions of counts-as axiomatically characterized by Gelati *et al.* and Jones *et al.*

First, the notion of counts-as statements as conditional counterparts of preferential reasoning ($\Rightarrow$) represents a defeasible form of contextual classification ($\Rightarrow_c^{cl}$), since the only properties distinguishing the two notions are strengthening of the antecedent (B) and transitivity (C), which in the presence of reflexivity (A) and cut (I) are actually equivalent (see [Kraus *et al.*, 1990]). From a semantic point of view, this constitutes a very interesting fact. In a way, it allows us to attach a precise meaning to the notion of counts-as axiomatized by Gelati *et al.* : if the statement "X counts-as Y in context C", intended as contextual classification, means "X is classified as Y in C", then the same statement read in the fashion of Gelati *et al.* would mean "X is classified as Y in C, *modulo exceptions*", or "it *normally* follows from C that X is classified as Y".

Second, the notions of proper contextual classification ($\Rightarrow_c^{cl+}$) and of acceptance-based counts-as ($\triangleright_c$) both appear to correspond to a slightly weaker version of the counts-as conditional proposed by Jones *et al.* ($\Rightarrow_c$) where transitivity (C) is substituted by the weaker property of cumulative transitivity (I).

Relative strength of modal counts-as

A final comparison is worth making, which focuses on those approaches to counts-as that are based on standard modal logic. Among all the logics of counts-as, the ones studied by Grossi *et al.* (see Section 3.3) and Lorini *et al.* (see Section 3.4) are the ones bearing the most similarities. In fact, [Lorini *et al.*, 2009] has proven an embedding of the logic of proper classificatory counts-as in a version of AL strengthened with suitable axioms.

Figure 4 displays the relative logical strength of the counts-as conditionals investigated by Grossi *et al.* and Lorini *et al.*, and relates them to the ones proposed by Jones *et al.* and by Gelati *et al.*. Proper contextual classification $\Rightarrow_c^{cl+}$ can be viewed as an extension of both contextual classification ($\Rightarrow_c^{cl}$) and acceptance-based counts-as ($\triangleright_c$). In turn, it is strengthened independently by ascriptive counts-as ($\Rightarrow_c^{as}$) and constitutive counts-as ($\Rightarrow_c^{co}$).

6 Some open problems

We conclude the chapter by pointing to a few open research questions in the formal analysis of counts-as and constitutive norms, which we consider most interesting and urgent.

First, the relation of logics of counts-as with logics of action. The chapter has not touched upon this issue which, although extensively highlighted by Jones *et al.*, has not been thus far systematically addressed by the litera-

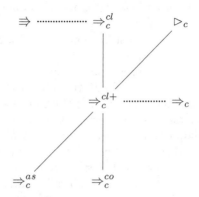

Figure 4: Diagram of all modal counts-as operators. Lines link weaker conditionals (top) to stronger ones (bottom). Dotted lines link the counts-as conditionals defined in the logics of Grossi et al. and Lorini et al. with conditionals defined in non-standard modal logics (in this case by Gelati et al., $\Rightarrow_c$ and Jones et al., $\Rightarrow_c$) to which they bear structural similarities (recall Table 3).

ture in the field. As a matter of fact, in Jones et al. the thrust towards a formal theory of counts-as comes from an attempt to provide a comprehensive formal theory of institutional action, accounting for key phenomena of institutional reality such as institutional power.

A recent interesting attempt to tackle this issue has been made in [Herzig et al., 2011b] which proposes a formal analysis of the theory of institutional action as delineated in [Goldman, 1976] (cf. the earlier Section 2). The crux of this proposal consists in embedding the logic of counts-as statements within a logic of agency[22], thereby making it possible to express the notion of *conventional generation* of actions from other actions.

Second, a key aspect of constitutive norms which still awaits a thorough formal analysis is the aspect which might be called 'language creation'. This aspect has only partially been addressed in [Grossi, 2011] (see the earlier Section 3.3) and concerns the capacity that constitutive norms have of, literally, creating new words. Constitutive norms not only define terms but, in a way, they really expand the current language of the normative

---

[22]More precisely the logic, which was first developed in [Herzig et al., 2011a], is a so-called logic of *propositional control* [Gerbrandy, 2006] which extends a logic known as Coalition Logic of Propositional Control [van der Hoek and Wooldridge, 2005] with Propositional Dynamic Logic [Harel et al., 2000] constructs.

system at issue. From a logical point of view, this poses several interesting technical questions whose answers would bear definite relevance for a full understanding of constitutive norms.

Third, the comparison of the properties of counts-as conditionals and causal conditionals. It has been claimed in, e.g., [Searle, 1969] and [Jones and Parent, 2007] that counts-as conditionals are central to understanding the foundations of interpersonal communication. However, as we see for instance in [Grice, 1957], a contrast is often drawn between signalling mechanisms that exploit causal connections (Grice's so-called natural meaning), and those that are characteristic of human communication. The latter were those for which Grice developed his intention-based account of non-natural meaning, but arguably the contrast is better understood in terms of the distinction between non-conventional and conventional meaning—see, e.g., [Lewis, 1969] and the recent work on signalling—and the natural processes that create conventions—by Brian Skyrms [Skyrms, 2010]. It may be that some further light can be thrown on this distinction by a comparison of the respective properties of counts-as and causal conditionals.

Finally, on the wave of attention to normative change that is characterizing much of the recent research in deontic logic[23], some works have been focusing on the dynamics of constitutive norms. In particular: [Aucher et al., 2009] has provided an analysis, based on a dynamic variant of the classificatory logic of counts-as [Grossi et al., 2006a], of how counts-as conditionals can be introduced or removed from the specification of a context, thereby offering a very stylized formal model of the enactment and abrogation of constitutive norms; [Boella et al., 2010] has interestingly argued for the formal study of the dynamics of constitutive norms as a privileged means to provide a solid understanding of the key judicial phenomenon of legal interpretation. Although these works provide some important first steps in the formal understanding of the dynamics of constitutive rules, much still remains unexplored.

7 Conclusions

The chapter has provided a detailed overview of that branch of deontic logic which, in the last fifteen years, has addressed the issue of the formal analysis of constitutive norms and counts-as conditionals.

The chapter has addressed some of the philosophical/conceptual features of the various approaches available in the literature and their key technical aspects. The resulting overview has then been used—in the key section of the chapter (Section 5)—to provide a systematic comparison of the various

---

[23]See dedicated chapter in this Handbook.

approaches along three different lines: the different aspects addressed in the formal analysis, the methodology chosen for delivering the analysis, and the formal relationships between the different logical systems.

Acknowledgements

Davide Grossi has been partly supported by the *Nederlandse Organisatie voor Wetenschappelijk Onderzoek* (Netherlands Organization for Scientific Research) under the NWO VENI grant 639.021.816.

BIBLIOGRAPHY

[Alchourrón and Bulygin, 1971] C. E. Alchourrón and E. Bulygin. *Normative Systems*. Springer Verlag, 1971.

[Anscombe, 1958] G. E. M. Anscombe. On brute facts. *Analysis*, 18:69–72, 1958.

[Artosi et al., 2004] A. Artosi, A. Rotolo, and S. Vida. On the logical nature of counts-as conditionals. In C. Cevenini, editor, *The Law and Electronic Agents: Proceedings of the LEA 04 workshop*, pages 9–34, Bologna, Italy, 2004.

[Aucher et al., 2009] G. Aucher, D. Grossi, A. Herzig, and E. Lorini. Dynamic context logic. In X. He, J. Horty, and E. Pacuit, editors, *Proceedings of LORI 2009*, volume 5834 of *LNAI*. Springer, 2009.

[Blackburn et al., 2001] P. Blackburn, M. de Rijke, and Y. Venema. *Modal Logic*. Cambridge University Press, Cambridge, 2001.

[Boella and van der Torre, 2003] G. Boella and L. van der Torre. Attributing mental attitudes to normative systems. In *AAMAS '03: Proceedings of the second international joint conference on Autonomous agents and multiagent systems*, pages 942–943, New York, NY, USA, 2003. ACM Press.

[Boella and Van der Torre, 2004] G. Boella and L. Van der Torre. Regulative and constitutive norms in normative multiagent systems. In D. Dubois, C. A. Christopher A. Welty, and M. Williams, editors, *Proceedings of KR2004, Whistler, Canada*, pages 255–266, 2004.

[Boella and van der Torre, 2006] G. Boella and L. van der Torre. A logical architecture of a normative system. In L. Goble and J.-J.Ch. Meyer, editors, *Deontic Logic and Artificial Normative Systems: 8th International Workshop on Deontic Logic in Computer Science (DEON 2006)*, volume 4048 of *LNCS*, pages 24–35, 2006.

[Boella et al., 2010] G. Boella, G. Governatori, A. Rotolo, and L. van der Torre. A logical understanding of legal interpretation. In L. Fangzhen, U. Sattler, and M. Truszczynski, editors, *Proceedings of the 12th International Conference on the Principles of Knowledge Representation and Reasoning (KR 2010)*, pages 563–565, 2010.

[Bulygin, 1992] E. Bulygin. On norms of competence. *Law and Philosophy* 11, pages 201–216, 1992.

[Chellas, 1980] B. F. Chellas. *Modal Logic. An Introduction*. Cambridge University Press, Cambridge, 1980.

[Cherry, 1973] C. Cherry. Regulative rules and constitutive rules. *The Philosophical Quarterly*, 23(93):301–315, 1973.

[Fornara et al., 2007] N. Fornara, F. Viganò, and M. Colombetti. Agent communication and artificial institutions. *Journal of Autonomous Agents and Multi-Agent Systems*, 14:121–142, 2007.

[Gaudou et al., 2008] B. Gaudou, D. Longin, E. Lorini, and L. Tummolini. Anchoring institutions in agents' attitudes: Towards a logical framework for autonomous multi-agent systems. In L. Padgham, D. Parkes, J. Müller, and S. Parsons, editors, *Proceedings of the 7th International Conference on Autonomous Agents and Multiagent Systems (AAMAS 2008)*, pages 728–735. ACM, 2008.

[Gelati et al., 2002] J. Gelati, G. Governatori, A. Rotolo, and G. Sartor. Actions, institutions, powers. Preliminary notes. In *International Workshop on Regulated Agent-Based Social Systems: Theories and Applications (RASTA'02)*, pages 131–147, 2002.

[Gelati et al., 2004] J. Gelati, A. Rotolo, G. Sartor, and G. Governatori. Normative autonomy and normative co-ordination: Declarative power, representation, and mandate. *Artif. Intell. Law*, 12(1-2):53–81, 2004.

[Gerbrandy, 2006] J. Gerbrandy. Logics of propositional control. In *Proceedings of the 5th International Conference on Autonomous Agents and Multiagent Systems (AAMAS 2006)*. ACM, 2006.

[Goldman, 1976] A. I. Goldman. *A Theory of Human Action*. Princeton University Press, Princeton, 1976.

[Governatori and Rotolo, 2008] G. Governatori and A. Rotolo. A computational framework for institutional agency. *Artificial Intelligenc and law*, 16:25–52, 2008.

[Grice, 1957] H. P. Grice. Meaning. *Philosophical Review*, 66:377–388, 1957.

[Grossi et al., 2005] D. Grossi, J.-J.Ch. Meyer, and F. Dignum. Modal logic investigations in the semantics of counts-as. In *Proceedings of the Tenth International Conference on Artificial Intelligence and Law (ICAIL'05)*, pages 1–9. ACM, June 2005.

[Grossi et al., 2006a] D. Grossi, J.-J.Ch. Meyer, and F. Dignum. Classificatory aspects of counts-as: An analysis in modal logic. *Journal of Logic and Computation*, 16(5):613–643, 2006. Oxford University Press.

[Grossi et al., 2006b] D. Grossi, J.-J.Ch. Meyer, and F. Dignum. Counts-as: Classification or constitution? An answer using modal logic. In L. Goble and J.-J.Ch. Meyer, editors, *Proceedings of DEON 2006*, LNAI 4048, pages 115–130, 2006.

[Grossi et al., 2008] D. Grossi, J.-J.Ch. Meyer, and F. Dignum. The many faces of counts-as: A formal analysis of constitutive-rules. *Journal of Applied Logic*, 6(2):192–217, 2008.

[Grossi, 2007] D. Grossi. *Designing Invisible Handcuffs. Formal Investigations in Institutions and Organizations for Multi-agent Systems*. PhD thesis, Utrecht University, SIKS, 2007.

[Grossi, 2008] D. Grossi. Pushing Anderson's envelope: The modal logic of ascription. In R. van der Meyden and L. van der Torre, editors, *Proceedings of the 9th International Conference on Deontic Logic in Computer Science (DEON 2008)*, number 5076/2008 in LNAI, pages 263–277. Springer, 2008.

[Grossi, 2011] D. Grossi. Norms as ascriptions of violations: An analysis in modal logic. *Journal of Applied Logic*, 9(2):95–112, 2011.

[Harel et al., 2000] D. Harel, D. Kozen, and J. Tiuryn. *Dynamic Logic*. MIT Press, 2000.

[Herzig et al., 2011a] A. Herzig, E. Lorini, F. Moisan, and N. Troquard. A dynamic logic of normative systems. In *Proceedings of the 22nd International Joint Conference on Artificial Intelligence (IJCAI'11)*, 2011.

[Herzig et al., 2011b] A. Herzig, E. Lorini, and N. Troquard. A dynamic logic of institutional actions. In *Proceedings of 12th International Workshop on Computational Logic in Multi-Agent Systems (CLIMA XII)*, 2011.

[Hindriks, 2009] F. Hindriks. Constitutive rules, language, and ontology. *Erkentniss*, 71:235–275, 2009.

[Jones and Parent, 2004] A. J. I. Jones and X. Parent. Convention signaling acts and conversation. In F. Dignum, editor, *Proceedings of ACL 2003*, pages 1–17. Springer, 2004.

[Jones and Parent, 2007] A. J. I. Jones and X. Parent. A convention-based approach to agent communication languages. *Group Decision and Negotiation*, 16:101–141, 2007.

[Jones and Sergot, 1996] A. J. I. Jones and M. Sergot. A formal characterization of institutionalised power. *Journal of the IGPL*, 3:427–443, 1996.

[Krabbendam and Meyer, 2000] J. Krabbendam and J.-J.Ch. Meyer. Release logics for temporalizing dynamic logic, orthogonalising modal logics. In M. Barringer, M. Fisher, D. Gabbay, and G. Gough, editors, *Advances in Temporal Logic*, pages 21–45. Kluwer Academic Publisher, 2000.

[Krabbendam and Meyer, 2003] J. Krabbendam and J.-J. Ch. Meyer. Contextual deontic logics. In P. McNamara and H. Prakken, editors, *Norms, Logics and Information Systems*, pages 347–362, Amsterdam, 2003. IOS Press.

[Kraus et al., 1990] S. Kraus, D. J. Lehmann, and M. Magidor. Nonmonotonic reasoning, preferential models and cumulative logics. *Artificial Intelligence*, 44(1-2):167–207, 1990.

[Lewis, 1969] D. Lewis. *Convention*. Harvard University Press, Cambridge, MA, 1969.

[Lindahl and Odelstad, 2000] L. Lindahl and J. Odelstad. An algebraic analysis of normative systems. *Ratio Juris*, 13:261–278, 2000.

[Lindahl and Odelstad, 2006] L. Lindahl and J. Odelstad. Open and closed intermediaries in normative systems. In T.M. van Engers, editor, *Proceedings of the Nineteenth JURIX Conference on Legal Knowledge and Information Systems (JURIX 2006)*, pages 91–100, 2006.

[Lindahl and Odelstad, 2008a] L. Lindahl and J. Odelstad. Intermediaries and intervenients in normative systems. *Journal of Applied Logic*, 6(2):229–258, 2008.

[Lindahl and Odelstad, 2008b] L. Lindahl and J. Odelstad. Strata of intervenient concepts in normative systems. In R. van der Meyden and L. van der Torre, editors, *Deontic Logic in Computer Science, 9th International Conference, Proceedings of DEON 2008*, volume 5076 of *LNCS*, pages 203–217. Springer, 2008.

[Lorini and Longin, 2008] E. Lorini and D. Longin. A logical account of institutions: from acceptances to norms via legislators. In G. Brewka and J. Lang, editors, *International Conference on Principles of Knowledge Representation and Reasoning (KR), Sidney, Australia, 16/09/08-19/09/08*, pages 38–48. AAAI Press, 2008.

[Lorini et al., 2009] E. Lorini, D. Longin, B. Gaudou, and A. Herzig. The logic of acceptance: Grounding institutions on agents' attitudes. *Journal of Logic and Computation*, To appear., 2009.

[Makinson and van der Torre, 2000] D. Makinson and L. van der Torre. Input-output logics. *Journal of Philosophical Logic*, 29(4):383–408, 2000.

[Makinson, 1999] D. Makinson. On a fundamental problem of deontic logic. In P. McNamara and H. Prakken, editors, *Norms, Logics and Information Systems. New Studies in Deontic Logic and Computer Science*, volume 49 of *Frontiers in Artificial Intelligence and Applications*, pages 29–53. IOS Press, Amsterdam, 1999.

[Nute, 1987] D. Nute. Defeasible logic. In *Handbook of Logic in Artificial Intelligence and Logic Programming*, volume 3, pages 353–395. Oxford University Press, 1987.

[Peczenik, 1989] A. Peczenik. *On Law and Reason*. Kluwer, Dordrecht, 1989.

[Pufendorf, 1688] S. Pufendorf. *De Jure Naturae et Gentium*. Clarendon Press, 1934 edition, 1688.

[Rawls, 1955] J. Rawls. Two concepts of rules. *The Philosophical Review*, 64(1):3–32, 1955.

[Ricciardi, 1997] M. Ricciardi. Constitutive rules and institutions. Paper presented at the meeting of the Irish Philosophical Club, Ballymascanlon, February 1997.

[Ross, 1957] A. Ross. Tû-tû. *Harvard Law Review*, 70:812–825, 1957.

[Searle, 1969] J. Searle. *Speech Acts. An Essay in the Philosophy of Language*. Cambridge University Press, Cambridge, 1969.

[Searle, 1995] J. Searle. *The Construction of Social Reality*. Free Press, 1995.

[Skyrms, 2010] B Skyrms. *Signals - Evolution, Learning and Information*. Oxford University Press, Oxford, UK, 2010.

[Tarski, 1944] A. Tarski. The semantic conception of truth and the foundations of semantics. *Philosophy and Phenomenological Research*, 4(3):341–376, 1944.

[Tarski, 1983] A. Tarski. The establishment of scientific semantics. In *Logic, Semantics, Metamathematics*, pages 401–408. Hackett, Indianapolis, 1983.

[van der Hoek and Wooldridge, 2005] W. van der Hoek and M. Wooldridge. On the logic of cooperation and propositional control. *Artificial Intelligence*, 164(1-2):81–119, 2005.

[Von Wright, 1963] G. H. Von Wright. *Norm and Action. A Logical Inquiry*. Routledge, London, 1963.

Davide Grossi
Department of Computer Science
University of Liverpool
Email: d.grossi@liverpool.ac.uk

Andrew J. I. Jones
Department of Informatics
King's College London
Email: andrewji.jones@kcl.ac.uk

PART III

NEW FRAMEWORKS

7
Alternative Semantics for Deontic Logic
SVEN OVE HANSSON

ABSTRACT. The major message of this chapter is that we have choices in the construction of deontic semantics. There are many different ways to build semantic models for deontic logic, and these models correspond to different views on how the subject matter of moral philosophy is structured. Therefore, the study of alternative deontic semantics can help us clarify the implications of moral theories and standpoints. The chapter has its focus on constructions that employ possible worlds and orderings as their main building blocks. These tools are surprisingly versatile and can give rise to widely divergent logical systems. Based on an analysis of the criticism of traditional deontic semantics, two major alternative lines of development are identified, namely those that remove agglomeration ($Op \& Oq \rightarrow O(p\&q)$) and those that give up necessitation (If $\vdash p \rightarrow q$ then $\vdash Op \rightarrow Oq$). Both these options are explored in some detail. The chapter also reviews some important classifications of (monadic and dyadic) normative statements that have bearing on the choice of an appropriate semantic model.

1 Introduction 446
2 Delimitations 448
    2.1 The deontic language 448
    2.2 Degrees of obligation 450
    2.3 Normative rules and veritable obligations 452
    2.4 The ideal ought and action representation 454
    2.5 Possible worlds and other holistic alternatives 456
3 Standard deontic logic 457
    3.1 The origins 457
    3.2 The standard system 458
    3.3 The Anderson–Kanger model 460
    3.4 An ordering of worlds 461
    3.5 Dyadic versions of standard deontic logic 462
4 Criticism of standard deontic logic 464
    4.1 Semantic criticism 465
    4.2 Syntactic criticism: Inconsistencies 468
    4.3 Syntactic criticism: Necessitation 472

		4.4	A summary of the criticism 474
5			Giving up agglomeration 479
6			Giving up necessitation 480
		6.1	The uses of orderings 480
		6.2	Preference-based deontic logic 482
		6.3	Reintroducing holistic alternatives 489
7			Conclusion 493

1 Introduction

There are two major ways to assess the plausibility of proposed principles in deontic logic. In the *syntactic* approach, deontic statements such as $Op \to Pp$ and $O(p\&q) \to Pp$ are directly compared to our intuitions about normative inferences. Is everything obligatory also permitted? And is every part of an obligation permitted?

In the *semantic* approach, a class of models for logical validity is constructed. A sentence in deontic language is assumed to be valid if and only if it holds in all such models. The commonly used semantic models in deontic logic are based on possible worlds in a manner well-known from modal logic. If a class of models is based on intuitively sound principles, then this may be a good reason for accepting deontic principles that are valid in all these models.

But why should we worry about semantics? Why is it not sufficient to directly investigate the plausibility of deontic principles, without constructing models for their validity? There are at least two good answers to that question, one general and one more specific.

The general answer is that neither of these two ways to tap our intuitions is perfect. When we lack a perfect method of investigation, and the methods at our disposal are associated with different weaknesses and strengths, then it is often useful to combine two or several of these methods. In the social sciences, such a combination of different methods is called "methodological pluralism" or "triangulation"; in philosophy the term "reflective equilibrium" is more common. [Hansson, 2010]

The more specific answer is that investigations that remain exclusively on the syntactic level have a most irritating shortcoming: In such an investigation we can easily make a "positive" list of deontic principles that should be valid, and we can also make a corresponding "negative" list of principles that should not be valid. However, we cannot in this way construct a list of valid deontic principles that is in any sense exhaustive. In a combined syntactic and semantic approach we can construct a "positive list" that is exhaustive in the sense that all semantically valid deontic principles are logically derivable from the principles on that list. In this way we obtain a complete characterization of the valid deontic inference principles.

Deontic logic has indeed a strong semantic tradition. Using possible worlds and orderings (preference relations) as their major building-blocks, deontic logicians have constructed models that can be used to determine the validity of deontic sentences. There are connections between the ways in which these models are constructed and the types of basic principles that are used in the construction of moral theories such as the various brands of utilitarianism and deontology. These connections, of course, add to the relevance of deontic logic for moral inquiry.

But, unfortunately, deontic logic also has a long history of getting stuck in semantic principles that support blatantly implausible deontic postulates. This is probably a major reason why the influence of deontic logic in moral philosophy has been rather weak.[1] It remains to develop the full potential that deontic logic has as a tool for moral philosophy. In order to achieve this, semantic tools and models are needed that provide us with a more plausible logic than those that have dominated the field.

It is the purpose of this chapter to give indications for such a development. Most importantly, the chapter will show that we have choices in the construction of deontic semantics. There are many different ways to build semantic models for deontic logic, and these models correspond to different views on how the subject-matter of moral philosophy is structured. Therefore, the study of alternative deontic semantics can help us clarify the implications of moral theories and standpoints.

The focus will be on semantical constructions that follow the general tradition in deontic logic in employing possible worlds and orderings.[2] These tools are surprisingly versatile, and can give rise to widely divergent logical systems.

After some delimitations have been made in Section 2 and traditional deontic semantics has been introduced in Section 3, the criticism of traditional semantics that has been used to justify alternative semantics is summarized in Section 4. This criticism goes in two different directions, and it has therefore given rise to two major lines in the development of alternative semantics. These lines of development are followed in Sections 5 and 6. Some final conclusions are offered in Section 7.

---

[1] The lack of influence of deontic logic in moral philosophy has often been blamed on reluctance among moral philosophers to embrace formal methods. But that can only be part of the explanation. Game theory that is not mathematically less demanding than deontic logic has had considerable influence in recent moral philosophy.

[2] For other approaches, see Chapters 8 and 9.

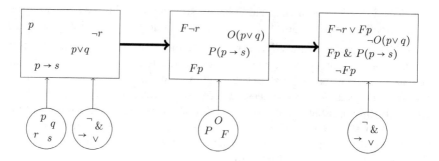

Figure 1: The three steps in the construction of the core of deontic languages

2 Delimitations

Several delimitations of the language and the subject-matter of deontic logic need to be made before we can turn to the construction and evaluation of deontic semantics.

2.1 The deontic language

Accounts of deontic logic differ in what the language is considered to consist in. The common core of deontic languages can be obtained recursively in three steps that are illustrated in Figure 1. First, we have as a starting-point a non-deontic language. It consists exclusively of statements of fact. These facts are the objects of deontic statements, such as "I pay €5.000 to you" (that is the object of the obligation "I am morally required to pay €5.000 to you"). This language is usually taken to be closed under truth-functional operators. (We will return in Section 2.4 to what types of facts these sentences represent.)

Secondly, atomic deontic sentences are formed by applying a deontic operator such as O (obligation), P (permission), and F (prohibition) to expressions in the factual language. This gives rise to expressions such as Op, $P(q \vee r)$ and $F(p \rightarrow r)$.

The third step consists in forming truth-functional combinations of the atomic deontic sentences from the second step. This gives rise to expressions such as $Op \rightarrow P(p \vee q)$, $Fp \vee P(p \& \neg r)$ etc. The expressions obtained in this third step form the common core of deontic languages. It is important to observe that the elements formed in the second step (such as $O(p \vee q)$) are retained in the third step, whereas the expressions from the first step (such as $\neg p \rightarrow q \vee r$) have been lost.

Three types of extensions of this common core of deontic languages are widespread.

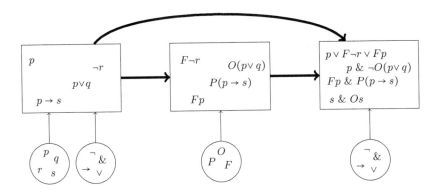

Figure 2: The inclusion of mixed formulas in a deontic language

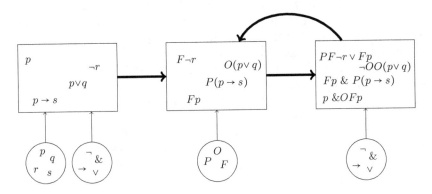

Figure 3: The inclusion of nested operators in a deontic language

First, formulas from the first step can be included in the third step, as shown in Figure 2. This will result in the inclusion of "mixed" formulas such as $p \,\&\, Op$ and $p \to Oq$ that are formed truth-functionally from both deontic and factual sentences. However, as we will see in Section 3.5, mixed formulas are not in general well suited to express the interconnections between norms and factual conditions. Therefore, mixed formulas will not be treated systematically in this chapter.

Secondly, the second and third step can be cyclically repeated, as shown in Figure 3. This gives rise to sentences with iterated deontic operators such as OOp and $FPFp$. More generally, it gives rise to sentences in which deontic operators are nested, i.e. they appear within the scope of other

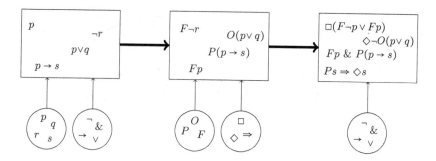

Figure 4: The inclusion of non-deontic operators in a deontic language

deontic operators. This results in expressions such as $O(Pp \vee O\neg p)$ and $O\neg(Op \, \& \, O\neg p)$. Such formulas have sometimes been interpreted as norms of higher order. Two formulas that have often been referred to in this way are $Op \rightarrow OOp$ and $O(Op \rightarrow p)$. However, there are difficulties in this interpretation.[3] It can also be argued that our understanding of unnested deontic operators needs to be improved before the additional problems connected with nested operators can be successfully attacked. In what follows, the focus will be on unnested operators, and nested operators will not be systematically discussed.[4]

The third extension is illustrated in Figure 4. It consists in including, at the second stage, operators representing other non-truthfunctional concepts than deontic ones, most commonly operators for necessity ($\Box$), possibility ($\Diamond$), and conditionality ($\Rightarrow$). Of these conditionality is the most important one, since a large part of the normative statements that we make are conditional. Conditional norms and their representation will be discussed below, especially in Sections 2.3 and 3.5.

2.2 Degrees of obligation

Moral requirements differ in stringency (strength). This is fairly obvious on an intuitive level. Every child knows the difference in stringency between the two requirements "do not speak with food in your mouth" and "do

---

[3] Using the terminology to be introduced in Section 2.4, a plausible interpretation of iterated deontic operators is available if we employ an ought-to-be interpretation of O. It can meaningfully be said that the world ought to be such that whatever occurs is not forbidden ($O(p \rightarrow \neg Fp)$). With an ought-to-do interpretation of the deontic operator it is much less clear what such sentences can mean.

[4] For a brief but clarifying discussion on nested operators, see [Chellas, 1980, pp. 193-4]. For further discussions on nested operators, see for instance [Hanson, 1965], [Føllesdal and Hilpinen, 1970, pp. 15-9], [Vorobej, 1982], and [von Wright, 1999b].

not erase the hard disk on mom's computer". There are various ways to explicate or codify the difference: the more stringent requirement is more important, its violation is worse, it is less easily overridden etc. [Chisholm, 1964, p. 151]

Not only do our moral requirements differ in stringency; so do the words and phrases that we use to express them. "Must" is more stringent than "ought", and "ought" is more stringent than "should". [Guendling, 1974][5] In order to express this relationship in the logical language, we can say that an operator O_1 includes another operator O_2 if and only if it is the case for all sentences p that if $O_2 p$, then $O_1 p$. Furthermore, O_1 properly includes O_2 if and only if O_1 includes O_2 but O_2 does not include O_1. Two operators of obligation differ in strength if one of them properly includes the other.[6] A set $\mathcal{O}$ of operators is inclusion-complete if it holds for all $O_1, O_2 \in \mathcal{O}$ that either O_1 includes O_2 or O_2 includes O_1. [Hansson, 2004b]

Most deontic systems contain only one operator of obligatoriness, and therefore they cannot express differences in strength.[7] However, some authors have introduced two degrees of obligatoriness, usually said to represent "must" and "ought".[8] Proposals have also been put forward that allow for a whole series of O operators, representing different degrees of strength.[9] Following tradition in deontic logic, in what follows the focus will mostly be on a single ought operator, but some results referring to operators with different degrees of strength will be reported.

In ordinary language, prescriptive expressions such as "ought", "must", "should", "has a duty to", "has an obligation to", etc. differ not only in strength but also in connotation. As an example of this, "obligations" typically derive from promises or agreements, whereas "duties" are associated with roles and offices in organizations and institutions. [Brandt, 1965][10] It does not seem possible to do justice to such variations in connotation in a formal language without making it much too complex. Therefore, deontic

---

[5]On differences in strength between moral requirements, see also [Ladd, 1957, p. 125], [Sloman, 1970, p. 391], [Harman, 1977, pp. 117-8], [Jones and Pörn, 1985], [Meyer, 1987b, p. 87], [Garcia, 1989], [Brown, 1996], [Hansson, 1999] and [Hansson, 2001].

[6]The relation "at least as strong as" is not necessarily complete. If there are p and q such that $O_1 p$, $\neg O_1 q$, $\neg O_2 p$, and $O_2 q$, then it does not hold in either direction.

[7]In principle, iterations of a single operator could be used to express strength. Hence $OOp \,\&\, Op$ may express a higher degree of stringency than $Op \,\&\, \neg OOp$. But this is a crude and interpretatively dubious way to express strength.

[8]This was done by Jones and Pörn [Jones and Pörn, 1986]. For discussions of their proposal, see [Hansson, 1989; Jones and Pörn, 1989]. Another system that contains separate representations of "must" and "ought" was put forward by Paul McNamara [McNamara, 1996].

[9]See for instance [Hansson, 1999; Hansson, 2001; Dellunde and Godo, 2008].

[10]Cf. [Mish'alani, 1969; Forrester, 1975].

logicians have almost invariably disregarded differences in connotation, a practice that will be followed here. With respect to connotations, the prescriptive operators of the formal language can be seen as representing that which is common between the various prescriptive expressions of natural language.

2.3 Normative rules and veritable obligations

Consider the following two sentences, said to someone who beats a cat:

1. "You must stop beating Puss."

2. "You are not allowed to be cruel to animals."

The first sentence differs from the second in offering a norm for only one situation, namely the present one. The second sentence refers to several situations. It exemplifies the most common type of norms referring to several situations, namely normative rules. The fundamental distinction that we need to make is that between situation-specific norms (as in sentence 1) and trans-situational ones (as in sentence 2). The situation-specific norms will be called *veritable norms*.[11] For most practical purposes the trans-situational norms will have the form of *normative rules*, and the crucial distinction can then be expressed as that between veritable norms and normative rules.[12] Unfortunately, this distinction is not easily made in ordinary language. English (like many other languages) employs the same linguistic forms for both purposes. Deontic logic follows natural language in this respect, and uses the same symbolic form (Op) to express that something (p) is obligatory in the present situation and to express that it is obligatory in general. In what follows, this convention will be followed, but it will often be important to keep the distinction between the two interpretations in mind.

Two other distinctions have often been confused with that between veritable norms and normative rules. One of these is that between over-all norms (non-overridden norms) and prima facie norms (norms that may or may not be overridden). However, veritable norms can be either over-all or prima facie. Example 1 above exemplifies the former, whereas the following sentence expresses a veritable prima facie norm:

---

[11] It would perhaps be more natural to call these norms "actual", but this terminology would be confusing due to the widespread use of "actual obligation" as a synonym of "over-all obligation". See for instance [Ross, 1930, p.20].

[12] This distinction was made in [Hansson, 1988]. Similarly, Carlos Alchourrón [Alchourrón, 1993] distinguished between "a norm for a single possible circumstance (which may be the actual circumstance)" and a norm for "all possible circumstances", and David Makinson [Makinson, 1999] distinguished between norms "in all circumstances" and norms "in present circumstances".

3. "You have an obligation to be there in time, but given the circumstances you are excused."

Example 2 is a prima facie normative rule. It is contestable whether there are any overall normative rules. (Kantians would say that there are, but proponents of other moral theories may object.)

The other confusable distinction is that between conditional and non-conditional norms.[13] Here, all four combinations are prevalent. Example 1 is a non-conditional veritable norm. The following is a conditional veritable norm:

4. "If you own ticket number 175, then you may collect the prize."

Example 2 is a non-conditional normative rule whereas the following is a conditional normative rule:

5. "If you own the winning ticket in a lottery, then you may collect the prize."

The application of the distinction between normative rules and veritable norms to conditional normative expressions can be examplified with the following two sentences:

6. "If your daughter had not been a victim of drunk driving, then you would not have had to pay these large hospital bills."

7. "If you borrow money, then you must pay it back."

Example 6 refers counterfactually to a particular situation, but one that is different from the present one. The norm in question is veritable, i.e. it is situation-specific and not a rule. The counterfactual "if ... then" in this sentence is not specific to the normative content; it has essentially the same meaning as in other counterfactual conditionals with the same antecedent, such as:

8. "If your daughter had not been a victim of drunk driving, then she would have been a successful football player."

Example 7 is of a quite different nature. This is a sentence that represents a normative rule saying that in a class of situations (those in which you have borrowed money) a particular norm applies (you have to pay it back). Conditionals of this type are major building-blocks of moral codes.

---

[13]The term "non-conditional" is preferable to "unconditional" that has a too strong connotation of absoluteness or indefeasibility.

This is one of several distinctions in deontic logic that are often overlooked because they are not made in everyday language. We use the same linguistic forms ("if... then") for both purposes. But that is not reason enough to assume that these two types of conditionality should have the same logical structure or even the same representation in formal languages.

All these three distinctions:

> normative rule vs. veritable norm

> prima facie norm vs. overall norm

> conditional norm vs. non-conditional norm

have implications on the logic. It is not necessary for an account of deontic logic to cover all the combinations of these three distinctions, but it is important to keep track of what it covers. It is also often useful to pay special attention to veritable, non-conditional norms. They form a basic fraction of normative discourse that we need to get right before introducing the further complications of trans-situational and conditional norms.

2.4 The ideal ought and action representation

As mentioned in Section 2.2, normative requirements can be expressed with many different words, including "ought", "obligatory", "must", "duty", "should", "have to", and several others. The term most often referred to in deontic logic is "ought". This is arguably unfortunate since "ought" is ambiguous between two meanings. We can distinguish between the *normative ought* (ought-to-do, Tunsollen) that prescribes or recommends actions and the *ideal ought* (ought-to-be, Seinsollen) that expresses wishes about the state of the world. "You ought to help your destitute brother" is an example of the former meaning, whereas "There ought to be no injustice in the world" exemplifies the latter. As Richard Robinson observed, statements of the latter type "are not prescriptive at all, either prudentially or morally, but express valuations. Such is 'Everybody ought to be happy'. This is not a prescription or command to anybody to act or to refrain." [Robinson, 1971, p. 195]

It is important to recognize that this double usage is specific for "ought", and does not apply to other prescriptive predicates. It would not make much sense to say that there is a duty for the world not to contain any injustice or that it is obligatory that everyone be happy. Since deontic logic is concerned with prescriptions in general, not only those that are expressed by the English word "ought", it would not be unreasonable to exclude the ideal ought from further consideration. This is how we treat non-prescriptive meanings of other prescriptive predicates, such as the "must" in "You must be wrong!". There is no reason to believe that the two meanings of "must" can be covered by one and the same logical representation. Similarly, there is no reason to assume that the normative and the ideal ought can both be adequately captured in one and the same logical representation. [Dayton, 1981]

However, it is not uncommon in deontic logic to try to unify the two notions of "ought". This is usually done by reconstructing the normative ought as an ideal ought referring to actions, according to the formula:

Person i ought to do X. = It ought to be the case that person i does X.

Hence, "You ought to sing in tune" is explicated (approximately) as meaning "The world ought to be such that you sing in tune." However, these two statements have quite different meanings. The former indicates that the person in question is able to sing more in tune, whereas the latter indicates that nature has not endowed her with that ability. As this example shows, the logical unification of the normative and ideal ought can at most be a very rough approximation.

In what follows I will assume that the O of deontic logic is intended to capture the normative ought (ought-to-do). However, after resolving this ambiguity we still have a problem with the representation of agency. In the deontic formula Op, neither O nor p refers to the agent who performs the action. So where is the agent?

The usual answer to that question is that p in Op is an action description that will, when spelt out, specify the agent. Hence, if Op represents the sentence "Eve ought to leave the house", then O represents "ought" and p represents "Eve leaves the house". To make this more precise, we can replace p by a sentence containing a Do operator (see-to-it-that operator) D and thus write OD_iq. Here, i represents Eve and q that she is outside of the house. The whole phrase D_iq represents "Eve sees to it that she is outside of the house". [Kanger, 1957]

In the sentence "Eve ought to leave the house" Eve is not only an agent; she is also the person who is subject to the obligation. These are two different roles. In principle we should therefore put an agent index not only on the action representation but also on the deontic operator. This would give rise to formulas such as O_iD_ip. However, in the cases usually discussed in deontic logic, the person under obligation coincides with the (one and only) person whose action(s) are represented in the argument of the O operator. The standard solution (to be followed here) is therefore to suppress the person index of deontic operators. In most cases this is unproblematic, but in representations of multi-agent actions or action complexes it may be necessary to introduce the index. We may for instance wish to distinguish between (i) a situation in which it is obligatory for each of i and j to see to it that one of them performs a certain type of action and (ii) a situation in which it is obligatory only for i to see to it that one of them performs the action. This difference can, at least approximately, be represented as a distinction between $O_{\{i,j\}}(D_i p \vee D_j p)$ and $O_i(D_i p \vee D_j p)$.

2.5 Possible worlds and other holistic alternatives

Possible worlds have traditionally a central role in the semantics of deontic logic. However, it has not always been made sufficiently clear what is meant by a possible world in this context. According to a metaphysical interpretation, a possible world is a possible state of the world, specified in full detail. The best representation of a (metaphysical) possible world in a language is the set of all sentences expressible in the language that are true in that world. Provided that every sentence in the language is either true or false in the world, such a set of sentences will be a maximally consistent subset of the language. In the logician's parlance, by a possible world is usually meant such a set. The metaphysical and the logical meanings of the term should be carefully kept apart.

A possible world in deontic logic can be a representation of a possible world in the metaphysical sense, i.e. one of the alternative ways in which the whole world could be. But it can also be a representation of a much more restricted alternative. If we discuss which to choose among the alternative courses of actions that are available to you, then the relevant holistic alternatives are complete action alternatives, i.e. each of them describes completely one way in which you can act. Such "small worlds" can be represented by maximal consistent subsets of a language or a subset of a language; they will then be possible worlds in the logical but not in the metaphysical sense. In order to avoid confusion, it is often advisable to use terms such as "holistic alternative" or "complete alternative" rather than "possible world" about such entities.

In standard uses of possible worlds in deontic logic it is taken for granted that the components of the possible worlds coincide with the admissible arguments of the deontic operators. Hence, if p is included in possible worlds, then Op is taken to be a well-formed formula. This is a formally convenient practice, but it requires careful consideration of what to include into the possible worlds. As was noted by Hector-Neri Castañeda, a clear distinction must be drawn between on the one hand sentences representing events and actions that are taken to be beyond the agent's control, and on the other hand sentences representing events and actions that are "practically considered", i.e. possible for the agent to bring about. [Castañeda, 1968; Castañeda, 1977; Castañeda, 1989] Only the latter are appropriate arguments of the deontic operators. If the holistic alternatives are so restricted that they only contain the practically considered sentences, then the usual practice of identifying these admissible arguments with the elements of possible worlds will not cause any problem. If they are not so restricted, then expressions with improper arguments will have to be excluded in some other way.

Sentences not referring to human actions, such as "The weather will be fine tomorrow", provide the most obvious examples of that which is not practically considered, in Castañeda's terminology. However, there may also be sentences that are excluded although the actions that they represent would be physically possible to carry through. When discussing what course of action to choose we employ a possibility perspective, i.e. a set of assumptions about what is possible or not possible to do. In doing so we tend to exclude certain options from further consideration. Such exclusions are often expressed in terms of (im)possibility, and give rise to "the frequent claim that one *cannot* do a certain thing because one has already decided to do something else, and not because one's will would not be efficacious as regards that act." [Kapitan, 1986, p. 246] As one example, I can say of a drug addict that he ought to use sterile needles. In saying this, I take his drug abuse for given. Soon afterwards, I may say that he ought not to take drugs at all, thus shifting the boundary between given and practically considered components of his behaviour. In a semantic account of these statements, the first statement seems to require a model containing no alternative in which he refrains from injecting, whereas the second would require a model in which such an alternative is included. Such shifts in perspective are an interesting but seemingly not much explored topic for studies in deontic logic.

3 Standard deontic logic

This section is devoted to the deontic logics that alternative semantics are alternatives to.

3.1 The origins

Although he had many forerunners, G.H. von Wright can rightly be said to have founded modern deontic logic in his famous 1951 paper.[14] Possible world semantics for deontic logic was introduced soon afterwards. Contrary to what has often been believed, the introduction of possible world semantics for deontic logic did not take place as a later addition to well-established possible world semantics for modal concepts (necessity and possibility). In-

---

[14][von Wright, 1951]. On the origins of deontic logic, see [Føllesdal and Hilpinen, 1970; von Wright, 1999a].

stead, deontic semantics arose in the same process of parallel inventions that gave rise to general modal semantics. The pioneers of possible world semantics all discussed deontic logic in their early writings. They were also all aware of the crucial difference between modal and deontic possible world semantics, namely that the accessibility relation should be reflexive in the former but not the latter type of semantics (so that $\Box p \to p$ holds but not $Op \to p$). [Wolenski, 1989] In 1957, Stig Kanger discussed deontic logic and introduced a system that contains what was later called standard deontic logic. [Kanger, 1957] In the same year, Hintikka expressed the idea that Pp holds if p can obtain "without violating any obligations", i.e. if p takes place in some possible world in which all obligations are satisfied. [Hintikka, 1957, p. 12] In a similar vein, Kripke wrote in 1963:

> "If we were to drop the condition that R be reflexive, this would be equivalent to abandoning the modal axiom $\Box p \to p$. In this way we could obtain systems of the type required for deontic logic." [Kripke, 1963, p. 95]

A paper by William Hanson [Hanson, 1965] seems to be the first in which possible world semantics for deontic logic was fully developed. Like his predecessors, he constructed deontic semantics to differ from modal semantics only in one important respect, namely that the accessibility relation was not reflexive. This elegant construction yields the logical principles that von Wright had proposed for deontic logic, but with one exception. In his 1951 paper, von Wright had proclaimed a principle of "deontic contingency" for tautologies, namely: "A tautologous act is not necessarily obligatory", i.e., $\neg O(p \vee \neg p)$. [von Wright, 1951] Subsequent authors have in most cases accepted the opposite principle:

$$O(p \vee \neg p)$$

that may be called the axiom of the empty duty. The reason why this postulate was so readily accepted is its formal convenience. It holds in the type of possible world semantics that was briefly described by Kanger, Hintikka, and Kripke, and further developed by William Hanson. However, in terms of intuitive plausibility it is not easily defended. A sentence such as "You are morally required to either drink a glass of water or not drink a glass of water" does not make much sense, which is a good reason to avoid the assumption that it is necessarily true. (Cf. [Jackson, 1985, p. 191] and [Lenk, 1978, p. 31].)

3.2 The standard system

As was observed in Section 2.1, it is reasonable to restrict the language of deontic logic to sentences in which no instance of O can occur within the scope of another instance of that operator. (This means that we exclude sentences such as OOp and $O(p \vee Oq)$, but we may still have sentences such as $O(p\vee q)$ and $Op \& \neg Oq$.) Under this assumption, the accessibility relation of modal-style deontic logic is equivalent with a much simpler semantic construction. Assuming a given set $\mathcal{W}$ of possible worlds (maximal consistent sets of sentences), it can be expressed as follows:

Ideal Worlds Intersection (IWI)

There is a subset $\mathcal{I}$ of the set $\mathcal{W}$ of possible worlds[15], such that:

For all p, Op holds if and only if $p \in \bigcap \mathcal{I}$.

$\mathcal{I}$ is called the set of "ideal" worlds (or "deontically ideal" or "(deontically) perfect" worlds). It is easy to show that the sentences that are valid in this simple model coincide with those that are derivable from the following three axioms:

$Op \to \neg O \neg p$,

$Op \& Oq \leftrightarrow O(p\&q)$, and

$O(p \vee \neg p)$

The first two of these axioms were present in von Wright's 1951 paper, whereas the third is the axiom of the empty duty that replaced its negation in most early developments from von Wright's system. The second axiom is logically equivalent to the combination of the following two postulates:[16]

$Op \& Oq \to O(p\&q)$ (agglomeration)[17]

If $\vdash p \to q$ then $\vdash Op \to Oq$ (necessitation)[18]

In 1969 Bengt Hansson introduced the term "standard deontic logic" (SDL) to denote the deontic logic that can be characterized either by these three axioms or by the semantic principle of Ideal Worlds Intersection. [Hansson, 1969; von Wright, 1981, p. 5]

The next two sections are devoted to two models that are equivalent with the one described above.

---

[15] We can leave it open whether $\mathcal{I}$ is the same for all (actual) worlds or a function of the world for which the evaluation is made.

[16] This holds only under the assumption of intersubstitutivity of logically equivalent sentences.

[17] The name "agglomeration" was apparently introduced by Bernard Williams [Williams, 1965, p. 118]. Another common name is "aggregation" [Schotch and Jennings, 1981, p. 152].

[18] Necessitation has many other names, including "the consequence principle" [Hilpinen, 1985, p. 191], "the inheritance principle" [Vermazen, 1977, p. 14], "Becker's law" [McArthur, 1981, p. 149], "transmission" [Routley and Plumwood, 1984, p. 4], and "entailment" [Jackson, 1985, p. 178].

3.3 The Anderson–Kanger model

In 1956 Alan Ross Anderson proposed that a deontic operator can be introduced into a modal language through the addition of a constant S denoting sanction, sanctionability or violation of morality. [Anderson, 1956] The deontic operator O can then be defined as follows:

$$Op \leftrightarrow \Box(\neg p \rightarrow S)$$

With the usual interdefinability between O, P, and F (i.e. $Pp \leftrightarrow \neg O \neg p$ and $Fp \leftrightarrow O \neg p$), it follows directly that:

$$Pp \leftrightarrow \Diamond(p \,\&\, \neg S)$$
$$Fp \leftrightarrow \Box(p \rightarrow S)$$

In 1957 Stig Kanger independently proposed another reconstruction of deontic logic within modal logic, namely by the introduction of a constant G denoting "what morality requires". [Kanger, 1957] He defined the prescriptive operator as follows:

$$Op \leftrightarrow \Box(G \rightarrow p)$$

With the usual interdefinability between O, P, and F, it follows directly that:

$$Pp \leftrightarrow \Diamond(G \,\&\, p)$$
$$Fp \leftrightarrow \Box(G \rightarrow \neg p)$$

These two approaches are easily seen to be equivalent, just let

$$G \leftrightarrow \neg S.$$

The connection between Anderson–Kanger semantics and the standard semantics (IWI) comes out particularly clearly in Kanger's version. We can regard G as a marker that is present in all ideal worlds and absent from all other worlds. Then the connection with IWI follows from the following series of equivalences:

Op

$\Leftrightarrow$

p holds in all ideal worlds

$\Leftrightarrow$

p holds in all worlds in which G holds

$\Leftrightarrow$ [19]

$G \rightarrow p$ holds in all worlds

$\Leftrightarrow$

$\Box(G \rightarrow p)$

With a fairly wide range of logics for $\Box$, the Anderson-Kanger approach gives rise exactly to the SDL logic. If nested modalities (O operators within the scope of other O operators) are allowed, then the choice of a logic for $\Box$ will have influence on the logic of O.

3.4 An ordering of worlds

From a viewpoint of intuitive interpretation it would seem natural to identify $\mathcal{I}$ with the worlds that are best according to some value standard that can be expressed with a (presumably transitive and complete) preference relation $\geq$:

$$\mathcal{I} = \{W \in \mathcal{W} \mid (\forall W' \in \mathcal{W})(W \geq W')\}$$

hence:

Op if and only if $p \in \bigcap\{W \in \mathcal{W} \mid (\forall W' \in \mathcal{W})(W \geq W')\}$

Technically, this does not work if there are infinitely many possible worlds, and among them an unending series of better and better worlds (and thus no best worlds). There are two major ways to deal with this problem. One is to exclude such structures by the introduction of a limit assumption, i.e. a condition saying that there is no infinite sequence of better and better worlds. [Lewis, 1974, p. 5] [von Kutschera, 1975, p. 204] The other is to modify the semantic evaluation principle for O so that Op holds whenever p holds in a top fragment of the set of possible worlds. A top fragment can be identified as consisting of all worlds that are at least as good as some particular world W): [Goldman, 1977, p. 243]

Op if and only if $(\exists W \in \mathcal{W})(\forall W' \in \mathcal{W})(W' \geq W \Rightarrow p \in W')$

The introduction of an ordering that underlies the set of ideal worlds may appear to be just an unnecessary complication. However, as we will soon see, it provides extra resources that can be used to account for conditional obligations and to distinguish between different degrees of obligatoriness.

---

[19]Since $G \rightarrow p$ holds in all worlds not containing G.

3.5 Dyadic versions of standard deontic logic

Some of the earliest developments in deontic logic were extensions of the standard system to express conditional deontic sentences. In his 1951 paper von Wright proposed that a conditional obligation such as

> If p, then q is morally required.

should be written:

$$O(p \to q).$$

But it was not long before serious problems with this proposal were discovered. In SDL, the following formula is provable:

$$O \neg p \to O(p \to q).$$

In combination with von Wright's interpretation of $O(p \to q)$, this gives rise to the implausible conclusion that if you do what is forbidden, then you are required to do anything whatsoever. "If it is forbidden for you to steal this car, then if you steal it you ought to run over a pedestrian." (The paradox of commitment.) This was pointed out by A.N. Prior [Prior, 1954]. To solve the problem Prior brought up an alternative originally proposed to him by G.E. Hughes, namely to represent "If p, then q is morally required" as follows: [Prior, 1962, p. 224]

$$p \to Oq$$

This proposal saves us from the paradox of commitment, but unfortunately it gives rise to another, equally serious problem. The following formula is provable:

$$\neg p \to (p \to Oq)$$

It has instances such as "If you do not visit your friend, then if you visit her you ought to burn down her house." The source of this problem is the interpretation of material implication ($\to$), not that of the operator of moral requirement (O). This can be seen by replacing Oq for instance by a sentence r meaning "she will paint your face red". It is similarly provable that if you do not visit your friend, then if you visit her she will paint your face red.

To sum up, there are two obvious ways to represent conditional obligation in SDL, namely $O(p \to q)$ and $p \to Oq$, but they both give rise to absurd conclusions. It seems as if the resources of the SDL language (the language of propositional logic reinforced with the additional operator O) are insufficient to express conditional obligations. One rather immediate option would be to include a modal necessity operator ($\Box$) to the language and define conditional obligation as follows:

$\Box(p \to Oq)$

In this way we avoid the problems just mentioned for the other two proposals. But unfortunately this proposal leads to other implausible conclusions. Let p denote that Robert has been rightfully sentenced to prison, q that his prison guard Edward keeps him locked in his prison cell at night, and r that Robert is in acute need of hospital care. We then have $\Box(p \to Oq)$, but from this it follows logically that $\Box(p \& r \to Oq)$. Hence, according to this representation of conditional obligation, if Edward has to keep Robert in his cell, then he has to do so even if this would cost Robert his life.

The addition of $\Box$ seems to be insufficient to solve the problem. The situation has been aptly summarized by Brian Chellas: The representation $p \to Oq$ makes "If p, then q is morally required' true whenever p is false or Oq is true. The representation $O(p \to q)$ makes the same conditional sentence true whenever either $O\neg p$ or Oq is true. All three representations make "If $p \& r$, then q is morally required" true whenever "If p, then q is morally required" is true. [Chellas, 1980, p. 201]

It was against this background that Bengt Hansson in 1969 investigated how SDL could be extended to give a more reasonable account of conditional obligations. He decided to introduce a primitive notation for "If p, then q is morally required", namely $O(q \mid p)$. In order to determine the validity of instances of this formula in a possible world model, he made use of an ordering of worlds that represents their relative value, or degree of ideality (as introduced in Section 3.4). In order to determine whether $O(q \mid p)$ holds, we restrict our attention to worlds in which p is true. $O(q \mid p)$ holds if and only if q holds in all the best (most ideal) p-worlds. [Hansson, 1969] This can be expressed as an extension of the IWI principle:

Dyadic Ideal Worlds Intersection (dIWI)

There is a relation $\geq$ on the set $\mathcal{W}$ of possible worlds, such that for all p and q, $O(q \mid p)$ holds if and only if:

$q \in \bigcap \{W \in \mathcal{W}_p \mid (\forall W' \in \mathcal{W}_p)(W \geq W')\}$,

where $\mathcal{W}_p$ is the set of elements of $\mathcal{W}$ that contain p.

Hence, according to SDL "You ought to offer her a loan" holds if and only if you offer her a loan in all the best (most ideal) worlds. According to Bengt

Hansson's conditional extension of SDL, "You ought to offer her a loan if she loses her job" holds if and only if you offer her a loan in all the best (most ideal) among those worlds in which she loses her job. For technical reasons this solution does not work if there are no best p-worlds but instead an infinite series of deontically better and better p-worlds. In such cases we can apply the top fragment approach from Section 3.4, and use the following definition:

$$O(q \mid p) \text{ iff } (\exists W \in \mathcal{W}_p)(\forall W' \in \mathcal{W}_p)(W' \geq W \Rightarrow q \in W')$$

Bengt Hansson proposed three *d*yadic extensions of *s*tandard *d*eontic *l*ogic, DSDL1, DSDL2, and DSDL3. They are all based dIWI but differ in the properties of the relation $\geq$.

> In *DSDL1*, $\geq$ is reflexive.
>
> In *DSDL2*, $\geq$ is reflexive, and for every consistent sentence p there is at least one $\geq$-maximal p-world.
>
> In *DSDL3*, $\geq$ has the same properties as in DSDL2, and in addition $\geq$ is transitive and complete.

DSDL3 was axiomatically characterized by Wolfgang Spohn [Spohn, 1975]. He transferred the SDL axioms to dyadic deontic logic in the following way:

$$O(p \mid r) \;\&\; O(q \mid r) \leftrightarrow O(p\&q \mid r)$$

In addition to the transferred axioms, only one more axiom was needed for the axiomatization, namely:

$$P(p \mid q) \rightarrow (O(r \mid p\&q) \leftrightarrow O(p \rightarrow r \mid q))$$

Additional results on DSDL3 have been obtained by Xavier Parent [Parent, 2008], who has also axiomatized DSDL2 [Parent, 2010]. The axiomatization of DSDL1 is still an open question.

4 Criticism of standard deontic logic

The argumentation for alternative semantics for deontic logic is largely based on critical discussions of SDL. As we have seen there is both a semantical and a syntactical presentation of SDL. They have both been subject to extensive criticism. Most of this criticism aims at properties that SDL has in common with modal logic; indeed von Wright himself conceded that SDL "may be said to stretch the analogy between modal and deontic logic to its utmost limit". [von Wright, 1981, p. 6] Therefore the criticism points in the direction of developing new systems of deontic logic that do more justice to the differences between modal and deontic concepts.

A large part of the criticism against SDL applies both to its basic monadic form and its dyadic extensions. In what follows, the emphasis will be on the monadic version.

4.1 Semantic criticism

Existence of ideal worlds. In principle it is possible to treat ideal worlds as a convenient construction that yields the right results but need not coincide with non-formal notions of "perfect" or "best" worlds. However, at least some deontic logicians seem to assign much more meaning to the notion. Von Wright wrote:

> "Generally speaking: a legal order and, similarly, any coherent code or system of norms may be said to envisage what I propose to call an ideal state of things when no obligation is ever neglected... Deontic logic, to put it in a nutshell, is the study of logical relations in deontically perfect worlds." [von Wright, 1986]

Critics have pointed out that our value judgments about possible worlds may not support the notion of best or ideal worlds. Even if all possible worlds are ordered by a transitive and complete preference relation, there may be no worlds that are better than all the other worlds. If the number of possible worlds is infinite, then there may instead be an infinite series of better and better worlds, so that no world is best. [Reichenbach, 1980] However, this is not a very strong argument against SDL, and this for two reasons. First, it is possible to interpret the ideal worlds of deontic logic as ideal only with respect to obligation-fulfilment, not necessarily with respect to other objects of evaluation. There may be possible worlds that are perfect from the viewpoint of obligation-fulfilment even if none of them is perfect from a more general moral point of view. Secondly, as we saw in Sections 3.4 and 3.5, if there is an infinite series of deontically better and better worlds, then the ideal worlds construction can be replaced by a definition according to which Op holds if and only p holds in a top fragment of the set of worlds. Therefore this argument can lead us to modify, but not to reject ideal worlds semantics.

Another argument against the notion of ideality was put forward by Eric Dayton, who pointed out that perhaps there are no possible worlds in which all desirable features are simultaneously realized. Whether such worlds are impossible in the sense of implying a contradiction "is an open question and should not be answered with an axiom". [Dayton, 1981, p. 139] However, this argument is based on the assumption that the ideal worlds are worlds in which each desideratum is maximally satisfied. Alternatively, ideal worlds could be identified as those worlds that are so good that they cannot be improved.

Too restricted information. According to SDL semantics, our obligations are completely determined by the ideal possible worlds. However, it is easy to show that the ideal worlds do not always supply sufficient information for determining what our obligations are. Information about non-ideal worlds may also be needed. Consider the case of an alcoholic who has driven drunk on previous occasions. It is not unreasonable to claim that he ought to have an alcohol lock installed in his car in order to prevent himself from driving drunk again. However, in an ideal world he would certainly control himself and never try to drive when he is drunk (if he is at all drunk in an ideal world).[20] Therefore, this obligation cannot be derived from information inherent in the ideal worlds. We derive it from information about causality and human agency that is obtainable from our knowledge of non-ideal rather than ideal worlds.

Compensatory and preventive actions. Obligations to perform compensatory actions are difficult to account for in SDL. Suppose that John sees a small child fall into the pool in front of him. It would be easy for him to save the child's life. Does he have an obligation to do so? According to SDL semantics we have to consider what his actions would have been in an ideal world. In an ideal world, the child would presumably not have fallen into the water. (This may apply even if we consider the ideal worlds to be ideal only in terms of obligation-fulfilment. If the child's parents had fulfilled their obligations then the accident would not have happened.) Hence, in the ideal worlds the child would not have been in danger, and John could not have saved it. It follows that John is under no moral obligation to save the child. This example is due to Holly Goldman, according to whom SDL "ignores the fact that particular obligations flow from abstract principles *together with* contingent features of the world", and these features "do not appear in *all* the morally best worlds". [Goldman, 1977, p. 244]

A similar argument can be raised against the dyadic versions of SDL. Suppose that I am bringing back a book that I borrowed from you two years ago, promising to give it back within a month. The best worlds in which I return the book so late are worlds in which you have released me from my promise to bring it back earlier. In these worlds, there would be no

---

[20] This feature of the ideal worlds follows from the assumption that he has, in the actual world, an obligation not to try to start the car while drunk. Hence, in this argument we do not need to assume that the ideal worlds are ideal in any other sense than that of obligation-fulfilment.

reason to apologize for the delay. This is the criterion according to which we are supposed to judge whether I have, in the present world, an obligation to apologize for the delay. It follows that I have no such obligation. [Goldman, 1977, p. 247]

Preventive actions are almost as difficult as compensatory actions to account for in SDL semantics. In an ideal world there will be no acts of violence or racism, since such acts would violate obvious moral obligations. Therefore, in ideal worlds no one will act to prevent such misdeeds. According to the IWI principle, it would follow that there can be no obligation in the actual world to act against violence or racism.

These examples illustrate a fundamental problem with IWI (and its dyadic variant dIWI): Since our obligations are identified with how we would act in an ideal world, these semantic principles do in fact recommend us to act as if we already lived in such a world. But generally speaking, that is bad advice. Acting as one would have done in an ideal world is the behaviour that we can expect to follow from wishful thinking, not from well-considered moral deliberation. [Hansson, 2006]

Agent identity. SDL also has difficulties to deal with individuals who would not exist in all the ideal worlds.

> "But perhaps one of my male ancestors raped one of my female ancestors and I would not exist except for this rape. But in a deontically perfect world surely there are no rapes. Thus how could I exist in such a world on that supposition?" [Purtill, 1973, p. 431]

If a person does not exist in all the ideal worlds, then there is nothing that this person does in all the ideal worlds. According to SDL semantics she is therefore subject to no obligations at all. Each of us owes our existence to a large number of ancestral acts of procreation (probably not much less than 200 such acts only in the last two centuries before the individual's birth). If only one of these resulted from a breach of an obligation then we would not exist in all the ideal worlds, and hence we cannot have any obligation to do something in the actual world. It seems doubtful whether anyone would have obligations if this analysis is taken seriously.

A partial solution: smaller holistic alternatives. Some of the above-mentioned problems with SDL semantics can be resolved if we replace the possible worlds of traditional SDL by much smaller holistic alternatives. (Cf. Section 2.5.) In particular, the holistic alternatives can represent the combinations of actions that are open to an individual, rather than the worlds that she might inhabit. It is necessary to distinguish between those properties of SDL that depend on the choice of an alternative set (traditionally: set of possible worlds) and those that depend on the method used to assign normative status to the sentences represented in the chosen alternative set (IWI).

Many of the problems with compensatory and preventive actions can be solved in this way. That the child fell into the pond was beyond John's control. If we apply SDL semantics to the set of options that are open for him to choose between, then the absurd result obtained with possible worlds will disappear. We can reasonably assume that in all the best action alternatives available to him (contrary to all the best possible worlds that he may inhabit) he will save the child; hence he has an obligation to do so.

Furthermore, this move also solves the problems related to agent identity. In all the alternatives open to an agent, that agent exists. The problems connected with possible worlds in which the agent does not exist disappear when the scope of the semantics is reduced to actions open to the agent.

However, some of the problems referred to above cannot be solved in this way. This applies for instance to the example of the alcoholic and the alcohol lock. In the best of the options available to him, he will never be drunk again. Considerations of these worlds provide no reason to assign to him an obligation to have an alcohol lock installed. That obligation only comes up when we consider what happens in the non-ideal options – which is exactly what SDL semantics prevents us from taking into account.

4.2 Syntactic criticism: Inconsistencies

The syntactic criticism of SDL has followed two major lines that have also determined the ways in which alternative semantics have been developed. The first line of criticism concerns the consequences of inserting moral dilemmas into the framework. The derivations are very simple indeed. See Figure 5. Suppose that we have a moral dilemma, so that both p and its negation are obligatory, i.e. Op & $O\neg p$. We can then apply the SDL axiom of agglomeration (cf. Section 3.2) and obtain $O(p\&\neg p)$. This is an inconsistent duty, and therefore even more troublesome than the inconsistency among duties represented by Op & $O\neg p$. It is one thing to have one moral requirement to be at the pub and another moral requirement not to be at the pub. This is a conundrum we can understand, and we can deliberate on various (partial and imperfect) solutions. It is something else to have a moral requirement to both be and not be at the pub. Such an obligation (if we at all take it seriously) is impossible to do anything at all about. We cannot even imagine an action that would in any way take us closer to complying with it. We can call this the problem of *self-inconsistent obligation*. When we have inconsistency among obligations, SDL produces an obligation to something that is inconsistent in itself.

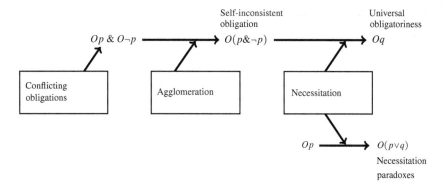

Figure 5: The most discussed implausible results in deontic logic

But it becomes even worse. We can continue by applying the SDL postulate of necessitation to $O(p\&\neg p)$. We can then derive Oq, for any sentence q. This is of course an absurd conclusion. The student in Sartre's famous example had both an obligation (Op) to join the Forces Françaises Libres in Britain and an obligation $(O\neg p)$ not to do so (in order to take care of his mother). [Sartre, 1946, pp. 39-40] These two obligations did not combine to put him under an obligation (Oq) to become a Nazi collaborator. We can call this the problem of *universal obligatoriness*. When we have a self-inconsistent obligation, SDL proclaims that everything is obligatory.

Both these problems, self-inconsistent obligation and universal obligatoriness, are artefacts of the logical system, i.e. they do not correspond to problems that we encounter in informal discourse about the subject-matter that deontic logic is intended to represent. Such artefacts may be more or less serious. It can be argued that self-inconsistent obligations are fairly harmless. Contrary to q in Oq the sentence $p\&\neg p$ in $O(p\&\neg p)$ does not describe an action, and its obligatoriness can therefore be regarded as an innocuous peculiarity of the logical system, devoid of meaningful interpretation. It would not be unreasonable to see the presence of self-inconsistent sentences as a price we have pay for other, more desirable features of a logical system (such as simplicity). But there can be no doubt that universal obligatoriness is an intolerable defect of a deontic system. We therefore have a choice: Either we disallow inconsistent combinations of obligations (such as $Op \& O\neg p$), or we have to weaken the logic so that it does not contain both agglomeration and necessitation.

The first of these options, disallowing inconsistent combinations of obligations, can be expressed as an axiom of consistency. A first approximation of such an axiom would be:

(1) $\neg(Op \,\&\, O\neg p)$

(which is equivalent with $\neg O\bot$ if both agglomeration and necessitation hold).[21] However, deontic inconsistencies can also arise between two sentences that are not each other's negations:

> You must pay at least 500 dollars. You are not allowed to pay more than 300 dollars.

In order to cover such cases, the consistency requirement can be generalized as follows:

(2) If $\vdash \neg(p\&q)$ then $\vdash \neg(Op \,\&\, Oq)$.

But this is not yet a sufficiently general consistency postulate. Consider the following example:

> You must report this incident either to the general or to the colonel. You are not allowed to report the incident to the general. You are not allowed to report the incident to the colonel.

This can be formalized as an inconsistency arising from the three sentences $O(p \lor q)$, $O\neg p$, and $O\neg q$. Unless some other postulate for O is added (such as agglomeration), this set of three prescriptions does not violate (2). Nevertheless, it is clearly an inconsistent set of prescriptions. (Cf. [Royakkers, 1996, p. 158]) Therefore, (2) should be generalized:

(3) $\{p \mid Op\} \nvdash \bot$

In the presence of agglomeration and necessitation, (1), (2), and (3) are all equivalent. They are also all valid in SDL. It has therefore often been taken for granted that (1) is a sufficiently general representation of deontic consistency. However, in studies of logics that are weaker than SDL it can lead us wrong. Therefore, (3) is a more adequate general-purpose postulate for deontic consistency.

Two complications have to be mentioned that may lead to modifications of (3) in some contexts. First, we have tacitly assumed that logical consequence exhausts the resources we have to express relations among sentences. In real life, options may be incompatible for contingent, non-logical reasons. Suppose that your only two options with respect to feeding your child are to steal food for the child and to let it starve. Then the following set of prescriptions is impossible to comply with, although it is consistent:

---

[21] $\bot$ denotes falsum, an arbitrary inconsistent sentence.

You should not steal. You should not let your child starve.

If we wish deontic logic to exclude the combinations of requirements that are practically impossible to comply with ("ought implies can") then we need additional resources in the language. With a modal possibility operator we can generalize (3) to:

(4) $\Diamond \& \{p \mid Op\}$

($\&A$ denotes the conjunction of all elements of the set A of sentences.[22])

The other complication concerns models including several agents. In such models we have to distinguish between agent-specific and universal requirements of consistency. If the O operator satisfies agent-specific consistency, then the obligations of each agent are consistent, but obligations pertaining to different agents may be in conflict. Thus $OD_i p$ & $OD_k \neg p$ is compatible with agent-specific consistency, but $OD_i p$ & $OD_i \neg p$ is not. According to the stronger principle of universal consistency, the performance of the overall duties of all agents should be consistent; hence neither $OD_i p$ & $OD_k \neg p$ nor $OD_i p$ & $OD_i \neg p$ can be accepted. Universal consistency is plausible according to classical utilitarianism, but it is implausible according to moral theories that allow for agent-specific commitments.

For a concrete example, suppose that I have promised my child to buy a certain unique object at an auction, and you have made a similar promise to your child to buy the same object. Arguably, both these promises give rise to obligations of fulfilment. We cannot both satisfy our obligations, but does the fact that you made this promise relieve me of my obligation to fulfil mine? If we wish to accept conflicts between obligations of different persons, but not between obligations pertaining to one and the same person, then we can restrict the consistency postulate to agent-specific consistency:

(5) If $A \subseteq \{p \mid Op\}$, all elements of A have the same agent, and $A \vdash q$, then $\Diamond q$.

Having elaborated on how a consistency postulate should be expressed, we also need to consider whether or not such a postulate should hold in deontic logic. The answer will depend on what notion of moral requirement the logic is intended to represent.[23] If the desired notion is that of obligatoriness all things considered ("overall ought"), and it is furthermore assumed

---

[22] For some infinite sets of sentences, $\&A$ is undefined. If $\{p \mid Op\}$ is such a set, then the following postulate can be used instead: If $\{p \mid Op\} \vdash q$ then $\Diamond q$.

[23] Dayton noted that a consistency postulate should hold for the so-called "ought-to-be". [Dayton, 1981, p. 138] This is quite plausible, but for reasons given in Section 2.4 the semantics of "ought-to-be" will not be further discussed here.

that moral dilemmas do not exist, then a consistency postulate may be reasonable. However, if we allow for moral dilemmas in an account of obligatoriness all things considered, then the best way to express such dilemmas is to allow for the simultaneous validity of obligations Op and Oq such that p and q are incompatible.[24] Likewise, in an account of prima facie obligations, incompatible obligations will have to be allowed.

The general conclusion to be drawn from all this is that if we wish deontic logic to be a versatile tool for moral reflection, adaptable to different notions of moral requirement, then we need to develop formal systems that are compatible with both the presence and the absence of conflicting obligations. In order to capture mechanisms such as the derivation of overall duties from prima facie duties or the choice of one of the horns of a moral dilemma we may also have to include, in one and the same logic, both operators that satisfy the consistency requirements referred to above and operators that violate them. [Hansson, 1999] But even in logics that allow for the simultaneous validity of obligations that cannot all be fulfilled, absurd results such as universal obligatoriness have to be avoided. In order achieve this, while allowing for inconsistent sets of obligations, we need a logic that differs from SDL in that agglomeration and necessitation do not both hold.

4.3 Syntactic criticism: Necessitation

The other major line of syntactic criticism of SDL is directed at the implications of the following rule of inference that is valid in SDL:

$$\text{If } \vdash p \rightarrow q \text{ then } \vdash Op \rightarrow Oq \text{ (necessitation)}$$

Necessitation implies that duties are arbitrarily divisible so that every part of a duty is itself a duty. It is not difficult to find examples in which this is contrary to common intuitions.

> "If I am obliged to pay for some goods and carry them away, but fail for some reason to pay for them, I can hardly carry the goods away, claiming that I am keeping at least one of my obligations!" [Purtill, 1975, p. 487]

---

[24] The absence of moral dilemmas in hypothetical ideal worlds may perhaps induce a deontic logician to believe that moral dilemmas should always be avoided. Since there are no dilemmas in a morally perfect world, and we want to come as close as possible to moral perfection, dilemmas should be avoided. But this is a fallacy. The imitation of some aspect of a far-away ideal does not necessarily lead us closer to that ideal. In the actual world, with all its moral imperfections, strategies that reduce the incidence of moral dilemmas may have side-effects that are not worth the price. See [Hansson, 1998] for an argument to the effect that moral dilemmas are an unavoidable component of a satisfactory moral life.

I work as a janitor at a bank. Usually I have no access to the money handled by the bank, but one day my boss orders me to fetch a box containing €10,000,000 and carry it to another bank office a couple of blocks away. Unknown to him, I suffer from weakness of will. I know that once I have the money in my hands it is in fact quite improbable that I will be able to resist the temptation to elope with it. Therefore, if I pick up the money (p) it is most unlikely that I will also hand it over to the other bank (q). Since it is part of my job to run errands for the bank, I certainly have the obligation represented by $O(p\&q)$. But do I have the obligation represented by Op?

Another way to express this criticism is that a deontic logic that satisfies the necessitation postulate conflates the distinction between our obligations and that which we have to do in order to fulfil them. These two categories do not necessarily coincide. "[T]he fact that we can't help but bring about the necessary consequences of our action does not mean we have an *obligation* to bring them about." [Sayre-McCord, 1986, p. 188]

The problems associated with necessitation have been expressed in terms of paradoxes; in fact most of the major deontic paradoxes depend on necessitation. We can call them necessitation paradoxes.

The most famous of them is Ross's paradox that is based on the instance $Op \to O(p \vee q)$ of necessitation. ("If you ought to mail the letter, then you ought to either mail or burn it.") [Ross, 1941, p. 62][25] The Good Samaritan operates on two sentences p and q, such that q denotes some atrocity and p some good act that can only take place if q has taken place. We then have $\vdash p \to q$, and it follows by necessitation that if Op then Oq. ("You ought to call an ambulance for the policeman you assaulted. Therefore, you ought to assault the policeman.") [Prior, 1958, p. 144][26] Åqvist's Knower paradox makes use of the epistemic principle that only that which is true can be known. Here, q denotes some wrongful action, and p denotes that q is

---

[25] Arguably, the "or" of the consequent is not truth-functional, but rather of the free-choice variant. (See Chapter 3, Section 5.) Therefore, it is tempting to believe that this paradox depends only on the non-truth-functional properties of the "or" of English and other natural languages. However, as can be seen from Purtill's example, quoted above, a similar problem arises if $p \vee q$ is replaced by a non-disjunctive logical consequence of p.

[26] With a similar argument, necessitation has also been put into question for the so-called ought-to-be: "The problem with this is that if we agree that:

(S) We feed the starving poor

ought to [be] true, then we seem to invite the consequence that the sentence

($\exists S$) There are starving poor

which is logically implied by (S) ought also to be true." [Schotch and Jennings, 1981, p. 151] known by someone who is required to know it. Again, we have $\vdash p \to q$ and Op, and it follows by necessitation that Oq. ("If the police officer ought to know that Smith robbed Jones, then Smith ought to rob Jones.") [Åqvist, 1967]

These paradoxes all follow directly from necessitation, and they do not seem to be solvable by plausible reinterpretations of the formalism. It has also since long been recognized that necessitation is the major source of deontic paradoxes. Both Purtill and Stranzinger pointed out that the paradoxes can be solved by changing deontic logic so that necessitation is given up while agglomeration is retained. [Purtill, 1975; Stranzinger, 1978] In a discussion of the paradoxes von Wright concluded that "in a deontic logic which rejects the implication from left to right in the equivalence $O(p\&q) \leftrightarrow Op \& Oq$ while retaining the implication from right to left, the 'paradoxes' would not appear." [von Wright, 1981, p. 7] (Given intersubstitutivity of logically equivalent sentences, the left-to-right direction of that equivalence is logically equivalent with necessitation.)

Explicit defences of necessitation are not common in the literature. Possibly the best defence of Ross's paradox, and with it necessitation, was provided by Bengt Hansson:

> "What we often call descriptions of acts are not descriptions of *how the act is performed* but of *the result of the act*. It is in general possible to proceed in several different ways to achieve the same goal and it is certainly so if the result is described as a disjunction. That somebody asserts an obligation does not mean that he approves of every way of making the obligatory formula true. Specifically, to fulfil the obligation to help or kill Mr. A by killing Mr. A would be an unacceptable way, but doing it by helping him is all right. In fact $O(h \vee k)$ is then not more paradoxical than Oh, because even this obligation can be fulfilled in unacceptable ways, e.g. by helping the crippled Mr. A downstairs by kicking him. Let s stand for 'you kick Mr. A down the stairs'. Already the fact that Oh is equivalent to $O((h\&s) \vee (h\&\neg s))$ constitutes Ross' paradox, which then is not specific to von Wright type deontic logics."[Hansson, 1969, p. 384]

4.4 A summary of the criticism

The syntactic criticism of SDL outlined in Sections 4.2-4.3 constitutes only part of the criticism that has been waged against the language and the logical principles of SDL. A somewhat more extensive list of such criticism can be found in Table 1. The terminology used in the table is based on statistical nomenclature: Type 1 denotes false positives and type 2 false negatives. Furthermore, each of these two types is subdivided into two subtypes: those concerning what can be expressed (linguistic divergencies) and those concerning what can be inferred (logical divergencies).

Type 1–language: Normative statements that can be expressed, or distinctions that can be made, in the formal language but not in ordinary language.
$O\bot$
$O\top$

Type 1–logic: Derivations that are valid in the formal language but not in ordinary language (paradoxes).
$Op \to O(p \vee q)$ (Ross's paradox)
$O\neg p \to O(p \to q)$ (the paradox of commitment)
$Op\ \&\ O\neg p \to O(p\ \&\ \neg p)$ (self-inconsistency)
$Op\ \&\ O\neg p \to Oq$ (universal obligatoriness)

Type 2–language: Normative statements that can be expressed, or distinctions that can be made, in ordinary language but not in the formal language.
Degrees of obligatoriness.
Deontic dilemmas (that do not lead to universal obligatoriness)

Type 2–logic: Derivations that are valid in ordinary language but not in the formal language.
$Pp \to P\neg p$ (See Chapter 3, Section 5)
$P(p \vee q) \to Pp\ \&\ Pq$ (free choice permission, see Chapter 3, Section 5)

Table 1: Examples of proposed inadequacies of traditional deontic logic, divided into four major categories

Although the table is far from complete it suffices to show that there is a large number of divergences between deontic logic and informal accounts of norms. Indeed, one deontic logician reached the pessimistic conclusion that "[u]nlike modal logic, which has been at least somewhat useful in dealing with substantive philosophical problems (the Ontological Argument, Determinism), deontic logic has so far created more problems than it has solved." [Purtill, 1980]

However, when assessing SDL and the needs to modify it, we should keep in mind that formalized treatments of philosophical subject-matter can never be expected to have a perfect fit with informal discourse on the same subject-matter. To formalize means to reduce to a simplified form in which we can get a clear view of some (structural) aspects at the expense of others. [Merrill, 1978, pp. 305-11] [Hansson, 2000] Therefore, formalization always involves a trade-off between simplicity and faithfulness to the original. If the subject-matter is complex, then a reasonably simple model will usually have to leave out some of its philosophically relevant features. This is certainly true in deontic logic.

It follows that a philosophically uncriticizable deontic logic is impossible. Since simplifications are necessary, it will always be possible to devise a counter-argument against a proposed deontic logic – typically in the form of a counter-example – that seemingly invalidates the model. However, even if such a counter-argument convincingly discloses an imperfection in the model, this is not necessarily a sufficient reason to give it up. If the counter-argument cannot be neutralized without substantial losses of simplicity, then an appropriate response may be to continue using the model, bearing in mind its weaknesses (and perhaps supplementing it with other models that have other strengths and weaknesses).

The following is an example of a problematic inference in SDL that it would probably be too costly to remove: SDL is an extensional logic, i.e. logically equivalent sentences are intersubstitutable. (The same applies to the vast majority of alternative deontic logics that have been proposed.) Therefore it satisfies the following postulate:

$$\text{If } \vdash p \leftrightarrow q \text{ then } \vdash Op \leftrightarrow Oq.^{27}$$

Let a_1 signify that John kills his wife's murderer, a_2 that he kills only other persons than his wife's murderer, and b that he does not kill anybody at all. We then have:

$$O(\neg a_1) \rightarrow O(a_2 \vee b)$$

In words: If John ought not to kill his wife's murderer, then he ought to kill either only other persons than his wife's murderer, or no one at all. This is the revenger's paradox. [Hansson, 1991b] It can be avoided by giving up intersubstitutivity. However, this would be a far-reaching weakening of deontic logic. The revenger's paradox shows that if we wish to avoid such a drastic weakening, then we will have to accept as valid seemingly absurd expressions such as "John is obliged to either kill only other persons than his wife's murderer, or no one at all". This can be justified by treating $a_2 \vee b$ as an unnatural or misleading way to express the equivalent sentence a_1. If we follow this line of reasoning (as we probably should), then we also have

---

[27] Equivalently (in a form that more clearly shows that this is a weakening of necessitation): if $\vdash p \leftrightarrow q$ then $\vdash Op \rightarrow Oq$.

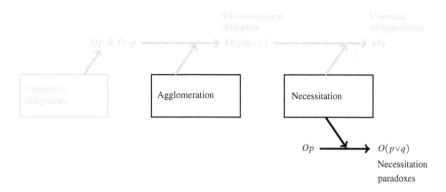

Figure 6: The effects on the implausible derivations of excluding the possibility of conflicting obligations

to ask ourselves which if any of the other implausible divergences between SDL and natural language we should accept.

The three SDL inferences that have most often been taken to be intolerable are those presented in Sections 4.2-4.3, namely self-inconsistent obligation, universal obligatoriness, and the necessitation paradoxes. Figure 5 summarizes how they depend on three major assumptions in deontic logic, namely that conflicting obligations (more precisely: inconsistent combinations of obligations) are possible and that the agglomeration and necessitation postulates hold. In Figures 6 and 7 we see what happens if we give up the possibility of conflicting obligations, respectively agglomeration. The result is the same, with respect to the three implausible derivations. In both cases we get rid of both self-inconsistent obligation and universal obligatoriness, but the necessitation paradoxes are not affected. This, however, does not mean that it is unimportant which of the two assumptions we give up. If we give up the possibility of conflicting obligations, then we cannot express moral dilemmas in our logic, but that is still possible if we instead give up agglomeration. This can be taken as a reason for further investigations of the latter approach.

Figure 8 shows what happens if we instead give up necessitation. In this way as well we get rid of universal obligatoriness. Self-inconsistent obligation is retained, but instead we remove the necessitation paradoxes.

It is of course possible to give up both necessitation and agglomeration (or both necessitation and the possibility of conflicting obligations). In this way we will get rid of all three problematic inferences. However, this would also make the resulting logic very weak. As already mentioned, the con-

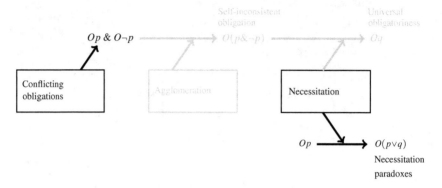

Figure 7: The effects of giving up agglomeration

struction of a deontic logic has to involve a trade-off between the exclusion of implausible expressions and the inclusion of plausible ones. (This can also be expressed as a balance between avoidance of divergencies of types 1–language and 1–logic respectively types 2–language and 2–logic, see Table 1.) Therefore it is no surprise that the development of alternative semantics for deontic logic has followed two major lines: giving up agglomeration but retaining necessitation and the other way around. In the next two sections, these two approaches will be investigated.

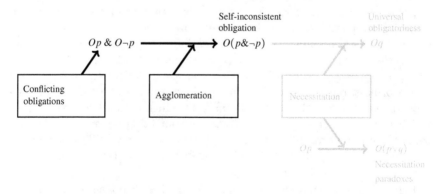

Figure 8: The effects of giving up necessitation

5 Giving up agglomeration

Of these two approaches, giving up agglomeration and retaining necessitation is by far the technically easiest one to implement. In 1974 Brian Chellas showed that agglomeration can be eliminated with a simple move that still retains the basic idea of IWI, namely that the obligatory acts are those that are performed in certain worlds. The trick is to have not one but several sets of ideal possible worlds. He called this system "minimal monadic deontic logic". [Chellas, 1974] It has subsequently often been called "minimal deontic logic" (MDL).[28] To introduce it, let $\mathcal{I}_1, ...\mathcal{I}_n$ be sets of possible worlds. Intuitively we can think of each of them as the collection of those worlds that are ideal (perfect, best) according to some particular standard. Furthermore, let Op hold if and only if p is true in all worlds in one of these sets. We can express this as follows:

Multiple Ideal Worlds Intersection (mIWI)

There is a set $\mathfrak{J}$ of subsets of the set $\mathcal{W}$ of possible worlds, such that:

For all p, Op holds if and only if $(\exists \mathcal{I} \in \mathfrak{J})(p \in \bigcap \mathcal{I})$.

Chellas summarized the intuition behind this construction as follows:

"Thus our account of the meaning of O is that a sentence of the form Op is true at a possible world just in case the world has a non-empty class of deontic alternatives throughout which p is true. The picture is one of possibly empty collections of non-empty classes of worlds functioning as moral standards: What ought to be true is what is entailed by one of these moral standards." [Chellas, 1974, p. 24]

To see that agglomeration does not hold in general according to mIWI, let $\mathfrak{J}$ consist of two sets $\mathcal{I}_1$ and $\mathcal{I}_2$ of possible worlds, such that $p \in \bigcap \mathcal{I}_1$, $q \notin \bigcap \mathcal{I}_1$, $p \notin \bigcap \mathcal{I}_2$, and $q \in \bigcap \mathcal{I}_2$. It then follows from $p \in \bigcap \mathcal{I}_1$ that Op and from $q \in \bigcap \mathcal{I}_2$ that Oq. However, since $p\&q \notin \bigcap \mathcal{I}_1$ and $p\&q \notin \bigcap \mathcal{I}_2$ we also have $\neg O(p\&q)$, contrary to agglomeration.

To see that necessitation holds according to mIWI. let $\vdash p \rightarrow q$ and Op. It follows from Op that there is some $\mathcal{I} \in \mathfrak{J}$ such that $p \in \bigcap \mathcal{I}$. It follows from $p \in \bigcap \mathcal{I}$ and $\vdash p \rightarrow q$ that $q \in \bigcap \mathcal{I}$, hence Oq.

---

[28]The following alternative readings of the abbreviations are proposed:

SDL = single-ideal deontic logic, and MDL = multiple-ideals deontic logic.

MDL was called "multiplex deontic logic" by Goble [Goble, 2000].

The axiomatic characterization of MDL is quite simple. As Chellas observed, it is characterized by the following two postulates: [Chellas, 1974, p. 24]

If $\vdash p \to q$ then $\vdash Op \to Oq$ (necessitation)

$\neg O\bot$

In its treatment of inconsistencies, MDL differs favourably from the proposal discussed in Section 4.2, namely the exclusion of conflicting obligations. Just like SDL, MDL disallows self-inconsistent obligations, i.e. $O\bot$ or $O(p\&\neg p)$ cannot hold. But MDL differs from SDL in allowing for conflicting obligations, i.e. Op and $O\neg p$ can both hold. This is what will happen if $\mathfrak{I}$ contains two sets $\mathcal{I}_1$ and $\mathcal{I}_2$ of ideal worlds such that $p \in \bigcap \mathcal{I}_1$ and $\neg p \bigcap \mathcal{I}_2$.

If we wish to exclude conflicting obligations in MDL, then that can be achieved with a simple semantic criterion, namely that there is some world that is an element of all elements of $\mathfrak{I}$ (more succinctly: $\bigcap \mathfrak{I} \neq \varnothing$).

Essentially the same construction has been advocated by other authors. [Schotch and Jennings, 1981; Goble, 2000]

Since MDL satisfies necessitation, it is not difficult to find counterexamples to its logic – the common necessitation paradoxes apply to MDL and they are equally problematic here as in SDL. In summary, MDL and its semantic principle (mIWI) are attractive alternatives if we wish to allow conflicting obligations but disallow self-inconstenct obligations, and consider necessitation to be an acceptable deontic principle.

6 Giving up necessitation

The other major approach consists in giving up necessitation while retaining agglomeration. From a formal point of view it turns out to require a more thorough revision of the semantics.

6.1 The uses of orderings

According to the approach introduced in Section 3.4, in order to construct a deontic logic we begin with an ordering over the set of worlds. This ordering is used to select a set of possible worlds, the ideal worlds. From these worlds we obtain a deontic operator O. See Figure 9.

The alternative approach that we are now going to investigate also uses an ordering (preference relation) to determine the deontic operator O. The difference is that we employ an ordering over the same objects that O refers to, namely sentences, rather than over possible worlds, see Figure 10.

This approach can be called "preference-based deontic logic" (PDL). [Hansson, 1990b] It is the subject of Section 6.2.

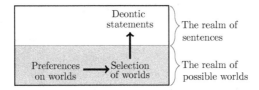

Figure 9: The structure of SDL semantics

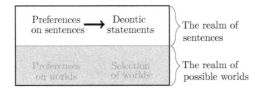

Figure 10: The structure of PDL semantics

Even if we choose this approach we can still use possible world semantics. The reason for this is that the preference relation over sentences on which we base our deontic operator can in its turn be derivable from a preference relation over worlds (holistic alternatives). This is illustrated in Figure 11.

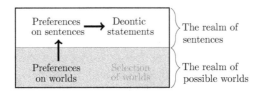

Figure 11: Reintroducing possible worlds into PDL semantics

This approach is supported by a tradition in preference logic that achieves coherence by requiring that our preferences over smaller things (represented by sentences) should be reconstructible in terms of an (underlying) preference relation over complete alternatives (possible worlds). This should of course not be seen as a faithful representation of actual deliberative or evaluative processes. Instead, the holistic preference relation can be conceived as a reconstruction used to effectuate a coherence requirement on preferences. This approach will be investigated in Section 6.3.

6.2 Preference-based deontic logic

In this section we are going to investigate the approach shown in Figure 10. In other words we are going to base deontic operators on preference relations that refer to sentences. But before doing so we need to consider the counter-argument that any such construction would be a category mistake. Arguably a norm sentence and a value sentence cannot have the same meaning, for the simple reason that they represent different conceptual categories. Norms (in the sense explained in Section 2.4) are action-guiding in a way that values are not. However, in order to include norm and value sentences into the same structure it is not necessary for them to have the same meaning. It is sufficient that they have the same extension, or – since this is model-building – approximatively the same extension. G.E. Moore may have been the first to point out that a norm sentence and a value sentence can be equivalent in an extensional but not an intensional sense. [Moore, 1912, pp. 172-3] [Hansson, 1991a]

In what follows, the underlying preference relation $\geq$ is assumed to be transitive and complete.[29] Furthermore, we will assume that a disjunction either has the same value as one of is disjuncts or is intermediate in value between them. We will call a relation $\geq$ *interpolative* if and only if this holds, i.e. if and only if it satisfies the condition $(p \geq (p \vee q) \geq q) \vee (q \geq (p \vee q) \geq p)$. A wide range of preference relations are interpolative. [Hansson, 2001, ch. 7] It can easily be shown that every transitive and interpolative relation is also complete. Therefore, the latter condition is redundant, and we can characterize the underlying preference relation $\geq$ as transitive and interpolative.

Deontic operators such as O, P (permitted), and F (forbidden) are monadic. There are also monadic value predicates such as those representing "good", "bad", and "best". Before inserting monadic deontic operators into the structure provided by a preference relation it is useful to consider how value predicates fit into such a structure. From this point of view there are three major types of monadic value predicates: [Hansson, 2001]

1 $\geq$-*positive predicates*

   Definition: Hp & $q \geq p \rightarrow Hq$

   Examples: Good, best, not bad, not worst, very good, not very bad.

2 $\geq$-*negative predicates*

   Definition: Hp & $p \geq q \rightarrow Hq$

---

[29]See [Hansson, 2001] for a treatment that covers cases when these assumptions are not satisfied.

Examples: Bad, worst, very bad, worst, not good, not very good, not best.

2 $\geq$-circumscriptive predicates

Definition: There is a $\geq$-positive predicate H^+ and a $\geq$-negative predicate H^- such that for all p: $Hp \leftrightarrow H^+p \,\&\, H^-p$.

Examples: Almost worst, neither good nor bad, fairly good.

More informally, we can say that a monadic predicate H is positive if it is possible for something to be too bad to be H, but impossible for something to be too good to be H. Similarly, H is negative if it is impossible for something to be too bad to be H, but possible for something to be too good to be H. Finally, H is circumscriptive if it possible for something to be too bad to be H, and it is also possible for something to be too good to be H.

Can the deontic operators be included in any of these categories? It does not seem plausible for any of them to be circumscriptive, but at first sight it would seem as if prescriptive (O) and permissive (P) operators can be $\geq$-positive and prohibitive (F) operators $\geq$-negative. However, this is not an issue that can be determined separately for each of the three types of operators. They are interconnected, and indeed even interdefinable through the conditions $Pp \leftrightarrow \neg O\neg p$ and $Fp \leftrightarrow O\neg p$. Given these definitions their interconnections are as follows:[30]

Observation 6.1 *Let O, P, and F be interdefinable through the equivalences $Pp \leftrightarrow \neg O\neg p$ and $Fp \leftrightarrow O\neg p$. Then:*

(1) The following three conditions are equivalent:

  (a) O is $\geq$-contranegative (If Op and $\neg p \geq \neg q$ then Oq.)

  (b) P is $\geq$-positive

  (c) F is $\geq$-negative

(2) The following three conditions are equivalent:

  (a) O is $\geq$- positive

  (b) P is $\geq$-contranegative (If Pp and $\neg p \geq \neg q$ then Pq.)

  (c) F is $\geq$-contrapositive ((If Fp and $\neg q \geq \neg p$ then Fq.)

---

[30] The first part of the observation can be found in [Hansson, 2001, p. 144]. The proofs are straightforward.

The observation is perhaps at first sight puzzling. It connects two of the above-mentioned first impressions, namely $\geq$-positivity of P and $\geq$-negativity of F, with each other but puts the third, $\geq$-positivity of O, in another category. But the puzzle is relatively easily solved. We can show that *prescriptive operators are not $\geq$-positive*. In more colloquial language: What is better than something morally required is not necessarily morally required.[31] At least two classes of counterexamples can be used to show this.

The first class of counterexamples compares a morally required act with a supererogatory version of that same act. For instance, let p denote that you return a borrowed motorcar in time to its owner and q that you return it in time to its owner after first having washed it and filled the petrol tank. It is quite plausible to value q higher than p but nevertheless maintain that p but not q is morally required. Or similarly, let p denote that you give your hungry visitor something to eat and q that you serve her a gourmet meal.

In the second class of counterexamples, we compare a morally required action with a variant of it that is specified in some morally irrelevant way. For instance, let p denote that I visit my sick aunt, and q that I do so, entering her flat with my left foot first. Then q is at least as good as (in fact equally good as) p, but it may nevertheless be the case that p but not q is morally required.

On the other hand, studies of these and other examples will support the idea that O is $\geq$-contranegative. If you ought to work hard, and it is worse to be drunk at work than not to work hard, then you ought not to be drunk at work. Additional confirming examples are easily found.

However, we also need to consider whether there are any counterarguments against the $\geq$-contranegativity of O. Attempts to find such counterexamples tend to end up in two types of examples. The first are those that equivocate either between different moral standards or between a moral and a legal standard. It is arguably (morally) worse to cut off all contacts with one's parents than to steal their bicycle pump. Nevertheless it is (legally) required that you do not steal the pump, but it is not (legally) required that you keep in contact with your parents. However, we should not expect legal obligatoriness to be contranegative with respect to moral betterness; these are two different standards.[32]

---

[31] Surprisingly many proposals made in moral philosophy and deontic logic imply $\geq$-positivity of O. A few examples: G.E. Moore identified the assertion "I am morally bound to perform this action" with the assertion "This action will produce the greatest possible amount of good in the Universe" [Moore, 1903, p. 174]. Gupta and von Kutschera both claim that "good" and "ought" coincide, and "good" is clearly $\geq$-positive. [Gupta, 1959; von Kutschera, 1975] Goble identifies Op with $p > \neg p$. [Goble, 1993, p. 149]

[32] The present analysis of obligatoriness in terms of preference is intended for moral, not legal obligatoriness. Legal obligations emanate from legal rules and therefore seem

The other type of alleged counterexamples consists of those that are based on comparisons between the outcomes of actions, rather than comparisons between the actions themselves. Consider Karen who has run into economic problems. She currently pays a monthly contribution to Oxfam (p). She also has to pay back the money that she borrowed from a rich relative (q). The consequences of $\neg p$ are expectedly worse than those of $\neg q$. However, we may nevertheless say that she has a moral obligation to pay back the loan (Oq) but not to continue paying to Oxfam ($\neg Op$). This may seem to be in conflict with $\geq$-contranegativity, but in fact it is not. The relevant comparison is between actions, not between consequences of actions. We would not hesitate to say, in retrospect, "it was bad of her not to pay back the loan", but we would probably not say "it was bad of her not to continue paying to Oxfam".

The $\geq$-positivity of P and the $\geq$-negativity of F are both immediately plausible, and it is equally difficult to construct counterexamples against them as against the $\geq$-contranegativity of O. This gives us a good reason to construct deontic logics in which these three principles all hold. As we saw in Observation 6.1 the three principles are equivalent.

This construction has the further advantage that there are surprisingly simple connections between the properties of $\geq$ and those of the $\geq$-contranegative operators (O). Some of these connections are summarized in Table 2.

The following result provides us with a plausible, semantically characterized, deontic logic that satisfies agglomeration but not necessitation.

Observation 6.2 ([Hansson, 2004a]) *Let $\mathcal{O}$ be a set of monadic operators. Then the following two conditions on $\mathcal{O}$ are equivalent:*

(1) $\mathcal{O}$ is finite and inclusion-complete[33], and each element $O \in \mathcal{O}$ satisfies:

  (a) Op & $Oq \to O(p\&q)$ (agglomeration),

  (b) $O(p\&q) \to Op \vee Oq$ (disjunctive division), and

  (c) There is some p such that $\neg Op$ (non-universality).

(2) $\mathcal{O}$ is the set of non-universal monadic operators that are contranegative with respect to some transitive and interpolative relation $\geq$ that has a finite number of equivalence classes.

---

to obey a logic different from that of moral obligations. [Hansson and Makinson, 1997]

[33] As was noted in Section 2.2, $\mathcal{O}$ is inclusion-complete if and only if it holds for all $O_1, O_2 \in \mathcal{O}$ either that $O_1 p \to O_2 p$ for all p or that $O_2 p \to O_1 p$ for all p.

Disjunctive division says that if one ought to do two things, then one ought to do at least one of them. It can be seen a much weakened variant of necessitation.

Additional properties from Table 2 can be added to obtain a stronger deontic logic. It is interesting to note that necessitation, the property that this exercise is intended to avoid, corresponds to a highly implausible property of $\geq$, namely that if $\vdash q \to p$, then $p \geq q$. (To see why it is implausible, let p denote that you steal and q that you steal in order to save your child from starving to death.)

A transitive and complete relation $\geq$ satisfies ...	if and only if every $\geq$-contranegative operator O satisfies ...
$(p \geq (p \vee q)) \vee (q \geq (p \vee q))$	$Op \ \& \ Oq \to O(p \& q)$ (agglomeration)
$((p \vee q) \geq p) \vee ((p \vee q) \geq q)$	$O(p \& q) \to Op \vee Oq$ (disjunctive division)
$(p \geq (p \& q)) \vee (p \geq (p \& \neg q))$	$P(p \& q) \ \& \ P(p \& \neg q) \to Pp$ (permissive cancellation)
If $\vdash q \to p$, then $p \geq q$.	If $\vdash p \to q$, then $Op \to Oq$. (necessitation)
$(p \geq (p \& q)) \vee (q \geq (p \& q))$	$Op \ \& \ Oq \to O(p \vee q)$ (disjunctive closure)
$(p \geq q) \vee ((\neg p \& q) \geq q)$	$Op \ \& \ O(p \to q) \to Oq$ (deontic detachment)
$p \geq (p \& \neg p)$	$Op \to O(p \vee \neg p)$

Table 2: The connections between properties of a transitive and complete preference relation $\geq$ over sentences and the properties of the $\geq$-contranegative operators. (For details, see [Hansson, 2001, pp. 149-59 and pp. 263-72].)

This construction provides us not only with a single prescriptive operator O but with a structure that can contain many prescriptive operators, corresponding to different degrees of moral requiredness. If we wish to fix one specific such operator, then there are at least three ways to do so.

First, we can couple O to some $\geq$-negative value predicate, perhaps most naturally "bad", denoted B. Under the assumption that B is $\geq$-negative, if we define:

$$Op \leftrightarrow B\neg p$$

then Op will be $\geq$-contranegative.[34] This is equivalent with treating "bad" and "forbidden" as (extensionally) equivalent. Unfortunately these connections are only – at best – rough approximations. Probably, "forbidden" is in most contexts somewhat stronger than "bad". As was noted by Chisholm and Sosa, there are actions of "permissive ill-doing", i.e. "minor acts of discourtesy which most of us feel we have a right to perform (e.g. taking too long in the restaurant when others are known to be waiting)." [Chisholm and Sosa, 1966, p. 326] Arguably, such acts are morally bad but not morally forbidden.

Secondly, we can define an O operator that is as demanding as it can be without making inconsistent demands. This is the *maxiconsistent* contranegative operator. [Hansson, 2001, pp. 166-9] Thirdly, we can define a whole cluster of contranegative operators by assigning to each of them a particular sentence that delimits its strength. For each sentence f we can define a contranegative operator O such that for all p:

$$Op \leftrightarrow f \geq \neg p$$

Under the assumption that $\geq$ is transitive and complete, this construction provides us with an inclusion-complete set of O operators, i.e. a linearly ordered series of O operators that differ only in strength.[35] In this way, we can account for the existence of moral requirements of different strengths. (Cf. Section 2.2.) This is needed to make sense of important features of ordinary deontic discourse. Consider the following dialogue:

"MORALIST: You have a large debt that is due today. You should pay it.

---

[34] B can in its turn be defined in terms of $\geq$. The most common such definition is that which equates "bad" with "worse than its negation", i.e. $Bp \leftrightarrow \neg p > p$. For an overview of such definitions, see [Hansson, 1990a] and [Hansson, 2001, ch. 8]. On the logic of "ought" that follows from defining it as "bad not", see [Hansson, 2001, pp. 164-5].

[35] Such a series is of course also implicit in Bengt Hansson's semantics, as described in Section 3.5. For each world W in his model we can use the set of worlds that are at least as good as it, $\{W' \in \mathcal{W} \mid W' \geq W\}$ as a set of (weakly) ideal worlds and use IWI to define an O operator from it. However, the operators obtained in that way will all satisfy necessitation.

SPENDTHRIFT: It is impossible for me to do that. I do not have the money.

MORALIST: I know that.

SPENDTHRIFT: Yes, and I already know what my obligations are. Please, as a moralist, tell me instead what I should do.

MORALIST: I have already told you. You should pay your debt." [Hansson, 1999]

Here, the "should" of "You should pay your debt" is unsuitable for action-guidance since it requires something that Spendthrift cannot do. The shift in focus that Spendthrift asks for but Moralist refuses to make can be described as a shift to a prescriptive operator that is weak enough to prescribe only doable actions. To make this more precise, let p denote that Spendthrift pays off her debts and q that she pays her creditors at least as much as she can without losing her means of subsistence. Our Moralist persists in speaking in terms of an "ought" predicate O such that Op (and of course Oq). An action-guiding "ought" would in this case have to be represented by a weaker predicate O', presumably such that $O'q \ \& \ \neg O'p$.[36]

Next, consider the following case:

"ADULTERER: I have put myself in a terrible situation. I have promised Anne to get a divorce and then marry her. She has waited for me more than five years. Now she is pregnant with my child and she entreats me to take the decisive step. But I also still love my wife, and I have promised never to leave her. What should I do?

MORALIST: Since you can only be married to one person you should not have promised two persons to be married to them. That is what is wrong.

ADULTERER: I know that. But please tell me what I should do."

Contrary to the previous example, this is not a conflict between ought and can but a conflict between two oughts, i.e. a moral dilemma. But Moralist is unhelpful in very much the same way as in the previous conversation, namely by sticking to a notion of moral requirement that is so strict that all available courses of action are impermissible. Such a notion may be adequate for parts of the deliberations that Adulterer should engage in,

---

[36] This example also illustrates the shifts in possibility perspectives discussed in Section 2.5.

for instance to learn from his mistakes and to determine what duties of reparation he may have. But when deciding which of his two major options to choose he will need a weaker notion of moral requirement that does not demand the impossible. This weaker notion must be such that staying with his wife (w) and marrying his mistress (m) are not both required (but for our present purpose we can leave it unsettled whether $Ow \ \& \ \neg Om$, $\neg Ow \ \& \ Om$, or $\neg Ow \ \& \ \neg Om$ holds for such a notion of moral requirement in this case).

Under the assumption that marriage is by definition monogamous, we have $\vdash \neg(w \& m)$. It therefore follows from the axiom of agglomeration that $Ow \ \& \ Om$ implies the unsatisfiable obligation $O(w \& m)$, which expresses why an O operator such that $Ow \ \& \ Om$ cannot be used for directly action-guiding deliberations. Hence the agglomeration postulate adequately reflects our reasoning about moral dilemmas in cases like this. It is also an important advantage of the approach presented in this section that necessitation does not hold for the deontic operators. As we saw in Section 4.2, if both agglomeration and necessitation hold then it would follow from $Ow \ \& \ Om$ that Os holds for any s (such as "flee the country" or "find an additional mistress"). These are absurdities of a type that an adequate deontic logic should avoid.

6.3 Reintroducing holistic alternatives

As we saw in Section 6.1 and Figure 11, holistic structures such as possible worlds or complete action descriptions can be reintroduced as a means to obtain the preference relation on sentences from which the (contranegative) deontic operators are constructed. The need to base preferences over sentences on preferences over larger, holistic structures can be seen from the following dialogue:

> A: Do you prefer coffee or tea?
>
> B: I am an inveterate tea-drinker, so I certainly prefer tea to coffee.
>
> A: Do you never drink coffee?
>
> B: Yes, like most Swedes I usually take coffee with a sweet dessert.
>
> A: So you prefer coffee with a sweet dessert to tea with a sweet dessert.
>
> B: Yes, I do.
>
> A: Do you never take tea with a sweet dessert?
>
> B: Well, there is one exception. In Chinese restaurants, when tea is served with the main dish, I do not switch to coffee.

A: So you prefer tea with a sweet dessert after being served tea with the main course to coffee with a sweet dessert after being served tea with the main course?

B: Yes, that is true.

As this example shows, even seemingly trivial preferences can be subject to exceptions and counter-exceptions. In this case we have:

$t > c$

$c\&s > t\&s$

$t\&s\&m > c\&s\&m$

where:

$t =$ I drink tea.

$c =$ I drink coffee.

$s =$ I eat a sweet dessert.

$m =$ I had tea with the main dish.

Hence, preferences tend to change when the alternatives are extended with additional information. The only preferences that are stable against such additions are those that refer to maximal alternatives, i.e. alternatives so large that no sentences can be consistently added to them. In logical parlance, such alternatives are called possible worlds. As was noted in Section 2.5 they can represent much smaller entities than (metaphysical) possible worlds, such as the complete action alternatives open to a particular agent in a particular situation. For preferences to be coherent, they should be extendible to preferences over such holistic structures. From the viewpoint of formal logic, we can then reconstruct preferences over sentences from the holistic preferences. This construction yields what is best described as a semantic account for preferences over sentences. It has turned out to be quite helpful in dealing with issues that arise for preferences over sentences but not for preferences over primitive entities, such as how to evaluate disjunctive and negated sentences. (Does $p > p \vee q$ imply $p \vee q \geq q$? Is $p \geq q$ equivalent with $p\&\neg q \geq q\&\neg p$? Etc.)

Perhaps the most obvious way to base preferences over sentences on preferences over holistic alternatives would be to use weighted averages of values assigned to each complete alternative. The idea is very simple: Assign to each alternative $W \in \mathcal{W}$ a value $u(W)$ and a weight $w(W)$. Then we can assign to each sentence a value that is equal to the weighted average of the values of the alternatives in which it holds. Hence, suppose that W_1 and W_2 are the (only) worlds in which p is true, and that:

$u(W_1) = 10$,

$u(W_2) = 22$,

$w(W_1) = 0.05$, and

$w(W_2) = 0.15$,

Then the weighted value of p is:

$$u(p) = \frac{w(W_1)}{w(W_1)+w(W_2)} u(W_1) + \frac{w(W_2)}{w(W_1)+w(W_2)} u(W_2) = 19$$

$p \geq q$ holds if and only if $u(p)$ is not smaller than $u(q)$. If the weights are interpreted as probabilities, then this construction corresponds to a notion of "probably better".[37] However, this is not a suitable explication of preferences for the purposes of deontic logic. Suppose that you are deliberating on whether to keep an extra income for yourself (s) or donate it to a charity (c). The probability that you will do one or the other (or the probabilities of holistic alternatives containing one or the other) should not influence your choice since if it did, then you would not really treat both alternatives as fully open. This argument speaks generally against the use of weighted-average preferences for deliberative purposes. In addition there is a counter-argument of a more formal nature: Weighted-average preferences are not in general interpolative and they do not even satisfy the highly plausible property that if $p \simeq q$ then $p \simeq (p \vee q)$, where $\simeq$ denotes indifference.[38]

Instead of weighted averages, another class of preference relations can be used that have a long tradition in decision theory, namely extremal preferences. [Barbera et al., 1984] [Hansson, 2001, pp. 102-13] These are the preferences over sentences such that the position of a sentence p in the derived preference relation is completely determined by the positions of the best and the worst p-worlds (complete alternatives containing p) in the underlying ordering. For every sentence p, let min(p) be one of the lowest-ranked elements of $\mathcal{W}$ that contain p and let max(p) be one of the highest-ranked elements of $\mathcal{W}$ that contain p. We can then derive several types of preference relations over sentences in terms of the underlying preference relation over worlds (complete alternatives). Some major such types are listed in Table 3. The deontic logics based on them have all been axiomatically characterized. [Hansson, 2001; Hansson, 2004b]. Possibly the

---

[37]More precisely, it corresponds to what is called "news value" in decision theory. [Gibbard and Harper, 1978]

[38]To see this, let there be three complete alternatives W_1, W_2, and W_3, such that $p\&\neg q \in W_1$, $p\&q \in W_2$, and $\neg p\&q \in W_3$. Let $u(W_1) = u(W_3) = 6$, $u(W_2) = 0$, and $w(W_1) = w(W_2) = w(W_3) = \frac{1}{3}$. Then $u(p) = u(q) = 3$ and $u(p \vee q) = 4$.

Maximin preferences:	
$p \geq_i q$ iff	$\min(p) \geq \min(q)$
Maximax preferences:	
$p \geq_x q$ iff	$\max(p) \geq \max(q)$
Interval maximin preferences:	
$p \geq_{ix} q$ iff	either $\min(p) > \min(q)$
	or both $\min(p) \simeq \min(q)$ and $\max(p) \geq \max(q)$
Interval maximax preferences:	
$p \geq_{xi} q$ iff	either $\max(p) > \max(q)$
	or both $\max(p) \simeq \max(q)$ and $\min(p) \geq \min(q)$
Doubly maximizing preferences	
$p \geq_{\ddagger} q$ iff	$\max(p) \geq \max(q)$ and $\min(p) \geq \min(q)$

Table 3: Five types of extremal preference relations

most plausible of them makes use of the contranegative predicate that is based on doubly maximizing preferences. It can be characterized by five axioms as follows:

Observation 6.3 ([Hansson, 2004b]) *O is $\geq_{\ddagger}$-contranegative if and only if it satisfies:*

(i) $Op \& Oq \to O(p\&q)$ (agglomeration)

(ii) $O(p\&q) \to Op \vee Oq$ (disjunctive division)

(iii) If $\vdash p \to q$, $\vdash q \to r$, Op and Or, then Oq (bilateral necessitation)

(iv) $\neg O\bot$ (self-consistency)

(v) There is some p such that Op (non-vacuity)

(i) and (ii) were also used in Observation 6.2. (iii) is a much weakened form of necessitation, (iv) is a consistency postulate, and (v) ensures that there is something that is morally required. It remains an open question whether there is a derivation principle by which we can obtain, from any complete ordering over holistic alternatives (possible worlds), a preference relation over sentences that is characterized only by axioms listed in Observation 6.2.

Generally speaking, the use of an underlying preference relation over holistic alternatives (as in Figure 11 above) is a considerable strength for an account of deontic logic. However, this advantage should not be bought at the price of implausible logical properties. As we saw in Observation 6.2, deontic logic can credibly be based on a preference relation over sentences that is not in its turn based on a preference relation over holistic alternatives.

7 Conclusion

In this chapter we have penetrated the preconditions and justifications for investigations of alternative deontic semantics. The focus has been on constructions that follow the general tradition in deontic logic in employing possible worlds (holistic alternatives) and/or preference relations as semantic devices. Many other devices have been used in deontic logic, such as non-monotonic inference [Horty, 1994; Asher and Bonevac, 1996; Asher and Bonevac, 1997], dynamic logic [Meyer, 1987a], fuzzy logic [Dellunde and Godo, 2008], modal preference logic [van der Torre and Tan, 1999] and input/output logic [Makinson and van der Torre, 2000].

The central message that has hopefully been conveyed in this chapter is that the choice of semantic and syntactic principles for deontic logic depends on which inferences involving normative sentences we take for valid, and this in its turn is closely related to issues in moral philosophy. The application of Ideal Worlds Intersection (IWI) to complete possible worlds (in a metaphysical sense) is easier to combine with consequentialist ethics than with ethical theories that emphasize moral considerations of actions *per se*. The latter type of ethics is more easily combined with semantic principles that are based on orderings of action-representing sentences, such as the contranegative logics introduced in Section 6.2. As was shown in Section 4.2, different approaches to normative inconsistencies also have an impact on deontic logic, both in the semantic and the syntactic perspective. In a multi-agent deontic logic views on co-ordination and the moral status of collective action can be a further decisive factor. There is an urgent need for a *rapprochement* between deontic logic and moral philosophy.

Acknowledgements

I would like to thank Ron van der Meyden for most valuable comments on an earlier version of this chapter.

BIBLIOGRAPHY

[Alchourrón, 1993] C. Alchourrón. Philosophical foundations of deontic logic and the logic of defeasible conditionals. In J.-J. Ch. Meyer and R. J. Wieringa, editors, *Deontic Logic in Computer Science*, pages 43–84. John Wiley & Son, Chichester, 1993.

[Anderson, 1956] A. R. Anderson. The formal analysis of normative systems. Technical report, No. 2, U.S. Office of Naval Research Contract No. SAR/Nonr-609 (16), 1956. Reprinted in N. Rescher (ed.) *The Logic of Decision and Action*. Pittsburgh: University of Pittsburgh Press, 1967, pp. 147-213.

[Åqvist, 1967] L. Åqvist. Good samaritans, contrary-to-duty imperatives, and epistemic obligations. *Noûs*, 1:361–379, 1967.

[Asher and Bonevac, 1996] N. Asher and D. Bonevac. Prima facie obligation. *Studia Logica*, 57:19–45, 1996.

[Asher and Bonevac, 1997] N. Asher and D. Bonevac. Common sense obligation. In D. Nute, editor, *Defeasible Deontic Logic*, pages 159–203. Kluwer, Dordrecht, 1997.

[Barbera et al., 1984] S. Barbera, C. R. Barrett, and P. K. Pattanaik. On some axioms for ranking sets of alternatives. *Journal of Economic Theory*, 33:301–308, 1984.

[Brandt, 1965] R. B. Brandt. The concepts of obligation and duty. *Mind*, 73:374–393, 1965.

[Brown, 1996] M. A. Brown. A logic of comparative obligation. *Studia Logica*, 57:117–137, 1996.

[Castañeda, 1968] H.-N. Castañeda. Acts, the logic of obligation, and deontic calculi. *Philosophical Studies*, 19:13–26, 1968.

[Castañeda, 1977] H.-N. Castañeda. Ought, time, and deontic paradoxes. *Journal of Philosophy*, 74:775–791, 1977.

[Castañeda, 1989] H.-N. Castañeda. Paradoxes of moral reparation: Deontic foci vs. circumstances. *Philosophical Studies*, 57:1–21, 1989.

[Chellas, 1974] B. F. Chellas. Conditional obligation. In S. Stenlund, editor, *Logical Theory and Semantic Analysis*, pages 23–33. Reidel, Dordrecht, 1974.

[Chellas, 1980] B. F. Chellas. *Modal Logic. An Introduction*. Cambridge University Press, Cambridge, 1980.

[Chisholm and Sosa, 1966] R. M. Chisholm and E. Sosa. Intrinsic preferability and the problem of supererogation. *Synthese*, 16:321–331, 1966.

[Chisholm, 1964] R. M. Chisholm. The ethics of requirement. *American Philosophical Quarterly*, 1:147–153, 1964.

[Dayton, 1981] E. Dayton. Two approaches to deontic logic. *Journal of Value Inquiry*, 15:137–147, 1981.

[Dellunde and Godo, 2008] P. Dellunde and L. Godo. Introducing grades in deontic logic. In R. van der Meyden and L. van der Torre, editors, *Deontic Logic in Computer Science, 9th International Conference (DEON 2008)*, number 5076 in Lecture Notes in Computer Science, pages 248–262. 2008.

[Føllesdal and Hilpinen, 1970] D. Føllesdal and R. Hilpinen. Deontic logic: An introduction. In R. Hilpinen, editor, *Deontic Logic: Introductory and Systematic Readings*, pages 1–35. Synthese Library, Dordrecht: Reidel, 1970.

[Forrester, 1975] M. Forrester. Some remarks on obligation, permission, and supererogation. *Ethics*, 85:219–226, 1975.

[Garcia, 1989] J. L. A. Garcia. The problem of comparative value. *Mind*, 98:277–283, 1989.

[Gibbard and Harper, 1978] A. Gibbard and W. L. Harper. Counterfactuals and two kinds of expected utility. *Foundations and Applications of Decision Theory*, 1:125–162, 1978.

[Goble, 1993] L. Goble. The logic of obligation, better and worse. *Philosophical Studies*, 70:133–163, 1993.

[Goble, 2000] L. Goble. Multiplex semantics for deontic logic. *Nordic Journal of Philosophical Logic*, 5:113–134, 2000.

[Goldman, 1977] H. S. Goldman. David Lewis's semantics for deontic logic. *Mind*, 86:242–248, 1977.

[Guendling, 1974] J. E. Guendling. Modal verbs and the grading of obligations. *Modern Schoolman*, 51:117–138, 1974.

[Gupta, 1959] R. K. Gupta. Good, duty and imperatives. *Methodos*, 11:161–167, 1959.

[Hanson, 1965] W. H. Hanson. Semantics for deontic logic. *Logique et Analyse*, 31:177–190, 1965.
[Hansson and Makinson, 1997] S. O. Hansson and D. Makinson. Applying normative rules with restraint. In M. L. Dalla Chiara, K. Doets, D. Mundici, and J. van Benthem, editors, *Logic and Scientific Methods*, volume 1 of *Tenth International Congress of Logic, Methodology and Philosophy of Science*, pages 313–332. Kluwer, Boston, Mass., 1997.
[Hansson, 1969] B. Hansson. An analysis of some deontic logics. *Noûs*, 3:373–398, 1969.
[Hansson, 1988] S. O. Hansson. Deontic logic without misleading alethic analogies, parts I-II. *Logique et Analyse*, 31:337–370, 1988.
[Hansson, 1989] S. O. Hansson. A note on the deontic system DL of Jones and Pörn. *Synthese*, 80:427–428, 1989.
[Hansson, 1990a] S. O. Hansson. Defining good and bad in terms of better. *Notre Dame Journal of Formal Logic*, 31:136–149, 1990.
[Hansson, 1990b] S. O. Hansson. Preference-based deontic logic (PDL). *Journal of Philosophical Logic*, 19:75–93, 1990.
[Hansson, 1991a] S. O. Hansson. Norms and values. *Critica*, 23:3–13, 1991.
[Hansson, 1991b] S. O. Hansson. The revenger's paradox. *Philosophical Studies*, 61:301–305, 1991.
[Hansson, 1998] S. O. Hansson. Should we avoid moral dilemmas? *Journal of Value Inquiry*, 32:407–416, 1998.
[Hansson, 1999] S. O. Hansson. But what should I do? *Philosophia*, 27:433–440, 1999.
[Hansson, 2000] S. O. Hansson. Formalization in philosophy. *Bulletin of Symbolic Logic*, 6:162–175, 2000.
[Hansson, 2001] S. O. Hansson. *The Structure of Values and Norms*. Cambridge University Press, Cambridge, 2001.
[Hansson, 2004a] S. O. Hansson. A new representation theorem for contranegative deontic logic. *Studia Logica*, 77:1–7, 2004.
[Hansson, 2004b] S. O. Hansson. Semantics for more plausible deontic logics. *Journal of Applied Logic*, 2:3–18, 2004.
[Hansson, 2006] S. O. Hansson. Ideal worlds: Wishful thinking in deontic logic. *Studia Logica*, 82:329–336, 2006.
[Hansson, 2010] S. O. Hansson. Methodological pluralism in philosophy. *Theoria*, 76:189–191, 2010.
[Harman, 1977] G. Harman. *The Nature of Morality*. Oxford U.P, New York, 1977.
[Hilpinen, 1985] R. Hilpinen. Normative conflicts and legal reasoning. In E. Bulygin, J.L. Gardies, and I. Niiniluoto, editors, *Man, Law and Modern Forms of Life*, pages 191–208. Reidel, Dordrecht, 1985.
[Hintikka, 1957] J. Hintikka. Quantifiers in deontic logic. *Societas Scientiarum Fennica, Commentationes Humanarum Literarum*, 23(4), 1957.
[Horty, 1994] J. Horty. Moral dilemmas and nonmonotonic logic. *Journal of Philosophical Logic*, 23:35–65, 1994.
[Jackson, 1985] F. Jackson. On the semantics and logic of obligation. *Mind*, 94:177–195, 1985.
[Jones and Pörn, 1985] A. J. I. Jones and I. Pörn. Ideality, sub-ideality and deontic logic. *Synthese*, 65:275–290, 1985.
[Jones and Pörn, 1986] A. J. I. Jones and I. Pörn. 'Ought' and 'must'. *Synthese*, 66:89–93, 1986.
[Jones and Pörn, 1989] A. J. I. Jones and I. Pörn. A rejoinder to Hansson. *Synthese*, 80:429–432, 1989.
[Kanger, 1957] S. Kanger. *New Foundations for Ethical Theory*. Almqvist and Wiksell, Stockholm, 1957. Reprinted in R. Hilpinen (ed.), *Deontic Logic: Introductory and Systematic Readings*, Synthese Library, Dordrecht, 1970, pp. 36-58.
[Kapitan, 1986] T. Kapitan. Deliberation and the presumption of open alternatives. *Philosophical Quarterly*, 36:230–251, 1986.

[Kripke, 1963] S. Kripke. Semantic analysis of modal logic I. Normal modal propositional calculi. *Zeitschrift für mathematische Logik und Grundlagen der Mathematik*, 9:67–96, 1963.

[Ladd, 1957] J. Ladd. *The Structure of a Moral Code*. Harvard U.P, Cambridge, Mass., 1957.

[Lenk, 1978] H. Lenk. Varieties of commitment. *Theory and Decision*, 9:17–37, 1978.

[Lewis, 1974] D. Lewis. Semantic analyses for dyadic deontic logic. In S. Stenlund, editor, *Logical Theory and Semantic Analysis*, pages 1–14. Reidel, Dordrecht, 1974.

[Makinson and van der Torre, 2000] D. Makinson and L. van der Torre. Input/output logics. *Journal of Philosophical Logic*, 29:383–408, 2000.

[Makinson, 1999] D. Makinson. On a fundamental problem of deontic logic. In P. McNamara and H. Prakken, editors, *Norms and Information Systems. New Studies on Deontic Logic and Computer Science*, pages 29–53. IOS Press, Amsterdam, 1999.

[McArthur, 1981] R. P. McArthur. Anderson's deontic logic and relevant implication. *Notre Dame Journal of Formal Logic*, 22:145–154, 1981.

[McNamara, 1996] P. McNamara. Must I do what I ought? (or will the least I can do do?). In M. A. Brown and J. Carmo, editors, *Deontic Logic, Agency and Normative systems*, pages 154–173. Springer, Berlin, 1996. DEON'96: Third International Workshop on Deontic Logic in Computer Science, Sesimbra, Portugal, 11-13 Jan 96.

[Merrill, 1978] G. H. Merrill. Formalization, possible worlds and the foundations of modal logic. *Erkenntnis*, 12:305–327, 1978.

[Meyer, 1987a] J.-J. Ch. Meyer. A different approach to deontic logic: Deontic logic viewed as a variant of dynamic logic. *Notre Dame Journal of Formal Logic*, 29:109–136, 1987.

[Meyer, 1987b] J.-J. Ch. Meyer. A simple solution to the deepest paradox in deontic logic. *Logique et Analysis*, 117-118:81–90, 1987.

[Mish'alani, 1969] J. K. Mish'alani. Duty, obligation and ought. *Analysis*, 30:33–40, 1969.

[Moore, 1903] G. E. Moore. *Principia Ethica*. Cambridge University Press, 1903.

[Moore, 1912] G. E. Moore. *Ethics*. Oxford University Press, London, 1912.

[Parent, 2008] X. Parent. On the strong completeness of Åqvist's dyadic deontic logic G. In R. van der Meyden and L. van der Torre, editors, *Deontic Logic in Computer Science, 9th International Conference (DEON 2008)*, volume 5076 of *Lecture Notes in Computer Science*, pages 189–202. 2008.

[Parent, 2010] X. Parent. A complete axiom set for Hansson's deontic logic DSDL2. *Logic Journal of IGPL*, 18:422–429, 2010.

[Prior, 1954] A. N. Prior. The paradoxes of derived obligation. *Mind*, 63:64–65, 1954.

[Prior, 1958] A. N. Prior. Escapism. In A. I. Melden, editor, *Essays in Moral Philosophy*, pages 135–146. University of Washington Press, Seattle, 1958.

[Prior, 1962] A. N. Prior. *Formal Logic*. Clarendon Press, Oxford, 2nd edition, 1962.

[Purtill, 1973] R. L. Purtill. Deontically perfect worlds and *prima facie* obligations. *Philosophia*, 3:429–438, 1973.

[Purtill, 1975] R. L. Purtill. Paradox-free deontic logics. *Notre Dame Journal of Formal Logic*, 16:483–490, 1975.

[Purtill, 1980] R. L. Purtill. Review of al-Hibri, Deontic Logic. *Southwestern Journal of Philosophy*, 11:171–174, 1980.

[Reichenbach, 1980] B. Reichenbach. Basinger on Reichenbach and the best possible world. *International Philosophical Quarterly*, 20:343–346, 1980.

[Robinson, 1971] R. Robinson. Ought and ought not. *Philosophy*, 46:193–202, 1971.

[Ross, 1930] D. Ross. *The Right and the Good*. Clarendon Press, Oxford, 1930.

[Ross, 1941] A. Ross. Imperatives and logic. *Theoria*, 7:53–71, 1941.

[Routley and Plumwood, 1984] R. Routley and V. Plumwood. Moral dilemmas and the logic of deontic notions. Discussion papers in environmental philosophy. Department of Philosophy, Australian National University, 1984.

[Royakkers, 1996] L. Royakkers. *Representing Legal Rules in Deontic Logic.* PhD thesis, Katholieke Universiteit Brabant, 1996.
[Sartre, 1946] J. P. Sartre. *L'Existentialisme est un Humanisme.* Les Editions Nagel, Paris, 1946.
[Sayre-McCord, 1986] G. Sayre-McCord. Deontic logic and the priority of moral theory. *Noûs*, 20:179–197, 1986.
[Schotch and Jennings, 1981] P. K. Schotch and R. E. Jennings. Non-kripkean deontic logic. In R. Hilpinen, editor, *New Studies in Deontic Logic*, pages 149–162. Reidel, Dordrecht, 1981.
[Sloman, 1970] A. Sloman. 'Ought' and 'better'. *Mind*, 79:385–394, 1970.
[Spohn, 1975] W. Spohn. An analysis of Hansson's dyadic deontic logic. *Journal of Philosophical Logic*, 4:237–252, 1975.
[Stranzinger, 1978] R. Stranzinger. Ein paradoxienfreies deontisches System. In I. Tammelo and H. Schreiner, editors, *Strukturierungen und Entscheidungen im Rechtsdenken: Notation, Terminologie und Datenverarbeitung in der Rechtslogik*, pages 183–192. Springer, Wien, 1978.
[van der Torre and Tan, 1999] L. van der Torre and Y.-H. Tan. Contrary-to-duty reasoning with preference-based dyadic obligations. *Annals of Mathematics and Artificial Intelligence*, 27:49–78, 1999.
[Vermazen, 1977] B. Vermazen. The logic of practical ought-sentences. *Philosophical Studies*, 32:1–71, 1977.
[von Kutschera, 1975] F. von Kutschera. Semantic analyses of normative concepts. *Erkenntnis*, 9:195–218, 1975.
[von Wright, 1951] G. H. von Wright. Deontic logic. *Mind*, 60:1–15, 1951.
[von Wright, 1981] G. H. von Wright. On the logic of norms and actions. In R. Hilpinen, editor, *New Studies in Deontic Logic*, pages 3–35. Reidel, Dordrecht, 1981.
[von Wright, 1986] G. H. von Wright. Is and ought. In M. C. Doeser and J. N. Kraay, editors, *Facts and Values*, pages 31–48. Martinus Nijhoff Publishers, Dordrecht, 1986.
[von Wright, 1999a] G. H. von Wright. Deontic logic – as I see it. In P. McNamara and H. Prakken, editors, *Norms, Logics and Information Systems. New Studies in Deontic Logic and Computer Science*, pages 15–25. IOS Press, Amsterdam, 1999.
[von Wright, 1999b] G. H. von Wright. Deontic logic: A personal view. *Ratio Juris*, 12:26–38, 1999.
[Vorobej, 1982] M. Vorobej. Deontic accessibility. *Philosophical Studies*, 41:317–319, 1982.
[Williams, 1965] B. Williams. Ethical consistency. *Aristotelian Society Supplementary Volume*, 39:103–124, 1965.
[Wolenski, 1989] J. Wolenski. Deontic logic and possible worlds semantics: A historical sketch. *Studia Logica*, 49:273–282, 1989.

Sven Ove Hansson
Division of Philosophy
Royal Institute of Technology, Stockholm
Email: soh@kth.se

8
Input/output Logic
XAVIER PARENT AND LEENDERT VAN DER TORRE

ABSTRACT. The chapter provides an overview of input/output logic as a framework for reasoning about conditional norms. First, we take a bird-eye view, and offer a general perspective on the latter framework. The key idea is to treat detachment as the central mechanism underlying normative reasoning. Next, we present the mathematical foundations of the framework. It comes in two levels, an unconstrained one and a constrained one. Both levels are explained in detail.

1 Introduction 500
   1.1 Objectives 501
   1.2 Detachment as a core mechanism 502
   1.3 Makinson's third way 505
   1.4 Secretarial assistant 506
   1.5 What counts as an input/output logic? 507
2 Obligations from an I/O perspective 508
   2.1 Basic representation 508
   2.2 Comparison with conceptual implicative structures . . 512
   2.3 Modal embedding 513
   2.4 A possible-worlds semantics 514
   2.5 Input/output without WO 516
   2.6 Reasoning about norm violation 517
   2.7 Conflicts amongst obligations 519
3 Input/output approaches to permission 522
   3.1 Different concepts of permission 523
   3.2 The MvT account of permission 524
   3.3 Permission as exception 527
4 Proof theory 532
   4.1 Obligation 532
   4.2 Syntactic characterization of permission 539
5 Conclusion and issues for future research 541
   5.1 Go top-down 541
   5.2 Equivalence and redundancy 541
   5.3 Changing the base logic 541

5.4	Lions	542
5.5	Proof theory for constrained I/O logic and permission	542
5.6	Moving towards implementation	542

1 Introduction

Makinson and van der Torre [2000; 2001; 2003a] introduce input/output logic (I/O logic) as a general framework for reasoning about conditional norms. The term "modal logic" may be used more broadly for a family of related systems known as K, KD, S4, S5,.... Much the same can be said of the term "input/output logic". Throughout the chapter, we will refer to the family itself as the input/output framework. And we will refer to its individual members by the names that were given to them in the three seminal papers mentioned above. The proposed framework has been applied to domains other than normative reasoning, for example causal reasoning, argumentation, logic programming and nonmonotonic logic (see [Bochman, 2005]). Here we restrict the discussion to normative reasoning. For a gentle introduction to input/output logic, the reader is referred to [Makinson and van der Torre, 2003b].

The aim of this chapter is twofold. Our first aim is to give an overview of input/output logic. We will discuss the objectives, the methodology and the mathematical foundations of the framework. We will also evaluate the results obtained thus far, and identify where the main challenges are. The basic idea is to consider detachment as the central mechanism of the semantics of normative reasoning, and define proof systems for it. The traditional interpretation of "x is obligatory if a" as "the preferred a's are x" is replaced by "x can be detached in context a". For example, if the obligation for $x \wedge y$ implies the obligation for x, then this means that if $x \wedge y$ can be detached from the normative system, then x can be detached too.

The second aim of the chapter is to correct a number of misunderstandings frequently made about I/O logic. These will be cleared up as we go along:

- I/O logic should not be reduced to a proof system. It has a well-defined semantics based on detachment. Soundness and completeness results link the two.

- I/O logic is not just an extension of classical logic. It can be built on top of, e.g., intuitionistic logic.

- So-called constrained I/O logic (which imposes further constraints on the process of detachment) provides a more satisfactory and sensitive analysis of normative reasoning than unconstrained I/O logic. Therefore, input/output logic should not be reduced to its unconstrained version.

- I/O logic should not be viewed as an account of obligation only. The framework can be used to reason about other kinds of regulative norms like permissions, and so-called constitutive norms. The I/O approach to constitutive norms is described in the Chapter "Constitutive norms and counts as conditionals" by D. Grossi and A. J. I. Jones. The present chapter will focus on the notion of permission.

Our aim is not to give an historical overview of the development of input/output logic. The interested reader can obtain the original papers, and reconstruct the short history of the field.

The layout of this chapter is as follows. In the remainder of this introductory section, we take a bird-eye view, and offer a general perspective on I/O logic. We emphasize the main problems the framework is meant to address. Next, in Sections 2-4, we present the mathematical foundations of I/O logic. We begin with the semantics for obligation and permission, then move to the proof theory. The framework comes in two levels, an unconstrained one and a constrained one. Both levels are explained in detail. Section 5 ends with some suggestions as to useful ways forward.

1.1 Objectives

Based on the seminal papers by Makinson and van der Torre, and the subsequent papers on the topic, we believe that the following objectives of the input/output logic framework can be identified:

1. Define a framework where detachment is the central mechanism of the semantics, together with a proof theory for the semantics thus conceived. On this basis, define and compare individual I/O logics, for example logics either validating or blocking the chaining of obligations.

2. Give an improved analysis of phenomena considered problematic in the deontic logic literature. These include:

    - Jørgensen [1937]'s dilemma. It roughly says that a proper logic of norms is impossible because norms do not have truth-values;
    - Contrary-to-duty (CTD) reasoning. It is the problem of reasoning about norm violation. This issue has not disappeared from the stage of deontic logic, since SDL was criticized for not being able to deal with it.

- Reflexivity law $O(x|x)$ ("if x is the case, then x is obligatory"). Such a law is validated by the semantics for dyadic deontic logic (DDL) devised by Hansson [1969] and Lewis [1973] in order to model contrary-to-duties. The intuitive standing of such a law is debatable, and casts doubts on the suitability of DDL for modelling normative reasoning.[1]

- The varieties of permission, and their interplay with the notion of obligation. Different notions of permission can be distinguished. A central question is how they interact, and how (when combined with obligations) they generate permissions that — in some sense — follow.

- Moral dilemmas. A core issue in deontic logic is accommodating their existence, and defining a mechanism for resolving conflicts amongst norms.

3. Provide theoretical foundations for the study of new challenges to normative reasoning. These concern logical architectures (also known as logical input/output nets), norm change, norm creation, epistemic norms, game theoretic norms, and the formal relation to other kinds of reasoning (like, for example, belief revision, default reasoning, decision making, and case based reasoning).

4. Provide a theoretical framework for the use of norms and normative reasoning in applications in computer science, law, linguistics, ethics, and other domains. For example, it can be useful for the development of norm programming languages, the specification and verification of normative multiagent systems, or the development of algorithms for the checking of compliance.

Some aspects of these objectives are explained in the remainder of this section.

1.2 Detachment as a core mechanism

The first objective states that detachment is viewed as the core mechanism of the semantics of normative reasoning. A few comments are in order.

As we shall see, in I/O logic, a conditional obligation is represented as a pair (a, x) of boolean formulae, where a and x are the antecedent and the consequent, respectively. The notation remains neutral on the question of whether "ought" takes wide scope or narrow scope over the conditional. The pair (a, x) may indifferently be read as "If a, then it ought to be that

---

[1] For a presentation of DDL, see Chapters 1 and 7 in this Handbook.

x", or "It ought to be that if a then x". In the chapter we will switch rather casually between the two readings.

The law of detachment or modus-ponens is well-known from propositional logic: from a conditional statement and its antecedent, the consequent of the conditional statement is inferred. For instance, from "If John loves Mary, Mary is happy" and "John loves Mary", "Mary is happy" is inferred. The case when the consequent of the conditional statement is the content of an obligation is covered by such a rule. In the deontic logic literature, this is referred to as "factual" detachment. The rule may be given the form:

(1) Factual detachment
  (i) If a is the case, then x is obligatory
  (ii) a is the case
  (iii) So, x is obligatory

In the I/O notation the rule may be expressed as:

$$\text{If } (a, x) \in N \text{ then } x \in out(N, a)$$

Here $out(N, a)$ denotes the output of a under some set N of conditional norms. This is shown diagrammatically in Figure 1.

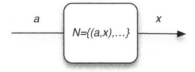

Figure 1: Factual detachment

Note that the normative status of output x is that of an unconditional obligation. As we shall see in Subsection 2.1, there are different ways to detach a conclusion about what is obligatory. The idea is to spell them out, and study them.

Detachment is accepted as valid in many systems. What makes I/O logic different is that it is the only assumption made. Other approaches to normative reasoning make some extra assumptions that are potentially controversial.

In particular, the Lewis-Hansson logic for conditional obligation mentioned above shares with the so-called classical theory of rational choice[2]

---

[2]It is also known as the theory of "revealed preferences". For a systematic survey of the field, see [Suzumura, 1983].

the assumption that an individual has (well-defined) preferences, and that a normative judgment is based on a maximization process. Advocates of (as [Simon, 1957] terms it) "bounded rationality" have argued that an approach to rationality in terms of maximization is not realistic, because human beings lack the cognitive resources to optimize. Usually we do not know the relevant probabilities of outcomes, we can rarely evaluate all outcomes with sufficient precision, and our memories are weak and unreliable.

The best way to avoid potential objections is to make as few assumptions as possible. We believe that the assumption that conditionals obey the detachment rule is one that can hardly be challenged. Obligations and permissions are contextual and vary based on the setting. Consequently, a norm always takes the form of a conditional statement. Some philosophers like [Boghossian, 2000] think (rightly, in our view) that the disposition to reason according to detachment is constitutive of the possession of the concept of conditional, and thus of the concept of norm. The idea is that, if some agent says "if a then x", and if he truly means it, then he commits himself to detaching x given a. If this agent refuses to acknowledge that he is justified in employing detachment, this will be good evidence that he fails to understand what is meant by "if ... then". Accepting detachment and acquiring an implication are simply two sides of the same coin.

As a matter of facts, there is more to detachment than what propositional logic can reveal. This becomes clear when we consider the full picture of detachment in the deontic logic literature. Another form of detachment that figures prominently in the discussions on CTDs is so-called "deontic" detachment. The rule ratifies the detachment of the obligatoriness of the consequent from the obligatoriness of the antecedent. It has the form:

(2)   Deontic detachment
    (i) If a is obligatory, then x is obligatory
    (ii) a is obligatory
    (iii) So, x is obligatory

In the I/O notation the rule may be expressed as:

If $a \in out(N, \top)$ and $(a, x) \in N$ then $x \in out(N, \top)$

This translates iteration of successive detachments as illustrated in Figure 2.

Deontic logic has traditionally been described as facing a choice between these two forms of detachment. The reason why can be illustrated with the well-known example from [Chisholm, 1963]. Suppose we have: a is obligatory; if a, then x is obligatory; if $\neg a$, then $\neg x$ is obligatory; $\neg a$. If both factual and deontic detachment are allowed, then "x is obligatory" and "$\neg x$ is obligatory" both follow, which seems counter-intuitive. What

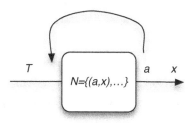

Figure 2: Iterated detachment

is so special about the input/output logic framework is that it allows both deontic and factual detachment yet it handles them coherently. The idea is to assume that deontic detachment does not unrestrictively: in case of conflict, factual detachment supersedes deontic detachment.

1.3 Makinson's third way

Makinson's notion of iterative detachment [Makinson, 1999], and the subsequent work on input/output logics [Makinson and van der Torre, 2000], has been driven by the need to give a precise formulation to the new reading of deontic formulae as reporting the norms that hold "according to" a particular code. This is the I/O way to handle Jørgensen [1937]'s dilemma. The idea is that, although norms are neither true nor false, one may state that (according to the norms), something ought to be done: the statement "John ought to leave the room" is, then, a true or false description of a normative situation. Such a statement is usually called a normative proposition, as distinguished from a norm.

The problem is to identify the norms that are implicit in a code on the basis of those that it presents explicitly, without appealing to some already given deontic logic. According to Makinson, this is not just one problem among many, but it is a fundamental challenge, because it presents an alternative to the traditional axiomatic and possible-worlds approaches to deontic logic. Most deontic logicians would agree that there is "no logic of norms without attention to a system of which they form part" [1999, p. 29]. However, Makinson is the first to observe that this observation leads to a new framework for doing deontic logic, posing an alternative to the traditional formal frameworks for studying normative reasoning.

> "Deontic logic has fallen into ruts. The older rut is the axiomatic approach, with its succession of propositional and occasionally quantified calculi. The newer one is possible-worlds semantics,

with endless minor variations in the details. We depart from these confines, with an approach which, whilst syntactic rather than semantic, is not at all axiomatic in character. It could be called *iterative*." [Makinson, 1999, p. 31]

1.4 Secretarial assistant

The traditional picture of logic is that of an inference motor. This is illustrated with Figure 3.

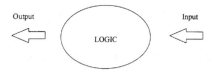

Figure 3: Logic as an inference motor

The view of logic underpinning the I/O framework is very different. Its role is not to create or determine a distinguished set of norms, but rather to prepare information before it goes in as input to such a normative code, to unpack output as it emerges and, if needed, coordinate the two in certain ways. A set of conditional norms is, thus, seen as a transformation device, and the task of logic is to act as its "secretarial assistant" [Makinson and van der Torre, 2000, p. 384]. This is illustrated with Figure 4.

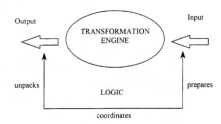

Figure 4: Logic as a secretarial assistant

More precisely, a normative code is a set N of conditional norms. These are ordered pairs of the form (a, x). For each such pair, the body a is thought of as an input, representing some condition or situation, and the head x is thought of as an output, representing what the norm tells us to be obligatory. Moreover, given any universe L such that $N \subseteq L \times L$ and an input $A \subset L$, the output of A under N may be understood as

$$N(A) = \{x : (a,x) \in N \text{ for some } a \in A\}$$

$N(A)$ can be viewed as the I/O implementation of the operation of detachment mentioned in Subsection 1.2. To obtain $N(A)$, you go through the elements in A. Each time one appears as body of a rule (a, x), you detach the head x, and put it in $N(A)$.

Input/output logic investigates what happens to this basic picture when we pass to the logical level, i.e. when L is a propositional language, closed under at least the usual truth-functional connectives, and N a set of ordered pairs (a, x) of formulae in L. These are referred to as "generators". It is worth stressing that the generators in N are not treated as formulae themselves. If they were, then we would commit ourselves to the view that norms bear truth-values.

In its full generality the detachment problem as studied in input/output logic can be stated as follows. Suppose that we are given a set A of formulae. How may we reasonably define the set of propositions x making up the output of A under N, or one might also say, of N given A, which we write $out(N, A)$? The task of logic must, thus, be seen as a modest one.

1.5 What counts as an input/output logic?

Makinson and van der Torre go bottom-up, starting from examples of input/output logics. However, they do not have much to say on what counts as an input/output logic. Since the full range of all individual input/output logics is yet to be discovered, this may be seen as restricting the scope beforehand unnecessarily. The definition we give in this section is tentative and quite general.

Definition 1.1 *An input/output logic consists of:*

1. *A definition of a normative system;*

2. *A semantics telling us how to way to detach obligations, permissions and institutional facts from the normative system in a given context;*

3. *A proof system for the detachment of obligations, permissions and institutional facts;*

4. *Soundness and completeness results relating the semantics and the proof system.*

The definition of a normative system may include deadlines, bearers and counterparts of obligations, so-called constitutive norms, and much more.

All these extra aspects will be put aside in this chapter. The only criterion is that, in a context, deontic statements can be detached from the system. It may be thought that the study of normative reasoning calls for the use of a much more expressive language than the one we will be using here. However, like [Makinson, 1999, p. 45], we believe that more complex machinery should be introduced only after the simpler framework is well-understood.

2 Obligations from an I/O perspective

We now turn to the mathematical foundations of I/O logic. For ease and conciseness of exposition all the proofs are omitted.

Sections 2 and 3 are devoted to the semantics for input/output logic. Here the term "semantics" is taken in a very broad sense. It can be called "operational semantics" to distinguish it from the usual truth-functional semantics for deontic logic. Roughly speaking, the meaning of deontic concepts is given in terms of a set of mechanisms yielding outputs for inputs.

This provides an opportunity to clear up the first common misunderstanding about input/output logic mentioned in the introductory section. It is a mistake to think that input/output logic is just syntax. The semantics for input/output logic is well-defined, and to some extent better understood than its proof theory.

From now onwards we use the letter G rather than N to denote the set of norms or generators.

2.1 Basic representation

The basic intuition is that input and output are both under the sway of the operation Cn of classical consequence. The simplest approach is to put

$$out(G, A) = Cn(G(Cn(A)))$$

where Cn (alias $\vdash$) is classical consequence, and the function $G(.)$ is defined by

$$G(X) = \{x : (a, x) \in G \text{ for some } a \in X\}$$

In other words, given a set A of formulae as input, we first collect all of its consequences, then apply G to them, and finally consider all of the consequences of what is thus obtained. It is possible to define various variants to deal with disjunctive inputs intelligently, and make outputs available for recycling as inputs.

Below we recall some well-known properties of Cn. They are needed for the completeness theorems to obtain. These are:

(inclusion) $A \subseteq Cn(A)$

(monotony) $A \subseteq B \to Cn(A) \subseteq Cn(B)$
(idempotence) $Cn(A) = CnCn(A)$
(compactness) $x \in Cn(A) \to Cn(A')$ for some finite $A' \subseteq A$

Definition 2.1 *Let L be the set of all Boolean formulae, and let G be a set of ordered pairs of L. Each pair (a, x) is called a generator, and is read as 'if input a then output x'. The following logical systems can be defined, where a complete set is one that is either maximal consistent[3] or equal to L:*

$$out_1(G, A) = Cn(G(Cn(A)))$$
$$out_2(G, A) = \cap\{Cn(G(V)) : A \subseteq V, V\, complete\}$$
$$out_3(G, A) = \cap\{Cn(G(B)) : A \subseteq B \supseteq Cn(B) \supseteq G(B)\}$$
$$out_4(G, A) = \cap\{Cn(G(V)) : A \subseteq V \supseteq G(V), V\, complete\}$$

out_1, out_2, out_3 and out_4 are called *simple-minded output, basic output, simple-minded reusable output* and *basic reusable output*, respectively. As this terminology suggests, out_3 is a variant of out_1, and out_4 is a variant of out_2.

The operation out_1 has already been explained. The operation of detachment is performed on the logical consequences of the input. The definition may be reformulated in terms of $\vdash$ as follows. For $x \in out_1(G, A)$ to be the case, the following must hold:

$$A \vdash a_1 \wedge ... \wedge a_n$$
$$\text{for } (a_1, x_1), ..., (a_n, x_n) \in G$$
$$\text{and } x_1 \wedge ... \wedge x_n \vdash x$$

Compactness for $\vdash$ guarantees that the numbers of norms (or generators) whose heads conjointly prove x is finite.

The operation out_2 works differently. Roughly speaking, the detachment operation is performed on the maximal consistent extensions of the input set. This is needed to allow reasoning by cases (see Example 2.6). We go through the list of all the supersets of A, and remove those that are not maximal under inclusion. Let $V_1, V_2, ...$ be the sets we are left with. Each V_i is either L, or a maximal consistent extension of A. We apply G to each such V_i (viz. we detach the head of all the rules in G whose antecedent is

---

[3] A set is maximal consistent if it is consistent, and any proper superset of it is inconsistent.

in V_i), and then take the logical closure of the set of heads thus detached. Having done this for each V_i, we take the intersection of the family of all the consequence sets thus obtained. This is illustrated with Figure 5.

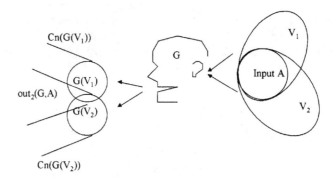

Figure 5: Basic output (out_2)

out_4 is much alike out_2 except that the choice of complete sets is restricted to those containing their own image under G.

[Stolpe, 2008b] has shown how to give an inductive characterization of out_3. It clarifies how out_3 connects with out_1, and it also makes out_3 look like much more intuitive. Roughly speaking, out_3 is just the iterative version of out_1. Once some output has been delivered, it is recycled as input in a circular motion, whilst accumulating the results as we go. In the notation out_3^b, the superscript b is mnemonic for "bulk increment".

Definition 2.2 $out_3^b(G, A) = \bigcup_{i=0}^{\omega} A_i$ where

- $A_0 = out_1(G, A)$
- $A_{n+1} = Cn(A_n \cup out_1(G, A_n \cup A))$

The following applies:

Theorem 2.3 $out_3(G, A) = out_3^b(G, A)$.

Proof. This is [Stolpe, 2008b, th. 4.3.12 and 4.3.13] ∎

It would be interesting to know if out_4 too can be given an inductive characterization in terms of out_2.

The differences between these four I/O operations appear more clearly at the proof-theoretical level. We postpone discussion of the proof theory until Section 4.1.1, and confine ourselves to comparing the four I/O logics on a number of examples.

The most characteristic property is that inputs are not in general outputs; that is, we do not have $A \subseteq out(G, A)$. The principle $A \subseteq out(G, A)$ is the I/O analog of the principle known as reflexivity or identity, $\bigcirc(x/x)$.

Example 2.4 (Reflexivity) *Put $G = \{(a, x)\}$ and $A = \{a\}$. For $i \in \{1, 2, 3, 4\}$, $out_i(G, A) = Cn(x)$. Thus $a \notin out_i(G, a)$. This gives a counterexample to the law $A \subseteq out(G, A)$.*

Example 2.5 (Contraposition) *Put $G = \{(a, x)\}$ again. For all $i \in \{1, 2, 3, 4\}$, $out_i(G, a) = Cn(x)$, and $out_i(G, \neg x) = Cn(\emptyset)$. This gives a counterexample to the law of contraposition. That is, $x \in out(G, a)$ does not imply $\neg a \in out(G, \neg x)$.*

Example 2.6 (Reasoning by cases) *Put $G = \{(a, x), (b, x)\}$.*

For $i \in \{1, 3\}$, $out_i(G, a) = Cn(x) = out_i(G, b)$, but $out_i(G, a \vee b) = Cn(\emptyset)$. Thus, neither out_1 nor out_3 support reasoning by cases. That is, $x \in out_i(G, a)$, and $x \in out_i(G, b)$ do not imply $x \in out_i(G, a \vee b)$,

For $i \in \{2, 4\}$, $out_i(G, a) = out_i(G, b) = out_i(G, a \vee b) = Cn(x)$. Thus, both out_2 and out_4 support reasoning by cases.

Example 2.7 (Transitivity) *Put $G = \{(a, x), (x, y)\}$.*

For $i \in \{1, 2\}$, $out_i(G, a) = Cn(x)$, and $out_i(G, x) = Cn(y)$. Thus, plain transitivity fails for out_1 and out_2. That is, $x \in out(G, a)$ and $y \in out(G, x)$ do not imply $y \in out(G, a)$.

For $i \in \{3, 4\}$, $out_i(G, a) = Cn(x, y)$, and $out_i(G, x) = Cn(y)$. Thus, plain transitivity holds for out_3 and out_4.

These four operations have four counterparts that also allow *throughput*. Intuitively, this amounts to requiring $A \subseteq out G(A)$. In terms of the definitions, it is to require that G is expanded to contain the diagonal, *i.e.*, all pairs (a, a).

All eight systems are distinct, with one exception: basic throughput, which we write as out_2^+, authorizes reusability, so that $out_2^+ = out_4^+$. This may be shown directly in terms of the definitions.

Classical propositional logic is used as the base logic for the sake of simplicity only. As shown in [Parent et al., 201x], one can use weaker logics like intuitionistic logic. This provides an opportunity to clear up another frequent misunderstanding about I/O logic. Just as it is a mistake to think that input/output logic is just proof theory, so also it is a mistake to think that it is just an "extension" of classical propositional logic. We use inverted commas, because even when classical logic is used as the base logic not all inference patterns that are classically valid are valid in I/O logic. In this respect, I/O logic is not so much an extension of classical logic, but rather a generalization of it. This can be made more precise using Theorem 2.8 below, due to [Makinson and van der Torre, 2000]:

Theorem 2.8 out_4^+ *collapses into classical consequence, in the sense that* $out_4^+(G, A) = Cn(m(G) \cup A)$ *where* $m(G)$ *is the materialization of* G, *i.e., the set of all formulae* $a \to x$ *where* $(a, x) \in G$.

Proof. This is [Makinson and van der Torre, 2000, obs. 16]. ∎

2.2 Comparison with conceptual implicative structures

It is a good place to show the analogy between I/O logic and the Lindahl-Odelstad algebraic approach to normative systems described in the next chapter.

We shall focus on (as they call it) the "cis-model", where "cis" abbreviates conceptual implicative structures. The theory they define in their chapter, which they call the "Theory of Joining Systems" (or TJS, for short), is more general.

According to the cis-model, a normative system consists of a system B_1 of potential grounds (or descriptive conditions), a system B_2 of potential consequences (or normative effects) and a set J of links or joinings from the system of grounds to the system of consequences. The set J is a set of norms. The cis-model is illustrated with Figure 6, where a norm is represented by an arrow from one system to the other. There is here an

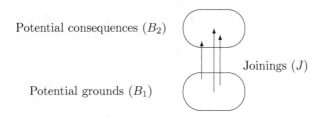

Figure 6: The cis-model

obvious connection with out_1. Recall the calculation of out_1 involves three main steps. First, input set A is expanded to its classical closure $Cn(A)$. This corresponds to the fact that grounds are ordered by an implication relation in the cis-model. Next, the set obtained at step 1 is passed into a "black box", which delivers the corresponding deontic output. This is done by taking the image of $Cn(A)$ under G. This is exactly what Lindahl and Odelstad call "joining". Finally, the set of outputs obtained at step 2 is expanded to its classical closure again. This last step corresponds to the fact that deontic consequences are also ordered by an implicative relation. It is interesting to remark that although the two accounts use different intuitions the results seem very close. Nevertheless, it is unclear whether the analogy still holds for the other input/output operations.

2.3 Modal embedding

out_2 and out_4 can be embedded within modal logic. The basic idea is to prefix heads with boxes, and use some appropriate modal system S as the target of the embedding. Let $G^{\square}$ denote the set of modal formulae $x \to \square y$ such that $(x, y) \in G$. Let $Z \vdash_S z$ mean that $\wedge Y \to z$ for some finite $Y \subseteq Z$. The modal systems of interest here are:

- System K_0: it is much alike the familiar modal system K except for the fact that it has a restricted form of necessitation:

$$\text{If } x \text{ is a tautology, then } \vdash \square x$$

- System $K45$: it is K supplemented with

$$4 : \square x \to \square\square x$$
$$5 : x \to \square \Diamond x$$

- System $K_0 T$: it is K_0 supplemented with

$$T : \square x \to x$$

- System $KT45$: it is $K45$ supplemented with T

We have:

Theorem 2.9 $x \in out_2(G, A)$ iff $x \in Cn(G(L))$ and $G^{\square} \cup A \vdash_S \square x$ for any S such that $K_0 \subseteq S \subseteq K45$.

Proof. This is [Makinson and van der Torre, 2000, obs. 4]. ∎

Not surprisingly, an analog of Theorem 2.9 is available for out_4.

Theorem 2.10 $x \in out_4(G, A)$ iff $x \in Cn(G(L))$ and

$$G^\Box \cup A \vdash_S \Box x \text{ for any } S \text{ such that } K_0 T \subseteq S \subseteq KT45$$

Proof. This is [Makinson and van der Torre, 2000, obs. 15]. ∎

The question may be raised if this modal embedding can be used to transfer decidability (and eventually complexity) results from modal logic back to I/O logic.

2.4 A possible-worlds semantics

Here we describe a possible-worlds semantics put forth by [Bochman, 2005]. Its main interest lies in putting the semantics of the I/O systems in the logical tradition of modal logic, which some might be more familiar with. Such a semantics does not come out of the blue. Bochman takes advantage of the modal reformulation of some of the I/O operations described in Subsection 2.3.

However, a word of caution is in order. Bochman's prime interest is in modelling production and causal reasoning rather than normative reasoning. Causal reasoning is the ability to identify relationships between causes — events or forces in the environment — and the effects they produce. I/O logic seems to be particularly well-suited for modelling causal reasoning. Indeed, there is a widespread agreement amongst philosophers that a causal inference relation should not satisfy the Reflexivity postulate of consequence relations. As mentioned, the distinctive feature of I/O logic is that it rejects such a postulate.

This change of application area leads Bochman to depart from the initial account in a significant way:

- Under Bochman's account, an input/output operation is called a production rule, and is represented as a propositional connective $\Rightarrow$ in the object language, alongside the truth-functional ones. Rules of the form $a \Rightarrow x$ form a basis for production and causal inference relations. The I/O view is quite different: the key relations and operations of the metalanguage are not internalized to become connectives of the object language.

- Correlatively, under Bochman's account, a production rule bears a truth-value, or kind of. A production rule is assigned the value "valid" or "invalid".

Below we will speak of such and such possible-worlds account as providing a semantics close to (or in the neighborhood of) such and such I/O operation. As a matter of facts, there is no direct (formal) connection between the semantics Bochman proposes and the operational semantics for I/O logic. The linkage between the two is established through the axiomatic characterization: both the possible-worlds semantics and the operational semantics give rise to almost the same axiom system.[4] This finding has an interest in its own, and is worth a mention.

Bochman provides a semantics close to out_1 and out_3 based on pairs of deductively closed theories called bimodels. A bimodel can be thought of as representing an initial state (input) and a possible final state (output) of a production process.

Definition 2.11 *A pair of classically consistent and deductively closed sets of propositions $\langle u, v \rangle$ is called a classical bimodel. A set of classical bimodels is called a classical binary semantics, and is denoted by $\mathcal{B}$.*

The evaluation rule for $\Rightarrow$ runs as follows.

Definition 2.12 *A rule $a \Rightarrow x$ is valid in a classical binary semantics $\mathcal{B}$ if, for any bimodel $\langle u, v \rangle$ from $\mathcal{B}$, $a \in v$ only if $x \in u$.*

The class of all models thus defined provides a semantics in the neighborhood of out_1. A semantics close to out_3 can be obtained by considering only bimodels $\langle u, v \rangle$ such that $u \subseteq v$. Such bimodels (and corresponding semantics) are called consistent.

For out_2 and out_4, the semantics is a little bit different. Here we restrict the set of bimodels to bimodels of the form (α, β), where α and β are worlds. The corresponding semantics can, then, be recast in terms of a possible-worlds semantics. A model $\mathbb{W}$ is a triplet (W, R, V), where W is a set of possible worlds, R a binary accessibility relation on W, and V a function assigning to each propositional letter a the set of possible worlds where a is true. Validity of rules are defined as follows:

Definition 2.13 *A rule $a \Rightarrow x$ is valid in a relational model $\mathbb{W} = (W, R, V)$ if, for any $\alpha, \beta \in W$ such that $R\alpha\beta$, if a holds at α then x holds at β.*

The class of all relational models thus conceived provides a semantics in the neighborhood of out_2. Let us call a relational model $\mathbb{W} = (W, R, V)$ quasi-reflexive if $R\alpha\beta$ implies $R\alpha\alpha$. The class of quasi-reflexive models provides a semantics in the neighborhood of out_4. As we will see in Subsection 4.1.1, the main difference between out_2 and out_4 is that the latter, but not

---

[4]We say "almost the same" because all the systems studied by Bochman contain an additional axiom, which is not available in I/O logic. It is $\bot \Rightarrow \bot$.

the former, validates the principle of cumulative transitivity. This is the principle:

$$\text{From } a \Rightarrow x \text{ and } a \wedge x \Rightarrow y \text{ infer } a \Rightarrow y$$

It is not difficult to see this principle may fail unless the model is required be quasi-reflexive.

2.5 Input/output without WO

The correctness of the rule "weakening of the output" (WO) has been called into question by a number of deontic logicians. For instance, some deontic paradoxes like Ross's Paradox, the Good Samaritan Paradox and the Paradox of Epistemic Obligation appear to depend on this rule. A natural reaction is to resolve the latter paradoxes by rejecting WO itself. This is the approach taken by e.g. [Jackson, 1985; Goble, 1991].

Two new I/O logics, which do not contain WO, have been proposed by [Stolpe, 2008a; Stolpe, 2008b]. These are variants of out_1 and out_3. Let us represent them by out_1^s and out_3^s, respectively. A wff x is said to be equivalent to some set X of wffs iff $Cn(x) = Cn(X)$.

Definition 2.14 $x \in out_1^s(G, A)$ iff x is equivalent to a subset of $G(Cn(A))$

For a counter-example to WO, put $G = \{(\top, a)\}$. We have $G(Cn(\top)) = \{a\}$. So $a \in out_1^s(G, \top)$ but $a \vee \neg a \notin out_1^s(G, \top)$ because $Cn(a) \nsubseteq Cn(\emptyset)$. It is not hard to establish that, more generally,

Theorem 2.15 $out_1(G, A) = Cn(out_1^s(G, A))$

Thus, out_1^s is alike the original out_1 except that the output is no longer closed under logical consequence.

The definition of out_3^s is recursive. It mimics the definition for out_3^b in the obvious way.

Definition 2.16 $x \in out_3^s(G, A)$ iff x is equivalent to a subset of $\bigcup_{i=0}^{\omega} A_i$ where

- $A_0 = G(Cn(A))$
- $A_{n+1} = A_n \cup G(Cn(A_n \cup A))$

2.6 Reasoning about norm violation

One of the most frequently discussed issues in deontic logic is the problem of how to reason about norm violation. This is known as the problem of contrary-to-duty reasoning. It has been discussed in the context of the notorious contrary-to-duty paradoxes like Chisholm's and Forrester's paradox. In input/output logic, it has led to the use of constraints [Makinson and van der Torre, 2001].

2.6.1 Threshold idea

The strategy is to cut back the set of norms to just below the threshold of yielding excess. To do that, we look at the maximal non-excessive subsets, i.e. the maximal $G' \subseteq G$ such that $out(G', A)$ is consistent with input A. In [Makinson and van der Torre, 2001], the family of such G' is called the *maxfamily* of (G, A), and the family of outputs $out(G', A)$ for G' in the maxfamily, is called the *outfamily* of (G, A). The formal definition below is general, covering as special case both inconsistency of the output and its inconsistency with the input.

Definition 2.17 (Constraints) *Let G be a set of generators and out be an input/output logic. Let C be an arbitrary set of formulae, which we may call 'consistency constraints'. We define:*

- *maxfamily(G, A, C) is the set of $\subseteq$-maximal subsets G' of G such that $out(G', A)$ is consistent with C.*
- *outfamily$(G, A, C) = \{out(G', A) \mid G' \in $ maxfamily$(G, A, C)\}$.*

Putting $C = \emptyset$ amounts to requiring the output be consistent. Putting $C = A$ amounts to requiring it be consistent with the input.

We stress that a set of generators and an input do not have a set of propositions as output, but a set of sets of propositions. So, like in the logics of belief change and non-monotonic inference, we can infer a set of propositions by taking either a credulous or a skeptical approach. In [Makinson and van der Torre, 2001], the two resulting operations are called full meet and full join constrained output, and they are noted $\cap outfamily(G, A, C)$ and $\cup outfamily(G, A, C)$, respectively. Variants are possible, but these need not concern us here. We use $out^\cap(G, A)$ as a shorthand for $\cap outfamily(G, A, A)$, and $out^\cup(G, A)$ as a shorthand for $\cup outfamily(G, A, A)$.

These notions are illustrated below using three examples from the literature on CTDs. In these examples, the credulous and sceptical approaches yield the same outcome.

Example 2.18 (Chisholm) *Put* $G = \{(\top, a), (a, t), (\neg a, \neg t)\}$, *where* $(\top, a)$ *means the norm that the man must go to the assistance of his neighbors,* (a, t) *means the norm that it ought to be that if he goes he ought to tell them he is coming, and* $(\neg a, \neg t)$ *means the norm that if he does not go he ought not to tell them he is coming. Put* $A = \{\neg a\} = C$. *Then:*

$$\text{maxfamily}(G, A, A) = \{\{(a, t), (\neg a, \neg t)\}\}$$
$$\text{outfamily}(G, A, A) = \{Cn(\neg t)\}$$

So

$$\text{out}^{\cap/\cup}(G, A) = Cn(\neg t)$$

This agrees with the intuitive assessment of the example: given the norms in G *and input* $A = \{\neg a\}$, *the man must not tell his neighbors he is coming.*

Example 2.19 (Forrester) *Put* $G = \{(\top, \neg k), (k, k \wedge g)\}$, *where* $(\top, \neg k)$ *means that you should not kill, and* $(k, k \wedge g)$ *means that if you kill you should do it gently. Put* $A = \{k\} = C$. *Then:*

$$\text{maxfamily}(G, A, A) = \{\{(k, k \wedge g)\}\}$$
$$\text{outfamily}(G, A, A) = \{Cn(k \wedge g)\}$$

So

$$\text{out}^{\cap/\cup}(G, A) = Cn(k \wedge g)$$

This agrees with the intuitive assessment of the example: given the norms in G *and input* $A = \{k\}$, *you must kill gently.*

Example 2.20 (Multiple levels of violation) *Let* a, x, y *be read as 'you break your promise', 'you apologize' and 'you are ashamed', respectively. Put* $G = \{(\top, \neg a), (a, x), (a \wedge \neg x, y)\}$. *Put* $A = \{a \wedge \neg x\} = C$. *Then:*

$$\text{maxfamily}(G, A, A) = \{(a \wedge \neg x, y)\}$$
$$\text{outfamily}(G, A, A) = \{Cn(y)\}$$

So

$$\text{out}^{\cap/\cup}(G, A) = Cn(y)$$

As described above, the approach has at least two salient features compared to other treatments. First, we stay mono-modal, in the sense that no distinction is made between different senses of 'ought' (see [Carmo and Jones, 2002; Cholvy and Garion, 2001]). Next, instead of restricting the application of the inference rules, we cut back on the set of generators itself. Here lies perhaps the main difference between the I/O account and the non-monotonic treatment based on defeasible logic (see [Nute, 1997]).

2.6.2 Nonmonotonic reasoning

Although the outfamily strategy is designed to deal with contrary-to-duty norms, its application turns out to be closely related to belief revision and nonmonotonic reasoning when the underlying input/output operation authorizes throughput.

When all elements of G are of the form $(\top, x)$, then for the degenerate input/output operation $out_2^+(G, a) = out_4^+(G, a) = Cn(m(G) \cup \{a\})$, the elements of $outfamily(G, a)$ are just the maxichoice revisions of $m(G)$ by a, in the sense of [Alchourrón et al., 1985]. These coincide, in turn, with the extensions of the default system $(m(G), a, \emptyset)$ of [Poole, 1988].

More surprisingly, there are close connections with the default logic of [Reiter, 1980]. Read elements (a, x) of G as normal default rules $a : x/x$ in the sense of Reiter, and write $extfamily(G, A)$ for the set of extensions of (G, A). Then, for reusable simple-minded throughput out_3^+, it can be shown that $extfamily(G, A) \subseteq outfamily(G, A)$ and indeed that $extfamily(G, A)$ consists of exactly the maximal elements (under set inclusion) of $outfamily(G, A)$.

But care should be taken here. The most that comes from the previous result is that the two accounts coincide under the credulous approach:

$$\cup outfamily(G, A) = \cup extfamily(G, A)$$

The equality may fail under the sceptical approach, which is more or less the standard approach in nonmonotonic logic.

These results are proven in [Makinson and van der Torre, 2001].

2.7 Conflicts amongst obligations

The aim of this section is two-fold. First, we show that I/O logic can accommodate the existence of conflicts amongst obligations. Next, we provide a procedure for resolving the latter conflicts using a priority relation. This section uses some material from [Parent, 2011]. We confine ourselves to the basic definitions, and the motivation. For a more comprehensive analysis, the reader is referred to the aforementioned paper.

In order to accommodate conflicts, one may use the same trick as in the treatment of CTDs: cut back the set of norms to just below the threshold of yielding excess, and consider the resulting output. The way it works is best explained by considering the simplest form of conflict between obligations, where the head of one rule is the negation of the head of the other. This form of conflict may be called "strict" or logical.

Example 2.21 (Logical conflict) *Put $G = \{(a,x),(a,\neg x)\}$, and consider input a. The reader may easily verify that, for all the (unconstrained) output operations, $out(G,a) = L$. This shows that none of the operations considered so far are conflict-tolerant. Take Definition 2.17, and put $C = \emptyset$. The maxfamily has two elements $\{(a,x)\}$ and $\{(a,\neg x)\}$, and thus the outfamily has two elements $Cn(x)$ and $Cn(\neg x)$. Let the final output be calculated using the full meet operation. We have*

$$\cap outfamily(G,A,C) = Cn(x) \cap Cn(\neg x) = Cn(x \vee \neg x)$$

Since no rule has priority over the other, the most that comes is that the disjunction of x and $\neg x$ is obligatory.

The same point can be made about "natural" (or non-strict) conflicts. These are of the form: (a,x) and (a,y), with x incompatible with y in the sense of natural or physical necessity, broadly conceived. In the treatment of this type of conflict, the set C of integrity constraints mentioned in Definition 2.17 is put into the foreground. In the above example, $C = \{x \to \neg y\}$. This is meant to indicate that x and y cannot be simultaneously true, given the agent's present physical and psychical capabilities, and so-on. The output is never allowed to contradict C. So, compared to the treatment of a strict conflict, the main difference is that we take the maximal $G' \subseteq G$ such that $out(G',A)$ is consistent with C.

Example 2.22 (Natural conflict) *Put $G = \{(a,x),(a,y)\}$, $C = \{x \to \neg y\}$ and $A = \{a\}$. For all the output operations, $out(G,A) = Cn(x,y)$, which is inconsistent with C. The maxfamily has two elements $\{(a,x)\}$ and $\{(a,y)\}$, and thus the outfamily has two elements $Cn(x)$ and $Cn(y)$. We have*

$$\cap outfamily(G,A,C) = Cn(x) \cap Cn(y) = Cn(x \vee y)$$

Again, the most that comes is that the disjunction of x and y is obligatory.

Now we bring priorities to the picture. The idea is to assume that conflicts are resolved based on a priority ordering on the powerset of G, so that only a "preferred" element of the maxfamily is used to generate the output. This is implemented using a relation on set of rules. However, in practical applications, one uses a relation on rules, not a relation on sets of rules. So the question is: given a relation on rules, how can it be lifted to a relation on sets of rules? The definition of lifting given below is taken from [Brass,

1991].[5] The relation $\geq$ is read "at least as strong as", and the superscript s (mnemonic for "set") is used to distinguish between the two relations.

We assume that $\geq$ is a pre-order, i.e., the relation is reflexive and transitive. $>$ denotes the strengthened complement of $\geq$, defined by putting $a > b$ whenever $a \geq b$ and $b \not\geq a$. $\sim$ is the equivalence relation generated by $\geq$, defined by putting $a \sim b$ whenever $a \geq b$ and $b \geq a$. Each of the latter two notions has a counterpart in terms of ordering on sets – the definition is analogous. And $a \in S$ is called a $\geq^s$-maximal element of S if, for all $b \in S$, $b \geq^s a$ implies $a \geq^s b$.

Definition 2.23 (Lifting) *Let G be a set of generators equipped with a pre-order $\geq$. For any $G_1, G_2 \subseteq G$, we define $G_1 \geq^s G_2$ to hold if for every $\delta_2 \in G_2 \setminus G_1$ there is $\delta_1 \in G_1 \setminus G_2$ with $\delta_1 \geq \delta_2$.*

Now comes the main construction. This is a lightweight version of the account described in [Parent, 2011].

Definition 2.24 (Outfamily with priorities) *Let G, A and C be a pre-ordered set of generators, an input set, and a set of integrity constraints, respectively. Let $\geq^s$ be taken as in Definition 2.23. Put*

- *maxfamily(G, A, C) is the set of $\subseteq$-maximal subsets G' of G such that $out(G', A)$ is consistent with C*

- *preffamily(G, A, C) is the set of $\geq^s$-maximal elements of maxfamily(G, A, C)*

There are two steps involved in the construction of the preffamily. We start by determining the maxfamily. It gathers all the maximal $G' \subseteq G$ such that $out(G', A)$ is consistent with C. This step is mandatory to guard against possible contradictions when applying rules to the input set. The preffamily, then, determines the preferred element in the maxfamily taking priorities amongst rules into account. This step is needed to resolve conflicts amongst norms.

Below we introduce the associated output operation. The subscript p is short for "preferred".

Definition 2.25 (Preferred output) *Let G, A and C be a pre-ordered set of generators, an input set, and a set of integrity constraints, respectively. We define*

$$out_p(G, A) = \cap \{out(G', A) \mid G' \in \text{preffamily}(G, A, C)\}$$

---

[5]The same proposal was repeated in a perhaps wider known paper by [Sakama and Inoue, 1996], and some equivalents to this definition are discussed in [Hansen, 2006, p. 9].

Example 2.26 provides an illustration.

Example 2.26 (The book) *Let b, s, y and l represent the respective propositions that I borrowed the book from you, that you stole it from the library, that I return the book to you, and that I return it to the library. Assume $C = \{y \to \neg l\}$, meaning that I cannot simultaneously return the book to you and the library. Let $G = \{(b,y),(s,l)\}$ with $(s,l) > (b,y)$.*

Case 1 *Assume $A = \{b\}$. For all output operations, $out(G, A)$ equates $Cn(y)$, which is consistent with C. In this case, the outfamily/preffamily has one element $\{(b,y),(s,l)\}$. And $out_p(G,A) = Cn(y)$.*

Case 2 *Assume $A = \{b,s\}$. For all output operations, $out(G,A)$ equates $Cn(y,l)$, and thus it is inconsistent with C. The maxfamily has two elements, $\{(b,y)\}$ and $\{(s,l)\}$. Furthermore, $\{(s,l)\} >^s \{(b,y)\}$ since $(s,l) > (b,y)$. So the preffamily has only one element $\{(s,l)\}$. Hence $out_p(G,A) = Cn(y)$.*

Example 2.27 (Order puzzle, [Horty, 2007]) *Let a be a shorthand for putting the heating on, and b be a shorthand for opening the window. Put $G = \{(\top,a),(\top,b),(a,\neg b)\}$. Assume these express orders issued by a priest, a bishop and a cardinal, respectively. So, $(a, \neg b) > (\top, b) > (\top, a)$. Let us assume that out is out_3 or out_4, require the output be consistent, viz $C = \emptyset$, and put $A = \emptyset$. The maxfamily have three elements*

$$\underbrace{\{(\top,a),(\top,b)\}}_{G_1} \quad \underbrace{\{(\top,a),(a,\neg b)\}}_{G_2} \quad \underbrace{\{(\top,b),(a,\neg b)\}}_{G_3}$$

And
$$G_3 >^s G_2 >^s G_1$$

Therefore, the preffamily has only one element, G_3, and $out_p(G,A) = Cn(b)$.

For more on the theme of conflicts amongst obligations, see the dedicated chapter by L. Goble in this Handbook and [Hansen, 2008; Horty, 2007; Horty, 2012]. For more on the I/O approach itself, see [Parent, 2011].

3 Input/output approaches to permission

Another frequent misunderstanding about input/output logic is that it is restricted to obligations and prohibitions. The aim of this section is to correct this misunderstanding by presenting the different notions of permission that can be formally articulated within the I/O set-up.

3.1 Different concepts of permission

We start with an informal discussion on the notion of permission. This one is notoriously ambiguous. Consider:

(1) It is permitted to drive at a speed of 95 km/h on a motorway

(2) An injured employee is allowed one change of treating doctor to another on the panel without prior authorization from the board

(3) Everyone has the right to liberty and security of person

(4) Personal data may be disclosed to a third party only if the data subject has unambiguously given his consent

(1) expresses what is usually called a "negative" permission: the act of driving at a speed of 95 km/h on a motorway is permitted by the normative system (a traffic regulation), in the sense that amongst the consequences of the system there is no norm which prohibits it. You will not find any road sign from which such a prohibition can be inferred.

The kind of permission expressed by (2) is stronger. It is usually called a "positive" permission: the act of changing doctors is permitted by the normative system (a workers' compensation scheme) in the sense that such a permission has been explicitly granted.

(3) is from the European Convention for the Protection of Human Rights and Fundamental Freedoms of 1950. (3) is a positive permission of a somewhat special kind. It expresses a so-called constitutional right, and thus it sets limits on what can be explicitly prohibited by a legal code.

(4) is from the European directive on Data Protection of 1995.[6] (4) is a positive permission too. The main difference with (2) and (3) is that such a permission provides an exception to a pre-existing prohibition. From a legal perspective, the obligation not to disclose any personal data is the general case. The right to freedom of expression, which is firmly embedded in the European Convention on Human Rights and Fundamental Freedoms, seems prima facie to imply the right to disclose any personal data. But the right to privacy, which is also recognized as fundamental in the 1950 Convention, takes precedence over the right to freedom of expression.[7]

In the next sections we show how these different concepts of permission can be clarified and precisely articulated using I/O logic.

---

[6] Directive 95/46/EC, Official Journal L281, 23/11/1995 pp. 31-50.

[7] For an analysis of the EU directive on Data Protection, see [Jones and Parent, 2008].

3.2 The MvT account of permission

We start with the [Makinson and van der Torre, 2003a] account, which will be referred to as the MvT one.

From now onwards we will switch rather casually between the notations $x \in out(G, a)$ and $(a, x) \in out(G)$. Both notations are equivalent.

Definition 3.1 defines a concept of negative permission in the line of the classic approach.

Definition 3.1 (Negative permission) *Let G be a set of generators, and out an input/output logic. Put*

$$(a, x) \in negperm(G) \text{ iff } (a, \neg x) \notin out(G)$$

So something is permitted by a code iff its negation is not obligatory according to the code and in the given situation. This is example (1) in Subsection 3.1.

The relationship between *out* and *negperm* may be expressed as follows. When A and B are sets of pairs, we say that A is almost included in B (notation: $A \subseteq^c B$) if $(a, x) \in B$ whenever $(a, x) \in A$ and a is classically consistent. We call G internally coherent if there is no classically consistent a such that $(a, x), (a, \neg x) \in out(G)$. We have

Theorem 3.2 $out(G) \subseteq^c negperm(G)$ *iff G is internally coherent.*

Proof. Immediate from definitions. ■

To formalize the kind of permission exemplified by sentence (2) in Subsection 3.1, we need a set P of explicit permissive norms, along with the set G of explicit obligations. As a first approximation, one may say that something is positively permitted by a code iff the code explicitly presents it as such. But this leaves a central logical question unanswered as to how explicitly given permissive and obligating norms may generate permissions that – in some sense – follow from the explicitly given norms. In the line of von Wright's later approach, we may use the following definition

Definition 3.3 (Static positive permission) *Let G and P be two sets of generators, and let out be an input/output logic. Put*

$$(a, x) \in statperm(P, G) \text{ iff } (a, x) \in out(G \cup Q)$$
$$\text{for some singleton or empty } Q \subseteq P$$

So there is a permission to realize x in conditions a if x is generated under these conditions either by the norms in G alone, or the norms in G together with some explicit permission (b, y) in P. We call this the "static" version of positive permission.

Example 3.4 *Put $G = \{(work, tax)\}$ and $P = \{(18y, vote)\}$ (all adults may take part in political elections). Then*

$$(work, tax) \in statperm(P, G)$$
$$(18y, vote) \in statperm(P, G)$$
$$(work \wedge male, tax) \in statperm(P, G)$$

and also

(#) $\qquad\qquad (\neg work \wedge 18y, vote) \in statperm(P, G)$

(So even unemployed adults are permitted to vote.)

Where negative permission is liberal, in the sense that anything is permitted that does not conflict with one's obligations, the concept of static permission is quite strict, as nothing is permitted that does not explicitly occur in the norms.

The relationship between *statperm* and *out/negperm* may be expressed as follows. Call G cross-coherent with P if there is no pair (c, z) with c classically consistent and $(c, \neg z) \in out(G)$ whilst $(c, z) \in statperm(P, G)$. Intuitively, G does not contain a prohibition that directly conflicts with a static permission. We have:

Theorem 3.5 *i) $out(G) \subseteq statperm(P, G)$*
ii) $statperm(P, G) \subseteq^c negperm(P, G)$ iff G is cross-coherent with P.

Proof. Immediate from the definitions involved. ∎

In between, one may define a concept of "dynamic permission" that defines something as permitted in some conditions a if forbidding it for these conditions would prevent an agent from making use of some explicit (static) permission. This is example (3) in Subsection 3.1. The formal definition may be phrased thus:

Definition 3.6 (Dynamic positive permission)

$(a, x) \in dynperm(P, G)$ iff $(c, \neg z) \in out(G \cup \{(a, \neg x)\})$
$\qquad\qquad$ for $(c, z) \in statperm(P, G)$ with c consistent

As Stolpe puts it,

"The paradigm example - but by no means the only one - is an action protected by constitutional law. Freedom of expression, for instance, is recognized as a human right under Article 19 of the Universal Declaration of Human Rights and is recognized in international human rights law in the International Covenant 4 on Civil and Political Rights. [...] An example that comes to mind is the Jyllands-Posten incident of 2005, when Muslim organizations led a complaint with the Danish police, following the publication of twelve cartoons depicting the Islamic prophet Mohammad. The investigation was discontinued by the Regional Prosecutor in Viborg, who concluded that Jyllands-Posten must be reckoned protected by the freedom of expression. The Director of Public Prosecutors in Denmark later agreed. One may say, therefore, that the printing of the cartoons was deemed [dynamically] permitted by the Danish authorities." [Stolpe, 2010b]

Stolpe's example may be analyzed as follows.

Example 3.7 *Let p and f represent the respective propositions that the cartoons are printed, and that I express myself freely. Put $P = \{(\top, f)\}$ and $G = \{(\neg p, \neg f)\}$. Here the pair $(\neg p, \neg f)$ expresses the fact that my not printing the cartoons would be the same as my not expressing myself freely. We have*

$$(\top, f) \in statperm(P, G)$$

and also

$$(\top, \neg f) \in out(G \cup \{(\top, \neg p)\})$$

Therefore, the printing of the cartoons is dynamically permitted by the code, i.e.,

$$(\top, p) \in dynperm(P, G)$$

The relationship between *dynperm* and *statperm* is as follows.

Theorem 3.8 $statperm(P, G) \subseteq^c dynperm(P, G)$

Proof. Follows from the definitions involved. ∎

Theorem 3.9 shows how *dynperm* connects with *negperm*. Intuitively, it says that in cross-coherence mode dynamic permission is a strengthened negative permission.

Theorem 3.9 $dynperm(P, G) \subseteq negperm(G)$ iff G is cross-coherent with P.

Proof. This is [Makinson and van der Torre, 2003a, coroll. 6]. ∎

From the above observations, it (very roughly) emerges that there are cases where the logical relationship amongst these notions boils down to set-theoretic subset ordering. In particular this will happen if

- G does not prescribe two incompatible actions in a consistent context (internal coherence);
- G does not contain a prohibition that directly conflicts with a static permission (cross-coherence).

The following figure shows what is included in what. A square represents the set of all pairs (a, x) subsumed under a given normative concept.

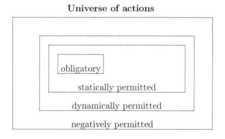

3.3 Permission as exception

The MvT account of permission cannot handle a permission statement like (4), Subsection 3.1, which provides an exception to a pre-existing prohibition. To see why, put $G = \{(\top, \neg x)\}$ and $P = \{(a, x)\}$. Obviously, this leads to deontic explosion w.r.t. permissions. Indeed, for any output operation, $(a, y) \in out(G \cup P)$ for an arbitrarily chosen y, and thus $(a, y) \in statperm(G, P)$.

Two ways around have been proposed in the literature. One is the theory of permission based on derogation due to [Stolpe, 2010b]. The other is the theory of permission based on constraints due to [Boella and van der Torre, 2008]. We will discuss them in turn.

3.3.1 Permission based on derogation

[Stolpe, 2010b] refers to the type of permission expressed by (4) as an "exemption". His account draws inspiration from the AGM analysis of belief change, and appeals to the notion of derogation (the AGM term for it is "contraction") defined in terms of remainder set. It is a straightforward matter to adapt the notion of remainder to the present set-up. The outcome of contracting G by (a, x) should be a subset of G that does not deliver (a, x) as output, but from which no elements of G have been unnecessarily removed. This is Definition 3.10 below. For $G \perp (a, x)$, read "G remainder (a, x)". The elements of $G \perp (a, x)$ are called "remainders". They are calculated by considering the maximal (under set-theoretic inclusion) subsets of G that do not deliver (a, x) as output. It is understood that out_1 is the underlying output operation.

Definition 3.10 (Remainder set) $out(G) \perp (a, x)$ *is the set of all H such that*

$H \subseteq out(G)$

$(a, x) \notin out(H)$

If $H \subset I \subseteq G$ *then* $(a, x) \in out(I)$

Note that we are taking remainders of the closure $out(G)$ rather than of the base G.

The contraction of $out(G)$ by a pair (a, x), i.e. the operation that just removes (a, x) from $out(G)$, is denoted $out(G) - (a, x)$, and is defined in terms of full meet. This, Stolpe calls it derogation.

Definition 3.11 (Derogation) $out(G) - (a, x) = \bigcap (out(G) \perp (a, x))$

Full meet derogation satisfies the following properties:

Theorem 3.12 *We have*

Closure: $out(G) - (a, x) = out(out(G) - (a, x))$
Vacuity: $out(G) \subseteq out(G) - (a, x)$ *if* $(a, x) \notin out(G)$
Failure: $out(G) \subseteq out(G) - (a, x)$ *if* $\vdash x$
Inclusion: $out(G) - (a, x) \subseteq out(G)$
Success: $(a, x) \notin out(G) - (a, x)$ *if* $(a, x) \in out(G)$ *and* $\nvdash x$
Local recovery: $(a, x) \in out((out(G) - (a, y)) \cup \{(a, y)\})$ *if* $(a, x) \in out(G)$

Proof. This is [Stolpe, 2010b, lem. 5.4]. ∎

Now comes the definition of exemption:

Definition 3.13 (Exemption I) (a, x) *is an exemption according to code* $\langle G, P \rangle$ *iff* $(a, \neg x) \in out(G) \setminus out(G) - (b, \neg y)$ *for some* $(b, y) \in P$.

Here the backslash symbol $\setminus$ denotes set-theoretic difference. The *definiens* contains two clauses:

i) $(a, \neg x) \in out(G)$

ii) $(a, \neg x) \notin out(G) - (b, \neg y)$ for some $(b, y) \in P$.

The meaning of i) is obvious: it says that the prohibition $(a, \neg x)$ can be derived from the code. Obviously, an exemption presupposes a pre-existing prohibition, of which it is an exception. What of ii)? Stolpe interprets it as meaning that "unless it [i.e., the prohibition $(a, \neg x)$] is removed, the code will contradict an implicit permission in P"[8], alias (b, y).

The following variant may be used to ensure that weakening of the input fails for conditional permission. The latter property is counterintuitive. It would allow us to move from a permission conditional upon the occurrence of some state of affairs to an unconditional permission.

Definition 3.14 (Exemption II) (a, x) *is an exemption according to code* $\langle G, P \rangle$ *iff* $(a, \neg x) \in out(G) \backslash out(G) - (b, \neg y)$ *for some* $(b, y) \in P$ *such that* $b \equiv a$.

Example 3.15 provides an illustration.

Example 3.15 *Put* $G = \{(\top, \neg x)\}$, *and* $P = \{(a, x)\}$. *On the one hand,* $(a, \neg x) \in out(G)$ *by SI. But* $(a, x) \in P$, *and by success for full meet contraction* $(a, \neg x) \notin out(G) - (a, \neg x)$. *So* (a, x) *is an exemption according to* $\langle G, P \rangle$.

The deontic explosion problem mentioned above is avoided, due to condition i) in Definition 3.13. That is, (a, y) is *not* an exemption according to $\langle G, P \rangle$, simply because $(a, \neg y) \notin out(G)$.

Theorem 3.16 shows that the notion of exemption can be captured without appealing to the notion of derogation at all. Both formulations are equivalent.

Theorem 3.16 (a, x) *is an exemption according to code* $\langle G, P \rangle$ *iff* $(a, \neg x) \in out(G)$ *and* $(a, \neg x \to \neg d) \in out(G)$ *for some* $(c, d) \in P$ *with* $c \equiv a$.

Proof. This is [Stolpe, 2010b, th. 5.27]. ∎

Intuitively, Theorem 3.16 says that (a, x) is an exemption if the prohibition $(a, \neg x)$, which follows from the code, "normatively" contradicts a positive permission (c, d), in the sense that complying with $(a, \neg x)$ makes d forbidden.

So far we have only covered permission in the static sense. It is a straightforward matter to extend the account to the dynamic version of permission. Stolpe calls it "antithetic permission".

---

[8][Stolpe, 2010b, p. 106]

Definition 3.17 (Antithetic permission) (a, x) *is antithetically permitted according to* $\langle G, P \rangle$ *iff* $(b, \neg y) \in out(G \cup \{(a, \neg x)\})$ *where* (b, y) *is an exemption or an explicit permission according to the same code, and* $a \equiv b$.

Definition 3.17 is much alike Definition 3.6 except that it uses a different concept of positive permission, and replaces the requirement "b consistent" with "$a \equiv b$".

The above account is only a first step in the right direction. In [Stolpe, 2010a], a triviality result is reported, which at first seriously undermines the value of the approach. This is Theorem 3.18 below.

Theorem 3.18 (Trivialization) *If* (a, x) *is an exemption according to code* $\langle G, P \rangle$, *then* (a, y) *is antithetically permitted according to the same code, for an arbitrarily chosen* y.

Proof. This is [Stolpe, 2010a, th. 4]. ∎

In order to avoid such a result, the author suggests incorporating a relevance requirement based on the theory of relevance through propositional letter sharing as adapted to belief revision theory by [Makinson, 2009]. The reader interested in these developments is referred to [Stolpe, 2010a]. We shall just point out that (even in the more elaborated framework described in the latter paper) the underlying concept of obligation remains unchanged, and validates the rule (SI) among others. This hints at an alternative way to handle the deontic explosion problem encountered in the MvT set-up when it comes to permission as exception. It consists in keeping the definition of the notion of static permission as it is, but assuming that the underlying output operation is in constrained rather than unconstrained mode. This is approach taken by [Boella and van der Torre, 2008], to which we now turn.

3.3.2 Permission based on constraints

We shall confine ourselves with a simplified version of the Boella and van der Torre account. They draw a distinction between a generator and (as they call it) a generator pointer, because the same generator may occur several times in the ordering. In fact, the same generator can be the object of norms enacted by different authorities: however, all these instances of the generator may have different priorities. So it is necessary to consider each norm, i.e., each instance of a generator, as a different generator pointer. For ease of exposition we do not make the distinction here.

Even with this simplification, the details turn out to be fiddle to state concisely. The notions of maxfamily and preffamily must be extended to take permissions into account. This is Definition 3.19 below.

Definition 3.19 (Maxfamily/Preffamily) *Let G and P be disjoint sets of generators, and $\geq$ a partial pre-order on the pairs in $G \cup P$. Assume $\geq$ is lifted to a relation $\geq^s$ on sets of pairs as in Definition 2.23. Put*

- *maxfamily(G, P, A) is the set of $\subseteq$-maximal $G' \cup P'$ such that $G' \subseteq G$, $P' \subseteq P$ and $out(G' \cup Q, A)$ is consistent with A for every singleton or empty $Q \subseteq P'$.*

- *preffamily$(G, P, \geq, A)$ is the set of $\geq^s$-maximal elements of maxfamily(G, P, A).*

The notion of permission as exception may be defined as follows:

Definition 3.20 (Permission as exception) *statpermfamily$(G, P, \geq, A)$ is the set of $out(G' \cup Q, A)$ such that $G' \cup P' \in$ preffamily$(G, P, \geq, A)$, $G' \subseteq G$, $Q \subseteq P' \subseteq P$, and Q is a singleton or empty.*

Example 3.21 shows the account avoids the deontic explosion problem.

Example 3.21 *Put $G = \{(\top, \neg x)\}$, and $P = \{(a, x)\}$ with $(a, x) > (\top, \neg x)$. Consider input a. The maxfamily has two elements, namely*

- $G_1 \cup P_1$ *with* $G_1 = \{(\top, \neg x)\}$ *and* $P_1 = \emptyset$
- $G_2 \cup P_2$ *with* $G_2 = \emptyset$ *and* $P_2 = \{(a, x)\}$

Since $(a, x) > (\top, \neg x)$, the preffamily has $G_2 \cup P_2$ as sole element. Thus, x is permitted in context a, but y is not, since $out(G_2 \cup P_2, a) = Cn(x)$.

A permission in turn leaves room for exception. It is a straightforward matter to define an output operation based on the above new notion of preffamily so we can talk and reason about obligations as exceptions to permissions.

Definition 3.22 (Obligation as exception) *Put*

$$outfamily(G, P, \geq, A) = \{out(G', A) : G' \cup P' \in preffamily(G, P, \geq, A)\}$$

And

$$out_p(G, P, \geq, A) = \cap\, outfamily(G, P, \geq, A)$$

The following examples illustrate permissions as exceptions, and obligations as exceptions to permissions.

Example 3.23 Let $G = \{(\top, \neg k), (s \wedge p, \neg k)\}$, and $P = \{(s, k)\}$, with $(s \wedge p, \neg k) > (s, k) > (\top, \neg k)$. Intuitively: It is forbidden to kill, but it is permitted to kill in case of self-defense, unless it is a policeman. Let $A = \{\top\}$. We have $\neg k \in out_p(G, P, \geq, \top)$, so k is forbidden in context $\top$.

Example 3.24 G, P and $>$ are as before, but $A = \{s\}$. We have

$maxfamily(G, P, s) = \{\{(\top, \neg k), (s \wedge p, \neg k)\}, \{(s \wedge p, \neg k), (s, k)\}\}$
$preffamily(G, P, \geq, s) = \{(s \wedge p, \neg k), (s, k)\}$
$outfamily(G, P, \geq, s) = \{Cn(\emptyset)\}$
$out_p(G, P, \geq, s) = Cn(\emptyset)$
$statpermfamily(G, P, \geq, s) = \{Cn(k)\}$

$\neg k \notin out_p(G, P, \geq, s)$, so k is no longer forbidden in context s. Instead $k \in statpermfamily(G, P, \geq, s)$, viz k is permitted in context s.

Example 3.25 G, P and $>$ are as before, but $A = \{s, p\}$. We have

$maxfamily(G, P, s \wedge p) = \{\{(\top, \neg k), (s \wedge p, \neg k)\}, \{(s, k)\}\}$
$preffamily(G, P, \geq, s \wedge p) = \{\{(\top, \neg k), (s \wedge p, \neg k)\}\}$
$outfamily(G, P, \geq, s \wedge p) = \{Cn(\neg k)\}$
$out_p(G, P, \geq, s \wedge p) = Cn(\neg k)$
$statpermfamily(G, P, \geq, s \wedge p) = \{Cn(\neg k)\}$

$\neg k \in out_p(G, P, \geq, s \wedge p)$, so in context $s \wedge p$ the prohibition of k is "in" again, and correlatively the permission of k is "out".

The next step would be to extend the account to cover the dynamic version of permission.

4 Proof theory

In this section we turn to the proof theory for I/O logic. We start with the unconstrained I/O logic for obligation whose proof theory is well understood. We, then, discuss a more syntactical way to characterize the constrained I/O logic for obligation, and also permission (without constraints).

4.1 Obligation

4.1.1 Basic

Unconstrained I/O logic for obligation is axiomatized as a kind of conditional logic. Here we use again the equivalence: $x \in out(G, A)$ iff $(A, x) \in out(G)$.

The specific rules of interest are shown below.

(SI) $$\frac{(a,x) \quad b \vdash a}{(b,x)}$$

(WO) $$\frac{(a,x) \quad x \vdash y}{(a,y)}$$

(AND) $$\frac{(a,x) \quad (a,y)}{(a, x \wedge y)}$$

(OR) $$\frac{(a,x) \quad (b,x)}{(a \vee b, x)}$$

(CT) $$\frac{(a,x) \quad (a \wedge x, y)}{(a,y)}$$

In general, for any set of rules, we say that a pair (a, x) of formulae is derivable using those rules from a set G of such pairs iff (a, x) is in the least set that includes G, contains the pair $(\top, \top)$, and is closed under the rules. In the systems studied here, it will make no difference which tautology $\top$ is chosen. Our notations are $(a, x) \in deriv(G)$ or equivalently $x \in deriv(G, a)$, with a subscript to indicate the set of rules employed.

When A is a set of formulae, derivability of (A, x) from G is defined as derivability of (a, x) from G for some conjunction $a = a_1 \wedge ... \wedge a_n$ of elements of A. We understand the conjunction of zero formulae to be a tautology, so that $(\emptyset, a)$ is derivable from G iff $(\top, a)$ is for some tautology $\top$.

In the particular case of simple-minded output, we use the three core rules (SI), (WO) and (AND). The system is called $deriv_1$. In the case of basic output, we add (OR) to the latter triplet, obtaining the system $deriv_2$. In the case of reusable simple-minded output, we add (CT) instead of (OR). This is $deriv_3$. In the case of reusable basic output, we add both (OR) and (CT) to the latter triplet, obtaining $deriv_4$. This is summarized in Table 1.

Output operation	Rules
Simple-minded (out_1)	{SI, WO, AND}
Basic (out_2)	{SI, WO, AND}+{OR}
Reusable simple-minded (out_3)	{SI, WO, AND}+{CT}
Reusable basic (out_4)	{SI, WO, AND}+{OR,CT}

Table 1: IOL systems

Taken together, (SI) and (CT) give plain transitivity ("From (a, x) and (x, y), infer (a, y)"). Therefore, the last two output operations can be described as obtained from the first two, by just allowing obligations be chained together.

The following applies:

Theorem 4.1 (Completeness)

$$out_1(G, A) = deriv_1(G, A)$$
$$out_2(G, A) = deriv_2(G, A)$$
$$out_3(G, A) = deriv_3(G, A)$$
$$out_4(G, A) = deriv_4(G, A)$$

Proof. We give the proof for out_1 only. The argument for the others can be found in [Makinson and van der Torre, 2000].

For the right in left inclusion (soundness), assume $x \in deriv_1(G, A)$. By definition $x \in deriv_1(G, a)$ for some conjunction $a = a_1 \wedge ... \wedge a_n$ of elements of A. We need to show $x \in out_1(G, a)$. (This is equivalent to $x \in out_1(G, \{a_1, ..., a_n\})$, from which $x \in out_1(G, A)$ follows by monotony in A.) The proof is by induction on the length n of the derivation. If $n = 1$, then either $(a, x) \in G$ or (a, x) is the pair $(\top, \top)$. Suppose $(a, x) \in G$. Since $a \in Cn(a)$, $x \in G(Cn(a))$, and thus $x \in out_1(G, a)$. Suppose (a, x) is the pair $(\top, \top)$. By definition, $\top$ is a logical consequence of every set of sentences, including $G(Cn(\top))$. In other words $\top \in Cn(G(Cn(\top))) = out_1(G, \top)$. In both cases, we are done.

Let the derivation of (a, x) be of length $n + 1$. Suppose the result holds for any derivation of length $k \leq n$. We break the argument into cases depending on the inference rule used to get (a, x).

Suppose it is (SI). Then there is a derivation of (a', x) of length $k \leq n$, and $a \vdash a'$. By the induction hypothesis, $x \in Cn(G(Cn(a')))$. By monotony and idempotence for Cn, $\{a'\} \subseteq Cn(a)$ yields $Cn(a') \subseteq CnCn(a) = Cn(a)$. By monotony for G, $G(Cn(a')) \subseteq G(Cn(a))$. By monotony for Cn again, $x \in Cn(G(Cn(a'))) \subseteq Cn(G(Cn(a)))$, so $x \in out_1(G, a)$ as required.

Suppose (a, x) is obtained using (WO). Then there is a derivation of (a, y) of length $k \leq n$, and $y \vdash x$. By the induction hypothesis, $y \in Cn(G(Cn(a)))$. Since $y \vdash x$, $x \in Cn(G(Cn(a))) = out_1(G, a)$ as required.

Suppose (a, x) is obtained using (AND). Put $x := y \wedge z$. Then there is a derivation of (a, y) of length $k \leq n$, and a derivation of (a, z) of length $k' \leq n$. By the induction hypothesis, $y \in Cn(G(Cn(a)))$ and $z \in Cn(G(Cn(a)))$, so $y \wedge z \in Cn(G(Cn(a))) = out_1(G, a)$ as required.

This completes the verification of the right in left inclusion.

For the left in right inclusion (completeness), let $x \in Cn(G(Cn(A)))$. We break the argument into cases depending on whether some elements of G are "triggered" or not.

Suppose $G(Cn(A)) = \emptyset$. In that case, $x \in Cn(\emptyset)$, and thus x is $\top$. But $(\top, \top)$ is derivable from G, and $a \vdash \top$ for any conjunction $a = a_1 \wedge ... \wedge a_n$ of elements of A. By (SI), $(a, \top)$ is derivable from G, and thus so is $(A, \top)$ as required.

Suppose $G(Cn(A)) \neq \emptyset$. By compactness of Cn, there are $x_1, ..., x_n$ ($n > 0$) in $G(Cn(A))$ such that $x_1 \wedge ... \wedge x_n \vdash x$. So G contains the pairs (a_1, x_1), ..., and (a_n, x_n) with $A \vdash a_1$, ..., and $A \vdash a_n$. By compactness and monotony for $\vdash$, $a \vdash a_1$, ..., and $a \vdash a_n$ for some conjunction a of elements of A. Since $G \subseteq deriv_1(G)$, each (a_i, x_i) is derivable from G. Based on this it is easy to obtain a derivation of (a, x) from G:

$$\cfrac{\cfrac{\cfrac{(a_1, x_1)}{(a, x_1)}\text{SI} \quad ...... \quad \cfrac{(a_n, x_n)}{(a, x_n)}\text{SI}}{(a, x_1 \wedge ... \wedge x_n)}\text{AND}}{(a, x)}\text{WO}$$

Since a is a conjunction of elements of A, (A, x) is derivable from G.

This completes the proof. ∎

The *throughput* counterparts out^+ of the output operations have a proof theory too. It is obtained by allowing arbitrary pairs of the form (a, a) to appear as leaves of a derivation; this is called the zero-premise identity rule:

(ID)   From no premise infer (a, a)

4.1.2 Universal orders of derivation

We say that a derivation respects an order R_1, ..., R_n of those rules if a rule R_j is never applied before a rule R_i ($i < j$). We say that an order is universal iff whenever $(a, x) \in out(G)$ then there is a derivation of (a, x) from G respecting that order. Theorem 4.2 below tells us how many universal orders there are for each output operation. The parenthesis indicate that every arrangement within them is counted.

Theorem 4.2 *The following applies:*

a) For simple-minded output, there are (at least) three universal orders of derivation: SI, AND, WO, and (SI,WO), AND.

b) For basic output, there are at least six universal orders: SI, AND, WO, OR, and (SI,WO), (AND,OR) and WO, OR, SI, AND.

c) For simple-minded reusable output, there are (at least) eight universal orders: SI, (WO,CT,AND) and WO,SI,(CT,AND).

d) For reusable output, there are (at least) eleven orders: SI, (WO,CT,AND),OR and SI, (WO,CT),OR,AND and WO,SI,(CT,AND),OR and WO,SI,CT,OR,AND.

Proof. See [Makinson and van der Torre, 2000, §7 and §8]. ∎

4.1.3 Bochman's systems

Bochman provides four completeness results with respect to the possible-worlds semantics described in Section 2.4. Completeness is established using a canonical model argument.

The counterpart of an I/O derivation rule R in terms of $\Rightarrow$ is denoted by $R_\Rightarrow$. For instance, $\top_\Rightarrow$ is the axiom $\top \Rightarrow \top$.

Definition 4.3 $\Rightarrow$ *is called a production inference relation if it satisfies* $SI_\Rightarrow$, $WO_\Rightarrow$, $AND_\Rightarrow$, $\top_\Rightarrow$, *and*

$$(\bot_\Rightarrow) \quad \bot \Rightarrow \bot$$

Production inference relations almost coincide with $deriv_1$, except for the presence of the last postulate, $(\bot_\Rightarrow)$. The latter makes the production relations inconsistency-preserving.

Definition 4.4 *A production inference relation $\Rightarrow$ is called regular if it satisfies* $CT_\Rightarrow$.

The following applies.

Theorem 4.5 $\Rightarrow$ *is a production inference relation if and only if it is determined by a classical binary semantics.*

Proof. This is [Bochman, 2005, coroll. 8.5, p. 235]. Hint on proof: given some production relation $\Rightarrow$, define its canonical semantics as the set of all classical binary models of the form $\langle C(w), w \rangle$, where w is some consistent and deductively closed theory, and $C(w) = \{x \mid \wedge a \Rightarrow x \text{ for some finite } a \subseteq w\}$. ∎

Theorem 4.6 $\Rightarrow$ *is a regular production inference relation if and only if it is generated by a consistent classical binary semantics.*

Proof. This is [Bochman, 2005, th. 8.9, p. 239]. Hint on proof: given some regular production relation $\Rightarrow$, use the same construction as in the proof for Theorem 4.5, but restrict it to the set of models for which $C(w) \subsetneq w$. ∎

Definition 4.7 *A production inference relation* $\Rightarrow$ *is called basic if it satisfies* $\text{OR}_\Rightarrow$.

Theorem 4.8 $\Rightarrow$ *is a basic production inference relation if and only if it has a relational possible worlds model* $\mathbb{W} = (W, R, V)$, *where the evaluation rule for* $\Rightarrow$ *is as in Definition 2.13.*

Proof. This is [Bochman, 2005, coroll. 8.43, p. 262]. Hint on proof: Define R in the canonical model by putting $R\alpha\beta$ whenever $C(\alpha) \subseteq \beta$. ∎

Definition 4.9 *A production inference relation* $\Rightarrow$ *is called causal if it satisfies* $\text{OR}_\Rightarrow$ *and* $\text{CM}_\Rightarrow$.

Theorem 4.10 $\Rightarrow$ *is a causal production inference relation if and only if it has a quasi-reflexive relational possible worlds model* $\mathbb{W} = (W, R, V)$.

Proof. This is [Bochman, 2005, th. 8.53, p. 269]. Hint on proof: Define R in the canonical model by putting $R\alpha\beta$ whenever $C(\alpha) \subseteq \alpha \cap \beta$. ∎

4.1.4 I/O logic without WO

We mentioned the attempt made by Stolpe to define two output operations out_1^s and out_3^s that do not validate the rule WO. Their axiomatic counterparts may be written as $deriv_1^s$ and $deriv_3^s$, respectively.

Definition 4.11 $(a,x) \in deriv_1^s(G)$ iff there is derivation of (a,x) from G using the rules of inference (SI), (AND), and

$$(Eq) \quad \frac{(a,x') \quad x' \equiv x}{(a,x)}$$

The following applies.

Theorem 4.12 $deriv_1^s(G) = out_1^s(G)$

Proof. The proof is a straightforward modification of the completeness proof for out_1 w.r.t. $deriv_1$. ∎

Definition 4.13 $(a,x) \in deriv_3^s(G)$ iff there is a derivation of (a,x) from G using the rules (SI), (AND), (Eq) and the rule of "mediated cumulative transitivity" (MCT), i.e.,

$$(MCT) \quad \frac{(a,x') \quad (a \wedge x, y) \quad x' \vdash x}{(a,y)}$$

Theorem 4.14 $deriv_3^s(G) = out_3^s(G)$

Proof. The proof proceeds via a series of lemmas, for which we refer the reader to [Stolpe, 2008a; Stolpe, 2008b]. ∎

We have chosen to stick with the syntactic characterization given by Stolpe. The distinctive axiom of $deriv_3^s$ is the rule (MCT). It shows that the question is not so much how to drop out WO but how to restrict it. Indeed MCT restricts weakening of the output to the chaining of obligations in the sense that it itself makes logical entailment sufficient for chaining.

As a matter of facts, $deriv_3^s$ may equally be axiomatized by (SI), (AND), (Eq) and (CT). Given the other axioms of the systems, (CT) and (MCT) are equivalent. On the one hand, given reflexivity for $\vdash$, MCT entails CT. For assume (a,x) and $(a \wedge x, y)$. Since $x \vdash x$, a direct application of MCT yields (a,y). On the other hand, given SI, CT entails MCT:

$$\text{CT} \; \frac{(a,x') \quad \dfrac{(a \wedge x, y) \quad \dfrac{x' \vdash x}{a \wedge x' \vdash a \wedge x} \, \text{SI}}{(a \wedge x', y)}}{(a,y)}$$

4.1.5 Constrained or safe derivations

A more syntactical way to characterize full join constrained output is given in [Makinson and van der Torre, 2001, §6].

We say that a derivation Δ is constrained or safe iff the body of the root is consistent with its own derivability set. To be precise, let Δ be a derivation of (a,x) from G, given a set R of rules. Let $L \subseteq \Delta$ be the set of the leaves of Δ. We say that Δ is constrained or safe with respect to rule-set R iff $(a, \neg a) \notin deriv(L)$ where $deriv(R)$ is derivability using only rules in R. Intuitively, the set L of leaves corresponds to the part of G the derivation Δ makes use of. What the definition says is that this part of G will not generate any inconsistency no matter the input.

We say that (a,x) is derivable with constraint from G given rule-set R iff there is some derivation of (a,x) from G given R that is constrained with respect to R. Theorem 4.15 shows this is equivalent with full join constrained output, where the constraint is the same as the input.

Theorem 4.15 *Let out be any one of the operations out_i or out_i^+ ($i = 1,...,4$), and let R be the corresponding set of derivation rules. Then $x \in \cup outfamily(G, a, a)$ iff there is a derivation of (a,x) from G (given rules from R) that is constrained (with respect to R).*

Proof. This is [Makinson and van der Torre, 2001, obs. 9]. ■

Roughly speaking, this means that (a,x) can be derived using only the part of G that is "safe" for the derivation rules in question, in the sense of not causing any trouble.

4.2 Syntactic characterization of permission

A syntactic characterization of permission is given in [Makinson and van der Torre, 2003a]. It is based on the notions of inverse and subverse of a Horn rule.

Consider any Horn rule for output of the following form:

$$\text{HR} \ \frac{(a_i, x_i) \in out(G)\ (0 \leq i \leq n) \qquad y_i \in Cn(b_j)\ (0 \leq j \leq m)}{(c, z) \in out(G)}$$

We call $(a_i, x_i) \in out(G)$ the substantive premises, and $y_i \in Cn(b_j)$ the auxiliary ones. Each of the rules used to characterize $out_i (i \in \{1,2,3,4\})$, and their extensions by throughput, is of this form.

The inverse of (HR) is used for negative permission and dynamic permission. The negative permission version of the rule has the form

$$\text{HR}^{-1} \frac{(a_i, x_i) \in \mathit{out}(G) \ (i < n) \quad (c, \neg z) \in \mathit{negperm}(G) \quad y_i \in Cn(b_j) \ (j \leq m)}{(a_n, \neg x_n) \in \mathit{negperm}(G)}$$

The conclusion of (HR) is swapped with one of the substantive premises, negating both their heads in the process and rewriting their *out* as *negperm*. The dynamic positive permission version of the rule is obtained in the straightforward way, by replacing *negperm* with *dynperm*.

The subverse of (HR) is used for static permission. The rule has the form

$$\text{HR}^{\downarrow} \frac{(a_i, x_i) \in \mathit{out}(G) \ (i < n) \quad (a_n, x_n) \in \mathit{statperm}(G, P) \quad y_i \in Cn(b_j) \ (j \leq m)}{(c, z) \in \mathit{statperm}(G, P)}$$

The subverse rule is obtained by downgrading to permission status one of the substantive premises, and also the conclusion of the rule. In the limiting case there are no substantive premises, the conclusion alone is downgraded.

Table 2 shows the inverse/subverse of the rules introduced in Subsection 4.1.1. There (a, x) is labelled with o or p depending on whether (a, x) is an obligation or a permission:

Horn rule	**Inverse**	**Subverse**
(SI)	$\dfrac{(a \wedge b, x)^p}{(a, x)^p}$	$\dfrac{(a, x)^p}{(a \wedge b, x)^p}$
(WO)	$\dfrac{(a, x)^p \ \ x \vdash y}{(a, y)^p}$	$\dfrac{(a, x)^p \ \ x \vdash y}{(a, y)^p}$
(AND)	$\dfrac{(a, x)^o \ \ (a, \neg(x \wedge y))^p}{(a, \neg y)^p}$	$\dfrac{(a, x)^o \ \ (a, y)^p}{(a, x \wedge y)^p}$
(OR)	$\dfrac{(a, x)^o \ \ (a \vee b, \neg x)^p}{(b, \neg x)^p}$	$\dfrac{(a, x)^o \ \ (b, x)^p}{(a \vee b, x)^p}$
(CT)	$\dfrac{(a, x)^o \ \ (a, \neg y)^p}{(a \wedge x, \neg y)^p}$ $\dfrac{(a \wedge x, y)^o \ \ (a, \neg y)^p}{(a, \neg x)^p}$	$\dfrac{(a, x)^o \ \ (a \wedge x, y)^p}{(a, y)^p}$ $\dfrac{(a, x)^p \ \ (a \wedge x, y)^o}{(a, y)^p}$

Table 2: Inverse/subverse of a Horn rule

The following applies.

Theorem 4.16 *Let out be any output operation. If out satisfies a rule of the form (HR), then the corresponding negperm operation satisfies the inverse(s) (HR^{-1}).*

Proof. This is [Makinson and van der Torre, 2003a, obs. 1]. ∎

Theorem 4.17 *Let out be any output operation. If out satisfies a rule of the form (HR), then the corresponding statperm operation satisfies the subverse(s) (HR$^{\downarrow}$).*

Proof. This is [Makinson and van der Torre, 2003a, obs. 2]. ∎

Theorem 4.18 *Let out be any output operation. If out satisfies a rule of the form (HR), then the corresponding dynperm operation satisfies the inverse(s) (HR^{-1}).*

Proof. This is [Makinson and van der Torre, 2003a, obs. 8]. ∎

5 Conclusion and issues for future research

In this chapter we have defined a framework where detachment is the central mechanism of the semantics, and a proof theory for the system. On this basis, we have introduced and compared several I/O logics, depending on the inference patterns they validate. We have also shown how to refine the account in order to give a more fine-grained analysis of a number of concepts playing a fundamental role in normative reasoning: permission, CTDs and conflicts. Below we discuss a number of topics for future research.

5.1 Go top-down

The input/output logic framework has been introduced bottom-up, starting from representative samples of I/O logics. It is natural to ask if the notion of input/output can be treated as a first-class citizen of the framework. This could yield new insights on the question of what counts as an input/output logic.

5.2 Equivalence and redundancy

One might wish to shift the emphasis from detachment to redundancy. One can use redundancy to simplify normative systems, which tend to grow quickly and be difficult to understand. The simplest approach is this. Treat two normative systems N and N' as equivalent if for all propositional formulae a, we have that $out(N, a) = out(N', a)$. Call a norm $(a, x) \in N$ redundant whenever N is equivalent to $N - \{(a, x)\}$. The implications of this shift of emphasis for deontic logic are discussed in [van der Torre, 2010].

5.3 Changing the base logic

Another natural next step is to investigate what happens if another logic (viz. other than classical propositional logic) is used as the base logic in the I/O framework. Some first findings are reported in [Parent et al., 201x]. Completeness results are given for (unconstrained) I/O logics based on intuitionistic logic. On the semantical side, the key idea is to work with so-called saturated sets instead of maximal consistent sets.

5.4 Lions

The notion of intermediary (see Chapters 6 and 9 in this Handbook) calls for the use of structured assemblies of I/O operations. Such structures, called "logical input/output nets" (or *lions* for short), are graphs, with the nodes labelled by pairs (G, out), where G is a normative code and out is an I/O operation (or recursively, by other lions). A relation may be used to indicate which nodes have access to others, providing passage for the transmission of local outputs as local inputs. The graph may be further equipped with an entry point and an exit point, for global input and output. The formal study of such lions remains to be done.

5.5 Proof theory for constrained I/O logic and permission

Stricto sensu the syntactical characterizations of constrained I/O logic and permission described in Section 4 are not a proof theory. However, there are proof systems for Reiter's default logic. One such is the sequent calculus for default logic due to [Bonatti and Olivetti, 1997]. A conspicuous feature of the framework is the use of (as they call it) "antisequent" $\Gamma \not\vdash a$, where Γ is a set of wffs. It remains to be seen whether or not a similar calculus can be built for constrained I/O logic.

5.6 Moving towards implementation

Another fruitful avenue for future research would be to study the complexity and computational tractability of I/O logic. For practical purposes fully automated I/O systems would be very desirable.

Acknowledgements

We wish to thank David Makinson, Jan Odelstad and Loes Olde Loohuis for valuable comments. Two earlier versions were presented at the 9th and 10th de Morgan Workshops on Deontic Logic held on 1 - 4 Dec 2009 at Luxembourg, and on 5 & 6 Jul 2010 at Firenze in Italy, respectively.

BIBLIOGRAPHY

[Alchourrón et al., 1985] C. E. Alchourrón, P. Gärdenfors, and D. Makinson. On the logic of theory change: Partial meet contraction and revision functions. *Journal of Symbolic Logic*, 50(2):510–530, 1985.

[Bochman, 2005] A. Bochman. *Explanatory Nonmonotonic Reasoning*. World Scientific Publishing Company, New York, London, 2005.

[Boella and van der Torre, 2008] G. Boella and L. van der Torre. Institutions with a hierarchy of authorities in distributed dynamic environments. *Artificial Intelligence and Law*, 16(1):53–71, 2008.

[Boghossian, 2000] P. Boghossian. Knowledge of logic. In P. Boghossian and C. Peacocke, editors, *New Essays on the A Priori*, pages 229–254. Clarendon Press, Oxford, 2000.

[Bonatti and Olivetti, 1997] P. A. Bonatti and N. Olivetti. A sequent calculus for skeptical default logic. In *Proc. of the Int. Conf. on Automated Reasoning with Analytic Tableaux and Related Methods*, pages 107–121. Springer-Verlag, 1997.

[Brass, 1991] S. Brass. Deduction with supernormal defaults. In G. Brewka, K. Jantke, and P. Schmitt, editors, *Nonmonotonic and Inductive Logics*, volume 659 of *Lecture Notes in Computer Science*, pages 153–174. Springer, Berlin, 1991.

[Carmo and Jones, 2002] J. Carmo and A. Jones. Deontic logic and contrary-to-duties. In D. Gabbay and F. Guenthner, editors, *Handbook of Philosophical Logic*, volume 8, pages 265–343. Kluwer Academic Publishers, Dordrecht, Holland, 2nd edition, 2002.

[Chisholm, 1963] R.M. Chisholm. Contrary-to-duty imperatives and deontic logic. *Analysis*, 24:33–66, 1963.

[Cholvy and Garion, 2001] L. Cholvy and C. Garion. An attempt to adapt a logic of conditional preferences for reasoning with contrary-to-duties. *Fundamenta Informaticae*, 48(2-3):183–204, 2001.

[Goble, 1991] L. Goble. Murder most gentle: the paradox deepens. *Philosophical Studies*, 64:217–227, 1991.

[Governatori and Sartor, 2010] G. Governatori and G. Sartor, editors. *Deontic Logic in Computer Science, 10th International Conference, DEON 2010*, volume 6181 of *Lecture Notes in Computer Science*. Springer, 2010.

[Hansen, 2006] J. Hansen. Deontic logics for prioritized imperatives. *Artificial Intelligence and Law*, 14(1-2):1–34, 2006.

[Hansen, 2008] J. Hansen. Prioritized conditional imperatives: problems and a new proposal. *Journal of Autonomous Agents and Multi-Agent Systems*, 17(1):11–35, 2008.

[Hansson, 1969] B. Hansson. An analysis of some deontic logics. *Noûs*, 3:373–398, 1969. Reprinted in [Hilpinen, 1971, pp. 121-47].

[Hilpinen, 1971] R. Hilpinen, editor. *Deontic Logic: Introductory and Systematic Readings*. Reidel, Dordrecht, 1971.

[Horty, 2007] J. F. Horty. Defaults with priorities. *Journal of Philosophical Logic*, 36:367–413, 2007.

[Horty, 2012] J. F. Horty. *Reasons as Defaults*. Oxford University Press, USA, 2012.

[Jackson, 1985] F. Jackson. On the semantics and logic of obligation. *Mind*, 94(1):177–195, 1985.

[Jones and Parent, 2008] A. J. I. Jones and X. Parent. Normative-informational positions: a modal-logical approach. *Artificial Intelligence and Law*, 16(1):7–23, 2008.

[Jørgensen, 1937] J. Jørgensen. Imperatives and logic. *Erkenntnis*, 7(1):288–296, 1937.

[Lewis, 1973] D.K. Lewis. *Counterfactuals*. Blackwell, Oxford, 1973.

[Makinson and van der Torre, 2000] D. Makinson and L. van der Torre. Input/output logics. *Journal of Philosophical Logic*, 29(4):383–408, 2000.

[Makinson and van der Torre, 2001] D. Makinson and L. van der Torre. Constraints for input/output logics. *Journal of Philosophical Logic*, 30(2):155–185, 2001.

[Makinson and van der Torre, 2003a] D. Makinson and L. van der Torre. Permission from an input/output perspective. *Journal of Philosophical Logic*, 32(3):391–416, 2003.

[Makinson and van der Torre, 2003b] D. Makinson and L. van der Torre. What is input/output logic? In B. Löwe, W. Malzkorn, and T. Räsch, editors, *Foundations of the Formal Sciences II. Applications of Mathematical Logic in Philosophy and Linguistics*, Frontiers in Artificial Intelligence and Applications, pages 163–174. Springer-Verlag, Berlin, 2003.

[Makinson, 1999] D. Makinson. On a fundamental problem in deontic logic. In P. McNamara and H. Prakken, editors, *Norms, Logics and Information Systems*, pages 29–54. IOS Press, Amsterdam, 1999.

[Makinson, 2009] D. Makinson. Propositional relevance through letter-sharing. *Journal of Applied Logic*, 7(4):377–387, 2009.

[Nute, 1997] D. Nute. Apparent obligation. In D. Nute, editor, *Defeasible Deontic Logic*, volume 263 of *Synthese Library*, pages 287–316. Kluwer Academic Publishers, Dordrecht, Holland, 1997.

[Parent et al., 201x] X. Parent, D. Gabbay, and L. van der Torre. Intuitionistic basis for input/output logic. In S. O. Hansson (ed.), *Classical Methods for Nonclassical Problems: Essays in Honor of David Makinson*, Springer, 201x.

[Parent, 2011] X. Parent. Moral particularism in the light of deontic logic. *Artificial Intelligence and Law*, 19(2-3):75–98, 2011.

[Poole, 1988] D. Poole. A logical framework for default reasoning. *Artificial Intelligence*, 36(1):27–47, 1988.

[Reiter, 1980] R. Reiter. A logic for default reasoning. *Artificial Intelligence*, 13(1-2):81–132, 1980.

[Sakama and Inoue, 1996] C. Sakama and K. Inoue. Representing priorities in logic programs. In *Proc. 1996 Joint Int. Conf. and Symp. on Logic Programming*, pages 82–96, 1996.

[Simon, 1957] H. Simon. *Models of Man, Social and Rational*. John Wiley and Sons, New York, 1957.

[Stolpe, 2008a] A. Stolpe. Normative consequence: The problem of keeping it whilst giving it up. In R. van der Meyden and L. van der Torre, editors, *Deontic Logic in Computer Science, 9th International Conference, DEON 2008*, volume 5076 of *Lecture Notes in Computer Science*, pages 174–188. Springer, 2008.

[Stolpe, 2008b] A. Stolpe. *Norms and Norm-System Dynamics*. PhD thesis, Department of Philosophy, University of Bergen, Norway, 2008.

[Stolpe, 2010a] A. Stolpe. Relevance, derogation and permission. In Governatori and Sartor [2010], pages 98–115.

[Stolpe, 2010b] A. Stolpe. A theory of permission based on the notion of derogation. *Journal of Applied Logic*, 8(1):97–113, 2010.

[Suzumura, 1983] K. Suzumura. *Rational Choice, Collective Decisions, and Social Welfare*. Cambridge University Press, Cambridge, 1983.

[van der Torre, 2010] L. van der Torre. Deontic redundancy: a fundamental challenge for deontic logic. In Governatori and Sartor [2010], pages 11–31.

Xavier Parent
University of Luxembourg
6, rue Richard Coudenhove-Kalergi
L-1359 Luxembourg
Luxembourg
Email: xavier.parent@uni.lu

Leendert van der Torre
University of Luxembourg
6, rue Richard Coudenhove-Kalergi
L-1359 Luxembourg
Luxembourg
Email: leon.vandertorre@uni.lu

9
The Theory of Joining-Systems

LARS LINDAHL AND JAN ODELSTAD

ABSTRACT. The theory of joining-systems (TJS), as developed in this chapter, consists of three main parts, developed after the informal introduction and overview in Sections 1 and 2. One part (Section 3) is the abstract theory of joining-systems, providing the framework for the subsequent analysis. Two other parts introduce those concepts and results of the theory that are in focus for the representation of normative systems. The first of these parts (Section 4) presents the model of condition implication structures (cis's) as applied to well-known issues in legal theory. In the second part (Section 5), the cis model of TJS is applied to a comprehensive new field, namely the theory of "intervenients". In a developed normative system, intervenient concepts serve as vehicles of inference for going from ultimate descriptive grounds to ultimate deontic consequences. Among the issues dealt with are: Boolean compounds of intervenients, intervenients as organic wholes, narrowing or widening of intervenients, the typology of various kinds of intervenient minimality.

1   The field of research and its origins 546
     1.1   Cases and solutions in the theory of Alchourrón and Bulygin 546
     1.2   Input-output logic 548
     1.3   The theory of joining-systems TJS 549
     1.4   TJS for simple normative systems 549
     1.5   Normative positions in TJS 550
     1.6   Subtraction and addition of norms in TJS 551
     1.7   Intermediaries and intervenients 552
     1.8   Advice to readers 562
2   First introduction to TJS 562
     2.1   General TJS irrespective of intervenients 562
     2.2   Intervenients in TJS 564
3   Formal development of TJS 566
     3.1   Basic concepts 566
     3.2   Joining-systems 568
     3.3   Weakest grounds, strongest consequences and minimal joinings 575

	3.4	Connectivity 577
	3.5	Lowerness 582
	3.6	The structure on minimal joinings 586
	3.7	Networks of joining-systems 590
	3.8	Intervenients 591
4	TJS for Boolean joining-systems 593	
	4.1	Boolean quasi-orderings and Boolean joining-systems . 593
	4.2	The condition implication model (cis) 596
	4.3	Subtraction and addition of norms: an example 598
	4.4	The cis version of normative positions 604
5	Intervenients for Boolean joining-systems 612	
	5.1	Introductory remarks on intervenients in Bjs' 612
	5.2	cis' with intervenients 615
6	Related work 625	
	6.1	Previous work of ours 625
	6.2	Recent work of others 627

1 The field of research and its origins

In the analysis of normative systems, one of the approaches is to represent a normative system as a *deductive mechanism*, giving a normative output for an input of facts. In modern literature, the foremost origin of this approach is the work *Normative Systems* by the Argentinians Carlos E. Alchourrón and Eugenio Bulygin. To this tradition belongs as well the recent "input-output logic" by David Makinson and Leon van der Torre and the Theory of Joining-Systems (TJS) proposed by the present authors.

A theory of representation for normative systems will be incomplete unless attention is paid to the role of *intermediate concepts* within the system (for example, the role of legal concepts such as ownership). If a normative system is represented as a deductive mechanism, there will be an emphasis on the role of intermediate concepts as "vehicles of inference" within the system. In this respect, the origin of later developments comes from Scandinavian legal philosophy in the 1950's, in particular the work of Anders Wedberg and Alf Ross.

1.1 Cases and solutions in the theory of Alchourrón and Bulygin

Alchourrón and Bulygin introduce the idea of deductive mechanism by contrasting the Aristotelian conception of science with the idea of deductive system in modern theory [Alchourrón and Bulygin, 1971, pp. 43ff.]. The notion of deductive system is based on Tarski's notion of deductive consequence, satisfying the following four requirements [Alchourrón and Bulygin, 1971, pp. 48ff.]:

1. The set of the consequences of a set of sentences consists solely of sentences.

2. Every sentence belonging to a given set is to be regarded as a consequence of this set.

3. The consequences of the consequences are, in turn, consequences.

4. If a sentence of a conditional form $(y \supset z)$ is a consequence of the set of sentences X, then z is a consequence of the set of sentences resulting from adding to X the sentence y.

Adopting the Tarskian conception of deductive system, Alchourrón and Bulygin conceive of a normative system as a set of sentences deductively correlating pairs of sentences. A set α of sentences deductively correlates a pair $\langle p, q \rangle$ of sentences if q is a deductive consequence of $\{p\} \cup \alpha$, or, using the relation Cn of consequence, if $q \in Cn(\{p\} \cup \alpha)$. Moreover, the statement $q \in Cn(\{p\} \cup \alpha)$ is equivalent to $(p \supset q) \in Cn(\alpha)$ where $\supset$ is the symbol for truth-functional implication [Alchourrón and Bulygin, 1971, pp. 54ff.]

For a set α to be a normative system the additional requirement is made that there be at least one pair $\langle p, q \rangle$ where $q \in Cn(\{p\} \cup \alpha)$ such that p is a "case" and q is a "solution". A solution is a normative sentence expressed in terms of a descriptive sentence (deontic content) preceded by a deontic operator for command, prohibition or permission. So, the character of the system as normative depends on the deontic character of the solutions inferred in the system. In the words of [Alchourrón and Bulygin, 1971, p. 169]: "Justifying the deontic qualification of an action by means of a normative system consists in showing that the obligation, the prohibition or the permission of this action can be inferred from (i.e., is a consequence of) this system."

If propositional logic is used as a basis, it is usually presupposed that p, q are closed sentences with no free variables, i.e., for example, p is the sentence "Smith has promised to pay Jones \$100" and q is "Smith has an obligation to pay \$100 to Jones". In these sentences, individuals are referred to by individual constants (names). While it is true that a normative system may correlate sentences of this kind, a set of sentences containing individual names is not, however, an appropriate representation of a normative system. A normative system expresses general rules where no individual names occur. If the task is to represent a normative system this feature of generality has to be taken into account.

When Alchourrón and Bulygin speak of normative "solutions" being correlated to "cases", however, they have in mind correlation of "generic" cases to "generic" solutions. They emphasize the distinction between individual and generic cases, and an analogous distinction holds for solutions. An individual case is a situation or a state of affairs. As such, appropriately, it should be described by a closed sentence. On the other hand, a generic case is a property or a set of individual cases, defined by a property.[1] Therefore, a "case" in the generic sense relevant to Alchourrón and Bulygin is an object described by an open sentence. It can be argued that, when the expression $q \in Cn(\{p\} \cup \alpha)$ is said to express that α correlates q to p, q and p must be thought of as "open" sentences (like "x has promised to pay \$$y$ to z", "It shall be that x pays \$$y$ to z"), not prefixed by any universal quantifier.[2]

1.2 Input-output logic

In a series of papers, Makinson and van der Torre have developed a logic called "input-output logic", see for example [Makinson and van der Torre, 2000; Makinson and van der Torre, 2003]. If G is a generating set, then $x \in out(G, A)$, i.e., x belongs to the output of A under G, if and only if $(A, x) \in out(G)$. The principal out-operation in input-output logic does not require reflexivity or contraposition.

Input-output logic can, but need not, apply specifically to normative systems, where norms are represented as ordered pairs.[3] The construction of norms in input-output logic, however, is different from the construction in [Alchourrón and Bulygin, 1971]. In Alchourrón and Bulygin, if a is a case and x is a solution, it is assumed that x is a normative sentence (a solution, see above). In contrast, in input-output logic, a generating set G of ordered pairs $\langle a, x \rangle$ can be understood as a set of conditional obligations in spite of the fact that x, the consequence, is descriptive rather than normative. The normative character, in this case, depends on the specific character of the set G as a set of conditional obligations. (Similarly if $\langle a, x \rangle$ is a conditional permission.)

For further details, the reader is referred to the Chapter "Input/output logic" of the present Handbook. A remark on the interrelation between

---

[1] By an individual case is meant an element of the universe of discourse. See [Alchourrón and Bulygin, 1971, p. 28, and p. 10]. A generic case is described alternatively as a subset of the universe of discourse, defined by a property, or as this defining property itself. See [Alchourrón and Bulygin, 1971, p. 29].

[2] Cf. [Alchourrón and Bulygin, 1971, p. 49], and the comments in [Lindahl and Odelstad, 2004, sect. 1.1].

[3] [Makinson and van der Torre, 2000, p. 383 and p. 392].

input-output logic and TJS is given below, Section 6.2.2.

1.3 The theory of joining-systems TJS

In TJS, implications are seen as relations between two objects. Thus a statement "a implies b" expresses that an implicative relation holds from a to b. The specific character of the objects a and b is a matter of which model is chosen for the abstract theory.

A first view of TJS is as follows. A simple normative system contains three basic kinds of implicative relations:

- a relation R_1 over a set A_1 of grounds,

- a relation R_2 over a set A_2 of consequences,

- a relation J from the grounds in A_1 to the consequences in A_2 (expressing the norms of the system).

We note that, though each of R_1, R_2 and J is a binary implicative relation, the relation J is different in kind from R_1, and R_2. Thus while the point of the latter two relations is to order elements of A_1 and A_2, respectively, relation J is a "correspondence", with the purpose of assigning consequences in A_2 to grounds in A_1 and vice versa. (This is particularly perspicuous in the case where A_1 and A_2 are disjunct.)

A picture of a joining relation is shown in Figure 1.

The resulting structures or systems are: The structure $\mathcal{A}_1 = \langle A_1, R_1 \rangle$ of grounds, the structure $\mathcal{A}_2 = \langle A_2, R_2 \rangle$ of consequences, and the system $\langle \mathcal{A}_1, \mathcal{A}_2, J \rangle$, called a joining-system, where the elements of J are joinings from $\mathcal{A}_1$ to $\mathcal{A}_2$. (The elements of the joining relation J constitute a subset of $A_1 \times A_2$, representing the norms of the normative system.) For a joining-system $\langle \mathcal{A}_1, \mathcal{A}_2, J \rangle$, if $\langle a_1, a_2 \rangle \in J$ (where $a_1 \in A_1$ and $a_2 \in A_2$), we say that a_1 is a ground for a_2 and a_2 is a consequence of a_1.

To the three relations R_1, R_2, J will be added a fourth implicative ordering relation $\trianglelefteq$, called "narrowness", over the set of elements in J. These elements (i.e., the norms from $A_1 \times A_2$) can be more or less "narrow", and this is expressed by the relation $\trianglelefteq$. From another aspect, $\trianglelefteq$ expresses implication between the norms in J. Thus, the expression $\langle a_1, a_2 \rangle \trianglelefteq \langle b_1, b_2 \rangle$ means that $\langle a_1, a_2 \rangle$ is at least as narrow as $\langle b_1, b_2 \rangle$, and also that $\langle a_1, a_2 \rangle$ implies $\langle b_1, b_2 \rangle$.

1.4 TJS for simple normative systems

TJS has a wider range of application than the representation of normative systems. As will appear in Sections 2 and 3, the general theory of joining-systems can be applied to quasi-orderings of any kind. Within this range, a

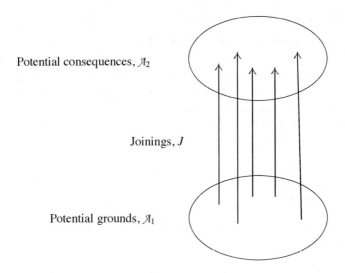

Figure 1

field of special interest is that of what may be called "Many-sorted implicative conceptual systems" (cf. [Odelstad, 2008]). From the perspective to be adopted here, a special area of this kind is the representation of normative systems with conditional norms. In TJS, this problem is dealt with in terms of joinings of normative consequences in A_2 to grounds in A_1.

If the sentence "a implies b" expresses a (conditional) norm, it is assumed that b, the consequence, is normative. In this respect, the representation of norms in TJS is akin to the theory of correlation of normative solutions to cases in the work of Alchourrón and Bulygin, but different from the representation of norms in input-output logic. The specific character of various normative consequences in TJS is dealt with in terms of so-called normative positions, made up by a combination of deontic concepts (constructed by "Shall", "May" for obligation and permission) and action concepts (constructed from "x sees to it that ...").

1.5 Normative positions in TJS

An important refinement of classical deontic logic is the theory of normative positions as the combination of a standard deontic operator *Shall*, expressing command (or *May*, expressing permission) with an action operator *Do* ("$\mathrm{Do}(x,F)$" for "x sees to it that F"), and exploiting the possibilities of external and internal negation of sentences where these operators are combined. See Chapter "The theory of normative positions " in the present Handbook.

As an illustration, imagine a normative system $\mathcal{N} = \langle \mathcal{A}_1, \mathcal{A}_2, J \rangle$ such that $\langle a_1, a_2, \rangle \in J$. Suppose F is the condition that the police is informed of which political party x sympathizes with. Let a_1, a_2 be as follows:

a_1 : x is not suspected of any crime, and y is a police authority.

a_2 the conjunction of (1)-(6) below:

(1) May Do(x,F)
(2) May Do($x, \neg F$),
(3) May ($\neg$Do(x,F) & $\neg$Do($x,\neg F$))
(4) $\neg$May Do(y,F)   (= Shall $\neg$ Do(y,F))
(5) May Do($y, \neg F$),
(6) May ($\neg$Do(y,F) & $\neg$Do($y,\neg F$))

Among these, (1)-(3), (5-6) express permissions, while (4) expresses a prohibition. (1) expresses that x may see to it that the police is informed of which political party x sympathizes with, (2) that x may see to it that the police is not so informed, (3) that x may be passive in this respect. (4) expresses that it shall be the case that y (a police authority) does not see to it that the police is informed, and so on. As will appear later, the conjunction of (1)-(3) exemplifies one-agent type T_1 of normative positions while the conjunction of (4)-(6) exemplifies one-agent type T_4.

As will be developed in Section 4.4 below, the TJS version of normative positions combines the TJS approach to joining-systems with an explicitly algebraic model of the theory of normative positions. In the system of grounds and consequences of a normative system, the algebraic version of normative positions is an algebra of normative consequences intended to handle the stratum $\mathcal{A}_2$ of a normative joining-system $\langle \mathcal{A}_1, \mathcal{A}_2, J \rangle$. In Section 4.4.1, we introduce an example of conditional norms concerning the normative positions of the owners of two adjacent estates.

1.6 Subtraction and addition of norms in TJS

An important issue within the representation of normative systems is the handling of changes, in the sense of subtracting and/or adding norms to the system. Section 4.3 below provides an example showing how TJS deals with these issues in terms of the lattice-like structure of so-called minimal joinings. The example concerns the legal effects of an illegal transfer of goods belonging to someone else. We illustrate the transition from an original normative system $\mathcal{S}_I$, satisfying specific requirements for minimal joinings, via an unsatisfactory system $\mathcal{S}_{II}$, to systems $\mathcal{S}_{III}$ and $\mathcal{S}_{IV}$, once more satisfying the requirements for joining-systems.

1.7 Intermediaries and intervenients

1.7.1 Facts, normative positions and intermediaries

Legal rules attach obligations, rights, normative positions to facts, i.e., the occurrence of actions and events, or the presence of circumstances. Normative positions are, so we might say, legal consequences of these facts. Facts and normative positions are objects of two different sorts; we might call them Is-objects and Ought-objects. In a legal system, when Ought-objects are said to be "attached to" or to be "consequences of" Is-objects, there is sense of direction. In a legal system, inferences and arguments go from Is-objects to Ought-objects, not vice versa.

In the Is-Ought partition, something very essential is missing, namely the great bulk of more specific legal concepts. A few examples are: property, tort, contract, trust, possession, guardianship, matrimony, citizenship, crime, responsibility, punishment. These concepts are links between grounds on the left hand side and normative consequences on the right hand side of the scheme below:

Facts	*Links*	*Normative positions*
Events	Ownership	Obligations
Actions	Valid contract	Claims
Circumstances	Citizenship (etc.)	Powers (etc.)

Using this three-column scheme, we might say that ownership, valid contract, citizenship etc. are attached to certain facts, and that normative positions, in turn, are attached to these legal positions.

As an example, Amendment XIV, Section 1, of the Constitution of the United States reads as follows:

> "All persons born or naturalized in the United States, and subject to the jurisdiction thereof, are citizens of the United States and of the State wherein they reside. No State shall make or enforce any law which shall abridge the privileges or immunities of citizens of the United States; nor shall any State deprive any person of life, liberty, or property, without due process of law; nor deny to any person within its jurisdiction the equal protection of the laws."

Two central terms in this constitutional rule are "citizen" and "person". The rule enumerates grounds for being a citizen of the United States and pronounces a number of legal consequences, expressed in terms of "shall", of this condition. It does not assert any grounds for being a "person", but it pronounces a number of legal consequences attached to personhood. Within the U.S. constitutional system, the article just referred to is supplemented by other rules established by the Constitution and by constitutional court decisions. These rules together, by specifying grounds and consequences, indicate the role of the term "citizen" or "person" within the system.

1.7.2 Wedberg and Ross on vehicles of inference

In the 1950's, each of the two Scandinavians Wedberg and Ross proposed the idea that a legal term such as "ownership", or "x is the owner of y at time t" is a syntactical tool serving the purpose of economy of expression of a set of legal rules.[4]

As an example, the function of the term "ownership" is illustrated as follows by [Ross, 1951], cf. [Ross, 1956 and 1957]:

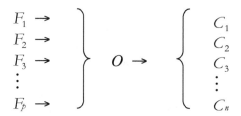

Figure 2

In the picture, the letters are to be interpreted as follows:

- $F_1 - F_p$ for: x has lawfully purchased y, x has inherited y, x has acquired y by prescription, and so on.

- $C_1 - C_n$ for: judgment for recovery shall be given in favor of x against other persons retaining y in their possession, judgment for damages shall be given in favor of x against other persons who culpably damage y, if x has raised a loan from z that it is not repaid at the proper time, z shall be given judgment for satisfaction out of y, and so on.

The letter "O" is a link between the left hand side and the right hand side. It can be read "x is the owner of y".

In Ross's scheme, the number of implications to ownership from the grounds for ownership is p (since the grounds are $F_1, ..., F_p$); similarly the

---

[4]In the same year 1951, when Ross published his well-known essay "Tû-Tû" in a Danish Festschrift [Ross, 1951] (English translation [Ross, 1956 and 1957]), Wedberg published an essay on the same theme in the Swedish journal *Theoria* [Wedberg, 1951]. Possibly, the two authors arrived at these ideas independently of each other. Cf. [Wedberg, 1951, p. 266, n. 15], and [Ross, 1956 and 1957, p. 822, n. 6].

number of implications from ownership to consequences of ownership is n (since there are n consequences). Therefore, the total number of implications in the scheme is $p + n$. On the other hand, if the rules were formulated by attaching each C_j among the consequences to each F_i among the grounds, the number of rules would be $p \cdot n$. Consequently, by the formulation in the scheme, the number of rules is reduced from $p \cdot n$ to $p + n$, a number that can be much smaller [Wedberg, 1951, pp. 273f.]. In this way, economy of expression is obtained.[5] (Cf., however, below, Section 1.7.4, on reductionism.)

1.7.3 Intermediaries and meaning

Both Wedberg and Ross emphasize that intermediaries like "ownership" fulfil their deductive purpose even if they are not defined. Ross claims that "ownership" is a meaningless word in legal language:

> "... the 'ownership' inserted between the conditioning facts and the conditioned consequences is in reality a meaningless word, a word without any semantic reference whatever, serving solely as a means of presentation." [Ross, 1956 and 1957, p. 820]

Already in 1944 (in a lecture in Uppsala), Anders Wedberg proposed the idea that the concept of a "right", as it appears in a normative system, is a syntactical tool for inferences, not a concept with "independent meaning".

> "In the normative rules, the concepts of rights function as syntactical tools, not as concepts with independent meaning." (See [Lindahl, 2004, p. 189, n. 16] for the reference.)

In his essay in 1951, [Wedberg, 1951], Wedberg, more cautiously, proposes this as a "third alternative", beside the alternatives of defining ownership in terms of grounds or in terms of consequences, respectively (alternatives one and two).

A plausible interpretation of Wedberg's idea of "not independent meaning" is that the rules stating the grounds and consequences of ownership (cf. Ross's figure above) are meaningful and that the sentence "O is the property of P at t" has a purposeful role as a component of these rules but that it has no meaning in abstraction from the rules where it functions as a vehicle of inference.

---

[5] The similarities between Wedberg's and Ross' ideas are striking. Both use the example of ownership. Central ideas propounded by both of them are: By use of the linking term, the number $p \cdot n$ of rules is reduced to $p + n$, and, the linking term has no independent meaning (Wedberg) or has no semantical reference (Ross).

"It may be shocking to unsophisticated common sense to admit such 'meaningless' expressions in the serious discourse of legal scientists. But, as a matter of fact, there is no reason why all expressions employed in a discourse, which as a whole is highly 'meaningful', should themselves have a 'meaning'." [Wedberg, 1951, p. 273]

[Sartor, 2009] contrasts the idea of vehicles of inference with the idea of legal concepts as "categories" in a domain ontology.[6] In the latter perspective, meaning inheres in words or terms, and the meaning of sentences results from the meaning of their lexical components. (See [Sartor, 2009, pp. 236f.].) In jurisprudential writing, systematization is sometimes achieved by the ordering of legal concepts in conceptual trees or pyramids.[7] (As a well-known analogue from natural science, we may think of the Linnaean system of plants, which influenced eighteenth century conceptual jurisprudence in Germany.) If such an ordering is to be congruent with an existing normative system, however, it should accord with the role the concepts have as vehicles of inference within the system. If A and B are subcategories of category C, then category C indicates some properties which members of A and B have in common.[8] As regards concepts in a normative system, these common properties may regard either grounds or consequences or both, according to the rules of the system in view.

Since there are many legal systems, there are (to take an example) many concepts of ownership, more or less similar. Thus one concept of ownership is ownership as a vehicle of inference in Swedish private law on January 1st 2010. This concept of ownership is determined by the particular normative system referred to; consequently, the concept is replaced by another whenever the grounds or consequences of ownership in the system are changed. We note that, when several different concepts (for example, ownership in actual Swedish law and ownership in Anglo-Saxon common law) are called "concepts of ownership", it is suggested that these varieties have properties in common, justifying that they are called "concepts of ownership". In particular, the concepts in view can have a common historical origin, and the "institution" that they are used for expressing (the institution of ownership) can have the same social purpose or function in the different systems.

---

[6] An earlier version of Sartor's paper is [Sartor, 2007].

[7] Cf. [Lindahl, 2000], in particular pp. 166f., on the reasoning of the German eighteenth century jurist Georg Friedrich Puchta. A systematization of concepts appears as well in the arrangement of norms in civil codes such as the German *Bürgerliches Gesetzbuch* and the French *Code Civil*.

[8] A recent development is the idea of semantic networks and inheritance, see [Horty et al., 1990], (referred to by [Sartor, 2009, p. 243, n. 27]. The focus in [Horty et al., 1990] is on defeasibility, in this case "multiple inheritance with exceptions".

Considerations of this kind are relevant for a critical assessment of the ownership rules of particular normative systems, and may cause assessment of what is the "essential content" of ownership.[9]

1.7.4 Reductionism

In the Ross-Wedberg example on ownership, the set of legal rules illustrated by the picture can be reformulated in two rules:

(1) $(F_1 \vee ... \vee F_m) \to O$.
(2) $O \to (S_1 \wedge ... \wedge S_n)$.

If the middle term M is eliminated, we get the single rule:

(3) $(F_1 \vee ... \vee F_m) \to (S_1 \wedge ... \wedge S_n)$.

The most economical way to express the rules of the two arrays above would seem to be by a single sentence like (3). By reductionism regarding intermediaries is meant the idea that legal reasoning might in general proceed directly from facts to normative consequences so as to dispense with intermediate concepts.

Concerning the accomplishment of reduction, two complications have to be born in mind. Firstly, the bulk of so-called "legal" concepts are intermediaries, and these intermediaries constitute complex networks. (Cf. [Lindahl and Odelstad, 2011]) Secondly, many legal intermediaries are vague or "open textured", so that power to decide on grounds and consequences for the intermediaries is conferred on judges and other persons who apply the law (see below, Section 5.2.2).

The question whether, in principle, it is possible to do away with the intermediaries is complex and will not be answered here. A formal theory for handling intermediaries, however, is needed both for any attempt to eliminate them and for representing the system as it is without reduction.

1.7.5 Open legal concepts

As mentioned, there are numerous cases where legal concepts are vague or "open textured", and power to interpret the concepts is conferred on judges and other persons who apply the law. Obvious examples are such concepts as "negligent" or "reasonable" but considerable openness also is a feature of such concepts as "public interest", "contract" and "ownership".

---

[9]To exemplify, in German constitutional law there is a guarantee of protection for the "essential content" (*Wesensgehalt*) of the basic rights of the German Constitution. In an essay by the Swedish philosopher Ingemar Hedenius, Max Weber's idea of "ideal types" is applied to the concept of ownership, where normative systems are represented as different alternatives of fulfilment on each of several dimensions. (See [Hedenius, 1977, pp. 130-55].) According to Hedenius' proposal, the concept of ownership in particular normative systems can be critically assessed according to their degree of fulfilment on the dimensions introduced.

An example might be the legal rule stipulating the ground for what, in Swedish law, is called "having a relationship similar to being married". If two persons are not married, nevertheless they can have a relationship similar to being married. From such a condition particular legal consequences follow by the law. First, if the relationship is dissolved, property acquired by one of the parties for use in common shall be partitioned between the parties according to rules similar to those applied when a marriage is dissolved. Secondly, if the relationship of the parties is dissolved, their dwelling can be allotted to that party who needs it most.

The law does not specify exactly which facts give rise to a "relationship similar to being married".[10] However, there are a number of criteria. Let us consider the following eleven criteria, calling them $F_1, F_2, ..., F_{11}$:

F_1 : cohabiting, F_2 : housekeeping in common, F_3 : having children in common, F_4 : having sexual intercourse, F_5 : having confirmed the relation by a contract, F_6 : living in emotional fellowship, F_7 : being faithful, F_8 : giving mutual support, F_9 : sharing economic assets and debts, F_{10} : having no legal impediments to marriage, F_{11} : having no similar relationship to another person.

If all of the criteria are satisfied by persons i and j, their relationship is "similar to being married". Conversely, if none of them is satisfied, their relationship is not "similar to being married". These two rules belong to established law.

However, the law does not say what is the result if some of the conditions are satisfied while others are not. This means that, in a sense, the set of grounds for having a relationship similar to being married is "open", and the grounds are not specified completely.

A great amount of legal concepts are "ground-open" like "relationship similar to being married". When such a concept occurs in a legal argument, there is room and need for decisions to be made by courts and other authorities applying the law. This task is an obstacle to reductionist efforts to do away with legal intermediaries in favor of rules attaching deontic consequences directly to factual events, actions, circumstances. In legal argument from facts to deontic consequences, the argument is a sequence of steps, passing through a number of stations involving legal concepts. Insofar as the concepts are open, decisions have to be made step by step.

---

[10] In 2003, a new statute (SFS 2003: 376) on cohabitant partners ("sambor") was enacted in Sweden. In article 1, paragraph 1, there is a definition of "cohabitant partners", intended to be a little more precise: "By cohabitant partners is meant two persons who live together permanently in a partner relationship and have their housekeeping in common." (Translated here.)

"Relationship similar to being married" is a concept that is ground-open, in the sense we have indicated. Similarly, a legal concept can be consequence-open. Taking a concept like "ownership", "citizenship" or "matrimony", for some deontic consequences it is established that they do follow, for others it is established that they do not follow. However, there are as well consequences for which it is not established whether they follow or not. Then the concept is consequence-open.

"Being the owner of" can serve as an example of a concept that is to some extent consequence-open. Thus it need not, for example, be entirely settled to what extent and by what means the owner of an estate may exclude others from entering on his/her ground.

The phenomenon of open concepts in a normative system is connected with the limits on what can be achieved by a legislator. If a legislator attempts to avoid openness, the probability increases that the norms enacted become oversimplified. As clearly understood already by Aristotle, it is not possible to create a complete legal code of "established law" without incurring into error by oversimplification:

> ".. all law is universal but about some things it is not possible to make a universal statement which shall be correct. In those cases, then, in which it is necessary to speak universally, but not possible to do so correctly, the law takes the usual case, though it is not ignorant of the possibility of error. And it is none the less correct; for the error is [not] in the law nor in the legislator but in the nature of the thing, since the matter of practical affairs is of this kind from the start. When the law speaks universally, then, and a case arises on it which is not covered by the universal statement, then it is right, where the legislator fails us and has erred by oversimplicity, to correct the omission - to say what the legislator himself would have said had he been present, and would have put into his law if he had known. Hence the equitable is just, and better than one kind of justice - not better than absolute justice but better than the error that arises from the absoluteness of the statement. And this is the nature of the equitable, a correction of law where it is defective owing to its universality. In fact this is the reason why all things are not determined by law, that about some things it is impossible to lay down a law, so that a decree is needed. For when the thing is indefinite the rule also is indefinite, like the leaden rule used in making the Lesbian moulding; the rule adapts itself to the shape of the stone and is not rigid, and so too the decree is adapted to the facts." [Aristotle, Nicomachean

Ethics, EN 1137b]

The issue of open legal concepts will be dealt with in Section 5.2.2 below.

1.7.6 Intermediaries outside the realm of legal systems

The idea of intermediaries is applicable outside the realm of legal systems. An example is Dummett's theory of language. Dummett distinguishes between the conditions for applying a term and the consequences of its application. According to Dummett both are parts of the meaning. Dummett exemplifies by the use of the term "Boche" as a pejorative term Cf. [Kremer, 1988; Lindahl and Odelstad, 2006a; Lindahl and Odelstad, 2008a; Sartor, 2007; Sartor, 2009]. (Since the example is interesting from a philosophical point of view, we use it even though it has the disagreeable feature of being offensive to German nationals.)

> "The condition for applying the term to someone is that he is of German nationality; the consequences of its application are that he is barbarous and more prone to cruelty than other Europeans. We should envisage the minimal joinings in both directions as sufficiently tight as to be involved in the very meaning of the word: neither could be severed without altering its meaning. Someone who rejects the word does so because he does not want to permit a transition from the grounds for applying the term to the consequences of doing so. The addition of the term 'Boche' to a language which did not previously contain it would produce a non-conservative extension, i.e., one in which certain statements which did not contain the term were inferable from other statements not containing it which were not previously inferable." [Dummett, 1973, p. 454]

Dummett's example illustrates how the use of a word is determined by two rules (1) and (2):

(1) Rule linking a concept a to an intermediary m : If $a(x,y)$ then $m(x,y)$,

(2) Rule linking intermediary m to a concept b : If $m(x,y)$ then $b(x,y)$.

The rules (1) and (2) can be compared to the rules of introduction and rules of elimination, respectively, in Gentzen's theory of natural deduction in [Gentzen, 1934]. If this comparison is made, (1) is regarded as an introduction rule and (2) as an elimination rule for m. (See [Lindahl and Odelstad, 2008a, sect. 1.2.3].)

In natural science, the idea of "intermediate" has been applied to the term "force" within physical theory. As is observed by [Wedberg, 1982, pp. 11ff.]

during the eighteenth century several thinkers thought of the forces spoken of in mechanics as a kind of mathematical fictions, useful for describing the movements of bodies in a convenient way. What exists in physical reality, according to this view, are configurations of mass, speeds, and accelerations. Forces are fictions, but they enable us to describe the interrelations of the former entities in a compact way. As Wedberg mentions, Berkeley is among the thinkers who held this opinion.

The position, held by Berkeley and others, that "force" is merely a device for compact expression, closely resembles the idea of intermediaries. This resemblance becomes even more obvious if the position in view is described in Wedberg's own words:

> "If a body k with mass m is in a particular (spatial and temporal) relation to certain other bodies, we say that a force of magnitude f affects k. If a force of magnitude f affects k, then k receives an acceleration a satisfying the equation:
>
> (i) $f = a \cdot m$
>
> Thus the force occurs as a middle term in the pair of hypothetical statements:
>
> (ii) Given a certain configuration of mass, a certain force exists.
>
> (iii) Given a certain force, a certain acceleration results.
>
> If the middle term is eliminated, we arrive at the conclusion:
>
> (iv) Given a certain configuration of mass, a certain acceleration results." [Wedberg, 1982, p. 11]

An objection to Berkeley's idea that forces are "fictions", however, is raised by Wedberg in pointing out that the term "force" can be defined in terms of such entities that Berkeley considers as real. Such a definition, in Wedberg's words, might be formulated as a definition of the entire statement (see [Wedberg, 1982, p. 12]):

The body k exerts a force f upon the body k'.

A definition of this statement, then, can read as follows:

> f is the product of the acceleration a, which k' receives from k and the mass of k'.

In connection with the possibility of defining "force" in terms of "real" entities, we recall the possibility of defining ownership, either in terms of grounds or in terms of consequences (Wedberg's "first" and "second" alternatives).

Another interesting example from physics is found in the work of Henri Poincaré. Poincaré proposed that "gravitation" can be regarded as an intermediary (*un intermédiaire*). According to Poincaré, the proposition "the stars obey Newton's laws" can be broken up into two others, namely (1) "gravitation obeys Newton's laws" and (2) "gravitation is the only force acting on the stars". Among these, proposition (1) is a definition and not subject to the test of experiment, while (2) is subject to such a test. "Gravitation", according to Poincaré, is an intermediary. Poincaré maintains that in science, when there is a relation between two facts A and B, an intermediary C is often introduced by the formulation of one relationship between A and C, and another between C and B. The relation between A and C, then, is often elevated to a principle, not subject to revision, while the relation between C and B is a law, subject to such revision. See [Poincaré, 1907, pp. 124f.], in the chapter "Is science artificial?" On the analogous question of definition and norm in a normative system, cf. [Lindahl, 1997, p. 298].

Still another example concerns probability (see [Lindahl and Odelstad, 1999a]). Consider statements of the kind "the probability of the event A equals m" (where m is a real number). Using the notion of conditions, introduced below in Section 4.2, page 596, one may speak of conditions on events, for example the condition of having probability m. Such a condition can be regarded as an intermediary between two conceptual structures, one concerning frequencies and symmetries, and the other concerning how one ought to choose between different games. It is a plausible idea that the so-called objective, or frequency, interpretation of probability deals with the structure of grounds for probability conditions, whereas the so-called subjective interpretation deals with the structure of consequences. This suggestion seems to assign a proper role to each of the two interpretations.

For a treatment of intermediate concepts in connection with weighing of interests in urban planning, see [Odelstad, 2002; Odelstad, 2009].

1.7.7 Counts-as-theory

When a rule r of a legal system $\mathcal{N}$ attaches an intermediary m, e.g., "x and y have made a contract to the effect that z", to a conjunction a of facts, the rule r can be expressed in different ways, e.g. "if a then m", or, sometimes, "*a counts as m*". A logical analysis of sentences of the kind "x counts-as y in s", where s is an institution (s can be a normative system), was proposed in [Jones and Sergot, 1996; Jones and Sergot, 1997].[11] The work of Jones and

---

[11] The original motivation of Jones and Sergot was, so it seems, to give a formal characterization of "institutionalized power", see [Jones and Sergot, 1997, pp. 349ff.]. For a comment on this matter, see [Lindahl and Odelstad, 2008a, sect. 3.5.3, n. 22].

Sergot on "Counts-as" has been continued by a number of other authors, in particular in the book-length study by Davide Grossi [Grossi, 2007]. For further details on Counts-as, the reader is referred to Chapter "Constitutive norms and counts as conditionals" of the present Handbook. A remark on the interrelationship between Counts-as and TJS, see below, Section 6.2.1.

1.7.8 "Intervenient" as a technical notion in TJS

An essential part of the theory of joining-systems is the theory of intervenients. Though this theory aims at providing tools for analyzing intermediaries as they appear in law, language, morals, and so on, "intervenient" is a technical notion defined (see Definition 5.2, below, Section 5.1) at the abstract algebraic level, used as a tool for analyzing different kinds of what, informally, is called intermediaries. The notion of intervenient is tied to the TJS approach, focusing on a normative system as a deductive mechanism and on intermediaries as vehicles of inference. Therefore, in the development of the theory of intervenients, the idea of economy of expression has a central role. This relates both to the effective representation of a normative system by intervenients and to changes in such a system accomplished by changing grounds and/or consequences of intervenients.

Special themes regarding intervenients dealt with in this Chapter are what we call "organic wholes" (Section 5.2.1), open concepts and "narrowing of intervenients" (Section 5.2.2), and the typology of intervenients (Section 5.2.4).

1.8 Advice to readers

Though a substantial part of the chapter is abstract and formal, there are as well several parts that are semi-formal. This holds for next Section 2, which is a first introduction to TJS, as well as for the subsections on *cis* applications in Sections 4 and 5. More exactly, these subsections are: Section 4.3 on subtraction and addition of norms, Section 4.4.1 on ownership to an estate, Section 5.2.1 on organic wholes of intervenients, Section 5.2.2 on open concepts and the "narrowing" of intervenients, and Section 5.2.3 on the legal example of grounds and consequences of ownership and trust.

2 First introduction to TJS

2.1 General TJS irrespective of intervenients

2.1.1 Strata and joining systems

The structure of grounds as well as the structure of consequences will be called a *stratum*. The word "stratum" is understood here in the sense of the result of arrangement of the parts or elements of something.[12] More precisely, in TJS, the general structure of a stratum is a set A of objects, ordered by an implicative relation R, which is binary, reflexive and transitive. It is not assumed that R is antisymmetric, nor that it is not. In other words, a stratum is conceived of as a quasi-ordering $\langle A, R \rangle$ of objects from a set A. (Another term for quasi-ordering is *preordering*.) The relation R is a relation ordering the objects within a stratum, and, therefore, is called an *intrastratum* relation.

In TJS, the relation J is an *interstrata* implicative relation from elements of a stratum of grounds to elements of a stratum of consequences. As will be made more explicit subsequently, the relation J (which, normally, is not a function) provides a "correspondence" between these two strata, depicting the set of grounds on the set of consequences and vice versa. In this respect, relation J differs from relation R which is an intrastratum ordering relation.

As mentioned (see Section 1.3), a *joining-system* $\langle \mathcal{A}_1, \mathcal{A}_2, J \rangle$ consists of two strata $\mathcal{A}_1, \mathcal{A}_2$ and a relation J. TJS leaves room for different kinds of structures over each of $\mathcal{A}_1, \mathcal{A}_2$. For $1 \leq i \leq 2$, a stratum can be a quasi-ordering $\langle A_i, R_i \rangle$, where A_i is (simply) a set, or it can be a "lattice-based quasi-ordering" $\langle L_i, \wedge, \vee, R_i \rangle$, where $\langle L_i, \wedge, \vee \rangle$ is a lattice, or it can be a "Boolean quasi-ordering", $\langle B_i, \wedge,' , R_i \rangle$, where $\langle B_i, \wedge,' \rangle$ is a Boolean algebra. A special case is where, for a lattice-based quasi-ordering $\langle L_i, \wedge, \vee, R_i \rangle$ or a Boolean quasi-ordering $\langle B_i, \wedge,' , R_i \rangle$, R_i is the relation $\leq$ of $\langle L_i, \wedge, \vee \rangle$ or of $\langle B_i, \wedge,' \rangle$, respectively.

As will appear, the definition of "joining-system" is the same, independently of which is the type of the strata connected in the joining-system, only provided that each stratum fulfills the minimum requirement of being a quasi-ordering. Thus while there is flexibility as regards the types of strata, the definition of joining-system gives stability to the theory: As we will see, a joining-system exhibits a number of important properties, relevant for the representation of a normative system.

While both the intrastratum R and the interstrata J express implication, an essential difference between R and J is that between "one-sort" objects and "two-sorts" objects. In TJS, the intrastratum R is a relation between objects conceived of as being of the same sort; in contrast, the interstrata relation J is a relation between objects thought of as being of two sorts. As regards normative systems, the idea of two sorts applies in particular to the difference between empirical/descriptive and normative. (In another area, consider the difference between physical and mental.)

---

[12]Cf. the online *Free Dictionary*: "One of a number of layers, levels, or divisions in an organized system." Note that "stratum" as used here is not to be understood in the sense of: "one of several parallel layers of material arranged one on top of another."

Norms are represented by ordered pairs $\langle a_1, a_2 \rangle$ where a_1, a_2 are of different sorts. The most general version of TJS is where the strata $\mathcal{A}_1, \mathcal{A}_2$ of a joining-system $\langle \mathcal{A}_1, \mathcal{A}_2, J \rangle$ are simply quasi-orderings. A substantial part of TJS will be developed within this general framework. As will appear, in this version, TJS yields a number of results for the formal representation of normative systems. In particular, by the relation $\trianglelefteq$ of narrowness (see above, end of Section 1.3), there is an implicative structure over the norms of the system, and the system can be expressed in an economic way by its set of "minimal joinings".

2.1.2 Minimal joinings

Suppose that a norm $\langle a_1, a_2 \rangle$ is a joining from a stratum $\mathcal{A}_1$ of grounds to a stratum $\mathcal{A}_2$ of consequences. Then, if (in a sense to be defined) a_1 is a "weakest ground" for a_2, and a_2 is a "strongest consequence" of a_1, the pair $\langle a_1, a_2 \rangle$ represents what in TJS is called a *minimal joining*. If a normative system fulfills a requirement called "connectivity", any norm in the system is always implied by a minimal joining.

In TJS, a normative system can be represented in a convenient way by its set of minimal joinings, and therefore, minimality is decisive for how economy of expression is accomplished and for how changes of a system can be effectively achieved. Furthermore, in a well-structured normative system, the set of minimal joinings has a number of perspicuous structural properties. Thus, firstly, the set of minimal joinings can be ordered in an interesting way as a lattice-like structure. Secondly, if $\langle a_1, a_2 \rangle$ belongs to the set J of joinings, let us call the ground a_1 the "bottom" of the joining $\langle a_1, a_2 \rangle$ and the consequence a_2 the "top" of this joining. Then, as we will see, there is a similarity between the set min J of minimal joinings and the set of bottoms of min J as well as to the set of tops of min J.

2.2 Intervenients in TJS

Suppose that we have in view three joining-systems $\mathcal{S}_1 = \langle \mathcal{A}_1, \mathcal{A}_2, J_{1,2} \rangle$, $\mathcal{S}_2 = \langle \mathcal{A}_2, \mathcal{A}_3, J_{2,3} \rangle$, $\mathcal{S}_3 = \langle \mathcal{A}_1, \mathcal{A}_3, J_{1,3} \rangle$ such that these systems constitute a chain in the sense that by $J_{1,2}$ you can go from $\mathcal{A}_1$ to $\mathcal{A}_2$, by $J_{2,3}$ you can go from $\mathcal{A}_2$ to $\mathcal{A}_3$, and by $J_{1,3}$ (using relative product) you can go directly from $\mathcal{A}_1$ to $\mathcal{A}_3$. In a sense, the stratum $\mathcal{A}_2$ is intermediate between $\mathcal{A}_1$ and $\mathcal{A}_3$. Certain elements in A_2 can be *intervenients* between elements in A_1 and elements in A_3.[13] (See Figure 3 on page 565.) If $a_1 \in A_1$, and $a_2 \in A_2$ and $a_3 \in A_3$, a_2 *corresponds* to the pair $\langle a_1, a_3 \rangle$ if, in a sense to be defined, later, a_1 is the weakest ground in $\mathcal{A}_1$ for a_2 and a_3 is the strongest

---

[13]Note that we use calligraphic letters $\mathcal{A}_1$, $\mathcal{A}_2$, $\mathcal{A}_3$ for the quasi-orderings $\langle A_1, R_1 \rangle$, $\langle A_2, R_2 \rangle$, $\langle A_3, R_3 \rangle$ and we use italics A_1, A_2, A_3 for the domains of these quasi-orderings.

consequence in $\mathcal{A}_3$ of a_2. The investigation of intervenients following in this

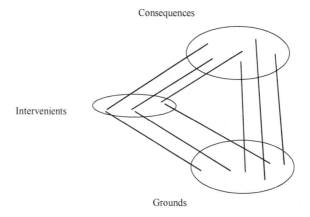

Figure 3

chapter has in view the structure and properties of the intervenients. To this subject-matter belongs a number of special issues. A few examples are as follows. If economy of expression is related to the notion of minimal joinings, what can be said about intervenients and minimality? Is there a typology of intervenients and minimality? Under what conditions can a normative system be represented by a base of intervenients? Furthermore, there is the issue of Boolean operations (conjunction, disjunction, negation) on intervenients. If a_2, b_2 are intervenients from $\mathcal{A}_1$ to $\mathcal{A}_3$, then what can be said about $a_2 \wedge b_2$, $a_2 \vee b_2$ and (the negations) a_2', b_2'? How do Boolean compounds of intervenients relate to corresponding compounds of grounds and of consequences? All of these questions are essential to the formal structure of intervenients and have a direct bearing on the formal representation of intermediaries in a normative system.

2.2.1 Subject-matter of sections 3-5

The following three main Sections 3-5 are organized as follows. (We recall what was said in Section 2.1.1 about joining as a relation between elements of two strata.) In Section 3, the basic theory of joining-systems is developed, while Section 4 is devoted to the theory of different kinds of strata. In Section 3, dealing with joining-systems in general, very little is presupposed about the structure of strata. In Section 4, on the other hand, the character of strata is the subject-matter of more differentiation. Here, what is in view is joining-systems where strata are Boolean-like structures or lattice-like structures. Since the development in Section 4 is intended for the representation of normative systems, the focus there will mainly be on so-called Boolean joining-systems. Section 5 is devoted to the theory of intervenients in Boolean joining-systems.

It should be observed that the general results regarding lattice-like structures in Section 3 are essential for the analysis of joining-systems, including the analysis of Boolean joining-systems (later pursued in Section 4) and the analysis of intervenients (in Section 5).

3 Formal development of TJS

3.1 Basic concepts

Much of the study of ordering relations in mathematics seems to have partial orderings as its basic structure. Lattices and Boolean algebras, for example, are partially ordered sets. In the study of norms and conceptual systems, it is more convenient to take quasi-orderings as the formal framework. The reason for choosing quasi-orderings instead of partial orderings is that in a quasi-ordering $\langle A, R \rangle$ two objects a and b can be similar with respect to R (for example, by having the same extension) without being identical. This feature is useful when dealing with concepts.

In the next subsection (Section 3.1.1), the notion of quasi-ordering is defined. After that, in the subsequent subsections, we generalize some well-known mathematical notions, so as to apply to quasi-orderings.

3.1.1 Quasi-orderings

First a note on terminology. Suppose that R is ν-ary relation on a set A and that X is a subset of A. Then $R \cap X^\nu$ is denoted R/X and is called the restriction of R to X.

Definition 3.1 *The binary relation R is a* quasi-ordering *on A if R is transitive and reflexive in A.*

(As mentioned, another name for quasi-ordering is preordering.)

Writing Q for the *equality* part of R we say that xQy holds iff xRy and yRx. Also, writing P for the *strict* part of R we put xPy iff xRy and not yRx.

A quasi-ordering is closely related to a partial ordering. If $\langle A, R \rangle$ is a quasi-ordering and Q is the equivalence part of R, then R generates a partial ordering on the set of Q-equivalence classes generated from A.

Definition 3.2 *Suppose that R is a quasi-ordering on A and that $X \subseteq A$ and $x \in X$. Then,*
(1) x is a minimal *element in X with respect to R iff there is no $y \in X$*

such that yPx,

(2) x is a maximal element *in X with respect to R iff there is no $y \in X$ such that xPy.*

(3) *The set of minimal elements in X with respect to R is denoted* $\min_R X$ *and the set of maximal elements of X with respect to R is denoted* $\max_R X$.

(4) x *is a* least element *in X with respect to R iff for all $y \in X$, xRy,*

(5) x *is a* greatest element *in X with respect to R iff for all $y \in X$, yRx.*

Note that in a quasi-ordering $\langle A, R \rangle$, a greatest and a least element in a set $X \subseteq A$ need not be unique. But if x and y are greatest elements (or least elements) in X with respect to R, then xQy.

3.1.2 Quasi-lattices and complete quasi-lattices

As will appear in Section 3.2.2, the notions of least upper bound and greatest lower bound are important in the definition of a joining-system. These notions are usually defined for partial orderings and not for quasi-orderings. Since quasi-ordering is a basic structure in TJS, we generalize the notions of least upper bound and greatest lower bound to quasi-orderings. We use ub and lb as abbreviations for upper bound and lower bound respectively, and lub and glb for least upper bound and greatest lower bound respectively. We note that (in contrast to what holds for partial orderings) a least upper bound or a greatest lower bound relative to a quasi-ordering $\langle A, R \rangle$ need not be unique.

Definition 3.3 *Let R be a quasi-ordering on a set A with $X \subseteq A$. Then*
$\text{ub}_R X = \{a \in A \mid \forall x \in X : xRa\}$
$\text{lb}_R X = \{a \in A \mid \forall x \in X : aRx\}$
$\text{lub}_R X = \{a \in A \mid a \in \text{ub}_R X \ \& \ \forall b \in \text{ub}_R X : aRb\}$
$\text{glb}_R X = \{a \in A \mid a \in \text{lb}_R X \ \& \ \forall b \in \text{lb}_R X : bRa\}$.

According to standard algebraic terminology, a partially ordered set $\langle L, \leq \rangle$ is a lattice if for all $a, b \in L$, $\sup_\leq \{a, b\}$ and $\inf_\leq \{a, b\}$ exist in L. (In connection with partial orderings, we prefer to use sup and inf instead of lub and glb respectively.) $\langle L, \leq \rangle$ is *complete* if $\inf_\leq X$ and $\sup_\leq X$ exist for all $X \subseteq L$. We generalize these notions to quasi-orderings.[14]

Definition 3.4 *If $\langle A, R \rangle$ is a quasi-ordering such that*

$$\text{lub}_R \{a, b\} \neq \varnothing \text{ and } \text{glb}_R \{a, b\} \neq \varnothing \text{ for all } a, b \in A,$$

---

[14] Note that the concept of completeness for lattices, quasi-lattices, and quasi-orderings should not be confounded with completeness in the sense that an ordering relation R on a set A is called complete if for all $x, y \in A$ it holds that xRy or yRx.

then $\langle A, R \rangle$ will be called a quasi-lattice. If $\mathrm{lub}_R X \neq \varnothing$ and $\mathrm{glb}_R X \neq \varnothing$ for all $X \subseteq A$, then $\langle A, R \rangle$ is a complete quasi-lattice.

If $\langle A, \leq \rangle$ is a partial order then $a \in \sup_\leq \varnothing$ iff a is the smallest element in A with respect to $\leq$ and $a \in \inf_\leq \varnothing$ iff a is the greatest element in A with respect to $\leq$. (See for example [Grätzer, 2011, p. 5].) Analogously, if $\langle A, R \rangle$ is a quasi-order then

(i) $a \in \mathrm{lub}_R \varnothing$ iff a is a smallest element in A with respect to R

(ii) $a \in \mathrm{glb}_R \varnothing$ iff a is a greatest element in A with respect to R.

We note that if a quasi-lattice is finite, then it is complete.

Theorem 3.5 *Suppose that $\langle A, R \rangle$ is a quasi-lattice, that Q is the indifference part of R, and that A_Q is the set of Q-equivalence classes generated by elements of A. Then $\langle A_Q, R^* \rangle$, where $[a]_Q R^* [b]_Q$ iff aRb, is a lattice. If $\langle A, R \rangle$ is a complete quasi-lattice then $\langle A_Q, R^* \rangle$ is a complete lattice.*

In analogy with what holds of complete lattices, see [Grätzer, 2011, p. 50], the following holds of a complete quasi-lattice.

Theorem 3.6 *Let $\langle A, R \rangle$ be a quasi-ordering in which $\mathrm{glb}_R X \neq \varnothing$ for all $X \subseteq A$. Then $\langle A, R \rangle$ is a complete quasi lattice.*

By duality, the theorem holds if instead $\mathrm{lub}_R X \neq \varnothing$ for all $X \subseteq A$.

In lattice theory the notion of a sublattice is introduced. Suppose $\langle L, \leq \rangle$ is a lattice and $\varnothing \neq M \subseteq L$. Let, furthermore, $\leq^* = \leq/M$. Then $\langle M, \leq^* \rangle$ is a sublattice of $\langle L, \leq \rangle$ if $a, b \in M$ implies that $\sup_{\leq^*} \{a, b\} = \sup_\leq \{a, b\}$ and $\inf_{\leq^*} \{a, b\} = \inf_\leq \{a, b\}$. We now generalize the notion of a sublattice to quasi-lattices and define the notion of a subquasi-lattice.

Definition 3.7 *Suppose that $\langle A, R \rangle$ is a quasi-lattice, $X \subseteq A$ and $S = R/X$. Then $\langle X, S \rangle$ is a subquasi-lattice of $\langle A, R \rangle$ if $x, y \in X$ implies that $\mathrm{lub}_R \{x, y\} \supseteq \mathrm{lub}_S \{x, y\} \neq \varnothing$ and $\mathrm{glb}_R \{x, y\} \supseteq \mathrm{glb}_S \{x, y\} \neq \varnothing$.*

Theorem 3.8 *If $\langle A, R \rangle$ is a quasi-lattice and $\langle X, S \rangle$ a subquasi-lattice of $\langle A, R \rangle$, then $\langle X_Q, S^* \rangle$ is a sublattice of $\langle A_Q, R^* \rangle$.*

(See the notation introduced in Theorem 3.5.)

3.2 Joining-systems

3.2.1 Narrowness

In TJS, the relation of "narrowness" is highly important. It is used in the definition of a joining-system, since it determines the relation of implication between norms and the set of minimal joinings (cf. above Section 2.1.2). The minimal joinings are essential in a normative system, since they serve as the tool for a succinct representation of the system.

Definition 3.9 (1) *The* narrowness relation determined by *the quasi-orderings* $\langle A_1, R_1 \rangle$ *and* $\langle A_2, R_2 \rangle$ *is the binary relation* $\trianglelefteq$ *on* $A_1 \times A_2$ *such that* $\langle a_1, a_2 \rangle \trianglelefteq \langle b_1, b_2 \rangle$ *iff* $b_1 R_1 a_1$ *and* $a_2 R_2 b_2$.
(2) $\langle x_1, x_2 \rangle$ *is a minimal element in* $X \subseteq A_1 \times A_2$ *with respect to* $\langle A_1, R_1 \rangle$ *and* $\langle A_2, R_2 \rangle$ *if* $\langle x_1, x_2 \rangle$ *is a minimal element in* X *with respect to* $\trianglelefteq$. *The set of minimal elements in* X *with respect to* $\trianglelefteq$ *is denoted* $\min_{R_1}^{R_2} X$. (*When there is no risk of ambiguity we write just* $\min X$.)

Note that $\trianglelefteq$ is a quasi-ordering, i.e. transitive and reflexive. Let $\simeq$ denote the equality part of $\trianglelefteq$ and $\triangleleft$ the strict part of $\trianglelefteq$. Then the following holds:

$\langle a_1, a_2 \rangle \simeq \langle b_1, b_2 \rangle$ iff $b_1 Q_1 a_1$ & $a_2 Q_2 b_2$
$\langle a_1, a_2 \rangle \triangleleft \langle b_1, b_2 \rangle$ iff $(b_1 P_1 a_1$ & $a_2 R_2 b_2)$ or $(b_1 R_1 a_1$ & $a_2 P_2 b_2)$

where Q_i is the equality-part of R_i and P_i is the strict part of R_i.

The notion of narrowness is illustrated in Figure 4. Note that $\langle x_1, x_2 \rangle$ is

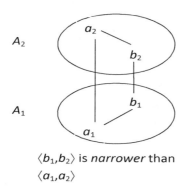

$\langle b_1, b_2 \rangle$ is *narrower* than $\langle a_1, a_2 \rangle$

Figure 4

a minimal element in $X \subseteq A_1 \times A_2$ with respect to $\langle A_1, R_1 \rangle$ and $\langle A_2, R_2 \rangle$ if there is no $\langle y_1, y_2 \rangle \in X$ such that $\langle y_1, y_2 \rangle \triangleleft \langle x_1, x_2 \rangle$, i.e. if there is no element $\langle y_1, y_2 \rangle \in X$ such that $x_1 R_1 y_1$ & $y_2 P_2 x_2$, or $x_1 P_1 y_1$ & $x_2 R_2 y_2$.

In TJS, *up-sets* with respect to the narrowness-relation will be of special interest. We give an explicit definition of up-set with respect to the narrowness-relation here.[15]

Definition 3.10 *Suppose that* $\mathcal{A}_1 = \langle A_1, R_1 \rangle$ *and* $\mathcal{A}_2 = \langle A_2, R_2 \rangle$ *are quasi-orderings and* $K \subseteq A_1 \times A_2$. *Then we say that* K *is an* up-set *with respect to* $\trianglelefteq$ *if the following holds: For all* $a_1, b_1 \in A_1$ *and* $a_2, b_2 \in A_2$, *if* $\langle a_1, a_2 \rangle \in K$ *and* $\langle a_1, a_2 \rangle \trianglelefteq \langle b_1, b_2 \rangle$, *then* $\langle b_1, b_2 \rangle \in K$.

3.2.2 The definition of a joining-system

As mentioned in Section 2.1.1, while TJS is flexible as regards the character of strata $\mathcal{A}_1$ and $\mathcal{A}_2$, in TJS the definition of "joining-system" is the same, independently of which is the type of the strata connected in the joining-system, only provided that each stratum fulfills the minimum requirement of being a quasi-ordering.

The definition of joining-system is as follows.

Definition 3.11 *A* joining-system *(Js), is an ordered triple* $\langle \mathcal{A}_1, \mathcal{A}_2, J \rangle$ *such that* $\mathcal{A}_1 = \langle A_1, R_1 \rangle$ *and* $\mathcal{A}_2 = \langle A_2, R_2 \rangle$ *are quasi-orderings, and* $J \subseteq A_1 \times A_2$, *and the following conditions are satisfied where* $\trianglelefteq$ *is the narrowness relation determined by* $\mathcal{A}_1$ *and* $\mathcal{A}_2$:
(1) *for all* $a_1, b_1 \in A_1$ *and* $a_2, b_2 \in A_2$, *if* $\langle a_1, a_2 \rangle \in J$ *and* $\langle a_1, a_2 \rangle \trianglelefteq \langle b_1, b_2 \rangle$, *then* $\langle b_1, b_2 \rangle \in J$,
(2) *for any* $X_1 \subseteq A_1$ *and* $a_2 \in A_2$, *if* $\langle a_1, a_2 \rangle \in J$ *for all* $a_1 \in X_1$, *then* $\langle b_1, a_2 \rangle \in J$ *for all* $b_1 \in \text{lub}_{R_1} X_1$,
(3) *for any* $X_2 \subseteq A_2$ *and* $a_1 \in A_1$, *if* $\langle a_1, a_2 \rangle \in J$ *for all* $a_2 \in X_2$, *then* $\langle a_1, b_2 \rangle \in J$ *for all* $b_2 \in \text{glb}_{R_2} X_2$.

(In what follows, when we use the expression $\langle \mathcal{A}_1, \mathcal{A}_2, J \rangle$, we presuppose that $\mathcal{A}_1 = \langle A_1, R_1 \rangle$ and $\mathcal{A}_2 = \langle A_2, R_2 \rangle$.)

If $\langle \mathcal{A}_1, \mathcal{A}_2, J \rangle$ is a joining-system, then the elements in J are called *joinings* from $\mathcal{A}_1$ to $\mathcal{A}_2$, and we call J the *joining-space* in $\langle \mathcal{A}_1, \mathcal{A}_2, J \rangle$. We call $\mathcal{A}_1$ the *bottom-structure* and $\mathcal{A}_2$ the *top-structure* in the Js $\langle \mathcal{A}_1, \mathcal{A}_2, J \rangle$.

Requirement (1) in the definition of a joining-system means that the joining-space J is an up-set with respect to the narrowness-relation. Note that from requirement (1) it follows, for example, that if $\mathcal{A}_1, \mathcal{A}_2$ are lattices such that $a_1, b_1 \in A_1$, $a_2, b_2 \in A_2$ and $\langle a_1, a_2 \rangle \in J$ then, $\langle a_1 \wedge b_1, a_2 \rangle \in J$ and $\langle a_1, a_2 \vee b_2 \rangle \in J$.

As an analogy, in propositional logic, for the implicative connective $\rightarrow$ it holds that from the conjunction of $p_1 \rightarrow q_1$ and $p_2 \rightarrow q_2$ it follows that

---

[15] For the notion of "up-set" in general, see for example [Davey and Priestley, 2002, p. 20].

if $q_1 \to p_2$ then $p_1 \to q_2$. Requirement (1) stipulates a similar result for a combination of the three implicative relations R_1, R_2 and J in a joining-system.

For a joining-system $\langle \mathcal{A}_1, \mathcal{A}_2, J \rangle$ conceived of as representing a normative system, let us interpret a formula $\langle x_1, x_2 \rangle \in J$ so as to mean that $\langle x_1, x_2 \rangle$ is a norm in $\langle \mathcal{A}_1, \mathcal{A}_2, J \rangle$. Then the import of requirement (1) is that if it holds that $\langle a_1, a_2 \rangle$ is a norm in $\langle \mathcal{A}_1, \mathcal{A}_2, J \rangle$ and $b_1 R a_1$ and $a_2 R b_2$ then $\langle b_1, b_2 \rangle$ as well is a norm in $\langle \mathcal{A}_1, \mathcal{A}_2, J \rangle$. This requirement is a corner-stone in the TJS approach to normative systems as deductive mechanisms. In a sense, a normative system $\langle \mathcal{A}_1, \mathcal{A}_2, J \rangle$ is represented by the quasi-ordering $\langle J, \trianglelefteq \rangle$. As we shall see, however, there are other representations that are more economical in expression.

The import of requirements (2) and (3) is easier to see if we suppose that $\langle \mathcal{A}_1, R_1 \rangle$ and $\langle \mathcal{A}_2, R_2 \rangle$ are lattices so that $\wedge$ and $\vee$ are defined for the elements in A_1 and A_2, respectively. In this case, from requirements (2) and (3) it follows: If $\langle a_1, a_2 \rangle \in J$ and $\langle b_1, a_2 \rangle \in J$ then $\langle a_1 \vee b_1, a_2 \rangle \in J$ (requirement (2)). And if $\langle a_1, a_2 \rangle \in J$ and $\langle a_1, b_2 \rangle \in J$ then $\langle a_1, a_2 \wedge b_2 \rangle \in J$ (requirement (3)).

We note that a joining-system as here defined gives rise to a closure system (see Section 3.2.5 below). Also, we note that in requirement (2) we do not presuppose that $\text{lub}_{R_1} X_1 \neq \varnothing$ and in requirement (3) we do not presuppose that $\text{glb}_{R_2} X_2 \neq \varnothing$. Furthermore note that $\langle \mathcal{A}_1, \mathcal{A}_2, \varnothing \rangle$ and $\langle \mathcal{A}_1, \mathcal{A}_2, A_1 \times A_2 \rangle$ are joining-systems, the *empty* joining-system and the *trivial* joining-system respectively. A joining-system that is not empty or trivial is called a *proper* joining-system.

In the definition of a joining-system, we do not presuppose that the domains in the quasi-orderings are disjunct sets. This is indeed the case in many intended applications, but in a large number of typical applications there is some overlap between the domains. The following remark will elucidate this situation.

Suppose that $\mathcal{B}_1 = \langle B_1, \wedge_1, \prime_1 \rangle$ and $\mathcal{B}_2 = \langle B_2, \wedge_2, \prime_2 \rangle$ are Boolean algebras and that $\leq_1$ and $\leq_2$ are the partial orderings determined by the Boolean algebras $\mathcal{B}_1$ and $\mathcal{B}_2$ respectively. Suppose further that $\langle \langle B_1, \leq_1 \rangle, \langle B_2, \leq_2 \rangle, J \rangle$ is a joining-system. From a formal point of view, it is possible that $\mathcal{B}_1$ and $\mathcal{B}_2$ are independent of each other, so that, for example the zero and unit elements in $\mathcal{B}_1$ are different from the zero and unit elements in $\mathcal{B}_2$.

In many applications, however, $\mathcal{B}_1$ and $\mathcal{B}_2$ are subalgebras of a common Boolean algebra $\mathcal{B} = \langle B, \wedge, \prime \rangle$, and if $\bot$ is the zero element in $\mathcal{B}$ and $\top$ is the unit element in $\mathcal{B}$, then this holds in $\mathcal{B}_1$ and $\mathcal{B}_2$ as well, and, hence, $\bot$ and $\top$ are elements in the intersection of B_1 and B_2. In this case it is also natural to denote $\wedge_1$ and $\wedge_2$ with $\wedge$ and, furthermore, $\prime_1$ and $\prime_2$ with $\prime$. In this chapter, when there is no risk of misunderstanding, we often use $\wedge$ and $\prime$ (without subscript) in various Boolean algebras even when the domains and operations are different.

3.2.3 Joinings as correspondences

For a joining-system $\langle \mathcal{A}_1, \mathcal{A}_2, J \rangle$ (where $\mathcal{A}_1 = \langle A_1, R_1 \rangle$ and $\mathcal{A}_2 = \langle A_2, R_2 \rangle$), the difference in kind between relations R_1, R_2 on one hand, and J on the other, becomes more perspicuous when we introduce the distinction between ordering relations and correspondences. Obviously, both relations R_1, R_2 and the relation J are sets of ordered pairs, i.e., relations in the sense of set theory. However, while the point of each of R_1 and R_2 is to order objects in a set, the point of J is to assign objects in one set A_2 to objects in another set A_1, or vice versa.[16] This idea of J as a correspondence between sets will prove to be useful in what follows. In particular, under some general conditions, by transition through equivalence classes, an "ordering preserving" correspondence will result in an isomorphism.

The triple $\langle X, Y, \gamma \rangle$ is a *correspondence* with X as domain and Y as codomain if X and Y are sets, γ is a binary relation, and $\gamma \subseteq X \times Y$.[17] Suppose that $\langle X, Y, \gamma \rangle$ is a correspondence. If $Z \subseteq X$ we define:

$$\gamma[Z] = \{y \in Y \mid \exists x \in Z : x\gamma y\}.$$

If $W \subseteq Y$ then

$$\gamma^{-1}[W] = \{x \in X \mid \exists y \in W : y\gamma^{-1}x\} = \{x \in X \mid \exists y \in W : x\gamma y\}.$$

The correspondence $\langle X, Y, \gamma \rangle$ is on X if $\gamma^{-1}[Y] = X$, onto Y if $\gamma[X] = Y$. If there is no risk of ambiguity, we denote $\gamma[\{a\}]$ with $\gamma[a]$ and $\gamma^{-1}[\{b\}]$ with $\gamma^{-1}[b]$.

If $\langle \mathcal{A}_1, \mathcal{A}_2, J \rangle$ is a Js then $\langle A_1, A_2, J \rangle$ is a correspondence with A_1 as domain and A_2 as codomain, and we can also say that J is a correspondence from A_1 to A_2.

Definition 3.12 *Suppose that $\langle A_1, A_2, \gamma \rangle$ is a correspondence from A_1 to A_2. If $\mathcal{A}_1 = \langle A_1, R_1 \rangle$ and $\mathcal{A}_2 = \langle A_2, R_2 \rangle$ are quasi-orderings, we say that $\Gamma = \langle \mathcal{A}_1, \mathcal{A}_2, \gamma \rangle$ is a quasi-ordering correspondence, abbreviated* qo-corr.

---

[16] Obviously, the idea of J as a correspondence should be distinguished from the fact that there are ordering relations over the set J of ordered pairs. As we have seen, in TJS the relation of narrowness is an ordering relation over the ordered pairs in J. Another ordering relation over J (to be introduced later on) is the relation "at least as low as".

[17] If the triple $\langle X, Y, \gamma \rangle$ is a correspondence, it is sometimes more convenient to say that γ is a correspondence from X to Y and that γ^{-1} is a correspondence from Y to X. If γ is a correspondence from X to Y, Y is often called the image of X by γ, or, shorter, the γ-image of X.

If $\langle \mathcal{A}_1, \mathcal{A}_2, J \rangle$ is a Js, then $\langle \mathcal{A}_1, \mathcal{A}_2, J \rangle$ is a qo-corr and $J[\mathcal{A}_1] \subseteq \mathcal{A}_2$, where $J[\mathcal{A}_1]$ contains the second components (belonging to $\mathcal{A}_2$) of the ordered pairs that are joinings from $\mathcal{A}_1$ to $\mathcal{A}_2$. Conversely, $J^{-1}[\mathcal{A}_2] \subseteq \mathcal{A}_1$, where $J^{-1}[\mathcal{A}_2]$ contains the first components (belonging to $\mathcal{A}_1$) of the joinings from $\mathcal{A}_1$ to $\mathcal{A}_2$.. Then $J^{-1}[\mathcal{A}_2]$ is the set of grounds and $J[\mathcal{A}_1]$ the set of consequences of the joinings in $\langle \mathcal{A}_1, \mathcal{A}_2, J \rangle$.

The relative product of two correspondences γ and δ is denoted $\gamma|\delta$. If $\langle \mathcal{A}_1, \mathcal{A}_2, J \rangle$ is a joining-system, then $R_1|J|R_2 = J$ and, therefore, J can be said to "absorb" R_1 and R_2. Note that $x_1(R_1|J|R_2)x_2$ iff $\exists y_1, y_2$: $x_1 R_1 y_1 \,\&\, y_1 J y_2 \,\&\, y_2 R_2 x_2$.

3.2.4 Order-preservation and order-similarity

The notion of qo-corr is a basis for the notions of "order-preservation" and "order-similarity". Suppose $\mathcal{A}_1 = \langle A_1, R_1 \rangle$ and $\mathcal{A}_2 = \langle A_2, R_2 \rangle$ are two strata, and that J is a qo-corr from $\mathcal{A}_1$ to $\mathcal{A}_2$. If $\langle \mathcal{A}_1, \mathcal{A}_2, J \rangle$ is order-preserving, Q_1-similar grounds in A_1 have the same consequences in A_2, Q_2-similar consequences in A_2 have the same grounds in A_1, and if $\langle a_1, a_2 \rangle$, $\langle b_1, b_2 \rangle$ are joinings from $\mathcal{A}_1$ to $\mathcal{A}_2$, then the R_1-structure on $\{a_1, b_1\}$ is similar to the R_2-structure on $\{a_2, b_2\}$ insofar as $a_1 R_1 b_1$ iff $a_2 R_2 b_2$. The general definition is as follows.

Definition 3.13 *Suppose that* $\Gamma = \langle \langle A_1, R_1 \rangle, \langle A_2, R_2 \rangle, \gamma \rangle$ *is a* qo-corr. *We say that* Γ *is order-preserving if the following holds for* $a_1, b_1 \in A_1$ *and* $a_2, b_2 \in A_2$:

(1) *If* $a_1 Q_1 b_1$ *then* $(a_1 \gamma a_2$ *iff* $b_1 \gamma a_2)$.
(2) *If* $a_2 Q_2 b_2$ *then* $(a_1 \gamma a_2$ *iff* $a_1 \gamma b_2)$.
(3) *If* $a_1 \gamma a_2$ *and* $b_1 \gamma b_2$ *then* $a_1 R_1 b_1$ *iff* $a_2 R_2 b_2$.

Definition 3.14 *Two quasi-orderings* $\langle A_1, R_1 \rangle$ *and* $\langle A_2, R_2 \rangle$ *are said to be order-similar if there is* $\gamma \subseteq A_1 \times A_2$ *such that* $\langle \langle A_1, R_1 \rangle, \langle A_2, R_2 \rangle, \gamma \rangle$ *is an order-preserving* qo-corr *on* A_1 *onto* A_2.

The notion of "order-preserving qo-corr" is elucidated by the fact that by transition from quasi-orderings to equivalence classes you get an isomorphism between the resulting structures; also, if there is an isomorphism between the equivalence classes, there is order-preservation between the quasi-orderings.

Theorem 3.15 *Suppose that* $\langle \langle A_1, R_1 \rangle, \langle A_2, R_2 \rangle, \gamma \rangle$ *is a* qo-corr *on* A_1 *onto* A_2. *Let* $[a]_i$ $[b]_i$ *be the equivalence-classes with respect to* Q_i *generated by* a *and* b, *respectively* $(i = 1, 2)$. *Let further* $A_1^* = \{[a]_1 \mid a \in \gamma^{-1}[A_2]\}$

and $A_2^* = \{[a]_2 \mid a \in \gamma [A_1]\}$ and let R_i^* be defined as follows: $[a]_i R_i^* [b]_i$ iff $a R_i b$.

(1) Suppose that $\langle\langle A_1, R_1\rangle, \langle A_2, R_2\rangle, \gamma\rangle$ is an order-preserving qo-corr and let γ^* be defined by $[a_1]_1 \gamma^* [a_2]_2$ iff $a_1 \gamma a_2$. Then γ^* is an isomorphism on $\langle A_1^*, R_1^*\rangle$ onto $\langle A_2^*, R_2^*\rangle$. If $\langle A_1, R_1\rangle$ and $\langle A_2, R_2\rangle$ are quasi-lattices (see Definition 3.4), then γ^* is an isomorphism on the lattice $\langle A_1^*, R_1^*\rangle$ onto the lattice $\langle A_2^*, R_2^*\rangle$.

(2) If φ is an isomorphism on $\langle A_1^*, R_1^*\rangle$ onto $\langle A_2^*, R_2^*\rangle$, then

$$\langle\langle A_1, R_1\rangle, \langle A_2, R_2\rangle, \gamma\rangle$$

is an order-preserving qo-corr on A_1 onto A_2, where γ is defined by $a_1 \gamma a_2$ iff $\varphi([a_1]_1) = [a_2]_2$.

3.2.5 Joining-closure and the generating of joining-spaces

An important aspect of TJS is that it gives a method (the forming of a "joining-closure") for representing an "elaborated" version of a set of "crude" conditional norms. Suppose that $\mathcal{A}_1$ is a quasi-ordering of grounds and $\mathcal{A}_2$ is a quasi-ordering of consequences. Let us suppose that K is a set of conditional norms with the antecedents taken from $\mathcal{A}_1$ and the consequences taken from $\mathcal{A}_2$. Hence, $K \subseteq A_1 \times A_2$ and K is a correspondence from A_1 to A_2. The set K can be thought of as a crude representation of a normative system $\mathcal{N}$. Then we can generate a set K^* by forming the "joining closure" of K such that $\langle \mathcal{A}_1, \mathcal{A}_2, K^*\rangle$ is a joining-system, which will be explained below.

The next theorem shows that if $\mathcal{A}_1$ and $\mathcal{A}_2$ are quasi-orderings and

$$\mathcal{J} = \{J \subseteq A_1 \times A_2 \mid \langle \mathcal{A}_1, \mathcal{A}_2, J\rangle \text{ is a } Js\},$$

then $\mathcal{J}$ is a closure system.[18] Note that $\mathcal{J}$ is the family of all joining-spaces from $\mathcal{A}_1$ to $\mathcal{A}_2$.

Theorem 3.16 If $\mathcal{J} = \{J \subseteq A_1 \times A_2 \mid \langle \mathcal{A}_1, \mathcal{A}_2, J\rangle \text{ is a } Js\}$ and $\mathcal{K} \subseteq \mathcal{J}$, then $\cap \mathcal{K} \in \mathcal{J}$.

Proof. If $\cap \mathcal{K} = \varnothing$, then $\langle \mathcal{A}_1, \mathcal{A}_2, \cap \mathcal{K}\rangle$ is the empty joining-system and hence $\cap \mathcal{K} \in \mathcal{J}$. Now suppose that $\cap \mathcal{K} \neq \varnothing$.

(I) Firstly, we prove that condition (1) in the definition of a joining-system is satisfied. Suppose therefore that $b_i, c_i \in A_i$ for $i = 1, 2$ and $\langle b_1, b_2\rangle \in \cap \mathcal{K}$ and $\langle b_1, b_2\rangle \trianglelefteq \langle c_1, c_2\rangle$. Let $K \in \mathcal{K}$. Then $\cap \mathcal{K} \subseteq K$ and thus $\langle b_1, b_2\rangle \in K$.

---

[18] For definition and results of closure systems, see for example [Grätzer, 1979, p. 23f.].

Since $K \in \mathcal{J}$ and $\langle b_1, b_2 \rangle \trianglelefteq \langle c_1, c_2 \rangle$ it follows that $\langle c_1, c_2 \rangle \in K$. Hence, for all $K \in \mathcal{K}$, $\langle c_1, c_2 \rangle \in K$ which implies $\langle c_1, c_2 \rangle \in \cap \mathcal{K}$.

(II) Secondly, we prove that condition (2) in the definition of a joining-system is satisfied. Suppose that $C_1 \subseteq A_1$, $b_2 \in A_2$, and $\langle c_1, b_2 \rangle \in \cap \mathcal{K}$ for all $c_1 \in C_1$. Then $\langle c_1, b_2 \rangle \in K$ for all $c_1 \in C_1$ and $K \in \mathcal{K}$. Since $K \in \mathcal{J}$ it follows that $\langle a_1, b_2 \rangle \in K$ for all $a_1 \in \mathrm{lub}_{R_1} C_1$. Hence, for all $K \in \mathcal{K}$, $\langle a_1, b_2 \rangle \in K$ for all $a_1 \in \mathrm{lub}_{R_1} C_1$, which implies $\langle a_1, b_2 \rangle \in \cap \mathcal{K}$ for all $a_1 \in \mathrm{lub}_{R_1} C_1$.

(III) Thirdly, we prove that condition (3) in the definition of a joining-system is satisfied. Suppose that $C_2 \subseteq A_2$, $b_1 \in A_1$, and $\langle b_1, c_2 \rangle \in \cap \mathcal{K}$ for all $c_2 \in C_2$. Then $\langle b_1, c_2 \rangle \in K$ for all $c_2 \in C_2$ and $K \in \mathcal{K}$. Since $K \in \mathcal{J}$ it follows that $\langle b_1, a_2 \rangle \in K$ for all $a_2 \in \mathrm{glb}_{R_2} C_2$. Hence, for all $K \in \mathcal{K}$, $\langle b_1, a_2 \rangle \in K$ for all $a_2 \in \mathrm{glb}_{R_2} C_2$, which implies $\langle b_1, a_2 \rangle \in \cap \mathcal{K}$ for all $a_2 \in \mathrm{glb}_{R_2} C_2$. ∎

From the theorem follows that if $K \subseteq A_1 \times A_2$ and

$$[K]_\mathcal{J} = \cap \{ J \mid J \in \mathcal{J},\ J \supseteq K \},$$

then $[K]_\mathcal{J}$ is the joining-space, here called the *joining-closure*, over $\mathcal{A}_1$ and $\mathcal{A}_2$ *generated* by K. (Note that since $A_1 \times A_2$ is a joining space, $\{ J \mid J \in \mathcal{J},\ J \supseteq K \} \neq \varnothing$.)

If J is the joining-closure from $\mathcal{A}_1$ to $\mathcal{A}_2$ generated by K but J is not generated by any proper subset of K, then we say that J is the joining-closure *non-redundantly generated* by K.

3.3 Weakest grounds, strongest consequences and minimal joinings

3.3.1 Weakest grounds and strongest consequences

Definition 3.17 *Suppose that $\mathcal{S} = \langle \mathcal{A}_1, \mathcal{A}_2, J \rangle$ is a joining-system, and that $C_1 \subseteq A_1$ and $C_2 \subseteq A_2$. Then,*

1. *$a_1 \in C_1 \subseteq A_1$ is one of the weakest grounds of $a_2 \in A_2$ in C_1 with respect to $\mathcal{S}$, which is denoted $\mathrm{WG}_\mathcal{S}(a_1, a_2, C_1)$, if*

   $\langle a_1, a_2 \rangle \in J$ *and, for any $b_1 \in C_1$,*
   it holds that $\langle b_1, a_2 \rangle \in J$ implies $b_1 R_1 a_1$.

2. *$a_2 \in C_2 \subseteq A_2$ is one of the strongest consequences of $a_1 \in A_1$ in C_2 with respect to $\mathcal{S}$, which is denoted $\mathrm{SC}_\mathcal{S}(a_2, a_1, C_2)$, if*

   $\langle a_1, a_2 \rangle \in J$, *and, for any $b_2 \in C_2$,*
   it holds that $\langle a_1, b_2 \rangle \in J$ implies $a_2 R_2 b_2$.

In Section 3.3.2, the interrelationship between minimal joinings and weakest grounds, strongest consequences will be further developed. Below, however, are some basic results. (Cf. [Lindahl and Odelstad, 2011, sect. 3.2].)

Theorem 3.18 *Let $\langle \mathcal{A}_1, \mathcal{A}_2, J \rangle$ be a joining-system.*
(1) *Suppose that* WG $(a_1, a_2, \mathcal{A}_1)$ *and* WG $(b_1, b_2, \mathcal{A}_1)$. *If $a_2 R_2 b_2$, then $a_1 R_1 b_1$.*
(2) *Suppose that* SC $(a_2, a_1, \mathcal{A}_2)$ *and* SC $(b_2, b_1, \mathcal{A}_2)$ *If $a_1 R_1 b_1$, then $a_2 R_2 b_2$.*
(3) *Suppose that* WG $(a_1, a_2, \mathcal{A}_1)$ *and* WG $(b_1, b_2, \mathcal{A}_1)$. *For all $c_1 \in \mathcal{A}_1$ and $c_2 \in \mathcal{A}_2$, if $c_1 \in \text{glb}_{R_1}\{a_1, b_1\}$ and $c_2 \in \text{glb}_{R_2}\{a_2, b_2\}$, then* WG $(c_1, c_2, \mathcal{A}_1)$.
(4) *Suppose that* SC $(a_2, a_1, \mathcal{A}_2)$ *and* SC $(b_2, b_1, \mathcal{A}_2)$. *For all $c_1 \in \mathcal{A}_1$ and $c_2 \in \mathcal{A}_2$, if $c_1 \in \text{lub}_{R_1}\{a_1, b_1\}$ and $c_2 \in \text{lub}_{R_2}\{a_2, b_2\}$, then* SC $(c_2, c_1, \mathcal{A}_2)$.

Proof. We prove (3). Note that $a_1 J a_2$ and $b_1 J b_2$. Suppose that $c_1 \in \text{glb}_{R_1}\{a_1, b_1\}$ and $c_2 \in \text{glb}_{R_2}\{a_2, b_2\}$. Hence, $c_1 J a_2$ and $c_1 J b_2$ and according to condition (3) in the definition of a joining-system, $c_1 J c_2$. Suppose that $d_1 J c_2$. Then $d_1 J a_2$ and $d_1 J b_2$, and since WG $(a_1, a_2, \mathcal{A}_1)$ and WG $(b_1, b_2, \mathcal{A}_1)$ it follows that $d_1 R_1 a_1$ and $d_1 R_1 b_1$ which implies that $d_1 R_1 c_1$. Thus WG $(c_1, c_2, \mathcal{A}_1)$. ∎

Item (1) in Theorem 3.18 is illustrated by Figure 5, and item (2) by Figure 6.

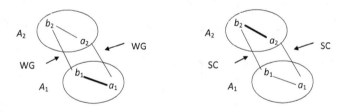

Thick line is conclusion   Thick line is conclusion

Figure 5   Figure 6

Theorem 3.19 *Let $\langle \mathcal{A}_1, \mathcal{A}_2, J \rangle$ be a joining-system.*
(1) *Suppose that $\mathcal{A}_1$ is a complete quasi-lattice (see Definition 3.4). Then* WG $(a_1, a_2, \mathcal{A}_1)$ *iff $a_1 \in \text{lub}_{R_1} J^{-1}[a_2]$.*
(2) *Suppose that $\mathcal{A}_2$ is a complete quasi-lattice. Then* SC $(a_2, a_1, \mathcal{A}_2)$ *iff $a_2 \in \text{glb}_{R_2} J[a_1]$.*

Proof. We prove (1) above. (i) Suppose that $\mathrm{WG}\,(a_1, a_2, A_1)$. Hence, $a_1 \in J^{-1}[a_2]$. Since $\mathcal{A}_1$ is a complete quasi-lattice it follows that there is $b_1 \in \mathrm{lub}_{R_1} J^{-1}[a_2]$ and $a_1 R_1 b_1$. From condition (2) of a joining-system it follows that $\langle b_1, a_2 \rangle \in J$. Since $\mathrm{WG}(a_1, a_2, A_1)$, it follows that $b_1 R_1 a_1$. Together with $a_1 R_1 b_1$, this implies $a_1 Q_1 b_1$. Thus $a_1 \in \mathrm{lub}_{R_1} J^{-1}[a_2]$. (ii) Suppose that $a_1 \in \mathrm{lub}_{R_1} J^{-1}[a_2]$. If $\langle b_1, a_2 \rangle \in J$ then $b_1 \in J^{-1}[a_2]$ and hence $b_1 R_1 a_1$. From this follows that $\mathrm{WG}\,(a_1, a_2, A_1)$. (Note that this part of the proof does not require that $\mathcal{A}_2$ is a complete quasi-lattice.) The proof of (2) is analogous. ∎

3.3.2 Minimal joinings

Minimal joinings in a Js will be a central theme in the subsequent presentation. The formal definition is as follows (we recall the definition of "minimal element" with respect to narrowness in Definition 3.9).

Definition 3.20 *Suppose that $\langle \mathcal{A}_1, \mathcal{A}_2, K \rangle$ is a qo-corr. A minimal element in $\langle \mathcal{A}_1, \mathcal{A}_2, K \rangle$ is a minimal element $\langle a_1, a_2 \rangle$ in K with respect to $\mathcal{A}_1$ and $\mathcal{A}_2$. The set of minimal elements in $\langle \mathcal{A}_1, \mathcal{A}_2, K \rangle$ is denoted $\min \langle \mathcal{A}_1, \mathcal{A}_2, K \rangle$ or just $\min K$.*

If $\langle \mathcal{A}_1, \mathcal{A}_2, J \rangle$ is a joining-system, then the elements in $\min J$ are often called minimal joinings. The connection between the notion of minimal joining on one hand and the notions of weakest ground and strongest consequence on the other side is made clear in the following theorem.

Theorem 3.21 *Suppose that $\langle \mathcal{A}_1, \mathcal{A}_2, J \rangle$ is a joining-system. Then $\langle a_1, a_2 \rangle \in \min J$ iff $\mathrm{WG}\,(a_1, a_2, A_1)$ and $\mathrm{SC}\,(a_2, a_1, A_2)$. See Figure 7.*

A proof of the theorem under the assumption that $\langle \mathcal{A}_1, \mathcal{A}_2, J \rangle$ is a Boolean joining-system is given in [Lindahl and Odelstad, 2011, theorem 36, p. 126], but it is easy to see that the theorem holds even if $\langle \mathcal{A}_1, \mathcal{A}_2, J \rangle$ is a mere joining-system.

3.4 Connectivity

As stated in the introductory Section 2.1.2, if a normative system fulfils a requirement called "connectivity", any norm in the system will always be implied by a minimal joining. Therefore, the idea of connectivity will be essential in the theory of minimal joinings to be developed in the next subsections. The definition of connectivity is given next.

Definition 3.22 *A qo-corr $\langle \mathcal{A}_1, \mathcal{A}_2, K \rangle$ such that K is an up-set with respect to $\trianglelefteq$ satisfies connectivity if whenever $\langle c_1, c_2 \rangle \in K$ there is $\langle b_1, b_2 \rangle \in K$ such that $\langle b_1, b_2 \rangle$ is a minimal element in K with respect to $\trianglelefteq$ and $\langle b_1, b_2 \rangle \trianglelefteq \langle c_1, c_2 \rangle$.*

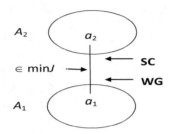

Thick character is conclusion

Figure 7

Definition 3.23 *Suppose that* $\langle \mathcal{A}_1, \mathcal{A}_2, K \rangle$ *is a qo-corr. Then the set*

$$\{\langle a_1, a_2 \rangle \in \mathcal{A}_1 \times \mathcal{A}_2 \mid \exists \langle b_1, b_2 \rangle \in K : \langle b_1, b_2 \rangle \trianglelefteq \langle a_1, a_2 \rangle\}$$

is called the enclosure of K *and is denoted* $\uparrow K$.

Note that $\uparrow K$ is an up-set (with respect to $\trianglelefteq$) and the smallest up-set containing K. (For the notion of up-set see Definition 3.10 in Section 3.2.1.) To use an expression from lattice theory, $\uparrow K$ is read 'up K' (with respect to $\trianglelefteq$). (See [Davey and Priestley, 2002, p. 20].) Note also that K is an up-set if and only if $K = \uparrow K$.

Theorem 3.24 *Suppose that* $\langle \mathcal{A}_1, \mathcal{A}_2, K \rangle$ *is a qo-corr such that K is an up-set with respect to $\trianglelefteq$. Then* $\langle \mathcal{A}_1, \mathcal{A}_2, K \rangle$ *satisfies connectivity iff* $K = \uparrow \min K$.

Proof. (I) Suppose $\langle \mathcal{A}_1, \mathcal{A}_2, K \rangle$ satisfies connectivity. (i) Suppose $\langle a_1, a_2 \rangle \in K$. Then there is $\langle b_1, b_2 \rangle \in \min K$ such that $\langle b_1, b_2 \rangle \trianglelefteq \langle a_1, a_2 \rangle$ and hence $\langle a_1, a_2 \rangle \in \uparrow \min K$. This shows that $K \subseteq \uparrow \min K$. (ii) Suppose $\langle a_1, a_2 \rangle \in \uparrow \min K$. Then there is $\langle b_1, b_2 \rangle \in \min K$ such that $\langle b_1, b_2 \rangle \trianglelefteq \langle a_1, a_2 \rangle$. Since $\langle \mathcal{A}_1, \mathcal{A}_2, K \rangle$ is a *qo-corr* such that K is an up-set with respect to $\trianglelefteq$, $\langle a_1, a_2 \rangle \in K$. Hence, $\uparrow \min K \subseteq K$.

(II) Suppose that $K = \uparrow \min K$ and that $\langle a_1, a_2 \rangle \in K$. Then $\langle a_1, a_2 \rangle \in \uparrow \min K$ and there is $\langle b_1, b_2 \rangle \in \min K$ such that $\langle b_1, b_2 \rangle \trianglelefteq \langle a_1, a_2 \rangle$. This shows that $\langle \mathcal{A}_1, \mathcal{A}_2, K \rangle$ satisfies connectivity. ∎

If a joining-system satisfies connectivity, then the set of minimal joinings determines the system in an interesting way, which will be explained below.

Corollary 3.25 *If the joining–system $\langle \mathcal{A}_1, \mathcal{A}_2, J \rangle$ satisfies connectivity, then $J = \uparrow \min J$, that is,*

$$J = \{\langle a_1, a_2\rangle \in A_1 \times A_2 \mid \exists \langle b_1, b_2\rangle \in \min J : \langle b_1, b_2\rangle \trianglelefteq \langle a_1, a_2\rangle\}.$$

The corollary shows that there is an interesting way of representing a normative system in terms of $\trianglelefteq$-minimal elements. This way of representing is different from the method of "joining-closure" presented above in Section 3.2.5 and we will here develop it a little further.

Note that we have not so far said anything about how to get a joining-system using the enclosure of a *qo-corr* (Definition 3.23). We will return to this problem in Section 3.6.

Theorem 3.26 *If $\mathcal{A}_1 = \langle A_1, R_1\rangle$ and $\mathcal{A}_2 = \langle A_2, R_2\rangle$ are complete quasi-lattices (see Definition 3.4, Section 3.1.2), and $\langle \mathcal{A}_1, \mathcal{A}_2, J\rangle$ is a joining-system, then $\langle \mathcal{A}_1, \mathcal{A}_2, J\rangle$ satisfies connectivity.*

Proof. Suppose $\langle c_1, c_2\rangle \in J$. Let $X_1 = \{x_1 \in A_1 \mid \langle x_1, c_2\rangle \in J\}$. Since $\mathcal{A}_1$ is a complete quasi-lattice it holds that $\mathrm{lub}\, X_1 \neq \varnothing$. Let $b_1 \in \mathrm{lub}\, X_1$. From (2) in the definition of a joining-system follows that $\langle b_1, c_2\rangle \in J$ and hence $b_1 \in X_1$. Let $X_2 = \{x_2 \in A_2 \mid \langle b_1, x_2\rangle \in J\}$. Since $\langle b_1, c_2\rangle \in J$, $X_2 \neq \varnothing$. $\mathcal{A}_2$ is a complete quasi-lattice and therefore it holds that $\mathrm{glb}\, X_2 \neq \varnothing$. Let $b_2 \in \mathrm{glb}\, X_2$. From (3) in the definition of a joining-system follows that $\langle b_1, b_2\rangle \in J$ and hence $b_2 \in X_2$. Since $c_1 \in X_1$ and $b_1 \in \mathrm{lub}\, X_1$ then $c_1 R_1 b_1$. And since $c_2 \in X_2$ and $b_2 \in \mathrm{glb}\, X_2$ then $b_2 R_2 c_2$. Hence, $\langle b_1, b_2\rangle \trianglelefteq \langle c_1, c_2\rangle$.

Suppose now that $\langle a_1, a_2\rangle \in J$ and $\langle a_1, a_2\rangle \trianglelefteq \langle b_1, b_2\rangle$. Thus $c_1 R_1 b_1 R_1 a_1$ and $a_2 R_2 b_2 R_2 c_2$, which implies that $\langle a_1, a_2\rangle \trianglelefteq \langle a_1, c_2\rangle$ and $\langle a_1, a_2\rangle \trianglelefteq \langle b_1, a_2\rangle$. According to condition (1) in the definition of a joining-system, it follows that $\langle a_1, c_2\rangle, \langle b_1, a_2\rangle \in J$ and thus $a_1 \in X_1$ and $a_2 \in X_2$. Since $b_1 \in \mathrm{ub}_{R_1} X_1$ it follows that $a_1 R_1 b_1$, and since $b_2 \in \mathrm{lb}_{R_2} X_2$ it follows that $b_2 R_2 a_2$. Hence, $a_1 Q_1 b_1$ and $a_2 Q_2 b_2$, and we conclude that $\langle b_1, b_2\rangle$ is a minimal element in $\langle \mathcal{A}_1, \mathcal{A}_2, J\rangle$. ∎

The next theorem states that if connectivity holds, then a weakest ground of an element is the bottom of a minimal joining and a strongest consequence of an element is the top of a minimal joining.

Theorem 3.27 *Suppose that $\langle \mathcal{A}_1, \mathcal{A}_2, J\rangle$ is a joining-system which satisfies connectivity (see Definition 3.22). Then:*

1. If $\mathrm{WG}(a_1, a_2, A_1)$ then there is $b_2 \in A_2$ such that $\langle a_1, b_2 \rangle \in \min J$ and $b_2 R_2 a_2$.

2. If $\mathrm{SC}(a_2, a_1, A_2)$ then there is $b_1 \in A_1$ such that $\langle b_1, a_2 \rangle \in \min J$ and $a_1 R_1 b_1$.

(For a proof, see [Odelstad, 2008, pp. 50f.].)

Considering a joining-system $\langle \mathcal{A}_1, \mathcal{A}_2, J \rangle$, a useful device is the introduction of projections $\pi_1[J] \subseteq A_1$ and $\pi_2[J] \subseteq A_2$, which implies that each $a_1 \in \pi_1[J]$ is a "ground" for some element a_2 of A_2 and, conversely, each $a_2 \in \pi_2[J]$ is a "consequence" of some element a_1 of A_1. The general definition is as follows.

Definition 3.28 *For sets A_1 and A_2, if $X \subseteq A_1 \times A_2$ then for $i = 1, 2$, $\pi_i : X \to A_i$ is such that $\pi_i(x_1, x_2) = x_i$ is the projection of X on the ith coordinate.*

Note that if $X \subseteq A_1 \times A_2$ then $\pi_1[X] = \{x_1 \in A_1 \mid \exists x_2 \in A_2 : \langle x_1, x_2 \rangle \in X\}$

$$\pi_2[X] = \{x_2 \in A_2 \mid \exists x_1 \in A_1 : \langle x_1, x_2 \rangle \in X\}$$

The subsequent Theorem 3.30 might be easier to grasp if we first consider the special case of a joining-system $\langle \mathcal{L}_1, \mathcal{L}_2, J \rangle$ where $\mathcal{L}_1 = \langle L_1, \wedge, \vee \rangle$, $\mathcal{L}_2 = \langle L_2, \wedge, \vee \rangle$ are lattices and $\leq_1$, $\leq_2$ are the partial orderings determined by these lattices. Then, according to Theorem 3.30, if $\langle a_1, a_2 \rangle, \langle b_1, b_2 \rangle \in \min J$, there is $c_2 \in L_2$, $d_1 \in L_1$ such that

(1) $\langle a_1 \wedge b_1, c_2 \rangle \in \min J$,

(2) $\langle d_1, a_2 \vee b_2 \rangle \in \min J$,

(3) $c_2 \leq_2 a_2 \wedge b_2$,

(4) $a_1 \vee b_1 \leq_1 d_1$.

The following theorem is used in the proof of Theorem 3.30.

Theorem 3.29 *Suppose that $\langle \mathcal{A}_1, \mathcal{A}_2, J \rangle$ is a joining-system that satisfies connectivity. Then the following holds:*

(i) *If $\langle a_1, a_2 \rangle \in \min J$, then $\langle a_1, b_2 \rangle \in J$ implies $a_2 R_2 b_2$ and $\langle b_1, a_2 \rangle \in J$ implies $b_1 R_1 a_1$. (See Figure 8 on page 581.)*

(ii) *If $\langle a_1, a_2 \rangle, \langle b_1, b_2 \rangle \in \min J$ then $a_1 R_1 b_1$ iff $a_2 R_2 b_2$.*

(iii) *If $\langle a_1, a_2 \rangle \in \min J$ then $\langle a_1, b_2 \rangle \in \min J$ implies $a_2 Q_2 b_2$ and $\langle b_1, a_2 \rangle \in \min J$ implies $a_1 Q_1 b_1$. (See Figure 9 on page 582.)*

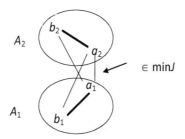

Thick line is conclusion

Figure 8

(For a proof, see [Odelstad, 2008, p. 51].)

Theorem 3.30 *Suppose that $\langle \mathcal{A}_1, \mathcal{A}_2, J \rangle$ is a joining-system and that $\mathcal{A}_1 = \langle A_1, R_1 \rangle$ and $\mathcal{A}_2 = \langle A_2, R_2 \rangle$ are complete quasi-lattices. If $X \subseteq \min J$ and $X \neq \emptyset$ then the following holds:*

(1) *There is $c_2 \in A_2$ such that for all $a_1 \in \mathrm{glb}_{R_1} \pi_1[X]$, $\langle a_1, c_2 \rangle \in \min J$, and, furthermore, it holds that $c_2 R_2 a_2$ for all $a_2 \in \mathrm{glb}_{R_2} \pi_2[X]$.*

(2) *There is $d_1 \in A_1$ such that for all $b_2 \in \mathrm{lub}_{R_2} \pi_2[X]$, $\langle d_1, b_2 \rangle \in \min J$, and, furthermore, it holds that $b_1 R_1 d_1$ for all $b_1 \in \mathrm{lub}_{R_1} \pi_1[X]$.*

Proof. Since $\mathcal{A}_1$ and $\mathcal{A}_2$ are complete quasi-lattices it follows from Theorem 3.26 that $\langle \mathcal{A}_1, \mathcal{A}_2, J \rangle$ satisfies connectivity.
(I) We prove (1). Since $\mathcal{A}_1$ is a complete quasi-lattice, it follows that there is $a_1 \in \mathrm{glb}_{R_1} \pi_1[X]$. Suppose that $x_2 \in \pi_2[X]$. Then there is $x_1 \in \pi_1[X]$ such that $\langle x_1, x_2 \rangle \in X$ and $\langle x_1, x_2 \rangle \trianglelefteq \langle a_1, x_2 \rangle$. Since $X \subseteq J$ it follows that $\langle a_1, x_2 \rangle \in J$ and this holds for all $x_2 \in \pi_2[X]$. Since $\mathcal{A}_2$ is a complete quasi-lattice, it follows that $\mathrm{glb}_{R_2} \pi_2[X] \neq \emptyset$. Let $a_2 \in \mathrm{glb}_{R_2} \pi_2[X]$. From condition (3) in the definition of a Js it follows that $\langle a_1, a_2 \rangle \in J$. Since J satisfies connectivity it follows that there is $\langle c_1, c_2 \rangle \in \min J$ such that $\langle c_1, c_2 \rangle \trianglelefteq \langle a_1, a_2 \rangle$. Let $\langle z_1, z_2 \rangle \in X$, which implies that $\langle z_1, z_2 \rangle \in \min J$ and since $z_2 \in \pi_2[X]$ and $a_2 \in \mathrm{glb}_{R_2} \pi_2[X]$ it follows that $a_2 R_2 z_2$. Furthermore, $c_2 R_2 a_2$ and thus $c_2 R_2 z_2$, which implies according to (ii) in theorem 3.29, that $c_1 R_1 z_1$. Hence, $c_1 \in \mathrm{lb}_{R_1} \pi_1[X]$. Since $a_1 \in \mathrm{glb}_{R_1} \pi_1[X]$ it follows that $c_1 R_1 a_1$, and since $a_1 R_1 c_1$ this implies $a_1 Q_1 c_1$. This shows that $\langle a_1, c_2 \rangle \in \min J$. Note that $c_2 R_2 a_2$.
(II) The proof of (2) is analogous. ∎

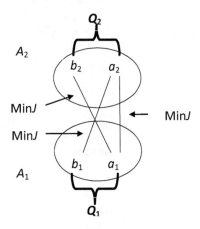

Figure 9

An illustration in a lattice framework of (1) and (2) in Theorem 3.30 is provided in Figures 10 on page 583 and Figure 11 on page 584, respectively.

3.5 Lowerness

In the literature on partial orderings, the notion "coordinatewise ordering" of a Cartesian product of partial ordered sets is introduced (see for example [Davey and Priestley, 2002, p. 18].) It is straight forward to generalize this notion to quasi-ordered sets. This is done in the definition below. With the interpretation of TJS in this chapter as a theory of normative systems, we call the relation "coordinatewise ordering" the *lowerness-relation*.

Definition 3.31 *The* lowerness relation determined by *the quasi-orderings* $\langle A_1, R_1 \rangle$ *and* $\langle A_2, R_2 \rangle$ *is the binary relation* $\precsim$ *on* $A_1 \times A_2$ *such that for all* $\langle a_1, a_2 \rangle, \langle b_1, b_2 \rangle \in A_1 \times A_2$

$$\langle a_1, a_2 \rangle \precsim \langle b_1, b_2 \rangle \quad \textit{iff} \quad a_1 R_1 b_1 \quad \textit{and} \quad a_2 R_2 b_2.$$

For elements in $A_1 \times A_2$ we read $\precsim$ as "at least as low as". If j_1 and j_2 are elements in $A_1 \times A_2$, then j_1 is at least as low as j_2, i.e. $j_1 \precsim j_2$, if the "bottom" of j_1 is at least as low as, i.e. stands in the relation R_1 to, the "bottom" of j_2, and the "top" of j_1 is at least as low as, i.e. stands in the relation R_2 to, the "top" of j_2. See Figure 12 on page 585. (As a contrast, see Figure 4 on page 569.) Note that $\precsim$ is a quasi-ordering, i.e. transitive

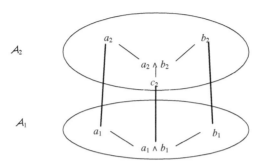

$\text{glb}_{R_1\pi_1}\{\langle a_1,a_2\rangle,\langle b_1,b_2\rangle\} = \text{glb}_{R_1}\{a_1,b_1\}=\{a_1 \wedge b_1\}$

$\text{glb}_{\leq/\min J}\{\langle a_1,a_2\rangle,\langle b_1,b_2\rangle\} = \{\langle a_1 \wedge b_1, c_2\rangle\},\ \pi_1\,[\text{glb}_{\leq/\min J}\{\langle a_1,a_2\rangle,\langle b_1,b_2\rangle\}] = \{a_1 \wedge b_1\}$

Figure 10

and reflexive. Let $\sim$ denote the equality part of $\precsim$ and $\prec$ the strict part of $\precsim$. Then the following holds:

$\langle a_1, a_2\rangle \sim \langle b_1, b_2\rangle$ iff $b_1 Q_1 a_1$ & $a_2 Q_2 b_2$
$\langle a_1, a_2\rangle \prec \langle b_1, b_2\rangle$ iff $(a_1 P_1 b_1$ & $a_2 R_2 b_2)$ or $(a_1 R_1 b_1$ & $a_2 P_2 b_2)$

where Q_i is the equality-part of R_i and P_i is the strict part of R_i.

The structure of the minimal joinings in a joining-system is similar to the structure of their "bottoms" and "tops". We recall the definition of projections π_i (Definition 3.28 in Section 3.4).

Theorem 3.32 *Suppose that $\langle \mathcal{A}_1, \mathcal{A}_2, J\rangle$ is a joining-system that satisfies connectivity (See Definition 3.22). Then for $i = 1, 2$, $\pi_i : \min J \longrightarrow \pi_i\,[\min J]$ is surjective, and the following holds:*

$$\text{for all } \alpha, \beta \in \min J,\ \alpha \precsim \beta \text{ iff } \pi_i\,(\alpha)\, R_i \pi_i\,(\beta).$$

Proof. Follows from Theorem 3.29, (ii). ∎

Corollary 3.33 *If $\langle \mathcal{A}_1, \mathcal{A}_2, J\rangle$ is a joining-system satisfying connectivity, then*

$$\langle\langle \pi_1\,[\min J], R_1\rangle, \langle \pi_2\,[\min J], R_2\rangle, \min J\rangle$$

is an order-preserving quasi-order correspondence (cf. Definitions 3.13 and 3.12).

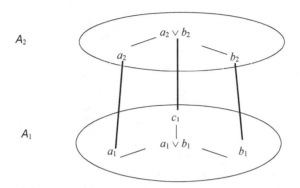

$\text{lub}_{R_2}\pi_2\{<a_1,a_2>,<b_1,b_2>\} = \text{lub}_{R_2}\{a_2,b_2\} = \{a_2 \vee b_2\}$
$[\text{lub}_{\leq/\text{minJ}}\{<a_1,a_2>,<b_1,b_2>\}] = \{<c_1, a_2 \vee b_2>\}, \pi_2[\text{lub}_{\leq/\text{minJ}}\{<a_1,a_2>,<b_1,b_2>\}] = \{a_2 \vee b_2\}$

Figure 11

The corollary says that in a joining-system $\langle \mathcal{A}_1, \mathcal{A}_2, J \rangle$, the R_1-structure of set of "bottoms" of min J is order similar to the R_2-structure of the set of "tops" of min J. (See Theorem 3.15 for how this result can be expressed in terms of the notion of isomorphism.)

3.5.1 A remark on the interrelation between narrowness and lowerness

Given the quasi-orderings $\langle A_1, R_1 \rangle$ and $\langle A_2, R_2 \rangle$, we have introduced two quasi-orderings on $A_1 \times A_2$, viz. the narrowness relation $\trianglelefteq$ and the lowerness relation $\precsim$. The interrelation between these two orderings is of great interest in the study of joining-systems.

How narrowness and lowerness are connected becomes more transparent if we if we restrict ourselves to consider lattices instead of quasi-orderings. Suppose that $\langle L_1, \leq_1 \rangle$ and $\langle L_2, \leq_2 \rangle$ are lattices. Let $\precsim$ be the lowerness-relation with respect to $\leq_1$ and $\leq_2$, i.e. for all $\langle a_1, a_2 \rangle, \langle b_1, b_2 \rangle \in L_1 \times L_2$

$$\langle a_1, a_2 \rangle \precsim \langle b_1, b_2 \rangle \text{ iff } a_1 \leq_1 b_1 \text{ and } a_2 \leq_2 b_2.$$

Then $\langle L_1 \times L_2, \precsim \rangle$ is a lattice and is the product of $\langle L_1, \leq_1 \rangle$ and $\langle L_2, \leq_2 \rangle$. Let $\langle L_1, \wedge_1, \vee_1 \rangle$ and $\langle L_2, \wedge_2, \vee_2 \rangle$ be the algebraic formulation of $\langle L_1, \leq_1 \rangle$ and $\langle L_2, \leq_2 \rangle$ respectively. Define

$$\begin{pmatrix} \wedge_2 \\ \wedge_1 \end{pmatrix} : L_1 \times L_2 \longrightarrow L_1 \times L_2$$

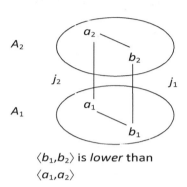

$\langle b_1, b_2 \rangle$ is *lower* than $\langle a_1, a_2 \rangle$

Figure 12

such that
$$\langle a_1, a_2 \rangle \begin{pmatrix} \wedge_2 \\ \wedge_1 \end{pmatrix} \langle b_1, b_2 \rangle = \langle a_1 \wedge_1 b_1, a_2 \wedge_2 b_2 \rangle.$$

And define
$$\begin{pmatrix} \vee_2 \\ \vee_1 \end{pmatrix} : L_1 \times L_2 \longrightarrow L_1 \times L_2$$

such that
$$\langle a_1, a_2 \rangle \begin{pmatrix} \vee_2 \\ \vee_1 \end{pmatrix} \langle b_1, b_2 \rangle = \langle a_1 \vee_1 b_1, a_2 \vee_2 b_2 \rangle.$$

Then
$$\left\langle L_1 \times L_2, \begin{pmatrix} \wedge_2 \\ \wedge_1 \end{pmatrix}, \begin{pmatrix} \vee_2 \\ \vee_1 \end{pmatrix} \right\rangle$$

is the coordinatewise product lattice of $\langle L_1, \wedge_1, \vee_1 \rangle$ and $\langle L_2, \wedge_2, \vee_2 \rangle$ and is the algebraic version of $\langle L_1 \times L_2, \precsim \rangle$, see [Davey and Priestley, 2002, p. 42].

Suppose as above that $\langle L_1, \leq_1 \rangle$ and $\langle L_2, \leq_2 \rangle$ are lattices. Let $\trianglelefteq$ be the narrowness-relation with respect to $\leq_1$ and $\leq_2$, i.e. for all $\langle a_1, a_2 \rangle, \langle b_1, b_2 \rangle \in L_1 \times L_2$

$$\langle a_1, a_2 \rangle \trianglelefteq \langle b_1, b_2 \rangle \quad \text{iff} \quad b_1 \leq_1 a_1 \quad \text{and} \quad a_2 \leq_2 b_2.$$

It can be shown that $\langle L_1 \times L_2, \trianglelefteq \rangle$ is a lattice. Let

$$\langle L_1, \wedge_1, \vee_1 \rangle \text{ and } \langle L_2, \wedge_2, \vee_2 \rangle$$

be the algebraic formulation of $\langle L_1, \leq_1 \rangle$ and $\langle L_2, \leq_2 \rangle$ respectively. Define

$$\begin{pmatrix} \wedge_2 \\ \vee_1 \end{pmatrix} : L_1 \times L_2 \longrightarrow L_1 \times L_2$$

such that

$$\langle a_1, a_2 \rangle \begin{pmatrix} \wedge_2 \\ \vee_1 \end{pmatrix} \langle b_1, b_2 \rangle = \langle a_1 \vee_1 b_1, a_2 \wedge_2 b_2 \rangle.$$

And define

$$\begin{pmatrix} \vee_2 \\ \wedge_1 \end{pmatrix} : L_1 \times L_2 \longrightarrow L_1 \times L_2$$

such that

$$\langle a_1, a_2 \rangle \begin{pmatrix} \vee_2 \\ \wedge_1 \end{pmatrix} \langle b_1, b_2 \rangle = \langle a_1 \wedge_1 b_1, a_2 \vee_2 b_2 \rangle.$$

Then

$$\left\langle L_1 \times L_2, \begin{pmatrix} \wedge_2 \\ \vee_1 \end{pmatrix}, \begin{pmatrix} \vee_2 \\ \wedge_1 \end{pmatrix} \right\rangle$$

is a lattice and is the algebraic version of $\langle L_1 \times L_2, \trianglelefteq \rangle$.

3.6 The structure on minimal joinings

The next theorem gives a characterization of the structure, with respect to the lowerness-relation, of the elements in a joining-space that are maximally narrow, i.e., those called minimal joinings. Note that with $\min J$ is meant $\min_\triangleleft J$.

We recall the definition 3.4 on page 567 of a complete quasi-lattice.

Theorem 3.34 *Suppose that $\langle \mathcal{A}_1, \mathcal{A}_2, J \rangle$ is a Js and that $\mathcal{A}_1$ and $\mathcal{A}_2$ are complete quasi-lattices and denote the relation $\precsim / \min J$ as $\precsim^*$. Let $X \subseteq \min J$. Then*

(i) $\mathrm{lub}_{\precsim^*} X \neq \emptyset$ and $\mathrm{glb}_{\precsim^*} X \neq \emptyset$

(ii) *if $X \neq \emptyset$ then $\pi_2 \left[\mathrm{lub}_{\precsim^*} X \right] \subseteq \mathrm{lub}_{R_2} \pi_2 [X]$*

(iii) *if $X \neq \emptyset$ then $\pi_1 \left[\mathrm{glb}_{\precsim^*} X \right] \subseteq \mathrm{glb}_{R_1} \pi_1 [X]$.*

Proof. Suppose that $X \subseteq \min J$. Note that since $\mathcal{A}_1 = \langle A_1, R_1 \rangle$ and $\mathcal{A}_2 = \langle A_2, R_2 \rangle$ are complete quasi-lattices, then $\mathrm{glb}_{R_1} \pi_1 [X] \neq \emptyset$ and $\mathrm{lub}_{R_2} \pi_2 [X] \neq \emptyset$.

(I) We prove (iii). Suppose that $X \neq \emptyset$. From (1) in Theorem 3.30 it follows that there is $c_2 \in A_2$ such that if $a_1 \in \mathrm{glb}_{R_1} \pi_1 [X]$, $\langle a_1, c_2 \rangle \in \min J$,

and, furthermore, it holds that $c_2 R_2 a_2$ for all $a_2 \in \mathrm{glb}_{R_2} \pi_2 [X]$. We shall now show that
$$\langle a_1, c_2 \rangle \in \mathrm{glb}_{\precsim^*} X.$$
Suppose that $\langle x_1, x_2 \rangle \in X$. Hence, $x_1 \in \pi_1 [X]$ and $x_2 \in \pi_2 [X]$. Since $a_1 \in \mathrm{glb}_{R_1} \pi_1 [X]$, it follows that $a_1 R_1 x_1$. Suppose that $a_2 \in \mathrm{glb}_{R_2} \pi_2 [X]$. Then $a_2 R_2 x_2$ and since $c_2 R_2 a_2$ it follows that $c_2 R_2 x_2$. From $a_1 R_1 x_1$ and $c_2 R_2 x_2$ follows that $\langle a_1, c_2 \rangle \precsim \langle x_1, x_2 \rangle$ and since $\langle a_1, c_2 \rangle, \langle x_1, x_2 \rangle \in \min J$ it follows that
$$\langle a_1, c_2 \rangle \precsim^* \langle x_1, x_2 \rangle.$$
Since $\langle x_1, x_2 \rangle$ is an arbitrary element in X, it follows that
$$\langle a_1, c_2 \rangle \in \mathrm{lb}_{\precsim^*} X.$$
Suppose now that $\langle y_1, y_2 \rangle \in \min J$ and $\langle y_1, y_2 \rangle \in \mathrm{lb}_{\precsim^*} X$. We shall prove that
$$\langle y_1, y_2 \rangle \precsim^* \langle a_1, c_2 \rangle.$$
Suppose $z_1 \in \pi_1 [X]$. Then there is $z_2 \in \pi_2 [X]$ such that $\langle z_1, z_2 \rangle \in X$ and hence $\langle y_1, y_2 \rangle \precsim^* \langle z_1, z_2 \rangle$, which implies that $y_1 R_1 z_1$. Thus $y_1 \in \mathrm{lb}_{R_1} \pi_1 [X]$ and since $a_1 \in \mathrm{glb}_{R_1} \pi_1 [X]$, it follows that $y_1 R_1 a_1$. Since
$$\langle a_1, c_2 \rangle, \langle y_1, y_2 \rangle \in \min J \quad \text{and} \quad y_1 R_1 a_1$$
it follows from (ii) in Theorem 3.29 that $y_2 R_2 c_2$, which implies that
$$\langle y_1, y_2 \rangle \precsim^* \langle a_1, c_2 \rangle.$$
This shows that $\langle a_1, c_2 \rangle \in \mathrm{glb}_{\precsim^*} X$ and hence $\mathrm{glb}_{\precsim^*} X \neq \varnothing$. Note that $a_1 \in \pi_1 \left[\mathrm{glb}_{\precsim^*} X \right]$ and $a_1 \in \mathrm{glb}_{R_1} \pi_1 [X]$. Suppose that $x_1 \in \pi_1 \left[\mathrm{glb}_{\precsim^*} X \right]$. Then there is x_2 such that $\langle x_1, x_2 \rangle \in \mathrm{glb}_{\precsim^*} X$. Since $\langle a_1, c_2 \rangle \in \mathrm{glb}_{\precsim^*} X$ it follows that
$$\langle x_1, x_2 \rangle \sim^* \langle a_1, c_2 \rangle$$
which implies $x_1 Q_1 a_1$. Since $a_1 \in \mathrm{glb}_{R_1} \pi_1 [X]$ it follows that $x_1 \in \mathrm{glb}_{R_1} \pi_1 [X]$. This shows that
$$\pi_1 \left[\mathrm{glb}_{\precsim^*} X \right] \subseteq \mathrm{glb}_{R_1} \pi_1 [X].$$

(II) The proof of (ii) is analogous with the proof of (iii).

(III) That (i) holds when $X \neq \varnothing$ follows from the proof of (ii) and (iii). The proof that $\mathrm{lub}_{\precsim^*} \varnothing \neq \varnothing$ and $\mathrm{glb}_{\precsim^*} \varnothing \neq \varnothing$ follows from the lemma below. (To see this, cf. as well the remark above Theorem 3.5.) ∎

Lemma 3.35 *Suppose that $\langle \mathcal{A}_1, \mathcal{A}_2, J \rangle$ is a non-empty joining-system and that $\mathcal{A}_1 = \langle A_1, R_1 \rangle$ and $\mathcal{A}_2 = \langle A_2, R_2 \rangle$ are complete quasi-lattices. Then*

(i) *there are $a_1 \in \mathrm{lub}_{R_1} \pi_1[J]$ and $a_2 \in \mathrm{glb}_{R_2} J[a_1]$ and the following holds: $\langle a_1, a_2 \rangle \in \min J$ and $\langle a_1, a_2 \rangle$ is a greatest element in $\min J$ with respect to $\precsim$.*

(ii) *there are $b_2 \in \mathrm{glb}_{R_2} \pi_2[J]$ and $b_1 \in \mathrm{lub}_{R_1} J^{-1}[b_2]$ and the following holds: $\langle b_1, b_2 \rangle \in \min J$ and $\langle b_1, b_2 \rangle$ is a least element in $\min J$ with respect to $\precsim$.*

Proof. (I) We prove (i). Since $\mathcal{A}_i$ ($i = 1, 2$) is a complete quasi-lattice, there is $g_i \in A_i$ such that g_i is a greatest element in $\mathcal{A}_i$ with respect to R_i and $l_i \in A_i$ such that l_i is a least element in $\mathcal{A}_i$. According to the assumption, $J \neq \varnothing$. Suppose that $\langle x_1, x_2 \rangle \in J$. Note that $x_2 R_2 g_2$ and from condition (1) in the definition of a joining-system follows $\langle x_1, g_2 \rangle \in J$. Since $\mathcal{A}_1$ is a complete quasi-lattice it follows that $\mathrm{lub}_{R_1} \pi_1[J] \neq \varnothing$. Suppose that $a_1 \in \mathrm{lub}_{R_1} \pi_1[J]$. From condition (2) of a joining-system follows that $\langle a_1, g_2 \rangle \in J$. Since $\mathcal{A}_2$ is a complete quasi-lattice $\mathrm{glb}_{R_2} J[a_1] \neq \varnothing$. Suppose that $a_2 \in \mathrm{glb}_{R_2} J[a_1]$. Then $\langle a_1, a_2 \rangle \in J$ according to condition (3) of a joining-system. Suppose that $\langle y_1, y_2 \rangle \in J$ and $\langle y_1, y_2 \rangle \lhd \langle a_1, a_2 \rangle$. Then

$$(*) \quad a_1 R_1 y_1 \& y_2 P_2 a_2$$

or

$$(**) \quad a_1 P_1 y_1 \& y_2 R_2 a_2$$

Since $y_1 \in \pi_1[J]$ and $a_1 \in \mathrm{lub}_{R_1} \pi_1[J]$ it follows that $y_1 R_1 a_1$ and therefore $(**)$ above does not hold. $a_1 R_1 y_1$ implies $y_1 Q_1 a_1$ and hence $y_2 \in J[a_1]$. Since $a_2 \in \mathrm{glb}_{R_2} J[a_1]$ it follows that $a_2 R_2 y_2$. This shows that $(*)$ above does not hold. Thus $\langle a_1, a_2 \rangle \in \min J$. Suppose that $\langle z_1, z_2 \rangle \in \min J$. Then $z_1 \in \pi_1[J]$ and since $a_1 \in \mathrm{lub}_{R_1} \pi_1[J]$ it follows that $z_1 R_1 a_1$ and thus $\langle z_1, z_2 \rangle \precsim \langle a_1, a_2 \rangle$.

(II) The proof of (ii) is analogous with the proof of (i). ∎

Corollary 3.36 *Given the assumption in Theorem 3.34, $\langle \min J, \precsim^* \rangle$ is a complete quasi-lattice.*

The theorem 3.37 below is a kind of converse of the theorem 3.34 above. We recall that $\uparrow K$ is the enclosure of K (see definition 3.23 above on page 578).

Theorem 3.37 *Suppose that $\mathcal{A}_1 = \langle A_1, R_1 \rangle$ and $\mathcal{A}_2 = \langle A_2, R_2 \rangle$ are quasi-orderings and $K \subseteq A_1 \times A_2$ is such that for all $\langle a_1, a_2 \rangle \in K$, $\langle a_1, a_2 \rangle$ is a minimal element in K with respect to $\trianglelefteq$. Suppose further that $\precsim_K$ is the relation $\precsim$ on $A_1 \times A_2$ restricted to K and that $\langle K, \precsim_K \rangle$ is a complete quasi-lattice and the following two conditions hold:*

(i) *For all $X \subseteq K$, $\pi_2 \left[\mathrm{lub}_{\precsim_K} X \right] \subseteq \mathrm{lub}_{R_2} \pi_2 [X]$.*

(ii) *For all $X \subseteq K$, $\pi_1 \left[\mathrm{glb}_{\precsim_K} X \right] \subseteq \mathrm{glb}_{R_1} \pi_1 [X]$.*

Then $\langle \mathcal{A}_1, \mathcal{A}_2, \uparrow K \rangle$ is a joining-system and $\min \uparrow K = K$.

Proof. (I) Proof of condition (1) in the definition of a joining-system. Suppose that $\langle a_1, a_2 \rangle \in \uparrow K$ and $\langle a_1, a_2 \rangle \trianglelefteq \langle b_1, b_2 \rangle$. Then there is $\langle c_1, c_2 \rangle \in K$ such that $\langle c_1, c_2 \rangle \trianglelefteq \langle a_1, a_2 \rangle$, and it follows that $\langle c_1, c_2 \rangle \trianglelefteq \langle b_1, b_2 \rangle$, which implies that $\langle b_1, b_2 \rangle \in \uparrow K$.

(II) Proof of condition (2) in the definition of a joining-system. Suppose that $C_1 \subseteq A_1$, $b_2 \in A_2$ and that $a_1 \in \mathrm{lub}_{R_1} C_1$. Suppose further that for all $c_1 \in C_1$, $\langle c_1, b_2 \rangle \in \uparrow K$. We show that $\langle a_1, b_2 \rangle \in \uparrow K$. For all $c_1 \in C_1$, there is an element $\langle c_1^*, b_2^{c_1} \rangle \in K$ such that $\langle c_1^*, b_2^{c_1} \rangle \trianglelefteq \langle c_1, b_2 \rangle$. Since $\langle K, \precsim_K \rangle$ is a complete quasi-lattice it follows that there is $\langle x_1, x_2 \rangle \in K$ such that

$$(***) \quad \langle x_1, x_2 \rangle \in \mathrm{lub}_{\precsim_K} \{ \langle c_1^*, b_2^{c_1} \rangle \mid c_1 \in C_1 \}.$$

Hence,
$$x_2 \in \pi_2 \left[\mathrm{lub}_{\precsim_K} \{ \langle c_1^*, b_2^{c_1} \rangle \mid c_1 \in C_1 \} \right].$$

From the assumption (i) follows that
$$x_2 \in \mathrm{lub}_{R_2} \pi_2 [\{ \langle c_1^*, b_2^{c_1} \rangle \mid c_1 \in C_1 \}]$$

and hence
$$x_2 \in \mathrm{lub}_{R_2} \{ b_2^{c_1} \mid c_1 \in C_1 \}.$$

Note that
$$b_2 \in \mathrm{ub}_{R_2} \{ b_2^{c_1} \mid c_1 \in C_1 \}$$

which implies that $x_2 R_2 b_2$.

From $(***)$ above it follows that for all $c_1 \in C_1$
$$\langle c_1^*, b_2^{c_1} \rangle \precsim_K \langle x_1, x_2 \rangle$$

and hence $c_1^* R_1 x_1$. For all $c_1 \in C_1$
$$\langle c_1^*, b_2^{c_1} \rangle \trianglelefteq \langle c_1, b_2 \rangle$$

which implies $c_1 R_1 c_1^*$ and hence $c_1 R_1 x_1$. Thus $x_1 \in \mathrm{ub}_{R_1} C_1$ and since $a_1 \in \mathrm{lub}_{R_1} C_1$ it follows that $a_1 R_1 x_1$. This together with $\langle x_1, x_2 \rangle \in K$ and $x_2 R_2 b_2$ implies (see part (I) in this proof) $\langle a_1, b_2 \rangle \in {\uparrow}K$.

(III) Proof of condition (3) in the definition of a joining-system is analogous to the proof of condition (2) in (II).

(IV) Proof of $\min{\uparrow}K = K$. Suppose that $\langle a_1, a_2 \rangle \in K$ and show that $\langle a_1, a_2 \rangle \in \min{\uparrow}K$. Suppose that $\langle b_1, b_2 \rangle \in {\uparrow}K$ such that $\langle b_1, b_2 \rangle \trianglelefteq \langle a_1, a_2 \rangle$. Since $\langle b_1, b_2 \rangle \in {\uparrow}K$ there is $\langle c_1, c_2 \rangle \in K$ such that $\langle c_1, c_2 \rangle \trianglelefteq \langle b_1, b_2 \rangle$. Hence, $\langle c_1, c_2 \rangle \trianglelefteq \langle a_1, a_2 \rangle$ and since $\langle a_1, a_2 \rangle, \langle c_1, c_2 \rangle \in K$ and all elements in K are minimal elements in K with respect to $\trianglelefteq$, it follows that $\langle a_1, a_2 \rangle \simeq \langle c_1, c_2 \rangle$, which implies that $\langle a_1, a_2 \rangle \simeq \langle b_1, b_2 \rangle$ and $\langle a_1, a_2 \rangle \in \min{\uparrow}K$.

Suppose that $\langle a_1, a_2 \rangle \in \min{\uparrow}K$. Then $\langle a_1, a_2 \rangle \in {\uparrow}K$ and there is $\langle b_1, b_2 \rangle \in K$ such that $\langle b_1, b_2 \rangle \trianglelefteq \langle a_1, a_2 \rangle$. According to what have just been proven, from $\langle b_1, b_2 \rangle \in K$ follows that $\langle b_1, b_2 \rangle \in \min{\uparrow}K$. This implies that $\langle b_1, b_2 \rangle \simeq \langle a_1, a_2 \rangle$, and thus $\langle a_1, a_2 \rangle \in K$. ∎

3.7 Networks of joining-systems

A normative system is not always represented by just one joining-system. More complex normative systems are usually represented by a network of joining-systems. (A rudimentary network is shown in Section 5.2.3.) In such representations, the relative product of joining spaces is an important operation for the construction of new joining-systems. The theorem below describes the situation.

Note that, when more than two joining-systems are involved, the sign J for a set of joinings will be annexed with two indices. Thus, the set of joinings from a quasi-ordering $\mathcal{A}_i$ to a quasi-ordering $\mathcal{A}_j$ will be denoted $J_{i,j}$. Accordingly, the joining-system from $\mathcal{A}_i$ to $\mathcal{A}_j$ is denoted $\langle \mathcal{A}_i, \mathcal{A}_j, J_{i,j} \rangle$.

Theorem 3.38 *Suppose that* $\langle \mathcal{A}_1, \mathcal{A}_2, J_{1,2} \rangle$ *and* $\langle \mathcal{A}_2, \mathcal{A}_3, J_{2,3} \rangle$ *are joining-systems and that* $\mathcal{A}_2$ *is a complete quasi-lattice. Then* $\langle \mathcal{A}_1, \mathcal{A}_3, J_{1,2}|J_{2,3} \rangle$ *is a joining-system and is called the relative product of* $\langle \mathcal{A}_1, \mathcal{A}_2, J_{1,2} \rangle$ *and* $\langle \mathcal{A}_2, \mathcal{A}_3, J_{2,3} \rangle$.

Proof. We begin by proving condition (1) in the definition of a Js (Definition 3.11 in Section 3.2.2). Suppose that $\langle a_1, a_3 \rangle \in J_{1,2}|J_{2,3}$ and $\langle a_1, a_3 \rangle \trianglelefteq \langle b_1, b_3 \rangle$. From $\langle a_1, a_3 \rangle \in J_{1,2}|J_{2,3}$ follows that there is $a_2 \in \mathcal{A}_2$ such that $\langle a_1, a_2 \rangle \in J_{1,2}$ and $\langle a_2, a_3 \rangle \in J_{2,3}$. From $\langle a_1, a_3 \rangle \trianglelefteq \langle b_1, b_3 \rangle$ follows that $b_1 R_1 a_1$ and $a_3 R_3 b_3$. Since $\langle \mathcal{A}_1, \mathcal{A}_2, J_{1,2} \rangle$ is a joining-system, $b_1 R_1 a_1$ and $\langle a_1, a_2 \rangle \in J_{1,2}$ implies that $\langle b_1, a_2 \rangle \in J_{1,2}$. And $a_3 R_3 b_3$ and $\langle a_2, a_3 \rangle \in J_{2,3}$ implies that $\langle a_2, b_3 \rangle \in J_{2,3}$, since $\langle \mathcal{A}_2, \mathcal{A}_3, J_{2,3} \rangle$ is a joining-system. From $\langle b_1, a_2 \rangle \in J_{1,2}$ and $\langle a_2, b_3 \rangle \in J_{2,3}$ follows that $\langle b_1, b_3 \rangle \in J_{1,2}|J_{2,3}$.

We now prove condition (2) in the definition of a *Js*. Suppose that $C_1 \subseteq A_1$ and $C_1 \neq \varnothing$ such that for all $c_1 \in C_1$, $\langle c_1, b_3 \rangle \in J_{1,2}|J_{2,3}$ and suppose $a_1 \in \text{lub}_{R_1} C_1$. Let

$$C_1^{(2)} = \{c_2 \in A_2 \mid \exists c_1 \in C_1 : \langle c_1, c_2 \rangle \in J_{1,2} \,\&\, \langle c_2, b_3 \rangle \in J_{2,3}\}$$

Hence, for all $c_2 \in C_1^{(2)}$, $\langle c_2, b_3 \rangle \in J_{2,3}$. Since $\mathcal{A}_2$ is a complete quasi-lattice (Definition 3.4), it follows that $\text{lub}_{R_2} C_1^{(2)} \neq \varnothing$. Suppose that $a_2 \in \text{lub}_{R_2} C_1^{(2)}$. Since $\langle \mathcal{A}_2, \mathcal{A}_3, J_{2,3} \rangle$ is a *Js* it follows that $\langle a_2, b_3 \rangle \in J_{2,3}$. For all $c_1 \in C_1$, there is $c_1^{(2)} \in C_1^{(2)}$ such that $\left\langle c_1, c_1^{(2)} \right\rangle \in J_{1,2}$. Since $\langle \mathcal{A}_1, \mathcal{A}_2, J_{1,2} \rangle$ is a *Js*, this implies that $\langle c_1, a_2 \rangle \in J_{1,2}$ for all $c_1 \in C_1$, and, consequently, $\langle a_1, a_2 \rangle \in J_{1,2}$. Since $\langle a_2, b_3 \rangle \in J_{2,3}$ it follows that $\langle a_1, b_3 \rangle \in J_{1,2}|J_{2,3}$.
The proof of condition (3) is analogous and is omitted. ∎

Note that from the assumption $J_{1,2}|J_{2,3} = J_{1,3}$ and the requirement of connectivity it follows that $\min J_{1,2}|\min J_{2,3} \subseteq \min J_{1,3}$. Also, however, note that $\subseteq$ cannot generally be strengthened to $=$ (Cf. [Lindahl and Odelstad, 2011, sect. 3.3.2]).

3.8 Intervenients

The notion of "intervenient" (cf. above, Section 2.2) will be treated in detail in Section 5, in connection with Boolean quasi-orderings and Boolean joining-systems. As a general notion, it is, however, introduced here.

Let us consider three joining-systems

$$\mathcal{S}_1 = \langle \mathcal{A}_1, \mathcal{A}_2, J_{1,2} \rangle, \mathcal{S}_2 = \langle \mathcal{A}_2, \mathcal{A}_3, J_{2,3} \rangle, \mathcal{S}_3 = \langle \mathcal{A}_1, \mathcal{A}_3, J_{1,3} \rangle,$$

where $\mathcal{A}_i = \langle A_i, R_i \rangle$. There can be $a_1 \in A_1$, $a_2 \in A_2$, and $a_3 \in A_3$ such that $\langle a_1, a_2 \rangle \in J_{1,2}$, $\langle a_2, a_3 \rangle \in J_{2,3}$, and $\langle a_1, a_3 \rangle \in J_{1,3}$. A case of special interest then, is when $\text{WG}_{\mathcal{S}_1}(a_1, a_2, A_1)$ and $\text{SC}_{\mathcal{S}_2}(a_3, a_2, A_3)$, i.e., when, in $\mathcal{S}_1$, a_1 is among the weakest grounds in A_1 for a_2, and a_3 is among the strongest consequences in A_3 of a_2. (Cf. above, Section 3.3). In this case, a_2, in a sense, is "intermediate" between a_1 and a_3 and "mediates" the joining $\langle a_1, a_3 \rangle$. Therefore, in this case we call a_2 an *intervenient*.

In order to give a more detailed formal exposition of what is said above, we first give the following definition of a *simple Js-triple*.

Definition 3.39 Suppose that $\mathcal{S}_1 = \langle \mathcal{A}_1, \mathcal{A}_2, J_{1,2} \rangle$, $\mathcal{S}_2 = \langle \mathcal{A}_2, \mathcal{A}_3, J_{2,3} \rangle$ and $\mathcal{S}_3 = \langle \mathcal{A}_1, \mathcal{A}_3, J_{1,3} \rangle$ are joining-systems where $\mathcal{A}_i = \langle A_i, R_i \rangle$. $\langle \mathcal{S}_1, \mathcal{S}_2, \mathcal{S}_3 \rangle$

is a simple Js-triple if A_1, A_2 and A_3 are pair-wise disjunct, and, for the relative product $J_{1,2}|J_{2,3}$ it holds that $J_{1,3} = J_{1,2}|J_{2,3}$.[19]

(For *Bjs-triples* of Boolean joining-systems, cf. Section 5.1.)

Then the notion of *intervenient* in a simple *Js*-triple is defined as follows.

Definition 3.40 *In a simple* Js*-triple* $\langle S_1, S_2, S_3 \rangle$, *the element* $a_2 \in A_2$, *is an* intervenient *from* A_1 *to* A_3 *corresponding to the joining* $\langle a_1, a_3 \rangle \in J_{1,3}$, *denoted* $a_2 \curvearrowright \langle a_1, a_3 \rangle$, *if* a_1 *is a weakest ground of* a_2 *in* S_1 *and* a_3 *is a strongest consequence of* a_2 *in* S_3.

Since weakest grounds and strongest consequences are related to minimal joinings, the same holds for intervenients. If a_2 is an intervenient corresponding to $\langle a_1, a_3 \rangle$, there is $b_2 \in A_2$ such that $\langle a_1, b_2 \rangle$ is a minimal joining and $b_2 R_2 a_2$. And, further, there is $c_2 \in A_2$ such that $\langle c_2, a_3 \rangle$ is a minimal joining and $a_2 R_2 c_2$. If $\langle a_1, a_2 \rangle$ is a minimal element, then, since a_2 is minimal with respect to the ground a_1, a_2 is called ground-minimal. If $\langle a_2, a_3 \rangle$ is a minimal element, then, since a_2 is minimal with respect to the consequence a_3, a_2 is called consequence-minimal. A very convenient way of representing a normative system is if all intervenients are ground- and consequence-minimal and the operation relative product is used. Changes of the normative system are then simplified and the notion of open intermediate concepts is elucidated.

A step towards analyzing more general structures in the law is taking into account chains of four or more quasi-orderings. Let us pay regard to joining-systems involving four quasi-orderings A_1, A_2, A_3, A_4 such that $a_2 \curvearrowright \langle a_1, a_3 \rangle$ and $a_3 \curvearrowright \langle a_2, a_4 \rangle$. (See Figure 13.) From this follows that WG (a_2, a_3, A_2) & SC (a_3, a_2, A_3). This conjunction is equivalent to $\langle a_2, a_3 \rangle \in \min J_{2,3}$, see Theorem 3.21. (This is illustrated by the thick line in Figure 13.) Note that a chain of four quasi-orderings can be continued at any length by adding A_5, A_6, and so on. The notion of intervenient is of particular interest when the three joining-systems are Boolean joining-systems. This will be the subject-matter of the subsequent Section 5, where conjunctions, disjunctions and negations of intervenients are studied, organic wholes of intervenients discussed and a typology of intervenients presented. Also, section 5 will contain several examples of legal intervenients.

---

[19]The triple is simple in the following sense. The presupposition of disjunct strata will make it possible in the present section to disregard the problem with "degenerated" weakest grounds and/or strongest consequences. This problem will be dealt with in connection with intervenients in Boolean joining systems.

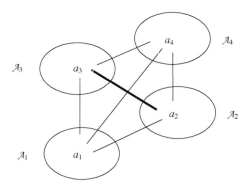

Figure 13

4 TJS for Boolean joining-systems

In the representation of a normative system, the connectives "and", "or" and "not" are often essential. This is neatly illustrated in the example of Amendment XIV in the U.S. Constitution, quoted above (Section 1.7.1):

> "All persons born *or* naturalized in the United States, *and* subject to the jurisdiction thereof, are citizens of the United States *and* of the State wherein they reside. *No* State shall make *or* enforce any law which shall abridge the privileges *or* immunities of citizens of the United States; *nor* shall any State deprive any person of life, liberty, *or* property, without due process of law; *nor* deny to any person within its jurisdiction the equal protection of the laws."

With a view to the connectives referred to, in the present Section 4 and the subsequent Section 5, we consider strata of *Boolean quasi-orderings* (*Bqo*'s) and joining-systems that are *Boolean joining-systems* (*Bjs*'). As mentioned, the development of TJS for *Bqo*'s and *Bjs*'s in this chapter of the Handbook relies much on earlier papers by the present authors and the reader will often be referred to these papers for further details and for proofs of the results.

4.1 Boolean quasi-orderings and Boolean joining-systems

4.1.1 Boolean quasi-orderings

The notion of Boolean quasi-ordering is defined as follows.

Definition 4.1 *The relational structure* $\mathcal{B} = \langle B, \wedge, ', R \rangle$ *is a Boolean quasi-ordering* (Bqo) *if* $\langle B, \wedge, ' \rangle$ *is a Boolean algebra and* R *is a quasi-ordering,* $\bot$ *is the zero element and* $\top$ *is the unit element, such that* R *satisfies the additional requirements:*

(1) aRb *and* aRc *implies* $aR(b \wedge c)$,

(2) aRb *implies* $b'Ra'$,

(3) $(a \wedge b)Ra$,

(4) not $\top R \bot$.

Note that if $\leq$ is the partial ordering determined by $\langle B, \wedge, ' \rangle$, from requirement (3) it follows that $a \leq b$ implies aRb. As usual, $\leq$ is defined by $a \leq b$ if and only if $a \wedge b = a$.

Requirements (3) and (4) can be expressed equivalently by saying that R is a non-total super-relation of the Boolean ordering $\leq$. More exactly, suppose that $\langle B, \wedge, ' \rangle$ is a Boolean algebra, that $\leq$ is the partial ordering determined by the algebra, and that R is a transitive relation on B. Then the conjunction of (3) and (4) is equivalent to the conjunction of (i) $\leq$ is a subset of R, and (ii) R is a proper subset of $B \times B$.

Some general notions relating to Bqo's are as follows (see [Lindahl and Odelstad, 2004, sect. 2.1]):

If $\langle B, \wedge, ', R \rangle$ is a Bqo then we say that the Boolean algebra $\langle B, \wedge, ' \rangle$ is the *reduct* of $\langle B, \wedge, ', R \rangle$. In what follows, the reduct $\langle B, \wedge, ' \rangle$ of a Bqo $\mathcal{B}$ will be denoted $\mathcal{B}^{red}$. Suppose that $\mathcal{B} = \langle B, \wedge, ', R \rangle$ is a Bqo and Q is the indifference part of R. The *quotient algebra* of $\mathcal{B}$ with respect to Q is a structure $\langle B/Q, \cap, -, \leq_Q \rangle$ such that $\langle B/Q, \cap, - \rangle$ is a Boolean algebra and $\leq_Q$ is the partial ordering determined by this algebra. The natural mapping of $\langle B, \wedge, ' \rangle$ onto $\langle B/Q, \cap, - \rangle$ is a homomorphism (cf. [Odelstad and Lindahl, 2000]). We call $\langle B/Q, \cap, - \rangle$ the *quotient reduction* of $\mathcal{B}$. Thus there are two Boolean algebras which should be kept apart, namely $\mathcal{B}^{red}$, i.e. the reduct of $\mathcal{B}$, and the quotient reduction of $\mathcal{B}$. If the quotient reduction of $\mathcal{B}$ is isomorphic to $\mathcal{B}^{red}$, $R = \leq$, and we say that $\mathcal{B}$ is *conservatively reducible*.

As just mentioned, the transition to the quotient algebra of $\langle B, \wedge, ', R \rangle$ with respect to the equality part Q of R will result in a new Boolean algebra. In what follows we will not make this transition. The point is that, in the models we have in mind, even though, for a and b it holds that aQb (and therefore a and b belong to the same Q-equivalence class), we may want to distinguish a and b because they can have different meaning. We get possibilities of finer divisions when we can distinguish the three possibilities: 1. $a = b$, 2. $a \neq b$ and aQb, 3. $a \neq b$ and not aQb. Therefore, there is a point in remaining within the framework of Boolean quasi-orderings as defined above.

Note that if $\mathcal{B} = \langle B, \wedge, ', R \rangle$ is a Bqo, then

$$(a \vee b) \in \mathrm{lub}_R \{a, b\},$$
$$(a \wedge b) \in \mathrm{glb}_R \{a, b\}.$$

If $\mathcal{B} = \langle B, \wedge, ', R \rangle$ is a Bqo, then $\langle B, R \rangle$ is a quasi-ordering and, of course, what is said about quasi-orderings in section 3 is applicable to $\mathcal{B}$. We say that the Bqo $\langle B, \wedge, ', R \rangle$ is *complete* if the quasi-ordering $\langle B, R \rangle$ is a complete quasi-lattice.

4.1.2 Boolean joining systems

A fundamental construction for the representation of a normative system is that of a Boolean joining-system. If $\mathcal{N}$ is a two-strata system of conditional norms, then $\mathcal{N}$ can be represented by a Bjs $\langle \mathcal{B}_1, \mathcal{B}_2, J \rangle$ where J is a set of conditional norms, where $\mathcal{B}_1$ is a Bqo of grounds, and $\mathcal{B}_2$ is a Bqo of normative consequences.

Definition 4.2 $\langle \mathcal{B}_1, \mathcal{B}_2, J \rangle$ *is a* Boolean joining system (Bjs) *if*

$$\mathcal{B}_1 = \langle B_1, \wedge, ', R_1 \rangle, \; \mathcal{B}_2 = \langle B_2, \wedge, ', R_2 \rangle$$

are Boolean quasi-orderings and $\langle \langle B_1, R_1 \rangle, \langle B_2, R_2 \rangle, J \rangle$ *is a joining-system.*

With the definition of a Bjs now given it is clear that the results for joining-systems in Section 3 apply to the Bjs version of joining-systems. This holds e.g., for the notions of weakest ground, strongest consequence, minimal joinings and connectivity.

In the study of Bjs's, structures that are not Bqo's play an essential role. This is exemplified by the following theorem, which is proved in [Lindahl and Odelstad, 2011, p. 128].

Theorem 4.3 *Suppose that* $\langle \mathcal{B}_1, \mathcal{B}_2, J \rangle$ *is a Bjs that satisfies connectivity. Then* $\langle \min J, \precsim \rangle$ *is a quasi-lattice.*

Cf. Corollary 3.36 above.

If $\langle \mathcal{B}_1, \mathcal{B}_2, J \rangle$ is a Boolean joining system, it is often reasonable that falsum in $\mathcal{B}_1$ and in $\mathcal{B}_2$ are the same element $\bot$ and that the same holds for verum $\top$. From this follows that in J there are joinings, which are degenerated in the sense that they do not seem to fulfill the intuitive idea behind the notion of a joining, for example $\langle \bot, \bot \rangle$ and $\langle \top, \top \rangle$.

Referring to a Bjs $\langle \mathcal{B}_1, \mathcal{B}_2, J \rangle$, however, we introduce a distinction between "degenerated" and "non-degenerated" for weakest ground, strongest consequences and joining.

(1) If WG $(\bot, a_2, \mathcal{B}_1)$, the weakest ground in B_1 for a_2 is degenerated; similarly, if SC$(\top, a_1, \mathcal{B}_2)$, the strongest consequence in B_2 of a_1 is degenerated.

(2) As joinings from $\mathcal{B}_1$ to $\mathcal{B}_2$, the elements in

$$\{\langle \bot, \bot \rangle, \langle \top, \top \rangle, \langle b_1, \top \rangle, \langle \bot, b_2 \rangle\}$$

are degenerated joinings.

Note that $\langle \bot, \bot \rangle, \langle \top, \top \rangle \in J$, and even $\langle \bot, \bot \rangle, \langle \top, \top \rangle \in \min J$. Note further that if $b_2 \in B_2$ and there is no $b_1 \in B_1 \setminus \{\bot\}$ such that $\langle b_1, b_2 \rangle \in J$, then $\langle \bot, b_2 \rangle \in \min J$. Analogously, if $b_1 \in B_1$ and there is no $b_2 \in B_2 \setminus \{\top\}$ such that $\langle b_1, b_2 \rangle \in J$, then $\langle b_1, \top \rangle \in \min J$.

4.2 The condition implication model (cis)

We recall the statement by [Alchourrón and Bulygin, 1971] (referred to in the introductory Section 1), that a set α of sentences deductively correlates a pair $\langle p, q \rangle$ of sentences if q is a deductive consequence of $\{p\} \cup \alpha$, (or, using the relation Cn of consequence, if $q \in Cn(\{p\} \cup \alpha)$.) Also, we recall our remark that if propositional logic is used as a basis, it is usually presupposed that p, q are closed sentences with no free variables,(i.e., for example, p is the sentence "Smith has promised to pay Jones \$100" and q is "Smith has an obligation to pay \$100 to Jones"). Thus, in such sentences, individuals are referred to by individual constants (names).

A sentence such as "Smith has an obligation to pay \$100 to Jones" is often said to express an "individual norm". Owing to its general character, the *Bjs* theory can be used for representing correlations of conditional individual norms and derivation of individual norms.

As mentioned in Section 1, however, a normative system usually expresses general rules where no individual names occur. If the task is to represent a normative system of this ordinary kind, the feature of generality has to be taken into account. What will here be called the theory of *condition implication structures* (*cis*'s) is a special variety of the *Bjs* theory where the elements of B in a *Bqo* $\langle B, \wedge,', R \rangle$ are *conditions*.

In general terms, a *cis* is a structure $\langle C, \rightarrow \rangle$ where C is a set of *conditions* and $\rightarrow$ is an implicative relation. In what follows we have in view especially the case of a *cis-Bqo* $\langle B, \wedge,', R \rangle$, where B is a set of conditions and R is the implicative relation. A *cis-Bjs* is a *Bjs* $\langle \mathcal{B}_1, \mathcal{B}_2, J \rangle$ where the *Bqo*'s $\mathcal{B}_1$ and $\mathcal{B}_2$ are *cis*'. Part of a normative system can often be represented by a *cis-Bjs* $\langle \mathcal{B}_1, \mathcal{B}_2, J \rangle$ where $\mathcal{B}_1, \mathcal{B}_2$ are *cis*', and J is a correspondence from the set B_1 of conditions to the set B_2 of conditions.

In simple cases, conditions can be denoted by expressions using the sign of the infinitive, such as "to be 21 years old", "to be a citizen of the U.S.", "to be a child of", "to be entitled to inherit", or by corresponding expressions in the ing-form, like "being 21 years old" etc. Often, however, conditions should appropriately be expressed by open sentences, like "x promises to pay \$$y$ to z", "x is a citizen of state y", "x is entitled to inherit y".

When a condition is expressed by an open sentence, free variables like $x, y, z, ...$ occurring in the sentence merely are place-holders for expressing the condition in a convenient way and keeping track of the order of the places. In simple cases like, "committing murder implies being liable to imprisonment", place-holders are not needed. For details about Boolean operations on conditions, the reader is referred to [Lindahl and Odelstad, 2004, sect. 3].

In a cis-Bqo $\langle B, \wedge, ', R \rangle$, a condition a in B, such as "x promises to pay \$$y$ to z ", is said to be *fulfilled* or *non-fulfilled* by a particular triple, like $\langle$Smith,100,Jones$\rangle$. The fulfillment of a condition by a particular $n-$tuple of individuals is expressed by a closed sentence naming the individuals of the $n-$tuple.

A framework with implication between conditions seems to accord with the presupposed ontology of legal language, where terms such as "citizenship", "inheritance", "ownership", denote conditions that are treated as objects between which there is an implicative relation of "ground-consequence", often expressed in terms of "gives rise to" or "causes", or "implies". Thus inheritance is said to give rise to ownership, and ownership is said to imply a bundle of liberties, claims, and immunities.

Let us recall the remark after Definition 3.12 that if $\langle \mathcal{A}_1, \mathcal{A}_2, J \rangle$ is a joining-system, then $R_1 | J | R_2 = J$ and, therefore, J can be said to "absorb" R_1 and R_2. From this it follows that if we have in view a cis-Bjs $\langle \mathcal{B}_1, \mathcal{B}_2, J \rangle$, where a_1, b_1, a_2, b_2 are conditions such that $a_1, b_1 \in B_1$ and $a_2, b_2 \in B_2$, we can use the following schema of derivation:

(1) $a_1 R_1 b_1$
(2) $\langle b_1, a_2 \rangle \in J$
(3) $a_2 R_2 b_2$
___
(4) $\langle a_1, b_2 \rangle \in J$

In this schema, the joining (4) of two conditions is derived from the joining (2) together with implications (1) and (3).

4.2.1 A note on cis models with lattice-based quasi-orderings

Some kinds of conditions do not constitute Boolean algebras. One example is equality-relations. The term "equality-relation" here refer to a relation of equality with respect to some aspect α, and it is presupposed in this context that an equality-relation is always an equivalence-relation, i.e. a reflexive, transitive and symmetric relation. Let A be a non-empty set and let $E(A)$ be the set of equivalence relations on A. Define the binary relation $\leq$ on $E(A)$ in the following way: For all $\varepsilon_1, \varepsilon_2 \in E(A)$

$$\varepsilon_1 \leq \varepsilon_2 \text{ iff } x\varepsilon_1 y \text{ implies } x\varepsilon_2 y.$$

The reader should be reminded of the fact that $\mathcal{E}(A) = \langle E(A), \leq \rangle$ is a complete lattice. Note that the negation ε' of an equivalence relation $\varepsilon \in E(A)$ is not an equivalence relation, i.e. $\varepsilon' \notin E(A)$. $\langle E(A), \leq \rangle$, therefore, does not constitute a Boolean algebra. (Cf. [Odelstad, 2008, pp. 38f.].)

As appear from the foregoing, a Boolean quasi-ordering is a Boolean algebra extended with a quasi-ordering satisfying certain conditions. We can define an analogous structure based on a lattice instead of a Boolean algebra.

Definition 4.4 *The relational structure $\langle L, \wedge, \vee, R \rangle$ is a* lattice-based quasi-ordering (Lqo) *if $\langle L, \wedge, \vee \rangle$ is a lattice and R is a quasi-ordering such that R satisfies the additional requirements:*

(1) aRb and aRc implies $aR(b \wedge c)$,

(2) aRc and bRc implies $(a \vee b)Rc$,

(3) $(a \wedge b)Ra$,

(4) $aR(a \vee b)$.

The transition to the quotient algebra of $\langle L, \wedge, \vee \rangle$ with respect to the equality part of R will result in a lattice. (Cf. [Lindahl and Odelstad, 1999a, p. 171].) Let $\leq$ be the partial ordering determined by the lattice-based quasi-ordering $\langle L, \wedge, \vee, R \rangle$.[20] From requirement (3) for lattice-based quasi-orderings it follows that $a \leq b$ implies aRb. If $\langle A, \wedge, \vee, R \rangle$ is a lattice-based quasi-ordering then $\langle L, R \rangle$ is a quasi-lattice. Note that a *Bqo* determines a *Lqo*.

4.3 Subtraction and addition of norms: an example

In Section 1.6 above, we mentioned that TJS deals with subtraction and addition of norms in terms of the structure of the set min J of minimal joinings. In the present subsection we illustrate this issue by a *cis* concerning the legal effects of an illegal transfer of goods belonging to someone else. (Cf. [Lindahl and Odelstad, 2003].)

---

[20] As usual, $\leq$ is defined by $a \leq b$ if and only if $a \wedge b = a$.

Consider the following example. Goods belonging to *owner* have been sold without owner's consent by *transferrer* to *transferee* by a contract. (We can suppose that transferrer has stolen or hired the goods from owner and had it in possession at the time of the contract with transferee.) The normative problem is: Under what conditions is there an obligation (denoted O1) for transferrer to deliver the goods to owner? Under what conditions is there an obligation (denoted O2) for transferee to deliver the goods to owner?

We consider four systems and for all of them we assume that the stratum of grounds coincides with its reduct and similarly for the stratum of consequences, i.e. R_i coincides with $\leq_i$.

The example is a *cis*-application representing four normative systems with general norms where descriptive conditions imply normative conditions. For convenience, the conditions involved will be referred to in an abbreviated way. So, for example, condition P below ("Transferee has the goods in possession") refers to a complex condition $C(x_1, ..., x_n)$ fulfilled or not fulfilled by an n-tuple of individuals $\langle i_1, ..., i_n \rangle$ in a situation s. For details on conditions in the *cis* of the present example, the reader is referred to [Lindahl and Odelstad, 2003, pp. 86ff.].

The conditions dealt with in this example are the following (where / signifies negation):

Grounds
   P = Transferee has (= the transferrer has not) the goods in possession.
   F = Transferee was in good faith at the time of the transfer.
   R = the owner offers to pay ransom to transferee for the goods.
Normative consequences
   O1 = Transferrer has an obligation to deliver the goods to owner.
   O2 = Transferee has an obligation to deliver the goods to owner.
Verum and falsum
   $\bot$ falsum
   $\top$ verum

To simplify the example, we stipulate that it is assumed that the goods are either in the possession of transferrer or in the possession of transferee (no third possibility).

The example is intended to illustrate that, by means of Theorems 3.34 and 3.37, we get a test for whether a legal system is a joining-system, useful in situations of subtraction of norms from a system and addition of norms to a system.

We consider four systems, $\mathcal{S}_I, \mathcal{S}_{II}, \mathcal{S}_{III}, \mathcal{S}_{IV}$, where

- $\mathcal{S}_I$ is a joining-system,

- S_{II}, the result of subtraction from S_I, is not a joining-system,
- S_{III}, the result of a more comprehensive subtraction from S_I, is a joining-system, and,
- S_{IV}, the result of an addition to S_{III}, is a joining-system.

We make the following assumptions concerning the Bqo's involved in the example:

1. The Bqo
$$\mathcal{B}_1 = \langle B_1, \wedge, ', R_1 \rangle, \text{ where } R_1 = \leq_1,$$
of grounds is the same for the systems S_I, S_{II}, S_{III}; B_1 consists of the Boolean combinations of F and P.
(The Bqo of grounds in S_{IV} will be indicated later).

2. The Bqo
$$\mathcal{B}_2 = \langle B_2, \wedge, ', R_2 \rangle, \text{ where } R_2 = \leq_2,$$
of consequences is the same for all of $S_I, S_{II}, S_{III}, S_{IV}$; B_2 consists of the Boolean combinations of O1 and O2;

We introduce the following names for some of the norms in $S_I - S_{III}$:

a = $\langle$F/∧P,O2$\rangle$
b = $\langle$P,O1/$\rangle$
c = $\langle$F∧P,O1/∧O2/$\rangle$
d = $\langle$F/∨P/,O1∨O2$\rangle$
e = $\langle$F∨P/,O2/$\rangle$
f = $\langle$P/,O1$\rangle$
$\langle \bot, \bot \rangle$
$\langle \top, \top \rangle$

In System S_I (which is a qo-$corr$ but, at this stage, not assumed to be a Js) the answer to the normative problem stated above depends on whether transferee has possession of the goods (denoted P) and whether transferee was in good faith at the time of the contract (denoted F). Let

$$K_I = \{\mathbf{a}, \mathbf{b}, \mathbf{c}, \mathbf{d}, \mathbf{e}, \mathbf{f}, \langle \bot, \bot \rangle, \langle \top, \top \rangle\}$$

be the set of norms in S_I that are minimal with respect to $\trianglelefteq$. Figure 14 on page 601 shows the six minimal, non-degenerated norms and their interrelation in system S_I: $\langle K_I, \precsim /K_I \rangle$ is a lattice, see Figure 15 on page 602. The

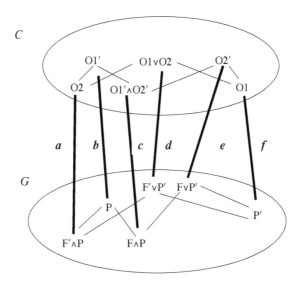

Figure 14

assumptions in Theorem 3.37 are satisfied. From Theorem 3.37 it follows that $\mathcal{S}_I = \langle \mathcal{B}_1, \mathcal{B}_2, \uparrow K_I \rangle$ is a Bjs and that $\min \uparrow K_I = K$.

We note that, for some $X \subseteq K_I$, $\langle \bot, \bot \rangle \in \mathrm{glb}_{\precsim} X$. Thus, for example, $\langle \bot, \bot \rangle \in \mathrm{glb}_{\precsim}\{\mathbf{a},\mathbf{c}\}$. Similarly, for some $X \subseteq K_I$, $\langle \top, \top \rangle \in \mathrm{lub}_{\precsim} X$. Thus, $\langle \top, \top \rangle \in \mathrm{lub}_{\precsim}\{\mathbf{b},\mathbf{d},\mathbf{e}\}$.

From the point of view of legal justice, System $\mathcal{S}_I$ may be thought to be unreasonable since it does not attach relevance to the possibility that owner can be willing to pay a ransom to transferee for getting the goods back. System $\mathcal{S}_{II}$ takes this consideration into account by elimination of some norms in the system. Suppose that the legislator in the set K_I of minimal joinings subtracts the minimal joining $\mathbf{c} = \langle F \wedge P, O1\prime \wedge O2\prime \rangle$, while $\mathbf{a}, \mathbf{b}, \mathbf{d}, \mathbf{e}, \mathbf{f}, \langle \bot, \bot \rangle$ and $\langle \top, \top \rangle$ are left.

System $\mathcal{S}_{II}$, where the set of minimal norms is

$$K_{II} = \{\mathbf{a},\mathbf{b},\mathbf{d},\mathbf{e},\mathbf{f},\langle \bot, \bot \rangle, \langle \top, \top \rangle\}$$

is a qo-$corr$ but not a Js. Indeed, $\langle K_{II}, \precsim /K_{II} \rangle$ is a lattice, see Figure 16. Greatest lower bound of $\mathbf{b}$ and $\mathbf{e}$ in this lattice is $\langle \bot, \bot \rangle$, i.e. $\langle \bot, \bot \rangle \in \mathrm{glb}_{\precsim/K_{II}}\{\mathbf{b},\mathbf{e}\}$. Note, however, that $\mathbf{c} \in \mathrm{glb}_{\precsim}\{\mathbf{b},\mathbf{e}\}$. Hence, $\bot \in \pi_1\left[\mathrm{glb}_{\precsim/K_{II}}\{\mathbf{b},\mathbf{e}\}\right]$ but $(F \wedge P) \in \mathrm{glb}_{R_1} \pi_1[\{\mathbf{b},\mathbf{e}\}]$. And so, though

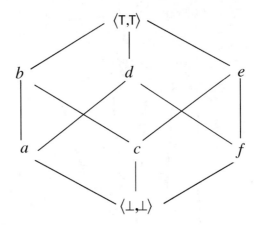

Figure 15

$\langle K_{II}, \precsim /K_{II}\rangle$ is a lattice (and complete since it is finite), it does not satisfy requirement (iii) in Theorem 3.34. Therefore, $\langle \mathcal{B}_1, \mathcal{B}_2, \uparrow K_{II}\rangle$ is not a Js.

If **c** is subtracted, in order to obtain a joining-system, the legislator has to subtract either **b** or **e**, or both, as well. Since elimination of **b** would seem unreasonable from a legal point of view, the appropriate choice would be to eliminate **e**. The resulting system will here be called *System* $\mathcal{S}_{III}$.

System $\mathcal{S}_{III}$ (which is a *qo-corr*, but, at this stage, is not assumed to be a joining-system) is such that

$$K_{III} = \{\mathbf{a}, \mathbf{b}, \mathbf{d}, \mathbf{f}, \langle \bot, \bot \rangle, \langle \top, \top \rangle\}$$

See Figure 17.

$\langle K_{III}, \precsim_{K_{III}}\rangle$ is a lattice. See Figure 18. Moreover, the assumptions in Theorem 3.37 are satisfied. Hence, it follows that $\mathcal{S}_{III} = \langle \mathcal{B}_1, \mathcal{B}_2, \uparrow K_{III}\rangle$ is a *Bjs*.

$\mathcal{S}_{III}$, however, is legally unsatisfactory, since it is merely the result of subtraction, without positively stipulating anything about the relevance of owner's offering/not offering to pay ransom for the goods. The next system to be considered, therefore, is *System* $\mathcal{S}_{IV}$, where "Ransom" is introduced. The *Bqo* of grounds in $\mathcal{S}_{IV}$ is

$$\mathcal{B}_3 = \langle B_3, \wedge, ', R_3\rangle \text{ with } R_3 = \precsim_3;$$

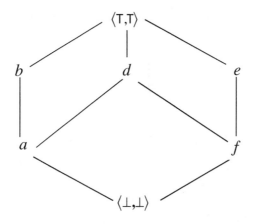

Figure 16

where B_3 consists of Boolean combinations of F, P and R. In S_{IV} the following norms are added:

⟨P∧R,O2⟩. If transferee has the goods in possession and owner pays ransom for the good, then transferee has the obligation to deliver the good to owner.

⟨F∧P∧R/,O2/⟩. If transferee has the good in possession and fulfills the good faith condition, and owner does not pay ransom, then transferee has no obligation to deliver the good back to owner. These added norms however, are not minimal elements.

In S_{IV} (which is assumed to be a *qo-corr* but not a *Js*) the set of minimal norms is

$$K_{IV} = \{\mathbf{b, f, g, h, i, j}, \langle \bot, \bot \rangle, \langle \top, \top \rangle\}$$

where
g = ⟨P∧(F/∨R),O2⟩
h = ⟨F∧P∧R/,O1/∧O2/⟩
i = ⟨F/∨P/∨R,O1∨O2⟩
j = ⟨P/ ∨ (F∧R/),O2/⟩

We note that, of the non-degenerated minimal norms in the original system S_I, only **b** and **f** remain unchanged in S_{IV}, while, due to the relevance of ransom, **g, h, i, j** are new minimal norms in S_{IV}.

The set of non-degenerated norms in K_{IV} and their interrelations is depicted in Figure 19. $\langle K_{IV}, \precsim /K_{IV} \rangle$ is a lattice, and hence complete, since it is finite. See Figure 20 on page 607. Moreover, the assumptions in Theorem

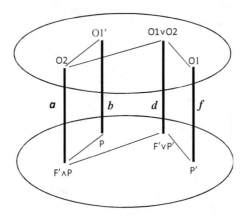

Figure 17

3.37 are satisfied and hence, it follows that $\langle \mathcal{B}_3, \mathcal{B}_2, \uparrow K_{IV} \rangle$ is a joining-system.[21] For further details on the example, cf. [Lindahl and Odelstad, 2003], developed within a slightly different framework (cf. Section 6.1 below).

4.4 The cis version of normative positions

The Kanger-Lindahl theory A natural approach to formulate normative concepts such as obligation and permission is to do so in terms of so-called *normative positions,* constructed by a combination of deontic logic and action logic. As is further developed in Marek Sergot's chapter "The theory of normative positions" of the present Handbook, the first version of the theory of normative positions, in its modern logical form, was developed by the Swedish logician Stig Kanger ([Kanger, 1957; Kanger, 1963]). Kanger's theory was inspired by the system of "fundamental jural relations" proposed by the American jurist W.N. Hohfeld in 1913. As realized by Kanger, standard deontic logic, with a deontic operator applied to sentences, is not adequate for expressing the Hohfeldian distinctions. The improvement proposed by Kanger was to combine a standard deontic operator *Shall* with an action operator *Do* (for "sees to it that") and to exploit the possibilities of external and internal negation of sentences where these operators are combined. Originally, Kanger's theory was conceived as a theory

---

[21] Basically, this was the system of Swedish legislation before 2003. That year, the law was changed so that, when the original owner has lost possession by theft, no ransom is required för getting the goods back.

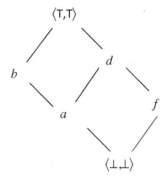

Figure 18

of *rights* (see [Lindahl, 1994]). As a theory of "legal" or "normative" positions, Kanger's theory was further developed by Lars Lindahl in [Lindahl, 1977]. Additional refinements of the so-called Kanger-Lindahl theory have been made by Andrew J.I. Jones and Marek Sergot ([Jones and Sergot, 1993; Jones and Sergot, 1996; Sergot, 1999; Sergot, 2001]). A special feature of the work of Jones and Sergot is that applications in computer science are in view.

A natural approach to the fine-grained structure of a *cis-Bjs* $\langle \mathcal{B}_1, \mathcal{B}_2, J \rangle$ where the stratum $\mathcal{B}_2$ is normative, is to formulate $\mathcal{B}_2$ in terms of an algebraic version of the Kanger-Lindahl theory of normative positions. (On this theory, see Sergot's chapter "The theory of normative positions" in the present Handbook.) The system of normative positions dealt with in what follows below is the system of *one-agent* types of normative position, in the sense of [Lindahl, 1977, ch. 3]. This system, chosen here since it is relatively simple, can easily be generalized to *n-agent* types, see Sergot's chapter and cf. Talja in [Talja, 1980].

To the Boolean connectives of negation, conjunction etc., are added the modal expressions "Shall" and "Do". If F is a state of affairs and x is an agent,[22] Shall F is to be read "It shall be the case that F" and $\mathrm{Do}(x, F)$ should be read "x sees to it that F". The expression May$F$ is an abbreviation for $\neg$Shall$\neg F$.

The basic idea in the Kanger-Lindahl theory is to exploit the possibilities of combining the deontic operator Shall with the action operator Do. One example is Shall $\mathrm{Do}(x, F)$ which means that it shall be that x sees to it that

---

[22] A state of affairs in Kanger's sense might be, for example, that Mr. Smith gets back the money lent by him to Mr. Black, or that Mr. Smith walks outside Mr. Black's shop.

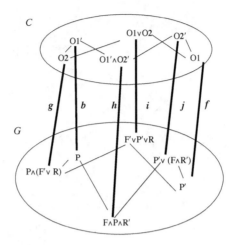

Figure 19

F; another is $\neg$ Shall Do$(y, \neg F)$ which means that it is not the case that it shall be that y sees to it that not F.

The logical postulates for Shall and Do assumed in the construction of one-agent types are as follows (cf. [Lindahl, 1977, p. 68]):

Rules for Do

RI. If $\vdash (A \longleftrightarrow B)$, then $\vdash (\text{Do}(s, A) \longleftrightarrow \text{Do}(s, B))$.
A1. Do$(s, A) \to A$.

Rules for Shall

RII. If $\vdash A$, then $\vdash$ Shall$A$.
A2. Shall $(A \to B) \to$ (Shall$A \to$ Shall$B$).
A3. Shall$A \to \neg$Shall$\neg A$.

The systems of normative positions can serve as a tools for describing the normative positions of different agents $x, y, z...$ with regard to states of affairs $F, G, H, ...$. For example, if x is the Swedish Government and F is the state of affairs that a paper on normative positions by Sergot is published in Sweden, the position, according to Swedish law, of x with regard to F can be described by Shall($\neg$Do(x, F) & $\neg$Do$(x, \neg F)$), expressing that the Government is not allowed either to bring about or prevent the publication.

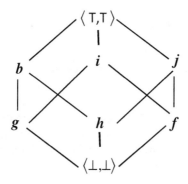

Figure 20

If x is an agent and F is a state of affairs, the seven one-agent types of position are as follows (see [Lindahl, 1977, p. 92]), where $\text{Pass}(x, F)$ is an abbreviation for $\neg \text{Do}(x, F)\ \&\ \neg \text{Do}(x, \neg F)$:

$T_1(x, F) : \text{MayDo}(x, F)\ \&\ \text{MayPass}(x, F)\ \&\ \text{MayDo}(x, \neg F)$.
$T_2(x, F) : \text{MayDo}(x, F)\ \&\ \text{MayPass}(x, F)\ \&\ \neg \text{MayDo}(x, \neg F)$.
$T_3(x, F) : \text{MayDo}(x, F)\ \&\ \neg \text{MayPass}(x, F)\ \&\ \text{MayDo}(x, \neg F)$.
$T_4(x, F) : \neg \text{MayDo}(x, F)\ \&\ \text{MayPass}(x, F)\ \&\ \text{MayDo}(x, \neg F)$.
$T_5(x, F) : \text{MayDo}(x, F)\ \&\ \neg \text{MayPass}(x, F)\ \&\ \neg \text{MayDo}(x, \neg F)$.
$T_6(x, F) : \neg \text{MayDo}(x, F)\ \&\ \text{MayPass}(x, F)\ \&\ \neg \text{MayDo}(x, \neg F)$.
$T_7(x, F) : \neg \text{MayDo}(x, F)\ \&\ \neg \text{MayPass}(x, F)\ \&\ \text{MayDo}(x, \neg F)$.

The numbering of the T_i conforms to the numbering of the corresponding one-agent types of normative position in [Lindahl, 1977]. The numbering suits the representation of the types in a Hasse diagram, exhibiting how the types are partially ordered by the relation "less free than" (see [Lindahl 1977, pp. 105 ff]).

The simplest way to combine the TJS approach with an algebraic version of the theory of one-agent normative positions is to transform the one-agent formulas $T_1(x, F), ..., T_7(x, F)$ into seven *conditions* $T_1 q, ..., T_7 q$. Thus T_i, when occurring in $T_i q$, is an operator on conditions, and the result is a normative condition, defined in terms of one-agent type T_i. A set $\{T_1 q, ..., T_7 q\}$ of seven normative conditions is obtained, and Boolean compounds of these seven conditions are formed by $\wedge, ', \vee$.

Next we construct a *normative position cis*. Let $\mathcal{B} = \langle B, \wedge, ', R \rangle$ be a *cis-Bqo* with a domain B of descriptive conditions $q_1, q_2, ...$. Furthermore,

let
$$T_\mathcal{B} = \{T_i q \mid q \in B - \{\bot, \top\}, 1 \leq i \leq 7\},$$
i.e., $T_\mathcal{B}$ is the set of all normative positions with regard to the descriptive conditions in B. Next, let $T_\mathcal{B}^*$ be the closure of $T_\mathcal{B}$ under $\wedge,'$. Then $\mathcal{T} = \langle T_\mathcal{B}^*, \wedge,' \rangle$ is a Boolean algebra, called a *Boolean normative position algebra*.

Finally, from $\mathcal{T}$ we construct a *cis-Bqo* $\langle T_\mathcal{B}^*, \wedge,', R \rangle$, called a *normative position cis*. Such as *cis* is to fulfil the requirements of deontic logic and action logic described in the theory of one-agent normative positions. These requirements are incorporated in the following definition.

Definition 4.5 A cis $\langle T_\mathcal{B}^*, \wedge,', R \rangle$ *is a* normative position cis *with regard to* $\mathcal{B}$ *if for any* $q, r \in \mathcal{B}$ *it holds that*
(1) *if* $i \neq j$, *then* $T_i q \wedge T_j q \, R \perp$ *(for* $i, j \in \{1, ..., 7\})$,
(2) $\top R (T_1 q \vee ... \vee T_7 q)$,
(3) $T_1 q \, Q \, T_1 q'$, $T_3 q \, Q \, T_3 q'$, $T_6 q \, Q \, T_6 q'$, $T_2 q \, Q \, T_4 q'$, $T_5 q \, Q \, T_7 q'$,
(4) *if* $q \, Q \, r$, *then* $T_i q \, Q \, T_i r$,
(5) *if* $i = 1, 3, 4, 7$, *then* $T_i \top Q \perp$, *and*,
(6) *if* $i = 1, 2, 3, 5$, *then* $T_i \perp Q \perp$.

Requirements (1)-(4) in the definition express restrictions on the relation R in a normative position algebra and correspond to three features of one agent types in the Kanger-Lindahl theory. Thus requirement (1) expresses that $T_1 q, ..., T_7 q$ are mutually incompatible, (2) that they are jointly exhaustive, and (3) that T_1, T_3, T_6 are neutral, while T_4 is the converse of T_2 and T_7 the converse of T_5. Requirements (4)-(6), finally, follow from the logic of Shall and Do, where (4) corresponds to the "extensionality" feature for combinations of operators Shall and Do in the Kanger-Lindahl theory, and (5) and (6) follow from the theorem $\neg \text{MayDo}(x, \perp)$. (See [Lindahl and Odelstad, 2004, sect. 1.2, 4 and 6] for details.)

Liberty conditions For seeing more clearly what various conditions in a normative position *cis* amount to in deontic terms, the notion of *liberty conditions* can be introduced (cf. Lindahl 1977, pp. 106 ff.). This device is available since each normative position condition equals a Boolean compound of liberty conditions.

There are three liberty operators L_1, L_2 and L_3. These can be called action permissibility, passivity permissibility and counter-action permissibility, respectively. In terms of May and Do we can read non-negated liberty conditions as follows.

Action permissibility: L_1
$L_1 q(x_1, ..., x_\nu, x_{\nu+1})$ iff May $\text{Do}(x_{\nu+1}, q(x_1, ..., x_\nu))$

Passivity permissibility: L_2
$L_2q(x_1,...,x_\nu,x_{\nu+1})$ iff May Pass$(x_{\nu+1},q(x_1,...,x_\nu))$

Counter-action permissibility: L_3
$L_3q(x_1,...,x_\nu,x_{\nu+1})$ iff May Do$(x_{\nu+1},q(x_1,...,x_\nu)')$

Liberty conditions L_1, L_2, L_3 can be defined in terms of disjunctions of basic *np*-conditions.

Definition 4.6 L_1, L_2, L_3 *are operators on conditions such that, if q is a condition:*
(1) L_1q is defined as: $T_1q \vee T_2q \vee T_3q \vee T_5q$.
(2) L_2q is defined as: $T_1q \vee T_2q \vee T_4q \vee T_6q$.
(3) L_3q is defined as: $T_1q \vee T_3q \vee T_4q \vee T_7q$.

Accordingly, it holds that (where ' signifies negation),
$T_1q \; Q \; L_1q \wedge L_2q \wedge L_3q$,
$T_2q \; Q \; L_1q \wedge L_2q \wedge (L_3q)'$,
$T_3q \; Q \; L_1q \wedge (L_2q)' \wedge L_3q$,
$T_4q \; Q \; (L_1q)' \wedge L_2q \wedge L_3q$,
$T_5q \; Q \; L_1q \wedge (L_2q)' \wedge (L_3q)'$,
$T_6q \; Q \; (L_1q)' \wedge L_2q \wedge (L_3q)'$,
$T_7q \; Q \; (L_1q)' \wedge (L_2q)' \wedge L_3q$.

Accordingly, if L_{iq} is denoted by 1 and $(L_{iq})'$ by 0, the basic *np*-conditions can be represented by the semi-lattice in Figure 21 (cf. [Lindahl, 1977, p. 105] and [Talja, 1980]).

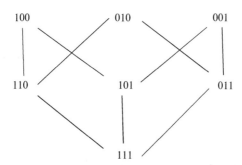

Figure 21

4.4.1 An example: ownership to an estate

Suppose we represent a normative system by a *cis* model of a joining-system with two strata one of which is a descriptive *cis*, and the other is a *normative position-cis*. We illustrate this representation by a simple example concerning the normative position of owners of real property in a legal system S. We consider a *cis* model of a Boolean joining-system $\langle \mathcal{B}_1, \mathcal{B}_2, J \rangle$ where $\mathcal{B}_1 = \langle B_1, \wedge, ', R_1 \rangle$ is descriptive, while $\mathcal{B}_2 = \langle B_2, \wedge, ', R_2 \rangle$ is a *normative position-cis*.

The two strata considered

The descriptive stratum $\mathcal{B}_1$.

We assume that conditions a_1 and b_1, appearing in the descriptive lower stratum $\mathcal{B}_1$ are as follows:

a_1: Being the owner of an estate E.[23]

b_1: Being the owner of an estate adjacent to estate E.

We furthermore assume that $\mathcal{B}_1$ is as depicted in the following diagram (where $\alpha \bowtie \beta$ is an abbreviation for $(\alpha \wedge \beta) \vee (\alpha' \wedge \beta')$ and where lines representing R_1 (implication) are omitted as being evident):

$\mathcal{B}_1$ $\qquad\qquad\qquad\qquad$ $\top$

$\qquad\qquad a_1 \vee b_1 \quad a_1 \vee b_1' \qquad a_1' \vee b_1 \quad a_1' \vee b_1'$

$\qquad a_1 \quad b_1 \quad a_1 \bowtie b_1 \quad a_1 \bowtie b_1' \quad b_1' \quad a_1'$

$\qquad\qquad a_1 \wedge b_1 \quad a_1 \wedge b_1' \qquad a_1' \wedge b_1 \quad a_1' \wedge b_1'$

$\qquad\qquad\qquad\qquad \bot$

We note that $\mathcal{B}_1$ coincides with its reduct $\langle B_1, \wedge, ' \rangle$ and that, therefore, in $\mathcal{B}_1$, R_1 coincides with $\leq_1$. As appears from the diagram, it is assumed that conditions $a_1 \wedge b_1, a_1 \wedge b_1', a_1' \wedge b_1, a_1' \wedge b_1'$ are atoms in $\mathcal{B}_1$.

---

[23] Letter E is to be regarded as a parameter, in the sense of a a quantity which is constant in a particular case considered, but which varies in different cases.

The normative stratum $\mathcal{B}_2$

Let conditions $q_1, ..., q_4$ be as follows:

q_1: Main building of estate E being painted white,
q_2: Main building on estate adjacent to E being painted white,
q_3: Cows of estate E entering land of adjacent estate,
q_4: Erecting a fence, going around estate E and adjacent estate.

Let $\mathcal{B} = \langle B, \wedge,' R \rangle$ be a *cis* such that the descriptive conditions q_1, q_2, q_3, q_4 are among the elements of its domain. Furthermore, as in Section 4.4, let $T_\mathcal{B} = \{T_i q \mid q \in B - \{\bot, \top\}, 1 \leq i \leq 7\}$, let $T_\mathcal{B}^*$ be the closure of $T_\mathcal{B}$ under $\wedge,'$ and let $\mathcal{T} = \langle T_\mathcal{B}^*, \wedge,' \rangle$ be a Boolean normative position algebra with regard to $\mathcal{B}$. Finally, let $\mathcal{B}_2 = \langle T_\mathcal{B}^*, \wedge,', R_2 \rangle$ be a normative position *cis* with regard to $\mathcal{B}$ (see above definition 4.5). Since $\mathcal{T}$ is the reduct of $\mathcal{B}_2$, the Boolean relation $\leq_T$ of $\mathcal{T}$ is a subset of the relation R_2 of $\mathcal{B}_2$.

Joining assumptions

We assume that in the Boolean joining-system $\langle \mathcal{B}_1, \mathcal{B}_2, J \rangle$, when referring to non-degenerated joinings, the following holds:

(i) $(a_1 \wedge b_1) \, J \, (T_1 q_1 \wedge T_1 q_2 \wedge T_1 q_3 \wedge T_1 q_4)$,

(ii) $(a_1 \wedge b_1') \, J \, (T_1 q_1 \wedge T_6 q_2 \wedge T_7 q_3 \wedge T_4 q_4)$,

(iii) $(a_1' \wedge b_1) \, J \, (T_6 q_1 \wedge T_1 q_2 \wedge T_4 q_3 \wedge T_4 q_4)$,

(iv) $(a_1' \wedge b_1') \, J \, (T_6 q_1 \wedge T_6 q_2 \wedge T_6 q_3 \wedge T_6 q_4)$.

Given the intended interpretation of conditions $T_i q_j$ in terms of Shall, May and Do, the joinings (i)-(iv) are plausible for a legal system. This can be seen by inspection of the different grounds and consequences correlated. For this purpose, the notion of liberty conditions is useful (on liberty conditions, see above Section 4.4). To exemplify, $a_1 \wedge b_1$ means being the owner of both estate E and adjacent estate. This condition is a ground for $T_1 q_1 \wedge T_1 q_2 \wedge T_1 q_3 \wedge T_1 q_4$, which is the normative position-condition denoting full freedom (operator T_1) with regard to all of $q_1, ..., q_4$ (painting the two buildings, letting the cows move around, erecting a surrounding fence). In contrast, $a_1 \wedge b_1'$ means owning estate E but not adjacent estate. This condition is ground for $T_1 q_1 \wedge T_6 q_2 \wedge T_7 q_3 \wedge T_4 q_4$. This condition denotes full freedom regarding the painting of building on estate E, no freedom to bring about or prevent painting of building on adjacent estate, obligation to see to it that cows from estate E do not enter land of adjacent estate, and, finally, freedom to prevent erection of the fence surrounding the estates and freedom to be passive about the matter, but no freedom to bring about the fence's being erected.

For further development of the example, see [Lindahl and Odelstad, 2004, sect. 6].

5 Intervenients for Boolean joining-systems

5.1 Introductory remarks on intervenients in Bjs'

In the present main section (Section 5) we will investigate the structure of a stratum $\langle \mathcal{B}_2, R_2 \rangle$ with intervenients, between one stratum $\langle \mathcal{B}_1, R_1 \rangle$ of grounds and one stratum $\langle \mathcal{B}_3, R_3 \rangle$ of consequences. In the present first subsection (Section 5.1), we introduce some notation and some basic results, in particular as regards Boolean operations on intervenients. Since these remarks have been dealt with extensively in [Lindahl and Odelstad, 2011], the general remarks are kept brief, and the reader is referred to [Lindahl and Odelstad, 2011] for proofs and further details.

One possible use of intervenients, not dealt with in the present chapter, is for characterizing a Boolean joining-system. Intervenients from B_1 to B_3 can be used for defining or characterizing the Boolean joining-system $\langle \mathcal{B}_1, \mathcal{B}_3, J_{1,3} \rangle$. Cf. [Lindahl and Odelstad, 2008a, sect. 2.3.5 and 4], on *gic*-systems, proto-intervenients and the methodology of intermediate concepts.

After these remarks, attention will be paid in particular to *cis* applications regarding some important issues. In particular, networks of strata with intervenients, organic wholes of intervenients and narrowing of intervenients will be dealt with.

In Section 3.8, the notion of an intervenient was defined with respect to simple *Js*-triples presupposing that the joinings of the strata are disjunct sets. This presupposition is not appropriate when it comes to intervenients in systems of *Bjs*'s, which can be seen in the following way. Suppose that $S_1 = \langle \mathcal{B}_1, \mathcal{B}_2, J_{1,2} \rangle$, $S_2 = \langle \mathcal{B}_2, \mathcal{B}_3, J_{2,3} \rangle$ and $S_3 = \langle \mathcal{B}_1, \mathcal{B}_3, J_{1,3} \rangle$, where $\mathcal{B}_i = \langle B_i, \wedge, ', R_i \rangle$, are *Bjs*'s and that $B_i \cap B_j = \{\bot, \top\}$ if $i \neq j$, $1 \leq i, j \leq 3$. Then it can be the case that for some $a_2 \in B_2$, $\bot$ is the weakest ground of a_2 or $\top$ is the strongest consequence of a_2. In either case, a_2 is not a proper intervenient since $\langle \bot, a_2 \rangle$ and $\langle a_2, \top \rangle$ are degenerated joinings (cf. Section 4.1.2). We say that a_2 is a non-degenerated intervenient if a_2 is an intervenient and $a_2 \curvearrowright \langle a_1, a_3 \rangle$, where $\langle a_1, a_3 \rangle$ is a non-degenerated joining.

Definition 5.1 *Suppose that* $S_1 = \langle \mathcal{B}_1, \mathcal{B}_2, J_{1,2} \rangle$, $S_2 = \langle \mathcal{B}_2, \mathcal{B}_3, J_{2,3} \rangle$ *and* $S_3 = \langle \mathcal{B}_1, \mathcal{B}_3, J_{1,3} \rangle$ *are joining-systems where* $\mathcal{B}_i = \langle B_i, \wedge, ', R_i \rangle$ *are complete and* $B_i \cap B_j = \{\bot, \top\}$ *for* $i \neq j$, $1 \leq i, j \leq 3$. *If* $J_{1,3} \supseteq J_{1,2} | J_{2,3}$ *we say that* $\langle S_1, S_2, S_3 \rangle$ *is a* Bjs-triple.

(Concerning completeness, see Section 4.1.1.)

Definition 5.2 *In a Bjs-triple $\langle S_1, S_2, S_3 \rangle$, the element $a_2 \in B_2$, is a non-degenerated intervenient from B_1 to B_3 corresponding to the joining $\langle a_1, a_3 \rangle \in J_{1,3}$, denoted $a_2 \curvearrowright \langle a_1, a_3 \rangle$, if a_1 is a non-degenerated weakest ground of a_2 in S_1 and a_3 is a non-degenerated strongest consequence of a_3 in S_2.*

Suppose that $\Phi = \langle S_1, S_2, S_3 \rangle$ is a *Bjs*-triple. Note that if $a_2 \in B_2$ is an intervenient in Φ from B_1 to B_3 then there is $a_1 \in B_1$ and $a_3 \in B_3$ such that a_2 is situated between B_1 and B_3 in S in the sense that $\langle a_1, a_2 \rangle \in J_{1,2}$, $\langle a_2, a_3 \rangle \in J_{2,3}$ and $\langle a_1, a_3 \rangle \in J_{1,3}$. Now, let us look at the converse of this statement. Suppose that $\langle a_1, a_2 \rangle \in J_{1,2}$, $\langle a_2, a_3 \rangle \in J_{2,3}$ and $\langle a_1, a_3 \rangle \in J_{1,3}$. Then, if a_1 is not similar to falsum and a_3 not similar to verum, then a_2 is an intervenient from B_1 to B_3. However, it is important to notice that, even though a_2 is an intervenient from B_1 to B_3 in Φ, it is not guaranteed that $a_2 \curvearrowright \langle a_1, a_3 \rangle$, i.e., that a_2 corresponds to $\langle a_1, a_3 \rangle$. But if $\langle a_1, a_2 \rangle \in \min J_{1,2}$, and $\langle a_2, a_3 \rangle \in \min J_{2,3}$, this holds. Note also that if $\langle a_1, a_3 \rangle \in \min J_{1,3}$ then there is $b_2 \in B_2$ such that b_2 is an intervenient in Φ from B_1 to B_3 and $b_2 \curvearrowright \langle a_1, a_3 \rangle$. (See [Lindahl and Odelstad, 2004, sect. 4] for details.)

5.1.1 Conjunction, disjunction and negation of intervenients

If we apply the Boolean operations conjunction, disjunction and negation on intervenients, will the result be intervenients as well? Which is the relationship between the conjunction of the weakest grounds of two intervenients and the weakest ground of their conjunction, and similarly for disjunction and negation? The same question arises with regard to strongest consequences. We will here consider conjunction and disjunction of pairs of intervenients. Of special interest is Boolean operations in connection with minimality.

Conjunction and disjunction of intervenients

In a *Bjs*-triple $\Phi = \langle S_1, S_2, S_3 \rangle$, we let $\mathrm{Iv}(B_2, B_1, B_3)$ denote the set of elements in B_2 which are intervenients from B_1 to B_3 in Φ. We state some results presented in [Lindahl and Odelstad, 2011, sect. 4.2].

The following theorem states a necessary and sufficient condition for a conjunction of intervenients being an intervenient, and similarly for a disjunction of intervenients.

Theorem 5.3 *Suppose that B_1 and B_3 are complete and that $a_2 \curvearrowright \langle a_1, a_3 \rangle$ and $b_2 \curvearrowright \langle b_1, b_3 \rangle$. Then*

1. $\perp P_1(a_1 \wedge b_1)$ *iff* $(a_2 \wedge b_2) \in \mathrm{Iv}(B_2, B_1, B_3)$, *and*

2. $(a_3 \vee b_3) P_3 \top$ iff $(a_2 \vee b_2) \in \text{Iv}(B_2, B_1, B_3)$.

The following theorem states the relationships between the Boolean operations on intervenients and the corresponding operations on grounds and consequences, respectively. These relationships are important for the discussion of organic wholes of intervenients in the Section 5.2.1.

Theorem 5.4 *Suppose that $\mathcal{B}_1$ and $\mathcal{B}_3$ are complete and that $a_2 \curvearrowright \langle a_1, a_3 \rangle$, $b_2 \curvearrowright \langle b_1, b_3 \rangle$. Then,*

1. *If $(a_2 \wedge b_2) \in \text{Iv}(B_2, B_1, B_3)$ then there is $c_3 \in B_3$ such that $a_2 \wedge b_2 \curvearrowright \langle a_1 \wedge b_1, c_3 \rangle$.*

2. *If $(a_2 \vee b_2) \in \text{Iv}(B_2, B_1, B_3)$ then there is $c_1 \in B_1$ such that $a_2 \vee b_2 \curvearrowright \langle c_1, a_3 \vee b_3 \rangle$.*

The following theorems connect Boolean operations of intervenients to minimality.

Theorem 5.5 *Suppose that $a_2 \curvearrowright \langle a_1, a_3 \rangle \in \min J_{1,3}$ and $b_2 \curvearrowright \langle b_1, b_3 \rangle \in \min J_{1,3}$ and not $a_1 \wedge b_1 R_1 \bot$ and not $\top R_3 a_3 \vee b_3$. Then the following holds:*

1. *If $\langle a_1 \wedge b_1, a_3 \wedge b_3 \rangle \in \min J_{1,3}$, then $a_2 \wedge b_2 \curvearrowright \langle a_1 \wedge b_1, a_3 \wedge b_3 \rangle$.*

2. *If $\langle a_1 \vee b_1, a_3 \vee b_3 \rangle \in \min J_{1,3}$, then $a_2 \vee b_2 \curvearrowright \langle a_1 \vee b_1, a_3 \vee b_3 \rangle$.*

Theorem 5.6 *Suppose that $a_2 \curvearrowright \langle a_1, a_3 \rangle \in \min J_{1,3}$ and $b_2 \curvearrowright \langle b_1, b_3 \rangle \in \min J_{1,3}$ and, furthermore, not $a_1 \wedge b_1 R_1 \bot$ and not $\top R_3 a_3 \vee b_3$. Then there are $c_2, d_2 \in B_2$, $c_3 \in B_3$ and $d_1 \in B_1$ such that*

1. *$c_2 \curvearrowright \langle a_1 \wedge b_1, c_3 \rangle \in \min J_{1,3}$, where $c_3 R_3 (a_3 \wedge b_3)$, and*

2. *$d_2 \curvearrowright \langle d_1, a_3 \vee b_3 \rangle \in \min J_{1,3}$, where $(a_1 \vee b_1) R_1 d_1$.*

Negations of intervenients

Negations of intervenients is an interesting subject. We will here give an overview. (For details and proofs, see [Lindahl and Odelstad, 2008a]). Suppose that a_2 is an intervenient from $\mathcal{B}_1$ to $\mathcal{B}_3$ corresponding to the joining $\langle a_1, a_3 \rangle \in J_{1,3}$ in the *Bjs*-triple $\Psi = \langle \mathcal{S}_1, \mathcal{S}_2, \mathcal{S}_3 \rangle$. Then there are two possibilities with regard to the negation a_2' of a_2:

1. a_2' is an intervenient from $\mathcal{B}_1$ to $\mathcal{B}_3$ in the *Bjs*-triple Ψ.

2. a_2' is not an intervenient from $\mathcal{B}_1$ to $\mathcal{B}_3$ in the *Bjs*-triple Ψ.

If a_2' is *not* an intervenient we can distinguish between three possibilities:

(i) a_2' has a non-degenerated weakest ground in B_1 but no non-degenerated strongest consequence in B_3.

(ii) a_2' has no non-degenerated weakest ground in B_1 but a non-degenerated strongest consequence in B_3.

(iii) a_2' has neither a non-degenerated weakest ground in B_1 nor a non-degenerated strongest consequence in B_3.

If a_2' *is* an intervenient it is important to note the relation between the joining corresponding to a_2 and to a_2'. Suppose that $a_2 \frown \langle a_1, a_3 \rangle$ and $a_2' \frown \langle b_1, b_3 \rangle$. Then:

(I) $\langle a_1', a_3' \rangle \trianglelefteq \langle b_1, b_3 \rangle$.

(II) If $\langle a_1, a_3 \rangle \in \min J_{1,3}$, then $\langle a_1', a_3' \rangle \simeq \langle b_1, b_3 \rangle$.

(III) If $\langle a_1', a_3' \rangle, \langle b_1', b_3' \rangle \in J_{1,3}$, then $\langle a_1', a_3' \rangle \simeq \langle b_1, b_3 \rangle$.

Note that if a_2' is an intervenient this constitutes a restriction on the possibility of narrowing a_2 (see Section 5.2.2 below), since a narrowing of a_2 implies a widening of $\langle a_1', a_3' \rangle$, and (I) above gives a restriction of how wide $\langle a_1', a_3' \rangle$ can be. If $a_2 \frown \langle a_1, a_3 \rangle$ and $\langle a_1, a_3 \rangle \in \min J_{1,3}$ and a_2' is an intervenient, then a_2 cannot be narrowed. The same holds if $a_2 \frown \langle a_1, a_3 \rangle$, $a_2' \frown \langle b_1, b_3 \rangle$ and $\langle a_1', a_3' \rangle, \langle b_1', b_3' \rangle \in J$. The subject of negations of intervenients is important in connection with open intermediaries (see Section 5.2.2 below).

5.2 cis' with intervenients

As appears from the foregoing, in TJS for intervenients, "intervenient" is a technical notion defined at the abstract algebraic level. The notion is intended as a tool for analyzing different kinds of what, informally, is called "intermediaries" and the aim is to provide tools for analyzing intermediaries as they appear in law, language, morals, and so on. For this reason cis' with intervenients is an important part of the chapter.

In the present Section 5.2, we assume that intervenients referred in the text are non-degenerated intervenients (see Definition 5.2).

5.2.1 Organic wholes

Attention should be drawn to the possible occurrence in normative systems of a phenomenon analogous to what G.E. Moore in *Principia Ethica* (first published in 1903) called an "organic unity" or "organic whole". Characteristic of an organic unity, according to Moore, is "that the value of such a whole bears no regular proportion to the sum of the values of its parts"[Moore, 1971, p. 27]. Using another terminology, the phenomenon can be called "synergy". In a context of norms, and within our algebraic framework of Boolean joining-systems, the idea of organic wholes refers to the normative impact of a Boolean compound of conditions rather than to "values" in Moore's sense. In the present section, this theme is dealt with as regards the normative impact of conjunction and disjunction of intervenients.

Definition 5.7 *Let* $a_2 \frown \langle a_1, a_3 \rangle$, $b_2 \frown \langle b_1, b_3 \rangle$, *and* $(a_2 \wedge b_2), (a_2 \vee b_2) \in \mathrm{Iv}(B_2, B_1, B_3)$.
(i) If there is $c_3 \in B_3$ *such that* $a_2 \wedge b_2 J_{2,3} c_3$ *and* $c_3 P_3 a_3 \wedge b_3$, *we say that* $a_2 \wedge b_2$ *is a conjunctive organic whole of* a_2 *and* b_2,
(ii) If there is $c_1 \in B_1$ *such that* $c_1 J_{1,2} a_2 \vee b_2$ *and* $a_1 \vee b_1 P_1 c_1$, *we say that* $a_2 \vee b_2$ *is a disjunctive organic whole of* a_2 *and* b_2.

Note that a disjunctive organic whole is constructed as the dual of a conjunctive organic whole.

A *cis* example of a conjunctive organic whole is a follows (cf. [Lindahl and Odelstad, 2003, sect. 5.1, p. 101]):

We imagine an athletic competition, where there are two events, running and high jumping. We consider three Bqo's where $\mathcal{B}_1$ (with $a_1, b_1, ...$) concerns competition *results* in the two events, where $\mathcal{B}_2$ (with $a_2, b_2, ...$) concerns winner's *titles*, and where $\mathcal{B}_3$ (with $a_3, b_3, c_3, ...$) concerns rights to competition *prizes*.

a_1 is to be the fastest runner, b_1 is to jump the highest,

a_2 is to be "master of running", b_2 is to be "master of jumping", $a_2 \wedge b_2$ is to be "twofold master".

a_3 is to have the right of the running prize, b_3 is to have the right of the jumping prize, $c_3 = a_3 \wedge b_3 \wedge d_3$ is to have the right of the excellence prize, namely (a_3) the right of the running prize, and (b_3) the right of the jumping prize, and, in addition, (d_3) the right of a special bonus prize for the twofold master. The example is illustrated in Figure 22.

In the example we have: $a_2 \frown \langle a_1, a_3 \rangle$, $b_2 \frown \langle b_1, b_3 \rangle$, $a_2 \wedge b_2 \frown \langle a_1 \wedge b_1, c_3 \rangle$, where $c_3 P_3 (a_3 \wedge b_3)$. Since we have $c_3 P_3 (a_3 \wedge b_3)$, it holds in the Bjs-triple $\langle \langle \mathcal{B}_1, \mathcal{B}_2, J_{1,2} \rangle, \langle \mathcal{B}_2, \mathcal{B}_3, J_{2,3} \rangle, \langle \mathcal{B}_1, \mathcal{B}_3, J_{1,3} \rangle \rangle$ that the intervenient $a_2 \wedge b_2$ is an organic whole in relation to $\mathcal{B}_3$. In other words: $a_2 \wedge b_2$ is an organic whole since the consequence $c_3 = a_3 \wedge b_3 \wedge d_3$ of the intervenient $a_2 \wedge b_2$ is "stronger" (P_3) than the "sum" $a_3 \wedge b_3$ of the consequence a_3 of a_2 and the consequence b_3 of b_2.

A subset of the minimal joinings from B_2 to B_3 is depicted by the thick lines in Figure 22.

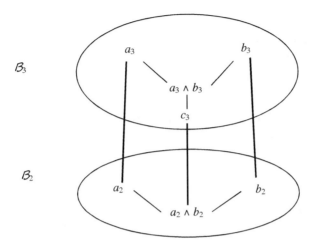

Figure 22

We observe that, in the sense of Theorem 3.34,

$$\mathrm{glb}_{R_2} \pi_1\{\langle a_2, a_3\rangle, \langle b_2, b_3\rangle\} = \mathrm{glb}_{R_2}\{a_2, b_2\} = \{a_2 \wedge b_2\} =$$
$$\pi_1[\mathrm{glb}_{\preceq/\min J}\{\langle a_2, a_3\rangle, \langle b_2, b_3\rangle\}].$$

For a legal example concerning citizenship, see [Lindahl and Odelstad, 2003, sect. 5.1].

5.2.2 Open concepts and the narrowing of intervenients

Ground-narrowing We recall the issue of open legal concepts and the example of "relationship similar to being married" (Section 1.7.5 above). Let $\Psi = \langle S_1, S_2, S_3\rangle$ be a *Bjs-triple* with

$$S_1 = \langle \mathcal{B}_1, \mathcal{B}_2, J_{1,2}\rangle, \ S_2 = \langle \mathcal{B}_2, \mathcal{B}_3, J_{2,3}\rangle, \ S_3 = \langle \mathcal{B}_1, \mathcal{B}_3, J_{1,3}\rangle.$$

Condition $a_2 \in B_2$ (where B_2 is the domain of stratum $\mathcal{B}_2$) is the condition of having a relationship similar to being married. The grounds for a_2 are among the elements of the domain B_1 of stratum $\mathcal{B}_1$ which includes Boolean combinations of the following conditions $a_{1_1}, a_{1_2}, ..., a_{1_{11}}$:

a_{1_1} : cohabiting, a_{1_2} : housekeeping in common, a_{1_3} : having children in common, a_{1_4} : having sexual intercourse, a_{1_5} : having confirmed the relation by a contract, a_{1_6} : living in emotional fellowship, a_{1_7} : being faithful, a_{1_8} : giving mutual support, a_{1_9} : sharing economic assets and debts, $a_{1_{10}}$: having no legal impediments to marriage, $a_{1_{11}}$: having no similar relationship to another person.

Let us suppose that the consequences of having a relationship similar to being married are among the Boolean combinations of conditions $a_{3_1}, ..., a_{3_5}$ belonging to the domain B_3 of stratum $\mathcal{B}_3$.

We assume that in the *Bjs-triple* Ψ, $a_2 \in B_2$ is an intervenient between $(a_{1_1} \wedge a_{1_2} \wedge ... \wedge a_{1_{11}}) \in B_1$ and $(a_{3_1} \wedge ... \wedge a_{3_5}) \in B_3$, i.e.,

$$a_2 \curvearrowright \langle (a_{1_1} \wedge a_{1_2} \wedge ... \wedge a_{1_{11}}), (a_{3_1} \wedge ... \wedge a_{3_5}) \rangle.$$

Thus we assume that in the *Bjs-triple* Ψ, the conjunction $a_{1_1} \wedge a_{1_2} \wedge ... \wedge a_{1_{11}}$ is the weakest ground in B_1 for a_2 and $a_{3_1} \wedge ... \wedge a_{3_5}$ is the strongest consequence in B_3 of a_2.

Next we consider a *Bjs-triple* $\Psi^* = \langle \mathcal{S}_1^*, \mathcal{S}_2, \mathcal{S}_3^* \rangle$ where $\mathcal{B}_1, \mathcal{B}_2, \mathcal{B}_3$ are the same as in Ψ and where $\mathcal{S}_2$ remains unchanged but where $J_{1,2}^*$ and $J_{1,3}^*$ in Ψ^* are different from $J_{1,2}$ and $J_{1,3}$ in Ψ. We assume that $\mathcal{S}_1^* = \langle \mathcal{B}_1, \mathcal{B}_2, J_{1,2}^* \rangle$ and $\mathcal{S}_3^* = \langle \mathcal{B}_1, \mathcal{B}_3, J_{1,3}^* \rangle$ in Ψ^* are different from $\mathcal{S}_1$ and $\mathcal{S}_3$ in Ψ since in Ψ^*,

$$a_2 \curvearrowright \langle (a_{1_1} \wedge a_{1_2} \wedge a_{1_9} \wedge a_{1_{11}}), (a_{3_1} \wedge ... \wedge a_{3_5}) \rangle.$$

Thus in Ψ^*, the conjunction of $a_{1_1} \wedge a_{1_2} \wedge a_{1_9} \wedge a_{1_{11}}$ is the weakest ground for a_2. This means that in Ψ^*, the weakest ground for a_2 is the conjunction of:

a_{1_1} : cohabiting, a_{1_2} : housekeeping in common, a_{1_9} : sharing economic assets and debts, $a_{1_{11}}$: having no similar relationship to another person.

Obviously, in both Ψ and Ψ^* it holds that $(a_{1_1} \wedge a_{1_2} \wedge ... \wedge a_{1_{11}}) R_1 (a_{1_1} \wedge a_{1_2} \wedge a_{1_9} \wedge a_{1_{11}})$. Therefore, the joining $\langle (a_{1_1} \wedge a_{1_2} \wedge a_{1_9} \wedge a_{1_{11}}), a_2 \rangle$ in $J_{1,2}^*$ is narrower than the joining $\langle (a_{1_1} \wedge a_{1_2} \wedge ... \wedge a_{1_{11}}), a_2 \rangle$ in $J_{1,2}$. Accordingly, it also holds that the joining $\langle (a_{1_1} \wedge a_{1_2} \wedge a_{1_9} \wedge a_{1_{11}}), (a_{3_1} \wedge ... \wedge a_{3_5}) \rangle$ in $J_{1,3}^*$ is narrower than the joining $\langle (a_{1_1} \wedge a_{1_2} \wedge ... \wedge a_{1_{11}}), (a_{3_1} \wedge ... \wedge a_{3_5}) \rangle$ in $J_{1,3}$. We describe the situation by saying that the intervenient a_2 is ground-narrower in Ψ^* than in Ψ. This means that the weakest ground for a_2 in Ψ^* is less restricted than in Ψ.

In general terms we can say: If $\Psi = \langle \mathcal{S}_1, \mathcal{S}_2, \mathcal{S}_3 \rangle$, $\Psi^* = \langle \mathcal{S}_1, \mathcal{S}_2^*, \mathcal{S}_3^* \rangle$ are *Bjs-triples* with $\mathcal{B}_i = \mathcal{B}_i^*$ ($1 \leq i \leq 3$) and $a_2 \curvearrowright \langle a_1, a_3 \rangle$ in Ψ, $a_2 \curvearrowright \langle b_1, a_3 \rangle$ in Ψ^* and $\langle b_1, a_3 \rangle \lhd \langle a_1, a_3 \rangle$, then a_2 is ground-narrower in Ψ^* than in Ψ.[24]

---

[24]In [Lindahl and Odelstad, 2008a, sect. 3.5.1], we discuss the narrowing of "relationship similar to being married" with a different framework and terminology.

As an illustrative elaboration of the example, let us consider a normative system such as "Swedish law" as a class of *Bjs-triples* Ψ, Ψ^*, Ψ^{**}... etc. Then we might think of Ψ as a representation of "established Swedish law" and of Ψ^*, Ψ^{**}... etc. as developments of Ψ, made by new authoritative court decisions. Referring to the example, a new court decision resulting in Ψ^* still respects the established law in Ψ insofar as the joining $\langle (a_{1_1} \wedge a_{1_2} \wedge a_{1_9} \wedge a_{1_{11}}), a_2 \rangle$ in established law Ψ still remains in system Ψ^*.

The possibility of narrowing an intervenient while respecting established law Ψ can be barred by a stipulation in Ψ that a certain combination of elements in $\mathcal{B}_1$ is *not* a ground for the intervenient a_2. As regards the handling of this case, see [Odelstad and Lindahl, 2002, sect. 3.4] (cf. [Lindahl and Odelstad, 1999b]).

If we say that "relationship similar to being married" is an "open" concept in Swedish law, this might be taken to mean that established law in Ψ represents only a part of what is considered to count as Swedish law, and that Ψ^* is a development of the first regulative step taken by establishing Ψ.

Consequence-narrowing

What has been said about ground-narrowing has an analogous application in consequence-narrowing. The outlines of an example might regard the consequences of the intervenient *being the owner of an estate*. Let $\Psi = \langle \mathcal{S}_1, \mathcal{S}_2, \mathcal{S}_3 \rangle$ be a *Bjs-triple* with

$$\mathcal{S}_1 = \langle \mathcal{B}_1, \mathcal{B}_2, J_{1,2} \rangle, \quad \mathcal{S}_2 = \langle \mathcal{B}_2, \mathcal{B}_3, J_{2,3} \rangle, \quad \mathcal{S}_3 = \langle \mathcal{B}_1, \mathcal{B}_3, J_{1,3} \rangle,$$

with

a_2: x is the owner of an estate,

and where in Ψ it holds that a_2 is an intervenient between the disjunction $(a_{1_1} \vee a_{1_2} \vee ... \vee a_{1_m})$ of grounds for ownership and the conjunction $(a_{3_1} \wedge a_{3_2} \wedge ... \wedge a_{3_n})$ of consequences of ownership, i.e., where in Ψ it holds that

$$a_2 \curvearrowright \langle (a_{1_1} \vee a_{1_2} \vee ... \vee a_{1_m}), (a_{3_1} \wedge a_{3_2} \wedge ... \wedge a_{3_n}) \rangle$$

Let $a_{3_{n+1}}$ be a consequence that is not a conjunct in the conjunction $(a_{3_1} \wedge a_{3_2} \wedge ... \wedge a_{3_n})$; for example let $a_{3_{n+1}}$ be the condition

$a_{3_{n+1}}$: x is permitted to erect a barbed-wire fence around the entire estate preventing others from entering.

In Ψ^* we have instead

$$a_2 \curvearrowright \langle (a_{1_1} \vee a_{1_2} \vee ... \vee a_{1_m}), (a_{3_1} \wedge a_{3_2} \wedge ... \wedge a_{3_n} \wedge a_{3_{n+1}}) \rangle$$

where $a_{3_{n+1}}$ is a conjunct in the conjunction of consequences.

Since $(a_{3_1} \wedge a_{3_2} \wedge ... \wedge a_{3_n} \wedge a_{3_{n+1}}) R_3 (a_{3_1} \wedge a_{3_2} \wedge ... \wedge a_{3_n})$, it follows that the joining $\langle a_2, (a_{3_1} \wedge a_{3_2} \wedge ... \wedge a_{3_n} \wedge a_{3_{n+1}}) \rangle$ which is narrowest in Ψ^* for the consequences of a_2 is narrower than the joining $\langle a_2, (a_{3_1} \wedge a_{3_2} \wedge ... \wedge a_{3_n}) \rangle$ which is narrowest in Ψ. In this sense, the intervenient a_2 is consequence-narrower in Ψ^* than in Ψ. This means that the strongest consequence of a_2 in Ψ^* is richer than in Ψ.

In general terms: If $\Psi = \langle \mathcal{S}_1, \mathcal{S}_2, \mathcal{S}_3 \rangle$, $\Psi^* = \langle \mathcal{S}_1, \mathcal{S}_2^*, \mathcal{S}_3^* \rangle$ are Bjs-triples with $\mathcal{B}_i = \mathcal{B}_i^*$ ($1 \leq i \leq 3$) and the joinings in Ψ, Ψ^* differ insofar as $a_2 \curvearrowright \langle a_1, a_3 \rangle$ in Ψ, $a_2 \curvearrowright \langle a_1, b_3 \rangle$ in Ψ^* where $\langle a_1, b_3 \rangle \triangleleft \langle a_1, a_3 \rangle$, then a_2 is consequence-narrower in Ψ^* than in Ψ.

What was said in the previous subsection of a normative system such as "Swedish law" as a class of Bjs-triples $\Psi, \Psi^*, \Psi^{**}...$ and of developing established law by narrowing an intervenient applies to consequence-narrowing in an analogous way.

5.2.3 A legal illustration of a network of strata

The present subsection (with Figure 23 on page 621) presents a *cis* example of joining-systems with intervenients for a network of Bqo strata. (Cf. [Lindahl and Odelstad, 2011]) The example is legal and concerns *ownership* and *trust* as intervenients. The legal rules in this example are expressed in terms of joinings between Bqo's $\mathcal{B}_1, \mathcal{B}_2, \mathcal{B}_4, \mathcal{B}_5$ for ownership, and between $\mathcal{B}_3, \mathcal{B}_4$ and $\mathcal{B}_5$ for trusteeship.[25] Both of $\mathcal{B}_2$ and $\mathcal{B}_4$ are intermediate structures, where $\mathcal{B}_4$ is supposed to contain the intervenients ownership and trusteeship and $\mathcal{B}_2$ the intervenients *purchase, barter, inheritance, occupation, specification, expropriation* (for public purposes or for other reasons), which are grounds for ownership. $\mathcal{B}_1$ contains grounds for the conditions in $\mathcal{B}_2$, such as making a contract for purchase or barter respectively, having particular kinship relationship to a deceased person, appropriating something not owned, creating a valuable thing out of worthless material, getting a verdict on disappropriation of property, either for public purposes or for other reasons. $\mathcal{B}_3$ contains different grounds for trusteeship. $\mathcal{B}_5$ contains the legal consequences of ownership and trusteeship, respectively, in terms of powers, permissions and obligations.

The example is a useful illustration in several ways. Thus it illustrates a TJS representation of a fairly complex normative system. Also, as will be shown in the nest subsection, it illustrates various properties of intervenients in terms of minimality.

---

[25]Trust is where a person (trustee) is made the nominal owner of property to be held or used for the benefit of another. Trusteeship is the legal position of a trustee.

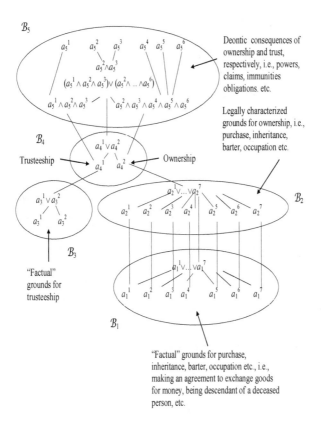

Figure 23

5.2.4 The typology of intervenient-minimality

The previous sections illustrate the role of intervenient concepts in the representation of a normative system. Of special interest is where intervenients exhibit different kinds of *minimality*. (To the following, see [Lindahl and Odelstad, 2011, pp. 132ff.]) Above, we have underlined the central role of minimal joinings and the formal structure of the set of minimal joinings. The previous sections provide tools for distinguishing between different kinds of intervenient minimality. We presuppose a *Bjs-triple* $\langle \mathcal{S}_1, \mathcal{S}_2, \mathcal{S}_3 \rangle$ in the sense of Definition 5.1 and non-degenerated intervenients in the sense of Definition 5.2.

If $a_2 \in \mathrm{Iv}\,(B_2, B_1, B_3)$ and $a_2 \curvearrowright \langle a_1, a_3 \rangle$, we say that,

a_2 is *correspondence-minimal* if $\langle a_1, a_3 \rangle \in \min J_{1,3}$,
a_2 is *ground-minimal* if $\langle a_1, a_2 \rangle \in \min J_{1,2}$,
a_2 is *consequence-minimal* if $\langle a_2, a_3 \rangle \in \min J_{2,3}$.

Combining the three cases,

(1) $\langle a_1, a_3 \rangle \in \min J_{1,3}$,

(2) $\langle a_1, a_2 \rangle \in \min J_{1,2}$,

(3) $\langle a_2, a_3 \rangle \in \min J_{2,3}$,

with their negations $(1'), (2'), (3')$, eight (2^3) cases are obtained. In the case $(1')\&(2')\&(3')$, the intervenient a_2 will be called *non-minimal*.

Not all eight cases are possible to realize. If $J_{1,3} = J_{1,2}|J_{2,3}$, then (1) is implied by (2)&(3). Hence, under this supposition, the case $(1')\&(2)\&(3)$ is impossible to realize.

As regards the importance of minimality emphasized above, note that the following holds: Suppose that $X_2 \subseteq B_2$ is such that for any $\langle x_1, x_3 \rangle \in \min J_{1,3}$ there is $x_2 \in X_2$ such that $x_2 \curvearrowright \langle x_1, x_3 \rangle$. Then

$$J_{1,3} = \{\langle a_1, a_3 \rangle \in B_1 \times B_3 \mid \exists b_2 \in X_2 : \langle a_1, b_2 \rangle \in J_{1,2} \text{ and } \langle b_2, a_3 \rangle \in J_{2,3}\}.$$

Hence, a set of correspondence-minimal intervenients can be a convenient way for characterizing a set of joinings.

However, intervenients can be useful even if they are not correspondence-minimal. A type worth considering is $(1')\&(2)\&(3')$, i.e., where a_2 is ground-minimal but neither correspondence-minimal nor consequence-minimal. For instance, murder and high treason can have the same legal consequence (life imprisonment) notwithstanding that these crimes have different grounds.[26] Thus let

a_1 : grounds for murder, b_1: grounds for high treason

a_2 : murder, b_2 : high treason,

a_3 : life imprisonment

The example is illustrated by Figure 24.

We have $a_2 \curvearrowright \langle a_1, a_3 \rangle$, $b_2 \curvearrowright \langle b_1, a_3 \rangle$, $a_2 \vee b_2 \curvearrowright \langle a_1 \vee b_1, a_3 \rangle$. The intervenient $a_2 \vee b_2$ is correspondence-minimal as well as ground- and consequence-minimal. Each of the intervenients a_2 and b_2, however, though ground-minimal, is neither consequence-minimal nor correspondence-minimal.

---

[26] See also [Lindahl and Odelstad, 2008a, sect. 3.2], for the case of "Boche" in the "Boche-Berserk" example. "Boche" and "Berserk" have different grounds but the same consequence.

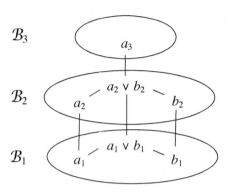

Figure 24

Types of intervenient minimality in the ownership/trust example

The ownership/trust example (Figure 23 on page 621) can be used for illustrating some types of intervenient minimality.

1. a_2^i ($1 \leq i \leq 7$) is an intervenient from B_1 to B_4. This holds since WG (a_1^i, a_2^i, B_1) and SC (a_4^2, a_2^i, B_4) and hence $a_2^i \curvearrowright \langle a_1^i, a_4^2 \rangle$. Note that (it is assumed that) $\langle a_1^i, a_2^i \rangle \in \min J_{1,2}$. Hence, the intervenient a_2^i is ground-minimal. However, a_2^i is neither correspondence-minimal (since $\langle a_1^i, a_4^2 \rangle \notin \min J_{1,4}$), nor consequence-minimal (since $\langle a_2^i, a_4^2 \rangle \notin \min J_{2,4}$).

2. $a_2^1 \vee ... \vee a_2^7$ is an intervenient from B_1 to B_4. This holds since

$$\text{WG}\left(a_1^1 \vee ... \vee a_1^7, a_2^1 \vee ... \vee a_2^7, B_1\right)$$

and

$$\text{SC}\left(a_4^2, a_2^1 \vee ... \vee a_2^7, B_4\right)$$

and hence

$$a_2^1 \vee ... \vee a_2^7 \curvearrowright \langle a_1^1 \vee ... \vee a_1^7, a_4^2 \rangle.$$

It is assumed that

$$\langle a_1^1 \vee ... \vee a_1^7, a_2^1 \vee ... \vee a_2^7 \rangle \in \min J_{1,2}$$

and that $\langle a_2^1 \vee ... \vee a_2^7, a_4^2 \rangle \in \min J_{2,4}$. It then follows that $\langle a_1^1 \vee ... \vee a_1^7, \min J_{1,4}$. (See the remark at the end of Section 3.7.) Hence, $a_2^1 \vee ... \vee a_2^7$ is ground-, consequence- and correspondence-minimal.

3. a_4^2 (being owner) is an intervenient from B_2 to B_5. This holds since
$$\text{WG}\left(a_2^1 \vee ... \vee a_2^7, a_4^2, B_2\right)$$
and SC $\left(a_5^2 \wedge ... \wedge a_5^6, a_4^2, B_5\right)$ and hence
$$a_4^2 \curvearrowright \langle a_2^1 \vee ... \vee a_2^7, a_5^2 \wedge ... \wedge a_5^6 \rangle.$$
It is assumed that $\langle a_2^1 \vee ... \vee a_2^7, a_4^2 \rangle \in \min J_{2,4}$ and
$$\langle a_4^2, a_5^2 \wedge ... \wedge a_5^6 \rangle \in \min J_{4,5}.$$
It follows that
$$\langle a_2^1 \vee ... \vee a_2^7, a_5^2 \wedge ... \wedge a_5^6 \rangle \in \min J_{2,5}.$$
Hence, the intervenient a_4^2 is ground-, consequence- and correspondence-minimal.

4. a_4^1 (being trustee) is an intervenient from B_3 to B_5. This holds since
$$\text{WG}\left(a_3^1 \vee a_3^2, a_4^1, B_3\right)$$
and SC $\left(a_5^1 \wedge a_5^2 \wedge a_5^3, a_4^1, B_5\right)$ and hence
$$a_4^1 \curvearrowright \langle a_3^1 \vee a_3^2, a_5^1 \wedge a_5^2 \wedge a_5^3 \rangle.$$
It is assumed that $\langle a_3^1 \vee a_3^2, a_4^1 \rangle \in \min J_{3,4}$ and that
$$\langle a_4^1, a_5^1 \wedge a_5^2 \wedge a_5^3 \rangle \in \min J_{4,5}.$$
Once more it follows that
$$\langle a_3^1 \vee a_3^2, a_5^1 \wedge a_5^2 \wedge a_5^3 \rangle \in \min J_{3,5}.$$
Hence, a_4^1 is ground-, consequence- and correspondence-minimal. On the other hand, since
$$\langle a_4^1 \vee a_4^2, a_5^2 \wedge a_5^3 \rangle \in J_{4,5}$$
it follows that $\langle a_4^1, a_5^2 \wedge a_5^3 \rangle \notin \min J_{4,5}$.

5. a_4^2 (being an owner) is an intervenient from B_1 to B_5. (Cf. 3 above.) This holds since
$$\text{WG}\left(a_1^1 \vee ... \vee a_1^7, a_4^2, B_1\right)$$

and
$$\text{SC}\left(a_5^2 \wedge \ldots \wedge a_5^6, a_4^2, B_5\right)$$
and hence
$$a_4^2 \curvearrowright \langle a_1^1 \vee \ldots \vee a_1^7, a_5^2 \wedge \ldots \wedge a_5^6 \rangle.$$
Here, it is assumed that (i)
$$\langle a_1^1 \vee \ldots \vee a_1^7, a_2^1 \vee \ldots \vee a_2^7 \rangle \in \min J_{1,2},$$
that (ii)
$$\langle a_2^1 \vee \ldots \vee a_2^7, a_4^2 \rangle \in \min J_{2,4}$$
and that (iii)
$$\langle a_4^2, a_5^2 \wedge \ldots \wedge a_5^6 \rangle \in \min J_{4,5}.$$
From (iii) it follows that a_4^2 is consequence minimal. From (i)-(iii) and (once more) the remark in Section 3.7 it follows that $\langle a_1^1 \vee \ldots \vee a_1^7, a_4^2 \rangle \in \min J_{1,4}$ (ground minimality for a_4^2), and that
$$\langle a_1^1 \vee \ldots \vee a_1^7, a_5^2 \wedge \ldots \wedge a_5^6 \rangle \in \min J_{1,5}$$
(correspondence minimality for a_4^2). Hence, a_4^2 is ground-, consequence- and correspondence minimal.

6 Related work

6.1 Previous work of ours

In our first major joint work on the subject of intermediate concepts, viz. [Lindahl and Odelstad, 1999a], we presented a number of ideas to be further developed in subsequent papers of ours.[27] Our concern with intermediaries originally was inspired by the Scandinavian legal and philosophical discussion of intermediate concepts in the law, a discussion started in the 1940's by Ekelöf and Wedberg. Other sources of inspiration were Dummett's theory of language, Gentzen's theory of natural deduction and the theory of normative systems of Alchourrón and Bulygin. (See Section 1.7 above.)

Our aim in [Lindahl and Odelstad, 1999a] was to provide tools for a rational reconstruction of a legal system with intermediaries; the formal framework was that of a *lattice* $\langle L, \leq \rangle$ of conditions and an implicative relation $\wp$ over L such that $\langle L_\wp, \leq_\wp \rangle$ is generated by the equivalence relation

---

[27][Lindahl and Odelstad, 1999a] was based on our presentation at DEON'98 in Bologna. Our joint theory was presented for the first time in 1996 at the workshop (a cura di V. A. A. Martino) *Logica, Informatica, Diritto*, Pisa, 1996, in honor of Carlos Alchourrón. For references to another preparatory joint work in 1996 see [Lindahl and Odelstad, 1999a]. An early paper in Swedish by Lindahl on intermediate concepts is [Lindahl, 1985].

$\approx_\wp$. Within this framework, we defined the notion of a lattice joining-system $\langle \mathbf{A}, \mathbf{B}, \mathbf{C} \rangle$, with $\mathbf{A}$ the under-lattice, $\mathbf{B}$ the over-lattice and $\mathbf{C}$ the background lattice. We defined two kinds of linking relations between $\mathbf{A}$ and $\mathbf{C}$, viz. the relations of "connection" and "coupling". We treated themes such as couplings satisfying a constraint, the relations "narrower" and "wider" for couplings, and the interrelation between coupling conditions and the notion of "intermediary".

In subsequent papers, we exchanged the main framework of lattices for a framework of *Boolean quasi-orderings* (*Bqo*'s, cf. Section 4.1.1 above.)[28] Connections and couplings now were thought of as relations between what we called "fragments" of a *Bqo*. A *Bqo* $\langle B, \wedge, ', R \rangle$ was thought of as the "closure" of a supplemented Boolean algebra $\langle B, \wedge, ', \rho \rangle$.[29] Also, the algebraic framework was made more abstract, so as to consider "condition implication structures" as models of the more abstract framework. Within this framework, the theory was further developed in various respects. In [Lindahl and Odelstad, 1999b], we introduced the idea of a normative system as a set of *Bqo*'s, among which a "core" and a number of "amplifications"; in [Lindahl and Odelstad, 2000], we treated the problem of intermediate legal concepts that (like disposition concepts) express hypothetical consequences; in [Odelstad and Lindahl, 2002], we further developed the theory of connections; in [Lindahl and Odelstad, 2003], we treated the idea of subtraction and addition of norms; in [Lindahl and Odelstad, 2004], we proposed a model for normative positions within the algebraic framework; and, in [Lindahl and Odelstad, 2006b], we dealt summarily with open and closed intermediaries.

A third stage of development with regard to the general framework appeared with the introduction of *Boolean joining-systems* (*Bjs*'s, cf. above Section 4), first presented in [Odelstad and Boman, 2004]. Instead of considering connections and couplings between two fragments of one single *Bqo*, we now introduced the idea of a *Bjs* $\langle \mathcal{B}_1, \mathcal{B}_2, J \rangle$ with a joining relation J from one *Bqo* $\mathcal{B}_1$ to another *Bqo* $\mathcal{B}_2$. We adjusted the analyses of the issues mentioned above to this framework and developed new themes. In particular, in [Lindahl and Odelstad, 2006a], we introduced the notion of "intervenient" as a formal tool for analyzing intermediaries in normative systems and began the development of a formal theory of intervenients. The theory of intervenients was further developed in [Lindahl and Odelstad, 2008a] and included topics such as "bases of intervenients", "extendable and non-extendable intervenients", and negations of intervenients. The formal

---

[28] The idea of Boolean quasi-orderings and fragments was first presented already in 1998, see references in [Odelstad and Lindahl, 2000].

[29] Cf. [Lindahl and Odelstad, 1999b].

analysis of intervenients was continued in [Lindahl and Odelstad, 2008b; Lindahl and Odelstad, 2011]. The focus of the latter paper is on intervenient minimality, conjunctions and disjunctions of intervenients, organic wholes of intervenients, and a typology of different kinds of intervenients. Also [Lindahl and Odelstad, 2011] pays attention to the properties of intervenients in a network of several *Bjs*'s, with "strata" of *Bqo*'s $\mathcal{B}_1, \mathcal{B}_2, \mathcal{B}_3, ...$.

6.2 Recent work of others

6.2.1 A remark on the "Counts-as" theory

A logical analysis of external sentences of the kind "x counts-as y in s", where s is an institution (s can be a normative system), was proposed by Jones and Sergot in [Jones and Sergot, 1996; Jones and Sergot, 1997]. The work of Jones and Sergot on "Counts-as" has been continued by a number of other authors. This subsequent work has many facets, developed over the past ten years. The book-length study [Grossi, 2007] by Grossi provides axiomatization and semantics of the different counts-as operators.

When a rule r of a legal system $\mathcal{N}$ attaches an intermediary m, e.g., "x and y have made a contract to the effect that z", to a conjunction a of facts, the rule r can be expressed in different ways, e.g. "if a then m", "a is a ground for m" or, sometimes, "a counts as m".

As appears from the foregoing, in our formal representation of $\mathcal{N}$ by a *cis* model of *Bjs-triples* $\langle \mathcal{S}_i, \mathcal{S}_j, \mathcal{S}_k \rangle$ we represent such a statement by $a_i R_i b_i$, or (if different sorts of objects are in view) $a_i J_{i,j} a_j$, which statements are read "a_i implies b_i" and "a_i is a ground for a_j" respectively. In TJS, no counts-as operator is introduced, and in the present chapter we do not examine the question in which cases the counts-as vocabulary might be appropriate. Rather, referring to the joint paper [Grossi et al., 2007] by Grossi, Meyer and Dignum, we will be content, by an example, merely to suggest how some of the material dealt with in the Counts-as theory might be represented in our theory. (Cf. [Lindahl and Odelstad, 2008a, sect. 3.5.3].)

In [Grossi et al., 2007, p. 2], the following example is given of three kinds of Counts-as:

> "It is a rule of normative system Γ that conveyances transporting people or goods count as vehicles; it is always the case that bikes count as conveyances transporting people or goods but not that bikes count as vehicles; therefore, in the context of normative system Γ, bikes count as vehicles."

According to [Grossi et al., 2007, p. 2], the first premise states a rule of Γ and is a constitutive Counts-as, the second premise states a generally acknowledged classification, thus states a general classificatory Counts-as, and the conclusion states a classification that holds in Γ and is a Counts-as brought about by Γ though it is not a constitutive Counts-as.

The example can be further developed by the assumption that in Γ vehicles are not admitted in public parks (cf. [Grossi et al., 2006, p. 615]).

If counts-as sentences are seen as internal to a normative system Γ, a representation of the example might be made in terms of Figure 25 on page 628. We can conceive of the example in such a way that "being a

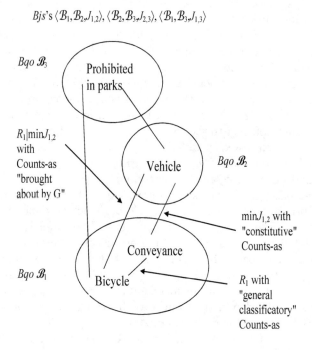

Figure 25

vehicle" is an intervenient from B_1 to B_3 corresponding to the pair ⟨being a conveyance, being prohibited in parks⟩ in $B_1 \times B_3$.

In this chapter, there is no room for going into possible developments of the example. A brief comment should be made, however, on how we might represent something similar to the distinction between three kinds of Counts-as made by Grossi, Meyer and Dignum. We can assume that relation R_1 (a subset of $B_1 \times B_1$) represents implications that hold in an uncontroversial way independently of the instituted rules of Γ. In contrast, the set of minimal joinings $\min J_{1,2}$ (a subset of $B_1 \times B_2$) can be seen as expressing implications that are instituted by the rules in Γ. If this view is taken, distinctions can be made as follows. (We write b, c, v for "bicycle", "conveyance", "vehicle".) Firstly, the general classification of bicycles as conveyances is due to $\langle b, c \rangle \in R_1$ ("bikes always count as conveyances"). Secondly, the classification of conveyances as vehicles is due to $\langle c, v \rangle \in \min J_{1,2}$ ("Conveyances ... are to count as vehicles"). Thirdly, the classification of bicycles as vehicles is due to $\langle b, v \rangle \in R_1 | \min J_{1,2}$ (the relative product).

6.2.2 Input-output logic

In a series of papers, Makinson and van der Torre have developed a theory called input-output logic, see for example [Makinson and van der Torre, 2000; Makinson and van der Torre, 2003]. Important similarities between input-output logic and our approach are that we study normative systems as deductive mechanisms yielding outputs for inputs and that norms are represented as ordered pairs.[30] Other similarities worth mentioning are that neither the principal output operation in input-output logic, nor the relation J in a *Bjs*, requires reflexivity or contraposition.

TJS, however, differs from input-output logic, as developed in [Makinson and van der Torre, 2000; Makinson and van der Torre, 2003], in a number of respects. Thus, in TJS,

1. if a pair $\langle a_1, a_2 \rangle$ represents a norm, this is due to the normative character of a_2 (see Sections 1 and 4.4);

2. a central theme is "intermediaries" (intermediate concepts) in the system;

3. a normative system is represented as a network of subsystems and relations between them; the study comprises stratification of a normative system with structures ("strata") that are intermediate;

4. emphasis is put on the analysis of minimality of joinings and of closeness between strata; representation by a base of minimal joinings is of special importance;

5. the strata of the kind of system called a Boolean joining-system are Boolean structures (*Bqo*'s to be more precise); however, the strata of joining-systems of other kinds need not in TJS be Boolean structures. Thus, in Section 3 of the present chapter, there is a general algebraic

---

[30]Cf. [Lindahl and Odelstad, 1999b, sect. 1.1], with a reference to the work of Alchourrón and Bulygin in [Alchourrón and Bulygin, 1971].

framework for joining-systems that need not be Boolean, for example joining-systems containing strata of lattice-like structures. (In input-output logic, the set of inputs constitutes a Boolean algebra and the same holds for the set of outputs.)

The following remark sheds some light on the relation between input-output logic and the theory of joining-systems. Suppose that $\langle \mathcal{B}_1, \mathcal{B}_2, J\rangle$ is a *Bjs* where $\mathcal{B}_1 = \langle B_1, \wedge,', R_1\rangle$ and $\mathcal{B}_2 = \langle B_2, \wedge,', R_2\rangle$. Makinson and van der Torre state a number of rules for the so-called "basic" output operator (called out_2) that they define. Translated to a *Bjs* these rules are as follows (cf. Definitions 3.11 in Section 3.2):

Strengthening Input: From $\langle a_1, a_2\rangle \in J$ to $\langle b_1, a_2\rangle \in J$ whenever $b_1 R_1 a_1$.
   Follows from condition (1) of a *Bjs*.
Conjoining Output: From $\langle a_1, a_2\rangle \in J$ and $\langle a_1, b_2\rangle \in J$ to $\langle a_1, a_2 \wedge b_2\rangle \in J$.
   Follows from condition (3) of a *Bjs*.
Weakening Output: From $\langle a_1, a_2\rangle \in J$ to $\langle a_1, b_2\rangle \in J$ whenever $a_2 R_2 b_2$.
   Follows from condition (1) of a *Bjs*.
Disjoining Input: From $\langle a_1, a_2\rangle \in J$ and $\langle b_1, a_2\rangle \in J$ to $\langle a_1 \vee b_1, a_2\rangle \in J$.
   Follows from condition (2) of a *Bjs*.

There are three conditions on a joining space in a Boolean joining-system. The comparison with input-output logic above shows that it could be of interest to define weaker kinds of systems characterized by, for example, conditions (1) and (3).

In TJS the notion of completeness plays an important role. If in a joining-system the quasi-orderings are complete quasi-lattices, then the joining-system satisfies connectivity, one of the key feature in TJS. Even in the definition of a joining-system itself, the notion of completeness is in some sense present although in a concealed form. To see this, we recall condition (2) and (3) in the definition of a joining-system. In these conditions, least upper bounds (lub's) and greatest lower bounds (glb's) of arbitrary sets are called for, although such bounds are not required to exist. Instead certain things must hold for those lub's or glb's of infinite sets that exist. Admittedly, however, this may in certain contexts be regarded as too demanding a requirement: if so, it may seem reasonable to restrict attention to lub's and glb's of pairs of objects. This reasoning leads to the following definition of a kind of systems called prejoining-systems.

Definition 6.1 *A prejoining-system, is an ordered triple* $\langle \mathcal{A}_1, \mathcal{A}_2, J\rangle$ *such that* $\mathcal{A}_1 = \langle A_1, R_1\rangle$ *and* $\mathcal{A}_2 = \langle A_2, R_2\rangle$ *are quasi-orderings and* $J \subseteq A_1 \times A_2$ *and the following conditions are satisfied where* $\trianglelefteq$ *is the narrowness relation*

determined by $\mathcal{A}_1$ *and* $\mathcal{A}_2$:
(1) *for all* $b_1, c_1 \in A_1$ *and* $b_2, c_2 \in A_2$, *if* $\langle b_1, b_2 \rangle \in J$ *and* $\langle b_1, b_2 \rangle \trianglelefteq \langle c_1, c_2 \rangle$, *then* $\langle c_1, c_2 \rangle \in J$,
(2) *for all* $b_1, c_1 \in A_1$ *and* $b_2 \in A_2$, *if* $\langle b_1, b_2 \rangle \in J$ *and* $\langle c_1, b_2 \rangle \in J$, *then* $\langle a_1, b_2 \rangle \in J$ *for all* $a_1 \in \mathrm{lub}_{R_1}\{b_1, c_1\}$,
(3) *for all* $b_2, c_2 \in A_2$ *and* $b_1 \in A_1$, *if* $\langle b_1, b_2 \rangle \in J$ *and* $\langle b_1, c_2 \rangle \in J$, *then* $\langle b_1, a_2 \rangle \in J$ *for all* $a_2 \in \mathrm{glb}_{R_2}\{b_2, c_2\}$.

Connectivity is not so firmly connected with prejoining-systems as with TJS joining-systems. The reason is roughly that the occurrence of lub's and glb's of infinite sets fits well with quasi-orderings satisfying completeness in the sense of being complete quasi-lattices. The importance of connectivity in TJS has been stressed several times.

A brief note on the role of the notion of closure system in TJS is in order. An important aspect of TJS is that it gives a method for representing a set of conditional norms in an elaborated way. Suppose that $\mathcal{B}_1$ is a *Bqo* of grounds and $\mathcal{B}_2$ is a *Bqo* of consequences. Let us suppose that K is a set of conditional norms with the antecedents taken from B_1 and the consequences taken from B_2. Hence, $K \subseteq B_1 \times B_2$ and K is a correspondence from B_1 to B_2. K can be thought of as a "crude" representation of a normative system $\mathcal{N}$. Then, a set K^* can be generated by forming the "joining closure" of K such that $\langle \mathcal{B}_1, \mathcal{B}_2, K^* \rangle$ is a *Bjs*. This is an "elaborated" representation of $\mathcal{N}$.

The *out*-operations introduced by Makinson and van der Torre also use a closure-operation, viz. classical consequence. With some simplification one can say that Makinson and van der Torre form the closure of the input and of the output but leave the set of norms as it is. However, it turns out that, regarded only as deductive mechanisms, input-output logic and the theory of joining-systems give rather similar results in spite of their use of different closure-operations in different ways. As a conjecture we suggest the following. Suppose that the *Bqo*'s $\mathcal{B}_1$ and $\mathcal{B}_2$ are Boolean algebras, i.e. for $i = 1, 2$, R_i is the partial ordering determined by the Boolean algebra $\langle B_i, \wedge,' \rangle$. Then $J = out_1(J)$. Furthermore, if $\mathcal{B}_1$ is a complete Boolean algebra and some general conditions are satisfied, then $J = out_2(J)$.

Acknowledgments

We want to thank the two referees for painstaking work och valuable comments. Financial support was given by the Swenson Foundation of the Faculty of Law, University of Lund, The Swedish Research Council, and the University of Gävle.[31]

---

[31]The chapter is the result of wholly joint work where the order of appearance of our author names has no significance.

BIBLIOGRAPHY

[Alchourrón and Bulygin, 1971] C. E. Alchourrón and E. Bulygin. *Normative Systems*. Springer, Wien, 1971.
[Aristotle, Nicomachean Ethics] Aristotle, *Nicomachean Ethics*. Written 350 BC, translated by W. D. Ross, The Internet Classic Archive.
[Davey and Priestley, 2002] B. A. Davey and H. A. Priestley. *Introduction to Lattices and Order*. Cambridge UP, Cambridge, 2nd edition, 2002.
[Dummett, 1973] M. Dummett. *Frege: Philosophy of Language*. Duckworth, London, 1973.
[Gentzen, 1934] G. Gentzen. Untersuchungen über das logische schließen, I. *Mathematische Zeitschrift*, 39:176–210, 1934.
[Grätzer, 1979] G. Grätzer. *Universal Algebra*. Springer-Verlag, New York, 2nd edition, 1979.
[Grätzer, 2011] G. Grätzer. *Lattice Theory: Foundation*. Birkhäuser, Basel, 2011.
[Grossi et al., 2006] D. Grossi, J-J. Ch. Meyer, and F. Dignum. Classificatory aspects of counts-as: An analysis in modal logic. *Journal of Logic and Computation*, 16:613–643, 2006.
[Grossi et al., 2007] D. Grossi, J.-J. Ch Meyer, and F. Dignum. On the logic of constitutive rules. Dagstuhl Seminar Proceedings 07122, Normative Multi-Agent Systems, http://drops.dagstuhl.de/opus/volltexte2007913, 2007.
[Grossi, 2007] D. Grossi. *Designing Invisible Handcuffs. Formal Investigations in Institutions and Organizations for Multi-agent Systems*. PhD thesis, Utrecht University, 2007. SIKS Dissertation Series No. 2007-16.
[Hedenius, 1977] I. Hedenius. *Filosofien i ett föränderligt samhälle*. Bonnier, Stockholm, 1977. (English title: Philosophy in a changing society).
[Hilpinen, 1971] R. Hilpinen, editor. *Deontic Logic: Introductory and Systematic Readings*. Reidel, Dordrecht, 1971.
[Holmström-Hintikka et al., 2001] G. Holmström-Hintikka, S. Lindström, and R. Sliwinski. *Collected Papers of Stig Kanger with Essays on His Life and Work,*, volume I-II. Kluwer, Dordrecht, 2001.
[Horty et al., 1990] J. F Horty, R. H. Thomason, and D. S. Touretzky. A skeptical theory of inheritance in nonmonotonic networks. *Artificial Intelligence*, 42:311–348, 1990.
[Jones and Sergot, 1993] A. J. I. Jones and M. Sergot. On the characterization of law and computer systems: The normative systems perspective. In J-J. Ch. Meyer and R. J Wieringa, editors, *Deontic Logic in computer Science*. Wiley, 1993.
[Jones and Sergot, 1996] A. J. I. Jones and M. Sergot. A formal characterisation of institutionalised power. *Journal of the IPGL*, 3:427–443, 1996. Reprinted in [Jones and Sergot, 1997].
[Jones and Sergot, 1997] A. J. I. Jones and M. Sergot. A formal characterisation of institutionalised power. In E.G. Valdés, W. Krawietz, G. H. von Wright, and R. Zimmerling, editors, *Normative Systems in Legal and Moral Theory. Festschrift for Carlos E. Alchourrón and Eugenio Bulygin*. Duncker and Humblot, Berlin, 1997.
[Kanger and Kanger, 1966] S. Kanger and H. Kanger. Rights and parliamentarism. *Theoria*, 32:85–116, 1966. Reprinted in [Holmström-Hintikka et al., 2001, Vol. 1, pp. 120-145].
[Kanger, 1957] S. Kanger. *New Foundations for Ethical Theory. Part 1*. Almqvist and Wiksell, Stockholm, 1957. Reprinted, first, in [Hilpinen, 1971, pp. 36–58], second, in [Holmström-Hintikka et al., 2001, Vol. 1, pp. 99–119].
[Kanger, 1963] S. Kanger. Rättighetsbegreppet. In *Sju Filosofiska Studier tillägnade Anders Wedberg den 30 mars 1963*, number 9 in Philosophical Studies. Department of Philosophy, University of Stockholm, Stockholm, 1963. (English title: The concept of a right). Reprinted, in English translation, as the first part of [Kanger and Kanger, 1966].

[Kremer, 1988] M. Kremer. The philosophical significance of the sequent calculus. *Mind*, 97:50–72, 1988.

[Lindahl and Odelstad, 1999a] L. Lindahl and J. Odelstad. Intermediate concepts as couplings of conceptual structures. In H. Prakken and P. McNamara, editors, *Norms, Logics and Informations Systems. New Studies on Deontic Logic and Computer Science*, pages 163–179. IOS Press, Amsterdam, 1999.

[Lindahl and Odelstad, 1999b] L. Lindahl and J. Odelstad. Normative systems: Core and amplifications. In R. Sliwinski, editor, *Philosophical Crumbs. Essays Dedicated to Ann-Mari Henschen-Dahlquist on the Occasion of Her Seventy-Fifth Birthday*, Uppsala Philosophical Studies. Department of Philosophy, Uppsala, 1999.

[Lindahl and Odelstad, 2000] L. Lindahl and J. Odelstad. An algebraic analysis of normative systems. *Ratio Juris*, 13:261–278, 2000.

[Lindahl and Odelstad, 2003] L. Lindahl and J. Odelstad. Normative systems and their revision: An algebraic approach. *Artificial Intelligence and Law*, 11:81–104, 2003.

[Lindahl and Odelstad, 2004] L. Lindahl and J. Odelstad. Normative positions within an algebraic approach to normative systems. *Journal Of Applied Logic*, 2:63–91, 2004.

[Lindahl and Odelstad, 2006a] L. Lindahl and J. Odelstad. Intermediate concepts in normative systems. In L. Goble and J-J. Ch. Meyer, editors, *Deontic Logic and Artificial Normative Systems (DEON 2006)*. Springer, Berlin, 2006.

[Lindahl and Odelstad, 2006b] L. Lindahl and J. Odelstad. Open and closed intermediaries in normative systems. In T. M. van Engers, editor, *Legal Knowledge and Information Systems (Jurix 2006)*, pages 91–99. IOS Press, Amsterdam, 2006.

[Lindahl and Odelstad, 2008a] L. Lindahl and J. Odelstad. Intermediaries and intervenients in normative systems. *Journal of Applied Logic*, 6:229–250, 2008.

[Lindahl and Odelstad, 2008b] L. Lindahl and J. Odelstad. Strata of intervenient concepts in normative systems. In R. van der Meyden and L. van der Torre, editors, *Deontic Logic in Computer Science - DEON 2008*. Springer, Berlin, 2008.

[Lindahl and Odelstad, 2011] L. Lindahl and J. Odelstad. Stratification of normative systems with intermediaries. *Journal of Applied Logic*, 9:113–136, 2011.

[Lindahl, 1977] L. Lindahl. *Position and Change: A Study in Law and Logic*. Reidel, Dordrecht, 1977.

[Lindahl, 1985] L. Lindahl. Definitioner, begreppsanalys och mellanbegrepp i juridiken. In *Rationalitet och Empiri i Rättsvetenskapen*, number 6 in Juridiska Fakultetens i Stockholm skriftserie. Juridiska Fakulteten, Stockholm, 1985. (English title: Definitions, conceptual analysis and intermediate concepts in the law).

[Lindahl, 1994] L. Lindahl. Stig Kanger's theory of rights. In D. Prawitz, B. Skyrms, and D. Westerståhl, editors, *Logic, Methodology, and Philosophy of Science IX*. Elsevier Science B.V., 1994. Reprinted, with minor changes, in [Holmström-Hintikka et al., 2001, Vol. II].

[Lindahl, 1997] L. Lindahl. Norms, meaning postulates, and legal predicates. In E.G. Valdés, W. Krawietz, G. H. von Wright, and R. Zimmerling, editors, *Festschrift for Carlos E. Alchourrón and Eugenio Bulygin*. Duncker and Humblot, Berlin, 1997.

[Lindahl, 2000] L. Lindahl. De juridiska begreppens Janusansikte. In A. Numhauser-Henning, editor, *Normativa perspektiv : Festskrift till Anna Christensen*. Juristförl, Lund, 2000. (English title: The Janus-face of legal concepts).

[Lindahl, 2004] L. Lindahl. Deduction and justification in the law: The role of legal terms and concepts. *Ratio Juris*, 17:182–202, 2004.

[Makinson and van der Torre, 2000] D. Makinson and L. van der Torre. Input-output logics. *Journal of Philosophical Logic*, 29:383–408, 2000.

[Makinson and van der Torre, 2003] D. Makinson and L. van der Torre. What is input/output logic? In B. Löwe, W. Malzkorn, and T. Räsch, editors, *Foundations of the Formal Sciences II: Applications of Mathematical Logic in Philosophy and Linguistics*, volume 17 of *Trends in Logic*, pages 163–174. Dordrecht, Kluwer, 2003.

[Moore, 1971] G. Moore. *Principia Ethica*. University Press, Cambridge, 1971.

[Odelstad and Boman, 2004] J. Odelstad and M. Boman. Algebras for agent norm-regulation. *Annals of Mathematics and Artificial Intelligence*, 42:141–166, 2004.

[Odelstad and Lindahl, 2000] J. Odelstad and L. Lindahl. Normative systems represented by Boolean quasi-orderings. *Nordic Journal of Philosophical Logic*, 5:161–174, 2000.

[Odelstad and Lindahl, 2002] J. Odelstad and L. Lindahl. The role of connections as minimal norms in normative systems. In T. Bench-Capon, A. Daskalopulu, and R. Winkels, editors, *Legal Knowledge and Information Systems (Jurix 2002)*. IOS Press, Amsterdam, 2002.

[Odelstad, 2002] J. Odelstad. *Intresseavvägning. En beslutsfilosofisk studie med tillämpning på planering*. Thales, Stockholm, 2002. (English title: Weighing of Interests. A Study in the Philosophy of Decision, with Applications in Planning).

[Odelstad, 2008] J. Odelstad. *Many-Sorted Implicative Conceptual Systems*. PhD thesis, KTH, School of Information and Communication Technology (ICT), Computer and Systems Sciences, DSV, Stockholm, 2008. http://urn.kb.se/resolve?urn=urn:nbn:se:kth:diva-9624.

[Odelstad, 2009] J. Odelstad. An essay on Msic-systems. In G. Boella, L. van der Torre, and H. Verhagen, editors, *Normative Multi-Agent Systems*, number 09121 in Dagstuhl Seminar Proceedings. Schloss Dagstuhl, 2009. http://drops.dagstuhl.de/opus/volltexte/2009/1914/.

[Poincaré, 1907] H. Poincaré. *The Value of Science*. Science Press, London, 1907. (English translation of *La Valeur de la Science*, Paris 1905).

[Ross, 1951] A. Ross. Tû-Tû. In O.A. Borum and K. Illum, editors, *Festskrift til Henry Ussing*. København, Juristforbundet, 1951.

[Ross, 1956 and 1957] A. Ross. Tû-Tû. *Harvard Law Review*, 70:812–825, 1956 and 1957. (English translation of [Ross, 1951]).

[Sartor, 2007] G. Sartor. The nature of legal concepts: Inferential nodes or ontological categories?, 2007.

[Sartor, 2009] G. Sartor. Legal concepts as inferential nodes and ontological categories. *Artificial Intelligence and the Law*, 17:217–251, 2009.

[Sergot, 1999] M. J. Sergot. Normative positions. In P. McNamara and H. Prakken, editors, *Norms, Logics and Information Systems*, pages 289–310. IOS Press, Amsterdam, 1999.

[Sergot, 2001] M. J. Sergot. A computational theory of normative positions. *ACM Transactions on Computational Logic (TOCL)*, 2:581–622, 2001.

[Talja, 1980] J. Talja. A technical note on Lars Lindahl's Position and Change. *Journal of Philosophical Logic*, 9:167–183, 1980.

[Wedberg, 1951] A. Wedberg. Some problems in the logical analysis of legal science. *Theoria*, 17:246–275, 1951.

[Wedberg, 1982] A. Wedberg. *A History of Philosophy. Vol. 2, The Modern Age to Romanticism*. Clarendon Press, Oxford, 1982.

Lars Lindahl
Lund University
Email: lars-lindahl@live.se

Jan Odelstad
University of Gävle
Email: jod@hig.se

CPSIA information can be obtained
at www.ICGtesting.com
Printed in the USA
BVHW041454071019
560248BV00011B/49/P